Veronique Fischer • Michael Ruzhansky

Quantization on Nilpotent Lie Groups

Veronique Fischer
Department of Mathematics
University of Bath
Bath, UK

Michael Ruzhansky
Department of Mathematics
Imperial College London
London, UK

DOI 10.1007/978-3-319-29558-9

Library of Congress Control Number: 2016932499

Mathematics Subject Classification (2010): 22C05, 22E25, 35A17, 35H10, 35K08, 35R03, 35S05, 43A15, 43A22, 43A77, 43A80, 46E35, 46L10, 47G30, 47L80

Ferran Sunyer i Balaguer (1912–1967) was a self-taught Catalan mathematician who, in spite of a serious physical disability, was very active in research in classical mathematical analysis, an area in which he acquired international recognition. His heirs created the Fundació Ferran Sunyer i Balaguer inside the Institut d'Estudis Catalans to honor the memory of Ferran Sunyer i Balaguer and to promote mathematical research.

Each year, the Fundació Ferran Sunyer i Balaguer and the Institut d'Estudis Catalans award an international research prize for a mathematical monograph of expository nature. The prize-winning monographs are published in this series. Details about the prize and the Fundació Ferran Sunyer i Balaguer can be found at

http://ffsb.espais.iec.cat/EN

This book has been awarded the Ferran Sunyer i Balaguer 2014 prize.

The members of the scientific commitee of the 2014 prize were:

Alejandro Adem
University of British Columbia

Hyman Bass
University of Michigan

Núria Fagella
Universitat de Barcelona

Eero Saksman
University of Helsinki

Yuri Tschinkel
Courant Institute of Mathematical Sciences, New York University

Ferran Sunyer i Balaguer Prize winners since 2004:

2004 Guy David
Singular Sets of Minimizers for the Mumford-Shah Functional, PM 233

2005 Antonio Ambrosetti and Andrea Malchiodi
Perturbation Methods and Semilinear Elliptic Problems on R^n, PM 240

José Seade
On the Topology of Isolated Singularities in Analytic Spaces, PM 241

2006 Xiaonan Ma and George Marinescu
Holomorphic Morse Inequalities and Bergman Kernels, PM 254

2007 Rosa Miró-Roig
Determinantal Ideals, PM 264

2008 Luis Barreira
Dimension and Recurrence in Hyperbolic Dynamics, PM 272

2009 Timothy D. Browning
Quantitative Arithmetic of Projective Varieties, PM 277

2010 Carlo Mantegazza
Lecture Notes on Mean Curvature Flow, PM 290

2011 Jayce Getz and Mark Goresky
Hilbert Modular Forms with Coefficients in Intersection Homology and Quadratic Base Change, PM 298

2012 Angel Cano, Juan Pablo Navarrete and José Seade
Complex Kleinian Groups, PM 303

2013 Xavier Tolsa
Analytic capacity, the Cauchy transform, and non-homogeneous Calderón–Zygmund theory, PM 307

Preface

The purpose of this monograph is to give an exposition of the global quantization of operators on nilpotent homogeneous Lie groups. We also present the background analysis on homogeneous and graded nilpotent Lie groups. The analysis on homogeneous nilpotent Lie groups drew a considerable attention from the 70's onwards. Research went in several directions, most notably in harmonic analysis and in the study of hypoellipticity and solvability of partial differential equations. Over the decades the subject has been developing on different levels with advances in the analysis on the Heisenberg group, stratified Lie groups, graded Lie groups, and general homogeneous Lie groups.

In the last years analysis on homogeneous Lie groups and also on other types of Lie groups has received another boost with newly found applications and further advances in many topics. Examples of this boost are subelliptic estimates, multiplier theorems, index formulae, nonlinear problems, potential theory, and symbolic calculi tracing full symbols of operators. In particular, the latter has produced further applications in the study of linear and nonlinear partial differential equations, requiring the knowledge of lower order terms of the operators.

Because of the current advances, it seems to us that a systematic exposition of the recently developed quantizations on Lie groups is now desirable. This requires bringing together various parts of the theory in the right generality, and extending notions and techniques known in particular cases, for instance on compact Lie groups or on the Heisenberg group.

In order to do so, we start with a review of the recent developments in the global quantization on compact Lie groups. In this, we follow mostly the development of this subject in the monograph [RT10a] by Turunen and the second author, as well as its further progress in subsequent papers. After a necessary exposition of the background analysis on graded and homogeneous Lie groups, we present the quantization on general graded Lie groups. As the final part of the monograph, we work out details of the general theory developed in this book in the particular case of the Heisenberg group.

In the introduction, we will provide a link between, on one hand, the symbolic calculus of matrix valued symbols on compact Lie groups with, on the other hand, different approaches to the symbolic calculus on the Heisenberg group for instance. We will also motivate further our choices of presentation from the point of view of the development of the theory and of its applications.

We would like to thank Fulvio Ricci for discussions and for useful comments on the historical overview of parts of the subject that we tried to present in the introduction. We would also like to thank Gerald Folland for comments leading to improvements of some parts of the monograph.

Finally, it is our pleasure to acknowledge the financial support by EPSRC (grant EP/K039407/1), Marie Curie FP7 (Project PseudodiffOperatorS - 301599), and by the Leverhulme Trust (grant RPG-2014-02) at different stages of preparing this monograph.

Véronique Fischer
Michael Ruzhansky

London, 2015

Contents

Introduction

Nilpotent Lie groups appear naturally in the analysis of manifolds and provide an abstract setting for many notions of Euclidean analysis. As is generally the case when studying analysis on nilpotent Lie groups, we restrict ourselves to the very large subclass of homogeneous (nilpotent) Lie groups, that is, Lie groups equipped with a family of dilations compatible with the group structure. They are the groups appearing 'in practice' in the applications (some of them are described below). From the point of view of general harmonic analysis, working in this setting also leads to the distillation of the results of the Euclidean harmonic analysis depending only on the group and dilation structures.

In order to motivate the work presented in this monograph, we focus our attention in this introduction on three aspects of the analysis on nilpotent Lie groups: the use of nilpotent Lie groups as local models for manifolds, questions regarding hypoellipticity of differential operators, and the development of pseudo-differential operators in this setting. We only outline the historical developments of ideas and results related to these topics, and on a number of occasions we refer to other sources for more complete descriptions. We end this introduction with the main topic of this monograph: the development of a pseudo-differential calculus on homogeneous Lie groups.

Nilpotent Lie groups by themselves and as local models

It has been realised for a long time that the analysis on nilpotent Lie groups can be effectively used to prove subelliptic estimates for operators such as 'sums of squares' of vector fields on manifolds. Such ideas started coming to light in the works on the construction of parametrices for the Kohn-Laplacian $\Box_b$ (the Laplacian associated to the tangential CR complex on the boundary X of a strictly pseudoconvex domain), which was shown earlier by J. J. Kohn to be hypoelliptic (see e.g. an exposition by Kohn [Koh73] on the analytic and smooth hypoelliptic-ities). Thus, the corresponding parametrices and subsequent subelliptic estimates have been obtained by Folland and Stein in [FS74] by first establishing a version of the results for a family of sub-Laplacians on the Heisenberg group, and then for the Kohn-Laplacian $\Box_b$ by replacing X locally by the Heisenberg group. These ideas soon led to powerful generalisations. The general techniques for approximating vector fields on a manifold by left-invariant operators on a nilpotent Lie group have been developed by Rothschild and Stein in [RS76]. Here the dimension of the nilpotent Lie group is normally larger than that of the manifold, and a first step of such a construction is to perform the 'lifting' of vector fields to the group. Consequently, this approach allowed one to produce parametrices for the original differential operator on the manifold by using the analysis on homogeneous Lie groups. A more geometric version of these constructions has been carried out by Folland in [Fol77b], see also Goodman [Goo76] for the presentation of nilpotent Lie algebras as tangent spaces (of sub-Riemannian manifolds). The functional analytic background for the analysis in the stratified setting was laid down by Folland in [Fol75]. A general approach to studying geometries appearing from systems of vector fields has been developed by Nigel, Stein and Wainger [NSW85].

Thus, one of the motivations for carrying out the analysis and the calculus of operators on nilpotent Lie groups comes from the study of differential operators on CR (Cauchy-Riemann) or contact manifolds, modelling locally the operators there on homogeneous invariant convolution operators on nilpotent groups. In 'practice' and from this motivation, only nilpotent Lie groups endowed with some compatible structure of dilations, i.e. homogeneous Lie groups, are considered. This will be also the setting of our present exposition.

The simplest example (apart from $\mathbb{R}^n$) of a nilpotent Lie group is the Heisenberg group, and the harmonic analysis there is a very well researched topic. We do not intend to make an overview of the subject here, but we refer to the books of Stein [Ste93] and Thangavelu [Tha98] for an introduction to the harmonic analysis on the Heisenberg group and for the historic development of the area. Elements of the harmonic analysis on different groups can be also found in Taylor's book [Tay86]. The Heisenberg group enters many applied areas, including various aspects of quantum mechanics, signal analysis, optics, thermodynamics; we refer to the recent book of Binz and Pods [BP08] for an overview of this subject. We mention another recent book by Calin, Chang and Greiner [CCG07] containing many explicit calculations related to the Heisenberg group and its sub-Riemannian geometry, as well as a sub-Riemannian treatment in Capogna, Danielli, Pauls and Tyson [CDPT07]. As such, in this monograph we will deal with the Heisenberg group almost exclusively in the context of pseudo-differential operators, and we refer to excellent surveys of Folland [Fol77a] and Howe [How80] on the role played by the Heisenberg group in the theory of partial differential equations and in harmonic analysis, as well as to Folland's book [Fol89] for its relation to the theory of pseudo-differential operators on $\mathbb{R}^n$ through the Weyl quantization. See also a more recent short survey by Semmes [Sem03] and a book by Krantz [Kra09].

Well-posedness questions for hyperbolic partial differential equations on the Heisenberg group have been considered parallel to their Euclidean counterparts. For example, the conditions for the well-posedness of the wave equation for the Laplacian associated to the $\bar{\partial}_b$ complex have been found by Nachman [Nac82], the L^p-estimates for the wave equation for the sub-Laplacian have been established by Müller and Stein [MS99], the smoothness of the Schrödinger kernel has been analysed by Sikora and Zienkiewicz [SZ02], a space-time estimate for the Schrödinger equation has been obtained by Zienkiewicz [Zie04], etc. Nonlinear wave and Schrödinger equations and Strichartz estimates have been analysed on the Heisenberg group as well, see e.g. Zuily [Zui93], Bahouri, Gérard and Xu [BGX00] and Furioli, Melzi and Veneruso [FMV07], as well as other equations, e.g. the Ginzburg-Landau equation by Birindelli and Valdinoci [BV08], quasilinear equations by Capogna [Cap99], etc.

The Hardy spaces on homogeneous Lie groups and the surrounding harmonic analysis have been investigated by Folland and Stein in their monograph [FS82]. In general, there are different machineries available depending on a degree of generality: the stratified Lie groups enjoy additional hypoellipticity techniques going back to Hörmander's celebrated sum of the squares theorem, while on the Heisenberg group explicit expressions from its representation theory can be used.

A typical example of such different degrees of generality within homogeneous Lie groups is, for instance, a problem of characterising the Hardy space H^1 in L^1 by families of singular integrals. Thus, in [CG84], Christ and Geller presented sufficient conditions for general homogeneous Lie groups, gave explicit examples of (generalised) Riesz transforms for such a family of integral operators on stratified Lie groups, and derived further necessary and sufficient conditions on the Heisenberg group in terms of its representation theory (see also further work by Christ [Chr84]).

A related aspect of harmonic analysis, the Calderón-Zygmund theory on homogeneous Lie groups, has a long history as well. Again, this started with the analysis of convolution operators (with earlier works e.g. by Korányi and Vági [KV71] in the nilpotent direction), but in this book we will adopt an utilitarian approach, and the setting of Coifman and Weiss [CW71a] of spaces of homogeneous type will be sufficient for our purposes (see Section 3.2.3 and Section A.4).

Proceeding with this part of the introduction on general homogeneous Lie groups, let us follow Folland and Stein [FS82] and mention another important occurrence of homogeneous Lie groups. If G is a non-compact real connected semi-simple Lie group, its Iwasawa decomposition $G = KAN$ contains the homogeneous Lie group N whose family of dilations comes from an appropriate one-parameter subgroup of the abelian group A (more precisely, if $\mathfrak{g} = \mathfrak{k} \oplus \mathfrak{k}^{\perp}$ is the Cartan decomposition of the Lie algebra $\mathfrak{g}$, the decomposition $G = KAN$ corresponds to the Iwasawa decomposition of the Lie algebra, $\mathfrak{g} = \mathfrak{k} + \mathfrak{a} + \mathfrak{n}$, where $\mathfrak{a}$ is the maximal abelian subalgebra of $\mathfrak{k}^{\perp}$, and the nilpotent Lie algebra $\mathfrak{n}$ is the sum of the positive root spaces corresponding to eigenvalues of $\mathfrak{a}$ acting on $\mathfrak{g}$). This decomposition generalises the decomposition of a real matrix as a product of an orthogonal, diagonal, and an upper triangular with 1 at the diagonal matrix. Furthermore, the the symmetric space G/K has the homogeneous nilpotent Lie group N as its 'boundary' in the sense that N may be identified with a dense subset of the maximal boundary of G/K. As we show in Section 6.1.1 for $n_o = 1$, if $G = \mathrm{SU}(n_o + 1, 1)$, G/K may be identified with the unit ball in $\mathbb{C}^{n_o+1}$ and the Heisenberg group $\mathbb{H}_{n_o}$ acts simply transitively on the complex sphere of $\mathbb{C}^{n_o+1}$

where one point has been excluded. This provides a link between the Heisenberg group $\mathbb{H}_{n_o}$, the analysis of the complex spheres, and the group $\mathrm{SU}(n_o+1,1)$ or, more generally, between general semi-simple Lie groups and homogeneous Lie groups as boundaries of their symmetric spaces. For example, harmonic functions on the symmetric space G/K can be represented by convolution operators on N (see e.g. the survey of Koranyi [Kor72]).

Our setting contains the realm of Carnot groups as this class of groups consists of the stratified Lie groups equipped with a specified metrics on the first layer, see e.g. Gromov [Gro96] for a survey on geometric analysis of Carnot groups. Our setting includes any class of stratified Lie groups, for instance H-groups, Heisenberg-Kaplan groups, Métivier-type groups [Mét80], filiform groups, as well as Kolmogorov-type groups appearing in the study of hypoelliptic ultraparabolic operators including the Kolmogorov-Fokker-Planck operator (see Kolmogorov [Kol34], Lanconelli and Polidoro [LP94]). We refer to the book [BLU07] by Bonfiglioli, Lanconelli and Uguzonni for a detailed consideration of these groups and of their sub-Laplacians as well as related operators.

Hypoellipticity and Rockland operators

On compact Lie groups, the Fourier analysis and the symbolic calculus developed in [RT10a] are based on the Laplacian and on the growth rate of its eigenvalues. While on compact Lie groups the Laplacians (or the Casimir element) are operators naturally associated to the group, it is no longer the case in the nilpotent setting. Thus, on nilpotent Lie groups it is natural to work with operators associated with the group through its Lie algebra structure. On stratified Lie groups these are the sub-Laplacians, and such operators are not elliptic but hypoelliptic. More generally, on graded Lie groups invariant hypoelliptic differential operators are the so-called Rockland operators.

Indeed, in [Roc78], Rockland showed that if T is a homogeneous left-invariant differential operators on the Heisenberg group, then the hypoellipticity of T and T^t is equivalent to a condition now called the Rockland condition (see Definition 4.1.1). He also asked whether this equivalence would be true for more general homogeneous Lie groups. Soon after, Beals showed in [Bea77b] that the hypoellipticity of a homogeneous left-invariant differential operator on any homogeneous Lie group implies the Rockland condition. In the same paper he also showed that the converse holds in some step-two cases. Eventually in [HN79], Helffer and Nourrigat settled what has become known as Rockland's conjecture by proving that the hypoellipticity is equivalent to the Rockland condition (see Section 4.1.3). At the same time, it was shown by Miller [Mil80] that in the setting of homogeneous Lie groups, the existence of an operator satisfying the Rockland condition (hence of an invariant hypoelliptic differential operator in view of Helffer and Nourrigat's result), implies that the group is graded, see also Section 4.1.1. This means, altogether, that the setting of graded Lie groups is the right generality for marrying the harmonic analysis techniques with those coming from the theory of partial differential equations.

A number of well-known functional inequalities can be extended to the graded setting, for example, see Bahouri, Fermanian-Kammerer and Gallagher [BFKG12b]. Also, there are many contributions to questions of solvability related to the hypoellipticity problem: for a good introduction to local and non-local solvability questions on nilpotent Lie groups see Corwin and Rothschild [CR81] and, missing to mention many contributions, for a more recent discussion of the topic see Müller, Peloso and Ricci [MPR99].

The hypoellipticity of second order operators is a very well researched subject. Its beginning may be traced to the 19th century with the diffusion problems in probability arising in Kolmogorov's work [Kol34]. Hörmander made a major contribution [Hör67b] to the subject which then developed rapidly after that (see e.g. the book of Oleinik and Radkevich [OR73]) until nowadays. We will not be concerned much with these nor with the solvability problems in this book, since one of topics of importance to us will be Rockland operators of an arbitrary degree, and we will be giving more relevant references as we go along.

Here we want to mention that the question of the analytic hypoellipticity turns out to be more involved than that in the smooth setting. In general, if a graded Lie group is not stratified, there are no homogeneous analytic hypoelliptic left-invariant differential operators, a result by Helffer [Hel82]. For stratified Lie groups, the situation is roughly as follows: for H-type groups the analytic hypoellipticity is equivalent to the smooth hypoellipticity, while for step ≥ 3 (and an additional assumption that the second stratum is one-dimensional) the sub-Laplacians are not analytic hypoelliptic, see Métivier [Mét80] and Helffer [Hel82], respectively, and the discussions therein. For the Kohn-Laplacian $\Box_b$ in the $\bar{\partial}$-Neumann problem as well as for higher order operators in this setting the analytic hypoellipticity was shown earlier by Tartakoff [Tar78, Tar80]. Below we will mention a few more facts concerning the analytic hypoellipticity in the framework of the analytic calculus of pseudo-differential operators.

Pseudo-differential operators

Several versions of the smooth calculi of pseudo-differential operators on the Heisenberg group have been considered over the years. An earlier attempt yielding the calculus of invariant operators with symbols on the dual $\mathfrak{g}'$ of the Lie algebra of the group was made by Strichartz [Str72]. A calculus for (right-invariant) operators has been also constructed by Melin [Mel81] yielding parametrices for operators elliptic in the so-called generating directions. In particular, the symbolic calculus for invariant operators on stratified and graded Lie groups developed by Melin further in [Mel83] provided a simpler proof of many of Helffer and Nourrigat's arguments.

The question of a general symbolic calculus for convolution operators on nilpotent Lie groups was raised by Howe in [How84], who also tackled questions related to the Calderón-Vaillancourt theorem. A more recent development of the calculus for invariant operators on homogeneous Lie groups and applications to the corresponding symbolic conditions for the L^2-boundedness of convolution operators was given by Głowacki in [Gło04] and [Gło07]. All this analysis applies to invariant operators and employs the Euclidean Fourier transform yielding a symbol on the dual $\mathfrak{g}'$. The symbol classes of such operators on the group are defined as coming from the usual Hörmander classes on the (Euclidean) vector space $\mathfrak{g}'$. They satisfy the spectral invariance properties and yield further useful generalisations of parametrix constructions, see Głowacki [Gło12]. An approach to Melin's operators on nilpotent Lie groups from the point of view of the Weyl calculus was done by Manchon, with further applications to the Weyl spectral asymptotics for the infinitesimal representations of elliptic operators in his calculus, see [Man91]. There exists also a calculus of left-invariant integral operators on the Heisenberg group, using Laguerre polynomials for its Fourier analysis, see Beals, Gaveau, Greiner and Vauthier [BGGV86], or using Leray's quadratic Fourier transform by Gaveau, Greiner and Vauthier [GGV86].

While these are mostly the calculi of invariant operators, the geometric considerations require one to also understand operators in the non-invariant setting. However, here the amount of knowledge is more limited and most of the symbolic calculus is restricted to the Heisenberg group. Dynin's construction of certain operators on the Heisenberg group in [Dyn76] (see also [Dyn78]), was also developed by Folland into considering meta-Heisenberg groups in [Fol94]. Beside this, a non-invariant pseudo-differential calculus on any homogeneous Lie group was developed by Christ, Geller, Głowacki and Polin in [CGGP92] but this is not symbolic since the operator classes are defined via properties of the kernel. In the revised version of [Tay84], Taylor described several (non-invariant) operator calculi and, in a different direction, he also noted a way to develop symbolic calculi: using the representations of the group, he defines a general quantization and symbols on any unimodular type I group (by quantization, we mean a procedure which associates an operator with a symbol). He illustrated this on the Heisenberg group and obtained there several important applications for, e.g., the study of hypoellipticity. He used the fact that, because of the properties of the Schrödinger representations of the Heisenberg group, a symbol is a family of operators in the Euclidean space, themselves given by symbols via the Weyl quantization. Recently, the definition of suitable classes of Shubin type for these Weyl-symbols led to another version of the calculus on the Heisenberg group by Bahouri, Fermanian-Kammerer and Gallagher [BFKG12a].

A calculus of pseudo-differential operators on the Heisenberg group in the analytic setting was developed by Geller [Gel90], with applications to the analytic hypoellipticity and further extensions of the calculus to real analytic CR manifolds. In particular, this implies the analytic hypoellipticity of the Kohn-Laplacian on q-forms on pseudoconvex real analytic manifolds. It also implies that the Szegö projection preserves analyticity, recovering earlier results on the relations between Szegö projections and $\bar{\partial}_b^*$-operator by Greiner, Kohn and Stein [GKS75], in turn related to the solvability of the Lewy equation. The analytic hypoellipticity of the complex boundary Kohn-Laplacian on the (p, q)-forms arising in the $\bar{\partial}$-Neumann problem was proved earlier by Tartakoff [Tar78] using L^2-methods and by Trèves [Trè78] using the calculus. Here we note that a corresponding Euclidean symbolic calculus with applications to the propagation of analytic singularities and corresponding version of Fourier integral operators has been developed by Sjöstrand [Sjö82], and the propagation of the analytic wave front set for the sub-Laplacian was studied by Grigis and Sjöstrand [GS85].

The analysis of the nilpotent setting can be extended to more general manifolds. Here, a typical application of the analysis on the Heisenberg groups is to questions on contact manifolds. Indeed, a contact structure defines a grading on the space of vector fields assigning them the degree of one or two. Locally, a contact manifold is diffeomorphic to the Heisenberg group, and the principal symbol of a differential operator on the contact manifold is its higher order terms, with the calculus on the contact manifold induced by that on the Heisenberg group, at least on the principal symbol level. The ellipticity condition for an operator on a contact manifold is thus replaced by the Rockland condition for the homogeneous principal part of the corresponding operator on the Heisenberg group. Such constructions can be carried out in more general settings, in particular on the so-called Heisenberg manifolds, which are smooth manifolds with a distinguished hyperplane bundle. The calculus of operators in this setting was carried out by Beals and Greiner in [BG88], in particular also generalising the calculus on CR manifolds needed for the construction of parametrices for the Kohn-Laplacian $\Box_b$. A recent advance in this direction mostly aimed at the second order operators on Heisenberg manifolds, with a more intrinsic notion of the principal symbol of such operators, was made by Ponge [Pon08]. Examples of such analysis include CR manifolds and contact manifolds, with applications to the Kohn-Laplacian, the Gover-Graham operators, the contact Laplacian associated to the Rumin complex, as well as to more general Rockland operators.

Moreover, such operators are subelliptic and their index may be calculated by the Atiyah-Singer index formula, see van Erp [vE10a]. The explicit knowledge of the Bargmann-Fock representations of the Heisenberg group allows one to construct the necessary Heisenberg calculus adapted to subelliplic operators in this setting, leading to the index formula also for subelliptic pseudo-differential operators on contact manifolds, see van Erp [vE10b].

The calculus of pseudo-differential operators on homogeneous Lie groups in terms of their kernels developed by Christ, Geller, Głowacki and Polin in [CGGP92] extended the parametrix construction of Helffer and Nourrigat in [HN79] and some properties of Taylor's calculus from [Tay84] in the case of the Heisenberg group. However, this calculus is not symbolic since it is based on the properties of the kernel. The same is true for the analysis of operators on unimodular Lie groups considered by Meladze and Shubin [MS87] where the operator classes were defined in terms of local properties of the kernels.

Recently in [MR15], Mantoiu and the second author developed a more general τ-quantization scheme on general locally compact unimodular type I groups, thus encompassing in particular the cases of compact Lie groups by the second author and Turunen [RT10a] and nilpotent Lie groups including the one developed in this monograph. Moreover, the τ-quantizations there allow one to deal with analogues of both Kohn-Nirenberg and Weyl quantizations. However, due to the generality the scope of available results at the moment is much more limited than the one presented in [RT10a] in the compact case, or in this book. The type I assumption is useful for having a rich machinery concerning the abstract Plancherel theorem (see Section 1.8.2), however it can be dropped for some questions: e.g. for an L^p-L^q Fourier multiplier theorem on general locally compact separable unimodular groups (without type I assumption) see Akylzhanov and Ruzhansky [AR15]. For nilpotent Lie groups, the relation between such quantizations and Melin's quantization described above has been also established in [MR15].

Quantization on homogeneous Lie groups and the book structure

Most of the above works that concern the non-invariant symbolic calculi of operators on nilpotent Lie groups, are restricted to the Heisenberg groups or to manifolds having the Heisenberg group as a local model (except for the calculi which are not symbolic). One of the reasons is that they rely in an essential way on the explicit formulae for representations of the Heisenberg group. However, in all the motivating aspects described above graded Lie groups appear as well as the Heisenberg group. Also, graded groups appear as local models once one is dealing with operators which are not in the form of 'sum of squares' even on manifolds such as the Heisenberg manifolds.

Recently, in [RT10a, RT13], the second author and Turunen developed a global symbolic calculus on any compact Lie group. They defined symbol classes so that the quantization procedure, analogous to the Kohn-Nirenberg quantization on $\mathbb{R}^n$, makes sense on compact Lie groups, and the resulting operators form an algebra with properties 'close enough' to the one enjoyed by the Euclidean Hörmander calculus. In particular, one can also recover the Hörmander classes of pseudo-differential operators on compact Lie groups viewed as compact manifolds through conditions imposed on global full matrix-valued symbols. This approach works for any compact Lie group and is intrinsic to the group in the sense that it does not depend on pseudo-differential calculus of (Euclidean) Hörmander classes on its Lie algebra. While relying on the representation theory of the group, the quantization and the calculus do not depend on explicit formulae for its representations. It does not depend either on the available Riemannian structure. This gives an advantage over the calculi expressed in terms of a fixed connection of a manifold, such as the one developed by Widom [Wid80], Safarov [Saf97], and Sharafutdinov [Sha05], see also the survey of McKeag and Safarov [MS11].

The crucial and new ingredient in the definition of symbol classes in [RT10a] was the introduction and systematic use of *difference operators* in order to replace the Euclidean derivatives in the Fourier variables by 'analogous' operators acting on the unitary dual of the group. These difference operators allow one to express the pseudo-differential behaviour directly on the group and are very natural from the point of view of the Calderón-Zygmund theory. These operators and their properties on compact Lie groups will be reviewed in Section 2.2.2.

It is not possible however to extend readily the results of the compact case developed in [RT10a] to the nilpotent context. Indeed, the global analysis on a non-compact setting is usually more challenging than in the case of a compact manifold. In the specific case of Lie groups, the dual of a non-compact group is no longer discrete and the unitary irreducible representations may be infinite dimensional, and are often so. More problematically there is no Laplacian and one expects to replace it by a sub-Laplacian on stratified Lie groups or, more generally, by a positive Rockland operator on graded Lie groups; such operators are not central.

Thus, in this book we study the global quantization of operators on graded Lie groups, in particular aiming at developing an intrinsic symbolic calculus of such operators. This is done in Chapter 5. As noted earlier, the graded Lie groups is a natural generality for such analysis since we can still make a full use of the Rockland operators as well as from the representation theory which is well understood e.g. by Kirillov's orbit method [Kir04]. The consequent Fourier analysis is then also well understood from earlier works on the Plancherel formula on nilpotent and even on more general locally compact unimodular type I groups; an overview of this topic is given in Section 1.7.

Summarising very briefly the results presented in Chapter 5, we introduce a global quantization on graded Lie groups and classes $S^m_{\rho,\delta}$ of symbols and of the corresponding operators in $\Psi^m_{\rho,\delta} = \operatorname{Op} S^m_{\rho,\delta}$ such that for each (ρ, δ) with $1 \geq \rho \geq \delta \geq 0$ and $\delta \neq 1$, we have an operator calculus, in the sense that the set $\bigcup_{m\in\mathbb{R}} \Psi^m_{\rho,\delta}$ forms an algebra of operators, stable under taking the adjoint, and acting on the Sobolev spaces in such a way that the loss of derivatives is controlled by the order of the operator. Moreover, the operators that are elliptic or hypoelliptic within these classes allow for a parametrix construction whose symbol can be obtained from the symbol of the original operator. Some applications of the constructed calculus are contained in Chapter 5, see also the authors' paper [FR13] for further applications to lower bounds of operators on graded Lie groups. A preliminary very brief outline of the constructions here was given in [FR14a].

To lay down the necessary foundation for the quantization of operators and symbols, we also make an exposition of the construction of the Sobolev spaces on graded Lie groups based on positive Rockland operators. Such construction has been previously done on stratified Lie groups by Folland [Fol75], for Sobolev spaces based on the (left-invariant) sub-Laplacian. Sub-Laplacians in this context are (up to a sign) a particular case of positive Rockland operators on stratified groups and our results coincide with Folland's in this case. However if we follow Folland's treatment but now in the more general context of graded Lie groups, beside the appearance of several technical problems, we would be led to make further assumptions. One of them would be that the degree of the positive Rockland operator ν must be less than the homogeneous dimension Q of the group, $\nu < Q$

(assuming $\nu < Q$ ensures the uniqueness of its homogeneous fundamental solution). In fact, Goodman makes such an assumption in his treatment of Sobolev spaces on graded Lie group in [Goo76]. In order to avoid this assumption and also to deal with other issues, we need to develop other arguments in the study of the powers of a general positive Rockland operator $\mathcal{R}$ and of $\mathrm{I}+\mathcal{R}$ and in the study of the associated Sobolev spaces. This is done under no assumptions on the relation between ν and Q in Sections 4.3 and 4.4.

The analysis of Sobolev spaces is based on the heat kernel associated with a positive Rockland operator. We make an exposition of this topic in Section 4.2.2. Our presentation there follows essentially the arguments of Folland and Stein [FS82]. The heat kernels are not necessarily positive functions and the heat semi-group does not necessarily correspond to a martingale as in the stratified case or more generally for sums of squares of vector fields with Hörmander's condition. Such sums of squares have been analysed in much more general settings. For example, we can refer to the book of Varopoulos, Saloff-Coste and Coulhon [VSCC92] for a treatment of the heat kernel on unimodular Lie groups of polynomial growth, and the usually associated to it estimates, such as Harnack and Sobolev inequalities. For another point of view, allowing dealing with heat kernels associated to more general subelliptic second order order differential operators we refer to Dungey, ter Elst and Robinson's book [DtER03].

Overall, the majority of the background material can be (sometimes even more easily) introduced in the setting of homogeneous Lie groups and we discuss these in Chapter 3. Our treatment in this chapter is inspired by those of Folland and Stein [FS82] and Ricci [Ric] but is slightly more general than that in the existing literature since we allow kernels and operators to have complex-valued homogeneity degrees. This allows us to treat complex powers of operators later on, e.g. in Section 4.3.3.

We assume that the reader is familiar with analysis at a graduate level, e.g. as presented in the books of Rudin [Rud87, Rud91], Reed and Simon [RS80, RS75], or Folland [Fol99]. Nevertheless, we make a brief exposition of topics, mostly from the representation theory of groups, to remind the reader of the necessary concepts used in later parts of the book and to fix the terminology and notation. This is done in Chapter 1, and references to more material are given throughout.

The exposition of the (matrix) quantization on compact Lie groups from [RT10a] and its related works is given in Chapter 2. This serves both as an introduction to the topic as well as provides motivation and examples for some of the concepts presented later in the book.

Chapter 6 is devoted to presenting an application of the general theory developed in Chapter 5 to the concrete setting of the Heisenberg groups. Some results from this chapter have been announced in the authors' paper [FR14b] and this chapter provides their proofs. We give the necessary preliminaries of the analysis on the Heisenberg group, including a description of its dual using Schrödinger representations, with further concrete expressions for the Plancherel measure and Plancherel formula. For the Heisenberg group $\mathbb{H}_n$, its Schrödinger representations

π_λ are acting on the space $L^2(\mathbb{R}^n)$, thus yielding symbols acting on the Schwartz space $\mathcal{S}(\mathbb{R}^n)$, the space of smooth vectors of π_λ. In turn, these symbols can be conveniently described using their Weyl quantization, giving a notion of scalar-valued λ-symbols. In this particular case of the Heisenberg group, the symbol classes of Chapter 5 can be characterised by the property that these λ-symbols belong to some Shubin spaces, more precisely, a semiclassical-λ-type version of the usual Shubin classes of symbols. Consequently, this is applied to giving criteria for ellipticity and hypoellipticity of operators on the Heisenberg group in terms of the invertibility properties of their λ-symbols. We provide a list of examples to show the applicability of these results in several settings.

Notation and conventions

$\mathbb{N} = \{1, 2, 3, \ldots\}$
$\mathbb{N}_0 = \mathbb{N} \cup \{0\}$
$\mathbb{R}_* = \mathbb{R}\backslash\{0\}$
$\mathbb{R}_+ = (0, \infty)$
$\mathbb{C}_+ = \{z \in \mathbb{C}, \operatorname{Re} z > 0\}$
$0! = 1$
For $\alpha = (\alpha_1, \ldots, \alpha_N)$, $\alpha! = \alpha_1! \cdots \alpha_N!$, $|\alpha| = \alpha_1 + \cdots + \alpha_N$
$[\alpha]$ is the homogeneous degree, defined in (3.12)
$\lceil x \rceil$ is the smallest $n \in \mathbb{Z}$ such that $n > x$
$\lfloor x \rfloor$ is the largest $n \in \mathbb{Z}$ such that $n \leq x$
$\lceil M \rfloor = \max\{|\alpha| : \alpha \in \mathbb{N}_0^n \text{ with } [\alpha] \leq M\}$, as defined in (3.35)
$A \asymp B$ means there is some $c > 0$ such that $c^{-1}A \leq B \leq cA$
$\overline{\sum_j} a_j := \sum_j c_j a_j$ denotes a (finite) linear combination with some (irrelevant) constants c_j
Unit elements: e on general groups, and 0 on nilpotent groups
δ_x is the delta-distribution at x: $\delta_x(\phi) = \phi(x)$
$\delta_{j,k}$ is the Kronecker delta: $\delta_{j,k} = 0$ for $j \neq k$, and $\delta_{j,j} = 1$
$\mathscr{L}(\mathcal{H}_1, \mathcal{H}_2)$ is the space of all linear continuous mappings from $\mathcal{H}_1$ to $\mathcal{H}_2$
$\mathscr{L}(\mathcal{H}) := \mathscr{L}(\mathcal{H}, \mathcal{H})$
$\mathcal{U}(\mathcal{H})$ is the space of unitary mappings in $\mathscr{L}(\mathcal{H})$
G is a Lie group, $\mathfrak{g}$ is its Lie algebra, and
$\mathfrak{U}(\mathfrak{g})$ is its universal enveloping algebra defined in Section 1.3
$\bar{T}$, T^* and T^t are defined by (1.8), (1.9), and (1.10) for an element $T \in \mathfrak{U}(\mathfrak{g})$ and in Definition 1.3.1 for an operator T on $L^2(G)$
$\operatorname{Rep} G$ is the set of all strongly continuous unitary irreducible representations of the group G
$\widehat{G}$ is the unitary dual of G, i.e. $\operatorname{Rep} G$ modulo the equivalence of representations
$L^2(\widehat{G})$ is the space of square integrable fields on $\widehat{G}$ with respect to the Plancherel measure, see (1.29)
$L^\infty(\widehat{G})$, $\mathscr{L}_L(L^2(G))$, and $\mathcal{K}(G)$ are the realisations of the von Neumann algebra of the group G as the spaces of the bounded fields of operators on $\widehat{G}$, of the left-

invariant operators in $\mathscr{L}(L^2(G))$, and of the convolution kernels corresponding to the latter, respectively, see Section 1.8.2

$L^\infty_{a,b}(\widehat{G})$, $\mathscr{L}_L(L^2_a(G), L^2_b(G))$, and $\mathcal{K}_{a,b}(G)$ are the Sobolev versions of the above, see Section 5.1.2

$\mathrm{Diff}^k(G)$ is the space of left-invariant differential operators of order k

$\mathrm{diff}^k(\widehat{G})$ is the space of difference operators on $\widehat{G}$ of order k

$C_c(G)$ is the space of continuous functions on G with compact support

$C_o(G)$ is the space of continuous functions on G vanishing at infinity (see Definition 3.1.57)

$\mathcal{D}(G) = C_c^\infty(G)$ or $\mathcal{D}(\mathbb{R}) = C_c^\infty(\mathbb{R})$ are spaces of smooth compactly supported functions

$C^\infty(G, F)$ denotes the set of smooth functions from G to a Fréchet space F

$L^p(G)$ or simply L^p for $1 \le p \le \infty$ is the usual Lebesgue space on G with norm $\|\cdot\|_{L^p} = \|\cdot\|_{L^p(G)} = \|\cdot\|_p$

$(\cdot,\cdot)_{L^2} = (\cdot,\cdot)_{L^2(G)}$ is the Hilbert sesquilinear form on $L^2(G)$ corresponding to the norm $\|\cdot\|_{L^2}$

$M(G)$ the Banach space of regular complex measures on G endowed with the total mass $\|\cdot\|_{M(G)}$

$\langle\cdot,\cdot\rangle$ denotes the distributional duality

Q denotes the homogeneous dimension of a homogeneous Lie group. After Section 6.4.2, it denotes the harmonic oscillator

sup always means the essential supremum with respect to the corresponding measure

Chapter 1

Preliminaries on Lie groups

In this chapter we provide the reader with basic preliminary facts about Lie groups that we will be using in the sequel. At the same time, it gives us a chance to fix the notation for the rest of the monograph. The topics presented here are all well-known and we decided to give a brief account without proofs referring the reader for more details to excellent sources where this material is treated from different points of view; for example, the monographs by Chevalley [Che99], Fegan [Feg91], Nomizu [Nom56], Pontryagin [Pon66], to mention only a few. Thus, this chapter can also serve as a quick and informal introduction to the subject, and we refer to monographs [RT10a] for an undergraduate level introduction to general Lie groups and their representation theory, and to Corwin and Greenleaf [CG90] or Goodman [Goo76] for a rather comprehensive treatment of nilpotent Lie groups. The groups that we are dealing with in the monograph are either compact or nilpotent Lie groups, so we can restrict our attention to unimodular Lie groups only.

The choice of material is adapted to our subsequent needs and, after giving basic definitions, we go straight to discussing convolutions, invariant differential operators, and elements of the representation theory. More information on compact or homogeneous nilpotent Lie groups will be given in relevant chapters at appropriate places. In particular Section 3.1.1 will provide examples and basic properties of graded nilpotent Lie groups. Relevant monographs to consult on invariant differential operators and related harmonic analysis may be Helgason's books [Hel84b, Hel01] or Wallach's [Wal73].

1.1 Lie groups, representations, and Fourier transform

A *Lie group* G is a smooth manifold endowed with the smooth mappings

$$G \times G \ni (x, y) \mapsto xy \in G \quad \text{and} \quad G \ni x \mapsto x^{-1} \in G$$

satisfying, for all $x, y, z \in G$, the properties

V. Fischer, M. Ruzhansky, *Quantization on Nilpotent Lie Groups*, Progress in Mathematics 314,
DOI 10.1007/978-3-319-29558-9_1

1. $x(yz) = (xy)z$;
2. $ex = xe = x$;
3. $xx^{-1} = x^{-1}x = e$,

where $e \in G$ is an element of the group called the *unit* element. To avoid unnecessary technicalities at a few places, we will always assume that G is connected, although we sometimes will emphasise it explicitly. A *compact Lie group* is a Lie group which is compact as a manifold.

Lie groups are naturally topological groups. Recall that a *topological group* G is a topological set G endowed with the continuous mappings

$$G \times G \ni (x, y) \mapsto xy \in G \quad \text{and} \quad G \ni x \mapsto x^{-1} \in G$$

satisfying, for all $x, y, z \in G$, the same properties 1., 2. and 3. as above. When the topology of a topological group is locally compact (i.e. every point has a compact neighbourhood), we say that the group is locally compact. Lie groups are (Hausdorff) locally compact.

Representations

A *representation* π of a group G on a Hilbert space $\mathcal{H}_\pi \neq \{0\}$ is a homomorphism π of G into the group of bounded linear operators on $\mathcal{H}_\pi$ with bounded inverse. This means that

- for every $x \in G$, the linear mapping $\pi(x) : \mathcal{H}_\pi \to \mathcal{H}_\pi$ is bounded and has bounded inverse,
- for any $x, y \in G$, we have $\pi(xy) = \pi(x)\pi(y)$.

A representation π of a group G is *unitary* when $\pi(x)$ is unitary for every $x \in G$. Hence a unitary representation π of a group G is a homomorphism $\pi \in \mathrm{Hom}(G, \mathcal{U}(\mathcal{H}_\pi))$, which means that

- for every $x \in G$, the linear mapping $\pi(x) : \mathcal{H}_\pi \to \mathcal{H}_\pi$ is unitary:
$$\pi(x)^{-1} = \pi(x)^*;$$
- for any $x, y \in G$, we have $\pi(xy) = \pi(x)\pi(y)$.

Here and everywhere, if $\mathcal{H}$ is a topological vector space, $\mathscr{L}(\mathcal{H})$ denotes the space of all continuous linear operators $\mathcal{H} \to \mathcal{H}$, and $\mathcal{U}(\mathcal{H})$ the space of unitary ones, with respect to the inner product on $\mathcal{H}$. For two different topological vector spaces $\mathcal{H}_1$ and $\mathcal{H}_2$, we denote by $\mathscr{L}(\mathcal{H}_1, \mathcal{H}_2)$ the space of all linear continuous mappings from $\mathcal{H}_1$ to $\mathcal{H}_2$.

An *invariant subspace* for a representation π is a vector subspace $W \subset \mathcal{H}_\pi$ such that $\pi(x)W \subset W$ holds for every $x \in G$. A representation π is called *irreducible* when it has no closed invariant subspaces.

Let us give the prototype example of a representation which is not irreducible. If $\pi_j \in \mathrm{Hom}(G,\mathcal{U}(\mathcal{H}_{\pi_j}))$ is a family of representations, then using the direct sum

$$\mathcal{H}_\pi := \bigoplus_j \mathcal{H}_{\pi_j}$$

with the induced inner product, we get a representation π which is the direct sum of π_j:

$$\pi = \bigoplus_j \pi_j \in \mathrm{Hom}(G,\mathcal{U}(\mathcal{H}_\pi)), \quad \pi(x)|_{\mathcal{H}_{\pi_j}} = \pi_j(x).$$

Naturally, a sum of several π_j's can not be irreducible as each $\mathcal{H}_{\pi_j}$ is a closed invariant subspace of $\mathcal{H}_\pi$.

If the space $\mathcal{H}_\pi$ is finite dimensional, the representation π is said to be *finite dimensional* and its *dimension/degree* is defined by

$$d_\pi := \dim \mathcal{H}_\pi.$$

The *trivial representation*, sometimes denoted by 1, is given by the group homomorphism $G \ni x \mapsto 1 \in \mathbb{C}$, and its dimension is one. If $\mathcal{H}_\pi$ is infinite dimensional, then the representation π is said to be *infinite dimensional.*

Two representations π_1 and π_2 are said to be *equivalent* if there exists a bounded linear mapping $A : \mathcal{H}_{\pi_1} \to \mathcal{H}_{\pi_2}$ between their representation spaces with a bounded inverse such that the relation

$$A\pi_1(x) = \pi_2(x)A \tag{1.1}$$

holds for all $x \in G$. In this case we write

$$\pi_1 \sim \pi_2 \quad \text{or, more precisely sometimes,} \quad \pi_1 \sim_A \pi_2$$

and denote their equivalence class by $[\pi_1] = [\pi_2]$. For unitary representations, A is assumed to be unitary as well. A bounded linear mapping with bounded inverse satisfying the relation (1.1) is sometimes called an intertwining operator or intertwiner. The set of bounded linear mappings A with bounded inverse satisfying the relation (1.1) is denoted by $\mathrm{Hom}(\pi_1, \pi_2)$.

Note that for any representation π, $\mathrm{Hom}(\pi, \pi)$ contains at least $\lambda \mathrm{I}_{\mathcal{H}_\pi}$, $\lambda \in \mathbb{C}$, where $\mathrm{I}_{\mathcal{H}_\pi}$ is the identity mapping on $\mathcal{H}_\pi$.

We now assume that the group G is topological. A representation π of G is *continuous* if the mapping

$$\begin{cases} G \times \mathcal{H}_\pi & \longrightarrow & \mathcal{H}_\pi \\ (x, v) & \longmapsto & \pi(x)v \end{cases}$$

is continuous. A representation π of G is called *strongly continuous* if the mapping $\pi : G \to \mathscr{L}(\mathcal{H}_\pi)$ is continuous for the strong operator topology in $\mathscr{L}(\mathcal{H}_\pi)$, that is, if the mapping

$$\begin{cases} G & \longrightarrow & \mathcal{H}_\pi \\ x & \longmapsto & \pi(x)v \end{cases}$$

is continuous for all $v \in \mathcal{H}_\pi$.

A continuous representation is strongly continuous. The converse is true for unitary representations. Indeed, if π is a unitary representation of G, then we have for any $x, x_0 \in G$ and $v, v_0 \in \mathcal{H}_\pi$,

$$\begin{aligned} \|\pi(x)v - \pi(x_0)v_0\|_{\mathcal{H}_\pi} &= \|\pi(x_0)(\pi(x_0^{-1}x)v - v_0)\|_{\mathcal{H}_\pi} = \|\pi(x_0^{-1}x)v - v_0\|_{\mathcal{H}_\pi} \\ &= \|\pi(x_0^{-1}x)(v - v_0) + (\pi(x_0^{-1}x)v_0 - v_0)\|_{\mathcal{H}_\pi} \\ &\leq \|\pi(x_0^{-1}x)(v - v_0)\|_{\mathcal{H}_\pi} + \|\pi(x_0^{-1}x)v_0 - v_0\|_{\mathcal{H}_\pi} \\ &= \|v - v_0\|_{\mathcal{H}_\pi} + \|\pi(x_0^{-1}x)v_0 - v_0\|_{\mathcal{H}_\pi}, \end{aligned}$$

having used only the unitarity of π and the triangle inequality. This shows that if a representation of G is unitary and strongly continuous then it is continuous.

Schur's lemma: Let π be a strongly continuous unitary representation of a topological group G on a Hilbert space $\mathcal{H}_\pi$. The representation π is irreducible if and only if the only bounded linear operators on $\mathcal{H}_\pi$ commuting with all $\pi(x)$, $x \in G$, are the scalar operators. Equivalently,

$$\pi \text{ irreducible} \iff \mathrm{Hom}(\pi, \pi) = \{\lambda \mathrm{I}_{\mathcal{H}_\pi} : \lambda \in \mathbb{C}\}.$$

The set of all equivalence classes of strongly continuous irreducible unitary representations of G is called the *unitary dual of* G or just dual of G and is denoted by $\widehat{G}$.

Later, we will give more details on representations of compact or nilpotent Lie groups and their dual.

The unitary dual of G is never a group unless G is commutative. However, if G is a commutative locally compact group, then $\widehat{G}$ has a natural structure of a commutative locally compact group and we have

Pontryagin duality: if G is a commutative locally compact group, then $\widehat{\widehat{G}} \simeq G$.

For most of the statements in the sequel, if they hold for one representation, they will also hold for all equivalent representations. That is why we may simplify the notation a little writing $\pi \in \widehat{G}$ instead of $[\pi] \in \widehat{G}$ for its equivalence class. In this case we can think of π as either any representative from its class or the equivalence class itself. If we need to work with a particular representation from an equivalence class (for example the one diagonalising certain operators in a particular choice of the basis in $\mathcal{H}_\pi$) we will specify this explicitly.

Haar measure

A fundamental fact, valid on general locally compact groups, is the existence of an invariant measure, called *Haar measure*:

Theorem 1.1.1. *Let G be a locally compact group. Then there exists a non-zero left-invariant measure on G, and it is unique up to a positive constant. More precisely, there exists a positive Radon measure on G satisfying*

$$|xA| = |A| \text{ for every Borel set } A \subset G \text{ and every } x \in G,$$

where $|A|$ denotes the measure of the set A with respect to this Radon measure. In the sequel, we denote this measure by dx, dy, etc., depending on the variable of integration. Then, for every $x \in G$ and every continuous compactly supported function f on G, we have

$$\int_G f(xy)dy = \int_G f(y)dy.$$

We fix one of such measures. In this monograph, we will be only dealing with either compact or nilpotent Lie groups, in which case it can be shown that the Haar measure is also right-invariant:

$$|Ax| = |A| \text{ for every Borel set } A \subset G \text{ and every } x \in G,$$

and also

$$\int_G f(yx)dy = \int_G f(y)dy;$$

such groups are called *unimodular*. Since the mapping $f \mapsto \int_G f(y^{-1})dy$ is positive, left-invariant, and normalised, by uniqueness we must also have

$$\int_G f(y^{-1})dy = \int_G f(y)dy.$$

For a more general definition of a modular function we can refer to Definition B.2.10. Here we can summarise a few properties of (unimodular) groups:

- Any Lie group is a locally compact (Hausdorff) group.
- Any compact (Hausdorff) group is a locally compact (Hausdorff) group and it is also unimodular.
- Any abelian locally compact (Hausdorff) group is unimodular.
- Any nilpotent or semi-simple Lie group is unimodular.

If $1 \leq p \leq \infty$, $L^p(G)$ or simply L^p denote the usual *Lebesgue space* on G with respect to the Haar measure, with the norm

$$\|\cdot\|_{L^p} = \|\cdot\|_{L^p(G)} = \|\cdot\|_p,$$

given for $p \in [1, \infty)$ by

$$\|f\|_p = \left(\int_G |f(x)|^p dx \right)^{1/p},$$

and for $p = \infty$ by

$$\|f\|_\infty = \sup_{x \in G} |f(x)|.$$

Here the supremum refers to the essential supremum with respect to the Haar measure.

The Hilbert sesquilinear form on $L^2(G)$ is denoted by

$$(f_1, f_2)_{L^2} = (f_1, f_2)_{L^2(G)} = \int_G f_1(x)\overline{f_2(x)}dx.$$

Example 1.1.2. Let us give an important example of so-called left and right regular representations leading to the notions of left- and right-invariant operators. We define the *left* and *right regular representations* of G on $L^2(G)$, $\pi_L, \pi_R : G \to \mathcal{U}(L^2(G))$, respectively, by

$$\pi_L(x)f(y) := f(x^{-1}y) \quad \text{and} \quad \pi_R(x)f(y) := f(yx).$$

Definition 1.1.3. An operator A is called *left* (*right*, resp.) *invariant* if it commutes with the left (right, resp.) regular representation of G.

Fourier analysis

For $f \in L^1(G)$ we define its *Fourier coefficient* or *group Fourier transform* at the strongly continuous unitary representation π as

$$\mathcal{F}_G f(\pi) \equiv \widehat{f}(\pi) \equiv \pi(f) := \int_G f(x)\pi(x)^* dx. \tag{1.2}$$

More precisely, we can write

$$(\widehat{f}(\pi)v_1, v_2)_{\mathcal{H}_\pi} = \int_G f(x)(\pi(x)^* v_1, v_2)_{\mathcal{H}_\pi} dx.$$

This gives a linear mapping $\widehat{f}(\pi) : \mathcal{H}_\pi \to \mathcal{H}_\pi$. If the representation π is finite dimensional, then after a choice of a basis in the representation space $\mathcal{H}_\pi$, the Fourier coefficient $\widehat{f}(\pi)$ can be also viewed as a matrix $\widehat{f}(\pi) \in \mathbb{C}^{d_\pi \times d_\pi}$.

Remark 1.1.4. The choice of taking the adjoint $\pi(x)^*$ in (1.2) is natural if we think of the unitary dual of the torus $\mathbb{T}^n = \mathbb{R}^n/\mathbb{Z}^n$ being $\widehat{\mathbb{T}^n} = \{\pi_\xi(x) = e^{2\pi i x \cdot \xi}\}_{\xi \in \mathbb{Z}^n} \simeq \mathbb{Z}^n$, and the Fourier transform on the torus defined by

$$\widehat{f}(\pi_\xi) \equiv \widehat{f}(\xi) = \int_{\mathbb{T}^n} e^{-2\pi i x \cdot \xi} f(x) dx = \int_{\mathbb{T}^n} f(x)\pi_\xi(x)^* dx.$$

In other contexts, the other choice, that is, integrating against $\pi(x)$ instead of $\pi(x)^*$, may be made. This is the case for instance in the study of C^*-algebras associated with groups.

Remark 1.1.5. We note that the Fourier coefficient $\widehat{f}(\pi)$ depends on the choice of the representation π from its equivalence class $[\pi]$. Namely, if $\pi_1 \sim \pi_2$, so that

$$\pi_2(x) = U^{-1}\pi_1(x)U$$

for some unitary U and all $x \in G$, then

$$\widehat{f}(\pi_2) = U^{-1}\widehat{f}(\pi_1)U.$$

This means that strictly speaking, we need to look at Fourier coefficients modulo conjugations induced by the equivalence of representations. This should, however, cause no problems, and we refer to Remark 2.2.1 for more discussion on this.

Recalling that the Fourier transform on $\mathbb{R}^n$ maps translations to modulations, here we have an analogous property, namely, if $\pi \in \widehat{G}$, $f \in L^1(G)$ and $x \in G$, then

$$\widehat{f(\cdot\, x)}(\pi) = \pi(x)\widehat{f}(\pi) \text{ and } \widehat{f(x\,\cdot)}(\pi) = \widehat{f}(\pi)\pi(x), \tag{1.3}$$

whenever the right hand side makes sense. Let us show these properties by a formal argument, which can be made rigorous on Lie groups, see the proof of Proposition 1.7.6, (iv). We have

$$\begin{aligned} \pi(x)\widehat{f}(\pi) &= \int_G f(y)\pi(x)\pi(y)^* dy \\ &= \int_G f(y)\pi(yx^{-1})^* dy \\ &= \int_G f(yx)\pi(y)^* dy \\ &= \widehat{f(\cdot\, x)}(\pi), \end{aligned} \tag{1.4}$$

as well as

$$\begin{aligned} \widehat{f(x\,\cdot)}(\pi) &= \int_G f(xy)\pi(y)^* dy \\ &= \int_G f(y)\pi(x^{-1}y)^* dy \\ &= \int_G f(y)\pi(y)^*\pi(x) dy \\ &= \widehat{f}(\pi)\pi(x). \end{aligned} \tag{1.5}$$

We will continue with a more detailed discussion of the Fourier transform on compact Lie groups in Section 2.1, on nilpotent Lie groups in Section 1.8.1 and, more generally, on a separable locally compact connected, unimodular, amenable group G of type I in Section 1.8.2.

1.2 Lie algebras and vector fields

A (real) *Lie algebra* is a real vector space V endowed with a bilinear mapping

$$V \times V \ni (a, b) \mapsto [a, b] \in V,$$

called *the commutator of a and b*, such that

- $[a, a] = 0$ for every $a \in V$;
- *Jacobi identity:* $[a, [b, c]] + [b, [c, a]] + [c, [a, b]] = 0$ for all $a, b, c \in V$.

By writing $[a+b, a+b] = [a, a] + [a, b] + [b, a] + [b, b]$ we see that the first property is equivalent to the condition that

$$\forall a, b \in V \quad [a, b] = -[b, a].$$

We now proceed to equip the tangent space of G (at every point) with a Lie algebra structure. A map $X_{(x)} : C^\infty(G) \to \mathbb{R}$ is called a *tangent vector* to G at $x \in G$ if

- $X_{(x)}(f + g) = X_{(x)}f + X_{(x)}g$;
- $X_{(x)}(fg) = X_{(x)}(f)g(x) + f(x)X_{(x)}(g)$.

The notation $X_{(x)}$ is used only in this section and the reason for its choice is that we want to reserve the notation X_x for derivatives, to be used later.

The space of all tangent vectors at x is a finite dimensional vector space of dimension equal to the dimension of G as a manifold; the finite dimensionality can be seen by passing to local coordinates. This vector space is denoted by T_xG. The disjoint union,

$$TG := \bigcup_{x \in G} T_xG$$

is a vector bundle over X, called *the tangent bundle.* The canonical projection $\mathrm{proj} : TG \to G$ is given by $\mathrm{proj}\, X_{(x)} := x$. If U_x is a (sufficiently small) open neighbourhood of x in G, we can trivialise the vector bundle TG by $\mathrm{proj}^{-1}(U_x) \simeq U_x \times E$ with a vector space E of dimension equal to that of G. This induces the manifold structure on TG.

A *(smooth) vector field* on G is a (smooth) section of TG, i.e. a (smooth) mapping $X : G \to TG$ such that $X(x) \equiv X_{(x)} \in T_xG$. It acts on $C^\infty(G)$ by

$$(Xf)(x) := (X_{(x)}f)(x), \quad f \in C^\infty(G).$$

There is a bracket structure on the space of vector fields acting on $C^\infty(G)$ given by

$$[X, Y](x)(f) := X(x)Yf - Y(x)Xf, \quad x \in G,$$

leading to the corresponding (smooth) vector field $[X, Y] : G \to TG$ given by $x \mapsto [X, Y](x)$. One can readily check that $[X, X] = 0$ for every vector field X and that the introduced bracket satisfies the Jacobi identity. This bracket $[\cdot,\cdot]$ is called the *commutator bracket* for vector fields.

We now recall that G is also a group, and relate vector fields to the group structure. First, we define the left and right translations by an element $y \in G$:

$$L_y, R_y : G \to G, \quad L_y(x) := yx, \; R_y(x) := xy.$$

Consequently, their derivatives are the mappings

$$dL_y, dR_y : TG \to TG \text{ such that } dL_y \in \mathscr{L}(T_xG, T_{yx}G), \; dR_y \in \mathscr{L}(T_xG, T_{xy}G).$$

Now, a vector field $X : G \to TG$ is called *left-invariant* if it commutes with the left translations, in the sense that

$$X \circ L_y = dL_y \circ X \quad \forall y \in G. \tag{1.6}$$

A similar construction leads to the notion of *right-invariant* vector fields, satisfying

$$X \circ R_y = dR_y \circ X$$

for all $y \in G$.

It follows that once a left-invariant vector field is defined at any one point, by the left-invariance it is uniquely determined at all points. Thus, the mapping $X \mapsto X_{(e)}$ is a one-to-one correspondence between left-invariant vector fields on G and the tangent space T_eG at the unit element $e \in G$. Conversely, given $X_{(e)} \in T_eG$, the vector field X defined by (1.6) is automatically smooth and, by definition, left-invariant. With this identification, we can now simplify the notation for left-invariant vector fields X, writing X also for its value $X_{(e)}$ at the unit element. It can be readily checked that if X and Y are left-invariant vector fields, so is also their commutator $[X, Y]$.

Definition 1.2.1. The *Lie algebra* $\mathfrak{g}$ *of the Lie group* G is the space T_eG equipped with the commutator $[\cdot,\cdot]$ induced by the commutator bracket of vector fields.

We now define the *exponential mapping* $\exp_G$. For $X \in \mathfrak{g}$, consider the initial value problem for a function $\gamma : [0, \epsilon) \to G$, $\epsilon > 0$, given by the ordinary differential equation determined by the left-invariant vector field associated with X:

$$\gamma'(t) = X_{(\gamma(t))}, \quad \gamma(0) = e.$$

From the theory of ordinary differential equations we know that this equation is uniquely solvable on some interval $[0, \epsilon)$ and the solution depends smoothly on $X_{(e)}$. Moreover, we notice that we can increase the interval of existence by taking smaller vectors $X_{(e)}$, in particular, in such a way that the solution exists on the interval $[0, 1]$. In this case we set $\exp_G X := \gamma(1)$. Altogether, it follows that the mapping $\exp_G$ is a smooth diffeomorphism from some open neighbourhood of $0 \in \mathfrak{g}$ to some open neighbourhood of $e \in G$.

Now, each vector $X \in \mathfrak{g}$ can be viewed as a left-invariant differential operator on $C^\infty(G)$ defined by

$$Xf(x) := \frac{d}{dt} f(x \ \exp_G(tX))|_{t=0}. \tag{1.7}$$

Indeed, it can be readily checked that $X\pi_L(y) = \pi_L(y)X$ for all $y \in G$. Analogously, the same vector $X \in \mathfrak{g}$ defines a right-invariant differential operator, which we denote by

$$\tilde{X}f(x) := \frac{d}{dt} f(\exp_G(tX)\, x)|_{t=0}.$$

Thus, throughout this book, we will be interpreting the Lie algebra $\mathfrak{g} = T_eG$ of G as the vector space of first order left-invariant partial differential operators on G. The space of all left-invariant vector fields will be sometimes denoted by $\mathbb{D}(G)$ or by $\mathrm{Diff}^1(G)$, and the space of all right-invariant vector fields by $\tilde{\mathbb{D}}(G)$.

1.3 Universal enveloping algebra and differential operators

Roughly speaking, the universal enveloping algebra of a Lie algebra $\mathfrak{g}$ is the natural non-commutative polynomial algebra on $\mathfrak{g}$. If $\mathfrak{g}$ is the Lie algebra of a Lie group G, then, similarly to the interpretation of $\mathfrak{g}$ as the space of left-invariant derivatives on G, the universal enveloping algebra $\mathfrak{U}(\mathfrak{g})$ of the Lie algebra of G will be also interpreted as the vector space of left-invariant partial differential operators on G of finite order. The associative algebra will be generated as a complex algebra over $\mathfrak{g}$, so that we could write $\mathfrak{U}(\mathfrak{g}^{\mathbb{C}})$ for it, where $\mathfrak{g}^{\mathbb{C}}$ denotes the complexification of $\mathfrak{g}$. However, we will simplify the notation writing $\mathfrak{U}(\mathfrak{g})$, and will later use the Poincaré-Birkhoff-Witt theorem to identify it with the left-invariant differential operators on G with complex coefficients. Let us now formalise these statements.

The following construction is algebraic and works for any real Lie algebra $\mathfrak{g}$. Let us denote the m-fold tensor product of $\mathfrak{g}^{\mathbb{C}}$ by $\otimes^m \mathfrak{g}^{\mathbb{C}} := \mathfrak{g}^{\mathbb{C}} \otimes \cdots \otimes \mathfrak{g}^{\mathbb{C}}$, and let

$$\mathcal{T} := \bigoplus_{m=0}^{\infty} \otimes^m \mathfrak{g}^{\mathbb{C}}$$

be the tensor product algebra of $\mathfrak{g}$, which means that $\mathcal{T}$ is the linear span of the elements of the form

$$\lambda_{00}\mathbf{1} + \sum_{m=1}^{M} \sum_{k=1}^{K_m} \lambda_{mk} X_{mk1} \otimes \cdots \otimes X_{mkm},$$

where $\mathbf{1}$ is the formal unit element of $\mathcal{T}$, $\lambda_{mk} \in \mathbb{C}$, $X_{mkj} \in \mathfrak{g}$, and $M, K_M \in \mathbb{N}$. This $\mathcal{T}$ becomes an associative algebra with the product

$$(X_1 \otimes \cdots \otimes X_p)(Y_1 \otimes \cdots \otimes Y_q) := X_1 \otimes \cdots \otimes X_p \otimes Y_1 \otimes \cdots \otimes Y_q$$

extended to a uniquely determined bilinear mapping $\mathcal{T} \times \mathcal{T} \to \mathcal{T}$. We now want to induce the commutator structure on $\mathcal{T}$: let $\mathcal{I}$ be the two-sided ideal in $\mathcal{T}$ spanned by the set

$$\mathcal{O} := \{X \otimes Y - Y \otimes X - [X,Y] : X, Y \in \mathfrak{g}\},$$

i.e. $\mathcal{I}$ is the smallest vector subspace of $\mathcal{T}$ such that

- $\mathcal{O} \subset \mathcal{I}$;
- for every $J \in \mathcal{I}$ and $T \in \mathcal{T}$ we have $JT, TJ \in \mathcal{I}$.

The quotient algebra

$$\mathfrak{U}(\mathfrak{g}) := \mathcal{T}/\mathcal{I}$$

is called the *universal enveloping algebra* of $\mathfrak{g}$; the quotient mapping

$$\iota : \mathcal{T} \ni T \mapsto T + \mathcal{I} \in \mathfrak{U}(\mathfrak{g}) = \mathcal{T}/\mathcal{I},$$

restricted to $\mathfrak{g}$, $\iota|_{\mathfrak{g}} : \mathfrak{g} \to \mathfrak{U}(\mathfrak{g})$, is called the canonical mapping of $\mathfrak{g}$. This gives the embedding of $\mathfrak{g}$ into $\mathfrak{U}(\mathfrak{g})$:

Ado-Iwasawa theorem: the canonical mapping $\iota|_{\mathfrak{g}} : \mathfrak{g} \to \mathfrak{U}(\mathfrak{g})$ is injective.

Let $n = \dim G$ and let $\{X_j\}_{j=1}^n$ be a basis of the Lie algebra $\mathfrak{g}$ of G. Regarded as first order left-invariant derivatives, they give rise to higher order left-invariant differential operators

$$X^\alpha = X_1^{\alpha_1} \dots X_n^{\alpha_n}, \quad \alpha = (\alpha_1, \dots, \alpha_n) \in \mathbb{N}_0^n.$$

The converse is also true (for a stronger version of this see e.g. [Bou98, Ch 1, Sec. 2.7]):

Poincaré-Birkhoff-Witt theorem: any left-invariant differential operator T on G can be written in a unique way as a finite sum

$$T = \sum_{\alpha \in \mathbb{N}_0^n} c_\alpha X^\alpha,$$

where all but a finite number of the coefficients $c_\alpha \in \mathbb{C}$ are zero. This gives an identification between the universal enveloping algebra $\mathfrak{U}(\mathfrak{g})$ and the space of left-invariant differential operators on G.

We denote the space of all left-invariant differential operators of order k by $\mathrm{Diff}^k(G)$.

If T is as above, we define three new elements $\bar{T}$, T^*, and T^t of $\mathfrak{U}(\mathfrak{g})$ via

$$\bar{T} := \sum_{\alpha \in \mathbb{N}_0^n} \bar{c}_\alpha (X_n)^{\alpha_n} \dots (X_1)^{\alpha_1}, \tag{1.8}$$

$$T^* := \sum_{\alpha \in \mathbb{N}_0^n} \bar{c}_\alpha (-X_n)^{\alpha_n} \dots (-X_1)^{\alpha_1}, \tag{1.9}$$

and

$$T^t := \sum_{\alpha \in \mathbb{N}_0^n} c_\alpha (-X_n)^{\alpha_n} \dots (-X_1)^{\alpha_1}. \tag{1.10}$$

These T^* and T^t are called the (formal) *adjoint* and *transpose* operators of T, respectively. Naturally, they coincide with the natural transpose and formal adjoint operators of their corresponding left-invariant vector fields. Recall that the latter operators are defined via:

Definition 1.3.1. Let T be an operator T on $L^2(G)$ with domain $\mathcal{D}(G)$ (T may be unbounded, $\mathcal{D}(G) \subset \operatorname{Dom} T$). The natural transpose and formal adjoint operators of T are the operators T^t and T^* on $L^2(G)$ defined via

$$\langle T\phi, \psi \rangle = \langle \phi, T^t \psi \rangle \quad \text{and} \quad (T\phi, \psi)_{L^2(G)} = (\phi, T^*\psi)_{L^2(G)}, \quad \phi, \psi \in \mathcal{D}(G).$$

We also define the operator $\bar{T}$ on $L^2(G)$ via

$$\bar{T}\phi := \overline{T\bar{\phi}},$$

for $\phi, \bar{\phi} \in \operatorname{Dom} T$.

Note that we also have, e.g.,

$$T^* = \overline{\{T^t\}} = \{\bar{T}\}^t$$

and so on. Denoting

$$\tilde{f}(x) := f(x^{-1}),$$

the left- and right- invariant differential operators are related by

$$\tilde{X} f(x) = -(X\tilde{f})(x^{-1}) \quad \text{and hence} \quad \tilde{X}^\alpha f(x) = (-1)^{|\alpha|}(X^\alpha \tilde{f})(x^{-1}). \tag{1.11}$$

Indeed, we can write

$$X\tilde{f}(x) = \frac{d}{dt} f((x \exp_G(tX))^{-1})|_{t=0} = \frac{d}{dt} f(\exp_G(-tX)x^{-1})|_{t=0} = -(\tilde{X}f)(x^{-1}),$$

implying (1.11).

For any $X \in \mathfrak{g}$ identified with a left-invariant vector field, we have

$$\tilde{X}_y\{f(xy)\} = \frac{d}{dt} f(xe^{tX}y)_{t=0} = X_x\{f(xy)\}.$$

Recursively, we obtain

$$\tilde{X}_y^\alpha\{f(xy)\} = X_x^\alpha\{f(xy)\}. \tag{1.12}$$

The first order differential operators are formally skew-symmetric:

$$\int_G (Xf_1)f_2 = -\int_G f_1(Xf_2) \quad \text{and} \quad \int_G (\tilde{X}f_1)f_2 = -\int_G f_1(\tilde{X}f_2),$$

so that from (1.11) we also have

$$\tilde{X}f(x) = -(X\tilde{f})(x^{-1}) = (X^t\tilde{f})(x^{-1}).$$

We now summarise several further notions and their properties that will be of use to us in the sequel:

- there is a natural representation of the Lie group G acting on its Lie algebra $\mathfrak{g}$, called the *adjoint representation*. To introduce it, first define the inner automorphism $I_x(y) := xyx^{-1}$. We have $I_x : G \to G$ and $I_{xy} = I_x I_y$. Its differential at e gives a linear mapping from T_eG to T_eG, and we denote it by
$$\mathrm{Ad}(x) := (dI_x)_e : \mathfrak{g} \to \mathfrak{g}.$$
We have $\mathrm{Ad}(e) = \mathrm{I}$ and $\mathrm{Ad}(xy) = \mathrm{Ad}(x)\mathrm{Ad}(y)$, so that $\mathrm{Ad} : G \to \mathscr{L}(\mathfrak{g})$ becomes a representation of G on $\mathfrak{g}$;
- the left and right multiplications on G are related by
$$x \exp_G X = \exp(\mathrm{Ad}(x)X)x, \quad x \in G,\ X \in \mathfrak{g};$$
- a Lie group G is called a *linear Lie group* if it is a closed subgroup of $\mathrm{GL}(n, \mathbb{C})$; the adjoint representation of such G is given by
$$\mathrm{Ad}(X)Y = XYX^{-1}$$
as multiplication of matrices;
- *universality of unitary groups*: any compact Lie group is isomorphic to a subgroup of $U(N)$, the group of $(N \times N)$-unitary matrices, for some $N \in \mathbb{N}$;
- let $\mathrm{ad} : \mathfrak{g} \to \mathcal{L}(\mathfrak{g})$ be the linear mapping defined by
$$\mathrm{ad}(X)Y := [X, Y];$$
then $d(\mathrm{Ad})_e = \mathrm{ad}$; see also Definition 1.7.4;
- the *Killing form* of the Lie algebra $\mathfrak{g}$ is the bilinear mapping $B : \mathfrak{g} \times \mathfrak{g} \to \mathbb{R}$ defined by
$$B(X, Y) := \mathrm{Tr}(\mathrm{ad}(X)\ \mathrm{ad}(Y));$$
it satisfies
$$B(X, Y) = B(Y, X) \text{ and } B(X, [Y, Z]) = B([X, Y], Z)$$
and is invariant under the adjoint representation of G, namely,
$$B(X, Y) = B(\mathrm{Ad}(x)(X), \mathrm{Ad}(x)(Y)) \text{ for all } x \in G,\ X, Y \in \mathfrak{g};$$

- A connected Lie group G is called *semi-simple* if B is non-degenerate; a connected semi-simple group G is compact if and only if B is negative definite.

The Ad-invariance of the Killing form has its consequences. On one hand, any bilinear form on $\mathfrak{g}$ can be extended to a bilinear (non-necessarily positive definite) metric on G by left translations. It is automatically left-invariant. On the other hand, if the form on $\mathfrak{g}$ is Ad-invariant, then the extended metric is also right-invariant. Thus, we can conclude that the Killing form induces a bi-invariant metric on G. By the last property above, if G is semi-simple, the Killing form is non-degenerate, and hence the corresponding metric is pseudo-Riemannian. Moreover, if G is a connected semi-simple compact Lie group, the positive-definite form $-B$ induces the bi-invariant Riemannian metric on G.

For the basis $\{X_j\}_{j=1}^n$ as above, let us define $R_{ij} := B(X_i, X_j)$. If the group G is semi-simple, the matrix (R_{ij}) is invertible, and we denote its inverse by R^{-1}. This leads to another vector space basis on $\mathfrak{g}$ given by

$$X^i := \sum_{j=1}^n (R^{-1})_{ij} X_j,$$

and to the so-called *Casimir element* of $\mathfrak{U}(\mathfrak{g})$ defined by

$$\Omega := \sum_{i=1}^n X_i X^i.$$

It has the crucial property: Ω is independent of the choice of the basis $\{X_j\}$, and

$$\Omega T = T\Omega \quad \text{for all} \quad T \in \mathfrak{U}(\mathfrak{g}).$$

We finish this section with the formula for the group product which will be useful for us, especially in the nilpotent case:

Theorem 1.3.2 (Baker-Campbell-Hausdorff formula). *Let G be a Lie group with Lie algebra $\mathfrak{g}$. There exists a neighbourhood V of 0 in $\mathfrak{g}$ such that for any $X, Y \in V$, we have*

$$\begin{aligned} \exp_G X \exp_G Y &= \exp_G \Big(\sum_{n>0} \frac{(-1)^{n+1}}{n} \sum_{\substack{p,q\in\mathbb{N}_0^n \\ p_i+q_i>0}} \frac{(\sum_{j=1}^n (p_j+q_j))^{-1}}{p_1!q_1!\dots p_n!q_n!} \\ &\qquad \times (\mathrm{ad}X)^{p_1}(\mathrm{ad}Y)^{q_1}\dots(\mathrm{ad}X)^{p_n}(\mathrm{ad}Y)^{q_n-1}Y\Big). \end{aligned}$$

The equality holds whenever the sum on the right-hand side is convergent.

Writing first few terms explicitly, we have

$$\begin{aligned} &\exp_G X \exp_G Y \\ &= \exp_G \left(X + Y + \frac{1}{2}[X,Y] + \frac{1}{12}[[X,Y],Y] - \frac{1}{12}[[X,Y],X] + \dots \right). \end{aligned}$$

1.4 Distributions and Schwartz kernel theorem

Here we fix the notation concerning distributions. For an extensive analysis of spaces of distributions and their properties on manifolds we refer to [Hör03].

The space of smooth functions compactly supported in a smooth manifold M will be denoted by $\mathcal{D}(M)$. Throughout the book, any smooth manifold is assumed to be paracompact (i.e. every open cover has an open refinement that is locally finite) and this allows us to consider the space of distributions $\mathcal{D}'(M)$ as the dual of $\mathcal{D}(M)$. Note that any Lie group is paracompact.

If $u \in \mathcal{D}'(M)$ and $\phi \in \mathcal{D}(M)$, we shall denote the evaluation of u on ϕ by $\langle u, \phi \rangle$, or even by $\langle u, \phi \rangle_M$ when we wish to be precise; however, we shall usually pretend that the distributions are functions and write

$$\langle u, \phi \rangle = \int_M u(x)\phi(x)dx, \quad u \in \mathcal{D}'(M),\ \phi \in \mathcal{D}(M).$$

The Schwartz space $\mathcal{S}(\mathbb{R}^n)$ of rapidly decreasing functions will be equipped with a family of seminorms defined by

$$\|f\|_{\mathcal{S}(\mathbb{R}^n),N} := \sup_{|\alpha|\leq N,\, x\in\mathbb{R}^n} (1+|x|)^N \left| \left(\frac{\partial}{\partial x}\right)^\alpha f(x) \right|. \tag{1.13}$$

Its dual, the space of tempered distributions, is denoted by $\mathcal{S}'(\mathbb{R}^n)$.

Theorem 1.4.1 (Schwartz kernel theorem). *We have the following statements:*

- *Let $T : \mathcal{S}(\mathbb{R}^n) \to \mathcal{S}'(\mathbb{R}^n)$ be a continuous linear operator. Then there exists a unique distribution $\kappa \in \mathcal{S}'(\mathbb{R}^n \times \mathbb{R}^n)$ such that*

  $$T\phi(x) = \int_{\mathbb{R}^n} \kappa(x,y)\phi(y)dy.$$

  In other words, T is an integral operator with kernel κ. The converse is also true.

- *Let M be a smooth connected manifold and let $T : \mathcal{D}(M) \to \mathcal{D}'(M)$ be a continuous linear operator. There exists a unique distribution $\kappa \in \mathcal{D}'(M\times M)$ such that*

  $$T\phi(x) = \int_M \kappa(x,y)\phi(y)dy.$$

  In other words, T is an integral operator with kernel κ. The converse also is true.

In both cases, the map $\kappa \mapsto T$ is an isomorphism of topological vector space.

We refer to e.g. [Tre67] for further details. We will also give a version of this theorem on Lie groups for left-invariant operators in Corollary 3.2.1.

Let Ω be an open set in $\mathbb{R}^n$ or in M. We say that $u \in \mathcal{D}'(\Omega)$ is supported in the set $K \subset \Omega$ if $\langle u, \phi \rangle = 0$ for all $\phi \in \mathcal{D}(\Omega)$ such that $\phi = 0$ on K. The smallest closed set in which u is supported is called the *support* of u and is denoted by $\operatorname{supp} u$. The space of compactly supported distributions on M is denoted by $\mathcal{E}'(M)$, and the duality between $\mathcal{E}'(M)$ and $C^\infty(M)$ will still be denoted by $\langle \cdot, \cdot \rangle$.

We write $u \in \mathcal{D}'_j(\Omega)$ for the space of distributions of order j on Ω, which means that for any compact subset K of Ω,

$$\exists C > 0 \qquad \forall \phi \in \mathcal{D}(K) \qquad |\langle u, \phi \rangle| \le C \|\phi\|_{C^j(K)},$$

but j does not depend on K. An important property of such distributions, useful for us, is the following

Proposition 1.4.2. *If a distribution $u \in \mathcal{D}'_j(\mathbb{R}^n)$ has support* $\operatorname{supp} u = \{0\}$, *then there exist constants $a_\alpha \in \mathbb{C}$ such that*

$$u = \sum_{|\alpha| \le j} a_\alpha \partial^\alpha \delta_0,$$

where $\delta_0(\phi) = \phi(0)$ is the delta-distribution at zero.

1.5 Convolutions

Let $f, g \in L^1(G)$ be integrable function on a locally compact group. The *convolution* $f * g$ is defined by

$$(f * g)(x) := \int_G f(y) g(y^{-1}x) dy.$$

In this monograph we consider only unimodular groups. This means that the Haar measure is both left- and right-invariant. Consequently we also have

$$(f * g)(x) = \int_G f(xy^{-1}) g(y) dy.$$

On a nilpotent or compact Lie group which is not abelian, the convolution is not commutative: in general, $f * g \neq g * f$. However, apart from the lack of commutativity, group convolution and the usual convolution on $\mathbb{R}^n$ share many properties. For example, we have

$$\begin{aligned} \langle f * g, h \rangle &= \int_G (f * g)(x)\, h(x)\, dx \\ &= \int_G \int_G f(y)\, g(y^{-1}x)\, h(x)\, dy\, dx \\ &= \langle f, h * \tilde{g} \rangle, \quad \text{with} \quad \tilde{g}(x) = g(x^{-1}). \end{aligned} \tag{1.14}$$

We also have

$$\begin{aligned}\langle f * g, h\rangle &= \int_G \int_G f(y)\, g(y^{-1}x)\, h(x)\, dy\, dx \\ &= \int_G \int_G f(y)\, g(z)\, h(yz)\, dy\, dz \\ &= \int_G \int_G f(wz^{-1})\, g(z)\, h(w)\, dz\, dw \\ &= \langle g, \tilde{f} * h\rangle. \qquad (1.15)\end{aligned}$$

With the notation $\tilde{\cdot}$ for the operation given by $\tilde{g}(x) = g(x^{-1})$, we also have

$$(f * g)\tilde{} = \tilde{g} * \tilde{f}. \qquad (1.16)$$

One can readily check the following simple properties:

- if $f, g \in L^1(G)$ then $f * g \in L^1(G)$, and we have $\|f * g\|_{L^1} \leq \|f\|_{L^1}\|g\|_{L^1}$;
- under the assumptions above, we have
$$(f * g)(x) = \int_G f(y^{-1})g(yx)dy = \int_G f(xy)g(y^{-1})dy$$
for almost every $x \in G$;
- if either f or g are continuous on G then $f * g$ is continuous on G;
- $\|f * g\|_{L^\infty} \leq \|f\|_{L^2}\|g\|_{L^2}$;
- the convolution is associative: $f * (g * h) = (f * g) * h$, for $f, g, h \in L^1(G)$;
- the convolution is commutative if and only if G is commutative;
- (if G is a Lie group and) if X is a left-invariant vector field, whenever it makes sense, we have
$$X(f * g) = f * (Xg) \quad \text{and} \quad \tilde{X}(f * g) = (\tilde{X} f) * g;$$
moreover, we also have
$$(Xf) * g = f * (\tilde{X} g);$$
- the right convolution operator $f \mapsto f * \kappa$ is left-invariant; the left convolution operator $f \mapsto \kappa * f$ is right-invariant.

To check the last statement, let us show that the right convolution operator given via $Af = f * \kappa$ is left-invariant:

$$\begin{aligned}\pi_L(z)Af(x) = (f * \kappa)(z^{-1}x) &= \int_G f(y)\, \kappa(y^{-1}z^{-1}x)dy \\ &= \int_G f(z^{-1}y)\, \kappa(y^{-1}x)dy = (\pi_L(z)f) * \kappa(x) = A\pi_L(z)f(x).\end{aligned}$$

Conversely, it follows from the Schwartz integral kernel theorem that if A is left-invariant, it can be written as a right convolution $Af = f * \kappa$, and if A is right-invariant, it can be written as a left convolution $Af = f * \kappa$, see Section 1.4 and later Corollary 3.2.1.

With our choice of the definition of the convolution and the Fourier transform in (1.2), one can readily check that for $f, g \in L^1(G)$, we have

$$\widehat{f * g}(\pi) = \widehat{g}(\pi)\widehat{f}(\pi) \tag{1.17}$$

or, in the other notation,

$$\pi(f * g) = \pi(g)\pi(f).$$

We say that an operator A is of *weak type* (p,p) if there is a constant $C > 0$ such that for every $\lambda > 0$ we have

$$|\{x \in G : |Af(x)| > \lambda\}| \leq C \frac{\|f\|^p_{L^p(G)}}{\lambda^p},$$

where $|\{\cdot\}|$ denotes the Haar measure of a set in G.

Proposition 1.5.1 (Marcinkiewicz interpolation theorem). *Let $r < q$ and assume that operator A is of weak types (r,r) and (q,q). Then A is bounded on $L^p(G)$ for all $r < p < q$.*

An important fact, the Young inequality, relates convolution to L^p-spaces:

Proposition 1.5.2 (Young's inequality). *Suppose*

$$1 \leq p, q, r \leq \infty \quad \textit{and} \quad \frac{1}{p} + \frac{1}{q} = \frac{1}{r} + 1.$$

If $f_1 \in L^p(G)$ and $f_2 \in L^q(G)$ then $f_1 * f_2 \in L^r(G)$ and

$$\|f_1 * f_2\|_r \leq \|f_1\|_p \|f_2\|_q.$$

If $p, q \in (1, \infty)$ are such that $\frac{1}{p} + \frac{1}{q} > 1$, $f_1 \in L^p(G)$, and f_2 satisfies the weak-$L^q(G)$ condition:

$$\sup_{s>0} s^q \, |\{x \ : \ |f_2(x)| > s\}| \ =: \|f_2\|^q_{w-L^q(G)} < \infty,$$

then $f_1 * f_2 \in L^r$ with r as above and

$$\|f_1 * f_2\|_r \leq \|f_1\|_p \|f_2\|_{w-L^q(G)}.$$

The proof is an easy adaptation of the Euclidean case which can be found e.g., in [SW71] or, in the nilpotent case, in [FS82, Proposition 1.18] and [Fol75, Proposition 1.10].

Convolution of distributions

We now define the convolution of distributions on a Lie group G. For $\phi \in C^\infty(G)$, we recall that

$$\tilde{\phi}(x) = \phi(x^{-1})$$

and

$$\pi_L(x)\phi(y) = \phi(x^{-1}y).$$

Consequently, we note that

$$(\pi_L(x)\tilde{\phi})(y) = \tilde{\phi}(x^{-1}y) = \phi(y^{-1}x).$$

It follows that we can write the convolution as

$$(f * g)(x) = \langle f, \pi_L(x)\tilde{g}\rangle,$$

and hence it make sense to define

Definition 1.5.3. Let $v \in \mathcal{D}'(G)$ and $\phi \in \mathcal{D}(G)$. Then we define their convolution as

$$(v * \phi)(x) := \langle v, \pi_L(x)\tilde{\phi}\rangle \equiv \langle v, \tilde{\phi}(x^{-1}\,\cdot)\rangle.$$

We also define

$$(\phi * v)(x) := \langle v, \pi_R(x^{-1})\tilde{\phi}\rangle \equiv \langle v, \tilde{\phi}(\cdot\, x^{-1})\rangle,$$

where

$$\pi_R(x^{-1})\tilde{\phi}(y) = \tilde{\phi}(yx^{-1}),$$

and which is also consistent with the convolution of functions.

We note that this expression makes since since $\pi_L(x), \pi_R(x^{-1})$ and $\phi \mapsto \tilde{\phi}$ are continuous mappings from $\mathcal{D}(G)$ to $\mathcal{D}(G)$.

For example, for the delta-distribution δ_e at the unit element $e \in G$, it follows that

$$\delta_e * \phi = \phi \quad \text{for every} \quad \phi \in \mathcal{D}(G),$$

since we can calculate

$$(\delta_e * \phi)(x) = \langle \delta_e, \pi_L(x)\tilde{\phi}\rangle = \phi(y^{-1}x)|_{y=e} = \phi(x).$$

The following properties are easy to check using Definition 1.5.3:

- if $v \in \mathcal{D}'(G)$ and $\phi \in \mathcal{D}(G)$, then $v * \phi \in C^\infty(G)$;
- if $u, v, \phi \in \mathcal{D}(G)$, then $\langle u * v, \phi\rangle = \langle u, \phi * \tilde{v}\rangle$, in consistency with (1.14).

For $v \in \mathcal{D}'(G)$, we now define $\tilde{v} \in \mathcal{D}'(G)$ by

$$\langle \tilde{v}, \phi\rangle := \langle v, \tilde{\phi}\rangle.$$

In particular, if $v \in \mathcal{D}'(G)$ and $\phi \in \mathcal{D}(G)$, then $\phi * \tilde{v} \in C^\infty(G)$. This shows that the following convolution of distributions is correctly defined:

Definition 1.5.4. Let $u \in \mathcal{E}'(G)$ and $v \in \mathcal{D}'(G)$. Then we define their convolution as

$$\langle u * v, \phi \rangle := \langle u, \phi * \tilde{v} \rangle, \quad \forall \phi \in \mathcal{D}(G).$$

This gives $u * v \in \mathcal{D}'(G)$ which is consistent with the convolution of functions in view of (1.15). If we start with a compactly supported distribution $v \in \mathcal{E}'(G)$ in Definition 1.5.3, we arrive at the definition of the composition $u * v$ for $u \in \mathcal{D}'(G)$ and $v \in \mathcal{E}'(G)$, given by the same formula as in Definition 1.5.4.

A word of caution has to be said about convolution of distributions, namely, it is not in general associative for distributions, although it is associative for functions.

1.6 Nilpotent Lie groups and algebras

From now on, any Lie algebra $\mathfrak{g}$ is assumed to be real and finite dimensional.

Proposition 1.6.1. *The following are equivalent:*

- ad *is a nilpotent endomorphism over* $\mathfrak{g}$, *i.e.*

$$\exists k \in \mathbb{N} \quad \forall X \in \mathfrak{g} \qquad (\mathrm{ad}X)^k = 0;$$

- *the lower central series of* $\mathfrak{g}$, *defined inductively by*

$$\mathfrak{g}_{(1)} := \mathfrak{g}, \quad \mathfrak{g}_{(j)} := [\mathfrak{g}, \mathfrak{g}_{(j-1)}], \tag{1.18}$$

terminates at 0 in a finite number of steps.

Definition 1.6.2. (i) If a Lie algebra $\mathfrak{g}$ satisfies any of the equivalent conditions of Proposition 1.6.1, then it is called *nilpotent.*

(ii) Moreover, if $\mathfrak{g}_{(s+1)} = \{0\}$ and $\mathfrak{g}_{(s)} \neq \{0\}$, then $\mathfrak{g}$ is said to be nilpotent of *step s*.

(iii) A Lie group G is *nilpotent* (of step s) whenever its Lie algebra is nilpotent (of step s).

Here are some examples of nilpotent Lie groups and their Lie algebras.

Example 1.6.3. The abelian group $\mathbb{R}^n$ equipped with the usual addition is nilpotent. Its Lie algebra is $\mathbb{R}^n$ equipped with the trivial Lie bracket.

Example 1.6.4. If $n_o \in \mathbb{N}$, the Heisenberg group $\mathbb{H}_{n_o}$ is the Lie group whose underlying manifold is $\mathbb{R}^{2n_o+1}$ and whose law is

$$h_1 h_2 = \Big(x_1 + x_2, y_1 + y_2, t_1 + t_2 + \frac{1}{2}(x_1 y_2 - y_1 x_2)\Big), \tag{1.19}$$

for $h_1 = (x_1, y_1, t_1)$ and $h_2 = (x_2, y_2, t_2)$ in $\mathbb{R}^{n_o} \times \mathbb{R}^{n_o} \times \mathbb{R}$. Here, for vectors $x_1, y_1, x_2, y_2 \in \mathbb{R}^{n_o}$, we denote by $x_1 y_2$ and $y_1 x_2$ their usual inner products on $\mathbb{R}^{n_o}$.

Its Lie algebra $\mathfrak{h}_{n_o}$ is $\mathbb{R}^{2n_o+1}$ equipped with the Lie bracket given by the commutator relations of its canonical basis $\{X_1, \ldots, X_{n_o}, Y_1, \ldots, Y_{n_o}, T\}$:

$$[X_j, Y_j] = T \qquad \text{for } j = 1, \ldots, n_o,$$

and all the other Lie brackets (apart from those obtained by anti-symmetry) are trivial.

In the case $n_o = 1$, we will often simplify the notation and write X, Y, T for the basis of $\mathfrak{h}_1$, etc...

Example 1.6.5. Let T_{n_o} be the group of $n_o \times n_o$ matrices which are upper triangular with 1 on the diagonal. The matrix group T_{n_o} is a nilpotent Lie group.

It can be proved that any (connected simply connected) nilpotent Lie group can be realised as a subgroup of T_{n_o}.

Its Lie algebra $\mathfrak{t}_{n_o}$ is the space of $n_o \times n_o$ matrices which are upper triangle with 0 on the diagonal. A basis is $\{E_{i,j}, 1 \le i < j \le n_o\}$ where $E_{i,j}$ is the matrix with all zero entries except the i-th row and j-th column which is 1.

Proposition 1.6.6. *Let G be a connected simply connected nilpotent Lie group with Lie algebra $\mathfrak{g}$. Then*

(a) The exponential map $\exp_G$ is a diffeomorphism from $\mathfrak{g}$ onto G.

(b) If G is identified with $\mathfrak{g}$ via $\exp_G$, the group law $(x, y) \mapsto xy$ is a polynomial map.

(c) If $d\lambda_{\mathfrak{g}}$ denotes a Lebesgue measure on the vector space $\mathfrak{g}$, then $d\lambda_{\mathfrak{g}} \circ \exp_G^{-1}$ is a bi-invariant Haar measure on G.

This proposition can be found in, e.g. [FS82, Proposition 1.2] or [CG90, Sec. 1.2].

After the choice of a basis $\{X_1, \ldots, X_n\}$ for $\mathfrak{g}$, Proposition 1.6.6, Part (a), implies that the group G is identified with $\mathbb{R}^n$ via the exponential mapping; this means that a point $x = (x_1, \ldots, x_n) \in \mathbb{R}^n$ is identified with the point

$$\exp_G(x_1 X_1 + \ldots + x_n X_n)$$

of the group. Part (b) implies that the law can be written as

$$x \cdot y = (P_1(x, y), P_2(x, y), \ldots, P_n(x, y)), \tag{1.20}$$

where $P_j : \mathbb{R}^n \times \mathbb{R}^n \to \mathbb{R}$, $j = 1, \ldots, n$, are polynomial mappings given via the Baker-Campbell-Hausdorff formula (see Theorem 1.3.2). Indeed in the nilpotent case, since ad is nilpotent, the Baker-Campbell-Hausdorff formula is finite and holds for any two elements of the Lie algebra.

Remark 1.6.7. More is known.

1. Certain choices of bases, namely the so-called Jordan-Hölder or strong-Malcev bases ([Puk67, CG90]), lead to a 'triangular' shaped law, that is,

$$\begin{aligned} P_1(x,y) &= x_1 + y_1, \\ P_2(x,y) &= x_2 + y_2 + Q_2(x_1,y_1), \\ &\vdots \\ P_n(x,y) &= x_n + y_n + Q_n(x_1,\ldots,x_{n-1},y_1,\ldots,y_{n-1}), \end{aligned}$$

   with $Q_1,\ldots,Q_n$ polynomials.

   In Chapter 3 we will see that in the particular case of homogeneous Lie groups, with the choice of the basis made in Section 3.1.3, this fact together with some additional homogeneous properties is proved in Proposition 3.1.24.

2. The second type of exponential coordinates

$$\mathbb{R}^n \ni (x_1,\ldots,x_n) \longmapsto \exp_G(x_1X_1)\ldots\exp_G(x_nX_n) \in G,$$

   may be used to identify a nilpotent Lie group with $\mathbb{R}^n$ after the choice of a suitable basis as in Part 1.

   In the particular case of homogeneous Lie groups, with the choice of the basis made in Section 3.1.3, this fact together with some additional homogeneous properties is proved in Lemma 3.1.47.

3. The converse of (a) and (b) in Proposition 1.6.6 holds in the following sense: if a Lie group G can be identified with $\mathbb{R}^n$ such that

   (a) its law is a polynomial mapping (as in (1.20)),

   (b) and for any $s,t\in\mathbb{R}$, $x\in\mathbb{R}^n$, the product of the two points sx and tx is the point $(s+t)x$,

   then the Lie group G is nilpotent [Puk67, Part. II chap. I].

However, we will not use these general facts.

Setting aside the abelian case $(\mathbb{R}^n,+)$, we use the multiplicative notation for the group law of any other connected simply connected nilpotent Lie group G. The identification of G with $\mathfrak{g}$ leads to *consider the origin* 0 *as the unit element* (even if the equality $xx^{-1}=0$ may look surprising at first sight). Because of the Baker-Campbell-Hausdorff formula (see Theorem 1.3.2), the inverse of an element is in fact its opposite, that is, with the notation above,

$$x^{-1} = (-x_1,\ldots,-x_n).$$

The identification of G with $\mathfrak{g}$ allows us to define objects which usually live on a vector space, for example the Schwartz class:

Definition 1.6.8. A Schwartz function f on G is a function f such that $f \circ \exp_G$ is a Schwartz function on $\mathfrak{g}$. We denote by $\mathcal{S}(G)$ the class of Schwartz functions. It is naturally a Fréchet space and its dual space is the space of tempered distribution $\mathcal{S}'(G)$.

Formally a distribution $T \in \mathcal{D}'(G)$ is tempered when $T \circ \exp_G$ is a tempered distribution on $\mathfrak{g}$. The distribution duality is formally given by

$$\langle f, \phi \rangle = \int_G f(x)\phi(x)dx, \quad f \in \mathcal{S}'(G), \ \phi \in \mathcal{S}(G).$$

The Schwartz space and the tempered distributions on a nilpotent homogeneous Lie group will be studied more thoroughly in Section 3.1.9.

1.7 Smooth vectors and infinitesimal representations

In this section we describe the basics of the part of the representation theory of non-compact Lie groups that is relevant to our context. For most statements of this section we give proofs since understanding of these ideas will be important for the developments of pseudo-differential operators in Chapter 5. Thus, the setting that we have in mind is that of nilpotent Lie groups, although we do not need to make this assumption for the following discussion. For the general representation theory of locally compact groups we can refer to, for example, the books of Knapp [Kna01], Wallach [Wal92, Chapter 14] or Folland [Fol95].

Let us first recall some basic definitions about differentiability of a Banach space-valued function.

Definition 1.7.1. Let f be a function from on open subset Ω of $\mathbb{R}^n$ to a Banach space B with norm $|\cdot|_B$.

The function f is said to be *differentiable* at $x_o \in \Omega$ if there exists a (necessarily unique) linear map $f'(x_o) : \mathbb{R}^n \to B$ such that

$$\frac{1}{|x - x_o|_{\mathbb{R}^n}} \left| f(x) - f(x_o) - f'(x_o)(x - x_o) \right|_B \underset{x \to x_o}{\longrightarrow} 0.$$

We call $f'(x_o)$ the differential of f at x_o.

If f is differentiable at each point of Ω, then $x \mapsto f'(x)$ is a function from Ω to the Banach space $\mathscr{L}(\mathbb{R}^n, B)$ of linear mappings from $\mathbb{R}^n$ to B (recall that linear mappings from $\mathbb{R}^n$ to B are automatically bounded.) We say that f is *of class* C^1 if $x \mapsto f'(x)$ is continuous, and that f is *of class* C^2 if $x \mapsto f'(x)$ is of class C^1 and so on. We say that f is of *class* C^∞ if f is of class C^k for all $k \in \mathbb{N}$.

These definitions extend to any open set of any smooth manifold.

As in the case of functions valued in a finite dimensional Euclidean space, we have the basic properties for a function f as in Definition 1.7.1:

- The function f is of class C^k if and only if all of its partial derivatives of order $1, 2, \dots, k$ exist and are continuous.
- The chain rule holds for a composition $f \circ h$ where h is a mapping from an open subset of a finite dimensional Euclidean space into Ω.

We can now define the smooth vectors of a representation.

Definition 1.7.2. Let G be a Lie group and let π be a representation of G on a Hilbert space $\mathcal{H}_\pi$. A vector $v \in \mathcal{H}_\pi$ is said to be *smooth* or *of type* C^∞ if the function

$$G \ni x \mapsto \pi(x)v \in \mathcal{H}_\pi$$

is of class C^∞.

We denote by $\mathcal{H}_\pi^\infty$ the space of all smooth vectors of π.

The following is a necessary preparation to introduce the notion of the *infinitesimal representation* and of the operator $d\pi(X)$. This will be of fundamental importance in the sequel.

Proposition 1.7.3. *Let G be a Lie group with Lie algebra $\mathfrak{g}$. Let π be a strongly continuous representation of G on a Hilbert space $\mathcal{H}_\pi$. Then for any $X \in \mathfrak{g}$ and $v \in \mathcal{H}_\pi^\infty$, the limit*

$$\lim_{t \to 0} \frac{1}{t} \left(\pi(\exp_G(tX))v - v\right)$$

exists in the norm topology of $\mathcal{H}_\pi$ and is denoted by $d\pi(X)v$. Each $d\pi(X)$ leaves $\mathcal{H}_\pi^\infty$ invariant, and $d\pi$ is a representation of $\mathfrak{g}$ on $\mathcal{H}_\pi^\infty$ satisfying

$$\forall X, Y \in \mathfrak{g} \qquad d\pi(X)d\pi(Y) - d\pi(Y)d\pi(X) - d\pi\left([X, Y]\right) = 0. \tag{1.21}$$

Consequently, $d\pi$ extends to a representation of the Lie algebra $\mathfrak{U}(\mathfrak{g})$ on $\mathcal{H}_\pi^\infty$ with $d\pi(0) = 0$ and $d\pi(1) = 0$.

Recalling the derivative with respect to X in (1.7), we may formally abbreviate writing

$$d\pi(X)v = X(\pi(x)v)|_{x=e} \quad \text{or even} \quad d\pi(X) = X\pi(e). \tag{1.22}$$

Sketch of the proof of Proposition 1.7.3. Let $v \in \mathcal{H}_\pi^\infty$. The function $f : \mathfrak{g} \to \mathcal{H}_\pi$ defined by $f(X) := \pi(\exp X)v$ is of class C^∞, and for any $X \in \mathfrak{g}$ we have

$$f'(0)(X) = \lim_{t \to 0} \frac{1}{t} \left(\pi(\exp_G(tX))v - v\right).$$

By definition $f'(0)(X) = d\pi(X)$.

Since π is continuous we have, using the identification of $\mathfrak{g}$ with the space of left-invariant vector fields,

$$\begin{aligned} \pi(x)d\pi(X)v &= \lim_{t \to 0} \frac{1}{t}\pi(x)\left(\pi(\exp_G(tX))v - v\right) \\ &= \lim_{t \to 0} \frac{1}{t}\left(\pi(x\exp_G(tX))v - \pi(x)v\right) = XF(x), \end{aligned}$$

where $F : G \to \mathcal{H}$ is the function defined by $F(x) := \pi(x)v$. By assumption F is of type C^∞ thus $x \mapsto XF(x)$ is also of type C^∞ and the equality above says that $d\pi(X)v$ is smooth. Hence $d\pi(X)$ leaves $\mathcal{H}_\pi^\infty$ stable. Consequently $X \mapsto d\pi(X)$ can be extended to an algebra homomorphism $\mathfrak{U}(\mathfrak{g}) \to \mathcal{H}_\pi^\infty$ as in the statement.

It remains to prove (1.21), i.e. that

$$\forall X, Y \in \mathfrak{g} \qquad d\pi(X)d\pi(Y) - d\pi(Y)d\pi(X) - d\pi\left([X,Y]\right) = 0.$$

We fix $X, Y \in \mathfrak{g}$ and define a path c by

$$c(t) := \exp_G\left((-\mathrm{sgn}t)|t|^{\frac{1}{2}}X\right)\exp_G\left(-|t|^{\frac{1}{2}}Y\right)\exp_G\left((\mathrm{sgn}t)|t|^{\frac{1}{2}}X\right)\exp_G\left(|t|^{\frac{1}{2}}Y\right).$$

Clearly c is defined on a neighbourhood of 0 in $\mathbb{R}$ and valued in G, and is of class C^1 with $c'(0) = [X,Y]$. Let $v \in \mathcal{H}_\pi^\infty$. By the chain rule the map $t \mapsto \pi(c(t))v$ has differential $F'(e)([X,Y])$ at $t = 0$, where F is $F(x) = \pi(x)v$ as above and e is the neutral element. Thus

$$d\pi([X,Y]) = \lim_{t\to 0}\frac{1}{t}\left(\pi(c(t))v - v\right) = \lim_{t\to 0}\frac{1}{t^2}\left(\pi(c(t^2))v - v\right).$$

The strong continuity of π implies then

$$\begin{aligned}
&\lim_{t\to 0}\frac{1}{t^2}\left(\pi(\exp_G(tX)\exp_G(tY))v - \pi(\exp_G(tY)\exp_G(tX))v\right)\\
&\quad = \lim_{t\to 0}\pi(\exp_G(tY)\exp_G(tX))\frac{1}{t^2}\left(\pi(c(t^2))v - v\right)\\
&\quad = d\pi([X,Y])v. \qquad (1.23)
\end{aligned}$$

But we can also compute

$$\begin{aligned}
&(d\pi(X)d\pi(Y)v, u) = \partial_{s=0}\partial_{t=0}(\pi(\exp_G(sX)\exp_G(tY))v, u)\\
&= \lim_{t\to 0}(\frac{1}{t^2}\left\{\pi(\exp_G(tX)\exp_G(tY)) - \pi(\exp_G(tX)) - \pi(\exp_G(tY)) + \mathrm{I}\right\}v, u).
\end{aligned}$$

Interchanging X and Y and subtracting we find

$$\begin{aligned}
&((d\pi(X)d\pi(Y) - d\pi(Y)d\pi(X))v, u) \qquad (1.24)\\
&= \lim_{t\to 0}(\frac{1}{t^2}\left\{\pi(\exp_G(tX)\exp_G(tY)) - \pi(\exp_G(tY)\exp_G(tX))\right\}v, u).
\end{aligned}$$

Comparing this with (1.23), we obtain (1.21). This concludes the proof of Proposition 1.7.3. □

Definition 1.7.4. Let G be a Lie group with Lie algebra $\mathfrak{g}$ and let π be a strongly continuous representation of G on a Hilbert space $\mathcal{H}_\pi$. The representation $d\pi$ defined in Proposition 1.7.3 is called the *infinitesimal representation* associated to π. We will often denote it also by π. Consequently, for $T \in \mathfrak{U}(\mathfrak{g})$ or for its corresponding left-invariant differential operator, we write

$$\pi(T) := d\pi(T).$$

Example 1.7.5. For example, the infinitesimal representation of Ad is ad, see Section 1.3.

We now collect some properties of the infinitesimal representations.

Proposition 1.7.6. *Let G be a Lie group with Lie algebra $\mathfrak{g}$ and let π be a strongly continuous unitary representation of G on a Hilbert space $\mathcal{H}_\pi$. Then we have the following properties.*

(i) For the infinitesimal representation $d\pi$ of $\mathfrak{g}$ on $\mathcal{H}_\pi^\infty$ each $d\pi(X)$ for $X \in \mathfrak{g}$ is skew-hermitian: $d\pi(X)^* = -d\pi(X)$.

(ii) The space $\mathcal{H}_\pi^\infty$ of smooth vectors is invariant under $\pi(x)$ for every $x \in G$, and

$$\forall D \in \mathfrak{U}(\mathfrak{g})\ \forall v \in \mathcal{H}_\pi^\infty \qquad \pi(x)d\pi(D)\pi(x)^{-1}v = d\pi(\mathrm{Ad}(x)D)v.$$

(iii) If S is a vector subspace of $\mathcal{H}_\pi$ such that for all $v \in S$ and $X \in \mathfrak{g}$, the limits of $t^{-1}\{\pi(\exp_G(tX))v - v\}$ as $t \to 0$ exist, then $S \subset \mathcal{H}_\pi^\infty$.

(iv) Let $\phi \in \mathcal{D}(G)$. For any $X \in \mathfrak{g}$, viewed as a left-invariant vector field,

$$\forall v \in \mathcal{H}_\pi \qquad \pi(\phi)v \in \mathcal{H}_\pi^\infty \quad \text{and} \quad d\pi(X)\pi(\phi)v = \pi(X\phi)v,$$

and viewing X as a right-invariant vector field $\tilde{X}$,

$$\forall v \in \mathcal{H}_\pi^\infty \qquad \pi(\phi)d\pi(X)v = \pi(\tilde{X}\phi)v.$$

If G is a connected simply connected nilpotent Lie group, one can replace $\mathcal{D}(G)$ by the Schwartz space $\mathcal{S}(G)$.

Proof. Let us prove Part (i). Let $u, v \in \mathcal{H}_\pi^\infty$. The unitarity of π implies

$$\left(v, \frac{i}{t}\left(\pi(\exp_G(tX))u - u\right)\right) = \left(\frac{i}{-t}\left(\pi(\exp_G(-tX))v - v\right), u\right).$$

By definition of $d\pi(X)u$ and $d\pi(X)v$, the limits as $t \to 0$ of the left and right hand sides are $(v, id\pi(X)u)$ and $(id\pi(X)v, u)$, respectively. Hence they are equal and $d\pi(X)$ is skew-hermitian. This proves Part (i).

For (ii), we first observe that the map $x \mapsto \pi(x)\pi(x_o)v$ is the composition of $x \mapsto xx_o$ and $x \mapsto \pi(x)v$. Hence $\mathcal{H}_\pi^\infty$ is an invariant subspace for $\pi(x_o)$.

Now let $X \in \mathfrak{g}$, $x \in G$ and $v \in \mathcal{H}_\pi^\infty$. Then we compute easily

$$\begin{aligned}\frac{1}{t}\left(\pi(\exp_G(tX)) - \mathrm{I}\right)\pi(x)^{-1}v &= \pi(x)^{-1}\frac{1}{t}\left(\pi(x\exp_G(tX)x^{-1}) - \mathrm{I}\right)v \\ &= \pi(x)^{-1}\frac{1}{t}\left(\pi(\exp_G(\mathrm{Ad}(x)(tX)) - \mathrm{I}\right)v.\end{aligned}$$

Passing to the limit as $t \to 0$, we obtain

$$d\pi(X)\pi(x)^{-1}v = \pi(x)^{-1}d\pi(\mathrm{Ad}(x)(tX))v.$$

Hence

$$\pi(x)d\pi(X)\pi(x)^{-1} = d\pi(\mathrm{Ad}(x)(tX))$$

on $\mathcal{H}_\pi^\infty$. Using Proposition 1.7.3, we obtain a similar property for $D \in \mathfrak{U}(\mathfrak{g})$ instead of X. This shows (ii).

For (iii), by assumption for $v \in S$ the map $F_v : G \ni x \mapsto \pi(x)v$ is differentiable at the neutral element e, the partial derivative in the $X \in \mathfrak{g}$ direction being

$$XF_v(e) = \lim_{t\to 0} \frac{1}{t} \{\pi(\exp_G(tX))v - v\}.$$

More generally, since π is strongly continuous, we have for any $x \in G$,

$$\pi(x)XF_v(e) = \lim_{t\to 0} \frac{1}{t}\pi(x)\{\pi(\exp_G(tX))v - v\} = \lim_{t\to 0} \frac{1}{t} \{F_v(x\exp_G(tX)) - F_v(x)\}.$$

Thus F_v is also differentiable at $x \in G$ and

$$XF_v(x) = \pi(x)XF_v(e)$$

for any $X \in \mathfrak{g}$. This shows that the first derivatives of F_v are continuous, thus F_v must be of class C^1. Furthermore,

$$F_v'(x)(X) = \pi(x)XF_v(e).$$

If F_v is of class C^k for $k \in \mathbb{N}$, then the map $x \mapsto XF_v(x) = \pi(x)XF_v(e)$ is of class C^k and F_v must be of class C^{k+1}. Inductively this shows that F_v is of type C^∞. This shows Part (iii).

For (iv), for any $\phi \in L^1(G)$ and $x \in G$, recalling (1.3), we have

$$\pi(x)\pi(\phi) = \pi(\phi(\cdot\, x)).$$

Hence for any $\phi \in \mathcal{D}(G)$, $v \in \mathcal{H}_\pi$ and $X \in \mathfrak{g}$,

$$\frac{1}{t}\left(\pi(\exp_G(tX))\pi(\phi)v - \pi(\phi)v\right) = \pi\left(\frac{\phi(\cdot\,\exp_G(tX)) - \phi}{t}\right)v.$$

This last expression tends to $\pi(X\phi)v$ as $t \to 0$. Applying (iii) to $S = \pi(\phi)\mathcal{H}_\pi$, we see that $S \subset \mathcal{H}_\pi^\infty$. We also have

$$d\pi(X)\pi(\phi)v = \pi(X\phi)v.$$

For the right-invariant case, again by (1.3), we have

$$\pi(\phi)\pi(x) = \pi(\phi(x\,\cdot))$$

for any $\phi \in L^1(G)$ and $x \in G$. Hence for any $\phi \in \mathcal{D}(G)$, $v \in \mathcal{H}_\pi$ and $X \in \mathfrak{g}$,

$$\frac{1}{t}\left(\pi(\phi)\pi(\exp_G(tX))v - \pi(\phi)v\right) = \pi\left(\frac{\phi(\exp_G(tX)\,\cdot) - \phi}{t}\right)v.$$

This last expression tends to $\pi(\tilde{X}\phi)v$ as $t \to 0$ while the left-hand side tends to $\pi(\phi)d\pi(X)v$ if $v \in \mathcal{H}_\pi^\infty$. This proves Part (iv) in the general case. The changes for G connected simply connected nilpotent Lie group, and to replace $\mathcal{D}(G)$ by $\mathcal{S}(G)$ are straightforward. This concludes the proof of Proposition 1.7.6. □

In the following proposition, we show that the space of smooth vectors is dense in the space of a strongly continuous representation. The argument is famously due to Gårding.

Proposition 1.7.7. *Let G be a Lie group and let π be a strongly continuous representation of G on a Hilbert space $\mathcal{H}_\pi$. Then the subspace $\mathcal{H}_\pi^\infty$ of smooth vectors is dense in $\mathcal{H}_\pi$.*

Proof. Let $v \in \mathcal{H}_\pi$ and $\epsilon > 0$ be given. Since π is strongly continuous, the set

$$\Omega := \{x \in G \, : \, |\pi(x)^* v - v|_{\mathcal{H}_\pi} < \epsilon\}$$

is open. We can find a non-negative function $\phi \in \mathcal{D}(G)$ supported in Ω satisfying $\int_G \phi(x)dx = 1$. Then

$$\begin{aligned} |\pi(\phi)v - v|_{\mathcal{H}_\pi} &= \left| \int_G \phi(x)(\pi(x)^* v - v)dx \right|_{\mathcal{H}_\pi} \\ &\leq \int_\Omega \phi(x)|\pi(x)^* v - v|_{\mathcal{H}_\pi} dx \leq \int_G \phi(x)\epsilon dx = \epsilon. \end{aligned}$$

By Proposition 1.7.6, we know that $\pi(\phi)v$ is a smooth vector. This shows that $\mathcal{H}_\pi^\infty$ is dense in $\mathcal{H}_\pi$. □

In the proof above, we have in fact showed that the vectors $\pi(\phi)v$ for $v \in \mathcal{H}_\pi$ and $\phi \in \mathcal{D}(G)$ form a dense subspace of $\mathcal{H}_\pi$. If G is nilpotent connected simply connected, the same property holds with $\phi \in \mathcal{S}(G)$. The finite linear combinations of those vectors form a subspace called the *Gårding subspace*, which is included in $\mathcal{H}_\pi^\infty$ by Proposition 1.7.6 (iv).

It turns out that the Gårding subspace is not only included in the subspace $\mathcal{H}_\pi^\infty$ but is in fact equal to $\mathcal{H}_\pi^\infty$. This is a consequent of the following theorem, due to Dixmier and Malliavin [DM78]:

Theorem 1.7.8 (Dixmier-Malliavin). *Let G be a Lie group and let π be a strongly continuous representation of G on a Hilbert space $\mathcal{H}_\pi$.*

The space $\mathcal{H}_\pi^\infty$ of smooth vectors is spanned by all the vectors of the form $\pi(\phi)v$ for $v \in \mathcal{H}_\pi^\infty$ and $\phi \in \mathcal{D}(G)$. This means that any smooth vector can be written as a finite linear combination of vectors of the form $\pi(\phi)v$.

If G is a connected simply connected nilpotent Lie group, one can replace $\mathcal{D}(G)$ by the Schwartz space $\mathcal{S}(G)$.

1.8 Plancherel theorem

Here we discuss the Plancherel theorem for locally compact groups and for the special case of nilpotent Lie groups. Our presentation will be rather informal. One reason is that we decided not to present here in full detail the orbit method yielding the representations of the nilpotent Lie groups but to limit ourselves only to its consequences useful for our subsequent analysis. The reason behind this choice is that it could take quite much space to prove the general results for the orbit method and would lead us too much away from our main exposition also risking overwhelming the reader with technical discussions somewhat irrelevant for our purposes. In general, this subject is well-known and we can refer to books by Kirillov [Kir04] or by Corwin and Greenleaf [CG90] for excellent expositions of this topic. The same reasoning applies to the abstract Plancherel theorem: it is known in a much more general form, due to e.g. Dixmier [Dix77, Dix81], and we will limit ourselves to describing its implications for nilpotent Lie groups relevant to our subsequent work.

As we will see in Chapter 2, all the results of the abstract Plancherel theorem in the case of compact groups can be recaptured there thanks to the Peter-Weyl theorem (see Theorem 2.1.1). However, for nilpotent Lie groups, even if the orbit method provides a description of the dual of the group and of the Plancherel measure, in our analysis we will need to use the properties of the von Neumann algebra of the group provided by the general abstract Plancherel theorem. This will replace the use of the Fourier coefficients in the compact case.

Before we proceed, let us adopt two useful conventions. First, the set of all strongly continuous unitary irreducible representations of a locally compact group G will be denoted by $\operatorname{Rep} G$, i.e.

$$\operatorname{Rep} G = \{\text{all strongly continuous unitary irreducible representations of } G\}.$$

The equivalence of representations in $\operatorname{Rep} G$ leads to the unitary dual $\widehat{G}$. We have already agreed to write $\pi \in \widehat{G}$ meaning that the expressions, when dealing with Fourier transforms, may depend on π as described in Remark 1.1.5. However, in this section we will sometimes want to show that certain expressions do not depend on the equivalence class of π, and for this purpose we will be sometimes distinguishing between the sets $\operatorname{Rep} G$ and $\widehat{G}$.

The second useful convention that we will widely use especially in Chapter 5 is that we may denote the Fourier transform in three ways, namely, we have

$$\widehat{\phi}(\pi) \equiv \pi(\phi) \equiv \mathcal{F}_G(\phi)(\pi).$$

Although this may seem as too much notation for the same object, the reason for this is two-fold. Firstly, the notation $\pi(\phi)$ is widely adopted in the representation theory of C^*-algebra associated with groups. Secondly, it becomes handy for longer expressions as well as for expressing properties like

$$\pi(T\phi) = \pi(T)\pi(\phi)$$

where $\pi(T)$ is the infinitesimal representation given in Definition 1.7.4. The notation $\widehat{\phi}(\pi)$ is useful as an analogy for the Euclidean case and will be extensively used in the case of compact groups. When we want to write the Fourier transform as a mapping between different spaces, the notation $\mathcal{F}_G$ becomes useful.

1.8.1 Orbit method

In this section we briefly discuss the idea of the orbit method and its implications for our analysis. In general, we will not use the orbit method by itself in our analysis, but only the existence of a Plancherel measure and some Fourier analysis similar to the compact case as described in Section 2.1.

Let G be a connected, simply connected, nilpotent Lie group with Lie algebra $\mathfrak{g}$. The orbit method describes a way to associate to a given linear functional on $\mathfrak{g}$ a collection of unitary irreducible representations of G which are all unitarily equivalent between themselves. Consequently, to any element of the dual $\mathfrak{g}'$ of $\mathfrak{g}$, one can associate an equivalence class of unitary irreducible representations. It turns out that any such class is realised in this way. Furthermore, two elements $f_1, f_2 \in \mathfrak{g}'$ lead to the same class if and only if the two elements are in the same orbit under the natural action of G on $\mathfrak{g}'$; this natural action is the so-called *co-adjoint representation*: since the group G acts on $\mathfrak{g}$ by the adjoint representation Ad, it also acts on its dual $\mathfrak{g}'$ by

$$\text{co-Ad} : G \times \mathfrak{g}' \ni (g, f) \longmapsto f(\text{Ad}^{-1}g \,\cdot) \in \mathfrak{g}'.$$

This gives a one-to-one correspondence between

- on the one hand, the dual $\widehat{G}$ of the group, that is, the collection of unitary irreducible representations modulo unitary equivalence, and
- on the other hand, $\mathfrak{g}'/\text{co-Ad}(G)$, that is, the set of co-adjoint orbits.

Example 1.8.1. In the case of the Heisenberg group $\mathbb{H}_{n_o}$ presented in Example 1.6.4, a family of representatives of all co-adjoint orbits is

1. either of the form $\lambda T'$ if $\lambda \in \mathbb{R}\backslash\{0\}$,
2. or of the form $\sum_{j=1}^{n_o} \left(x'_j X'_j + y'_j Y'_j\right)$ with $x'_j, y'_j \in \mathbb{R}$,

where $\{X'_1, \ldots, X'_{n_o}, Y'_1, \ldots, Y'_{n_o}, T'\}$ is the dual basis to the canonical basis of $\mathfrak{h}_{n_o}$ given in Example 1.6.4. To $\lambda T'$ is associated the Schrödinger representation π_λ, and to $\sum_{j=1}^{n_o} x'_j X'_j + y'_j Y'_j$ is associated the 1-dimensional representation $(x, y, t) \mapsto \exp\left(i(xx' + yy')\right)$, where xx' and yy' denote the canonical scalar product on $\mathbb{R}^n$. See Section 6.2.

As for Schrödinger representations, the representations constructed via the orbit method can be realised as acting on some $L^2(\mathbb{R}^m)$ and the dual $\widehat{G}$ may be identified with $\mathfrak{g}'/\text{co-Ad}(G)$, or even with suitable representatives of this quotient.

Thus, by the orbit method the unitary dual $\widehat{G}$ is 'concretely' described as a subset of some Euclidean space. It is then possible to construct 'explicitly' a measure μ on $\widehat{G}$ such that we have the Fourier inversion theorem (where we recall once more the notation and conventions described in the beginning of Section 1.8):

Theorem 1.8.2. *Let G be a connected simply connected nilpotent Lie group. The dual $\widehat{G}$ is then equipped with a measure μ called* the Plancherel measure *satisfying the following property for any $\phi \in \mathcal{S}(G)$.*

The operator $\pi(\phi) \equiv \widehat{\phi}(\pi)$ is trace class for any strongly continuous unitary irreducible representation $\pi \in \operatorname{Rep} G$, and $\operatorname{Tr}(\pi(\phi))$ depends only on the class of π; the function $\widehat{G} \ni \pi \mapsto \operatorname{Tr}(\pi(\phi))$ is integrable against μ and the following formula holds:

$$\phi(0) = \int_{\widehat{G}} \operatorname{Tr}(\pi(\phi))\, d\mu(\pi). \tag{1.25}$$

For the explicit expression of the Plancherel measure μ, see, e.g., [CG90, Theorem 4.3.9].

Applying formula (1.25) to $\phi(\cdot) = f(\cdot\, x)$ and using $\pi(\phi) = \pi(x)\pi(f)$ in view of (1.3), we obtain:

Corollary 1.8.3 (Fourier inversion formula). *Let G be a connected simply connected nilpotent Lie group and let μ be the Plancherel measure on $\widehat{G}$.*

If $f \in \mathcal{S}(G)$, then $\pi(x)\pi(f)$ and $\pi(f)\pi(x)$ are trace class for every $x \in G$, the function $\widehat{G} \ni \pi \mapsto \operatorname{Tr}(\pi(x)\pi(f))$ is integrable against μ, and we have

$$f(x) = \int_{\widehat{G}} \operatorname{Tr}(\pi(x)\pi(f))\, d\mu(\pi) = \int_{\widehat{G}} \operatorname{Tr}(\pi(f)\pi(x))\, d\mu(\pi). \tag{1.26}$$

The latter equality can be seen by the same argument as above, applied to the function $f(x\,\cdot)$.

Example 1.8.4. In the case of the Heisenberg group $\mathbb{H}_{n_o}$, the Plancherel measure is given by integration over $\mathbb{R}\backslash\{0\}$ against $c_{n_0}|\lambda|^{n_o} d\lambda$, with a suitable constant c_{n_o} (depending on normalisations):

$$\phi(0) = c_{n_o} \int_{\mathbb{R}\backslash\{0\}} \operatorname{Tr}\big(\pi_\lambda(\phi)\big)|\lambda|^{n_o} d\lambda.$$

An orthonormal basis for $\mathcal{H}_{\pi_\lambda} = L^2(\mathbb{R}^{n_o})$ is given by the Hermite functions. The subset of $\widehat{G}$ formed by the 1-dimensional representations is negligible with respect to the Plancherel measure. We refer to Section 6.2.3 for a more detailed discussion as well as for the constant c_{n_o}.

Applying the inversion formula to $\phi * (\phi^*)$, where $\phi^*(x) = \bar{\phi}(x^{-1})$, one obtains:

Theorem 1.8.5 (Plancherel formula). *We keep the notation of Theorem 1.8.2. Let $\phi \in \mathcal{S}(G)$. Then the operator $\pi(\phi)$ is Hilbert-Schmidt, that is,*

$$\|\pi(\phi)\|_{\mathrm{HS}}^2 = \mathrm{Tr}\,(\pi(\phi)\pi(\phi)^*) < \infty$$

for any $\pi \in \mathrm{Rep}\, G$, and its Hilbert-Schmidt norm is constant on the equivalence class of π. The function $\widehat{G} \ni \pi \mapsto \|\pi(\phi)\|_{\mathrm{HS}}^2$ is integrable against μ and

$$\int_G |\phi(x)|^2 dx = \int_{\widehat{G}} \|\pi(\phi)\|_{\mathrm{HS}}^2 \, d\mu(\pi). \tag{1.27}$$

Formula (1.27) can be extended unitarily to hold for any $\phi \in L^2(G)$, permitting the definition of the group Fourier transform of a square integrable function on G.

Applying the inversion formula to $\phi * (\psi^*)$, or bilinearising the Plancherel formula, we also obtain:

Corollary 1.8.6. *Let $\phi, \psi \in \mathcal{S}(G)$. Then the operator $\pi(\phi)\pi(\psi)^*$ is trace class for any $\pi \in \mathrm{Rep}\, G$, and its trace is constant on the equivalence class of π. The function $\widehat{G} \ni \pi \mapsto \mathrm{Tr}\,(\pi(\phi)\pi(\psi)^*)$ is integrable against μ and*

$$(\phi, \psi)_{L^2(G)} = \int_G \phi(x)\overline{\psi(x)} dx = \int_{\widehat{G}} \mathrm{Tr}\,(\pi(\phi)\pi(\psi)^*) \, d\mu(\pi).$$

1.8.2 Plancherel theorem and group von Neumann algebras

In this section we describe the concept of the group von Neumann algebra that becomes handy in associating symbols with convolution kernels of invariant operators on G. For the details of the constructions described below we refer to Dixmier's books [Dix77, Dix81] and to Section B in the appendix of this monograph. For the Plancherel theorem on locally compact groups with emphasis on the decomposition of reducible representations in continuous Hilbert sums, see also Bruhat [Bru68]. A more extensive discussion of this subject is given in Appendix B.2, more precisely in Section B.2.5. An abstract version of the Plancherel theorem is also given in the appendix in Theorem B.2.32.

Our framework

The representation theory of a general locally compact group may be very wild. However, in favourable cases most of the traditional Fourier analysis on compact Lie groups (described in Section 2.1) remains valid under natural modifications; for instance, the sum over the discrete dual in the compact case is replaced by an integral. By favourable cases we mean the following hypothesis:

(H) The group G is separable locally compact, unimodular, and of type I.

(See e.g. Dixmier [Dix77]). For our purpose, it suffices to know that any Lie group which is either compact or nilpotent satisfies (H). Its unitary dual $\widehat{G}$ is a standard Borel space.

We will now present the abstract Plancherel theorem as obtained by Dixmier in [Dix77, §18.8] and stated in Theorem B.2.32. Here, we will formulate it neither in its logical order with the viewpoint of proving its statement nor in its full generality since this would require introducing a lot of additional notation. Instead, we present its consequences applicable to our setting, starting with the existence of the Plancherel measure.

The Plancherel formula

We start by describing the part of the Plancherel theorem dealing with the Plancherel formula. First if $\phi \in C_c(G)$ and $\pi \in \operatorname{Rep} G$, then $\widehat{\phi}(\pi)$ is a bounded operator on $\mathcal{H}_\pi$ (as the group Fourier transform of an integrable function) and one checks easily that its Hilbert-Schmidt norm is constant on the class of $\pi \in \operatorname{Rep} G$ in $\widehat{G}$. Hence $\|\widehat{\phi}(\pi)\|_{\mathtt{HS}(\mathcal{H}_\pi)}$ may be viewed as depending on $\pi \in \widehat{G}$. The Plancherel formula states that there exists a unique positive σ-finite measure μ, called the *Plancherel measure*, such that for any $\phi \in C_c(G)$ we have

$$\int_G |\phi(x)|^2 dx = \int_{\widehat{G}} \left\|\widehat{\phi}(\pi)\right\|^2_{\mathtt{HS}(\mathcal{H}_\pi)} d\mu(\pi). \tag{1.28}$$

In the compact or nilpotent case, the Plancherel measure can be described explicitly via the Peter-Weyl Theorem (see Theorem 2.1.1) or the orbit method (see Theorem 1.8.5), respectively.

The Plancherel formula in (1.28) may be reformulated in the following (more precise) way. The group Fourier transform is an isometry from $C_c(G)$ endowed with the $L^2(G)$-norm to the Hilbert space

$$L^2(\widehat{G}) := \int_{\widehat{G}}^{\oplus} \mathtt{HS}(\mathcal{H}_\pi) d\mu(\pi). \tag{1.29}$$

Hence the space $L^2(\widehat{G})$ is defined (see Section B.1 or, e.g., [Dix81, Part II ch. I]) as the space of μ-measurable fields of Hilbert-Schmidt operators $\{\sigma_\pi \in \mathtt{HS}(\mathcal{H}_\pi) : \pi \in \widehat{G}\}$ which are square integrable in the sense that

$$\|\sigma\|^2_{L^2(\widehat{G})} := \int_{\widehat{G}} \|\sigma_\pi\|^2_{\mathtt{HS}} d\mu(\pi) < \infty.$$

Here we use the usual identifications of a strongly continuous irreducible unitary representation from $\operatorname{Rep} G$ with its equivalence class in $\widehat{G}$, and of a field of operators on $\widehat{G}$ with its equivalence class with respect to the Plancherel measure μ. One can check that indeed, the properties above do not depend on a particular representative of π and of the field of operators. The Plancherel formula implies that $\mathcal{F}_G$ extends to an isometry on $L^2(G)$. We keep the same notation $\mathcal{F}_G$ for this map, allowing us to consider the Fourier transform of a square integrable function. The abstract Plancherel theorem states moreover that the isometry $\mathcal{F}_G : L^2(G) \to L^2(\widehat{G})$ is surjective. In other words, $\mathcal{F}_G$ maps $L^2(G)$ onto $L^2(\widehat{G})$ isometrically.

Note that for any $\phi, \psi \in L^2(G)$, the operator $\pi(\phi)\ \pi(\psi)^*$ is trace class on $\mathcal{H}_\pi$ for almost all $\pi \in \operatorname{Rep} G$ with

$$\operatorname{Tr}|\pi(\phi)\ \pi(\psi)^*| \leq \|\pi(\phi)\|_{\mathtt{HS}(\mathcal{H}_\pi)} \|\pi(\psi)^*\|_{\mathtt{HS}(\mathcal{H}_\pi)} = \|\pi(\phi)\|_{\mathtt{HS}(\mathcal{H}_\pi)} \|\pi(\psi)\|_{\mathtt{HS}(\mathcal{H}_\pi)},$$

and that $\operatorname{Tr}|\pi(\phi)\ \pi(\psi)^*|$ and $\operatorname{Tr}(\pi(\phi)\ \pi(\psi)^*)$ are constant on the class of $\pi \in \operatorname{Rep} G$ in $\widehat{G}$. Thus these traces can be viewed as being parametrised by $\pi \in \widehat{G}$. The bilinearisation of the Plancherel formula yields

$$\int_G \phi(x)\overline{\psi(x)}dx = \int_{\widehat{G}} \operatorname{Tr}\left(\pi(\phi)\ \pi(\psi)^*\right) d\mu(\pi). \tag{1.30}$$

One also checks easily, for example by density of $C_c(G)$ in $L^2(G)$, that Formula (1.17), that is,

$$\widehat{f * g}(\pi) = \widehat{g}(\pi)\widehat{f}(\pi) \tag{1.31}$$

or, in the other notation,

$$\pi(f * g) = \pi(g)\pi(f),$$

remains valid for $f \in L^1(G)$ and $g \in L^2(G)$ and also for $f \in L^2(G)$ and $g \in L^1(G)$.

We now present the parts of the Plancherel theorem (relevant for our subsequent analysis) regarding the description of the group von Neumann algebra.

Group von Neumann algebra

In this monograph, we realise the von Neumann algebra of a group G as the algebra denoted by $\mathscr{L}_L(L^2(G))$ and defined as follows.

Definition 1.8.7. Let $\mathscr{L}(L^2(G))$ denote the set of bounded linear operators $L^2(G) \to L^2(G)$, and let $\mathscr{L}_L(L^2(G))$ be the subset formed by the operators in $\mathscr{L}(L^2(G))$ which are left-invariant (in the sense of Definition 1.1.3).

Endowed with the operator norm and composition of operators, one checks easily that $\mathscr{L}_L(L^2(G))$ is a von Neumann algebra, see Section B.2.5 for the exposition of its general ideas.

Given a μ-measurable field of uniformly bounded operators $\sigma = \{\sigma_\pi\}$, the operator $T_\sigma \in \mathscr{L}_L(L^2(G))$ defined via

$$\widehat{T_\sigma \phi}(\pi) = \sigma_\pi \widehat{\phi}(\pi), \quad \phi \in L^2(G), \tag{1.32}$$

is in $\mathscr{L}_L(L^2(G))$. Using (1.30), this yields that the operator $T_\sigma : \mathcal{S}(G) \to \mathcal{S}'(G)$ can also be defined by

$$(T_\sigma \phi, \psi)_{L^2(G)} = \int_{\widehat{G}} \mathrm{Tr}\,(\sigma_\pi \; \pi(\phi) \; \pi(\psi)^*) \, d\mu(\pi), \quad \phi, \psi \in L^2(G). \tag{1.33}$$

This defines a map $\sigma \mapsto T_\sigma$ from $L^\infty(\widehat{G})$ to $\mathscr{L}_L(L^2(G))$ where the space $L^\infty(\widehat{G})$ is defined by

Definition 1.8.8. Let $L^\infty(\widehat{G})$ denote the space of μ-measurable fields on $\widehat{G}$ of uniformly bounded operators $\sigma = \{\sigma_\pi \in \mathscr{L}(\mathcal{H}_\pi), \; \pi \in \widehat{G}\}$, that is,

$$\sup_{\pi \in \widehat{G}} \|\sigma_\pi\|_{\mathscr{L}(\mathcal{H}_\pi)} < \infty. \tag{1.34}$$

Here we use the usual identifications of a strongly continuous irreducible unitary representation from $\mathrm{Rep}\, G$ with its equivalence class in $\widehat{G}$, and of a field of operators on $\widehat{G}$ with its equivalence class with respect to the Plancherel measure μ. One can check that indeed, being in $L^\infty(\widehat{G})$ does not depend on a particular representative of π and of the field of operators. In (1.34), the supremum is to be understood as the essential supremum with respect to the Plancherel measure μ.

We endow $L^\infty(\widehat{G})$ with the pointwise composition given by

$$\sigma\tau := \{\sigma_\pi \tau_\pi, \; \pi \in \widehat{G}\}, \quad \text{for} \quad \sigma = \{\sigma_\pi, \; \pi \in \widehat{G}\}, \tau = \{\tau_\pi, \; \pi \in \widehat{G}\} \in L^\infty(\widehat{G}),$$

and the essential supremum norm

$$\|\sigma\|_{L^\infty(\widehat{G})} := \sup_{\pi \in \widehat{G}} \|\sigma_\pi\|_{\mathscr{L}(\mathcal{H}_\pi)}. \tag{1.35}$$

We may sometimes abuse the notation and write $\|\sigma_\pi\|_{L^\infty(\widehat{G})}$ when no confusion is possible.

One checks easily that $L^\infty(\widehat{G})$ is a von Neumann algebra and that the map

$$L^\infty(\widehat{G}) \ni \sigma \longmapsto T_\sigma \in \mathscr{L}_L(L^2(G)),$$

is a morphism of von Neumann algebras. The Plancherel theorem implies that this map is in fact a bijection and an isometry, and hence a von Neumann algebra isomorphism. More precisely it yields that for any $T \in \mathscr{L}_L(L^2(G))$, there exists a μ-measurable field of uniformly bounded operators $\{\sigma_\pi^{(T)}\}$ such that for any $\phi \in L^2(G)$ the Hilbert-Schmidt operators $\widehat{T\phi}(\pi)$ and $\sigma_\pi^{(T)} \widehat{f}(\pi)$ are equal μ-almost everywhere; the field $\{\sigma_\pi^{(T)}\}$ is unique up to a μ-negligible set.

Note that by the Schwartz kernel theorem (see Corollary 3.2.1), an operator $T \in \mathscr{L}_L(L^2(G))$ is of convolution type with kernel $\kappa \in \mathcal{D}'(G)$,

$$Tf = f * \kappa, \quad f \in \mathcal{D}(G).$$

If $\kappa \in \mathcal{D}'(G)$ is such that the corresponding convolution operator $\mathcal{D}(G) \ni f \mapsto f*\kappa$ extends to a bounded operator T_κ on $L^2(G)$ then $T_\kappa \in \mathscr{L}_L(L^2(G))$ and we extend the definition of the group Fourier transform by setting

$$\sigma_\pi^{(T)} := \pi(\kappa) \equiv \widehat{\kappa}(\pi). \tag{1.36}$$

We denote by $\mathcal{K}(G)$ the set of such distributions κ:

Definition 1.8.9. Let $\mathcal{K}(G)$ denote the space of distributions $\kappa \in \mathcal{D}'(G)$ such that the corresponding convolution operator

$$\mathcal{D}(G) \ni f \mapsto f * \kappa$$

extends to a bounded operator on $L^2(G)$.

If G is a connected simply connected nilpotent Lie group, the Schwartz kernel theorem (see Corollary 3.2.1), implies in fact that the distributions in $\mathcal{K}(G)$ are tempered, i.e. $\mathcal{K}(G) \subset \mathcal{S}'(G)$.

If $\kappa \in \mathcal{K}(G)$, then κ^* defined via $\kappa^*(x) = \bar{\kappa}(x^{-1})$ is also in $\mathcal{K}(G)$. If $\kappa_1, \kappa_2 \in \mathcal{K}(G)$ and $T_{\kappa_1}, T_{\kappa_2} \in \mathscr{L}_L(L^2(G))$ denote the associated right-convolution operator, then $T_{\kappa_1} T_{\kappa_2} \in \mathscr{L}_L(L^2(G))$ and we denote by $\kappa_2 * \kappa_1$ its convolution kernel. One checks easily that this convolution product coincides or extends the already defined convolution products in Section 1.5. Furthermore $\mathcal{K}(G)$ equipped with this convolution product, the $*$-adjoint and the operator norm

$$\|\kappa\|_{\mathcal{K}(G)} := \|f \mapsto f * \kappa\|_{\mathscr{L}(L^2(G))} \tag{1.37}$$

is a von Neumann algebra. It is naturally isomorphic to $\mathscr{L}_L(L^2(G))$.

The part of the Plancherel theorem that we have already presented implies that the space $\mathcal{K}(G)$ is a von Neumann algebra isomorphic to $\mathscr{L}_L(L^2(G))$ and to $L^\infty(\widehat{G})$. Moreover, the group Fourier transform defined on $\mathcal{K}(G)$ gives the isomorphism between $\mathcal{K}(G)$ and $L^\infty(\widehat{G})$.

Naturally, $L^1(G)$ is embedded in $\mathcal{K}(G)$ since if $\kappa \in L^1(G)$, then the operator $\phi \mapsto \phi * \kappa$ is in $\mathscr{L}_L(L^2(G))$. Note that Young's inequality (see Proposition 1.5.2) implies

$$\|\widehat{\kappa}\|_{L^\infty(\widehat{G})} = \|\kappa\|_{\mathcal{K}} \leq \|\kappa\|_{L^1(G)}. \tag{1.38}$$

Furthermore, as $\mathcal{F}_G(\phi * \kappa) = \widehat{\kappa}\widehat{\phi}$ (see e.g. (1.31)), there is no conflict of notation between the group Fourier transforms defined first on $L^1(G)$ via (1.2) and then on $\mathcal{K}(G)$ in (1.36) as these group Fourier transforms coincide, since the field of operators associated to an operator in $\mathscr{L}_L(L^2(G))$ is unique.

More generally, the proof of Example 1.8.10 below shows that the space of complex Borel measures $M(G)$ (which contains $L^1(G)$) is contained in $\mathcal{K}(G)$, that is,

$$L^1(G) \subset M(G) \subset \mathcal{K}(G).$$

Moreover, their group Fourier transform may be defined directly via (1.39) below or as of an element of $\mathcal{K}(G)$ via Definition 1.36.

Example 1.8.10 (Complex Borel measures). Any complex Borel measure η on G is in $\mathcal{K}(G)$ and

$$\|\eta\|_{\mathcal{K}} \leq \|\eta\|_{M(G)},$$

where $\|\eta\|_{M(G)}$ denotes the total mass of η.

The group Fourier transform of a complex Borel measure η is given in the sense of Bochner by the integral

$$\mathcal{F}_G(\eta)(\pi) \equiv \widehat{\eta}(\pi) \equiv \pi(\eta) := \int_G \pi(x)^* d\eta(x). \tag{1.39}$$

In particular, the group Fourier transform of the Dirac measure δ_e at the neutral element is the identity operator

$$\widehat{\delta_e}(\pi) \equiv \pi(\delta_e) = \mathrm{I}_{\mathcal{H}_\pi}$$

on the representation space $\mathcal{H}_\pi$. More generally, the group Fourier transform of the Dirac measure δ_{x_o} at the element $x_o \in G$ is

$$\widehat{\delta_{x_o}}(\pi) = \pi(x_o).$$

Proof of Example 1.8.10. By Jensen's inequality, for $p = 1$ and 2 (in fact for any $p \in [1, \infty)$), the operator $T_\eta : \mathcal{D}(G) \ni \phi \mapsto \phi * \eta$ extends to an L^p-bounded operator with norm $\|\eta\|$.

If $\phi \in C_c(G)$, then $\phi * \eta \in L^1(G)$ (see Example 1.8.10) and we have in the sense of Bochner, using the change of variable $y = xz^{-1}$,

$$\begin{aligned}
\pi(\phi * \eta) &= \int_{G\times G} \phi(xz^{-1})\pi(x)^* dx d\eta(z) = \int_{G\times G} \phi(y)\pi(yz)^* dy d\eta(z) \\
&= \int_{G\times G} \phi(y)\pi(z)^*\pi(y)^* dy d\eta(z) = \int_G \pi(z)^* d\eta(z) \int_G \phi(y)\pi(y)^* dy \\
&= \pi(\eta)\,\pi(\phi),
\end{aligned}$$

confirming the formula for $\pi(\eta)$. Since the field of operators associated to an operator in $\mathscr{L}_L(L^2(G))$ is unique, the group Fourier transform of η as an element of $\mathcal{K}(G)$ is $\{\pi(\eta), \pi \in \widehat{G}\}$ defined in (1.39). □

The abstract Plancherel theorem

We now summarise the consequences of Dixmier's abstract Plancherel theorem, see Theorem B.2.32, that we will use:

Theorem 1.8.11 (Abstract Plancherel theorem). *Let G be a Lie group satisfying hypothesis (H). We denote by μ its Plancherel measure.*

The Fourier transform $\mathcal{F}_G$ extends to an isometry from $L^2(G)$ onto

$$L^2(\widehat{G}) := \int_{\widehat{G}}^{\oplus} \mathtt{HS}(\mathcal{H}_\pi) d\mu(\pi).$$

The Fourier transform of an element f of $\mathcal{K}(G)$, i.e. $f \in \mathcal{D}'(G)$ such that the operator $\mathcal{D}(G) \ni \phi \mapsto \phi * f$ extends boundedly to $L^2(G)$, has a meaning as a field of uniformly (μ-essentially) bounded operators

$$\{\widehat{f}(\pi) \equiv \pi(f) \, : \, \pi \in \widehat{G}\} \in L^\infty(\widehat{G})$$

satisfying

$$\pi(\phi * f) = \pi(f)\pi(\phi)$$

for any $\phi \in \mathcal{D}(G)$ and $\pi \in \widehat{G}$. Conversely, any field in $L^\infty(\widehat{G})$ leads to an element of $\mathcal{K}(G)$. Furthermore

$$\|f\|_{\mathcal{K}} = \|\phi \mapsto \phi * f\|_{\mathscr{L}(L^2(G))} = \sup_{\pi\in\widehat{G}} \|\widehat{f}(\pi)\|_{\mathscr{L}(\mathcal{H}_\pi)}. \tag{1.40}$$

The Fourier transform is a von Neumann algebra isomorphism from $\mathcal{K}(G)$ onto $L^\infty(\widehat{G})$. In particular, it is a bijection from $\mathcal{K}(G)$ onto $L^\infty(\widehat{G})$ and satisfies

$$\forall f_1, f_2, f \in \mathcal{K}(G) \quad \mathcal{F}_G(f_1 * f_2) = \mathcal{F}_G(f_2)\mathcal{F}_G(f_1) \quad \text{and} \quad \mathcal{F}_G(f^*) = \mathcal{F}_G(f)^*,$$

if $f^*(x) = \bar{f}(x^{-1})$. Moreover

$$\|\widehat{f}\|_{L^\infty(\widehat{G})} = \|f\|_{\mathcal{K}(G)}.$$

If G is a connected simply connected nilpotent Lie group, the elements of $\mathcal{K}(G)$ are tempered distributions.

Naturally the various definitions of group Fourier transforms on $L^1(G)$ or on the space $M(G)$ of regular complex measures on G, on $L^2(G)$ or on $\mathcal{K}(G)$, coincide on any intersection of these subspaces of $\mathcal{D}'(G)$. This can be seen easily using the abstract Plancherel theorem, especially the bijections $\mathcal{F}_G : L^2(G) \to L^2(\widehat{G})$ and $\mathcal{F}_G : \mathcal{K}(G) \to L^\infty(\widehat{G})$, together with the properties of the convolution and of the representations, especially (1.31).

1.8.3 Fields of operators acting on smooth vectors

Let us assume that the group G satisfies hypothesis (H) as in the previous section and is also a Lie group. This means that G is a unimodular Lie group of type I, for instance a compact or nilpotent Lie group.

In our subsequent analysis, we will need to consider fields of operators parametrised by $\widehat{G}$ but not necessarily bounded, for instance the fields given by the $\pi(X)^\alpha$'s.

The definition of fields of smooth vectors or of operators defined on smooth vectors will be a consequence of the following lemma. For a more general setting for measurable fields of operators see Section B.1.5.

Lemma 1.8.12. *Let $\pi_1, \pi_2 \in \operatorname{Rep} G$ with $\pi_1 \sim_T \pi_2$, that is, we assume that π_1 and π_2 are intertwined by the unitary operator T, i.e. $T\pi_1 = \pi_2 T$. Then T maps $\mathcal{H}_{\pi_1}^\infty$ onto $\mathcal{H}_{\pi_2}^\infty$ bijectively.*

Proof. This is an easy consequence of the Dixmier-Malliavin theorem, see Theorem 1.7.8. □

Lemma 1.8.12 allows us to define fields of operators not necessarily bounded but just defined on smooth vectors:

Definition 1.8.13. A *$\widehat{G}$-field of operators defined on smooth vectors* is a family of classes of operators $\{\sigma_\pi, \pi \in \widehat{G}\}$ where

$$\sigma_\pi := \{\sigma_{\pi_1} : \mathcal{H}_{\pi_1}^\infty \to \mathcal{H}_{\pi_1}, \pi_1 \in \pi\}$$

for each $\pi \in \widehat{G}$ viewed as a subset of $\operatorname{Rep} G$, satisfying for any two elements σ_{π_1} and σ_{π_2} in σ_π:

$$\pi_1 \sim_T \pi_2 \Longrightarrow \sigma_{\pi_2} T = T\sigma_{\pi_1}.$$

It is measurable when for one (and then any) choice of realisation π_1 and any vector $x_{\pi_1} \in \mathcal{H}_{\pi_1}^\infty$, as π runs over $\widehat{G}$, the resulting field $\{\sigma_{\pi_1} x_{\pi_1}, \pi \in \widehat{G}\}$ is μ-measurable whenever $\int_{\widehat{G}} \|x_{\pi_1}\|_{\mathcal{H}_{\pi_1}}^2 d\mu(\pi) < \infty$.

We will allow ourselves the shorthand notation

$$\sigma = \{\sigma_\pi : \mathcal{H}_\pi^\infty \to \mathcal{H}_\pi, \pi \in \widehat{G}\}$$

to indicate that the $\widehat{G}$-field of operators is defined on smooth vectors. Unless otherwise stated, all the $\widehat{G}$-fields of operators are assumed to be measurable and with operators defined on smooth vectors. We may allow ourselves to write $\sigma = \{\sigma_\pi, \pi \in \widehat{G}\}$. Note that we do not require the domain of each operator to be the whole representation space $\mathcal{H}_{\pi_1}$ but just the space of smooth vectors.

The next definition would allow us to compose such fields of operators.

Definition 1.8.14. A *measurable $\widehat{G}$-field of operators acting on the smooth vectors* is a measurable $\widehat{G}$-field of operators $\sigma = \{\sigma_\pi : \mathcal{H}_\pi^\infty \to \mathcal{H}_\pi, \pi \in \widehat{G}\}$ such that for any $\pi_1 \in \operatorname{Rep} G$, we have

$$\sigma_{\pi_1}(\mathcal{H}_{\pi_1}^\infty) \subset \mathcal{H}_{\pi_1}^\infty.$$

We will often abuse the notation and write

$$\{\sigma_\pi : \mathcal{H}_\pi^\infty \to \mathcal{H}_\pi^\infty, \pi \in \widehat{G}\}$$

to express the fact that the measurable $\widehat{G}$-field of operators act on smooth vectors.

Remark 1.8.15. Let $\sigma = \{\sigma_\pi : \mathcal{H}_\pi^\infty \to \mathcal{H}_\pi, \pi \in \widehat{G}\}$ be a $\widehat{G}$-field. If $\pi_1 \sim_T \pi_2$ that is, we assume that π_1 and π_2 are intertwined by the unitary operator T, then T maps $\sigma_{\pi_1}(\mathcal{H}_{\pi_1}^\infty)$ onto $\sigma_{\pi_2}(\mathcal{H}_{\pi_2}^\infty)$ bijectively. Thus the range $\sigma_\pi(\mathcal{H}_\pi^\infty)$ makes sense as the collection of the equivariant ranges $\sigma_{\pi_1}(\mathcal{H}_\pi^\infty)$ for $\pi_1 \in \pi \subset \operatorname{Rep} G$.

Consequently, in Definition 1.8.14, it suffices that $\sigma_{\pi_1}(\mathcal{H}_{\pi_1}^\infty) \subset \mathcal{H}_{\pi_1}^\infty$ for one representation $\pi_1 \in \pi$ for each $\pi \in \widehat{G}$.

Remark 1.8.16. We will often consider measurable field of operators $\sigma_{\pi,s}$ acting on smooth vectors and parametrised not only by $\widehat{G}$ but also by another set S. When this set S is a subset of some $\mathbb{R}^n$, we say that this parametrisation is smooth whenever the map appearing in Definition 1.8.14 above is not only measurable with respect to $\widehat{G}$ but also smooth with respect to the S-variable. Note that this hypothesis yields the existence of the fields of operators given by $D_s\sigma_{\pi,s}$ where D_s is a (smooth) differential operator on S.

It is clear that one can sum two fields $\sigma = \{\sigma_\pi : \mathcal{H}_\pi^\infty \to \mathcal{H}_\pi, \pi \in \widehat{G}\}$ and $\tau = \{\tau_\pi : \mathcal{H}_\pi^\infty \to \mathcal{H}_\pi, \pi \in \widehat{G}\}$ defined on smooth vectors. We may then write

$$\sigma + \tau = \{\sigma_\pi + \tau_\pi : \mathcal{H}_\pi^\infty \to \mathcal{H}_\pi, \pi \in \widehat{G}\}$$

for the resulting field. If σ and τ act on smooth vectors, then so does $\sigma + \tau$.

It is also clear that one can compose two fields $\sigma = \{\sigma_\pi : \mathcal{H}_\pi^\infty \to \mathcal{H}_\pi, \pi \in \widehat{G}\}$ and $\tau = \{\tau_\pi : \mathcal{H}_\pi^\infty \to \mathcal{H}_\pi^\infty, \pi \in \widehat{G}\}$ defined on smooth vectors if the first one acts on smooth vectors. We may then write

$$\sigma\tau = \{\sigma_\pi\tau_\pi : \mathcal{H}_\pi^\infty \to \mathcal{H}_\pi, \pi \in \widehat{G}\}$$

for the resulting field which is then defined on smooth vectors. Note that $\sigma\tau$ is not obtained as the composition of two unbounded operators on $\mathcal{H}_\pi$ as in Definition A.3.2 but as the composition of two operators acting on the same space $\mathcal{H}_\pi^\infty$.

Almost by definition of smooth vectors, we have the following example of measurable fields of operators acting on smooth vectors:

Example 1.8.17. If $T \in \mathfrak{U}(\mathfrak{g})$ then $\{\pi(T), \pi \in \widehat{G}\}$ yields a measurable field of operators acting on smooth vectors and parametrised by $\widehat{G}$ (see also Proposition 1.7.3).

If $T_1, T_2 \in \mathfrak{U}(\mathfrak{g})$ then the composition of $\{\pi(T_1), \pi \in \widehat{G}\}$ with $\{\pi(T_2), \pi \in \widehat{G}\}$ as field of operators acting on smooth vectors is $\{\pi(T_1T_2), \pi \in \widehat{G}\}$.

The definition of Fourier transform and Proposition 1.7.6 (iv) easily imply the next example of measurable fields of operators acting on smooth vectors:

Example 1.8.18. If $\phi \in \mathcal{D}(G)$, then $\widehat{\phi} = \{\pi(\phi) : \mathcal{H}_\pi^\infty \to \mathcal{H}_\pi^\infty, \pi \in \widehat{G}\}$ is a measurable $\widehat{G}$-field of operators acting on smooth vectors.

If $\phi_1, \phi_2 \in \mathcal{D}(G)$, then the composition of $\widehat{\phi_1}$ with $\widehat{\phi_2}$ as fields of operators acting on smooth vectors is $\widehat{\phi_2 * \phi_1}$.

If G is simply connected and nilpotent, the properties above also hold for Schwartz functions.

A field $\sigma = \{\sigma_\pi : \mathcal{H}_\pi \to \mathcal{H}_\pi, \pi \in \widehat{G}\}$ always gives by restriction operators that are defined on smooth vectors. If we start from a field of operators $\sigma = \{\sigma_\pi : \mathcal{H}_\pi^\infty \to \mathcal{H}_\pi, \pi \in \widehat{G}\}$ defined on smooth vectors, we can not always extend it to operators defined on every $\mathcal{H}_\pi$. However, since the space $\mathcal{H}_\pi^\infty$ of smooth vectors is dense in $\mathcal{H}_\pi$ (see Proposition 1.7.7), each operator $\sigma_{\pi_1} : \mathcal{H}_{\pi_1}^\infty \to \mathcal{H}_{\pi_1}$, $\pi_1 \in \operatorname{Rep} G$, has a unique extension to a bounded operator on $\mathcal{H}_{\pi_1}$ provided that such an extension exists. In this case, σ_{π_2} would have the same property if $\pi_1 \sim \pi_2$, and the operator norm $\|\sigma_{\pi_1}\|_{\mathscr{L}(\mathcal{H}_{\pi_1})}$ or the Hilbert-Schmidt norm $\|\sigma_{\pi_1}\|_{\mathtt{HS}(\mathcal{H}_{\pi_1})}$ of σ_{π_1} are constant (maybe infinite) for $\pi_1 \in \pi$. Hence we may regard these norms as being parametrised by $\pi \in \widehat{G}$. Furthermore, if $\|\sigma\|_{L^\infty(\widehat{G})}$ or $\|\sigma\|_{L^2(\widehat{G})}$ are finite, then the field of bounded operators in $L^\infty(\widehat{G})$ or $L^2(\widehat{G})$ (resp.) is unique and extends σ.

On a compact Lie group, any $\widehat{G}$-field of operators is measurable and the operators act on smooth vectors. This is because in this case $\widehat{G}$ is discrete and countable, and all the strongly continuous irreducible representations are finite dimensional and these have only smooth vectors, see the Peter-Weyl theorem in Theorem 2.1.1.

However on a non-compact Lie group, we can not restrict ourselves to the case of $\widehat{G}$-fields acting on smooth vectors in general since a non-compact Lie group may have infinite dimensional (strongly continuous irreducible) representations with non-smooth vectors and we then can find fields in $L^2(\widehat{G})$ which do not act on smooth vectors. Indeed, in this case, we can find a measurable field $\{v_\pi, \pi \in \widehat{G}\}$ of non-smooth vectors satisfying $\int_{\widehat{G}} \|v_\pi\|^2_{\mathtt{HS}(\mathcal{H}_\pi)} d\mu(\pi) < \infty$, and then construct the field of operators $\{v_\pi \otimes v_\pi^*, \pi \in \widehat{G}\}$ in $L^2(\widehat{G})$ which does not act on smooth vectors. Such field of vectors $\{v_\pi\}$ are easy to find for instance on the Heisenberg group $\mathbb{H}_n$ whose case is detailed in Chapter 6: in this case, almost all the representations in $\widehat{\mathbb{H}}_n$ may be realised on $L^2(\mathbb{R}^n)$ and the space of smooth vectors then coincides with the Schwartz space $\mathcal{S}(\mathbb{R}^n)$, see Section 6.2.1.

We can give a sufficient condition for a field to act on smooth vectors:

Lemma 1.8.19. *Let $\sigma = \{\sigma_\pi : \mathcal{H}_\pi^\infty \to \mathcal{H}_\pi\}$ be a field defined on smooth vectors. If for each $\phi \in \mathcal{D}(G)$, $\sigma\widehat{\phi}$ is a field of operators acting on smooth vectors, that is,*

$$\sigma\widehat{\phi} = \{\sigma_\pi \pi(\phi) : \mathcal{H}_\pi^\infty \to \mathcal{H}_\pi^\infty\},$$

then σ acts on smooth vectors.

Proof. Let us assume that $\sigma\widehat{\phi}$ is a field of operators acting on smooth vectors for every $\phi \in \mathcal{D}(G)$. Then, for each $\pi \in \widehat{G}$ realised as a representation and each smooth vector $v \in \mathcal{H}_\pi^\infty$, $\sigma_\pi \widehat{\phi}(\pi) v$ is smooth. By the Dixmier-Malliavin Theorem, see Theorem 1.7.8. the finite linear combination of the vectors of the form $\phi(\pi)v$ form $\mathcal{H}_\pi^\infty$. Therefore $\sigma_\pi : \mathcal{H}_\pi^\infty \to \mathcal{H}_\pi^\infty$, and the statement is proved. □

As an application of Lemma 1.8.19, we see that the field $\widehat{\delta_{x_c}}$ given at the end of Example 1.8.10 acts on smooth vectors:

Example 1.8.20. For any $x_o \in G$, the field $\widehat{\delta_{x_o}} = \{\pi(x_o) : \mathcal{H}_\pi^\infty \to \mathcal{H}_\pi^\infty\} \in L^\infty(\widehat{G})$ acts on smooth vectors.

Proof. Let $x_o \in G$. If $\phi \in \mathcal{D}(G)$, then by (1.4), $\pi(x_o)\pi(\phi) = \widehat{\phi(\cdot\, x_o)}(\pi)$ and $\phi(\cdot\, x_o) \in \mathcal{D}(G)$. Thus for any $v \in \mathcal{H}_\pi^\infty$, $\pi(x_o)\pi(\phi)v$ is smooth. We conclude using Lemma 1.8.19. □

To summarise, we will identify measurable $\widehat{G}$-fields $\sigma = \{\sigma_\pi : \mathcal{H}_\pi^\infty \to \mathcal{H}_\pi, \pi \in \widehat{G}\}$ defined on smooth vectors with their possible extensions whenever possible. If the group is non-compact, we can not restrict ourselves to fields acting on smooth vectors.

Chapter 2

Quantization on compact Lie groups

In this chapter we briefly review the global quantization of operators and symbols on compact Lie groups following [RT13] and [RT10a] as well as more recent developments of this subject in this direction. Especially the monograph [RT10a] can serve as a companion for the material presented here, so we limit ourselves to explaining the main ideas only. This quantization yields full (finite dimensional) matrix-valued symbols for operators due to the fact that the unitary irreducible representations of compact Lie groups are all finite dimensional. Here, in order to motivate the developments on nilpotent groups, which is the main subject of the present monograph, we briefly review key elements of this theory referring to [RT10a] or to other sources for proofs and further details.

Technically, the machinery for such global quantization of operators on compact Lie groups appears to be simpler than that on graded Lie groups that we deal with in subsequent chapters. Indeed, since the symbols can be viewed as matrices (more precisely, as linear transformations of finite dimensional representation spaces), we do not have to worry about their domains of definitions, extensions, and other functional analytical properties arising in the nilpotent counterpart of the theory. Also, we have the Laplacian at our disposal, which is elliptic and bi-invariant, simplifying the analysis compared to the analysis based on, for example, the sub-Laplacian on the Heisenberg group, or more general Rockland operators on graded Lie groups. On the other hand, the theory on graded Lie groups is greatly assisted by the homogeneous structure, significantly simplifying the analysis of appearing difference operators and providing additional tools such as the naturally defined dilations on the group.

When we will be talking about the quantization on graded Lie groups in Chapter 5 we will be mostly concerned, at least in the first stage, about assigning an operator to a given symbol. In fact, it will be a small challenge by itself to make

V. Fischer, M. Ruzhansky, *Quantization on Nilpotent Lie Groups*, Progress in Mathematics 314,
DOI 10.1007/978-3-319-29558-9_2

rigorous sense of a notion of a symbol there, but eventually we will show that the correspondence between symbols and operators is one-to-one. The situation on compact Lie groups is considerably simpler in this respect. Moreover, in (2.19) we will give a simple formula determining the symbol for a given operator. Thus, here we may talk about quantization of both symbols and operators, with the latter being often preferable from the point of view of applications, when we are concerned in establishing certain properties of a given operator and use its symbol as a tool for it.

Overall, this chapter is introductory, also serving as a motivation for the subsequent analysis, so we only sketch the ideas and refer for a thorough treatise with complete proofs to the monograph [RT10a] or to the papers that we point out in relevant places.

We do not discuss here all applications of this analysis in the compact setting. For example, we can refer to [DR14b] for applications of this analysis to Schatten classes, r-nuclearity, and trace formulae for operators on $L^2(G)$ and $L^p(G)$ for compact Lie groups G. For the functional calculus of matrix symbols and operators on G we refer to [RW14].

A related but different approach to the pseudo-differential calculus of [RT10a] has been also recently investigated in [Fis15]; there, a different notion of difference operators is defined intrinsically on each compact groups. This will not be discussed here.

2.1 Fourier analysis on compact Lie groups

Throughout this chapter G is always a compact Lie group. As in Chapter 1, we equip it with the uniquely determined probability Haar measure which is automatically bi-invariant by the compactness of G. We denote it by dx. We start by making a few remarks on the representation theory specific to compact Lie groups.

2.1.1 Characters and tensor products

An important first addition to Section 1.1 is that for a compact group G, every continuous irreducible unitary representation of G is finite dimensional. We denote by d_π the dimension of a finite dimensional representation π, $d_\pi = \dim \mathcal{H}_\pi$.

Another important property is the orthogonality of representation coefficients as follows. Let $\pi_1, \pi_2 \in \widehat{G}$ and let us choose some basis in the representation spaces so that we can view π_1, π_2 as matrices $\pi_1 = ((\pi_1)_{ij})_{i,j=1}^{d_{\pi_1}}$ and $\pi_2 = ((\pi_2)_{kl})_{k,l=1}^{d_{\pi_2}}$. Then:

- if $\pi_1 \neq \pi_2$, then $((\pi_1)_{ij}, (\pi_2)_{kl})_{L^2(G)} = 0$ for all i, j, k, l;
- if $\pi_1 = \pi_2$ but $(i, j) \neq (k, l)$, then $((\pi_1)_{ij}, (\pi_2)_{kl})_{L^2(G)} = 0$;

- if $\pi_1 = \pi_2$ and $(i,j) = (k,l)$, then

$$((\pi_1)_{ij}, (\pi_2)_{kl})_{L^2(G)} = \frac{1}{d_\pi}, \qquad \text{with } d_\pi = d_{\pi_1} = d_{\pi_2}.$$

For a finite dimensional continuous unitary representation $\pi : G \to \mathcal{U}(\mathcal{H}_\pi)$ we denote

$$\chi_\pi(x) := \mathrm{Tr}(\pi(x)),$$

the *character* of the representation π. Characters have a number of fundamental properties most of which follow from properties of the trace:

- $\chi_\pi(e) = d_\pi$;
- $\pi_1 \sim \pi_2$ if and only if $\chi_{\pi_1} = \chi_{\pi_2}$;
- consequently, the character χ_π does not depend on the choice of the basis in the representation space $\mathcal{H}_\pi$;
- $\chi(yxy^{-1}) = \chi_\pi(x)$ for any $x, y \in G$;
- $\chi_{\pi_1 \oplus \pi_2} = \chi_{\pi_1} + \chi_{\pi_2}$;
- $\chi_{\pi_1 \otimes \pi_2} = \chi_{\pi_1}\chi_{\pi_2}$, with the tensor product $\pi_1 \otimes \pi_2$ defined in (2.1);
- a finite dimensional continuous unitary representation π of G is irreducible if and only if $\|\chi_\pi\|_{L^2(G)} = 1$.
- for $\pi_1, \pi_2 \in \widehat{G}$, $(\chi_{\pi_1}, \chi_{\pi_2})_{L^2(G)} = 1$ if $\pi_1 \sim \pi_2$, and $(\chi_{\pi_1}, \chi_{\pi_2})_{L^2(G)} = 0$ if $\pi_1 \not\sim \pi_2$;
- for any $f \in L^2(G)$, there is the decomposition

$$f = \sum_{\pi \in \widehat{G}} d_\pi f * \chi_\pi,$$

given by the projections (2.7).

If we take $\pi_1 \in \mathrm{Hom}(G, \mathcal{U}(\mathcal{H}_1))$ and $\pi_2 \in \mathrm{Hom}(G, \mathcal{U}(\mathcal{H}_2))$ two finite dimensional representations of G on $\mathcal{H}_1$ and $\mathcal{H}_2$, respectively, their *tensor product* $\pi_1 \otimes \pi_2$ is the representation on $\mathcal{H}_1 \otimes \mathcal{H}_2$, $\pi_1 \otimes \pi_2 \in \mathrm{Hom}(G, \mathcal{U}(\mathcal{H}_1 \otimes \mathcal{H}_2))$, defined by

$$(\pi_1 \otimes \pi_2)(x)(v_1 \otimes v_2) := \pi_1(x)v_1 \otimes \pi_2(x)v_2. \tag{2.1}$$

Here the inner product on $\mathcal{H}_1 \otimes \mathcal{H}_2$ is induced from those on $\mathcal{H}_1$ and $\mathcal{H}_2$ by

$$(v_1 \otimes v_2, w_1 \otimes w_2)_{\mathcal{H}_1 \otimes \mathcal{H}_2} := (v_1, w_1)_{\mathcal{H}_1}(v_2, w_2)_{\mathcal{H}_2}.$$

In particular, it follows that

$$((\pi_1 \otimes \pi_2)(x)(v_1 \otimes v_2), w_1 \otimes w_2)_{\mathcal{H}_1 \otimes \mathcal{H}_2} = (\pi_1(x)v_1, w_1)_{\mathcal{H}_1}(\pi_2(x)v_2, w_2)_{\mathcal{H}_2}. \tag{2.2}$$

If $\pi_1, \pi_2 \in \widehat{G}$, the representation $\pi_1 \otimes \pi_2$ does not have to be irreducible, and we can decompose it into irreducible ones:

$$\pi_1 \otimes \pi_2 = \bigoplus_{\pi \in \widehat{G}} m_\pi \pi. \tag{2.3}$$

The constants $m_\pi = m_\pi(\pi_1, \pi_2)$ are called the *Clebsch-Gordan coefficients* and they determine the multiplicity of π in $\pi_1 \otimes \pi_2$,

$$m_\pi \pi \equiv \oplus_1^{m_\pi} \pi.$$

Also, we can observe that in view of the finite dimensionality only finitely many of m_π's are non-zero. Combining this with (2.2), we see that the product of any of the matrix coefficients of representations $\pi_1, \pi_2 \in \widehat{G}$ can be written as a finite linear combination of matrix coefficients of the representations from (2.3) with non-zero Clebsch-Gordan coefficients. In fact, this can be also seen on the level of characters providing more insight into the multiplicities m_π. First, for the tensor product of π_1 and π_2 we have $\chi_{\pi_1 \otimes \pi_2} = \chi_{\pi_1} \chi_{\pi_2}$. Consequently, equality (2.3) implies

$$\chi_{\pi_1} \chi_{\pi_2} = \chi_{\pi_1 \otimes \pi_2} = \sum_{\pi \in \widehat{G}} m_\pi \chi_\pi \tag{2.4}$$

with

$$m_\pi = m_\pi(\pi_1, \pi_2) = (\chi_{\pi_1} \chi_{\pi_2}, \chi_\pi)_{L^2(G)}.$$

This equality can be now reduced to the maximal torus of G, for which we recall

Cartan's maximal torus theorem: Let $\mathbb{T}^l \hookrightarrow G$ be an injective group homomorphism with the largest possible l. Then two representations of G are equivalent if and only if their restrictions to $\mathbb{T}^l$ are equivalent. In particular, the restriction $\chi_\pi|_{\mathbb{T}^l}$ of χ_π to $\mathbb{T}^l$ determines the equivalence class $[\pi]$.

Now, coming back to (2.4), we can conclude that we have

$$\chi_{\pi_1}|_{\mathbb{T}^l} \ \chi_{\pi_2}|_{\mathbb{T}^l} = \sum_{\pi \in \widehat{G}} m_\pi \ \chi_\pi|_{\mathbb{T}^l}.$$

For a compact connected Lie group G, the maximal torus is also called the Cartan subgroup, and its dimension is denoted by $\operatorname{rank} G$, the *rank of* G.

Explicit formulae for representations and the Clebsch-Gordan coefficients on a number of compact groups have been presented by Vilenkin [Vil68] or Zhelobenko [Žel73], with further updates in [VK91, VK93] by Vilenkin and Klimyk.

2.1.2 Peter-Weyl theorem

As discussed in Section 1.3, the Casimir element of the universal enveloping algebra $\mathfrak{U}(\mathfrak{g})$ can be viewed as an elliptic linear second order bi-invariant partial differential operator on G. If G is equipped with the uniquely determined (normalised) bi-invariant Riemannian metric, the Casimir element can be viewed as its (negative definite) Laplace-Beltrami operator, which we will denote by $\mathcal{L}_G$. Consequently, for any $D \in \mathfrak{U}(\mathfrak{g})$ we have

$$D\mathcal{L}_G = \mathcal{L}_G D.$$

The fundamental result on compact groups is the Peter-Weyl Theorem [PW27] giving a decomposition of $L^2(G)$ into eigenspaces of the Laplacian $\mathcal{L}_G$ on G, which we now sketch.

Theorem 2.1.1 (Peter-Weyl). *The space $L^2(G)$ can be decomposed as the orthogonal direct sum of bi-invariant subspaces parametrised by $\widehat{G}$,*

$$L^2(G) = \bigoplus_{\pi \in \widehat{G}} V_\pi, \quad V_\pi = \{x \mapsto \mathrm{Tr}(A\pi(x)) : A \in \mathbb{C}^{d_\pi \times d_\pi}\},$$

the decomposition given by the Fourier series

$$f(x) = \sum_{\pi \in \widehat{G}} d_\pi \, \mathrm{Tr}\left(\widehat{f}(\pi)\pi(x)\right). \tag{2.5}$$

After a choice of the orthonormal basis in each representation space $\mathcal{H}_\pi$, the set

$$\mathcal{B} := \left\{\sqrt{d_\pi}\ \pi_{ij} : \pi = (\pi_{ij})_{i,j=1}^{d_\pi}, \pi \in \widehat{G}\right\} \tag{2.6}$$

becomes an orthonormal basis for $L^2(G)$. For $f \in L^2(G)$, the convergence of the series in (2.5) holds for almost every $x \in G$, and also in $L^2(G)$.

One possible idea for the proof of the Peter-Weyl theorem is as follows. Let us take $\mathcal{B}$ as in (2.6). Finite linear combinations of elements of $\mathcal{B}$ are called the *trigonometric polynomials* on G, and we denote them by $\mathrm{span}(\mathcal{B})$. From the orthogonality of representations (see Section 2.1.1) we know that $\mathcal{B}$ is an orthonormal set in $L^2(G)$. It follows from (2.3) and the consequent discussion that $\mathrm{span}(\mathcal{B})$ is a subalgebra of $C(G)$, trivial representation is its identity, and it is involutive since $\pi^* \in \widehat{G}$ if $\pi \in \widehat{G}$. By invariance it is clear that $\mathcal{B}$ separates points of G. Consequently, by the Stone-Weierstrass theorem $\mathrm{span}(\mathcal{B})$ is dense in $C(G)$. Therefore, it is also dense in $L^2(G)$, giving the basis and implying the Peter-Weyl theorem.

For $f \in L^2(G)$, the decomposition

$$f = \sum_{\pi \in \widehat{G}} d_\pi f * \chi_\pi$$

given in Section 2.1.1 corresponds to the decomposition (2.5), the projections of $L^2(G)$ to V_π given by the convolution mappings

$$L^2(G) \ni f \mapsto f * \chi_\pi \in V_\pi. \tag{2.7}$$

The Peter-Weyl theorem can be also viewed as the decomposition of left or right regular representations of G on $L^2(G)$ into irreducible components. Indeed, from the homomorphism property of representations it follows that in the decomposition

$$L^2(G) = \bigoplus_{\pi\in\widehat{G}} \bigoplus_{j=1}^{d_\pi} \operatorname{span}\{\pi_{ij} : 1 \le i \le d_\pi\}, \tag{2.8}$$

the spans on the right hand side are π_L-invariant, and the restriction of π_L to each such space is equivalent to the representation π itself. This gives the decomposition of π_L into irreducible components as

$$\pi_L \sim \bigoplus_{\pi\in\widehat{G}} \bigoplus_{1}^{d_\pi} \pi.$$

The same is true for the decomposition of $L^2(G)$ into π_R-invariant subspaces $\operatorname{span}\{\pi_{ij} : 1 \le j \le d_\pi\}$, replacing the spans in (2.8).

It follows that the spaces V_π are bi-invariant subspaces of $L^2(G)$ and, therefore, they are eigenspaces of all bi-invariant operators. In particular, they are eigenspaces for the Laplacian $\mathcal{L}_G$ and, by varying the basis in the representation space $\mathcal{H}_\pi$, we see that V_π corresponds to the same eigenvalue of $\mathcal{L}_G$, which we denote by $-\lambda_\pi$, i.e.

$$-\mathcal{L}_G|_{V_\pi} = \lambda_\pi \mathrm{I}, \quad \lambda_\pi \ge 0. \tag{2.9}$$

It is useful to introduce also the quantity corresponding to the first order elliptic operator $(\mathrm{I} - \mathcal{L}_G)^{1/2}$,

$$\langle\pi\rangle := (1 + \lambda_\pi)^{1/2}, \tag{2.10}$$

so that we also have

$$(\mathrm{I} - \mathcal{L}_G)^{1/2}|_{V_\pi} = \langle\pi\rangle \mathrm{I}.$$

The quantity $\langle\pi\rangle$ and its powers become very useful in quantifying the growth/decay of Fourier coefficients, and eventually of symbols of pseudo-differential operators.

Using the Fourier series expression (2.5) and the orthogonality of matrix coefficients of representations, one can readily show that the Plancherel identity takes the form

$$(f, g)_{L^2(G)} = \sum_{\pi\in\widehat{G}} d_\pi \operatorname{Tr}\left(\widehat{f}(\pi)\widehat{g}(\pi)^*\right).$$

From this, it becomes natural to define the norm $\|\cdot\|_{\ell^2(\widehat{G})}$,

$$\|\widehat{f}\|_{\ell^2(\widehat{G})} = \left(\sum_{\pi\in\widehat{G}} d_\pi \|\widehat{f}(\pi)\|_{\mathtt{HS}}^2\right)^{1/2}, \tag{2.11}$$

with

$$\|\widehat{f}(\pi)\|_{\mathtt{HS}} = \sqrt{\mathrm{Tr}\left(\widehat{f}(\pi)\widehat{f}(\pi)^*\right)}.$$

This norm defines the Hilbert space $\ell^2(\widehat{G})$ with the inner product

$$(\sigma,\tau)_{\ell^2(\widehat{G})} := \sum_{\pi\in\widehat{G}} d_\pi \,\mathrm{Tr}\,(\sigma(\pi)\tau(\pi)^*), \qquad \sigma,\tau\in\ell^2(\widehat{G}), \tag{2.12}$$

and

$$\|\sigma\|_{\ell^2(\widehat{G})} = (\sigma,\sigma)^{1/2}_{\ell^2(\widehat{G})} = \left(\sum_{\pi\in\widehat{G}} d_\pi \|\sigma(\pi)\|^2_{\mathtt{HS}}\right)^{1/2}, \quad \sigma\in\ell^2(\widehat{G}),$$

so that the Plancherel identity yields

$$\|f\|_{L^2(G)} = \|\widehat{f}\|_{\ell^2(\widehat{G})}. \tag{2.13}$$

We conclude the preliminary part by recording some useful relations between the dimensions d_π and the eigenvalues $\langle\pi\rangle$ for representations $\pi\in\widehat{G}$: there exists $C>0$ such that

$$d_\pi \le C\langle\pi\rangle^{\frac{\dim G}{2}} \quad \text{and, even stronger,} \quad d_\pi \le C\langle\pi\rangle^{\frac{\dim G - \mathrm{rank} G}{2}}. \tag{2.14}$$

The first estimate follows immediately from the Weyl asymptotic formula for the eigenvalue counting function for the first order elliptic operator $(\mathrm{I}-\mathcal{L}_G)^{1/2}$ on the compact manifold G recalling that d_π^2 is the multiplicity of the eigenvalue $\langle\pi\rangle$, and the second one follows with a little bit more work from the Weyl character formula, with $\mathrm{rank}\, G$ denoting the rank of G. There is also a simple convergence criterion

$$\sum_{\pi\in\widehat{G}} d_\pi^2 \langle\pi\rangle^{-s} < \infty \quad \text{if and only if} \quad s > \dim G, \tag{2.15}$$

which follows from property (ii) in Section 2.1.3 applied to the delta-distribution δ_e at the unit element $e\in G$.

2.1.3 Spaces of functions and distributions on G

Different spaces of functions and distributions can be characterised in terms of the Fourier coefficients. For this, it is convenient to introduce the space of matrices taking into account the dimensions of representations. Thus, we set

$$\begin{aligned} \Sigma &:= \left\{\sigma = (\sigma(\pi))_{\pi\in\widehat{G}} : \sigma(\pi)\in\mathscr{L}(\mathcal{H}_\pi)\right\} \\ &\simeq \left\{\sigma = (\sigma(\pi))_{\pi\in\widehat{G}} : \sigma(\pi)\in\mathbb{C}^{d_\pi\times d_\pi}\right\}, \end{aligned}$$

the second line valid after a choice of basis in $\mathcal{H}_\pi$, and we are interested in the images of function spaces on G in Σ under the Fourier transform.

As it will be pointed out in Remark 2.2.1, we should rather consider the quotient space $\Sigma/\sim$ as the space of Fourier coefficients, with the equivalence in Σ induced by the equivalence of representations. However, in order to simplify the exposition, we will keep the notation Σ as above.

The set Σ can be considered as a special case of the direct sum of Hilbert spaces described in (1.29), with the corresponding interpretation in terms of von Neumann algebras. However, a lot of the general machinery can be simplified in the present setting since the Fourier coefficients allow the interpretation of matrices indexed over the discrete set $\widehat{G}$, with the dimension of each matrix equal to the dimension of the corresponding representation.

Distributions

For any distribution $u \in \mathcal{D}'(G)$, its matrix Fourier coefficient at $\pi \in \widehat{G}$ is defined by

$$\widehat{u}(\pi) := \langle u, \pi^* \rangle.$$

These are well-defined since $\pi(x)$ are smooth (even analytic). This gives rise to the Fourier transform of distributions on G but we will come to this after stating a few properties of several function spaces.

The following equivalences are easy to obtain for spaces defined initially via their localisations to coordinate charts, in terms of the quantity $\langle \pi \rangle$ introduced in (2.10):

(i) as we have already seen, $f \in L^2(G)$ if and only if $\widehat{f} \in \ell^2(\widehat{G})$, i.e. if

$$\sum_{\pi \in \widehat{G}} d_\pi \|\widehat{f}(\pi)\|_{\mathtt{HS}}^2 < \infty.$$

(ii) For any $s \in \mathbb{R}$, we have $f \in H^s(G)$ if and only if $\langle \pi \rangle^s \widehat{f} \in \ell^2(\widehat{G})$ if and only if

$$\sum_{\pi \in \widehat{G}} d_\pi \langle \pi \rangle^{2s} \|\widehat{f}(\pi)\|_{\mathtt{HS}}^2 < \infty.$$

(iii) $f \in C^\infty(G)$ if and only if for every $M > 0$ there exits $C_M > 0$ such that

$$\|\widehat{f}(\pi)\|_{\mathtt{HS}} \leq C_M \langle \pi \rangle^{-M}$$

holds for all $\pi \in \widehat{G}$.

(iv) $u \in \mathcal{D}'(G)$ if and only if there exist $M > 0$ and $C > 0$ such that

$$\|\widehat{u}(\pi)\|_{\mathtt{HS}} \leq C \langle \pi \rangle^{M}$$

holds for all $\pi \in \widehat{G}$.

The second characterisation (ii) follows from (i) if we observe that $f \in H^s(G)$ means that $(\mathrm{I} - \mathcal{L}_G)^{s/2} f \in L^2(G)$, and then pass to the Fourier transform side. The third characterisation (iii) follows if we observe that $\widehat{f}(\pi)$ must satisfy (ii) for all s and use estimates (2.14), and (iv) follows from (iii) by duality. The last two characterisations motivate to define spaces $\mathcal{S}(\widehat{G}), \mathcal{S}'(\widehat{G}) \subset \Sigma$ by

$$\mathcal{S}(\widehat{G}) := \left\{\sigma \in \Sigma : \forall M > 0\ \exists C_M > 0 \text{ such that } \|\sigma(\pi)\|_{\mathtt{HS}} \leq C_M \langle\pi\rangle^{-M}\right\}$$

and

$$\mathcal{S}'(\widehat{G}) := \left\{\sigma \in \Sigma : \exists M > 0, C > 0 \text{ such that } \|\sigma(\pi)\|_{\mathtt{HS}} \leq C\langle\pi\rangle^{M}\right\},$$

with the seminormed topology on $\mathcal{S}(\widehat{G})$ defined by family

$$p_k(\sigma) = \sum_{\pi \in \widehat{G}} d_\pi \langle\pi\rangle^k \|\sigma(\pi)\|_{\mathtt{HS}},$$

and the dual topology on $\mathcal{S}'(\widehat{G})$. It follows that the Fourier inversion formula (2.5) can be extended to the following: the Fourier transform $\mathcal{F}_G$ in (1.2) and its inverse, defined by

$$(\mathcal{F}_G^{-1}\sigma)(x) := \sum_{\pi \in \widehat{G}} d_\pi \operatorname{Tr}(\sigma(\pi)\pi(x)), \tag{2.16}$$

are continuous as $\mathcal{F}_G : C^\infty(G) \to \mathcal{S}(\widehat{G})$, $\mathcal{F}_G^{-1} : \mathcal{S}(\widehat{G}) \to C^\infty(G)$, and are inverse to each other on $C^\infty(G)$ and $\mathcal{S}(\widehat{G})$. In particular, this implies that $\mathcal{S}(\widehat{G})$ is a nuclear Montel space. The distributional duality between $\mathcal{S}'(\widehat{G})$ and $\mathcal{S}(\widehat{G})$ is given by

$$\langle\sigma_1, \sigma_2\rangle_{\widehat{G}} = \sum_{\pi \in \widehat{G}} d_\pi \operatorname{Tr}(\sigma_1(\pi)\sigma_2(\pi)), \quad \sigma_1 \in \mathcal{S}'(\widehat{G}),\ \sigma_2 \in \mathcal{S}(\widehat{G}).$$

The Fourier transform can be then extended to the space of distributions $\mathcal{D}'(G)$. Thus, for $u \in \mathcal{D}'(G)$, we define $\mathcal{F}_G u \equiv \widehat{u} \in \mathcal{S}'(\widehat{G})$ by

$$\langle\mathcal{F}_G u, \tau\rangle_{\widehat{G}} := \left\langle u, \iota \circ \mathcal{F}_G^{-1}\tau\right\rangle_G, \quad \tau \in \mathcal{S}(\widehat{G}),$$

where $(\iota \circ \varphi)(x) = \varphi(x^{-1})$ and $\langle\cdot,\cdot\rangle_G$ is the distributional duality between $\mathcal{D}'(G)$ and $C^\infty(G)$. Analogously, its inverse is given by

$$\left\langle\mathcal{F}_G^{-1}\sigma, \varphi\right\rangle_G := \langle\sigma, \mathcal{F}_G(\iota \circ \varphi)\rangle_{\widehat{G}}, \quad \sigma \in \mathcal{S}'(\widehat{G}),\ \varphi \in C^\infty(G),$$

and these extended mappings are continuous between $\mathcal{D}'(G)$ and $\mathcal{S}'(\widehat{G})$ and are inverse to each other. It can be readily checked that they agree with their restrictions to spaces of test functions, explaining the appearance of the inversion mapping ι.

Gevrey spaces and ultradistributions

Recently, Gevrey spaces of ultradifferentiable functions as well as spaces of corresponding ultradistributions have been characterised as well. We say that a function $\phi \in C^\infty(G)$ is a *Gevrey-Roumieu ultradifferentiable function*, $\phi \in \gamma_s(G)$, if in every local coordinate chart, its local representative $\psi \in C^\infty(\mathbb{R}^n)$ belongs to $\gamma_s(\mathbb{R}^n)$, that is, satisfies the condition that there exist constants $A > 0$ and $C > 0$ such that

$$|\partial^\alpha \psi(x)| \leq C A^{|\alpha|} (\alpha!)^s$$

holds for all $x \in \mathbb{R}^n$ and all multi-indices α. For $s = 1$ we obtain the space of analytic functions on G. As with other spaces before, $\gamma_s(G)$ is thus defined as having its localisations in $\gamma_s(\mathbb{R}^n)$, and a question of its characterisation in terms of its Fourier coefficients arises.

Analogously, we say that ϕ is a *Gevrey-Beurling ultradifferentiable function*, $\phi \in \gamma_{(s)}(G)$, if its local representatives ψ satisfy the condition that for every $A > 0$ there exists $C_A > 0$ such that

$$|\partial^\alpha \psi(x)| \leq C_A A^{|\alpha|} (\alpha!)^s$$

holds for all $x \in \mathbb{R}^n$ and all multi-indices α. For $1 \leq s < \infty$, these spaces do not depend on the choice of local coordinates on G in the definition, and can be characterised as follows:

Proposition 2.1.2. *Let* $1 \leq s < \infty$.
(1) *We have* $\phi \in \gamma_s(G)$ *if and only if there exist* $B > 0$ *and* $K > 0$ *such that*

$$||\widehat{\phi}(\pi)||_{\mathtt{HS}} \leq K e^{-B\langle\pi\rangle^{1/s}}$$

holds for all $\pi \in \widehat{G}$.
(2) *We have* $\phi \in \gamma_{(s)}(G)$ *if and only if for every* $B > 0$ *there exists* $K_B > 0$ *such that*

$$||\widehat{\phi}(\pi)||_{\mathtt{HS}} \leq K_B e^{-B\langle\pi\rangle^{1/s}}$$

holds for all $\pi \in \widehat{G}$.

The space of continuous linear functionals on $\gamma_s(G)$ $\big($or $\gamma_{(s)}(G)\big)$ is called the space of *ultradistributions* and is denoted by $\gamma_s'(G)$ $\left(\text{or } \gamma_{(s)}'(G)\right)$, respectively.

For any $v \in \gamma_s'(G)$ $\left(\text{or } \gamma_{(s)}'(G)\right)$, we note that its Fourier coefficient at $\pi \in \widehat{G}$ can be defined analogously to the case of distributions by

$$\widehat{v}(\pi) := \langle v, \pi^* \rangle \equiv v(\pi^*).$$

These are well-defined since G is compact and hence $\pi(x)$ are analytic.

Proposition 2.1.3. *Let $1 \leq s < \infty$.*
(1) *We have $v \in \gamma_s'(G)$ if and only if for every $B > 0$ there exists $K_B > 0$ such that*

$$\|\widehat{v}(\pi)\|_{\mathtt{HS}} \leq K_B e^{B\langle\pi\rangle^{1/s}}$$

holds for all $\pi \in \widehat{G}$.
(2) *We have $v \in \gamma_{(s)}'(G)$ if and only if there exist $B > 0$ and $K > 0$ such that*

$$\|\widehat{v}(\pi)\|_{\mathtt{HS}} \leq K e^{B\langle\pi\rangle^{1/s}}$$

holds for all $\pi \in \widehat{G}$.

Proposition 2.1.2 can be actually extended to hold for any $0 < s < \infty$, and we refer to [DR14a] for proofs and further details. This can be viewed also from the point of view of general eigenfunction expansions of function of compact manifolds, see [DR16] for the treatment of more general Komatsu-type classes of ultradifferentiable functions and ultradistributions, building on an analogous description for analytic functions by Seeley [See69].

For a review of the representation theory of compact Lie groups and further constructions using the Littlewood-Paley decomposition based on the heat kernel we refer to Stein's book [Ste70b].

2.1.4 ℓ^p-spaces on the unitary dual $\widehat{G}$

For a general theory of non-commutative integration on locally compact unimodular groups we refer to Dixmier [Dix53] and Segal [Seg50, Seg53]. In this framework, the Hausdorff-Young inequality has been established (see Kunze [Kun58]) for a version of ℓ^p-spaces on the unitary dual $\widehat{G}$ based on the Schatten classes, namely, an inequality of the type

$$\left(\sum_{\pi\in\widehat{G}} d_\pi \|\widehat{f}(\pi)\|^{p'}_{S^{p'}_{d_\pi}}\right)^{1/p'} \leq \|f\|_{L^p(G)} \text{ for } 1 < p \leq 2,$$

with an obvious modification for $p = 1$, where $\frac{1}{p} + \frac{1}{p'} = 1$, and $S^{p'}_{d_\pi}$ is the $(d_\pi \times d_\pi)$-dimensional Schatten p'-class. While the theory of the above spaces is well-known (see e.g. Hewitt and Ross [HR70, Section 31] or Edwards [Edw72, Section 2.14]), here we describe and develop a little further another class of ℓ^p-spaces on $\widehat{G}$ which was considered in [RT10a, Section 10.3.3], to which we refer for details and proofs of statement that we do not prove here.

For $1 \leq p < \infty$, we define the space $\ell^p(\widehat{G}) \subset \Sigma$ by the condition

$$\|\sigma\|_{\ell^p(\widehat{G})} := \left(\sum_{\pi\in\widehat{G}} d_\pi^{p\left(\frac{2}{p}-\frac{1}{2}\right)} \|\sigma(\pi)\|^p_{\mathtt{HS}}\right)^{1/p} < \infty.$$

For $p = \infty$, we define the space $\ell^\infty(\widehat{G}) \subset \Sigma$ by

$$\|\sigma\|_{\ell^\infty(\widehat{G})} := \sup_{\pi\in\widehat{G}} d_\pi^{-1/2} \|\sigma(\pi)\|_{\mathtt{HS}} < \infty.$$

For $p = 2$ we recover the space $\ell^2(\widehat{G})$ defined in (2.11), while the $\ell^1(\widehat{G})$-norm becomes

$$\|\sigma\|_{\ell^1(\widehat{G})} := \sum_{\pi\in\widehat{G}} d_\pi^{3/2} \|\sigma(\pi)\|_{\mathtt{HS}}.$$

This space and the Hausdorff-Young inequality for it become useful in, for example, proving Proposition 2.1.2. Also, it appears naturally in questions concerning the convergence of the Fourier series:

Remark 2.1.4. If $\sigma \in \ell^1(\widehat{G})$, then the (Fourier) series (2.16) converges absolutely and uniformly on G.

On the other hand, one can show that if $f \in C^k(G)$ with an even $k > \frac{1}{2}\dim G$, then $\widehat{f} \in \ell^1(\widehat{G})$ and the Fourier series (2.5) converges uniformly. Indeed, we can estimate

$$\begin{aligned} \|\widehat{f}\|_{\ell^1(\widehat{G})} &= \sum_{\pi\in\widehat{G}} \frac{d_\pi^{3/2}}{\langle\pi\rangle^k} \|\pi((\mathrm{I}-\mathcal{L}_G)^{k/2} f)\|_{\mathtt{HS}} \\ &\le \left(\sum_{\pi\in\widehat{G}} d_\xi^2 \langle\pi\rangle^{-2k}\right)^{1/2} \left(\sum_{\pi\in\widehat{G}} d_\pi \|\pi((\mathrm{I}-\mathcal{L}_G)^{k/2} f)\|_{\mathtt{HS}}^2\right)^{1/2} \\ &\le C\|(\mathrm{I}-\mathcal{L}_G)^{k/2} f\|_{L^2(G)} < \infty, \end{aligned}$$

in view of the Plancherel formula and (2.15), provided that $2k > \dim G$. In fact, the same argument shows the implication

$$f \in H^s(G), \quad s > \frac{1}{2}\dim G \qquad \Longrightarrow \qquad \widehat{f} \in \ell^1(\widehat{G}),$$

with the uniform convergence of the Fourier series (2.5) of f.

Regarding these $\ell^p(\widehat{G})$-spaces as weighted sequence spaces with weights given by powers of d_π, a general theory of interpolation spaces [BL76, Theorem 5.5.1] implies that they are interpolation spaces, namely, for any $1 \le p_0, p_1 < \infty$, we have

$$\left(\ell^{p_0}(\widehat{G}), \ell^{p_1}(\widehat{G})\right)_{\theta,p} = \ell^p(\widehat{G}),$$

where $0 < \theta < 1$ and $\frac{1}{p} = \frac{1-\theta}{p_0} + \frac{\theta}{p_1}$, see [RT10a, Proposition 10.3.40].

The Hausdorff-Young inequality holds for these spaces as well. Namely, if $1 \le p \le 2$ and $\frac{1}{p} + \frac{1}{p'} = 1$, we have

$$\|\widehat{f}\|_{\ell^{p'}(\widehat{G})} \leq \|f\|_{L^p(G)} \tag{2.17}$$

for all $f \in L^p(G)$, and

$$\|\mathcal{F}_G^{-1}\sigma\|_{L^{p'}(G)} \leq \|\sigma\|_{\ell^p(\widehat{G})}, \tag{2.18}$$

for all $\sigma \in \ell^p(\widehat{G})$.

We give a brief argument for these. To prove (2.18), on one hand we already have Plancherel's identity (2.13). On the other hand, from (2.16) we have

$$|(\mathcal{F}_G^{-1}\sigma)(x)| \leq \sum_{\pi\in\widehat{G}} d_\pi \|\sigma(\pi)\|_{\mathtt{HS}} \|\pi(x)\|_{\mathtt{HS}} = \sum_{\pi\in\widehat{G}} d_\pi^{3/2} \|\sigma(\pi)\|_{\mathtt{HS}} = \|\sigma\|_{\ell^1(\widehat{G})}.$$

Now the Stein-Weiss interpolation (see e.g. [BL76, Corollary 5.5.4]) implies (2.18). From this, (2.17) follows using the duality $\ell^p(\widehat{G})' = \ell^{p'}(\widehat{G})$, $1 \leq p < \infty$.

We remark that it is also possible to prove (2.17) directly by interpolation as well. However, one needs to employ an ℓ^∞-version of the interpolation theory with the change of measure, as e.g. in Lizorkin [Liz75].

Let us point out the continuous embeddings, similar to the usual ones:

Proposition 2.1.5. *We have*

$$\ell^p(\widehat{G}) \hookrightarrow \ell^q(\widehat{G}) \quad \text{and} \quad \|\sigma\|_{\ell^q(\widehat{G})} \leq \|\sigma\|_{\ell^p(\widehat{G})} \qquad \forall \sigma \in \Sigma, \quad 1 \leq p \leq q \leq \infty.$$

Proof. We can assume $p < q$. Then, in the case $1 \leq p < \infty$ and $q = \infty$, we can estimate

$$\|\sigma\|^p_{\ell^\infty(\widehat{G})} = \left(\sup_{\pi\in\widehat{G}} d_\pi^{-\frac{1}{2}} \|\sigma(\pi)\|_{\mathtt{HS}} \right)^p \leq \sum_{\pi\in\widehat{G}} d_\pi^{2-\frac{p}{2}} \|\sigma(\pi)\|^p_{\mathtt{HS}} = \|\sigma\|^p_{\ell^p(\widehat{G})}.$$

Let now $1 \leq p < q < \infty$. Denoting $a_\pi := d_\pi^{\frac{2}{q}-\frac{1}{2}} \|\sigma(\pi)\|_{\mathtt{HS}}$, we get

$$\begin{aligned}\|\sigma\|_{\ell^q(\widehat{G})} &= \left(\sum_{\pi\in\widehat{G}} a_\pi^q \right)^{\frac{1}{q}} \leq \left(\sum_{\pi\in\widehat{G}} a_\pi^p \right)^{\frac{1}{p}} = \left(\sum_{\pi\in\widehat{G}} d_\pi^{p(\frac{2}{q}-\frac{1}{2})} \|\sigma(\pi)\|^p_{\mathtt{HS}} \right)^{\frac{1}{p}} \\ &\leq \|\sigma\|_{\ell^p(\widehat{G})},\end{aligned}$$

completing the proof. ∎

Finally, we establish a relation between the family $\ell^p(\widehat{G})$ and the corresponding Schatten family of ℓ^p-spaces, which we denote by $\ell^p_{sch}(\widehat{G})$, defined by the norms

$$\|\sigma\|_{\ell^p_{sch}(\widehat{G})} := \left(\sum_{\pi\in\widehat{G}} d_\pi \|\sigma(\pi)\|^p_{S^p} \right)^{1/p}, \quad \sigma \in \Sigma,\ 1 \leq p < \infty,$$

where $S^p = S^p_{d_\pi}$ is the $(d_\pi \times d_\pi)$-dimensional Schatten p-class, and

$$\|\sigma\|_{\ell^\infty_{sch}(\widehat{G})} := \sup_{\pi\in\widehat{G}} \|\sigma(\pi)\|_{\mathscr{L}(\mathcal{H}_\pi)}, \quad \sigma \in \Sigma.$$

We have the following relations:

Proposition 2.1.6. *For $1 \le p \le 2$, we have continuous embeddings as well as the estimates*

$$\ell^p(\widehat{G}) \hookrightarrow \ell^p_{sch}(\widehat{G}) \quad \text{and} \quad \|\sigma\|_{\ell^p_{sch}(\widehat{G})} \le \|\sigma\|_{\ell^p(\widehat{G})} \qquad \forall \sigma \in \Sigma, \quad 1 \le p \le 2.$$

For $2 \le p \le \infty$, we have

$$\ell^p_{sch}(\widehat{G}) \hookrightarrow \ell^p(\widehat{G}) \quad \text{and} \quad \|\sigma\|_{\ell^p(\widehat{G})} \le \|\sigma\|_{\ell^p_{sch}(\widehat{G})} \qquad \forall \sigma \in \Sigma, \quad 2 \le p \le \infty.$$

Proof. For $p = 2$, the norms coincide since $S^2 = \mathtt{HS}$. Let first $1 \le p < 2$. Since $\sigma(\pi) \in \mathbb{C}^{d_\pi \times d_\pi}$, denoting by s_j its singular numbers, by the Hölder inequality we have

$$\|\sigma(\pi)\|^p_{S^p} = \sum_{j=1}^{d_\pi} s_j^p \le \left(\sum_{j=1}^{d_\pi} 1\right)^{\frac{2-p}{2}} \left(\sum_{j=1}^{d_\pi} s_j^{p\frac{2}{p}}\right)^{\frac{p}{2}} = d_\pi^{\frac{2-p}{2}} \|\sigma(\pi)\|^p_{\mathtt{HS}},$$

i.e.

$$\|\sigma(\pi)\|_{S^p} \le d_\pi^{\frac{2-p}{2p}} \|\sigma(\pi)\|_{\mathtt{HS}} \qquad (1 \le p \le 2).$$

Consequently, it follows that

$$\|\sigma\|^p_{\ell^p_{sch}(\widehat{G})} = \sum_{\pi\in\widehat{G}} d_\pi \|\sigma(\pi)\|^p_{S^p} \le \sum_{\pi\in\widehat{G}} d_\pi^{2-\frac{p}{2}} \|\sigma(\pi)\|^p_{\mathtt{HS}} = \|\sigma\|^p_{\ell^p(\widehat{G})},$$

proving the first claim. Conversely, for $2 < p < \infty$, we can estimate

$$\|\sigma(\pi)\|^2_{\mathtt{HS}} = \sum_{j=1}^{d_\pi} s_j^2 \le \left(\sum_{j=1}^{d_\pi} 1\right)^{\frac{p-2}{p}} \left(\sum_{j=1}^{d_\pi} s_j^{2\frac{p}{2}}\right)^{\frac{2}{p}} = d_\pi^{\frac{p-2}{p}} \|\sigma(\pi)\|^2_{S^p},$$

implying

$$\|\sigma(\pi)\|_{\mathtt{HS}} \le d_\pi^{\frac{p-2}{2p}} \|\sigma(\pi)\|_{S^p} \qquad (2 < p < \infty).$$

It follows that

$$\|\sigma\|^p_{\ell^p(\widehat{G})} = \sum_{\pi\in\widehat{G}} d_\pi^{2-\frac{p}{2}} \|\sigma(\pi)\|^p_{\mathtt{HS}} \le \sum_{\pi\in\widehat{G}} d_\pi \|\sigma(\pi)\|^p_{S^p} = \|\sigma\|^p_{\ell^p_{sch}(\widehat{G})},$$

proving the second claim for $2 < p < \infty$. Finally, for $p = \infty$, the inequality

$$\|\sigma(\pi)\|_{\mathtt{HS}} \le d_\pi^{1/2} \|\sigma(\pi)\|_{\mathscr{L}(\mathcal{H}_\pi)}$$

implies

$$\|\sigma\|_{\ell^\infty(\widehat{G})} = \sup_{\pi\in\widehat{G}} d_\pi^{-1/2}\|\sigma(\pi)\|_{\mathtt{HS}} \leq \sup_{\pi\in\widehat{G}} \|\sigma(\pi)\|_{\mathscr{L}(\mathcal{H}_\pi)} = \|\sigma\|_{\ell^\infty_{sch}(\widehat{G})},$$

completing the proof. □

2.2 Pseudo-differential operators on compact Lie groups

In this section we look at linear continuous operators $A : C^\infty(G) \to \mathcal{D}'(G)$ and a global quantization of A yielding its full matrix-valued symbol. By the Schwartz kernel theorem (Theorem 1.4.1) there exists a unique distribution $K_A \in \mathcal{D}'(G\times G)$ such that

$$Af(x) = \int_G K_A(x,y)f(y)dy,$$

interpreted in the distributional sense. We can rewrite this as a right-convolution kernel operator

$$Af(x) = \int_G R_A(x, y^{-1}x)f(y)dy,$$

with

$$R_A(x,y) = K_A(x, xy^{-1}),$$

so that

$$Af(x) = (f * R_A(x,\cdot))(x).$$

2.2.1 Symbols and quantization

The idea for the following construction is that we define the symbol of A as the Fourier transform of its right convolution kernel in the second variable. However, for the presentation purposes we now take a different route and, instead, we define the mapping $\sigma_A : G \times \widehat{G} \to \Sigma$ by

$$\sigma_A(x,\pi) := \pi(x)^*(A\pi)(x), \tag{2.19}$$

with $(A\pi)(x) \in \mathscr{L}(\mathcal{H}_\pi)$ defined by

$$(A\pi(x)u, v)_{\mathcal{H}_\pi} := A(\pi(x)u, v)_{\mathcal{H}_\pi}$$

for all $u, v \in \mathcal{H}_\pi$. After choosing a basis in the representation space $\mathcal{H}_\pi$, we can interpret this as a matrix $\sigma_A(x,\pi) \in \mathbb{C}^{d_\pi\times d_\pi}$ and $(A\pi)_{ij} = A(\pi_{ij})$, i.e. the operator A acts on the matrix $\pi(x)$ componentwise, so that

$$\sigma_A(x,\pi)_{ij} = \sum_{k=1}^{d_\pi} \overline{\pi_{ki}(x)} A\pi_{kj}(x).$$

We note that the symbol in (2.19) is well-defined since we can multiply the distribution $A\pi$ by a smooth (even analytic) matrix π.

Remark 2.2.1. We also observe that strictly speaking, the definition (2.19) depends on the choice of the representation π from its equivalence class $[\pi]$. Namely, if $\pi_1 \sim \pi_2$, so that

$$\pi_2(x) = U^{-1}\pi_1(x)U$$

for some unitary U and all $x \in G$, then

$$\widehat{f}(\pi_2) = U^{-1}\widehat{f}(\pi_1)U$$

and, therefore,

$$\sigma_A(x, \pi_2) = U^{-1}\sigma_A(x, \pi_1)U. \tag{2.20}$$

However, it can be readily checked that the quantization formula (2.22) below remains unchanged due to the presence of the trace. So, denoting by $\operatorname{Rep} G$ the set of all strongly continuous unitary irreducible representations of G, the symbol is well defined as a mapping

$$\sigma_A : G \times \operatorname{Rep} G \to \Sigma \qquad \text{or as} \qquad \sigma_A : G \times \widehat{G} \to \Sigma/\sim$$

where the equivalence on Σ is given by the equivalence of representations on $\operatorname{Rep} G$ inducing the equivalence on Σ by conjugations, as in formula (2.20). We will disregard this technicality in the current presentation to simplify the exposition, referring to [RT10a] for a more rigorous treatment. We note, however, that if $\pi_1 \sim \pi_2$, then

$$\operatorname{Tr}\left(\pi_1(x)\sigma_A(x, \pi_1)\widehat{f}(\pi_1)\right) = \operatorname{Tr}\left(\pi_2(x)\sigma_A(x, \pi_2)\widehat{f}(\pi_2)\right). \tag{2.21}$$

Using the symbol σ_A, it follows that the linear continuous operator $A : C^\infty(G) \to \mathcal{D}'(G)$ can be (de-)quantized as

$$Af(x) = \sum_{\pi \in \widehat{G}} d_\pi \operatorname{Tr}\left(\pi(x)\sigma_A(x, \pi)\widehat{f}(\pi)\right). \tag{2.22}$$

If the operator A maps $C^\infty(G)$ to itself and $f \in C^\infty(G)$, the formula (2.22) can be understood in the pointwise sense to hold for all $x \in G$, with the absolute convergence of the series. It can be shown that formulae (2.19) and (2.22) imply that σ_A is the Fourier transform of R_A, namely, we have

$$\sigma_A(x, \pi) = \int_G R_A(x, y)\pi(y)^* dy.$$

If the formula (2.22) holds, we will also write $A = \operatorname{Op}(\sigma_A)$.

In view of (2.21), the sum in (2.22) does not depend on the choice of a representation π from its equivalence class $[\pi]$.

Example 2.2.2. For the identity operator I we have its symbol

$$\sigma_{\mathrm{I}}(x,\pi) = \pi(x)^*\pi(x) = \mathrm{I}_{d_\pi}$$

is the identity matrix in $\mathbb{C}^{d_\pi \times d_\pi}$, by the unitarity of $\pi(x)$, so that (2.22) recovers the Fourier inversion formula (2.5) in this case. For the Laplacian $\mathcal{L}_G$ on G, we have

$$\sigma_{\mathcal{L}_G}(x,\pi) = \pi(x)^*\mathcal{L}_G\pi(x) = -\lambda_\pi \mathrm{I}_{d_\pi}$$

by the unitarity of π and (2.9), where $-\lambda_\pi$ are the eigenvalues of $\mathcal{L}_G$ corresponding to π. Consequently, we also have

$$\sigma_{(\mathrm{I}-\mathcal{L}_G)^{\mu/2}}(x,\pi) = \langle\pi\rangle^{\mu}\mathrm{I}_{d_\pi}.$$

Example 2.2.3. In the case of the torus $G = \mathbb{T}^n = \mathbb{R}^n/\mathbb{Z}^n$, and the representations $\{\pi_\xi\}_{\xi\in\mathbb{Z}^n}$ fixed as in Remark 1.1.4, we see that all $d_{\pi_\xi} = 1$. Hence

$$\sigma_A(x,\pi_\xi) \equiv \sigma_A(x,\xi) = e^{-2\pi i x\cdot\xi}A(e^{2\pi i x\cdot\xi}) \in \mathbb{C}, \quad (x,\xi) \in \mathbb{T}^n \times \mathbb{Z}^n,$$

with the quantization (2.22) becoming the *toroidal quantization*

$$Af(x) = \sum_{\xi\in\mathbb{Z}^n} e^{2\pi i x\cdot\xi}\,\sigma_A(x,\xi)\,\widehat{f}(\xi),$$

for a thorough analysis of which we refer to [RT10b] and [RT10a, Section 4].

Example 2.2.4. With our choices of definitions, the symbols of left-invariant operators on G become independent of x. As shown in Section 1.5, if

$$Af = f * \kappa$$

for some $\kappa \in L^1(G)$, then it is left-invariant. Consequently, the right convolution kernel of A is $R_A(x,y) = \kappa(y)$ and, therefore, its Fourier transform is

$$\sigma_A(x,\pi) = \widehat{\kappa}(\pi).$$

On the other hand, if

$$Af = \kappa * f$$

for some $\kappa \in L^1(G)$, then it is right-invariant. In this case its right convolution kernel is $R_A(x,y) = \kappa(xyx^{-1})$ and, therefore, its Fourier transform in y gives

$$\sigma_A(x,\pi) = \pi(x)^*\widehat{\kappa}(\pi)\pi(x).$$

The notion of the symbol σ_A becomes already useful in stating a criterion for the L^2-boundedness for an operator A. We recall from Section 1.3 that X^α denotes the left-invariant partial differential operators of order $|\alpha|$ corresponding to a basis of left-invariant vector fields $X_1,\cdots,X_n$, $n = \dim G$, of the Lie algebra

$\mathfrak{g}$ of G. As the derivatives with respect to these vector fields in general do not commute, in principle we have to take into account their order in forming partial differential operators of higher degrees. However, we note that the subsequent statements remain valid if we restrict our choice to

$$X^\alpha = X_1^{\alpha_1} \cdots X_n^{\alpha_n}.$$

We will sometimes write X_x^α to emphasise that the derivatives are taken with respect to the variable x.

Theorem 2.2.5. *Let G be a compact Lie group and let $A : C^\infty(G) \to C^\infty(G)$ be a linear continuous operator. Let k be an integer such that $k > \frac{1}{2}\dim G$. Assume that there is a constant $C > 0$ such that*

$$\|X_x^\alpha \sigma_A(x,\pi)\|_{\mathscr{L}(\mathcal{H}_\pi)} \le C$$

for all $(x,\pi) \in G \times \widehat{G}$, and all $|\alpha| \le k$. Then A extends to a bounded operator from $L^2(G)$ to $L^2(G)$.

In this theorem and elsewhere, $\|\cdot\|_{\mathscr{L}(\mathcal{H}_\pi)}$ denotes the operator norm of $\sigma_A(x,\pi) \in \mathscr{L}(\mathcal{H}_\pi)$ or, after a choice of the basis, the operator norm of the matrix multiplication by the matrix $\sigma_A(x,\pi) \in \mathbb{C}^{d_\pi \times d_\pi}$. The appearance of the operator norm is natural since for the convolution operators we have

$$\|f \mapsto f * h\|_{\mathscr{L}(L^2(G))} = \|f \mapsto h * f\|_{\mathscr{L}(L^2(G))} = \sup_{\pi \in \widehat{G}} \|\widehat{h}(\pi)\|_{\mathscr{L}(\mathcal{H}_\pi)}, \tag{2.23}$$

following from $\widehat{f * h}(\pi) = \widehat{h}(\pi)\widehat{f}(\pi)$ and Plancherel's theorem.

2.2.2 Difference operators and symbol classes

In order to describe the symbolic properties and to establish the symbolic calculus of operators we have to replace the derivatives in frequency, used in the symbolic calculus on $\mathbb{R}^n$, by suitable operations acting on the space Σ of Fourier coefficients. We call these operations *difference operators*. Roughly speaking, this corresponds to the idea that in the Calderón-Zygmund theory, the integral kernel K_A has singularities at the diagonal or, in other words, the right-convolution kernel $R_A(x,\cdot)$ has singularity at the unit element e of the group only. Therefore, if we form an operator with a new integral kernel $q(\cdot)R_A(x,\cdot)$ with a smooth $q \in C^\infty(G)$ satisfying $q(e) = 0$, the properties of this new operator should be better than those of the original operator A.

In [RT10a], the corresponding notion of difference operators has been introduced leading to the symbolic calculus of operators on G. However, we now follow the ideas of [RTW14] with a slightly more general treatment of difference operators.

Definition 2.2.6. Let $q \in C^\infty(G)$ vanish of order $k \in \mathbb{N}$ at the unit element $e \in G$, i.e. $(Dq)(e) = 0$ for all left-invariant differential operators $D \in \mathrm{Diff}^{k-1}(G)$ of order $k-1$. Then the *difference operator of order* k is an operator acting on the space Σ of Fourier coefficients by the formula

$$(\Delta_q \widehat{f})(\pi) := \widehat{qf}(\pi).$$

We denote the set of all difference operators of order k by $\mathrm{diff}^k(\widehat{G})$.

We now define families of first order difference operators replacing derivatives in the frequency variable in the Euclidean setting.

Definition 2.2.7. A collection of ℓ first order difference operators $\Delta_{q_1}, \dots, \Delta_{q_\ell} \in \mathrm{diff}^1(\widehat{G})$ is called *admissible*, if the corresponding functions $q_1, \dots, q_\ell \in C^\infty(G)$ satisfy

$$q_j(e) = 0, \ \ \mathrm{d}q_j(e) \neq 0, \ \ j = 1, \dots, \ell,$$

and, moreover,

$$\mathrm{rank}(\mathrm{d}q_1(e), \dots, \mathrm{d}q_\ell(e)) = \dim G.$$

It follows, in particular, that e is an isolated common zero of the family $\{q_j\}_{j=1}^{\ell}$. We call an admissible collection *strongly admissible*, if it is the only common zero, i.e. if

$$\bigcap_{j=1}^{\ell} \{x \in G : q_j(x) = 0\} = \{e\}.$$

We note that difference operators all commute with each other. For a given admissible collection of difference operators we use the multi-index notation

$$\Delta_\pi^\alpha := \Delta_{q_1}^{\alpha_1} \cdots \Delta_{q_\ell}^{\alpha_\ell} \ \text{ and } \ q^\alpha(x) := q_1(x)^{\alpha_1} \cdots q_\ell(x)^{\alpha_\ell},$$

the dimension of the multi-index $\alpha \in \mathbb{N}_0^\ell$ depending on the number ℓ of difference operators in the collection. Consequently, there exist corresponding differential operators $X^{(\alpha)} \in \mathrm{Diff}^{|\alpha|}(G)$ such that the Taylor expansion formula

$$f(x) = \sum_{|\alpha| \leq N-1} \frac{1}{\alpha!} q^\alpha(x^{-1}) X^{(\alpha)} f(e) + \mathcal{O}(h(x)^N), \qquad h(x) \to 0, \tag{2.24}$$

holds true for any smooth function $f \in C^\infty(G)$ and any N, with $h(x)$ the geodesic distance from x to the identity element e. An explicit construction of operators $X^{(\alpha)}$ in terms of $q^\alpha(x)$ can be found in [RT10a, Section 10.6]. Operators X^α and $X^{(\alpha)}$ can be expressed in terms of each other.

Example 2.2.8. In the case of the torus, $G = \mathbb{T}^n = \mathbb{R}^n/\mathbb{Z}^n$, let

$$q_j(x) = e^{-2\pi i x_j} - 1, \ \ j = 1, \dots, n.$$

The collection $\{q_j\}_{j=1}^n$ is strongly admissible, and the corresponding difference operators take the form

$$(\Delta_{q_j}\sigma)(\pi_\xi) \equiv (\Delta_{q_j}\sigma)(\xi) = \sigma(\xi + e_j) - \sigma(\xi), \quad j = 1, \dots, n,$$

with $\pi_\xi \in \widehat{\mathbb{T}^n}$ identified with $\xi \in \mathbb{Z}^n$, where e_j is the jth unit vector in $\mathbb{Z}^n$. The periodic Taylor expansion takes the following form (see [RT10a, Theorem 3.4.4]): for any $\phi \in C^\infty(\mathbb{T}^n)$ we have

$$\phi(x) = \sum_{|\alpha|<N} \frac{1}{\alpha!}(e^{2\pi i x} - 1)^\alpha X_z^{(\alpha)}\phi(z)|_{z=0} + \sum_{|\alpha|=N} \phi_\alpha(x)(e^{2\pi i x} - 1)^\alpha,$$

where $\phi_\alpha \in C^\infty(\mathbb{T}^n)$ and

$$(e^{2\pi i x} - 1)^\alpha := (e^{2\pi i x_1} - 1)^{\alpha_1} \cdots (e^{2\pi i x_n} - 1)^{\alpha_n}.$$

The operators $X_z^{(\alpha)}$ have the form

$$X_z^{(\alpha)} = X_{z_1}^{(\alpha_1)} \cdots X_{z_n}^{(\alpha_n)} \quad \text{with} \quad X_{z_k}^{(\alpha_k)} = \prod_{j=0}^{\alpha_k - 1} \left(\frac{1}{2\pi i} \frac{\partial}{\partial z_k} - j \right).$$

Example 2.2.9. For partial differential operators, it can be readily observed that the application of difference operators reduces the order of symbols. Thus, let

$$D = \sum_{|\alpha|\le N} c_\alpha(x) X_x^\alpha, \quad c_\alpha \in C^\infty(G).$$

Then it was shown in [RT10a, Proposition 10.7.4] that

$$\Delta_q \sigma_D(x,\pi) = \sum_{|\alpha|\le N} c_\alpha(x) \sum_{\beta\le\alpha} \binom{\alpha}{\beta} (-1)^{|\beta|} (X_x^\beta q)(e) \sigma_{X_x^{\alpha-\beta}}(x,\pi).$$

In particular, if q has zero of order M at $e \in G$ then $\mathrm{Op}(\Delta_q\sigma_D)$ is of order $N - M$.

Remark 2.2.10. We can estimate differences in terms of original symbols: assume that the symbol $\sigma \in \Sigma$ satisfies

$$\mu := \sup_\pi \langle \pi \rangle^{-m} \|\sigma(\pi)\|_{\mathscr{L}(\mathcal{H}_\pi)} < \infty$$

for some $m \in \mathbb{R}$. Then for any difference operator Δ_q defined in terms of a function $q \in C^\infty(G)$ we have the estimate

$$\|\Delta_q \sigma(\pi)\|_{\mathscr{L}(\mathcal{H}_\pi)} \le C\mu \|q\|_{C^{\varkappa + \lceil |m| \rceil}(G)} \langle \pi \rangle^m$$

with a constant C independent of σ and q, where $\varkappa = \lceil (\dim G)/2 \rceil$ is the smallest integer larger than half the dimension of G and $\lceil |m| \rceil$ is the smallest integer larger than $|m|$. We refer to [RW14, Lemma 7.1] for the proof. However, if q vanishes at the unit element e to some order, we can impose a much better behaviour.

The usual Hörmander classes $\Psi^m(G)$ of pseudo-differential operators on G viewed as a manifold can be characterised in terms of the matrix-valued symbols. Here we recall that $A \in \Psi^m(G)$ means that in every local coordinate chart $U \subset G$, the pullback of $A|_U$ to $\mathbb{R}^n$ is a pseudo-differential operator $A_U \in \Psi^m_{1,0}(\mathbb{R}^n)$, i.e. it can be written as

$$A_U f(x) = \int_{\mathbb{R}^n} e^{2\pi i x\cdot\xi} a(x,\xi)\widehat{f}(\xi)d\xi \text{ with } \widehat{f}(\xi) = \int_{\mathbb{R}^n} e^{-2\pi i x\cdot\xi} f(x)dx, \tag{2.25}$$

with symbol $a = a_U \in S^m_{1,0}(\mathbb{R}^n)$, i.e. satisfying

$$|\partial_x^\beta \partial_\xi^\alpha a(x,\xi)| \le C_{\alpha\beta}(1+|\xi|)^{m-|\alpha|}$$

for all multi-indices α, β, and all $x, \xi \in \mathbb{R}^n$.

The following characterisation was partly proved in [RT10a, RT13] (namely (A)$\Longleftrightarrow$(C)) and completed in [RTW14] (namely (B)$\Longleftrightarrow$(C)$\Longleftrightarrow$(D)) .

Theorem 2.2.11. *Let G be a compact Lie group of dimension n. Let A be a linear continuous operator from $C^\infty(G)$ to $\mathcal{D}'(G)$. Then the following statements are equivalent:*

(A) $A \in \Psi^m(G)$.

(B) For every left-invariant differential operator $D \in \mathrm{Diff}^k(G)$ of order k and every difference operator $\Delta_q \in \mathrm{diff}^l(\widehat{G})$ of order l the symbol estimate

$$\|\Delta_q D\sigma_A(x,\pi)\|_{\mathscr{L}(\mathcal{H}_\pi)} \le C_{qD}\langle\pi\rangle^{m-l}$$

is valid.

(C) For an admissible collection $\Delta_1, \dots, \Delta_\ell \in \mathrm{diff}^1(\widehat{G})$ we have

$$\|\Delta_\pi^\alpha X_x^\beta \sigma_A(x,\pi)\|_{\mathscr{L}(\mathcal{H}_\pi)} \le C_{\alpha\beta}\langle\pi\rangle^{m-|\alpha|}$$

for all multi-indices $\alpha \in \mathbb{N}_0^\ell$ and $\beta \in \mathbb{N}_0^n$. Moreover,

$$\mathrm{sing\,supp}\, R_A(x,\cdot) \subseteq \{e\}.$$

(D) For a strongly admissible collection $\Delta_1, \dots, \Delta_\ell \in \mathrm{diff}^1(\widehat{G})$ we have

$$\|\Delta_\pi^\alpha X_x^\beta \sigma_A(x,\pi)\|_{\mathscr{L}(\mathcal{H}_\pi)} \le C_{\alpha\beta}\langle\pi\rangle^{m-|\alpha|}$$

for all multi-indices $\alpha \in \mathbb{N}_0^\ell$ and $\beta \in \mathbb{N}_0^n$.

Motivated by Theorem 2.2.11, (D), we may define symbol classes $S^m_{\rho,\delta}(G)$. Fixing a strongly admissible collection of difference operators

$$\Delta_1, \dots, \Delta_\ell \in \mathrm{diff}^1(\widehat{G}),$$

we say that $\sigma_A \in S^m_{\rho,\delta}(G)$ if $\sigma_A(x,\cdot) \in \Sigma$ satisfies

$$\|\Delta_\pi^\alpha X_x^\beta \sigma_A(x,\pi)\|_{\mathscr{L}(\mathcal{H}_\pi)} \leq C_{\alpha\beta}\langle\pi\rangle^{m-\rho|\alpha|+\delta|\beta|} \tag{2.26}$$

for all $(x,\pi) \in G \times \widehat{G}$ and for all multi-indices $\alpha \in \mathbb{N}_0^\ell$ and $\beta \in \mathbb{N}_0^n$. If $\rho > \delta$, this definition is independent of the choice of a strongly admissible collection of difference operators. The equivalence (A)$\Longleftrightarrow$(D) in Theorem 2.2.11 can be rephrased as

$$A \in \Psi^m(G) \Longleftrightarrow \sigma_A \in S^m_{1,0}(G).$$

For any $0 \leq \delta < \rho \leq 1$, the equivalence (B)$\Longleftrightarrow$(C)$\Longleftrightarrow$(D) in Theorem 2.2.11 remains valid for the symbol class $S^m_{\rho,\delta}(G)$ if we replace the symbolic conditions there by the condition (2.26). As we shall see later, the class $S^m_{\rho,\delta}(G)$ with different values of ρ and δ becomes useful in a number of applications.

Theorem 2.2.5 has analogue for (ρ,δ) classes:

Theorem 2.2.12. *Let $0 \leq \delta < \rho \leq 1$ and let A be an operator with symbol in $S^m_{\rho,\delta}(G)$. Then A is a bounded from $H^s(G)$ to $H^{s-m}(G)$ for any $s \in \mathbb{R}$.*

See [RW14, Theorem 5.1] for the proof.

2.2.3 Symbolic calculus, ellipticity, hypoellipticity

We now give elements of the symbolic calculus on the compact Lie group G. Here, we fix some strongly admissible collection of difference operators, with corresponding operators $X_x^{(\alpha)}$ coming from the Taylor expansion formula (2.24). We refer to [RT10a, Section 10.7.3] for proofs and other variants of the calculus below. We start with the composition.

Theorem 2.2.13. *Let $m_1, m_2 \in \mathbb{R}$ and $0 \leq \delta < \rho$. Let $A, B : C^\infty(G) \to C^\infty(G)$ be linear continuous operators with symbols $\sigma_A \in S^{m_1}_{\rho,\delta}(G)$ and $\sigma_B \in S^{m_1}_{\rho,\delta}(G)$. Then $\sigma_{AB} \in S^{m_1+m_2}_{\rho,\delta}(G)$ and we have*

$$\sigma_{AB} \sim \sum_{\alpha\geq 0} \frac{1}{\alpha!}(\Delta_\pi^\alpha \sigma_A)(X^{(\alpha)}\sigma_B),$$

where the asymptotic expansion means that for every $N \in \mathbb{N}$ we have

$$\sigma_{AB}(x,\pi) - \sum_{|\alpha|<N} \frac{1}{\alpha!}(\Delta_\pi^\alpha \sigma_A)(x,\pi) X_x^{(\alpha)} \sigma_B(x,\pi) \in S^{m_1+m_2-(\rho-\delta)N}_{\rho,\delta}(G).$$

The composition formula together with Theorem 2.2.5 imply a criterion for the boundedness in L^2-Sobolev spaces.

Corollary 2.2.14. *Let G be a compact Lie group and let $A : C^\infty(G) \to C^\infty(G)$ be a linear continuous operator. Let $m \in \mathbb{R}$. Assume that the symbol σ_A satisfies*

$$\|X_x^\alpha \sigma_A(x,\pi)\|_{\mathscr{L}(\mathcal{H}_\pi)} \leq C_\alpha \langle \pi \rangle^m$$

for all $(x,\pi) \in G \times \widehat{G}$, and all multi-indices α. Then A extends to a bounded operator from $H^s(G)$ to $H^{s-m}(G)$, for all $s \in \mathbb{R}$.

Let us now present a construction of amplitude operators in our setting. Let $0 \leq \rho, \delta \leq 1$. We say that $a : G \times G \times \widehat{G} \to \Sigma$ is a *matrix-valued amplitude* in the class $\mathcal{A}^m_{\rho,\delta}(G)$ if for a strongly admissible collection of difference operators on $\widehat{G}$ we have the amplitude inequalities

$$\|\Delta_\pi^\alpha X_x^\beta X_y^\gamma a(x,y,\pi)\|_{\mathscr{L}(\mathcal{H}_\pi)} \leq C_{\alpha\beta\gamma} \langle \pi \rangle^{m-\rho|\alpha|+\delta|\beta+\gamma|},$$

for all multi-indices α, β, γ and for all $(x,y,\pi) \in G \times G \times \widehat{G}$. The corresponding *amplitude operator* $\mathrm{Op}(a) : C^\infty(G) \to \mathcal{D}'(G)$ is defined by

$$\mathrm{Op}(a)f(x) := \sum_{\pi \in \widehat{G}} d_\pi \operatorname{Tr}\left(\pi(x) \int_G a(x,y,\eta) f(y) \pi(y)^* dy\right). \tag{2.27}$$

In the case $a(x,y,\pi) = \sigma_A(x,\pi)$ independent of y, we recover the quantization (2.22), namely, we have $\mathrm{Op}(a) = A$.

Theorem 2.2.15. *Let $a \in \mathcal{A}^m_{\rho,\delta}(G)$. If $0 \leq \delta < 1$ and $0 \leq \rho \leq 1$ then* $\mathrm{Op}(a)$ *is a continuous linear operator from $C^\infty(G)$ to $C^\infty(G)$. Moreover, if $0 \leq \delta < \rho \leq 1$, then $A = \mathrm{Op}(a)$ is a pseudo-differential operator with a matrix-valued symbol $\sigma_A \in S^m_{\rho,\delta}(G)$, which has the asymptotic expansion*

$$\sigma_A(x,\pi) \sim \sum_{\alpha \geq 0} \frac{1}{\alpha!} \Delta_\pi^\alpha X_y^{(\alpha)} a(x,y,\pi)|_{y=x},$$

where the asymptotic expansion means that for every $N \in \mathbb{N}$ we have

$$\sigma_A(x,\pi) - \sum_{|\alpha|<N} \frac{1}{\alpha!} \Delta_\pi^\alpha X_y^{(\alpha)} a(x,y,\pi)|_{y=x} \in S^{m-(\rho-\delta)N}_{\rho,\delta}(G).$$

For the proof of this theorem we refer to [RT11]. Given the formula for the amplitude operators in Theorem 2.2.15, the symbol of the adjoint operator can be found as follows.

Theorem 2.2.16. *Let $m \in \mathbb{R}$ and $0 \leq \delta < \rho$. Let $A : C^\infty(G) \to C^\infty(G)$ be a linear continuous operator with symbol $\sigma_A \in S^m_{\rho,\delta}(G)$. Then the symbol σ_{A^*} of the adjoint operator A^* satisfies $\sigma_{A^*} \in S^m_{\rho,\delta}(G)$, and is given by*

$$\sigma_{A^*}(x,\pi) \sim \sum_{\alpha \geq 0} \frac{1}{\alpha!} \Delta_\pi^\alpha X_x^{(\alpha)} \sigma_A(x,\pi)^*,$$

where $\sigma_A(x,\pi)^*$ is the adjoint matrix to $\sigma_A(x,\pi)$, and the asymptotic expansion means that for every $N \in \mathbb{N}$ we have

$$\sigma_{A^*}(x,\pi) - \sum_{|\alpha|<N} \frac{1}{\alpha!}\Delta_\pi^\alpha X_x^{(\alpha)} \sigma_A(x,\pi)^* \in S_{\rho,\delta}^{m-(\rho-\delta)N}(G).$$

We recall that the operator $A \in \Psi^m(G)$ on G viewed as a manifold is elliptic if all of its localisations to coordinate charts are (locally) elliptic. This can be characterised in terms of the matrix-valued symbols. A combination of [RTW14, Theorem 4.1] and [RT10a, Theorem 10.9.10] yields

Theorem 2.2.17. *An operator $A \in \Psi^m(G)$ is elliptic if and only if its symbol $\sigma_A(x,\pi)$ is invertible for all but finitely many $\pi \in \widehat{G}$, and for all such π satisfies*

$$\|\sigma_A(x,\pi)^{-1}\|_{\mathscr{L}(\mathcal{H}_\pi)} \le C\langle\pi\rangle^{-m}$$

for all $x \in G$. Furthermore, in this case, assume that

$$\sigma_A \sim \sum_{j=0}^{\infty} \sigma_{A_j}, \quad A_j \in \Psi^{m-j}(G).$$

Let $\sigma_B \sim \sum_{k=0}^{\infty} \sigma_{B_k}$, where

$$\sigma_{B_0}(x,\pi) = \sigma_{A_0}(x,\pi)^{-1}$$

for large $\langle\pi\rangle$, and the symbols σ_{B_k} are defined recursively by

$$\sigma_{B_N} = -\sigma_{B_0} \sum_{k=0}^{N-1} \sum_{j=0}^{N-k} \sum_{|\gamma|=N-j-k} \frac{1}{\gamma!}(\Delta_\pi^\gamma \sigma_{B_k})(X_x^{(\gamma)} \sigma_{A_j}).$$

Then $\mathrm{Op}(\sigma_{B_k}) \in \Psi^{-m-k}(G)$, $B = \mathrm{Op}(\sigma_B) \in \Psi^{-m}(G)$, and the operators $AB - I$ and $BA - I$ are in $\Psi^{-\infty}(G)$.

One can also provide a criterion for the hypoellipticity in terms of matrix-valued symbols ([RTW14]), in analogy to the one on $\mathbb{R}^n$ given by Hörmander ([Hör67b]).

Theorem 2.2.18. *Let $m \ge m_0$ and $0 \le \delta < \rho \le 1$. Let $A \in \mathrm{Op}(S_{\rho,\delta}^m(G))$ be a pseudo-differential operator with symbol $\sigma_A \in S_{\rho,\delta}^m(G)$ which is invertible for all but finitely many $\pi \in \widehat{G}$, and for all such π satisfies*

$$\|\sigma_A(x,\pi)^{-1}\|_{\mathscr{L}(\mathcal{H}_\pi)} \le C\langle\pi\rangle^{-m_0}$$

for all $x \in G$. Assume also that (for a strongly admissible collection of difference operators) we have

$$\|\sigma_A(x,\pi)^{-1}\left[\Delta_\pi^\alpha X_x^\beta \sigma_A(x,\pi)\right]\|_{\mathscr{L}(\mathcal{H}_\pi)} \le C\langle\pi\rangle^{-\rho|\alpha|+\delta|\beta|}$$

for all multi-indices α, β, all $x \in G$, and all but finitely many $\pi \in \widehat{G}$. Then there exists an operator $B \in \mathrm{Op}(S^{-m_0}_{\rho,\delta}(G))$ such that $AB - I$ and $BA - I$ belong to $\Psi^{-\infty}(G)$. Consequently, we have

$$\textit{sing supp } Au = \textit{sing supp } u$$

for all $u \in \mathcal{D}'(G)$.

We finish this section with several results that are usually expected from the calculus. The following asymptotic expansion formula was established in [RW14].

Proposition 2.2.19. *Let $\sigma_j \in S^{m_j}_{\rho,\delta}(G)$, $j \in \mathbb{N}_0$, $0 \leq \delta < \rho \leq 1$, be a family of symbols with $m_j \searrow -\infty$. Then there exists a symbol $\sigma \in S^{m_0}_{\rho,\delta}(G)$ such that*

$$\sigma - \sum_{j=0}^{N-1} \sigma_j \in S^{m_N}_{\rho,\delta}(G)$$

for all $N \in \mathbb{N}_0$.

The functional calculus of matrix valued symbols and its operator counterpart have been also developed in [RW14]. A notable corollary of such functional calculus is the following

Corollary 2.2.20. *Let $0 \leq \delta < \rho \leq 1$ and let $m \geq 0$. Assume $\sigma_A \in S^{2m}_{\rho,\delta}(G)$ satisfies $\sigma_A(x,\pi) > 0$ and*

$$\|\sigma_A(x,\pi)^{-1}\|_{\mathscr{L}(\mathcal{H}_\pi)} \leq C\langle\pi\rangle^{-2m}$$

for all x and π. Then the square root

$$\sigma_B(x,\pi) = \sqrt{\sigma_A(x,\pi)}$$

in the sense of positive matrices is a symbol satisfying $\sigma_B \in S^m_{\rho,\delta}(G)$.

This is the corollary of the following more general result:

Theorem 2.2.21. *Let $0 \leq \delta \leq 1$ and $0 < \rho \leq 1$. Assume $\sigma_A \in S^m_{\rho,\delta}(G)$, $m \geq 0$, is positive definite, invertible, and satisfies*

$$\|\sigma_A(x,\pi)^{-1}\|_{\mathscr{L}(\mathcal{H}_\pi)} \leq C\langle\pi\rangle^{-m}$$

for all x and for all but finitely many π. Then for any number $s \in \mathbb{C}$,

$$\sigma_B(x,\pi) := \sigma_A(x,\pi)^s = \exp(s \log \sigma_A(x,\pi))$$

defines a symbol $\sigma_B \in S^{m'}_{\rho,\delta}(G)$, with $m' = \mathrm{Re}\,(ms)$.

In fact, the assumptions of Theorem 2.2.21 imply something stronger, namely, that the symbol $\sigma_A(x,\pi)$ is parameter-elliptic with respect to $\mathbb{R}_-$; we refer to [RW14] for the definition of parameter-ellipticity in this setting, and for a more general exposition and statements of the functional calculus on compact Lie groups.

2.2.4 Fourier multipliers and L^p-boundedness

Here we give an overview of the L^p-estimates for the Fourier multipliers and for non-invariant operators on compact Lie groups following [RW13, RW15]. We set aside the case of bi-invariant operators (or spectral multipliers) noting that there exist many results in this direction (see e.g. N. Weiss [Wei72], Coifman and G. Weiss [CW74], Stein [Ste70b], Cowling [Cow83], Alexopoulos [Ale94], to refer the reader to only a few). Instead, we concentrate on the case of left-invariant operators (or Fourier multipliers). To the best of our knowledge the literature in this case is much smaller, with a notable exception of a multiplier theorem for left-invariant operators on the group SU(2) treated by Coifman and Weiss [CW71b], Coifman and de Guzmán [CdG71], and appearing in more detail in the monograph by Coifman and Weiss [CW71a]. The conditions there are formulated using specific explicit expressions involving Clebsch-Gordan coefficients on SU(2), but they can be recast in a much shorter form using the concept of difference operators. It also allows one to treat the case of general compact Lie groups. Finally we note that there exist also results for the spectral multipliers in the sub-Laplacian, also on SU(2), for which we refer to Cowling and Sikora [CS01].

First, we discuss left-invariant operators $A : C^\infty(G) \to \mathcal{D}'(G)$, so that the matrix-valued symbol $\sigma_A(x,\pi) = \sigma_A(\pi)$ is independent of x and can be given as

$$\sigma_A(\pi) = \pi(x)^*(A\pi)(x) = (A\pi)(e).$$

The multiplier theorems that we will present can be said to be of *Mihlin-Hörmander type* in the sense that they provide analogues of famous multiplier theorems on $\mathbb{R}^n$ by Mihlin [Mih56, Mih57] and Hörmander [Hör60].

In order to formulate the results, we need to fix a particular collection of first order difference operators associated to the elements of the unitary dual $\widehat{G}$. Thus, for a fixed representation $\pi_0 \in \widehat{G}$, we notice that the $(d_{\pi_0} \times d_{\pi_0})$-matrix $\pi_0(x) - \mathrm{I}_{d_{\pi_0}}$ vanishes at $x = e$. Consequently, we define the difference operators ${}_{\pi_0}\mathbb{D} = ({}_{\pi_0}\mathbb{D}_{ij})_{i,j=1}^{d_{\pi_0}}$ associated with its elements,

$${}_{\pi_0}\mathbb{D}_{ij} := \Delta_{(\pi_0)_{ij}-\delta_{i,j}},$$

where $\delta_{i,j}$ is the Kronecker delta. For a family of difference operators of this type,

$$\mathbb{D}_1 = {}_{\pi_1}\mathbb{D}_{i_1j_1},\ \mathbb{D}_2 = {}_{\pi_2}\mathbb{D}_{i_2j_2},\ \ldots, \mathbb{D}_m = {}_{\pi_m}\mathbb{D}_{i_mj_m}, \tag{2.28}$$

with $\pi_k \in \widehat{G}$, $1 \le i_k, j_k \le d_{\pi_k}$, $1 \le k \le m$, we define

$$\mathbb{D}^\alpha := \mathbb{D}_1^{\alpha_1} \cdots \mathbb{D}_m^{\alpha_m}. \tag{2.29}$$

The described difference operators ${}_{\pi_0}\mathbb{D}$ have a number of useful properties. For example, they satisfy the *finite Leibniz formula* (while general difference operators satisfy only an asymptotic Leibniz formula, see [RT10a, Section 10.7.4]). Namely, for any fixed π_0, they satisfy

$$\mathbb{D}_{ij}(\sigma\tau) = (\mathbb{D}_{ij}\sigma)\tau + \sigma(\mathbb{D}_{ij}\tau) + \sum_{k=1}^{d_{\pi_0}} (\mathbb{D}_{ik}\sigma)(\mathbb{D}_{kj}\tau). \tag{2.30}$$

The collection of difference operators

$$\{{}_{\pi_0}\mathbb{D}_{ij} : \pi_0 \in \widehat{G}, 1 \le i,j \le d_{\pi_0}\}$$

is strongly admissible. Moreover, it has a finite strongly admissible sub-collection. Indeed, a homomorphic embedding of G into $\mathcal{U}(N)$ for some N is itself a representation of G. Decomposing it into irreducible components gives the desired finite family of π_0's.

We now formulate the first result on the L^p-boundedness of left-invariant operators.

Theorem 2.2.22. *Let $A : C^\infty(G) \to \mathcal{D}'(G)$ be a left-invariant linear continuos operator on a compact Lie group G, and let k denote the smallest even integer such that $k > \frac{1}{2}\dim G$. Assume that the symbol σ_A of A satisfies*

$$\|\mathbb{D}^\alpha \sigma_A(\pi)\|_{\mathscr{L}(\mathcal{H}_\pi)} \le C_\alpha \langle\pi\rangle^{-|\alpha|} \tag{2.31}$$

for all multi-indices $|\alpha| \le k$ and all $\pi \in \widehat{G}$. Then the operator A is of weak type (1,1) and is bounded on $L^p(G)$ for all $1 < p < \infty$.

We note that by Theorem 2.2.11, imposing conditions (2.31) for all multi-indices α would imply that A is a left-invariant pseudo-differential operator in Hörmander's class, $A \in \Psi^0(G)$, for which the L^p-boundedness would follow from the corresponding L^p-boundedness in $\mathbb{R}^n$ for its localisations. However, imposing conditions (2.31) for multi-indices $|\alpha| \le k$ still assures that the operator A is of Calderón-Zygmund type (in the sense of Coifman and Weiss, see Section A.4). The proof of the L^p-boundedness for $1 < p \le 2$ follows by Marcinkiewicz interpolation theorem (see Proposition 1.5.1) from the L^2-boundedness (and hence also weak (2,2) type) in Theorem 2.2.5, and from weak (1,1) type, which becomes, therefore, the main task.

For $2 < p < \infty$, the result follows by duality. Before we give an idea behind the proof of the weak (1,1) type, let us formulate several corollaries from Theorem 2.2.22. We recall that the Sobolev space $W^{p,s}(G)$ on G is the usual Sobolev space on G as a manifold defined by requiring all the localisations to belong to the Euclidean space $W^{p,s}(\mathbb{R}^n) = (\mathrm{I} - \mathcal{L}_{\mathbb{R}^n})^{-s/2}L^p(\mathbb{R}^n)$, where $\mathcal{L}_{\mathbb{R}^n}$ is the Laplacian on $\mathbb{R}^n$ and $s \in \mathbb{R}$.

Corollary 2.2.23. *Let $A : C^\infty(G) \to \mathcal{D}'(G)$ be a left-invariant linear continuous operator on a compact Lie group G. Let $0 \le \rho \le 1$ and let k denote the smallest* *even integer such that $k > \frac{1}{2}\dim G$. Assume that the symbol σ_A of A satisfies*

$$\|\mathbb{D}^\alpha \sigma_A(\pi)\|_{\mathscr{L}(\mathcal{H}_\pi)} \leq C_\alpha \langle \pi \rangle^{-\rho|\alpha|}$$

for all multi-indices $|\alpha| \leq k$ and all $\pi \in \widehat{G}$. Then the operator A extends to a bounded operator from the Sobolev space $W^{p,r}(G)$ to $L^p(G)$ for any $1 < p < \infty$, with

$$r = k(1-\rho)|\frac{1}{p} - \frac{1}{2}|.$$

Example 2.2.24. Let

$$\mathcal{L}_{sub} = X^2 + Y^2$$

be a sub-Laplacian on SU(2). Then it was shown in [RTW14] that it has a parametrix with the matrix-valued symbol in the class $S^{-1}_{\frac{1}{2},0}(\mathrm{SU}(2))$. Consequently, for any $1 < p < \infty$, Corollary 2.2.23 implies the subelliptic estimate

$$\|f\|_{W^{p,s+1-|\frac{1}{p}-\frac{1}{2}|}(\mathrm{SU}(2))} \leq C_p \|\mathcal{L}_{sub} f\|_{W^{p,s}(\mathrm{SU}(2))},$$

where the estimate is extended from $s = 0$ to any $s \in \mathbb{R}$ by the calculus. We refer to [RTW14] for the construction and discussion of parametrices for other operators, including the heat and the wave operator, d'Alambertian, and some higher order operators, on SU(2) and on $\mathbb{S}^3$, and to [RW13, RW15] for the corresponding L^p-estimates.

Example 2.2.25. Let (ϕ, θ, ψ) be the standard Euler angles on SU(2), see e.g. [RT10a, Chapter 11] for a detailed treatment of SU(2). Thus, we have $0 \leq \phi < 2\pi$, $0 \leq \theta \leq \pi$, and $-2\pi \leq \psi < 2\pi$, and every element

$$u = u(\phi,\theta,\psi) = \begin{pmatrix} a & b \\ -\bar{b} & \bar{a} \end{pmatrix} \in \mathrm{SU}(2)$$

is parametrised in such a way that

$$2a\bar{a} = 1 + \cos\theta, \; 2ab = ie^{i\phi}\sin\theta, \; -2a\bar{b} = ie^{i\psi}\sin\theta.$$

Conversely, we can also write

$$u(\phi,\theta,\psi) = \begin{pmatrix} \cos(\frac{\theta}{2})e^{i(\phi+\psi)/2} & i\sin(\frac{\theta}{2})e^{i(\phi-\psi)/2} \\ i\sin(\frac{\theta}{2})e^{-i(\phi-\psi)/2} & \cos(\frac{\theta}{2})e^{-i(\phi+\psi)/2} \end{pmatrix} \in \mathrm{SU}(2).$$

Let X be a left-invariant vector field on G normalised in such a way that $\|X\| = \|\partial/\partial\psi\|$ with respect to the Killing form. It was shown in [RTW14] that for $\gamma \in \mathbb{C}$,

$$\text{the operator } X + \gamma \text{ is invertible if and only if } i\gamma \notin \frac{1}{2}\mathbb{Z},$$

and, moreover, for such γ, the inverse $(X+\gamma)^{-1}$ has its matrix-valued symbol in the class $S^0_{0,0}(\mathrm{SU}(2))$. The same conclusion remains true if we replace SU(2) by $\mathbb{S}^3$, with the corresponding selection of Euler's angles. Then, Corollary 2.2.23 and the calculus imply the subelliptic estimate

$$\|f\|_{W^{p,s}(\mathbb{S}^3)} \leq C_p \|(X+\gamma)f\|_{W^{p,s+2|\frac{1}{p}-\frac{1}{2}|}(\mathbb{S}^3)}, \quad 1<p<\infty,\ s\in\mathbb{R}.$$

There is an analogue of this estimate on arbitrary compact Lie groups, see [RW15].

Let us briefly indicate an idea behind the proof of Theorem 2.2.22. In order to use the theory of singular integral operators (according to Coifman and Weiss, see Section A.4), we first define a suitable quasi-distance on G.

Let $\mathrm{Ad} : G \to \mathcal{U}(\mathfrak{g})$ be the adjoint representation of G. Then by the Peter-Weyl theorem it can be decomposed as a direct sum of irreducible representations,

$$\mathrm{Ad} = (\dim Z(G))1 \oplus \bigoplus_{\pi\in\Theta_0} \pi,$$

where $Z(G)$ is the centre of G, 1 is the trivial representation, and Θ_0 is an index set for the representations entering in this decomposition. Then we define a smooth non-negative function

$$r^2(x) := \dim G - \mathrm{Tr}\,\mathrm{Ad}(x) = \sum_{\pi\in\Theta_0} (d_\pi - \chi_\pi(x)), \tag{2.32}$$

which is central, non-degenerate, and vanishes of the second order at the unit element $e \in G$. It can be then checked that the function

$$d(x,y) := r(x^{-1}y)$$

is the quasi-distance in the sense of Section A.4. Consequently, one can check that the operator A satisfies Calderón-Zygmund conditions of spaces of homogeneous type, in terms of the quasi-distance above. Such a verification relies heavily on the developed symbolic calculus, Leibniz rules for difference operators, and criteria for the weak (1,1) type in terms of suitably defined mollifiers. We refer to [RW15] for further details of this construction.

Using the function $r(x)$, one can refine the statement of Theorem 2.2.22. Thus, let us define the difference operator associated with $r^2(x)$, namely,

$$⩓ := \Delta_{r^2} = \mathcal{F}_G\, r^2(x)\, \mathcal{F}_G^{-1},$$

and we have that $⩓ \in \mathrm{diff}^2(\widehat{G})$ is the second order difference operator.

Theorem 2.2.26. *Let $A : C^\infty(G) \to \mathcal{D}'(G)$ be a left-invariant linear continuous operator on a compact Lie group G, and let k denote the smallest even integer such that $k > \frac{1}{2}\dim G$. Assume that the symbol σ_A of A satisfies*

$$\|⩓^{k/2}\sigma_A(\pi)\|_{\mathscr{L}(\mathcal{H}_\pi)} \leq C\langle\pi\rangle^{-k} \tag{2.33}$$

as well as

$$\|\mathbb{D}^\alpha \sigma_A(\pi)\|_{\mathscr{L}(\mathcal{H}_\pi)} \leq C_\alpha \langle \pi \rangle^{-|\alpha|} \tag{2.34}$$

for all multi-indices $|\alpha| \leq k-1$ *and all* $\pi \in \widehat{G}$. *Then the operator* A *is of weak type (1,1) and is bounded on* $L^p(G)$ *for all* $1 < p < \infty$.

We note that, comparing (2.33) to the condition (2.31) in Theorem 2.2.22, only a single difference condition of order k is required in Theorem 2.2.26. This has interesting consequences, already in the case of the torus, as we will show in Example 2.2.27.

Moreover, the assumption (2.34) can be refined further: namely, to form a strongly admissible family of first order difference operators giving $\mathbb{D}^\alpha$ in (2.28) and (2.29), it is enough to take only $\pi_k \in \Theta_0$, the set of the irreducible components of the adjoint representation.

In all the theorems of this section an assumption that k is an even integer is present. This seems to be related to the technical part of the argument, namely, to the usage of the second order difference operator $\mathbb{A}$ that is naturally related to the quasi-metric on G as well as satisfies the finite Leibniz formula. The latter can be derived from (2.30) using the decomposition

$$\mathbb{A} = -\sum_{\pi \in \Theta_0} \sum_{i=1}^{d_\pi} {}_\pi\mathbb{D}_{ii},$$

which follows from the definition of $r^2(x)$ in (2.32). Thus, it satisfies

$$\mathbb{A}(\sigma\tau) = (\mathbb{A}\sigma)\tau + \sigma(\mathbb{A}\tau) - \sum_{\pi \in \Theta_0} \sum_{i,j=1}^{d_\pi} ({}_\pi\mathbb{D}_{ij}\sigma)({}_\pi\mathbb{D}_{ji}\tau),$$

and becomes instrumental in establishing the relation between assumption (2.33) and properties of the integral kernel of A in terms of the quasi-metric. However, we note also that the condition on the even number of analogous expressions appears already in the multiplier theorem for bi-invariant operators, established by rather different methods by N. Weiss [Wei72].

Example 2.2.27. Let us consider now the case of the torus, $G = \mathbb{T}^n$. In this case, the left-invariant operators take the form

$$Af(x) = \sum_{\xi \in \mathbb{Z}^n} e^{2\pi i x \cdot \xi} \sigma(\xi) \widehat{f}(\xi) \quad \text{with } \widehat{f}(\xi) = \int_{\mathbb{T}^n} e^{-2\pi i x \cdot \xi} f(x) dx.$$

or, in other words,

$$\widehat{Af}(\xi) = \sigma(\xi)\widehat{f}(\xi), \ \ \xi \in \mathbb{Z}^n.$$

We take

$$r^2(x) = 2n - \sum_{j=1}^n (e^{2\pi i x_j} + e^{-2\pi i x_j}),$$

so that

$$\mathbb{A}\sigma(\xi) = 2n\sigma(\xi) - \sum_{j=1}^{n}(\sigma(\xi + e_j) + \sigma(\xi - e_j)),$$

where $\xi \in \mathbb{Z}^n$ and e_j is the jth unit vector in $\mathbb{Z}^n$. The appearing operator $\mathbb{A}$ is rather curious since it replaces the assumptions usually imposed on all highest order difference conditions as, for example, in the suitably modified toroidal version of Hörmander's multiplier theorem [Hör60] (where one would need to make assumptions on all differences of order $\left[\frac{n}{2}\right] + 1$), or in Marcienkiewicz' version of multiplier theorem of Nikolskii [Nik77, Section 1.5.3] (where one imposes difference conditions up to order n). To clarify the nature of the operator $\mathbb{A}$, we give the examples for $\mathbb{T}^2$ and $\mathbb{T}^3$. As a consequence of Theorem 2.2.26 we get the following statements. Let $1 < p < \infty$. Assume that

$$|\sigma(\xi)| \leq C \text{ and } |\xi|\ |\sigma(\xi + e_j) - \sigma(\xi)| \leq C,$$

for all $\xi \in \mathbb{Z}^2$ and $j = 1, 2$, or $\xi \in \mathbb{Z}^3$ and $j = 1, 2, 3$, respectively. Furthermore, assume that

$$|\xi|^2\ |\sigma(\xi) - \frac{1}{4}\sum_{j=1}^{2}(\sigma(\xi + e_j) + \sigma(\xi - e_j))| \leq C \quad \text{for } \mathbb{T}^2,$$

or

$$|\xi|^2|\sigma(\xi) - \frac{1}{6}\sum_{j=1}^{3}(\sigma(\xi + e_j) + \sigma(\xi - e_j))| \leq C \quad \text{for } \mathbb{T}^3,$$

respectively. Then the operator A is bounded on $L^p(\mathbb{T}^2)$ or $L^p(\mathbb{T}^3)$, respectively.

We now drop the assumption of left-invariance and consider general linear continuous operators from $C^\infty(G)$ to $\mathcal{D}'(G)$. Then we can assure the L^p-boundedness provided we complement the differences in π with derivatives with respect to x.

Theorem 2.2.28. *Let $A : C^\infty(G) \to \mathcal{D}'(G)$ be a linear continuous operator on a compact Lie group G, and let k denote the smallest even integer such that $k > \frac{1}{2}\dim G$. Let $1 < p < \infty$ and let $l > \frac{\dim G}{p}$ be an integer. Assume that the symbol σ_A of A satisfies*

$$\|X_x^\beta \mathbb{A}^{k/2}\sigma_A(x,\pi)\|_{\mathscr{L}(\mathcal{H}_\pi)} \leq C\langle\pi\rangle^{-k} \tag{2.35}$$

as well as

$$\|X_x^\beta \mathbb{D}^\alpha \sigma_A(x,\pi)\|_{\mathscr{L}(\mathcal{H}_\pi)} \leq C_\alpha\langle\pi\rangle^{-|\alpha|} \tag{2.36}$$

for all $\pi \in \widehat{G}$ and for all multi-indices α and β with $|\alpha| \leq k - 1$ and $|\beta| \leq l$. Then the operator A is bounded on $L^p(G)$.

We refer to [RW15] for the detailed proofs of all the results in this section.

2.2.5 Sharp Gårding inequality

The sharp Gårding inequality on $\mathbb{R}^n$ is an important lower bound for operators with positive symbols, finding many applications in the theory of partial differential equations of elliptic, parabolic and hyperbolic types. The original Gårding inequality for elliptic operators has been established by Gårding in [Går53]. It says that if $p \in S^m_{\rho,\delta}(\mathbb{R}^n)$, $0 \leq \delta < \rho \leq 1$, is a symbol satisfying

$$\operatorname{Re} p(x,\xi) \geq c|\xi|^m,$$

$c > 0$, for all $x \in \mathbb{R}^n$ and ξ large enough, then the corresponding pseudo-differential operator

$$p(x,D)f(x) = \int_{\mathbb{R}^n} e^{2\pi i x\cdot\xi} p(x,\xi)\widehat{f}(\xi)d\xi$$

satisfies the following lower bound: for every $s \in \mathbb{R}$ and every compact set $K \subset \mathbb{R}^n$ there exist some constants c_0, c_1 such that

$$\operatorname{Re}\,(p(x,D)f,f)_{L^2(\mathbb{R}^n)} \geq c_0 \|f\|^2_{H^{m/2}(\mathbb{R}^n)} - c_1\|f\|^2_{H^s(\mathbb{R}^n)} \tag{2.37}$$

holds for all $f \in \mathcal{D}(K)$. Its improvement, the so-called *sharp Gårding inequality* was obtained by Hörmander in [Hör66]. It says that if $p \in S^m_{\rho,\delta}(\mathbb{R}^n)$, $0 \leq \delta < \rho \leq 1$, is a non-negative symbol, $p(x,\xi) \geq 0$ for all $x, \xi \in \mathbb{R}^n$, then the corresponding pseudo-differential operator satisfies the lower bound

$$\operatorname{Re}\,(p(x,D)f,f)_{L^2(\mathbb{R}^n)} \geq -c\|f\|^2_{H^{(m-(\rho-\delta))/2}(\mathbb{R}^n)} \tag{2.38}$$

for all $f \in \mathcal{D}(K)$. This inequality was further generalised to systems by Lax and Nirenberg [LN66], Kumano-go [Kg81], and Vaillancourt [Vai70]. It has been also extended to regain two derivatives for the class $S^2_{1,0}(\mathbb{R}^n)$ by Fefferman and Phong [FP78]. For expositions concerning sharp Gårding inequalities with different proofs we refer to monographs of Kumano-go [Kg81], Taylor [Tay81], Lerner [Ler10], or Friedrichs' notes [Fri70]. There is also an approach based on constructions in space variables rather than in frequency one, developed by Nagase [Nag77].

The situation with Gårding inequalities on manifolds is more complicated. The main problem is that the assumption that the symbol of a pseudo-differential operator is non-negative is harder to formulate since the full symbol is not invariantly defined. For second order pseudo-differential operators, under the non-negativity assumption on the principal symbol and certain non-degeneracy assumptions on the sub-principal symbol, a lower bound now known as Melin-Hörmander inequality has been obtained by Melin [Mel71] and Hörmander [Hör77]. The non-degeneracy conditions on the sub-principal symbol can be somehow relaxed, see [MPP07].

Nevertheless, in our setting we are assisted by the fact that the algebraic structure of a Lie group gives us the notion of the full symbol in (2.19). This symbol, however, is not needed for the standard Gårding inequality (2.37) since the ellipticity is determined by the principal symbol only. Thus, the standard Gårding inequality (2.37) on compact Lie groups has been established in [BGJR89] using Langlands' results for semi-groups on Lie groups [Lan60].

Let us first look at a possible assumption for the positivity of an operator in the invariant situation. If an operator A is given by the convolution $Af = \kappa * f$, we obtain

$$(Af, f)_{L^2(G)} = (\kappa * f, f)_{L^2(G)} = (\widehat{f}\,\widehat{\kappa}, \widehat{f})_{\ell^2(\widehat{G})} = \sum_{\pi\in\widehat{G}} d_\pi \operatorname{Tr}\left(\widehat{f}(\pi)\ \widehat{\kappa}(\pi)\ \widehat{f}(\pi)^*\right),$$

where we used the Plancherel identity (2.13). On the other hand, according to Section 1.5, A is right-invariant, and according to Example 2.2.4 its symbol is $\sigma_A(x,\pi) = \pi(x)^*\widehat{\kappa}(\xi)\pi(x)$. Thus, we get that A is a positive operator if and only if the matrix $\widehat{\kappa}(\pi)$ is positive for all $\pi \in \widehat{G}$, i.e. when $(\widehat{\kappa}(\pi)v, v)_{\mathcal{H}_\pi} \geq 0$ for all $v \in \mathcal{H}_\pi$. But this means that the symbol σ_A is positive, $\sigma_A(x,\pi) \geq 0$ for all $(x,\pi) \in G\times\widehat{G}$. Analogously, for left-invariant operators $Af = f * \kappa$, one sees that

$$(Af, f)_{L^2(G)} = (f * \kappa, f)_{L^2(G)} = (\widehat{\kappa}\,\widehat{f}, \widehat{f})_{\ell^2(\widehat{G})} = \sum_{\pi\in\widehat{G}} d_\pi \operatorname{Tr}\left(\widehat{f}(\pi)^*\ \widehat{\kappa}(\pi)\ \widehat{f}(\pi)\right).$$

So again, A is a positive operator if and only if its symbol $\sigma_A(\pi) = \widehat{\kappa}(\pi)$ is positive.

This motivates a hypothesis that the positivity of the matrix-valued symbol on G would be an analogue of the positivity of the Kohn-Nirenberg symbol on $\mathbb{R}^n$. Indeed, we have the following criterion, which for non-invariant operators becomes a sufficient condition:

Theorem 2.2.29. *Let $A \in \Psi^m(G)$ be such that its matrix-valued symbol σ_A is positive, i.e.*

$$\sigma_A(x,\pi) \geq 0 \quad \textit{for all} \quad (x,\pi) \in G\times\widehat{G}.$$

Then there exists a constant c such that

$$\operatorname{Re}(Af, f)_{L^2(G)} \geq -c\|f\|^2_{H^{(m-1)/2}(G)}$$

for all $f \in C^\infty(G)$.

The usual proofs of the sharp Gårding inequality on $\mathbb{R}^n$ (that is, the proofs not relying on the anti-Wick quantization) make use of a positive approximation of a pseudo-differential operator, the so-called Friedrichs symmetrisation, approximating an operator with non-negative symbol of order m by a positive operator modulo an error of order $m-1$. This construction, indeed, allows one to gain one derivative needed for the sharp Gårding inequality. Unfortunately, such an approximation in the frequency variable seems to be less useful on a Lie group G because the unitary dual $\widehat{G}$ is not well adapted for such purpose. However, one can carry out, instead, a symmetrisation in the space variables using the symbolic calculus of operators for the construction. In particular, it relies heavily on dealing with the symbol class $S^m_{1,\frac{1}{2}}(G)$ defined in Section 2.2.3.

As in the case of operators on $\mathbb{R}^n$, the sharp Gårding inequality leads to several further conclusions concerning the L^2-boundedness of operators. For example, pseudo-differential operators of the first order are bounded on $L^2(\mathbb{R}^n)$ provided their matrix-valued symbols are bounded:

Corollary 2.2.30. *Let $A \in \Psi^1(G)$ be such that its matrix-valued symbol σ_A satisfies*

$$\|\sigma_A(x,\pi)\|_{\mathscr{L}(\mathcal{H}_\pi)} \leq C$$

for all $(x,\pi) \in G \times \widehat{G}$. Then the operator A is bounded from $L^2(G)$ to $L^2(G)$.

It can be also used to determine constants as bounds for operator norms of mappings between L^2-Sobolev spaces. For the proofs of the statements in this section, as well as for further details we refer the reader to [RT11].

In the above, we concentrated on symbol classes $S^m_{1,0}(G)$ of type $(1,0)$. However, certain conclusions can be made also for operators with symbols of type (ρ,δ).

Proposition 2.2.31 (Gårding's inequality on G)**.** *Let $0 \leq \delta < \rho \leq 1$ and $m > 0$. Let $A \in \operatorname{Op} S^{2m}_{\rho,\delta}(G)$ be elliptic and such that $\sigma_A(x,\xi) \geq 0$ for all x and co-finitely many ξ. Then there are constants $c_1, c_2 > 0$ such that for any function $f \in H^m(G)$ the inequality*

$$\operatorname{Re}(Af,f)_{L^2} \geq c_1\|f\|^2_{H^m} - c_2\|f\|^2_{L^2}$$

holds true.

The statement follows by the calculus from its special case $m = \rho - \delta$. We refer to [RW14, Corollary 6.2] for the proofs.

Chapter 3

Homogeneous Lie groups

By definition a homogeneous Lie group is a Lie group equipped with a family of dilations compatible with the group law. The abelian group $(\mathbb{R}^n, +)$ is the very first example of homogeneous Lie group. Homogeneous Lie groups have proved to be a natural setting to generalise many questions of Euclidean harmonic analysis. Indeed, having both the group and dilation structures allows one to introduce many notions coming from the Euclidean harmonic analysis. There are several important differences between the Euclidean setting and the one of homogeneous Lie groups. For instance the operators appearing in the latter setting are usually more singular than their Euclidean counterparts. However it is possible to adapt the technique in harmonic analysis to still treat many questions in this more abstract setting.

As explained in the introduction (see also Chapter 4), we will in fact study operators on a subclass of the homogeneous Lie group, more precisely on graded Lie groups. A graded Lie group is a Lie group whose Lie algebra admits a $(\mathbb{N})$-gradation. Graded Lie groups are homogeneous and in fact the relevant structure for the analysis of graded Lie groups is their natural homogeneous structure and this justifies presenting the general setting of homogeneous Lie groups. From the point of view of applications, the class of graded Lie groups contains many interesting examples, in fact all the ones given in the introduction. Indeed these groups appear naturally in the geometry of certain symmetric domains and in some subelliptic partial differential equations. Moreover, they serve as local models for contact manifolds and CR manifolds, or for more general Heisenberg manifolds, see the discussion in the Introduction.

The references for this chapter of the monograph are [FS82, ch. I] and [Goo76], as well as Fulvio Ricci's lecture notes [Ric]. However, our conventions and notation do not always follow the ones of these references. The treatment in this chapter is, overall, more general than that in the above literature since we also consider distributions and kernels of complex homogeneous degrees and adapt our analysis for subsequent applications to Sobolev spaces and to the op-

V. Fischer, M. Ruzhansky, *Quantization on Nilpotent Lie Groups*, Progress in Mathematics 314,
DOI 10.1007/978-3-319-29558-9_3

erator quantization developed in the following chapters. Especially, our study of complex homogeneities allows us to deal with complex powers of operators (e.g. in Section 4.3.2).

3.1 Graded and homogeneous Lie groups

In this section we present the definition and the first properties of graded Lie groups. Since many of their properties can be explained in the more general setting of homogeneous Lie groups, we will also present these groups.

3.1.1 Definition and examples of graded Lie groups

We start with definitions and examples of graded and stratified Lie groups.

Definition 3.1.1. (i) A Lie algebra $\mathfrak{g}$ is *graded* when it is endowed with a vector space decomposition (where all but finitely many of the V_j's are $\{0\}$):

$$\mathfrak{g} = \bigoplus_{j=1}^{\infty} V_j \quad \text{such that} \quad [V_i, V_j] \subset V_{i+j}.$$

(ii) A Lie group is *graded* when it is a connected simply connected Lie group whose Lie algebra is graded.

The condition that the group is connected and simply connected is technical but important to ensure that the exponential mapping is a global diffeomorphism between the group and its Lie algebra.

The classical examples of graded Lie groups and algebras are the following.

Example 3.1.2 (Abelian case). The abelian group $(\mathbb{R}^n, +)$ is graded: its Lie algebra $\mathbb{R}^n$ is trivially graded, i.e. $V_1 = \mathbb{R}^n$.

Example 3.1.3 (Heisenberg group). The Heisenberg group $\mathbb{H}_{n_o}$ given in Example 1.6.4 is graded: its Lie algebra $\mathfrak{h}_{n_o}$ can be decomposed as

$$\mathfrak{h}_{n_o} = V_1 \oplus V_2 \quad \text{where} \quad V_1 = \oplus_{i=1}^{n_o} \mathbb{R}X_i \oplus \mathbb{R}Y_i \quad \text{and} \quad V_2 = \mathbb{R}T.$$

(For the notation, see Example 1.6.4 in Section 1.6.)

Example 3.1.4 (Upper triangular matrices). The group T_{n_o} of $n_o \times n_o$ matrices which are upper triangular with 1 on the diagonal is graded: its Lie algebra $\mathfrak{t}_{n_o}$ of $n_o \times n_o$ upper triangular matrices with 0 on the diagonal is graded by

$$\mathfrak{t}_{n_o} = V_1 \oplus \ldots \oplus V_{n_o - 1} \quad \text{where} \quad V_j = \oplus_{i=1}^{n_o - j} \mathbb{R}E_{i,i+j}.$$

(For the notation, see Example 1.6.5 in Section 1.6.) The vector space V_j is formed by the matrices with only non-zero coefficients on the j-th upper off-diagonal.

As we will show in Proposition 3.1.10, a graded Lie algebra (hence possessing a natural dilation structure) must be nilpotent. The converse is not true, see Remark 3.1.6, Part 2.

Examples 3.1.2–3.1.4 are stratified in the following sense:

Definition 3.1.5. (i) A Lie algebra $\mathfrak{g}$ is *stratified* when $\mathfrak{g}$ is graded, $\mathfrak{g} = \oplus_{j=1}^{\infty} V_j$, and the first stratum V_1 generates $\mathfrak{g}$ as an algebra. This means that every element of $\mathfrak{g}$ can be written as a linear combination of iterated Lie brackets of various elements of V_1.

(ii) A Lie group is *stratified* when it is a connected simply connected Lie group whose Lie algebra is stratified.

Remark 3.1.6. Let us make the following comments on existence and uniqueness of gradations.

1. A gradation over a Lie algebra is not unique: the same Lie algebra may admit different gradations. For example, any vector space decomposition of $\mathbb{R}^n$ yields a graded structure on the group $(\mathbb{R}^n, +)$. More convincingly, we can decompose the 3 dimensional Heisenberg Lie algebra $\mathfrak{h}_1$ as
$$\mathfrak{h}_1 = \bigoplus_{j=1}^{3} V_j \quad \text{with} \quad V_1 = \mathbb{R}X_1, \; V_2 = \mathbb{R}Y_1, \; V_3 = \mathbb{R}T.$$
This last example can be easily generalised to find several gradations on the Heisenberg groups $\mathbb{H}_{n_o}$, $n_o = 2, 3, \ldots$, which are not the classical ones given in Example 3.1.3. Another example would be
$$\mathfrak{h}_1 = \bigoplus_{j=1}^{8} V_j \quad \text{with} \quad V_3 = \mathbb{R}X_1, \; V_5 = \mathbb{R}Y_1, \; V_8 = \mathbb{R}T, \tag{3.1}$$
and all the other $V_j = \{0\}$.

2. A gradation may not even exist. The first obstruction is that the existence of a gradation implies nilpotency; in other words, a graded Lie group or a graded Lie algebra are nilpotent, as we shall see in the sequel (see Proposition 3.1.10). Even then, a gradation of a nilpotent Lie algebra may not exist. As a curiosity, let us mention that the (dimensionally) lowest nilpotent Lie algebra which is not graded is the seven dimensional Lie algebra given by the following commutator relations:
$$[X_1, X_j] = X_{j+1} \quad \text{for } j = 2, \ldots, 6, \qquad [X_2, X_3] = X_6,$$
$$[X_2, X_4] = [X_5, X_2] = [X_3, X_4] = X_7.$$
They define a seven dimensional nilpotent Lie algebra of step 6 (with basis $\{X_1, \ldots, X_7\}$). It is the (dimensionally) lowest nilpotent Lie algebra which is not graded. See, more generally, [Goo76, ch.I §3.2].

3. To go back to the problem of uniqueness, different gradations may lead to 'morally equivalent' decompositions. For instance, if a Lie algebra $\mathfrak{g}$ is graded by $\mathfrak{g} = \oplus_{j=1}^{\infty} V_j$ then it is also graded by $\mathfrak{g} = \oplus_{j=1}^{\infty} W_j$ where $W_{2j'+1} = \{0\}$ and $W_{2j'} = V_{j'}$. This last example motivates the presentation of homogeneous Lie groups: indeed graded Lie groups are homogeneous and the natural homogeneous structure for the graded Lie algebra

$$\mathfrak{g} = \oplus_{j=1}^{\infty} V_j = \oplus_{j=1}^{\infty} W_j$$

is the same for the two gradations.

Moreover, the relevant structure for the analysis of graded Lie groups is their natural homogeneous structure.

4. There are plenty of graded Lie groups which are not stratified, simply because the first vector subspace of the gradation may not generate the whole Lie algebra (it may be $\{0\}$ for example). This can also be seen in terms of dilations defined in Section 3.1.2. Moreover, a direct product of two stratified Lie groups is graded but may be not stratified as their stratification structures may not 'match'. We refer to Remark 3.1.13 for further comments on this topic.

3.1.2 Definition and examples of homogeneous Lie groups

We now deal with a more general subclass of Lie groups, namely the class of homogeneous Lie groups.

Definition 3.1.7. (i) A family of *dilations* of a Lie algebra $\mathfrak{g}$ is a family of linear mappings

$$\{D_r,\ r > 0\}$$

from $\mathfrak{g}$ to itself which satisfies:

- the mappings are of the form

$$D_r = \mathrm{Exp}(A \ln r) = \sum_{\ell=0}^{\infty} \frac{1}{\ell!} (\ln(r) A)^{\ell},$$

where A is a diagonalisable linear operator on $\mathfrak{g}$ with positive eigenvalues, Exp denotes the exponential of matrices and $\ln(r)$ the natural logarithm of $r > 0$,

- each D_r is a morphism of the Lie algebra $\mathfrak{g}$, that is, a linear mapping from $\mathfrak{g}$ to itself which respects the Lie bracket:

$$\forall X, Y \in \mathfrak{g},\ r > 0 \qquad [D_r X, D_r Y] = D_r [X, Y].$$

(ii) A *homogeneous* Lie group is a connected simply connected Lie group whose Lie algebra is equipped with dilations.

(iii) We call the eigenvalues of A the *dilations' weights* or weights. The set of dilations' weights, or in other worlds, the set of eigenvalues of A is denoted by $\mathcal{W}_A$.

We can realise the mappings A and D_r in a basis of A-eigenvectors as the diagonal matrices

$$A \equiv \begin{pmatrix} \upsilon_1 & & & \\ & \upsilon_2 & & \\ & & \ddots & \\ & & & \upsilon_n \end{pmatrix} \quad \text{and} \quad D_r \equiv \begin{pmatrix} r^{\upsilon_1} & & & \\ & r^{\upsilon_2} & & \\ & & \ddots & \\ & & & r^{\upsilon_n} \end{pmatrix}.$$

The dilations' weights are $\upsilon_1, \ldots, \upsilon_n$.

Remark 3.1.8. Note that if $\{D_r\}$ is a family of dilations of the Lie algebra $\mathfrak{g}$, then $\tilde{D}_r := D_{r^\alpha} := \operatorname{Exp}(\alpha A \ln r)$ defines a new family of dilations $\{\tilde{D}_r, \, r > 0\}$ for any $\alpha > 0$. By adjusting α if necessary, we may assume that the dilations' weights satisfy certain properties in order to compare different families of dilations and in order to fix one of such families. For example in [FS82], it is assumed that the minimum eigenvalue is 1.

Graded Lie algebras are naturally equipped with dilations: if the Lie algebra $\mathfrak{g}$ is graded by

$$\mathfrak{g} = \oplus_{j=1}^{\infty} V_j,$$

then we define the dilations

$$D_r := \operatorname{Exp}(A \ln r)$$

where A is the operator defined by $AX = jX$ for $X \in V_j$.

The converse is true:

Lemma 3.1.9. *If a Lie algebra $\mathfrak{g}$ has a family of dilations such that the weights are all rational, then $\mathfrak{g}$ has a natural gradation.*

Proof. By adjusting the weights (see Remark 3.1.8), we may assume that all the eigenvalues are positive integers. Then the decomposition in eigenspaces gives the the gradation of the Lie algebra. □

Before discussing the dilations in the examples given in Section 3.1.1 and other examples of homogeneous Lie groups, let us state the following crucial property.

Proposition 3.1.10. *The following holds:*

(i) A Lie algebra equipped with a family of dilations is nilpotent.

(ii) A homogeneous Lie group is a nilpotent Lie group.

Proof of Proposition 3.1.10. Let $\{D_r = \text{Exp}(A \ln r)\}$ be the family of dilations. By Remark 3.1.8, we may assume that the smallest weight is 1. For $\upsilon \in \mathcal{W}_A$ let $W_\upsilon \subset \mathfrak{g}$ be the corresponding eigenspace of A. If $\upsilon \in \mathbb{R}$ but $\upsilon \notin \mathcal{W}_A$ then we set $W_\upsilon := \{0\}$.

Thus $D_r X = r^\upsilon X$ for $X \in W_\upsilon$. Moreover, if $X \in W_\upsilon$ and $Y \in W_{\upsilon'}$ then

$$D_r[X,Y] = [D_r X, D_r Y] = r^{\upsilon+\upsilon'}[X,Y]$$

and hence

$$[W_\upsilon, W_{\upsilon'}] \subset W_{\upsilon+\upsilon'}.$$

In particular, since $\upsilon \geq 1$ for $\upsilon \in \mathcal{W}_A$, we see that the ideals in the lower series of $\mathfrak{g}$ (see (1.18)) satisfy

$$\mathfrak{g}_{(j)} \subset \oplus_{a \geq j} W_a.$$

Since the set $\mathcal{W}_A$ is finite, it follows that $\mathfrak{g}_{(j)} = \{0\}$ for j sufficiently large. Consequently the Lie algebra $\mathfrak{g}$ and its corresponding Lie group G are nilpotent. □

Let G be a homogeneous Lie group with Lie algebra $\mathfrak{g}$ endowed with dilations $\{D_r\}_{r>0}$. By Proposition 3.1.10, the connected simply connected Lie group G is nilpotent. We can transport the dilations to the group using the exponential mapping $\exp_G = \exp$ of G (see Proposition 1.6.6 (a)) in the following way: the maps

$$\exp_G \circ D_r \circ \exp_G^{-1}, \quad r > 0,$$

are automorphisms of the group G; we shall denote them also by D_r and call them *dilations on* G. This explains why homogeneous Lie groups are often presented as Lie groups endowed with dilations.

We may write

$$rx := D_r(x) \quad \text{for } r > 0 \text{ and } x \in G.$$

The dilations on the group or on the Lie algebra satisfy

$$D_{rs} = D_r D_s, \quad r, s > 0.$$

As explained above, Examples 3.1.2, 3.1.3 and, 3.1.4 are naturally homogeneous Lie groups:

In Example 3.1.2: The abelian group $(\mathbb{R}^n, +)$ is homogeneous when equipped with the usual dilations $D_r x = rx$, $r > 0$, $x \in \mathbb{R}^n$.

In Example 3.1.3: The Heisenberg group $\mathbb{H}_{n_o}$ is homogeneous when equipped with the dilations

$$rh = (rx, ry, r^2 t), \quad h = (x, y, t) \in \mathbb{R}^{n_o} \times \mathbb{R}^{n_o} \times \mathbb{R}.$$

The corresponding dilations on the Heisenberg Lie algebra $\mathfrak{h}_{n_o}$ are given by

$$D_r(X_j) = rX_j, \; D_r(Y_j) = rY_j, \; j = 1, \ldots, n_o, \text{ and } D_r(T) = r^2 T.$$

In Example 3.1.4: The group T_{n_o} is homogeneous when equipped with the dilations defined by

$$[D_r(M)]_{i,j} = r^{j-i}[M]_{i,j} \quad 1 \leq i < j \leq n_o, \; M \in T_{n_o}.$$

The corresponding dilations on the Lie algebra $\mathfrak{t}_{n_o}$ are given by

$$D_r(E_{i,j}) = r^{j-i}E_{i,j} \quad 1 \leq i < j \leq n_o.$$

As already seen for the graded Lie groups, the same homogeneous Lie group may admit various homogeneous structures, that is, a nilpotent Lie group or algebra may admit different families of dilations, even after renormalisation of the eigenvalues (see Remark 3.1.8). This can already be seen from the examples in the graded case (see Remark 3.1.6 part 1). These examples can be generalised as follows.

Example 3.1.11. On $\mathbb{R}^n$ we can define

$$D_r(x_1, \ldots, x_n) = (r^{\upsilon_1}x_1, \ldots, r^{\upsilon_n}x_n),$$

where $0 < \upsilon_1 \leq \ldots \leq \upsilon_n$, and on $\mathbb{H}_{n_o}$ we can define

$$D_r(x_1, \ldots, x_{n_o}, y_1, \ldots, y_{n_o}, t) = (r^{\upsilon_1}x_1, \ldots, r^{\upsilon_{n_o}}x_{n_o}, r^{\upsilon_1'}y_1, \ldots, r^{\upsilon_{n_o}'}y_{n_o}, r^{\upsilon''}t),$$

where $\upsilon_j > 0$, $\upsilon_j' > 0$ and $\upsilon_j + \upsilon_j' = \upsilon''$ for all $j = 1, \ldots, n_o$.

These families of dilations give graded structures whenever the weights υ_j for $\mathbb{R}^n$ and $\upsilon_j, \upsilon_j', \upsilon''$ for $\mathbb{H}_{n_o}$ are all rational or, more generally, all in $\alpha\mathbb{Q}^+$ for a fixed $\alpha \in \mathbb{R}_+$. From this remark it is not difficult to construct a homogeneous non-graded structure: on $\mathbb{R}^3$, consider the diagonal 3×3 matrix A with entries, e.g., 1 and π and $1 + \pi$.

Example 3.1.12. Continuing the example above, choosing the υ_j and υ_j''s rational in a certain way, it is also possible to find a homogeneous structure for $\mathbb{H}_{n_o}$ such that the corresponding gradation of $\mathfrak{h}_{n_o} = \oplus_{j=1}^{\infty} V_j$ does exist but is necessarily such that $V_1 = \{0\}$: we choose υ_j, υ_j' positive integers different from 1 but with 1 as greatest common divisor (for instance for $n_o = 2$, take $\upsilon_1 = 3, \upsilon_2 = 2, \upsilon_1' = 5, \upsilon_2' = 6$ and $\upsilon'' = 8$). As an illustration for Corollary 4.1.10 in the sequel, with this example, the homogeneous dimension is $Q = 3+2+5+6+8 = 24$ while the least common multiple is $\nu_o = 2 \times 3 \times 5 = 30$, so we have here $Q < \nu_o$.

If nothing is specified, we assume that the groups $(\mathbb{R}^n, +)$ and $\mathbb{H}_{n_o}$ are endowed with their classical structure of graded Lie groups as described in Examples 3.1.2 and 3.1.3.

Remark 3.1.13. We continue with several comments following those given in Remark 3.1.6.

1. The converse of Proposition 3.1.10 does not hold, namely, not every nilpotent Lie algebra or group admits a family of dilations. An example of a nine dimensional nilpotent Lie algebra which does not admit any family of dilations is due to Dyer [Dye70].

2. A direct product of two stratified Lie groups is graded but may be not stratified as their stratification structures may not 'match'. This can be also seen on the level of dilations defined in Section 3.1.2. Jumping ahead and using the notion of homogeneous operators, we see that this remark may be an advantage for example when considering the sub-Laplacian $\mathcal{L} = X^2 + Y^2$ on the Heisenberg group $\mathbb{H}_1$. Then the operator
$$-\mathcal{L} + \partial_t^k$$
for $k \in \mathbb{N}$ odd, becomes homogeneous on the direct product $\mathbb{H}_1 \times \mathbb{R}$ when it is equipped with the dilation structure which is not the one of a stratified Lie group, see Lemma 4.2.11 or, more generally, Remark 4.2.12.

3. In our definition of a homogeneous structure we started with dilations defined on the Lie algebra inducing dilations on the Lie group. If we start with a Lie group the situation may become slightly more involved. For example, $\mathbb{R}^3$ with the group law
$$xy = (\operatorname{arcsinh}(\sinh(x_1) + \sinh(y_1)), x_2 + y_2 + \sinh(x_1)y_3, x_3 + y_3)$$
is a 2-step nilpotent stratified Lie group, the first stratum given by
$$X = \cosh(x_1)^{-1}\partial_{x_1}, \quad Y = \sinh(x_1)\partial_{x_2} + \partial_{x_3},$$
and their commutator is
$$T = [X, Y] = \partial_{x_2}.$$
It may seem like there is no obvious homogeneous structure on this group but we can see it going to its Lie algebra which is isomorphic to the Lie algebra $\mathfrak{h}_1$ of the Heisenberg group $\mathbb{H}_1$. Consequently, the above group itself is isomorphic to $\mathbb{H}_1$ with the corresponding dilation structure.

4. In fact, the same argument as above shows that if we defined a stratified Lie group by saying that there is a collection of vector fields on it stratified with respect to their commutation relations, then for every such stratified Lie group there always exists a homogeneous stratified Lie group isomorphic to it. Indeed, since the Lie algebra is stratified and has a natural dilation structure with integer weights, we obtain the required homogeneous Lie group by exponentiating this Lie algebra. We refer to e.g. [BLU07, Theorem 2.2.18] for a detailed proof of this.

Refining the proof of Proposition 3.1.10, we can obtain the following technical result which gives the existence of an 'adapted' basis of eigenvectors for the dilations.

Lemma 3.1.14. *Let $\mathfrak{g}$ be a Lie algebra endowed with a family of dilations $\{D_r, r > 0\}$. Then there exists a basis $\{X_1, \ldots, X_n\}$ of $\mathfrak{g}$, positive numbers $\upsilon_1, \ldots, \upsilon_n > 0$, and an integer n' with $1 \le n' \le n$ such that*

$$\forall t > 0 \quad \forall j = 1, \ldots, n \qquad D_t(X_j) = t^{\upsilon_j} X_j, \tag{3.2}$$

and

$$[\mathfrak{g}, \mathfrak{g}] \subset \mathbb{R}X_{n'+1} \oplus \ldots \oplus \mathbb{R}X_n. \tag{3.3}$$

Moreover, $X_1, \ldots, X_{n'}$ generate the algebra $\mathfrak{g}$, that is, any element of $\mathfrak{g}$ can be written as a linear combination of these vectors together with all their iterated Lie brackets.

This result and its proof are due to ter Elst and Robinson (see [tER97, Lemma 2.2]). Condition (3.2) says that $\{X_j\}_{j=1}^n$ is a basis of eigenvectors for the mapping A given by

$$D_r = \text{Exp}(A \ln r).$$

Condition (3.3) says that this basis can be chosen so that the first n' vectors of this basis generate the whole Lie algebra and the others span (linearly) the derived algebra $[\mathfrak{g}, \mathfrak{g}]$.

Proof of Lemma 3.1.14. We continue with the notation of the proof of Proposition 3.1.10. For each weight $\upsilon \in \mathcal{W}_A$, we choose a basis

$$\{Y_{\upsilon,1}, \ldots, Y_{\upsilon,d'_\upsilon}, Y_{\upsilon,d'_\upsilon+1}, \ldots, Y_{\upsilon,d_\upsilon}\} \text{ of } W_\upsilon$$

such that $\{Y_{\upsilon,d'_\upsilon+1}, \ldots, Y_{\upsilon,d_\upsilon}\}$ is a basis of the subspace

$$W_\upsilon \bigcap \left(\text{Span} \bigcup_{\upsilon'+\upsilon''=\upsilon} [W_{\upsilon'}, W_{\upsilon''}] \right).$$

Since $\mathfrak{g} = \oplus_{\upsilon \in \mathcal{W}_A} W_\upsilon$, we have by construction that

$$[\mathfrak{g}, \mathfrak{g}] \subset \text{Span} \{Y_{\upsilon,j} \ : \ \upsilon \in \mathcal{W}_A, \ d'_\upsilon + 1 \le j \le d_\upsilon\}.$$

Let $\mathfrak{h}$ be the Lie algebra generated by

$$\{Y_{\upsilon,j} \ : \ \upsilon \in \mathcal{W}_A, \ 1 \le j \le d'_\upsilon\}. \tag{3.4}$$

We now label and order the weights, that is, we write

$$\mathcal{W}_A = \{\upsilon_1, \ldots, \upsilon_m\}$$

with $1 \le \upsilon_1 < \ldots < \upsilon_m$. It follows by induction on $N = 1, 2 \ldots, m$ that $\oplus_{j=1}^N W_{\upsilon_j}$ is contained in $\mathfrak{h}$ and hence $\mathfrak{h} = \mathfrak{g}$ and the set (3.4) generate (algebraically) $\mathfrak{g}$.

A basis with the required property is given by

$$Y_{\upsilon_1,1}, \ldots, Y_{\upsilon_1,d'_{\upsilon_1}}, \ldots, Y_{\upsilon_m,1}, \ldots, Y_{\upsilon_m,d'_{\upsilon_m}} \quad \text{for } X_1, \ldots, X_{n'},$$

and

$$Y_{\upsilon_1,d'_{\upsilon_1}+1}, \ldots, Y_{\upsilon_1,d_{\upsilon_1}}, \ldots, Y_{\upsilon_m,d'_{\upsilon_m}+1}, \ldots, Y_{\upsilon_m,d_{\upsilon_m}} \quad \text{for } X_{n'+1}, \ldots, X_n.$$

□

3.1.3 Homogeneous structure

In this section, we shall be working on a fixed homogeneous Lie group G of dimension n with dilations

$$\{D_r = \mathrm{Exp}(A \ln r)\}.$$

We denote by $\upsilon_1, \ldots, \upsilon_n$ the weights, listed in increasing order and with each value listed as many times as its multiplicity, and we assume without loss of generality (see Remark 3.1.8) that $\upsilon_1 \geq 1$. Thus,

$$1 \leq \upsilon_1 \leq \upsilon_2 \leq \ldots \leq \upsilon_n. \tag{3.5}$$

If the group G is graded, then the weights are also assumed to be integers with one as their greatest common divisor (again see Remark 3.1.8).

By Proposition 3.1.10 the Lie group G is nilpotent connected simply connected. Thus it may be identified with $\mathbb{R}^n$ equipped with a polynomial law, using the exponential mapping $\exp_G$ of the group (see Section 1.6). With this identification its unit element is $0 \in \mathbb{R}^n$ and it may also be denoted by 0_G or simply by 0.

We fix a basis $\{X_1, \ldots, X_n\}$ of $\mathfrak{g}$ such that

$$AX_j = \upsilon_j X_j$$

for each j. This yields a Lebesgue measure on $\mathfrak{g}$ and a Haar measure on G by Proposition 1.6.6. If x or g denotes a point in G the Haar measure is denoted by dx or dg. The Haar measure of a measurable subset S of G is denoted by $|S|$.

We easily check that

$$|D_r(S)| = r^Q |S|, \qquad \int_G f(rx)dx = r^{-Q} \int_G f(x)dx, \tag{3.6}$$

where

$$Q = \upsilon_1 + \ldots + \upsilon_n = \mathrm{Tr} A. \tag{3.7}$$

The number Q is larger (or equal) than the usual dimension of the group:

$$n = \dim G \leq Q,$$

and may replace it for certain questions of analysis. For this reason the number Q is called the *homogeneous dimension* of G.

Homogeneity

Any function defined on G or on $G\backslash\{0\}$ can be composed with the dilations D_r. Using property (3.6) of the Haar measure and the dilations, we have for any measurable functions f and ϕ on G, provided that the integrals exist,

$$\int_G (f \circ D_r)(x)\ \phi(x)\ dx = r^{-Q} \int_G f(x)\ (\phi \circ D_{\frac{1}{r}})(x)\ dx. \tag{3.8}$$

Therefore, we can extend the map $f \mapsto f \circ D_r$ to distributions via

$$\langle f \circ D_r, \phi \rangle := r^{-Q} \langle f, \phi \circ D_{\frac{1}{r}} \rangle, \quad f \in \mathcal{D}'(G), \ \phi \in \mathcal{D}(G). \tag{3.9}$$

We can now define the homogeneity of a function or a distribution in the same way:

Definition 3.1.15. Let $\nu \in \mathbb{C}$.

(i) A function f on $G \backslash \{0\}$ or a distribution $f \in \mathcal{D}'(G)$ is *homogeneous of degree* $\nu \in \mathbb{C}$ (or ν-homogeneous) when

$$f \circ D_r = r^\nu f \quad \text{for any } r > 0.$$

(ii) A linear operator $T : \mathcal{D}(G) \to \mathcal{D}'(G)$ is *homogeneous of degree* $\nu \in \mathbb{C}$ (or ν-homogeneous) when

$$T(\phi \circ D_r) = r^\nu (T\phi) \circ D_r \quad \text{for any } \phi \in \mathcal{D}(G), \, r > 0.$$

Remark 3.1.16. We will also say that a linear operator $T : E \to F$, where E is a Fréchet space containing $\mathcal{D}(G)$ as a dense subset, and F is a Fréchet space included in $\mathcal{D}'(G)$, is homogeneous of degree $\nu \in \mathbb{C}$ when its restriction as an operator from $\mathcal{D}(G)$ to $\mathcal{D}'(G)$ is. For example, it will apply to the situation when T is a linear operator from $L^p(G)$ to some $L^q(G)$.

Example 3.1.17 (Coordinate function). The coordinate function $x_j = [x]_j$ given by

$$G \ni x = (x_1, \ldots, x_n) \longmapsto x_j = [x]_j, \tag{3.10}$$

is homogeneous of degree υ_j.

Example 3.1.18 (Koranyi norm). The function defined on the Heisenberg group $\mathbb{H}_{n_o}$ by

$$\mathbb{H}_{n_o} \ni (x, y, t) \longmapsto \left(\left(|x|^2 + |y|^2 \right)^2 + t^2 \right)^{1/4},$$

where $|x|$ and $|y|$ denote the canonical norms of x and y in $\mathbb{R}^{n_o}$, is homogeneous of degree 1. It is sometimes called the Koranyi norm.

Example 3.1.19 (Haar measure). Equality (3.8) shows that the Haar measure, viewed as a tempered distribution, is a homogeneous distribution of degree Q (see (3.7)). We can write this informally as

$$d(rx) = r^Q dx,$$

see (3.6).

Example 3.1.20 (Dirac measure at 0). The Dirac measure at 0 is the probability measure δ_0 given by

$$\int_G f d\delta_0 = f(0).$$

It is homogeneous of degree $-Q$ since for any $\phi \in \mathcal{D}(G)$ and $r > 0$, we have

$$\langle \delta_0 \circ D_r, \phi \rangle = r^{-Q} \langle \delta_0, \phi \circ D_{\frac{1}{r}} \rangle = r^{-Q} \phi(\frac{1}{r} 0) = r^{-Q} \phi(0) = \langle r^{-Q} \delta_0, \phi \rangle.$$

Example 3.1.21 (Invariant vector fields). Let $X \in \mathfrak{g}$ be viewed as a left-invariant vector field X or a right-invariant vector field $\tilde{X}$ (cf. Section 1.3). We assume that X is in the υ_j-eigenspace of A. Then the left and right-invariant differential operators X and $\tilde{X}$ are homogeneous of degree υ_j. Indeed,

$$\begin{aligned} X(f \circ D_r)(x) &= \partial_{t=0} \{ f \circ D_r (x \exp_G(tX)) \} = \partial_{t=0} \{ f(rx \exp_G(r^{\upsilon_j} tX)) \} \\ &= r^{\upsilon_j} \partial_{t'=0} \{ f(rx \exp_G(t'X)) \} = r^{\upsilon_j} (Xf)(rx), \end{aligned}$$

and similarly for $\tilde{X}$.

The following properties are very easy to check:

Lemma 3.1.22. *(i) Whenever it makes sense, the product of two functions, distributions or operators of degrees ν_1 and ν_2 is homogeneous of degree $\nu_1\nu_2$.*

(ii) Let $T : \mathcal{D}(G) \to \mathcal{D}'(G)$ be a ν-homogeneous operator. Then its formal adjoint and transpose T^* and T^t, given by

$$\int_G (Tf)\bar{g} = \int_G f(\overline{T^* g}), \quad \int_G (Tf)g = \int_G f(T^t g), \quad f, g \in \mathcal{D}(G),$$

are also homogeneous with degree $\bar{\nu}$ and ν respectively.

Consequently for any non-zero multi-index $\alpha = (\alpha_1, \ldots, \alpha_n) \in \mathbb{N}_0^n \backslash \{0\}$, the function

$$x^\alpha := x_1^{\alpha_1} \ldots x_n^{\alpha_n}, \tag{3.11}$$

and the operators

$$\left(\frac{\partial}{\partial x}\right)^\alpha := \left(\frac{\partial}{\partial x_1}\right)^{\alpha_1} \cdots \left(\frac{\partial}{\partial x_n}\right)^{\alpha_n}, \ X^\alpha := X_1^{\alpha_1} \ldots X_n^{\alpha_n} \text{ and } \tilde{X}^\alpha := \tilde{X}_1^{\alpha_1} \ldots \tilde{X}_n^{\alpha_n},$$

are homogeneous of degree

$$[\alpha] := \upsilon_1 \alpha_1 + \ldots + \upsilon_n \alpha_n. \tag{3.12}$$

Formula (3.12) defines the *homogeneous degree* of the multi-index α. It is usually different from the length of α given by

$$|\alpha| := \alpha_1 + \ldots + \alpha_n.$$

For $\alpha = 0$, the function x^α and the operators $(\frac{\partial}{\partial x})^\alpha$, X^α, $\tilde{X}^\alpha$ are defined to be equal, respectively, to the constant function 1 and the identity operator I, which are of degree $[\alpha] := 0$.

With this convention for each $\alpha \in \mathbb{N}_0^n$, the differential operators $(\frac{\partial}{\partial x})^\alpha$, X^α and $\tilde{X}^\alpha$ are of order $|\alpha|$ but of homogeneous degree $[\alpha]$.

One easily checks for $\alpha_1, \alpha_2 \in \mathbb{N}_0^n$ that

$$[\alpha_1] + [\alpha_2] = [\alpha_1 + \alpha_2], \quad |\alpha_1| + |\alpha_2| = |\alpha_1 + \alpha_2| .$$

Proposition 3.1.23. *Let the operator T be homogeneous of degree ν_T and let f be a function or a distribution homogeneous of degree ν_f. Then, whenever Tf makes sense, the distribution Tf is homogeneous of degree $\nu_f - \nu_T$.*

In particular, if $f \in \mathcal{D}'(G)$ is homogeneous of degree ν, then

$$X^\alpha f, \tilde{X}^\alpha f, \partial^\alpha f$$

are homogeneous of degree $\nu - [\alpha]$.

Proof. The first claim follows from the formal calculation

$$(Tf) \circ D_r = r^{-\nu_T} T(f \circ D_r) = r^{-\nu_T} T(r^{\nu_f} f) = r^{-\nu_T + \nu_f} Tf.$$

The second claim follows from the first one since X^α, $\tilde{X}^\alpha f$ and $\partial^\alpha f$ are well defined on distributions and are homogeneous of the same degree $[\alpha]$ given by (3.12). □

3.1.4 Polynomials

By Propositions 3.1.10 and 1.6.6 we already know that the group law is polynomial. This means that each $[xy]_j$ is a polynomial in the coordinates of x and of y. The homogeneous structure implies certain additional properties of this polynomial.

Proposition 3.1.24. *For any $j = 1, \ldots, n$, we have*

$$[xy]_j = x_j + y_j + \sum_{\substack{\alpha,\beta \in \mathbb{N}_0^n \setminus \{0\} \\ [\alpha]+[\beta]=\upsilon_j}} c_{j,\alpha,\beta}\, x^\alpha y^\beta .$$

In particular, this sum over $[\alpha]$ and $[\beta]$ can involve only coordinates in x or y with degrees of homogeneity strictly less than υ_j.

For example,

$$\begin{aligned}
\text{for } \upsilon_1: \quad [xy]_1 &= x_1 + y_1, \\
\text{for } \upsilon_2: \quad [xy]_2 &= x_2 + y_2 + \sum_{[\alpha]=[\beta]=\upsilon_1} c_{2,\alpha,\beta}\, x^\alpha y^\beta, \\
\text{for } \upsilon_3: \quad [xy]_3 &= x_3 + y_3 + \sum_{\substack{[\alpha]=\upsilon_1,\, [\beta]=\upsilon_2 \\ \text{or } [\alpha]=\upsilon_2,\, [\beta]=\upsilon_1}} c_{3,\alpha,\beta}\, x^\alpha y^\beta,
\end{aligned}$$

and so on.

Proof. Let $j = 1, \ldots, n$. From the Baker-Campbell-Hausdorff formula (see Theorem 1.3.2) applied to the two vectors $X = x_1X_1 + \ldots + x_nX_n$ and $Y = y_1X_1 + \ldots + y_nX_n$ of $\mathfrak{g}$, we have with our notation that

$$[xy]_j = x_j + y_j + R_j(x, y)$$

where $R_j(x, y)$ is a polynomial in $x_1, y_1, \ldots, x_n, y_n$. Moreover, R_j must be a finite linear combination of monomials $x^\alpha y^\beta$ with $|\alpha| + |\beta| \geq 2$:

$$R_j(x, y) = \sum_{\substack{\alpha,\beta\in\mathbb{N}_0^n \\ |\alpha|+|\beta|\geq 2}} c_{j,\alpha,\beta} x^\alpha y^\beta.$$

We now use the dilations. Since the function x_j is homogeneous of degree υ_j, we easily check

$$R_j(rx, ry) = r^{\upsilon_j} R_j(x, y)$$

for any $r > 0$ and this forces all the coefficients $c_{j,\alpha,\beta}$ with $[\alpha] + [\beta] \neq \upsilon_j$ to be zero. The formula follows. □

Recursively using Proposition 3.1.24, we obtain for any $\alpha \in \mathbb{N}_0^n \backslash \{0\}$:

$$(xy)^\alpha = [xy]_1^{\alpha_1} \ldots [xy]_n^{\alpha_n} = \sum_{\substack{\beta_1,\beta_2\in\mathbb{N}_0^n \\ [\beta_1]+[\beta_2]=[\alpha]}} c_{\beta_1,\beta_2}(\alpha) x^{\beta_1} y^{\beta_2}, \tag{3.13}$$

with

$$c_{\beta_1,0}(\alpha) = \begin{cases} 0 & \text{if } \beta_1 \neq \alpha \\ 1 & \text{if } \beta_1 = \alpha \end{cases} \quad \text{and} \quad c_{0,\beta_2}(\alpha) = \begin{cases} 0 & \text{if } \beta_2 \neq \alpha \\ 1 & \text{if } \beta_2 = \alpha \end{cases}. \tag{3.14}$$

Definition 3.1.25. A function P on G is a *polynomial* if $P \circ \exp_G$ is a polynomial on $\mathfrak{g}$.

For example the coordinate functions $x_1, \ldots, x_n$ defined in (3.10) or, more generally, the monomials x^α defined in (3.11) are (homogeneous) polynomials on G.

It is clear that every polynomial P on G can be written as a unique finite linear combination of the monomials x^α, that is,

$$P = \sum_{\alpha\in\mathbb{N}_0^n} c_\alpha x^\alpha, \tag{3.15}$$

where all but finitely many of the coefficients $c_\alpha \in \mathbb{C}$ vanish. The *homogeneous degree* of a polynomial P written as (3.15) is

$$D^\circ P := \max\{[\alpha] : \alpha \in \mathbb{N}_0^n \text{ with } c_\alpha \neq 0\},$$

which is often different from its *isotropic degree*:

$$d^\circ P := \max\{|\alpha| : \alpha \in \mathbb{N}_0^n \text{ with } c_\alpha \neq 0\}.$$

For example on $\mathbb{H}_{n_o}$, $1+t$ is a polynomial of homogeneous degree 2 but isotropic degree 1.

Definition 3.1.26. We denote by $\mathcal{P}(G)$ the set of all polynomials on G. For any $M \geq 0$ we denote by $\mathcal{P}_{\leq M}$ the set of polynomials P on G such that $D^\circ P \leq M$ and by $\mathcal{P}^{iso}_{\leq M}$ the set of polynomials on G such that $d^\circ P \leq M$. We also define in the same way $\mathcal{P}_{<M}$, $\mathcal{P}_{=M}$, $\mathcal{P}_{\geq M}$ and so on, and similarly for $\mathcal{P}^{iso}$.

It is clear that $\mathcal{P}(G)$ is an algebra, for pointwise multiplication, which is generated by the x_j's.

It is not difficult to see:

Lemma 3.1.27. *The subspaces $\mathcal{P}_{\leq M}$ and $\mathcal{P}^{iso}_{\leq M}$ of $\mathcal{P}$ are finite dimensional with bases $\{x^\alpha : \alpha \in \mathbb{N}_0^n, [\alpha] \leq M\}$ and $\{x^\alpha : \alpha \in \mathbb{N}_0^n, |\alpha| \leq M\}$, respectively. Furthermore,*

$$\forall M \geq 0 \qquad \mathcal{P}_{\leq M} \subset \mathcal{P}^{iso}_{\leq M} \subset \mathcal{P}_{\leq \upsilon_n M}.$$

Proof. The first part of the lemma is clear. For the second, because of (3.5), we have

$$\forall \alpha \in \mathbb{N}_0^n \qquad |\alpha| \leq [\alpha] \leq \upsilon_n |\alpha|. \tag{3.16}$$

Therefore,

$$\forall P \in \mathcal{P} \qquad d^\circ P \leq D^\circ P \leq \upsilon_n d^\circ P,$$

and the inclusions follow. □

By Proposition 3.1.24, $[xy]_j$ is in $\mathcal{P}_{\leq \upsilon_j}$ as a function of x for each y, and also as a function of y for each x. Hence each subspace $\mathcal{P}_{\leq M}$ is invariant under left and right translation. This is not the case for $\mathcal{P}^{iso}_{\leq M}$ (unless $\mathcal{P}^{iso}_{\leq M} \sim \mathbb{C}$ or $G = (\mathbb{R}^n, +)$); consequently, it will not be of much use to us.

3.1.5 Invariant differential operators on homogeneous Lie groups

We now investigate expressions for left- and right-invariant operators on homogeneous Lie groups.

Proposition 3.1.28. *The left and right-invariant vector fields X_j and $\tilde{X}_j$, for any $j = 1, \ldots, n$, can be written as*

$$\begin{aligned} X_j &= \frac{\partial}{\partial x_j} + \sum_{\substack{1\leq k\leq n\\ \upsilon_j<\upsilon_k}} P_{j,k} \frac{\partial}{\partial x_k} = \frac{\partial}{\partial x_j} + \sum_{\substack{1\leq k\leq n\\ \upsilon_j<\upsilon_k}} \frac{\partial}{\partial x_k} P_{j,k} \\ \tilde{X}_j &= \frac{\partial}{\partial x_j} + \sum_{\substack{1\leq k\leq n\\ \upsilon_j<\upsilon_k}} Q_{j,k} \frac{\partial}{\partial x_k} = \frac{\partial}{\partial x_j} + \sum_{\substack{1\leq k\leq n\\ \upsilon_j<\upsilon_k}} \frac{\partial}{\partial x_k} Q_{j,k}, \end{aligned}$$

where $P_{j,k}$ *and* $Q_{j,k}$ *are homogeneous polynomials on* G *of homogeneous degree* $\upsilon_k - \upsilon_j > 0$.

Proof. For any $x \in G$, we denote by $L_x : G \to G$ the left-translation, i.e. $L_x(y) = xy$. Let $j = 1, \ldots, n$. Recall that X_j is the differential operator invariant under left-translation which agrees with $\frac{\partial}{\partial x_j}$ at 0, that is, for any $f \in C^\infty(G)$ and $x_o \in G$, we have

$$(X_j f) \circ L_{x_o}(0) = X_j(f \circ L_{x_o})(0) \quad \text{and} \quad X_j(f)(0) = \frac{\partial f}{\partial x_j}(0).$$

Thus

$$\begin{aligned}(X_j f)(x_o) &= (X_j f) \circ L_{x_o}(0) = X_j(f \circ L_{x_o})(0) = \frac{\partial}{\partial x_j}(f \circ L_{x_o})(0) \\ &= \sum_{k=1}^{n} \frac{\partial f}{\partial x_k}(x_o) \frac{\partial [x_o x]_k}{\partial x_j}(0),\end{aligned}$$

by the chain rule. But by Proposition 3.1.24,

$$\begin{aligned}\frac{\partial [x_o x]_k}{\partial x_j}(0) &= \frac{\partial}{\partial x_j}\left\{ [x_o]_k + x_k + \sum_{\substack{\alpha,\beta \in \mathbb{N}_0^n \setminus \{0\} \\ [\alpha]+[\beta]=\upsilon_k}} c_{k,\alpha,\beta} x_o^\alpha x^\beta \right\}(0) \\ &= \delta_{j,k} + \sum_{\substack{\beta = e_j,\, \alpha \in \mathbb{N}_0^n \setminus \{0\} \\ [\alpha]+[\beta]=\upsilon_k}} c_{k,\alpha,\beta} x_o^\alpha,\end{aligned}$$

where e_j is the multi-index with 1 in the j-th place and zeros elsewhere, and $\delta_{j,k}$ is the Kronecker delta. The assertion for X_j now follows immediately, and the assertion for $\tilde{X}_j$ is proved in the same way using right translations. □

Proposition 3.1.28 gives, in particular,

$$\begin{aligned}\text{for } \upsilon_n : \quad X_n &= \frac{\partial}{\partial x_n}, \\ \text{for } \upsilon_{n-1} : \quad X_{n-1} &= \frac{\partial}{\partial x_{n-1}} + P_{n-1,n}\frac{\partial}{\partial x_n}, \\ \text{for } \upsilon_{n-2} : \quad X_{n-2} &= \frac{\partial}{\partial x_{n-2}} + P_{n-2,n-1}\frac{\partial}{\partial x_{n-1}} + P_{n-2,n}\frac{\partial}{\partial x_n},\end{aligned}$$

so that

$$\begin{aligned}\frac{\partial}{\partial x_n} &= X_n, \\ \frac{\partial}{\partial x_{n-1}} &= X_{n-1} - P_{n-1,n}X_n, \\ \frac{\partial}{\partial x_{n-2}} &= X_{n-2} - P_{n-2,n-1}\left(X_{n-1} - P_{n-1,n}X_n\right) - P_{n-2,n}X_n,\end{aligned}$$

and so forth, with similar formulae for the right-invariant vector fields. This shows that there are formulas for the $\frac{\partial}{\partial x_j}$'s of the same sort as for the X_j's and $\tilde{X}_j$'s, that is,

$$\frac{\partial}{\partial x_j} = X_j + \sum_{\substack{1\leq k\leq n \\ \upsilon_j<\upsilon_k}} p_{j,k} X_k = \tilde{X}_j + \sum_{\substack{1\leq k\leq n \\ \upsilon_j<\upsilon_k}} q_{j,k} \tilde{X}_k, \tag{3.17}$$

where $p_{j,k}$ and $q_{j,k}$ are homogeneous polynomials on G of homogeneous degree $\upsilon_k - \upsilon_j > 0$.

Remark 3.1.29.
1. Given the formulae above and the condition on the degree, it is not difficult to see that the $P_{j,k}$ and $Q_{j,k}$ in Proposition 3.1.28 and the $p_{j,k}$ and $q_{j,k}$ in (3.17), with $\upsilon_k > \upsilon_j$, are polynomials in $(x_1, \ldots, x_{k-1})$ and commute with X_k, $\tilde{X}_k$ and $\frac{\partial}{\partial x_k}$ respectively.
2. The first part of Proposition 3.1.28 and its proof are valid for any nilpotent Lie group (see Remark 1.6.7, part (1)). In our setting here, the homogeneous structure implies the additional property that the $P_{j,k}$ and $Q_{j,k}$ are homogeneous.

Corollary 3.1.30. *For any $\alpha \in \mathbb{N}_0^n\backslash\{0\}$,*

$$\begin{aligned} X^\alpha &= \sum_{\substack{\beta\in\mathbb{N}_0^n,\,|\beta|\leq|\alpha| \\ [\beta]\geq[\alpha]}} P_{\alpha,\beta}\tilde{X}^\beta = \sum_{\substack{\beta\in\mathbb{N}_0^n,\,|\beta|\leq|\alpha| \\ [\beta]\geq[\alpha]}} \tilde{X}^\beta p_{\alpha,\beta}, \\ \tilde{X}^\alpha &= \sum_{\substack{\beta\in\mathbb{N}_0^n,\,|\beta|\leq|\alpha| \\ [\beta]\geq[\alpha]}} Q_{\alpha,\beta}X^\beta = \sum_{\substack{\beta\in\mathbb{N}_0^n,\,|\beta|\leq|\alpha| \\ [\beta]\geq[\alpha]}} X^\beta q_{\alpha,\beta}, \end{aligned}$$

where $P_{\alpha,\beta}, p_{\alpha,\beta}, Q_{\alpha,\beta}, q_{\alpha,\beta}$ are homogeneous polynomials of homogeneous degree $[\beta] - [\alpha]$.

Proof. By Proposition 3.1.28 we obtain recursively for any $\alpha \in \mathbb{N}_0^n\backslash\{0\}$ that

$$X^\alpha = \sum_{\substack{\beta\in\mathbb{N}_0^n,\,|\beta|\leq|\alpha| \\ [\beta]\geq[\alpha]}} P_{\alpha,\beta} \left(\frac{\partial}{\partial x}\right)^\beta, \tag{3.18}$$

with $P_{\alpha,\beta}$ homogeneous polynomial of degree $[\beta] - [\alpha]$. Similar formulae yield $\tilde{X}^\alpha$ in terms of the $\left(\frac{\partial}{\partial x}\right)^\beta$'s.

Recursively from (3.17), we also obtain similar formulae for $\left(\frac{\partial}{\partial x}\right)^\alpha$ in terms of the X^β or $\tilde{X}^\beta$.

The assertion comes form combining these formulae, with a similar argument for $p_{\alpha,\beta}$ and $q_{\alpha,\beta}$. □

Corollary 3.1.31. *For any $M \geq 0$, the maps*

$$\begin{aligned}
(i) \quad P &\longmapsto \left\{ \left(\frac{\partial}{\partial x}\right)^{\alpha} P(0) \right\}_{\alpha \in \mathbb{N}_0^n,\, [\alpha] \leq M},\\
(ii) \quad P &\longmapsto \{X^{\alpha} P(0)\}_{\alpha \in \mathbb{N}_0^n,\, [\alpha] \leq M},\\
(iii) \quad P &\longmapsto \left\{\tilde{X}^{\alpha} P(0)\right\}_{\alpha \in \mathbb{N}_0^n,\, [\alpha] \leq M},
\end{aligned}$$

are linear isomorphisms from $\mathcal{P}_{\leq M}$ to $\mathbb{C}^{\dim \mathcal{P}_{\leq M}}$. Also, the maps

$$\begin{aligned}
(i) \quad P &\longmapsto \left\{ \left(\frac{\partial}{\partial x}\right)^{\alpha} P(0) \right\}_{\alpha \in \mathbb{N}_0^n,\, [\alpha] = M},\\
(ii) \quad P &\longmapsto \{X^{\alpha} P(0)\}_{\alpha \in \mathbb{N}_0^n,\, [\alpha] = M},\\
(iii) \quad P &\longmapsto \left\{\tilde{X}^{\alpha} P(0)\right\}_{\alpha \in \mathbb{N}_0^n,\, [\alpha] = M},
\end{aligned}$$

are linear isomorphisms from $\mathcal{P}_{=M}$ to $\mathbb{C}^{\dim \mathcal{P}_{=M}}$.

Proof. By Lemma 3.1.27, the vector subspace $\mathcal{P}_{\leq M}$ of $\mathcal{P}$ is finite dimensional, with basis $\{x^{\alpha} : \alpha \in \mathbb{N}_0^n,\ [\alpha] \leq M\}$. Hence case (i) is a simple consequence of Taylor's Theorem on $\mathbb{R}^n$.

Note that in the formula (3.18), $P_{\alpha,\beta}$ is a constant function when $[\alpha] = [\beta]$ and $P_{\alpha,\beta}(0) = 0$ when $[\alpha] > [\beta]$. Hence

$$X^{\alpha}|_0 = \sum_{\substack{\beta \in \mathbb{N}_0^n,\, |\beta| \leq |\alpha| \\ [\beta] = [\alpha]}} P_{\alpha,\beta} \left(\frac{\partial}{\partial x}\right)^{\beta} \Bigg|_0 .$$

We have similar result from the other formulae relating X^{α}, $\tilde{X}^{\alpha}$ and $\left(\frac{\partial}{\partial x}\right)^{\alpha}$.

Cases (ii) and (iii) follow from these observations together with case (i). The case of the homogeneous polynomials of order M is similar. □

We may use the following property without referring to it.

Corollary 3.1.32. *Let $\alpha, \beta \in \mathbb{N}_0^n$. The differential operator $X^{\alpha} X^{\beta}$ is a linear combination of X^{γ} with $[\gamma] \in \mathbb{N}_0^n$, $[\gamma] = [\alpha] + [\beta]$:*

$$X^{\alpha} X^{\beta} = \sum_{\substack{\gamma \in \mathbb{N}_0^n,\, |\gamma| \leq |\alpha| + |\beta| \\ [\gamma] = [\alpha] + [\beta]}} c'_{\alpha,\beta,\gamma} X^{\gamma}. \tag{3.19}$$

The differential operator $\tilde{X}^{\alpha} \tilde{X}^{\beta}$ is a linear combination of $\tilde{X}^{\gamma}$ with $[\gamma] \in \mathbb{N}_0^n$, $|\gamma| \leq |\alpha| + |\beta|$ and $[\gamma] = [\alpha] + [\beta]$.

Proof. The differential operator $X^\alpha X^\beta$ is a left-invariant differential operator of order $|\alpha|+|\beta|$ by (3.18), and it is a linear combination of X^γ, $|\gamma|\leq|\alpha|+|\beta|$ (see Section 1.3),

$$X^\alpha X^\beta = \sum_{\gamma\in\mathbb{N}_0^n,\,|\gamma|\leq|\alpha|+|\beta|} c'_{\alpha,\beta,\gamma} X^\gamma.$$

By homogeneity, for any $r>0$ and any function $f\in C^\infty(G)$, we have on one hand,

$$X^\alpha X^\beta(f\circ D_r) = r^{[\alpha]+[\beta]}(X^\alpha X^\beta f)\circ D_r,$$

and on the other hand,

$$\begin{aligned} X^\alpha X^\beta(f\circ D_r) &= \sum_{\gamma\in\mathbb{N}_0^n,\,|\gamma|\leq|\alpha|+|\beta|} c'_{\alpha,\beta,\gamma} X^\gamma(f\circ D_r) \\ &= \sum_{\gamma\in\mathbb{N}_0^n,\,|\gamma|\leq|\alpha|+|\beta|} c'_{\alpha,\beta,\gamma} r^{[\gamma]}(X^\gamma f)\circ D_r. \end{aligned}$$

Choosing f suitably (for example f being polynomials of homogeneous degree at most $[\alpha]+[\beta]$, see Corollary 3.1.31), this implies that if $[\alpha]+[\beta]\neq[\gamma]$ then $c'_{\alpha,\beta,\gamma}=0$, showing (3.19).

The property for the right-invariant vector fields is similar. □

3.1.6 Homogeneous quasi-norms

We can define an Euclidean norm $|\cdot|_E$ on $\mathfrak{g}$ by declaring the X_j's to be orthonormal. We may also regard this norm as a function on G via the exponential mapping, that is,

$$|x|_E = |\exp_G^{-1} x|_E.$$

However, this norm is of limited use for our purposes, since it does not interact in a simple fashion with dilations. We therefore define:

Definition 3.1.33. A *homogeneous quasi-norm* is a continuous non-negative function

$$G\ni x\longmapsto |x|\in[0,\infty),$$

satisfying

(i) (symmetric) $|x^{-1}|=|x|$ for all $x\in G$,

(ii) (1-homogeneous) $|rx|=r|x|$ for all $x\in G$ and $r>0$,

(iii) (definite) $|x|=0$ if and only if $x=0$.

The $|\cdot|$-*ball centred at* $x\in G$ *with radius* $R>0$ is defined by

$$B(x,R) := \{y\in G:\ |x^{-1}y|<R\}.$$

Remark 3.1.34. With such definition, we have for any $x, x_o \in G$, $R > 0$,

$$x_o B(x,R) = B(x_o x, R), \tag{3.20}$$

since

$$z \in x_o B(x,R) \Longleftrightarrow x_o^{-1} z \in B(x,R) \Longleftrightarrow |x^{-1} x_o^{-1} z| < R \Longleftrightarrow z \in B(x_o x, R).$$

In particular, we see that

$$B(x,r) = xB(0,r).$$

It is also easy to check that

$$B(0,r) = D_r(B(0,1)).$$

Note that in our definition of quasi-balls, we choose to privilege the left translations. Indeed, the set $\{y \in G \ : \ |yx^{-1}| < R\}$ may also be defined as a quasi-ball but one would have to use the right translation instead of the left x_o-translation to have a similar property to (3.20).

An important example of a quasi-norm is given by Example 3.1.18 on the Heisenberg group $\mathbb{H}_{n_o}$. More generally, on any homogeneous Lie group, the following functions are homogeneous quasi-norms:

$$|(x_1,\ldots,x_n)|_p = \left(\sum_{j=1}^n |x_j|^{\frac{p}{\upsilon_j}}\right)^{\frac{1}{p}}, \tag{3.21}$$

for $0 < p < \infty$, and for $p = \infty$:

$$|(x_1,\ldots,x_n)|_\infty = \max_{1\leq j\leq n} |x_j|^{\frac{1}{\upsilon_j}}. \tag{3.22}$$

In Definition 3.1.33 we do not require a homogeneous quasi-norm to be smooth away from the origin but some authors do. Quasi-norms with added regularity always exist as well but, in fact, a distinction between different quasi-norms is usually irrelevant for many questions of analysis because of the following property:

Proposition 3.1.35. *(i) Every homogeneous Lie group G admits a homogeneous quasi-norm that is smooth away from the unit element.*

(ii) Any two homogeneous quasi-norms $|\cdot|$ and $|\cdot|'$ on G are mutually equivalent:

$$\|\cdot\| \asymp \|\cdot\|' \quad \text{in the sense that} \quad \exists a,b>0 \quad \forall x \in G \qquad a|x|' \leq |x| \leq b|x|'.$$

Proof. Let us consider the function

$$\Psi(r,x) = |D_r x|_E^2 = \sum_{j=1}^{n} r^{2\upsilon_j} x_j^2.$$

Let us fix $x \neq 0$. The function $\Psi(r,x)$ is continuous, strictly increasing in r and satisfies

$$\Psi(r,x) \underset{r\to 0}{\longrightarrow} 0 \quad \text{and} \quad \Psi(r,x) \underset{r\to+\infty}{\longrightarrow} +\infty.$$

Therefore, there is a unique $r > 0$ such that $|D_r x|_E = 1$. We set $|x|_o := r^{-1}$.

Hence we have defined a map

$$G\backslash\{0\} \ni x \mapsto |x|_o^{-1} \in (0,\infty)$$

which is the implicit function for $\Psi(r,x) = 1$. This map is smooth since the function $\Psi(r,x)$ is smooth from $(0,+\infty)\times G\backslash\{0\}$ to $(0,\infty)$ and $\partial_r \Psi(r,x)$ is always different from zero. Setting $|0_G|_o := 0$, the map $|\cdot|_o$ clearly satisfies the properties of Definition 3.1.33. This shows part (i).

For Part (ii), it is sufficient to prove that any homogeneous quasi-norm is equivalent to $|\cdot|_o$ constructed above. Before doing so, we observe that the unit spheres in the Euclidean norm and the homogeneous quasi-norm $|\cdot|_o$ coincide, that is,

$$\mathfrak{S} := \{x \in G : |x|_E = 1\} = \{x \in G : |x|_o = 1\}.$$

Let $|\cdot|$ be any other homogeneous norm. Since it is a definite function (see (iii) of Definition 3.1.33) its restriction to $\mathfrak{S}$ is never zero. By compactness of $\mathfrak{S}$ and continuity of $|\cdot|$, there are constants $a,b > 0$ such that

$$\forall x \in \mathfrak{S} \qquad a \leq |x| \leq b.$$

For any $x \in G\backslash\{0\}$, let $t > 0$ be given by $t^{-1} = |x|_o$. We have $D_t x \in \mathfrak{S}$, and thus

$$a \leq |D_t x| \leq b \quad \text{and} \quad a|x|_o = t^{-1}a \leq |x| \leq t^{-1}b = b|x|_o.$$

The conclusion of Part (ii) follows. $\square$

Remark 3.1.36. If G is graded, the formula (3.21) for $p = 2\upsilon_1 \ldots \upsilon_n$ gives another concrete example of a homogeneous quasi-norm smooth away from the origin since $x \mapsto |x|_p^p$ is then a polynomial in the coordinate functions $\{x_j\}$.

Proposition 3.1.35 and our examples of homogeneous quasi-norms show that the usual Euclidean topology coincides with the topology associated with any homogeneous quasi-norm:

Proposition 3.1.37. *If $|\cdot|$ is a homogeneous quasi-norm on $G \sim \mathbb{R}^n$, the topology induced by the $|\cdot|$-balls*

$$B(x,R) := \{y \in G : \ |x^{-1}y| < R\},$$

$x \in G$ and $R > 0$, coincides with the Euclidean topology of $\mathbb{R}^n$.

Any closed ball or sphere for any homogeneous quasi-norm is compact. It is also bounded with respect to any norm of the vector space $\mathbb{R}^n$ or any other homogeneous quasi-norm on G.

Proof of Proposition 3.1.37. It is a routine exercise of topology to check that the equivalence of norm given in Proposition 3.1.35 implies that the topology induced by the balls of two different homogeneous quasi-norms coincide. Hence we can choose the norm $|\cdot|_\infty$ given by (3.22) and the corresponding balls

$$B_\infty(x,R) := \{y \in G : \ |x^{-1}y|_\infty < R\}.$$

We also consider the supremum Euclidean norm given by

$$|(x_1,\ldots,x_n)|_{E,\infty} = \max_{1\leq j\leq n} |x_j|,$$

and its corresponding balls

$$B_{E,\infty}(x,R) := \{y \in G : \ |-x+y|_{E,\infty} < R\}.$$

That the topologies induced by the two families of balls

$$\{B_\infty(x,R)\}_{x\in G, R>0} \quad \text{and} \quad \{B_{E,\infty}(x,R)\}_{x\in G, R>0}$$

must coincide follows from the following two observations. Firstly it is easy to check for any $R \in (0,1)$

$$B_\infty(0, R^{\frac{1}{\upsilon_1}}) \subset B_{E,\infty}(0,R) \subset B_\infty(0, R^{\frac{1}{\upsilon_n}}).$$

Secondly for each $x \in G$, the mappings $\Psi_x : y \mapsto x^{-1}y$ and $\Psi_{E,x} : y \mapsto -x+y$ are two smooth diffeomorphisms of $\mathbb{R}^n$. Hence these mappings are continuous with continuous inverses (with respect to the Euclidean topology). Furthermore, by Remark 3.1.34, we have

$$\Psi_x(B_\infty(x,R)) = B_\infty(0,R) \quad \text{and} \quad \Psi_{E,x}(B_{E,\infty}(x,R)) = B_{E,\infty}(0,R).$$

The second part of the statement follows from the first and from the continuity of homogeneous quasi-norms. □

The next proposition justifies the terminology of 'quasi-norm' by stating that every homogeneous quasi-norm satisfies the triangle inequality up to a constant, the other properties of a norm being already satisfied.

Proposition 3.1.38. *If $|\cdot|$ is a homogeneous quasi-norm on G, there is a constant $C > 0$ such that*

$$|xy| \leq C\,(|x| + |y|) \qquad \forall x, y \in G.$$

Proof. Let $|\cdot|$ be a quasi-norm on G. Let $\bar{B} := \{x \,:\, |x| \leq 1\}$ be its associated closed unit ball. By Proposition 3.1.37, $\bar{B}$ is compact. As the product law is continuous (even polynomial), the set $\{xy : \ x, y \in \bar{B}\}$ is also compact. Therefore, there is a constant $C > 0$ such that

$$\forall x, y \in \bar{B} \qquad |xy| \leq C.$$

Let $x, y \in G$. If both of them are 0, there is nothing to prove. If not, let $t > 0$ be given by $t^{-1} = |x| + |y| > 0$. Then $D_t(x)$ and $D_t(y)$ are in $\bar{B}$, so that

$$t|xy| = |D_t(xy)| = |D_t(x)D_t(y)| \leq C,$$

and this concludes the proof. ◻

Note that the constant C in Proposition 3.1.38 satisfies necessarily $C \geq 1$ since $|0| = 0$ implies $|x| \leq C|x|$ for all $x \in G$. It is natural to ask whether a homogeneous Lie group G may admit a homogeneous quasi-norm $|\cdot|$ which is actually a norm or, equivalently, which satisfies the triangle inequality with constant $C = 1$. For instance, on the Heisenberg group $\mathbb{H}_{n_o}$, the homogeneous quasi-norm given in Example 3.1.18 turns out to be a norm (cf. [Cyg81]). In the stratified case, the norm built from the control distance of the sub-Laplacian, often called the Carnot-Caratheodory distance, is also 1-homogeneous (see, e.g., [Pan89] or [BLU07, Section 5.2]). This can be generalised to all homogeneous Lie groups.

Theorem 3.1.39. *Let G be a homogeneous Lie group. Then there exist a homogeneous quasi-norm on G which is a norm, that is, a homogeneous quasi-norm $|\cdot|$ which satisfies the triangle inequality*

$$|xy| \leq |x| + |y| \qquad \forall x, y \in G.$$

A proof of Theorem 3.1.39 by Hebisch and Sikora uses the correspondence between homogeneous norms and convex sets, see [HS90]. Here we sketch a different proof. Its idea may be viewed as an adaptation of a part of the proof that the control distance in the stratified case is a distance. Our proof may be simpler than the stratified case though, since we define a distance without using 'horizontal' curves.

Sketch of the proof of Theorem 3.1.39. If $\gamma : [0, T] \to G$ is a smooth curve, its tangent vector $\gamma'(t_o)$ at $\gamma(t_o)$ is usually defined as the element of the tangent space $T_{\gamma(t_o)}G$ at $\gamma(t_o)$ such that

$$\gamma'(t_o)(f) = \frac{d}{dt} f(\gamma(t))\Big|_{t=t_o}, \quad f \in C^\infty(G).$$

It is more convenient for us to identify the tangent vector of γ at $\gamma(t_o)$ with an element of the Lie algebra $\mathfrak{g} = T_0 G$. We therefore define $\tilde{\gamma}'(t_o) \in \mathfrak{g}$ via

$$\tilde{\gamma}'(t_o)(f) := \frac{d}{dt} f(\gamma(t_o)^{-1}\gamma(t))\Big|_{t=t_o}, \quad f \in C^\infty(G).$$

We now fix a basis $\{X_j\}_{j=1}^n$ of $\mathfrak{g}$ such that $D_r X_j = r^{\upsilon_j} X_j$. We also define the map $|\cdot|_\infty : \mathfrak{g} \to [0,\infty)$ by

$$|X|_\infty := \max_{j=1,\ldots,n} |x_j|^{\frac{1}{\upsilon_j}}, \quad X = \sum_{j=1}^n x_j X_j \in \mathfrak{g}.$$

Given a piecewise smooth curve $\gamma : [0,T] \to G$, we define its length adapted to the group structure by

$$\tilde{\ell}(\gamma) := \int_0^T |\tilde{\gamma}'(t)|_\infty dt.$$

If x and y are in G, we denote by $d(x,y)$ the infimum of the lengths $\tilde{\ell}(\gamma)$ of the piecewise smooth curves γ joining x and y. Since two points x and y can always be joined by a smooth compact curve, e.g. $\gamma(t) = ((1-t)x)\,ty$, the quantity $d(x,y)$ is always finite. Hence we have obtained a map $d : G \times G \to [0,\infty)$. It is a routine exercise to check that d is symmetric and satisfies the triangle inequality in the sense that we have for all $x, y, z \in G$, that

$$d(x,y) = d(y,x) \quad \text{and} \quad d(x,y) \leq d(x,z) + d(z,y).$$

Moreover, one can check easily that $\tilde{\ell}(D_r(\gamma)) = r\tilde{\ell}(\gamma)$ and $\tilde{\ell}(z\gamma) = \tilde{\ell}(\gamma)$, thus we also have for all $x, y, z \in G$ and $r > 0$, that

$$d(zx, zy) = d(x,y) \quad \text{and} \quad d(rx, ry) = rd(x,y). \tag{3.23}$$

Let us show that d is non-degenerate, that is, $d(x,y) = 0 \Rightarrow x = y$. First let $|\cdot|_E$ be the Euclidean norm on $\mathfrak{g} \sim \mathbb{R}^n$ such that the basis $\{X_j\}_{j=1}^n$ is orthonormal. We endow each tangent space $T_x G$ with the Euclidean norm obtained by left translation of the Euclidean norm $|\cdot|_E$. Hence we have for any smooth curve γ at any point t_o

$$|\gamma'(t_o)|_{T_{\gamma(t_o)}G} = |\tilde{\gamma}'(t_o)|_E.$$

Now we see that if $X = \sum_{j=1}^n x_j X_j \in \mathfrak{g}$ is such that

$$|X|_{E,\infty} := \max_{j=1,\ldots,n} |x_j| \leq 1,$$

then

$$|X|_E \asymp |X|_{E,\infty} \leq |X|_\infty.$$

This implies that if $\gamma : [0,T] \to G$ is a smooth curve satisfying

$$\forall t \in [0,T] \qquad |\gamma'(t)|_{T_{\gamma(t)}G} < 1, \tag{3.24}$$

then

$$\ell(\gamma) \leq C\tilde{\ell}(\gamma), \tag{3.25}$$

where ℓ is the usual length

$$\ell(\gamma) := \int_0^T |\gamma'(t)|_{T_{\gamma(t)}G} dt,$$

and $C > 0$ a positive constant independent of γ.

Let d_G be the Riemaniann distance induced by our choice of metric on the manifold G, that is, the infimum of the lengths $\ell(\gamma)$ of the piecewise smooth curves γ joining x and y. Very well known results in Riemaniann geometry imply that d_G induces the same topology as the Euclidean topology. Moreover, there exists a small open set Ω containing 0 such that any point in Ω may be joined to 0 by a smooth curve satisfying (3.24) at any point. Then (3.25) yields that we have $d_G(0,x) \leq Cd(0,x)$ for any $x \in \Omega$. This implies that d is non-degenerate since d is invariant under left-translation and is 1-homogeneous in the sense of (3.23),

Checking that the associated map $x \mapsto |x| = d(0,x)$ is a quasi-norm concludes the sketch of the proof of Theorem 3.1.39. □

Even if homogeneous norms do exist, it is often preferable to use homogeneous quasi-norms. Because the triangle inequality is up to a constant in this case, we do not necessarily have the inequality $||xy| - |x|| \leq C|y|$. However, the following lemma may help:

Proposition 3.1.40. *We fix a homogeneous quasi-norm $|\cdot|$ on G. For any $f \in C^1(G\backslash\{0\})$ homogeneous of degree $\nu \in \mathbb{C}$, for any $b \in (0,1)$ there is a constant $C = C_b > 0$ such that*

$$|f(xy) - f(x)| \leq C|y|\ |x|^{\operatorname{Re}\nu - 1} \quad \textit{whenever} \quad |y| \leq b|x|.$$

Indeed, applying it to a $C^1(G\backslash\{0\})$ homogeneous quasi-norm, we obtain

$$\forall b \in (0,1) \quad \exists C = C_b > 0 \quad \forall x,y \in G \quad |y| \leq b|x| \Longrightarrow \big||xy| - |x|\big| \leq C|y|. \tag{3.26}$$

Proof of Proposition 3.1.40. Let $f \in C^1(G\backslash\{0\})$. Both sides of the desired inequality are homogeneous of degree $\operatorname{Re}\nu$ so it suffices to assume that $|x| = 1$ and $|y| \leq b$. By Proposition 3.1.37 and the continuity of multiplication, the set $\{xy : |x| = 1 \text{ and } |y| \leq b\}$ is a compact which does not contain 0. So by the (Euclidean) mean value theorem on $\mathbb{R}^n$, we get

$$|f(xy) - f(x)| \leq C|y|_E.$$

We conclude using the next lemma. □

The next lemma shows that locally a homogeneous quasi-norm and the Euclidean norm are comparable:

Lemma 3.1.41. *We fix a homogeneous quasi-norm $|\cdot|$ on G. Then there exist $C_1, C_2 > 0$ such that*

$$C_1|x|_E \leq |x| \leq C_2|x|_E^{\frac{1}{\upsilon_n}} \quad \textit{whenever} \quad |x| \leq 1.$$

Proof of Lemma 3.1.41. By Proposition 3.1.37, the unit sphere $\{y : |y| = 1\}$ is compact and does not contain 0. Hence the Euclidean norm assumes a positive maximum C_1^{-1} and a positive minimum $C_2^{-\upsilon_n}$ on it, for some $C_1, C_2 > 0$.

Let $x \in G$. We may assume $x \neq 0$. Then we can write it as $x = ry$ with $|y| = 1$ and $r = |x|$. We observe that since

$$|ry|_E^2 = \sum_{j=1}^n y_j^2 r^{2\upsilon_j},$$

we have if $r \leq 1$

$$r^{\upsilon_n}|y|_E \leq |ry|_E \leq r|y|_E.$$

Hence for $r = |x| \leq 1$, we get

$$|x|_E = |ry|_E \leq r|y|_E \leq |x|C_1^{-1} \quad \text{and} \quad |x|_E = |ry|_E \geq r^{\upsilon_n}|y|_E \geq |x|^{\upsilon_n}C_2^{-\upsilon_n},$$

implying the statement. □

3.1.7 Polar coordinates

There is an analogue of polar coordinates on homogeneous Lie groups.

Proposition 3.1.42. *Let G be a homogeneous Lie group equipped with a homogeneous quasi-norm $|\cdot|$. Then there is a (unique) positive Borel measure σ on the unit sphere*

$$\mathfrak{S} := \{x \in G : |x| = 1\},$$

such that for all $f \in L^1(G)$, we have

$$\int_G f(x)dx = \int_0^\infty \int_{\mathfrak{S}} f(ry)r^{Q-1}d\sigma(y)dr. \tag{3.27}$$

In order to prove this claim, we start with the following averaging property:

Lemma 3.1.43. *Let G be a homogeneous Lie group equipped with a homogeneous quasi-norm $|\cdot|$. If f is a locally integrable function on $G\backslash\{0\}$, homogeneous of degree $-Q$, then there exists a constant $m_f \in \mathbb{C}$* (the average value of f) *such that for all $u \in L^1((0,\infty), r^{-1}dr)$, we have*

$$\int_G f(x)u(|x|)dx = m_f \int_0^\infty u(r)r^{-1}dr. \tag{3.28}$$

The proof of Lemma 3.1.43 yields the formula for m_f in terms of the homogeneous quasi-norm $|\cdot|$,

$$m_f = \int_{1\le|x|\le e} f(x)dx. \tag{3.29}$$

However, in Lemma 3.1.45 we will give an invariant meaning to this value.

Proof of Lemma 3.1.43. Let f be locally integrable function on $G\backslash\{0\}$, homogeneous of degree $-Q$. We set for any $r>0$,

$$\varphi(r) := \begin{cases} \int_{1\le|x|\le r} f(x)dx & \text{if } r\ge 1, \\ -\int_{r\le|x|\le 1} f(x)dx & \text{if } r<1. \end{cases}$$

The mapping $\varphi:(0,\infty)\to\mathbb{C}$ is continuous and one easily checks that

$$\varphi(rs)=\varphi(r)+\varphi(s) \qquad \text{for all } r,s>0,$$

by making the change of variable $x\mapsto sx$ and using the homogeneity of f. It follows that $\varphi(r)=\varphi(e)\ln r$ and we set

$$m_f := \varphi(e).$$

Then the equation (3.28) is easily satisfied when u is the characteristic function of an interval. By taking the linear combinations and limits of such functions, the equation (3.28) is also satisfied when $u\in L^1((0,\infty),r^{-1}dr)$. □

Proof of Proposition 3.1.42. For any continuous function f on the unit sphere $\mathfrak{S}$, we define the homogeneous function $\tilde f$ on $G\backslash\{0\}$ by

$$\tilde f(x) := |x|^{-Q} f(|x|^{-1}x).$$

Then $\tilde f$ satisfies the hypotheses of Lemma 3.1.43. The map $f\mapsto m_{\tilde f}$ is clearly a positive functional on the space of continuous functions on $\mathfrak{S}$. Hence it is given by integration against a regular positive measure σ (see, e.g. [Rud87, ch.VI]).

For $u\in L^1((0,\infty),r^{-1}dr)$, we have

$$\begin{aligned} \int f(|x|^{-1}x)u(|x|)dx &= \int \tilde f(x)|x|^Q u(|x|)dx = m_{\tilde f}\int_{r=0}^{\infty} r^{Q-1}u(r)dr \\ &= \int_0^\infty\int_{\mathfrak{S}} f(y)u(r)r^{Q-1}d\sigma(y)dr. \end{aligned}$$

Since linear combinations of functions of the form $f(|x|^{-1}x)u(|x|)$ are dense in $L^1(G)$, the proposition follows. □

We view the formula (3.27) as a change in polar coordinates.

Example 3.1.44. For $0 < a < b < \infty$ and $\alpha \in \mathbb{C}$, we have

$$\int_{a<|x|<b} |x|^{\alpha-Q} dx = C \begin{cases} \alpha^{-1}(b^\alpha - a^\alpha) & \text{if } \alpha \neq 0 \\ \ln\left(\frac{b}{a}\right) & \text{if } \alpha = 0 \end{cases} \qquad \text{with } C = \sigma(\mathfrak{S}).$$

And if $\alpha \in \mathbb{R}$ and f is a measurable function on G such that $f(x) = O(|x|^{\alpha-Q})$ then f is integrable either near ∞ if $\alpha < 0$, or near 0 if $\alpha > 0$.

The measure σ in the polar coordinates decomposition actually has a smooth density. We will not need this fact and will not prove it here, but refer to [FR66] and [Goo80].

Now, the polar change of coordinates depends on the choice of a homogeneous quasi-norm to fix the unit sphere. But it turns out that the average value of the $(-Q)$-homogeneous function considered in Lemma 3.1.43 does not. Let us prove this fact for the sake of completeness.

Lemma 3.1.45. *Let G be a homogeneous Lie group and let f be a locally integrable function on $G\backslash\{0\}$, homogeneous of degree $-Q$.*

Given a homogeneous quasi-norm, let σ be the Radon measure on the unit sphere $\mathfrak{S}$ giving the polar change of coordinate (3.27). Then the average value of f defined in (3.28) is given by

$$m_f = \int_{\mathfrak{S}} f d\sigma. \tag{3.30}$$

This average value m_f is independent of the choice of the homogeneous quasi-norm.

Proof of Lemma 3.1.45. For any homogeneous quasi-norm, using the polar change of coordinates (3.27), we obtain

$$\begin{aligned} \int_{a<|x|<b} f(x)dx &= \int_a^b \int_{\mathfrak{S}} f(rx)d\sigma(x) r^{Q-1} dr \\ &= \int_a^b \int_{\mathfrak{S}} f(x)d\sigma(x) r^{-1} dr = \int_a^b r^{-1} dr \int_{\mathfrak{S}} f(x)d\sigma(x) = \left(\ln \frac{b}{a}\right) m_f. \end{aligned}$$

This shows (3.30), taking $a = 1$ and $b = e$, see (3.29) and the proof of Lemma 3.1.43.

Let $|\cdot|$ and $|\cdot|'$ be two homogeneous quasi-norms on G. We denote by

$$\bar{B}_r := \{x \in G : |x| \leq r\} \quad \text{and} \quad \bar{B}'_r := \{x \in G : |x|' \leq r\},$$

the closed balls around 0 of radius r for $|\cdot|$ and $|\cdot|'$, respectively. By Proposition 3.1.35, Part (ii), there exists a constant $a > 0$ such that $\bar{B}'_a \subset \bar{B}_1$. We also have $\bar{B}'_a \subset \bar{B}'_{2a} \subset \bar{B}_2$ and, with the usual sign convention for integration, we have

$$\int_{\bar{B}_2\backslash\bar{B}_1} = \int_{\bar{B}_2\backslash\bar{B}'_a} - \int_{\bar{B}_1\backslash\bar{B}'_a} = \int_{\bar{B}_2\backslash\bar{B}'_{2a}} + \int_{\bar{B}'_{2a}\backslash\bar{B}'_a} - \int_{\bar{B}_1\backslash\bar{B}'_a}.$$

Using the homogeneities of f and of the Haar measure, we see, after the changes of variables $x = 2y$ and $x = az$, that

$$\int_{\bar{B}_2\setminus\bar{B}'_{2a}} f(x)dx = \int_{\bar{B}_1\setminus\bar{B}'_a} f(y)dy \quad \text{and} \quad \int_{\bar{B}'_{2a}\setminus\bar{B}'_a} f(x)dx = \int_{\bar{B}'_2\setminus\bar{B}'_1} f(z)dz.$$

Hence

$$\int_{\bar{B}_2\setminus\bar{B}_1} f = \int_{\bar{B}'_2\setminus\bar{B}'_1} f.$$

Using the first computations of this proof, the left and right hand sides are equal to $(\ln b/a)m_f$ and $(\ln b/a)m'_f$, respectively, where m_f and m'_f are the average values for $|\cdot|$ and $|\cdot|'$. Thus $m_f = m'_f$. □

3.1.8 Mean value theorem and Taylor expansion

Here we prove the mean value theorem and describe the Taylor series on homogeneous Lie groups. Naturally, the space $C^1(G)$ here is the space of functions f such that $X_j f$ are continuous on G for all j, etc. The following mean value theorem can be partly viewed as a refinement of Proposition 3.1.40.

Proposition 3.1.46. *We fix a homogeneous quasi-norm $|\cdot|$ on G. There exist group constants $C_0 > 0$ and $\eta > 1$ such that for all $f \in C^1(G)$ and all $x, y \in G$, we have*

$$|f(xy) - f(x)| \leq C_0 \sum_{j=1}^{n} |y|^{\upsilon_j} \sup_{|z|\leq \eta |y|} |(X_j f)(xz)|.$$

In order to prove this proposition, we first prove the following property.

Lemma 3.1.47. *The map $\phi : \mathbb{R}^n \to G$ defined by*

$$\phi(t_1, \ldots, t_n) = \exp_G(t_1 X_1) \exp_G(t_2 X_2) \ldots \exp_G(t_n X_n),$$

is a global diffeomorphism.

Moreover, fixing a homogeneous quasi-norm $|\cdot|$ on G, there is a constant $C_1 > 0$ such that

$$\forall (t_1, \ldots, t_n) \in \mathbb{R}^n,\ j = 1, \ldots, n, \qquad |t_j|^{\frac{1}{\upsilon_j}} \leq C_1 |\phi(t_1, \ldots, t_n)|.$$

The first part of the lemma is true for any nilpotent Lie group (see Remark 1.6.7 Part (ii)). But we will not use this fact here.

Proof. Clearly the map ϕ is smooth. By the Baker-Campbell-Hausdorff formula (see Theorem 1.3.2), the differential $d\phi(0) : \mathbb{R}^n \to T_0 G$ is the isomorphism

$$d\phi(0)(t_1, \ldots, t_n) = \sum_{j=1}^{n} t_j X_j|_0,$$

so that ϕ is a local diffeomorphism near 0 (this is true for any Lie group). More precisely, there exist $\delta, C' > 0$ such that ϕ is a diffeomorphism from U to the ball $B_\delta := \{x \in G : |x| < \delta\}$ with

$$\phi^{-1}(B_\delta) = U \subset \{(t_1, \dots, t_n) : \max_{j=1,\dots,n} |t_j|^{\frac{1}{v_j}} < C'\}.$$

We now use the dilations and for any $r > 0$, we see that

$$\begin{aligned} \phi(r^{v_1}t_1, \dots, r^{v_n}t_n) &= \exp_G(r^{v_1}t_1X_1)\dots\exp_G(r^{v_n}t_nX_n) \\ &= (r\exp_G(t_1X_1))\dots(r\exp_G(t_nX_n)) \\ &= r\left(\exp_G(t_1X_1)\dots\exp_G(t_nX_n)\right), \end{aligned}$$

hence

$$\phi(r^{v_1}t_1, \dots, r^{v_n}t_n) = r\phi(t_1, \dots, t_n). \tag{3.31}$$

If $\phi(t_1, \dots, t_n) = \phi(s_1, \dots, s_n)$, formula (3.31) implies that for all $r > 0$, we have

$$\phi(r^{v_1}t_1, \dots, r^{v_n}t_n) = \phi(r^{v_1}s_1, \dots, r^{v_n}s_n).$$

For r sufficiently small, this forces $t_j = s_j$ for all j since ϕ is a diffeomorphism on U. So the map $\phi : \mathbb{R}^n \to G$ is injective.

Moreover, any $x \in G\backslash\{0\}$ can be written as

$$x = ry \quad \text{with} \quad r := \frac{2}{\delta}|x| \quad \text{and} \quad y := r^{-1}x \in \overline{B_{\frac{\delta}{2}}} \subset \phi(U).$$

We may write $y = \phi(s_1, \dots s_n)$ with $|s_j|^{\frac{1}{v_j}} \leq C'$ and formula (3.31) then implies that $x = \phi(t_1, \dots, t_n)$ is in $\phi(\mathbb{R}^n)$ with $t_j := r^{v_j}s_j$ satisfying $|t_j|^{\frac{1}{v_j}} \leq C'r$. Setting $C_1 = 2C'/\delta$, the assertion follows. □

Proof of Proposition 3.1.46. First let us assume that $y = \exp_G(tX_j)$. Then

$$\begin{aligned} f(xy) - f(x) &= \int_0^t \partial_{s'=s}\{f(x\exp_G(s'X_j))\}\,ds \\ &= \int_0^t \partial_{s'=0}\{f(x\exp_G(sX_j)\exp_G(s'X_j))\}\,ds \\ &= \int_0^t X_jf(x\exp_G(sX_j))ds, \end{aligned}$$

and hence

$$\begin{aligned} |f(xy) - f(x)| &\leq |t| \sup_{0\leq s\leq t} |X_jf(x\exp_G(sX_j))| \\ &\leq |t| \sup_{|z|\leq|y|} |X_jf(xz)|. \end{aligned}$$

Since $|\exp_G(sX_j)| = |s|^{\frac{1}{\upsilon_j}} |\exp_G X_j|$ and hence $|y| = |t|^{\frac{1}{\upsilon_j}} |\exp_G X_j|$, setting

$$C_2 := \max_{k=1,\ldots,n} |\exp_G X_k|^{-\upsilon_k},$$

we obtain

$$|f(xy) - f(x)| \le C_2 |y|^{\upsilon_j} \sup_{|z|\le|y|} |X_j f(xz)|. \tag{3.32}$$

We now prove the general case, so let y be any point of G. By Lemma 3.1.47, it can be written uniquely as $y = y_1 y_2 \ldots y_n$ with $y_j = \exp_G(t_j X_j)$, and hence

$$|y_j| = |t|^{\frac{1}{\upsilon_j}} |\exp_G X_j| \le C_1 C_3 |y| \quad \text{where} \quad C_3 := \max_{k=1,\ldots,n} |\exp_G X_k|, \tag{3.33}$$

and C_1 is as in Lemma 3.1.47. We write

$$\begin{aligned} |f(xy) - f(x)| &\le |f(xy_1 \ldots y_n) - f(xy_1 \ldots y_{n-1})| \\ &\quad + |f(xy_1 \ldots y_{n-1}) - f(xy_1 \ldots y_{n-2})| + \ldots + |f(xy_1) - f(x)|, \end{aligned}$$

and applying (3.32) to each term, we obtain

$$|f(xy) - f(x)| \le \sum_{j=1}^{n} C_2 |y_j|^{\upsilon_j} \sup_{|z|\le|y_j|} |X_j f(xy_1 \ldots y_{j-1} z)|.$$

Let $C_4 \ge 1$ be the constant of the triangle inequality for $|\cdot|$ (see Proposition 3.1.38). If $|z| \le |y_j|$, then $z' = y_1 \ldots y_{j-1} z$ satisfies

$$\begin{aligned} |z'| &\le C_4(|y_1 \ldots y_{j-1}| + |y_j|) \le C_4\big(C_4(|y_1 \ldots y_{j-2}| + |y_{j-1}\,) + |y_j|\big) \\ &\le C_4^2\,(|y_1 \ldots y_{j-2}| + |y_{j-1}| + |y_j|) \le \ldots \le C_4^{j-1}(|y_1| + |y_2| + \ldots |y_j|) \\ &\le C_4^{j-1} j C_1 C_3 |y|, \end{aligned}$$

using (3.33). Therefore, setting $\eta := C_4^n n C_1 C_3$, using again (3.33), we have obtained

$$|f(xy) - f(x)| \le C_2 \sum_{j=1}^{n} (C_1 C_3 |y|)^{\upsilon_j} \sup_{|z'|\le\eta|y|} |X_j f(xz')|,$$

completing the proof. □

Remark 3.1.48. Let us make the following remarks.

1. In the same way, we can prove the following version of Proposition 3.1.46 for right-invariant vector fields: a homogeneous quasi-norm $|\cdot|$ being fixed on G, there exists group constants $C > 0$ and $b > 0$ such that for all $f \in C^1(G)$ and all $x, y \in G$, we have

$$|f(yx) - f(x)| \le C \sum_{j=1}^{n} |y|^{\upsilon_j} \sup_{|z|\le b|y|} |(\tilde{X}_j f)(zx)|.$$

2. If the homogeneous Lie group G is stratified, a more precise version of the mean value theorem exists involving only the vector fields of the first stratum, see Folland and Stein [FS82, (1.41)], but we will not use this fact here.
3. The statement and the proof of the mean value theorem can easily be adapted to hold for functions which are valued in a Banach space, the modulus being replaced by the Banach norm.

Taylor expansion

In view of Corollary 3.1.31, we can define Taylor polynomials:

Definition 3.1.49. The *Taylor polynomial* of a suitable function f at a point $x \in G$ of homogeneous degree $\leq M \in \mathbb{N}_0$ is the unique $P \in \mathcal{P}_{\leq M}$ such that

$$\forall \alpha \in \mathbb{N}_0^n,\ [\alpha] \leq M \qquad X^\alpha P(0) = X^\alpha f(x).$$

More precisely, we have defined the *left* Taylor polynomial, and a similar definition using the right-invariant differential operators $\tilde{X}^\alpha$ yields the right Taylor polynomial. However, in this monograph we will use only left Taylor polynomials.

We may use the following notation for the Taylor polynomial P of a function f at x and for its *remainder of order* M:

$$P_{x,M}^{(f)} := P \quad \text{and} \quad R_{x,M}^{(f)}(y) := f(xy) - P(y). \tag{3.34}$$

For instance, $P_{x,M}^{(f)}(0) = f(x)$. We will also extend the notation for negative M with

$$P_{x,M}^{(f)} := 0 \quad \text{and} \quad R_{x,M}^{(f)}(y) := f(xy) \quad \text{when } M < 0.$$

With this notation, we easily see (whenever it makes sense), the following properties.

Lemma 3.1.50. *For any $M \in \mathbb{N}_0$, $\alpha \in \mathbb{N}_0^n$ and suitable function f, we have*

$$X^\alpha P_{x,M}^{(f)} = P_{x,M-[\alpha]}^{(X^\alpha f)} \quad \text{and} \quad X^\alpha R_{x,M}^{(f)} = R_{x,M-[\alpha]}^{(X^\alpha f)}.$$

Proof. It is easy to check that the polynomial $P_o := X^\alpha P_{x,M}^{(f)}$ is homogeneous of degree $M - [\alpha]$. Furthermore, using (3.19), it satisfies for every $\beta \in \mathbb{N}_0^n$, such that $[\alpha] + [\beta] \leq M$, the equality

$$\begin{aligned} X^\beta P_o(0) &= X^\beta X^\alpha P_{x,M}^{(f)}(0) \\ &= \sum_{\substack{|\gamma| \leq |\alpha|+|\beta| \\ [\gamma]=[\alpha]+[\beta]}} c'_{\alpha,\beta,\gamma} X^\gamma P_{x,M}^{(f)}(0) = \sum_{\substack{|\gamma| \leq |\alpha|+|\beta| \\ [\gamma]=[\alpha]+[\beta]}} c'_{\alpha,\beta,\gamma} X^\gamma f(x) \\ &= X^\beta X^\alpha f(x). \end{aligned}$$

This shows the claim. □

In Definition 3.1.49 the suitable functions f are distributions on a neighbourhood of x in G whose (distributional) derivatives $X^\alpha f$ are continuous in a neighbourhood of x for $[\alpha] \leq M$. We will see in the sequel that in order to control (uniformly) a remainder of a function f of order M we would like f to be at least $(k+1)$ times continuously differentiable, i.e. $f \in C^{k+1}(G)$, where $k \in \mathbb{N}_0$ is equal to

$$\lceil M \rfloor := \max\{|\alpha| \, : \, \alpha \in \mathbb{N}_0^n \text{ with } [\alpha] \leq M\}; \tag{3.35}$$

this is indeed a maximum over a finite set because of (3.16).

We can now state and prove Taylor's inequality.

Theorem 3.1.51. *We fix a homogeneous quasi-norm $|\cdot|$ on G and obtain a corresponding constant η from the mean value theorem (see Proposition 3.1.46). For any $M \in \mathbb{N}_0$, there is a constant $C_M > 0$ such that for all functions $f \in C^{\lceil M \rfloor + 1}(G)$ and all $x, y \in G$, we have*

$$|R^{(f)}_{x,M}(y)| \leq C_M \sum_{\substack{|\alpha| \leq \lceil M \rfloor + 1 \\ [\alpha] > M}} |y|^{[\alpha]} \sup_{|z| \leq \eta^{\lceil M \rfloor + 1} |y|} |(X^\alpha f)(xz)|,$$

where $R^{(f)}_{x,M}$ and $\lceil M \rfloor$ are defined by (3.34) and (3.35).

Theorem 3.1.51 for $M = 0$ boils down exactly to the mean value theorem as stated in Proposition 3.1.46. Similar comments as in Remark 3.1.48 for the mean value theorem are also valid for Taylor's inequality.

Proof. Under the hypothesis of the theorem, a remainder $R^{(f)}_{x,M}$ is always C^1 and vanishes at 0. Let us apply the mean value theorem (see Proposition 3.1.46) at the point 0 to the remainders $R^{(f)}_{x,M}$, $R^{(X_{j_0} f)}_{x,M-\upsilon_{j_0}}$, $R^{(X_{\upsilon_{j_1}} X_{\upsilon_{j_0}} f)}_{x,M-(\upsilon_{j_0}+\upsilon_{j_1})}$, and so on as long as $M - (\upsilon_{j_0} + \ldots + \upsilon_{j_k}) \geq 0$; using this together with Lemma 3.1.50, we obtain

$$\begin{aligned}
\left| R^{(f)}_{x,M}(y_0) \right| &\leq C_0 \sum_{j_0=1}^{n} |y_0|^{\upsilon_{j_0}} \sup_{|y_1| \leq \eta |y_0|} \left| R^{(X_{\upsilon_{j_0}} f)}_{x,M-\upsilon_{j_0}}(y_1) \right|, \\
\left| R^{(X_{\upsilon_{j_0}} f)}_{x,M-\upsilon_{j_0}}(y_1) \right| &\leq C_0 \sum_{j_1=1}^{n} |y_1|^{\upsilon_{j_1}} \sup_{|y_2| \leq \eta |y_1|} \left| R^{(X_{\upsilon_{j_1}} X_{\upsilon_{j_0}} f)}_{x,M-(\upsilon_{j_0}+\upsilon_{j_1})}(y_2) \right|, \\
&\vdots \\
\left| R^{(X_{\upsilon_{j_k}} \ldots X_{\upsilon_{j_0}} f)}_{x,M-(\upsilon_{j_0}+\ldots+\upsilon_{j_k})}(y_k) \right| &\leq C_0 \sum_{j_k=1}^{n} |y_k|^{\upsilon_{j_k}} \sup_{|y_{k+1}| \leq \eta |y_k|} \left| R^{(X_{\upsilon_{j_{k+1}}} \ldots X_{\upsilon_{j_0}} f)}_{x,M-(\upsilon_{j_0}+\ldots+\upsilon_{j_{k+1}})}(y_k) \right|.
\end{aligned}$$

We combine these inequalities together, to obtain

$$\left| R^{(f)}_{x,M}(y_0) \right| \leq C_0^{k+1} \eta^k \sum_{\substack{j_i = 1, \ldots, n \\ i = 0, \ldots, k+1}} |y_0|^{\upsilon_{j_0} + \ldots + \upsilon_{j_k}} \sup_{|y_{k+1}| \leq \eta^{k+1} |y_0|} \left| R^{(X^{\upsilon_{j_{k+1}}} \ldots X^{\upsilon_{j_0}} f)}_{x,M-(\upsilon_{j_0}+\ldots+\upsilon_{j_{k+1}})}(y_k) \right|.$$

The process stops exactly for $k = \lceil M \rfloor$ by the very definition of $\lceil M \rfloor$. For this value of k, Corollary 3.1.32 and the change of discrete variable $\alpha := v_{j_0} e_{j_0} + \ldots v_{j_{k+1}} e_{j_{k+1}}$ (where e_j denotes the multi-index with 1 in the j-th place and zeros elsewhere) yield the result. □

Remark 3.1.52. 1. We can consider Taylor polynomials for right-invariant vector fields. The corresponding Taylor estimates would then approximate $f(yx)$ with a polynomial in y. See Part 1 of Remark 3.1.48, about the mean value theorem for the case of order 0. Note that in Theorem 3.1.51 we consider $f(xy)$ and its approximation by a polynomial in y.

2. If the homogeneous Lie group G is stratified, a more precise versions of Taylor's inequality exists involving only the vector fields of the first stratum, see Folland and Stein [FS82, (1.41)], but we will not use this fact here.

3. The statement and the proof of Theorem 3.1.51 can easily be adapted to hold for functions which are valued in a Banach space, the modulus being replaced by the Banach norm.

4. One can derive explicit formulae for Taylor's polynomials and the remainders on homogeneous Lie groups, see [Bon09] (see also [ACC05] for the case of Carnot groups), but we do not require these here.

As a corollary of Theorem 3.1.51 that will be useful to us later, the right-derivatives of Taylor polynomials and of the remainder will have the following properties, slightly different from those for the left derivatives in Lemma 3.1.50.

Corollary 3.1.53. *Let $f \in C^\infty(G)$. For any $M \in \mathbb{N}_0$ and $\alpha \in \mathbb{N}_0^n$, we have*

$$\tilde{X}^\alpha P^{(f)}_{x,M} = P^{(X_x^\alpha f(x\,\cdot))}_{0,M-[\alpha]} \quad \text{and} \quad \tilde{X}^\alpha R^{(f)}_{x,M} = R^{(X_x^\alpha f(x\,\cdot))}_{0,M-[\alpha]}.$$

Proof. Recall from (1.12) that for any $X \in \mathfrak{g}$ identified with a left-invariant vector field, we have

$$\tilde{X}_y\{f(xy)\} = \frac{d}{dt} f(xe^{tX}y)_{t=0} = X_x\{f(xy)\},$$

and recursively, we obtain

$$\tilde{X}_y^\alpha\{f(xy)\} = X_x^\alpha\{f(xy)\}. \tag{3.36}$$

Therefore, we have

$$\begin{aligned}
&\tilde{X}^\alpha P^{(f)}_{x,M}(y) - P^{(X_x^\alpha f(x\,\cdot))}_{0,M-[\alpha]}(y) \\
&= \tilde{X}_y^\alpha \left\{ f(xy) - R^{(f)}_{x,M}(y) \right\} - \left\{ X_x^\alpha f(xy) - R^{(X_x^\alpha f(x\,\cdot))}_{0,M-[\alpha]}(y) \right\} \\
&= -\tilde{X}^\alpha R^{(f)}_{x,M}(y) + R^{(X_x^\alpha f(x\,\cdot))}_{0,M-[\alpha]}(y).
\end{aligned} \tag{3.37}$$

By Corollary 3.1.30, we can write

$$\begin{aligned}\tilde{X}^\alpha R^{(f)}_{x,M}(y) &= \sum_{|\beta|\leq|\alpha|,\,[\beta]\geq[\alpha]} Q_{\alpha,\beta}(y) X^\beta R^{(f)}_{x,M}(y)\\ &= \sum_{|\beta|\leq|\alpha|,\,[\beta]\geq[\alpha]} Q_{\alpha,\beta}(y) R^{(X^\beta f)}_{x,M-[\beta]}(y),\end{aligned}$$

where each $Q_{\alpha,\beta}$ is a homogeneous polynomial of degree $[\beta]-[\alpha]$.

Fixing a homogeneous quasi-norm $|\cdot|$ on G, the Taylor inequality (Theorem 3.1.51) applied to $R^{(X^\alpha_x f(x\,\cdot))}_{0,M-[\alpha]}$ and $R^{(X^\beta f)}_{x,M-[\beta]}$ implies that, for $|y|\leq 1$,

$$|R^{(X^\alpha_x f(x\,\cdot))}_{0,M-[\alpha]}(y)| \leq C|y|^{M-[\alpha]+1} \quad \text{and} \quad |R^{(X^\beta f)}_{x,M-[\beta]}(y)| \leq C|y|^{M-[\beta]+1}.$$

Hence

$$|\tilde{X}^\alpha R^{(f)}_{x,M}(y)| \leq C|y|^{M-[\alpha]+1}.$$

Going back to (3.37), we have obtained that its left hand side can be estimated as

$$|\tilde{X}^\alpha P^{(f)}_{x,M}(y) - P^{(X^\alpha_x f(x\,\cdot))}_{0,M-[\alpha]}(y)| \leq C|y|^{M-[\alpha]+1}.$$

But $\tilde{X}^\alpha P^{(f)}_{x,M}(y) - P^{(X^\alpha_x f(x\,\cdot))}_{0,M-[\alpha]}(y)$ is a polynomial of homogeneous degree at most $M-[\alpha]$. Therefore, this polynomial is identically 0. This concludes the proof of Corollary 3.1.53. □

3.1.9 Schwartz space and tempered distributions

The Schwartz space on a homogeneous Lie group G is defined as the Schwartz space on any connected simply connected nilpotent Lie group, namely, by identifying G with the underlying vector space of its Lie algebra (see Definition 1.6.8). The vector space $\mathcal{S}(G)$ is naturally endowed with a Fréchet topology defined by any of a number of families of seminorms.

In the 'traditional' Schwartz seminorm on $\mathbb{R}^n$ (see (1.13)) we can replace (without changing anything for the Fréchet topology):

- $\left(\frac{\partial}{\partial x}\right)^\alpha$ and the isotropic degree $|\alpha|$ by X^α and the homogeneous degree $[\alpha]$, respectively, in view of Section 3.1.5,
- the Euclidean norm by the norm $|\cdot|_p$ given in (3.21), and then by any homogeneous norm since homogeneous quasi-norms are equivalent (cf. Proposition 3.1.35).

Hence we choose the following family of seminorms for $\mathcal{S}(G)$, where G is a homogeneous Lie group:

$$\|f\|_{\mathcal{S}(G),N} := \sup_{[\alpha]\leq N,\, x\in G} (1+|x|)^N |X^\alpha f(x)| \qquad (N\in\mathbb{N}_0),$$

after having fixed a homogeneous quasi-norm $|\cdot|$ on G.

Another equivalent family is given by a similar definition with the right-invariant vector fields $\tilde{X}^\alpha$ replacing X^α.

The following lemma proves, in particular, that translations, taking the inverse, and convolutions, are continuous operations on Schwartz functions.

Lemma 3.1.54. *Let $f \in \mathcal{S}(G)$ and $N \in \mathbb{N}$. Then we have*

$$\begin{aligned} \|f(y\,\cdot\,)\|_{\mathcal{S}(G),N} &\leq C_N(1+|y|)^N \|f\|_{\mathcal{S}(G),N} \qquad (y \in G), & (3.38)\\ \left\|\tilde{f}\right\|_{\mathcal{S}(G),N} &\leq C_N \|f\|_{\mathcal{S}(G),(\upsilon_n+1)N} \quad \text{where} \quad \tilde{f}(x) = f(x^{-1}), & (3.39)\\ \|f(\,\cdot\,y)\|_{\mathcal{S}(G),N} &\leq C_N(1+|y|)^{(\upsilon_n+1)N} \|f\|_{\mathcal{S}(G),(\upsilon_n+1)^2N} \quad (y \in G). & (3.40) \end{aligned}$$

Moreover,

$$\|f(y\,\cdot\,) - f\|_{\mathcal{S}(G),N} \longrightarrow_{y\to 0} 0 \quad \text{and} \quad \|f(\,\cdot\,y) - f\|_{\mathcal{S}(G),N} \longrightarrow_{y\to 0} 0. \tag{3.41}$$

The group convolution of two Schwartz functions $f_1, f_2 \in \mathcal{S}(G)$ satisfies

$$\|f_1 * f_2\|_{\mathcal{S}(G),N} \leq C_N \|f_1\|_{\mathcal{S}(G),N+Q+1} \|f_2\|_{\mathcal{S}(G),N}. \tag{3.42}$$

Proof. Let $C_o \geq 1$ be the constant of the triangle inequality, cf. Proposition 3.1.38. We have easily that

$$\forall x, y \in G \qquad (1+|x|) \leq C_o(1+|y|)(1+|yx|). \tag{3.43}$$

Thus,

$$\begin{aligned} \|f(y\,\cdot\,)\|_{\mathcal{S}(G),N} &\leq \sup_{[\alpha]\leq N,\, x\in G} (C_o(1+|y|)(1+|yx|))^N \,|X^\alpha f(yx)| \\ &\leq C_o^N (1+|y|)^N \|f\|_{\mathcal{S}(G),N}. \end{aligned}$$

This shows (3.38).

For (3.39), using (1.11) and Corollary 3.1.30, we have

$$\begin{aligned} \left\|\tilde{f}\right\|_{\mathcal{S}(G),N} &\leq \sup_{[\alpha]\leq N,\, x\in G} (1+|x|)^N |(\tilde{X}^\alpha f)(x^{-1})| \\ &\leq \sup_{[\alpha]\leq N,\, x\in G} \sum_{\substack{\beta\in\mathbb{N}_0^n,\, |\beta|\leq|\alpha| \\ [\beta]\geq[\alpha]}} (1+|x|)^N | \left(Q_{\alpha,\beta} X^\beta f\right)(x^{-1})| \\ &\leq C_N \sup_{[\beta]\leq \upsilon_n N,\, x\in G} (1+|x'|)^{N+[\beta]} |X^\beta f(x')| \end{aligned}$$

by homogeneity of the polynomials $Q_{\alpha,\beta}$ and (3.16).

Since $f(\,\cdot\,y) = (\tilde{f}(y^{-1}\,\cdot\,))\tilde{}$, we deduce (3.40) from (3.38) and (3.39).

By the mean value theorem (cf. Proposition 3.1.46),

$$\begin{aligned}\|f(y\,\cdot) - f\|_{\mathcal{S}(G),N} &= \sup_{[\alpha]\leq N,\, x\in G} (1+|x|)^N |X^\alpha f(yx) - X^\alpha f(x)| \\ &\leq C\sum_{j=1}^n |y|^{\upsilon_j} \sup_{\substack{[\alpha]\leq N \\ x\in G,\, |z|\leq \eta|y|}} (1+|x|)^N |(X_j X^\alpha f)(xz)| \\ &\leq C\sum_{j=1}^n |y|^{\upsilon_j} \|f\|_{\mathcal{S}(G),N+\upsilon_n}, \end{aligned} \tag{3.44}$$

and this proves (3.41) for the left invariance. The proof is similar for the right invariance and is left to the reader.

Since using (3.43) we have

$$\begin{aligned}(1+|x|)^N\, |X^\alpha(f_1 * f_2)(x)| &\leq \int_G (1+|x|)^N |f_1(y)|\, |X^\alpha f_2(y^{-1}x)| dy \\ \leq C_o^N \int_G (1+|y|)^N |f_1(y)| (1+|y^{-1}x|)^N |X^\alpha f_2(y^{-1}x)| dy \\ \leq C_o^N \sup_{z\in G} (1+|z|)^N |X^\alpha f_2(z)| \int_G (1+|y|)^N |f_1(y)| dy,\end{aligned}$$

we obtain (3.42) by the convergence in Example 3.1.44. □

The space of tempered distributions $\mathcal{S}'(G)$ is the (continuous) dual of $\mathcal{S}(G)$. Hence a linear form f on $\mathcal{S}(G)$ is in $\mathcal{S}'(G)$ if and only if

$$\exists N\in\mathbb{N}_0,\, C>0 \qquad \forall \phi\in\mathcal{S}(G) \quad |\langle f,\phi\rangle| \leq C\|\phi\|_{\mathcal{S}(G),N}. \tag{3.45}$$

The topology of $\mathcal{S}'(G)$ is given by the family of seminorms given by

$$\|f\|_{\mathcal{S}'(G),N} := \sup\{|\langle f,\phi\rangle|, \|\phi\|_{\mathcal{S}(G),N}\leq 1\}, \quad f\in\mathcal{S}'(G),\ N\in\mathbb{N}_0.$$

Now, with these definitions, we can repeat the construction in Section 1.5 and define convolution of a distribution in $\mathcal{S}'(G)$ with the test function in $\mathcal{S}(G)$. Then we have

Lemma 3.1.55. *For any $f\in\mathcal{S}'(G)$ there exist $N\in\mathbb{N}$ and $C>0$ such that*

$$\forall \phi\in\mathcal{S}(G) \qquad \forall x\in G \qquad |(\phi * f)(x)| \leq C(1+|x|)^N \|\phi\|_{\mathcal{S}(G),N}. \tag{3.46}$$

The constant C may be chosen of the form $C = C'\|f\|_{\mathcal{S}'(G),N'}$ for some C' and N' independent of f.

For any $f\in\mathcal{S}'(G)$ and $\phi\in\mathcal{S}(G)$, $\phi * f\in C^\infty(G)$. Moreover, if $f_\ell \longrightarrow_{\ell\to\infty} f$ in $\mathcal{S}'(G)$ then for any $\phi\in\mathcal{S}(G)$,

$$\phi * f_\ell \longrightarrow_{\ell\to\infty} \phi * f$$

in $C^\infty(G)$.

Furthermore, if $f \in \mathcal{S}'(G)$ *is compactly supported then* $\phi * f \in \mathcal{S}(G)$ *for any* $\phi \in \mathcal{S}(G)$.

Proof. Let $f \in \mathcal{S}'(G)$ and $\phi \in \mathcal{S}(G)$. By definition of the convolution in Definition 1.5.3 and continuity of f (see (3.45)) we have

$$\begin{aligned} |(\phi * f)(x)| &= |\langle f, \tilde{\phi}(\cdot\, x^{-1})\rangle| \leq C\|\tilde{\phi}(\cdot\, x^{-1})\|_{\mathcal{S}(G),N} \\ &\leq C(1+|x^{-1}|)^{(v_n+1)N}\|\tilde{\phi}\|_{\mathcal{S}(G),(v_n+1)^2N} \qquad \text{(by (3.40))} \\ &\leq C(1+|x|)^{(v_n+1)N}\|\phi\|_{\mathcal{S}(G),(v_n+1)^3N} \qquad \text{(by (3.39))}. \end{aligned}$$

This shows (3.46). Consequently

$$\tilde{X}^\alpha(\phi * f) = (\tilde{X}^\alpha\phi) * f$$

is also bounded for every $\alpha \in \mathbb{N}_0^n$ and hence $\phi * f$ is smooth. The convergence statement then follows from the definition of the convolution for distributions.

Let us now assume that the distribution f is compactly supported. Its support is included in the ball of radius R for R large enough. There exists $N \in \mathbb{N}_0$ such that

$$\begin{aligned} |(\phi * f)(x)| &= |\langle f, \tilde{\phi}(\cdot\, x^{-1})\rangle| \leq C \sup_{|y|\leq R,\, |\alpha|\leq N} \left| \left(\frac{\partial}{\partial y}\right)^\alpha (\phi(xy^{-1}))\right| \\ &\leq C_R \sup_{|y|\leq R,\, [\alpha]\leq v_n N} \left|\tilde{X}_y^\alpha\{\phi(xy^{-1})\}\right|, \end{aligned}$$

using (3.16) and (3.17). By (1.11), we have

$$\tilde{X}_y^\alpha\{\phi(xy^{-1})\} = (-1)^{|\alpha|}(X^\alpha\phi)(xy^{-1}),$$

and so for every $M \in \mathbb{N}_0$ with $M \geq [\alpha]$, we obtain

$$\left|\tilde{X}_y^\alpha\{\phi(xy^{-1})\}\right| = \left|X^\alpha\phi(xy^{-1})\right| \leq \|\phi\|_{\mathcal{S}(G),M}(1+|xy^{-1}|)^{-M}.$$

By (3.43), we have also

$$(1+|xy^{-1}|)^{-1} \leq C_o(1+|y|)(1+|x|)^{-1}.$$

Therefore, for every $M \in \mathbb{N}$ with $M \geq v_n N$ we get

$$\begin{aligned} |(\phi * f)(x)| &\leq C_R \sup_{|y|\leq R} C_o^M (1+|y|)^M(1+|x|)^{-M}\|\phi\|_{\mathcal{S}(G),M} \\ &\leq C_R'(1+|x|)^{-M}\|\phi\|_{\mathcal{S}(G),M}. \end{aligned}$$

This shows $\phi * f \in \mathcal{S}(G)$. □

We note that there are certainly different ways of introducing the topology of the Schwartz spaces by different choices of families of seminorms.

Lemma 3.1.56. *Other families of Schwartz seminorms defining the same Fréchet topology on $\mathcal{S}(G)$ are*

- $\phi \mapsto \max_{[\alpha],[\beta]\leq N} \|x^\alpha X^\beta \phi\|_p$
- $\phi \mapsto \max_{[\alpha],[\beta]\leq N} \|X^\beta x^\alpha \phi\|_p$
- $\phi \mapsto \max_{[\beta]\leq N} \|(1+|\cdot|)^N X^\beta \phi\|_p$

(for the first two we don't need a homogeneous quasi-norm) where $p \in [1,\infty]$.

Proof. The first two families with the usual Euclidean derivatives instead of left-invariant vector fields are known to give the Fréchet topologies. Therefore, by e.g. using Proposition 3.1.28, this is also the case for the first two families.

The last family would certainly be equivalent to the first one for the homogeneous quasi-norm $|\cdot|_p$ in (3.21), for p being a multiple of $\upsilon_1,\ldots,\upsilon_n$, since $|x|_p^p$ is a polynomial. Therefore, the last family also yields the Fréchet topology for any choice of homogeneous quasi-norm since any two homogeneous quasi-norms are equivalent by Proposition 3.1.35. □

3.1.10 Approximation of the identity

The family of dilations gives an easy way to define approximations to the identity.

If ϕ is a function on G and $t>0$, we define ϕ_t by

$$\phi_t := t^{-Q}\phi \circ D_{t^{-1}} \quad \text{i.e.} \quad \phi_t(x) = t^{-Q}\phi(t^{-1}x).$$

If ϕ is integrable then $\int \phi_t$ is independent of t.

We denote by $C_o(G)$ the space of continuous functions on G which vanish at infinity:

Definition 3.1.57. We denote by $C_o(G)$ the space of continuous function $f: G \to \mathbb{C}$ such that for every $\epsilon > 0$ there exists a compact set K outside which we have $|f| < \epsilon$.

Endowed with the supremum norm $\|\cdot\|_\infty = \|\cdot\|_{L^\infty(G)}$, $C_o(G)$ is a Banach space.

We also denote by $C_c(G)$ the space of continuous and compactly supported functions on G. It is easy to see that $C_c(G)$ is dense in $L^p(G)$ for $p\in[1,\infty)$ and in $C_o(G)$ (in which case we set $p=\infty$).

Lemma 3.1.58. *Let $\phi \in L^1(G)$ and $\int_G \phi = c$.*

(i) *For every $f \in L^p(G)$ with $1 \leq p < \infty$ or every $f \in C_o(G)$ with $p = \infty$, we have*

$$\phi_t * f \underset{t\to 0}{\longrightarrow} cf \quad \text{in } L^p(G) \text{ or } C_o(G), \text{ i.e. } \quad \|\phi_t * f - cf\|_{L^p(G)} \underset{t\to 0}{\longrightarrow} 0.$$

The same holds for $f * \phi_t$.

(ii) *If $\phi \in \mathcal{S}(G)$, then for any $\psi \in \mathcal{S}(G)$ and $f \in \mathcal{S}'(G)$, we have*

$$\phi_t * \psi \underset{t\to 0}{\longrightarrow} c\psi \quad \text{in } \mathcal{S}(G) \quad \text{and} \quad \phi_t * f \underset{t\to 0}{\longrightarrow} cf \quad \text{in } \mathcal{S}'(G).$$

The same holds for $\psi * \phi_t$ and $f * \psi_t$.

The proof is very similar to its Euclidean counterpart.

Proof. Let $\phi \in L^1(G)$ and $c = \int_G \phi$. If $f \in C_c(G)$ then

$$\begin{aligned}
(\phi_t * f)(x) - cf(x) &= \int_G t^{-Q}\phi(t^{-1}y)f(y^{-1}x)dy - cf(x) \\
&= \int_G \phi(z)f((tz)^{-1}x)dz - \int_G \phi(z)dzf(x) \\
&= \int_G \phi(z)\left(f((tz)^{-1}x) - f(x)\right)dz.
\end{aligned}$$

Hence by the Minkowski inequality we have

$$\|\phi_t * f - cf\|_p \leq \int_G |\phi(z)| \left\|f((tz)^{-1}\cdot\,) - f\right\|_p dz.$$

Since $\left\|f((tz)^{-1}\cdot\,) - f\right\|_p \leq 2\|f\|_p$, this shows (i) for any $f \in C_c(G)$ by the Lebesgue dominated convergence theorem. Let f be in $L^p(G)$ or $C_o(G)$ (in this case $p = \infty$). By density of $C_c(G)$, for any $\epsilon > 0$, we can find $f_\epsilon \in C_c(G)$ such that $\|f - f_\epsilon\|_p \leq \epsilon$. We have

$$\|\phi_t * (f - f_\epsilon)\|_p \leq \|\phi_t\|_1\|f - f_\epsilon\|_p \leq \|\phi\|_1\epsilon,$$

thus

$$\begin{aligned}
\|\phi_t * f - cf\|_p &\leq \|\phi_t * (f - f_\epsilon)\|_p + |c|\|f_\epsilon - f\|_p + \|\phi_t * f_\epsilon - cf_\epsilon\|_p \\
&\leq (\|\phi\|_1 + |c|)\epsilon + \|\phi_t * f_\epsilon - cf_\epsilon\|_p.
\end{aligned}$$

Since $\|\phi_t * f_\epsilon - cf_\epsilon\|_p \to 0$ as $t \to 0$, there exists $\eta > 0$ such that

$$\forall t \in (0, \eta) \qquad \|\phi_t * f_\epsilon - cf_\epsilon\|_p < \epsilon.$$

Hence if $0 < t < \eta$, we have

$$\|\phi_t * f - cf\|_p \leq (\|\phi\|_1 + |c| + 1)\epsilon.$$

This shows the convergence of $\phi_t * f - cf$ for any $f \in L^p(G)$ or $C_o(G)$.

With the notation $\tilde{\cdot}$ for the operation given by $\tilde{g}(x) = g(x^{-1})$, we also have

$$(f * g)\tilde{} = \tilde{g} * \tilde{f}.$$

Hence applying the previous result to $\tilde{f}$ and $\tilde{\phi}$, we obtain the convergence of $f * \phi_t - cf$.

Let us prove (ii) for $\phi, \psi \in \mathcal{S}(G)$. We have as above

$$(\phi_t * \psi)(x) - c\psi(x) = \int_G \phi(z) \left(\psi((tz)^{-1}x) - \psi(x)\right) dz,$$

thus

$$\begin{aligned} \|\phi_t * \psi - c\psi\|_{\mathcal{S}(G),N} &\leq \int_G |\phi(z)| \, \|\psi((tz)^{-1}\cdot) - \psi\|_{\mathcal{S}(G),N} \, dz \\ &\leq \int_G |\phi(z)| \, C \sum_{j=1}^n |(tz)^{-1}|^{\upsilon_j} \, \|\psi\|_{\mathcal{S}(G),N+\upsilon_n} \, dz \end{aligned}$$

by (3.44). And this shows

$$\|\phi_t * \psi - c\psi\|_{\mathcal{S}(G),N} \leq C \sum_{j=1}^n \|\phi\|_{\mathcal{S}(G),Q+1+\upsilon_j} \, \|\psi\|_{\mathcal{S}(G),N+\upsilon_n} \, t^{\upsilon_j} \underset{t\to 0}{\longrightarrow} 0.$$

Hence we have obtained the convergence of $\phi_t * \psi - c\psi$. As above, applying the previous result to $\tilde{\psi}$ and $\tilde{\phi}$, we obtain the convergence of $\psi * \phi_t$.

Let $f \in \mathcal{S}'(G)$. By (1.14) for distributions, we see for any $\psi \in \mathcal{S}(G)$, that

$$\langle f * \phi_t, \psi \rangle = \langle f, \psi * \tilde{\phi}_t \rangle \underset{t\to 0}{\longrightarrow} c\langle f, \psi \rangle$$

by the convergence just shown above. This shows that $f * \phi_t$ converges to f in $\mathcal{S}'(G)$. As above, applying the previous result to $\tilde{f}$ and $\tilde{\phi}$, we obtain the convergence of $f * \phi_t$. □

In the sequel we will need (only in the proof of Theorem 4.4.9) the following collection of technical results. Recall that a simple function is a measurable function which takes only a finite number of values.

Lemma 3.1.59. *Let $\mathscr{B}$ denote the space of simple and compactly supported functions on G. Then we have the following properties.*

(i) *The space $\mathscr{B}$ is dense in $L^p(G)$ for any $p \in [1, \infty)$.*

(ii) *If $\phi \in \mathcal{S}(G)$ and $f \in \mathscr{B}$, then $\phi * f$ and $f * \phi$ are in $\mathcal{S}(G)$.*

(iii) *For every $f \in \mathscr{B}$ and $p \in [1,\infty]$,*

$$\phi_t * f \underset{t\to 0}{\longrightarrow} (\int_G \phi) f$$

in $L^p(G)$. The same holds for $f * \phi_t$.

Proof. Part (i) is well-known (see, e.g., Rudin [Rud87, ch. 1]).

As a convolution of a Schwartz function ϕ with a compactly supported tempered distribution $f \in \mathscr{B}$, $f * \phi$ and $\phi * f$ are Schwartz by Lemma 3.1.55. This proves (ii).

Part (iii) follows from Lemma 3.1.58 (i) for $1 \le p < \infty$. For the case $p = \infty$, we proceed as in the first part of the proof of Lemma 3.1.58 (i) taking f not in $C_c(G)$ but a simple function with compact support. □

Remark 3.1.60. In Section 4.2.2 we will see that the heat semi-group associated to a positive Rockland operator gives an approximation of the identity h_t, $t > 0$, which is commutative:

$$h_t * h_s = h_s * h_t = h_{s+t}.$$

3.2 Operators on homogeneous Lie groups

In this section we analyse operators on a (fixed) homogeneous Lie group G. We first study sufficient conditions for a linear operator to extend boundedly from some L^p-space to an L^q-space. We will be particularly interested in the case of left-invariant homogeneous linear operators. In the last section, we will focus our attention on such operators which are furthermore differential and on the possible existence of their fundamental solutions. As an application, we will give a version of Liouville's Theorem which holds on homogeneous Lie groups. All these results have well-known Euclidean counterparts.

All the operators we consider here will be linear so we will not emphasise their linearity in every statement.

3.2.1 Left-invariant operators on homogeneous Lie groups

The Schwartz kernel theorem (see Theorem 1.4.1) says that, under very mild hypothesis, an operator on a smooth manifold has an integral representation. An easy consequence is that a left-invariant operator on a Lie group has a convolution kernel.

Corollary 3.2.1 (Kernel theorem on Lie groups). *We have the following statements.*

- *Let G be a connected Lie group and let $T : \mathcal{D}(G) \to \mathcal{D}'(G)$ be a continuous linear operator which is invariant under left-translations, i.e.*

$$\forall x_o \in G,\ f \in \mathcal{D}(G) \qquad T(f(x_o\,\cdot)) = (Tf)(x_o\,\cdot).$$

Then there exists a unique distribution $\kappa \in \mathcal{D}'(G)$ *such that*

$$Tf_1 : x \longmapsto f_1 * \kappa(x) = \int_G \kappa(y^{-1}x) f_1(y) dy.$$

In other words, T is a convolution operator with (right convolution) kernel κ. The converse is also true.

- *Let G be a connected simply connected nilpotent Lie group identified with $\mathbb{R}^n$ endowed with a polynomial law (see Proposition 1.6.6). Let $T : \mathcal{S}(G) \to \mathcal{S}'(G)$ be a continuous linear operator which is invariant under left translations, i.e.*

  $$\forall x_o \in G,\ f \in \mathcal{S}(G) \qquad T(f(x_o \cdot)) = (Tf)(x_o \cdot).$$

  Then there exists a unique distribution $\kappa \in \mathcal{S}'(\mathbb{R}^n)$ *such that*

  $$Tf_1 : x \longmapsto f_1 * \kappa(x) = \int_G \kappa(y^{-1}x) f_1(y) dy.$$

  In other words, T is a convolution operator with (right convolution) kernel κ. The converse is also true.

In both cases, for any test function f_1, the function Tf_1 is smooth. Furthermore, the map $\kappa \mapsto T$ is an isomorphism of topological vector spaces.

A similar statement holds for right-invariant operators.

We omit the proof: it relies on approaching the kernels $\kappa(x, y)$ by continuous functions for which the invariance forces them to be of the form $\kappa(y^{-1}x)$. The converses are much easier and have been shown in Section 1.5.

In this monograph, we will often use the following notation:

Definition 3.2.2. Let T be an operator on a connected Lie group G which is continuous as an operator $\mathcal{D}(G) \to \mathcal{D}'(G)$ or as $\mathcal{S}(G) \to \mathcal{S}'(G)$. Its right convolution kernel κ, as given in Corollary 3.2.1, is denoted by

$$T\delta_0 = \kappa.$$

In the case of left-invariant differential operators, we obtain easily the following properties.

Proposition 3.2.3. *If T is a left-invariant differential operator on a connected Lie group G, then its kernel is by definition the distribution $T\delta_0 \in \mathcal{D}'(G)$ such that*

$$\forall \phi \in \mathcal{D}(G) \qquad T\phi = \phi * T\delta_0.$$

The distribution $T\delta_0 \in \mathcal{S}'(G)$ is supported at the origin. The equality

$$f * T\delta_0 = Tf$$

holds for any $f \in \mathcal{E}'(G)$, the left-hand side being the group convolution of a distribution with a compactly supported distribution. The equality

$$T\delta_0 * f = \tilde{T} f$$

for the right-invariant differential operator corresponding to T also holds for any $f \in \mathcal{E}'(G)$.

The kernel of $T^t\delta_0$ is given formally by

$$T^t\delta_0(x) = T\delta_0(x^{-1}).$$

If $T = X^\ell$, for a left-invariant vector field X on G and $\ell \in \mathbb{N}$, then the distribution $(-1)^\ell X^\ell \delta_0(x^{-1})$ is the left convolution kernel of the right-invariant differential operator $\tilde{T}$.

We can also see from (1.14) and Definition 1.5.4 that the adjoint of the bounded on $L^2(G)$ operator $Tf = f * \kappa$ is the convolution operator $T^* f = f * \tilde{\kappa}$, well defined on $\mathcal{D}(G)$, with the right convolution kernel given by

$$\tilde{\kappa}(x) = \bar{\kappa}(x^{-1}). \tag{3.47}$$

The transpose operation is defined in Definition A.1.5, and for left-invariant differential operators it takes the form given by (1.10). Clearly the transpose of a left-invariant differential operator on G is a left-invariant differential operator on G.

Proof. A left-invariant differential operator is necessarily continuous as $\mathcal{D}(G) \to \mathcal{D}(G)$. Hence it admits the kernel $T\delta_0$. We have for $\phi \in \mathcal{D}(G)$ with $\tilde{\phi}(x) = \phi(x^{-1})$ that

$$\langle T\delta_0, \tilde{\phi} \rangle = (\phi * T\delta_0)(0) = T\phi(0).$$

So if $0 \notin \operatorname{supp} \phi$ then $\langle T\delta_0, \phi \rangle = 0$. This shows that $T\delta_0$ is supported at 0.

If $\phi, \psi \in \mathcal{D}(G)$, then

$$\langle \phi * T\delta_0, \psi \rangle = \langle T\phi, \psi \rangle = \langle \phi, T^t\psi \rangle = \langle \phi, \psi * T^t\delta_0 \rangle.$$

By (1.14) this shows that $T^t\delta_0 = (T\delta_0)\tilde{}$. Furthermore, if $f \in \mathcal{D}'(G)$, then

$$\langle Tf, \phi \rangle = \langle f, T^t\phi \rangle = \langle f, \phi * T^t\delta_0 \rangle = \langle f, \phi * (T\delta_0)\tilde{}\rangle = \langle f * T\delta_0, \phi \rangle.$$

This shows $Tf = f * T\delta_0$.

Now we can check easily (see (1.11)) that

$$\tilde{X} f = -(X\tilde{f})\tilde{}$$

and, more generally,

$$\tilde{X}^\ell f = (-1)^\ell (X^\ell \tilde{f})\tilde{}$$

for $\ell \in \mathbb{N}$. Since the equality $(f * g)\tilde{} = \tilde{g} * \tilde{f}$ holds as long as it makes sense, this shows that

$$(-1)^\ell (X^\ell \delta_0)\tilde{} * f = \tilde{T} f.$$

□

In fact, our primary concern will be to study operators of a different nature, and their possible extensions to some L^p-spaces. This (i.e. the L^p-boundedness) is certainly not the case for general differential operators.

Assuming that an operator is continuous as $\mathcal{S}(G) \to \mathcal{S}'(G)$ or as $\mathcal{D}(G) \to \mathcal{D}'(G)$ is in practice a very mild hypothesis. It ensures that a potential extension into a bounded operator $L^p(G) \to L^q(G)$ is necessarily unique, by density of $\mathcal{D}(G)$ in $L^p(G)$. Hence we may abuse the notation, and keep the same notation for an operator which is continuous as $\mathcal{S}(G) \to \mathcal{S}'(G)$ or as $\mathcal{D}(G) \to \mathcal{D}'(G)$ and its possible extension, once we have proved that it gives a bounded operator from $L^p(G)$ to $L^q(G)$.

We want to study in the context of homogeneous Lie groups the condition which implies that an operator as above extends to a bounded operator from $L^p(G)$ to $L^q(G)$.

As the next proposition shows, only the case $p \le q$ is interesting.

Proposition 3.2.4. *Let G be a homogeneous Lie group and let T be a linear left-invariant operator bounded from $L^p(G)$ to $L^q(G)$, for some (given) finite $p, q \in [1, \infty)$. If $p > q$ then $T = 0$.*

The proof is based on the following lemma:

Lemma 3.2.5. *Let $f \in L^p(G)$ with $1 \le p < \infty$. Then*

$$\lim_{x \to \infty} \|f - f(x\,\cdot)\|_{L^p(G)} = 2^{\frac{1}{p}} \|f\|_{L^p(G)}.$$

Proof of Lemma 3.2.5. First let us assume that the function f is continuous with compact support E. For $x_o \in G$, the function $f(x_o\,\cdot)$ is continuous and supported in $x_o^{-1}E$. Therefore, if x_o is not in $EE^{-1} = \{yz \,:\, y \in E,\ z \in E^{-1}\}$, then f and $f(x_o\,\cdot)$ have disjoint supports, and

$$\|f - f(x_o\,\cdot)\|_p^p = \int_E |f|^p + \int_{x_o^{-1}E} |f(x_o\,\cdot)|^p = 2\|f\|_p^p.$$

Now we assume that $f \in L^p(G)$. For each sufficiently small $\epsilon > 0$, let f_ϵ be a continuous function with compact support $E_\epsilon \subset \{|x| \le \epsilon^{-1}\}$ satisfying $\|f - f_\epsilon\|_p < \epsilon$. We claim that for any sufficiently small $\epsilon > 0$, we have

$$|x_o| > 2\epsilon^{-1} \Longrightarrow \left| \|f - f(x_o\,\cdot)\|_p - 2^{\frac{1}{p}} \|f\|_p \right| \le (2 + 2^{\frac{1}{p}})\epsilon. \tag{3.48}$$

Indeed, using the triangle inequality, we obtain

$$\left| \|f - f(x_o \cdot)\|_p - 2^{\frac{1}{p}} \|f\|_p \right| \le \left| \|f - f(x_o \cdot)\|_p - 2^{\frac{1}{p}} \|f_\epsilon\|_p \right| + 2^{\frac{1}{p}} \left| \|f_\epsilon\|_p - \|f\|_p \right|.$$

For the last term of the right-hand side we have

$$\left| \|f_\epsilon\|_p - \|f\|_p \right| \le \|f_\epsilon - f\|_p < \epsilon,$$

whereas for the first term, if $x_o \notin E_\epsilon E_\epsilon^{-1}$, using the first part of the proof and then the triangle inequality, we get

$$\begin{aligned} \left| \|f - f(x_o \cdot)\|_p - 2^{\frac{1}{p}} \|f_\epsilon\|_p \right| &= \left| \|f - f(x_o \cdot)\|_p - \|f_\epsilon - f_\epsilon(x_o \cdot)\|_p \right| \\ &\le \|(f - f(x_o \cdot)) - (f_\epsilon - f_\epsilon(x_o \cdot))\|_p \\ &\le \|f - f_\epsilon\|_p + \|f(x_o \cdot) - f_\epsilon(x_o \cdot)\|_p < 2\epsilon. \end{aligned}$$

This shows (3.48) and concludes the proof of Lemma 3.2.5. □

Proof of Proposition 3.2.4. Let $f \in \mathcal{D}(G)$. As T is left-invariant, we have

$$\|(Tf)(x_o \cdot) - Tf\|_q = \left\|T\big(f(x_o \cdot) - f\big)\right\|_q \le \|T\|_{\mathscr{L}(L^p(G),L^q(G))} \|f(x_o \cdot) - f\|_p .$$

Taking the limits as x_o tends to infinity, by Lemma 3.2.5, we get

$$2^{\frac{1}{q}} \|Tf\|_q \le \|T\|_{\mathscr{L}(L^p(G),L^q(G))} 2^{\frac{1}{p}} \|f\|_p .$$

But then

$$\|T\|_{\mathscr{L}(L^p(G),L^q(G))} \le 2^{\frac{1}{p} - \frac{1}{q}} \|T\|_{\mathscr{L}(L^p(G),L^q(G))}.$$

Hence $p > q$ implies $\|T\|_{\mathscr{L}(L^p(G),L^q(G))} = 0$ and $T = 0$. □

As in the Euclidean case, Proposition 3.2.4 is all that can be proved in the general framework of left-invariant bounded operators from $L^p(G)$ to $L^q(G)$. However, if we add the property of homogeneity more can be said and we now focus our attention on this case.

3.2.2 Left-invariant homogeneous operators

The next statement says that if the operator T is left-invariant, homogeneous and bounded from $L^p(G)$ to $L^q(G)$, then the indices p and q must be related in the same way as in the Euclidean case but with the topological dimension being replaced by the homogeneous dimension Q.

Proposition 3.2.6. *Let T be a left-invariant linear operator on G which is bounded from $L^p(G)$ to $L^q(G)$ for some (given) finite $p, q \in [1, \infty)$. If T is homogeneous of degree $\nu \in \mathbb{C}$ (and $T \neq 0$), then*

$$\frac{1}{q} - \frac{1}{p} = \frac{\operatorname{Re} \nu}{Q}.$$

Proof. We compute easily,

$$\|f \circ D_t\|_p = t^{-\frac{Q}{p}} \|f\|_p, \quad f \in L^p(G),\ t > 0.$$

Thus, since T is homogeneous of degree ν, we have

$$\begin{aligned} t^{\operatorname{Re}\nu - \frac{Q}{q}} \|Tf\|_q = \|t^\nu (Tf) \circ D_t\|_q = \|T(f \circ D_t)\|_q &\leq \|T\|_{\mathscr{L}(L^p(G),L^q(G))} \|f \circ D_t\|_p \\ &= \|T\|_{\mathscr{L}(L^p(G),L^q(G))} t^{-\frac{Q}{p}} \|f\|_p, \end{aligned}$$

so

$$\forall t > 0 \qquad \|T\|_{\mathscr{L}(L^p(G),L^q(G))} \leq t^{-\operatorname{Re}\nu + \frac{Q}{q} - \frac{Q}{p}} \|T\|_{\mathscr{L}(L^p(G),L^q(G))}.$$

Hence we must have

$$-\operatorname{Re}\nu + \frac{Q}{q} - \frac{Q}{p} = 0$$

as claimed. □

Combining together Propositions 3.2.4 and 3.2.6, we see that it makes sense to restrict one's attention to

$$\frac{\operatorname{Re}\nu}{Q} \in (-1, 0].$$

The case $\operatorname{Re}\nu = 0$ is the most delicate and we leave it aside for the moment (see Section 3.2.5). We shall discuss instead the case

$$-Q < \operatorname{Re}\nu < 0.$$

Let us observe that the homogeneity of the operator is equivalent to the homogeneity of its kernel:

Lemma 3.2.7. *Let T be a continuous left-invariant linear operator as $\mathcal{S}(G) \to \mathcal{S}'(G)$ or as $\mathcal{D}(G) \to \mathcal{D}'(G)$, where G is a homogeneous Lie group. Then T is ν-homogeneous if and only if its (right) convolution kernel is $-(Q+\nu)$-homogeneous.*

Proof. On one hand we have

$$T(f(r\,\cdot))(x) = \int_G f(ry)\kappa(y^{-1}x)dy,$$

and on the other hand,

$$\begin{aligned} Tf\,(rx) &= \int_G f(z)\kappa(z^{-1}rx)dz = \int_G f(ry)\kappa((ry)^{-1}rx)r^Q dy \\ &= r^Q \int_G f(ry)(\kappa \circ D_r)(y^{-1}x)dy. \end{aligned}$$

Now the statement follows from these and the uniqueness of the kernel. □

The following proposition gives a sufficient condition on the homogeneous kernel so that the corresponding left-invariant homogeneous operator extends to a bounded operator from $L^p(G)$ to $L^q(G)$.

Proposition 3.2.8. *Let T be a linear continuous operator as $\mathcal{S}(G) \to \mathcal{S}'(G)$ or as $\mathcal{D}(G) \to \mathcal{D}'(G)$ on a homogeneous Lie group G. We assume that the operator T is left-invariant and homogeneous of degree ν, that*

$$\operatorname{Re}\nu \in (-Q, 0),$$

and that the (right convolution) kernel κ of T is continuous away from the origin.

Then T extends to a bounded operator from $L^p(G)$ to $L^q(G)$ whenever $p, q \in (1, \infty)$ satisfy

$$\frac{1}{q} - \frac{1}{p} = \frac{\operatorname{Re}\nu}{Q}.$$

The integral kernel κ then can also be identified with a locally integrable function at the origin.

We observe that, by Corollary 3.2.1, κ is a distribution (in $\mathcal{S}'(G)$ or $\mathcal{D}'(G)$) on G. The hypothesis on κ says that its restriction to $G\backslash\{0\}$ coincides with a continuous function κ_o on $G\backslash\{0\}$.

Proof of Proposition 3.2.8. We fix a homogeneous norm $|\cdot|$ on G. We denote by $\bar{B}_R := \{x : |x| \leq R\}$ and $\mathfrak{S} := \{x : |x| = 1\}$ the ball of radius R and the unit sphere around 0. By Lemma 3.2.7, κ_o is a continuous homogeneous function of degree $-(Q+\nu)$ on $G\backslash\{0\}$. Denoting by C its maximum on the unit sphere, we have

$$\forall x \in G\backslash\{0\} \qquad |\kappa_o(x)| \leq \frac{C}{|x|^{Q+\operatorname{Re}\nu}}.$$

Hence κ_o defines a locally integrable function on G, even around 0, and we keep the same notation for this function. Therefore, the distribution $\kappa' = \kappa - \kappa_o$ on G is, in fact, supported at the origin. It is also homogeneous of degree $-Q-\nu$. Due to the compact support of κ', $|\langle \kappa', f\rangle|$ is controlled by some C^k norm of f on a fixed small neighbourhood of the origin. But, because of its homogeneity, and using (3.9), we get

$$\forall t > 0 \qquad \langle \kappa', f\rangle = t^{-Q-\nu}\langle \kappa' \circ D_{\frac{1}{t}}, f\rangle = t^{-\nu}\langle \kappa', f \circ D_t\rangle.$$

Letting t tend to 0, the C^k norms of $f \circ D_t$ remain bounded, so that $\langle \kappa', f\rangle = 0$ since $\operatorname{Re}\nu < 0$. This shows that $\kappa' = 0$ and so $\kappa = \kappa_o$.

Note that the weak $L^r(G)$-norm of κ is finite for $r = Q/(Q + \operatorname{Re}\nu)$. Indeed, if $s > 0$,

$$|\kappa_o(x)| > s \Longrightarrow |x|^{Q+\operatorname{Re}\nu} \leq \frac{C}{s},$$

so that

$$|\{x : |\kappa_o(x)| > s\}| \leq \left|B_{(C/s)^{\frac{1}{Q+\operatorname{Re}\nu}}}\right| \leq c\left(\frac{C}{s}\right)^{\frac{Q}{Q+\operatorname{Re}\nu}},$$

with $c = |B_1|$, and hence

$$\|\kappa_o\|_{w-L^r(G)} \leq c\,C^{\frac{Q}{Q+\operatorname{Re}\nu}} \quad \text{with } r = \frac{Q}{Q+\operatorname{Re}\nu}.$$

The proposition is now easy using the generalisation of Young's inequalities (see Proposition 1.5.2), so that we get that T is bounded from $L^p(G)$ to $L^q(G)$ for

$$\frac{1}{q} - \frac{1}{p} = \frac{1}{r} - 1 = \frac{\operatorname{Re}\nu}{Q},$$

as claimed. □

We may use the usual vocabulary for homogeneous kernels as in [Fol75] and [FS82]:

Definition 3.2.9. Let G be a homogeneous Lie group and let $\nu \in \mathbb{C}$.

A distribution $\kappa \in \mathcal{D}'(G)$ which is smooth away from the origin and homogeneous of degree $\nu - Q$ is called a *kernel of type* ν on G.

A (right) convolution operator $T : \mathcal{D}(G) \to \mathcal{D}'(G)$ whose convolution kernel is of type ν is called an *operator of type* ν. That is, T is given via

$$T(\phi) = \phi * \kappa,$$

where κ kernel of type ν.

Remark 3.2.10. We will mainly be interested in the $L^p \to L^q$-boundedness of operators of type ν. Thus, by Propositions 3.2.4 and 3.2.6, we will restrict ourselves to $\nu \in \mathbb{C}$ with $\operatorname{Re}\nu \in [0, Q)$.

If $\operatorname{Re}\nu \in (0, Q)$, then a $(\nu - Q)$-homogeneous function in $C^\infty(G\backslash\{0\})$ is integrable on a neighbourhood of 0 and hence extends to a distribution in $\mathcal{D}'(G)$, see the proof of Proposition 3.2.8. Hence, in the case $\operatorname{Re}\nu \in (0, Q)$, the restriction to $G\backslash\{0\}$ yields a one-to-one correspondence between the $(\nu - Q)$-homogeneous functions in $C^\infty(G\backslash\{0\})$ and the kernels of type ν.

We will see in Remark 3.2.29 that the case $\operatorname{Re}\nu = 0$ is more subtle.

In view of Lemma 3.2.7 and Proposition 3.2.8, we have the following statement for operators of type ν with $\operatorname{Re}\nu \in (0, Q)$.

Corollary 3.2.11. *Let G be a homogeneous Lie group and let $\nu \in \mathbb{C}$ with*

$$\operatorname{Re}\nu \in (0, Q).$$

Any operator of type ν is $(-\nu)$-homogeneous and extends to a bounded operator from $L^p(G)$ to $L^q(G)$ whenever $p, q \in (1, \infty)$ satisfy

$$\frac{1}{p} - \frac{1}{q} = \frac{\operatorname{Re}\nu}{Q}.$$

As we said earlier the case of a left-invariant operator which is homogeneous of degree 0 is more complicated and is postponed until the end of Section 3.2.4. In the meantime, we make a useful parenthesis about the Calderón-Zygmund theory in our context.

3.2.3 Singular integral operators on homogeneous Lie groups

In the case of $\mathbb{R}$, a famous example of a left-invariant 0-homogeneous operator is the Hilbert transform. This particular example has motivated the development of the theory of singular integrals in the Euclidean case as well as in other more general settings. In Section A.4, the interested reader will find a brief presentation of this theory in the setting of spaces of homogeneous type (due to Coifman and Weiss). In this section here, we check that homogeneous Lie groups are spaces of homogeneous type and we obtain the corresponding theorem of singular integrals together with some useful consequences for left-invariant operators. We also propose a definition of Calderón-Zygmund kernels on homogeneous Lie groups, thereby extending the one on Euclidean spaces (cf. Section A.4).

First let us check that homogeneous Lie groups equipped with a quasi-norm are spaces of homogeneous type in the sense of Definition A.4.2 and that the Haar measure is doubling (see Section A.4):

Lemma 3.2.12. *Let G be a homogeneous Lie groups and let $|\cdot|$ be a quasi-norm. Then the set G endowed with the usual Euclidean topology together with the quasi-distance*

$$d : (x, y) \mapsto |y^{-1}x|$$

is a space of homogeneous type and the Haar measure has the doubling property given in (A.5).

Proof of Lemma 3.2.12. We keep the notation of the statement. The defining properties of a quasi-norm and the fact that it satisfies the triangular inequality up to a constant (see Proposition 3.1.38) imply easily that d is indeed a quasi-distance on G in the sense of Definition A.4.1. By Proposition 3.1.37, the corresponding quasi-balls $B(x,r) := \{y \in G : d(x,y) < r\}$, $x \in G$, $r > 0$, generate the usual topology of the underlying Euclidean space. Hence the first property listed in Definition A.4.2 is satisfied.

By Remark 3.1.34, the quasi-balls satisfy $B(x,r) = xB(0,r)$ and $B(0,r) = D_r(B(0,1))$. By (3.6), the volume of $B(0,r)$ is $|B(0,r)| = r^Q|B(0,1)|$. Hence we have obtained that the volume of any open quasi-ball is $|B(x,r)| = r^Q|B(0,1)|$. This implies that the Haar measure satisfies the doubling condition given in (A.5). We can now conclude the proof of the statement with Lemma A.4.3. □

Lemma 3.2.12 implies that we can apply the theorem of singular integrals on spaces of homogeneous type recalled in Theorem A.4.4 and we obtain:

Theorem 3.2.13 (Singular integrals). *Let G be a homogeneous Lie group and let T be a bounded linear operator on $L^2(G)$, i.e.*

$$\exists C_o \qquad \forall f \in L^2 \quad \|Tf\|_2 \leq C_o \|f\|_2. \tag{3.49}$$

We assume that the integral kernel κ of T coincides with a locally integrable function away from the diagonal, that is, on $(G\times G)\backslash\{(x,y)\in G\times G : x=y\}$. We also assume that there exist $C_1, C_2 > 0$ satisfying

$$\forall y, y_o \in G \qquad \int_{|y_o^{-1}x|>C_1|y_o^{-1}y|} |\kappa(x,y)-\kappa(x,y_o)|dx \leq C_2. \tag{3.50}$$

for a quasi-norm $|\cdot|$.

Then for all p, $1 < p \leq 2$, T extends to a bounded operator on L^p because

$$\exists A_p > 0 \qquad \forall f \in L^2 \cap L^p \quad \|Tf\|_p \leq A_p \|f\|_p;$$

for $p = 1$, the operator T extends to a weak-type (1,1) operator since

$$\exists A_1 > 0 \qquad \forall f \in L^2 \cap L^1 \quad \mu\{x \ : \ |Tf(x)| > \alpha\} \leq A_1 \frac{\|f\|_1}{\alpha};$$

the constants A_p, $1 \leq p \leq 2$, depend only on C_o, C_1 and C_2.

Remark 3.2.14.

- The L^2-boundedness, that is, Condition (3.49), implies that the operator satisfies the Schwartz kernel theorem (see Theorem 1.4.1) and thus yields the existence of a distributional integral kernel. We still need to assume that this distribution is locally integrable away from the diagonal.
- Since any two quasi-norms on G are equivalent (see Proposition 3.1.35), if the kernel condition in (3.50) holds for one quasi-norm, it then holds for any quasi-norm (maybe with different constants C_1, C_2).

As recalled in Section A.4, the notion of Calderón-Zygmund kernels in the Euclidean setting appear naturally as sufficient conditions (often satisfied 'in practice') for (A.7) to be satisfied by the kernel of the operator and the kernel of its formal adjoint. This leads us to define the Calderón-Zygmund kernels in our setting as follows:

Definition 3.2.15. A *Calderón-Zygmund kernel on a homogeneous Lie group G* is a measurable function κ_o defined on $(G\times G)\backslash\{(x,y)\in G\times G : x=y\}$ satisfying for some γ, $0 < \gamma \leq 1$, $C_1 > 0$, $A > 0$, and a homogeneous quasi-norm $|\cdot|$ the inequalities

$$\begin{aligned}
|\kappa_o(x,y)| &\leq A|y^{-1}x|^{-Q}, \\
|\kappa_o(x,y)-\kappa_o(x',y)| &\leq A\frac{|x^{-1}x'|^\gamma}{|y^{-1}x|^{Q+\gamma}} \qquad \text{if } C_1|x^{-1}x'| \leq |y^{-1}x|, \\
|\kappa_o(x,y)-\kappa_o(x,y')| &\leq A\frac{|y^{-1}y'|^\gamma}{|y^{-1}x|^{Q+\gamma}} \qquad \text{if } C_1|y^{-1}y'| \leq |y^{-1}x|.
\end{aligned}$$

A linear continuous operator T as $\mathcal{D}(G) \to \mathcal{D}'(G)$ or as $\mathcal{S}(G) \to \mathcal{S}'(G)$ is called a *Calderón-Zygmund operator* if its integral kernel coincides with a Calderón-Zygmund kernel on $(G \times G)\backslash\{(x, y) \in G \times G : x = y\}$.

Remark 3.2.16. 1. In other words, we have modified the definition of a classical Calderón-Zygmund kernel (as in Section A.4)

- by replacing the Euclidean norm by a homogeneous quasi-norm
- and, more importantly, the topological (Euclidean) dimension of the underlying space n by the homogeneous dimension Q.

2. By equivalence of homogeneous quasi-norms, see Proposition 3.1.35, the definition does not depend on a particular choice of a homogeneous quasi-norm as we can change the constants C_1, A.

As in the Euclidean case, we have

Proposition 3.2.17. *Let G be a homogeneous Lie group and let T be a bounded linear operator on $L^2(G)$.*

If T is a Calderón-Zygmund operator on G (in the sense of Definition 3.2.15), then T is bounded on $L^p(G)$, $p \in (1, \infty)$, and weak-type (1,1).

Proof of Proposition 3.2.17. Let T be a bounded operator on $L^2(G)$ and $\kappa : (x, y) \mapsto \kappa(x, y)$ its distributional kernel. Then its formal adjoint T^* is also bounded on $L^2(G)$ with the same operator norm. Furthermore its distributional kernel is $\kappa^{(*)} : (x, y) \mapsto \bar{\kappa}(y, x)$. We assume that κ coincides with a Calderón-Zygmund kernel κ_o away from the diagonal. We fix a quasi-norm $|\cdot|$. The first inequality in Definition 3.2.15 shows that κ_o and $\kappa_o^{(*)}$ coincide with locally integrable functions away from the diagonal. Using the last inequality, we have for any $y, y_o \in G$,

$$\int_{|y_o^{-1}x| \geq C_1 |y_o^{-1}y|} |\kappa_o(x, y) - \kappa_o(x, y_o)| dx \leq A \int_{|y_o^{-1}x| \geq C_1 |y_o^{-1}y|} \frac{|y^{-1}y_o|^\gamma}{|y_o^{-1}x|^{Q+\gamma}} dx$$

and, using the change of variable $x' = y_o^{-1}x$, we have

$$\begin{aligned} \int_{|y_o^{-1}x| \geq C_1 |y_o^{-1}y|} \frac{1}{|y_o^{-1}x|^{Q+\gamma}} dx &= \int_{|x'| \geq C_1 |y_o^{-1}y|} |x'|^{-(Q+\gamma)} dx' \\ &\leq \int_{|x'| \geq C_1 |y_o^{-1}y|} |x'|^{-(Q+\gamma)} dx' \\ &= c \int_{r = C_1 |y_o^{-1}y|}^{+\infty} r^{-(Q+\gamma)} r^{Q-1} dr = c_1 |y_o^{-1}y|^{-\gamma}, \end{aligned}$$

having also used the polar coordinates (Proposition 3.1.42) with c denoting the mass of the Borel measure on the unit sphere, and c_1 a new constant (of C_1, γ and Q). Hence we have obtained

$$\int_{|y_o^{-1}x| \geq C_1 |y_o^{-1}y|} |\kappa_o(x, y) - \kappa_o(x, y_o)| dx \leq c_1 A.$$

Similarly for $\kappa_o^{(*)}$, we have

$$\int_{|y_o^{-1}x|\geq C_1|y_o^{-1}y|} |\kappa_o^{(*)}(x,y)-\kappa_o^{(*)}(x,y_o)|dx = \int_{|y_o^{-1}x|\geq C_1|y_o^{-1}y|} |\kappa_o(y,x)-\kappa_o(y_o,x)|dx$$
$$\leq A\int_{|y_o^{-1}x|\geq C_1|y_o^{-1}y|} \frac{|y_o^{-1}y|^\gamma}{|y_o^{-1}x|^{Q+\gamma}}dx,$$

having used the second inequality in Definition 3.2.15. The same computation as above shows that the last left-hand side is bounded by $c_1 A$. Hence κ_o and $\kappa_o^{(*)}$ satisfy (3.50). Proposition 3.2.17 now follows from Theorem 3.2.13. □

Remark 3.2.18. As in the Euclidean case, Calderón-Zygmund kernels do not necessarily satisfy the other condition of the L^2-boundedness (see (3.49)) and a condition of 'cancellation' is needed in addition to the Calderón-Zygmund condition to ensure the L^2-boundedness. Indeed, one can prove adapting the Euclidean case (see the proof of Proposition 1 in [Ste93, ch.VII §3]) that if κ_o is a Calderón-Zygmund kernel satisfying the inequality

$$\exists c>0 \qquad \forall x\neq y \qquad \kappa_o(x,y)\geq c|y^{-1}x|^{-Q},$$

then there does not exist an L^2-bounded operator T having κ_o as its kernel.

The following statement gives sufficient conditions for a kernel to be Calderón-Zygmund in terms of derivatives:

Lemma 3.2.19. *Let G be a homogeneous Lie group. If κ_o is a continuously differentiable function on $(G\times G)\backslash\{(x,y)\in G\times G : x=y\}$ satisfying the inequalities for any $x,y\in G$, $x\neq y$, $j=1,\ldots,n$,*

$$\begin{aligned}|\kappa_o(x,y)| &\leq A|y^{-1}x|^{-Q},\\ |(X_j)_x\kappa_o(x,y)| &\leq A|y^{-1}x|^{-(Q+\upsilon_j)},\\ |(X_j)_y\kappa_o(x,y)| &\leq A|y^{-1}x|^{-(Q+\upsilon_j)},\end{aligned}$$

for some constant $A>0$ and homogeneous quasi-norm $|\cdot|$, then κ_o is a Calderón-Zygmund kernel in the sense of Definition 3.2.15 with $\gamma=1$.

Again, if these inequalities are satisfied for one quasi-norm, then they are satisfied for all quasi-norms, maybe with different constants $A>0$.

Proof of Lemma 3.2.19. We fix a quasi-norm $|\cdot|$. We assume that it is a norm without loss of generality because of the remark just above and the existence of a homogeneous norm (Theorem 3.1.39); although we could give a proof without this hypothesis, it simplifies the constants below. Let κ_o be as in the statement. Using the Taylor expansion (Theorem 3.1.51) or the Mean Value Theorem (Proposition 3.1.46), we have

$$|\kappa_o(x',y)-\kappa_o(x,y)|\leq C_o\sum_{j=1}^n |x^{-1}x'|^{\upsilon_j} \sup_{|z|\leq\eta|x^{-1}x'|} |(X_j)_{x_1=xz}\kappa_o(x_1,y)|.$$

Using the second inequality in the statement, we have

$$\sup_{|z|\leq\eta|x^{-1}x'|} |(X_j)_{x_1=xz}\kappa_o(x_1,y)| \leq A \sup_{|z|\leq\eta|x^{-1}x'|} |y^{-1}xz|^{-(Q+\upsilon_j)}.$$

The reverse triangle inequality yields

$$|y^{-1}xz| \geq |y^{-1}x| - |z| \geq \frac{1}{2}|y^{-1}x| \qquad \text{if } |z| \leq \frac{1}{2}|y^{-1}x|.$$

Hence, if $2\eta|x^{-1}x'| \leq |y^{-1}x|$, then we have

$$\sup_{|z|\leq\eta|x^{-1}x'|} |y^{-1}xz|^{-(Q+\upsilon_j)} \leq 2^{Q+\upsilon_j}|y^{-1}x|^{-(Q+\upsilon_j)},$$

and we have obtained

$$\begin{aligned} |\kappa_o(x,y)-\kappa_o(x',y)| &\leq C_o\sum_{j=1}^n |x^{-1}x'|^{\upsilon_j}2^{Q+\upsilon_j}|y^{-1}x|^{-(Q+\upsilon_j)} \\ &\leq C_o\left(\sum_{j=1}^n (2\eta)^{-(\upsilon_j-1)}2^{Q+\upsilon_j}\right)|x^{-1}x'||y^{-1}x|^{-(Q-1)}. \end{aligned}$$

This shows the second inequality in Definition 3.2.15.

We proceed in a similar way to prove the third inequality in Definition 3.2.15: the Taylor expansion yields

$$|\kappa_o(x,y)-\kappa_o(x,y')| \leq C_o\sum_{j=1}^n |y^{-1}y'|^{\upsilon_j} \sup_{|z|\leq\eta|y^{-1}y'|} |(X_j)_{y_1=yz}\kappa_o(x,y_1)|$$

while one checks easily

$$\begin{aligned} \sup_{|z|\leq\eta|y^{-1}y'|} |(X_j)_{y_1=yz}\kappa_o(x,y_1)| &\leq A \sup_{|z|\leq\eta|y^{-1}y'|} |(yz)^{-1}x|^{-(Q+\upsilon_j)} \\ &\leq A2^{Q+\upsilon_j}|y^{-1}x|^{-(Q+\upsilon_j)}, \end{aligned}$$

when $2\eta|y^{-1}y'| \leq |y^{-1}x|$. We conclude in the same way as above and this shows that κ_o is a Calderón-Zygmund kernel. □

Corollary 3.2.20. *Let G be a homogeneous Lie group and let κ be a continuously differentiable function on $G\backslash\{0\}$. If κ satisfies for any $x \in G\backslash\{0\}$, $j = 1,\ldots,n$,*

$$\begin{aligned} |\kappa(x)| &\leq A|x|^{-Q}, \\ |X_j\kappa(x)| &\leq A|x|^{-(Q+\upsilon_j)}, \\ |\tilde{X}_j\kappa(x)| &\leq A|x|^{-(Q+\upsilon_j)}, \end{aligned}$$

for some constant $A > 0$ and homogeneous quasi-norm $|\cdot|$, then

$$\kappa_o : (x,y) \mapsto \kappa(y^{-1}x)$$

is a Calderón-Zygmund kernel in the sense of Definition 3.2.15 with $\gamma = 1$.

Corollary 3.2.20 will be useful when dealing with convolution kernels which are smooth away from the origin, in particular when they are also $(-Q)$-homogeneous, see Theorem 3.2.30.

Proof of Corollary 3.2.20. Keeping the notation of the statement, using properties (1.11) of left and right invariant vector fields, we have

$$\begin{aligned} (X_j)_x \kappa_o(x,y) &= (X_j\kappa)(y^{-1}x), \\ (X_j)_y \kappa_o(x,y) &= -(\tilde{X}_j\kappa)(y^{-1}x). \end{aligned}$$

The statement now follows easily from Lemma 3.2.19. □

Often, the convolution kernel decays quickly enough at infinity and the main singularity to deal with is about the origin. The next statement is an illustration of this idea:

Corollary 3.2.21. *Let G be a homogeneous Lie group and let T be a linear operator which is bounded on $L^2(G)$ and invariant under left translations.*

We assume that its distributional convolution kernel coincides on $G\backslash\{0\}$ with a continuously differentiable function κ which satisfies

$$\begin{aligned} \int_{|x|\geq 1/2} |\kappa(x)|dx &\leq A, \\ \sup_{0<|x|\leq 1} |x|^Q|\kappa(x)| &\leq A, \\ \sup_{0<|x|\leq 1} |x|^{Q+\upsilon_j}|X_j\kappa(x)| &\leq A, \quad j=1,\ldots,n, \end{aligned}$$

for some constant $A>0$ and a homogeneous quasi-norm $|\cdot|$. Then T is bounded on $L^p(G)$, $p\in(1,\infty)$, and is weak-type (1,1).

Proof. Let $\chi\in\mathcal{D}(G)$ be $[0,1]$-valued function such that $\chi\equiv 0$ on $\{|x|\geq 1\}$ and $\chi\equiv 1$ on $\{|x|\leq 1/2\}$. As $\int_{|x|\geq 1/2}|\kappa(x)|dx$ is finite, $(1-\chi)\kappa$ is integrable and the convolution operator with convolution kernel $(1-\chi)\kappa$ is bounded on $L^p(G)$ for $p\in[1,\infty]$. Hence it suffices to prove that the kernel κ_o given via $\kappa_o(x,y)=(\chi\kappa)(y^{-1}x)$ is Calderón-Zygmund.

From the estimates satisfied by κ, it is clear that the quantities

$$\sup_{x\in G\backslash\{0\}} |x|^Q|(\chi\kappa)(x)| \quad \text{and} \quad \sup_{x\in G\backslash\{0\}} |x|^{-(Q+\upsilon_j)}|X_j(\chi\kappa)(x)|$$

are finite. As each $\tilde{X}_j$ may be expressed as a combination of X_k with homogeneous polynomial coefficients, see Section 3.1.5, we have for any (regular enough) function f with compact support

$$\sup_{x\in G\backslash\{0\}} |x|^{-(Q+\upsilon_j)}|\tilde{X}_jf(x)| \leq C \sup_{\substack{x\in G\backslash\{0\}\\ k=1,\ldots,n}} |x|^{-(Q+\upsilon_k)}|X_kf(x)|.$$

Consequently, the quantities $\sup_{x\in G\backslash\{0\}} |x|^{-(Q+\upsilon_j)}|\tilde{X}_j(\chi\kappa)(x)|$ are also bounded. Applying Lemma 3.2.19 to κ_o defined above, one checks easily that it is a Calderón-Zygmund kernel. Applying Proposition 3.2.17 concludes the proof of Corollary 3.2.21. □

This closes our parenthesis about the Calderón-Zygmund theory in our context, and we can go back to the study of left-invariant homogeneous operators, this time of homogeneous degree 0.

3.2.4 Principal value distribution

As we will see in the sequel, many interesting operators for our analysis on a homogeneous Lie group G will be given by convolution operators with (right convolution distributional) kernels homogeneous of degree ν with $\operatorname{Re}\nu = -Q$. In most of the 'interesting' cases, the distribution κ will be given by a locally integrable function away from the origin; denoting by κ_o the restriction of κ to $G\backslash\{0\}$, one may wonder if there is a one-to-one correspondence between κ and κ_o. As in the Euclidean case, this leads to the notion of the principal value distribution and we adapt the ideas here to fit the homogeneous context; in particular, the topological (Euclidean) dimension is replaced by the homogeneous dimension Q.

So the question is: Considering a locally integrable function κ_o on $G\backslash\{0\}$ which is homogeneous of degree ν with $\operatorname{Re}\nu = -Q$, does there exist a distribution $\kappa \in \mathcal{D}'(G)$ on G, homogeneous of the same degree ν on G, whose restriction to $G\backslash\{0\}$ coincides with κ_o? that is,

$$\langle \kappa, f\rangle = \int_{G\backslash\{0\}} \kappa_o(x) f(x) dx,$$

whenever $f \in \mathcal{D}(G)$ and $0 \notin \operatorname{supp} f$. In other words, can the functional

$$\mathcal{D}(\mathbb{R}^n\backslash\{0\}) \ni f \longmapsto \int_{G\backslash\{0\}} \kappa_o(x) f(x) dx$$

be extended to a continuous functional on $\mathcal{D}(\mathbb{R}^n)$?

Remark 3.2.22. 1. We observe that if such an extension exists, it is not unique in general. For $\nu = -Q$, the reason is that the Dirac δ_0 at the origin is homogeneous of degree $-Q$ (see Example 3.1.20), so that if κ is a solution, then $\kappa + c\delta_0$ for any constant c is another solution. (However, see Proposition 3.2.27.)

2. The second observation is that the answer is negative in general:

Example 3.2.23. Let $|\cdot|$ be some fixed homogeneous quasi-norm on G smooth away from the origin. The function defined by $\kappa_o(x) = |x|^\nu$ with $\nu = -Q + i\tau$, $\tau \in \mathbb{R}$, is homogeneous of degree ν on $G\backslash\{0\}$ but can not be extended into a homogeneous distribution $\kappa \in \mathcal{D}'(G)$ of homogeneous degree ν.

Proof of Example 3.2.23. Indeed, let us assume that such a distribution κ exists for this κ_o. Homogeneity of degree $-Q+i\tau$ means that

$$\langle \kappa, \psi \circ D_t \rangle = t^{-i\tau} \langle \kappa, \psi \rangle, \quad t > 0, \ \psi \in \mathcal{D}(G).$$

Let $B_\delta := \{x \in G : |x| < \delta\}$ be the ball around 0 of radius δ. Let $\phi \in \mathcal{D}(G)$ be a real-valued function supported on $D_2(B_\delta)\backslash B_\delta$, such that

$$\int_G (\phi(x) - \phi(2x)) \, |x|^{-Q} dx \neq 0.$$

We now define

$$\psi(x) := |x|^{-i\tau} \phi(x) \quad \text{ and } \quad f := \psi - 2^{i\tau} (\psi \circ D_2), \ x \in G\backslash\{0\}.$$

Immediately we notice that

$$f(x) = |x|^{-i\tau} (\phi(x) - \phi(2x))$$

and, therefore, both ψ and f are supported inside $D_4(B_\delta)\backslash B_\delta$ and are smooth. We compute

$$\langle \kappa_o, f \rangle = \int_G (\phi(x) - \phi(2x)) \, |x|^{-Q} dx \neq 0$$

by the choice of ϕ. On the other hand,

$$\langle \kappa, f \rangle = \langle \kappa, \psi \rangle - 2^{i\tau} \langle \kappa, \psi \circ D_2 \rangle = 0.$$

We have obtained a contradiction. □

The next statement answers the question above under the assumption that κ_o is also continuous on $G\backslash\{0\}$.

Proposition 3.2.24. *Let G be a homogeneous Lie group and let κ_o be a continuous homogeneous function on $G\backslash\{0\}$ of degree ν with $\operatorname{Re} \nu = -Q$.*

Then κ_o extends to a homogeneous distribution in $\mathcal{D}'(G)$ if and only if its average value, defined in Lemmata 3.1.43 and 3.1.45, is $m_{\kappa_o} = 0$.

Proof. Let us fix a homogeneous quasi-norm $|\cdot|$. We denote by σ the measure on the unit sphere $\mathfrak{S} = \{x : |x| = 1\}$ which gives the polar change of coordinates (see Proposition 3.1.42) and $|\sigma|$ its total mass.

By Lemma 3.1.41, there exists $c > 0$ such that

$$|x| \leq 1 \Longrightarrow |x|_E \leq c|x|. \tag{3.51}$$

First let us assume $m_{\kappa_o} = 0$. Therefore, for any $a, b \in [0, \infty)$,

$$\int_{a<|x|<b} \kappa_o(x) dx = \int_{r=a}^{b} \int_{\mathfrak{S}} \kappa_o(rx) d\sigma(x) r^{Q-1} dr = m_{\kappa_o} \int_{r=a}^{b} r^{\nu} r^{Q-1} dr = 0,$$

see Section 3.1.7. We claim that, for each $f \in \mathcal{D}(G)$,

$$\exists \lim_{\epsilon\to 0} \int_{|x|>\epsilon} \kappa_o(x) f(x) dx \ < \infty. \tag{3.52}$$

Indeed, let us check the Cauchy condition for $0 < \epsilon < \epsilon'$. We see that

$$\begin{aligned}
\left| \int_{|x|>\epsilon} \kappa_o(x) f(x) dx - \int_{|x|>\epsilon'} \kappa_o(x) f(x) dx \right| &= \left| \int_{\epsilon<|x|<\epsilon'} \kappa_o(x) f(x) dx \right| \\
&= \left| \int_{\epsilon<|x|<\epsilon'} \kappa_o(x) \left(f(x) - f(0) \right) dx \right| \\
&\leq \int_{\epsilon<|x|<\epsilon'} |\kappa_o(x)| \, |f(x) - f(0)| \, dx.
\end{aligned}$$

The (Euclidean) mean value theorem and the estimate (3.51) imply

$$|f(x) - f(0)| \leq \|\nabla f\|_\infty |x|_E \leq \|\nabla f\|_\infty c|x| \quad \text{if } |x| < 1.$$

Since κ_o is ν-homogeneous with $\operatorname{Re}\nu = -Q$, denoting by C_o the maximum of $|\kappa_o|$ on the unit sphere $\{x : |x| = 1\}$, we have

$$\forall x \in G\backslash\{0\} \qquad |\kappa_o(x)| \leq C_o |x|^{-Q}.$$

Hence if $\epsilon' < 1$,

$$\begin{aligned}
\left| \int_{|x|>\epsilon} \kappa_o(x) f(x) dx - \int_{|x|>\epsilon'} \kappa_o(x) f(x) dx \right| &\leq \int_{\epsilon<|x|<\epsilon'} \|\nabla f\|_\infty c C_o |x|^{1-Q} dx \\
&= \|\nabla f\|_\infty c C_o (\epsilon' - \epsilon).
\end{aligned}$$

This implies the Cauchy condition. Therefore, Claim (3.52) is proved and we denote the limit by

$$\langle \kappa, f \rangle := \lim_{\epsilon\to 0} \int_{|x|>\epsilon} \kappa_o(x) f(x) dx, \quad f \in \mathcal{D}(G). \tag{3.53}$$

This clearly defines a linear functional. Moreover, this functional is continuous since if $f \in \mathcal{D}(G)$ is supported in a ball $\bar{B}_R = \{x : |x| \leq R\}$ for R large enough, then, for $\epsilon < 1$,

$$\begin{aligned}
\left| \int_{|x|>\epsilon} \kappa_o(x) f(x) dx \right| &\leq \left| \int_{\epsilon<|x|<1} \kappa_o(x) f(x) dx \right| + \left| \int_{1<|x|} \kappa_o(x) f(x) dx \right| \\
&\leq \|\nabla f\|_\infty c C_o (1 - \epsilon) + C_o \int_{1<|x|\leq R} |f(x)| dx \\
&\leq C_R (\|\nabla f\|_\infty + \|f\|_\infty).
\end{aligned}$$

For the converse, we proceed by contradiction: let us assume that κ exists and that $m_{\kappa_o} \neq 0$. Then

$$\kappa_o - \frac{m_{\kappa_o}}{|\sigma|}|x|^\nu$$

is a continuous homogeneous distribution of $G\backslash\{0\}$ of degree ν with mean average

$$\begin{aligned} \int_{\mathfrak{S}} \left(\kappa_o(x) - \frac{m_{\kappa_o}}{|\sigma|}|x|^\nu\right) d\sigma(x) &= \int_{\mathfrak{S}} \kappa_o(x) d\sigma(x) - \frac{m_{\kappa_o}}{|\sigma|}\int_{\mathfrak{S}} d\sigma(x) \\ &= m_{\kappa_o} - m_{\kappa_o} = 0. \end{aligned}$$

Hence it admits an extension into a homogeneous distribution by the first part of the proof. But this would imply that $|x|^\nu$ has such an extension and this is impossible by Example 3.2.23. □

Remark 3.2.25. (i) In view of the proof above, the hypothesis of continuity in Proposition 3.2.24 (and also in Proposition 3.2.27) can be relaxed into the following condition: κ_o is locally integrable and locally bounded on $G\backslash\{0\}$.

This ensures that all the computations make sense and, since the unit sphere of a given homogeneous quasi-norm is compact, $|\kappa_o|$ is bounded there.

We will not use this fact.

(ii) By Lemma 3.1.45 the condition $m_{\kappa_o} = 0$ is independent of the homogeneous quasi-norm. However, the distribution defined in (3.53) depends on the choice of a particular homogeneous quasi-norm. For instance, one can show that the function on $\mathbb{R}^2$ given in polar coordinates by

$$\kappa_o(re^{i\theta}) = \frac{\cos 4\theta}{r^2},$$

admits two different extensions κ via the procedure (3.53) when considering the Euclidean norm $(x, y) \mapsto (x^2 + y^2)^{1/2}$ and the ℓ^1-norm $(x, y) \mapsto |x| + |y|$.

Definition 3.2.26. The distribution given in (3.53) is called a *principal value distribution* denoted by

$$p.v.\ \kappa_o(x).$$

The notation is ambiguous unless a homogeneous norm is specified.

The next proposition states that, modulo a Dirac distribution at the origin, the only possible extension is the principal value distribution:

Proposition 3.2.27. *Let κ be a homogeneous distribution of degree ν with* $\operatorname{Re}\nu = -Q$ *on a homogeneous Lie group G. We assume that the restriction of κ to $G\backslash\{0\}$ coincides with a continuous function κ_o.*

Then κ_o is homogeneous of degree ν on $G\backslash\{0\}$ and $m_{\kappa_o} = 0$. Moreover, after the choice of a homogeneous norm,

$$\kappa(x) = p.v.\ \kappa_o(x) + c\delta_o,$$

for some constant $c \in \mathbb{C}$, with $c = 0$ if $\nu \neq -Q$.

Proof. By Proposition 3.2.24, $m_{\kappa_o} = 0$. Then

$$\kappa' := \kappa - p.v.\ \kappa_o$$

is also homogeneous of degree ν and supported at the origin.

Let $f \in \mathcal{D}(G)$ with $f(0) = 0$. Due to the compact support of κ', $|\langle \kappa', f\rangle|$ is controlled by some C^k norm of f on a fixed small neighbourhood of the origin. But, because of its homogeneity of degree ν with $\operatorname{Re}\nu = -Q$,

$$\forall t > 0 \qquad |\langle \kappa', f\rangle| = |\langle \kappa', f \circ D_t\rangle|.$$

Letting t tend to 0, the note that the C^k norms of $f \circ D_t$ remain bounded. Let us show that as $t \to 0$, we actually have $\langle \kappa', f \circ D_t\rangle \to 0$. We claim that $f \circ D_t \to 0$ in $C^k(U)$ for a neighbourhood U of 0. Indeed,

$$X^\alpha (f \circ D_t) = t^{[\alpha]} (X^\alpha f) \circ D_t \to 0 \quad \text{as } t \to 0,$$

provided that $\alpha \neq 0$. On the other hand, also $(f \circ D_t)(x) = f(tx) \to f(0) = 0$ as $t \to 0$, and same for the L^∞ norm over the set U. Thus, we have proved that $\langle \kappa', f\rangle = 0$ for any $f \in \mathcal{D}(G)$ vanishing at 0.

We now fix a function $\chi \in \mathcal{D}(G)$ with $\chi(0) = 1$. For any $f \in \mathcal{D}(G)$,

$$\langle \kappa', f\rangle = \langle \kappa', f - f(0)\chi\rangle + f(0)\langle \kappa', \chi\rangle = f(0)\langle \kappa', \chi\rangle,$$

since $f - f(0)\chi \in \mathcal{D}(G)$ vanishes at 0. This shows $\kappa' = c\delta_0$ where $c = \langle \kappa', \chi\rangle$. But δ_0 is homogeneous of degree $-Q$, see Example 3.1.20, whereas κ' is homogeneous of degree ν. So $c = 0$ if $\nu \neq -Q$.

Alternatively, we can also argue as follows. By Proposition 1.4.2 we must have

$$\kappa' = \kappa - p.v.\ \kappa_o = \sum_{|\alpha| \le j} a_\alpha \partial^\alpha \delta_0$$

for some j and some constants a_α. Now, we know by Example 3.1.20 that δ_0 is homogeneous of degree $-Q$, and by Proposition 3.1.23 that $\partial^\alpha \delta_0$ is homogeneous of degree $-Q - [\alpha]$. Since κ' is homogeneous of degree $-Q$, it follows that all $a_\alpha = 0$ for $-Q - [\alpha] \neq \nu$. The statement now follows since, if $\nu \neq -Q$, we must have all $a_\alpha = 0$, and if $\nu = -Q$, we take $c = a_0$. □

Using the vocabulary of kernels of type ν, see Definition 3.2.9, Proposition 3.2.24 implies easily:

Corollary 3.2.28. *Let G be a homogeneous Lie group and let κ_o be a smooth homogeneous function on $G \backslash \{0\}$ of degree ν with $\operatorname{Re}\nu = -Q$. Then κ_o extends to a homogeneous distribution in $\mathcal{D}'(G)$ if and only if its average value, defined in Lemmata 3.1.43 and 3.1.45, is $m_{\kappa_o} = 0$. In this case, the extension is a kernel of type ν.*

Remark 3.2.29. Remark 3.2.10 explained the correspondence between the kernels of type ν and their restriction to $G\backslash\{0\}$ in the case $\operatorname{Re}\nu \in (0, Q)$.

With Corollary 3.2.28, we obtain the case $\operatorname{Re}\nu = 0$: the restriction to $G\backslash\{0\}$ yields a correspondence between

- the $(\nu - Q)$-homogeneous functions in $C^\infty(G\backslash\{0\})$ with vanishing mean value
- and the kernels of type ν.

It is one-to-one if $\nu \neq 0$ but if $\nu = 0$, we have to consider the kernels of type ν modulo $\mathbb{C}\delta_0$.

3.2.5 Operators of type $\nu = 0$

We can now go back to our original motivation, that is, a condition on a left-invariant homogeneous operator of degree 0 to obtain continuity on every $L^p(G)$. Our condition here is that the operator is of type 0, or more generally of type ν, $\operatorname{Re}\nu = 0$.

Theorem 3.2.30. *Let G be a homogeneous Lie group and let $\nu \in \mathbb{C}$ with*

$$\operatorname{Re}\nu = 0.$$

Any operator of type ν on G is $(-\nu)$-homogeneous and extends to a bounded operator on $L^p(G)$, $p \in (1, \infty)$.

The proof consists in showing that the operator is Calderón-Zygmund (in the sense of Definition 3.2.15) and bounded on $L^2(G)$. Note that the cancellation condition (see Remark 3.2.18), is provided by $m_{\kappa_o} = 0$, see Proposition 3.2.27.

Proof. Let $\kappa \in \mathcal{D}'(G)$ be a kernel of type ν, $\operatorname{Re}\nu = 0$. We denote by κ_o its smooth restriction to $G\backslash\{0\}$. One checks easily that κ_o satisfies the hypotheses of Corollary 3.2.20. Consequently, κ_o is a Calderón-Zygmund kernel in the sense of Definition 3.2.15. By the Singular Integral Theorem, more precisely its form given in Proposition 3.2.17, to prove the L^p-boundedness for every $p \in (1, \infty)$, it suffices to prove the case $p = 2$.

Fixing a homogeneous norm $|\cdot|$ smooth away from the origin, by Proposition 3.2.27, we may assume that κ is the principal value distribution of κ_o (see Definition 3.2.26). We want to show that

$$f \mapsto f * p.v.\, \kappa_o$$

is bounded on $L^2(G)$. For this, we will apply the Cotlar-Stein lemma (see Theorem A.5.2) to the operators

$$T_j : f \mapsto f * K_j, \quad j \in \mathbb{Z},$$

where

$$K_j(x) = \kappa_o(x) 1_{2^{-j} \leq |x| \leq 2^{-j+1}}(x).$$

We claim that

$$\max\left(\|T_j^*T_k\|_{\mathscr{L}(L^2(G))}, \|T_jT_k^*\|_{\mathscr{L}(L^2(G))}\right) \leq C2^{-|j-k|}. \tag{3.54}$$

Assuming this claim, by the Cotlar-Stein lemma, $\sum_j T_j$ defines a bounded operator on $L^2(G)$ and its (right convolution) kernel is $\sum_j K_j$ which coincides, as a distribution, with $p.v.\,\kappa_o = \kappa$. This would conclude the proof.

Let us start to prove Claim (3.54). It is not difficult to see (see (3.47)) that the adjoint of the operator T_j on $L^2(G)$ is the convolution operator with right convolution kernel given by

$$K_j^*(x) = \bar{K}_j(x^{-1}),$$

which is compactly supported. Therefore, the operators $T_j^*T_k$ and $T_jT_k^*$ are convolution operators with kernels $K_k * K_j^*$ and $K_k^* * K_j$, respectively. We observe that, by homogeneity of κ_o, for any $j \in \mathbb{N}_0$,

$$|K_j(x)| = 2^{jQ}|K_0(2^jx)| \quad \text{and so} \quad \|K_j\|_{L^1(G)} = \|K_0\|_{L^1(G)}.$$

By the Young convolution inequality (see Proposition 1.5.2), the operators T_j, $T_j^*T_k$ and $T_jT_k^*$ are bounded on $L^2(G)$ with operator norms

$$\begin{aligned}
\|T_j\|_{\mathscr{L}(L^2(G))} &\leq \|K_j\|_1 = \|K_0\|_1,\\
\|T_j^*T_k\|_{\mathscr{L}(L^2(G))} &\leq \|K_k * K_j^*\|_1 \leq \|K_k\|_1\|K_j^*\|_1 = \|K_0\|_1^2,\\
\|T_jT_k^*\|_{\mathscr{L}(L^2(G))} &\leq \|K_k^* * K_j\|_1 \leq \|K_k^*\|_1\|K_j\|_1 = \|K_0\|_1^2.
\end{aligned}$$

In order to prove Claim (3.54) we need to obtain a better decay for $\|K_k * K_j^*\|_1$ and $\|K_k^* * K_j\|_1$ when j and k are 'far apart'. Since $\|K_k * K_j^*\|_1 = \|K_j * K_k^*\|_1$ and $\|K_k^* * K_j\|_1 = \|K_j^* * K_k\|_1$, we may assume $k > j$. Quantitatively we assume that $C_1 2^{j-k+1} < 1/2$ where $C_1 \geq 1$ is the constant appearing in (3.26) for $b = 1/2$.

We observe that the cancellation condition $m_{\kappa_o} = 0$ implies

$$\int_G K_k(x)dx = \int_{2^{-k}\leq|x|\leq 2^{-k+1}} \kappa_o(x)dx = m_{\kappa_o}\ln 2 = 0,$$

and so

$$\begin{aligned}
|K_k * K_j^*(x)| &= \left|\int_G K_k(y)K_j^*(y^{-1}x)dy\right| = \left|\int_G K_k(y)\left(K_j^*(y^{-1}x) - K_j^*(x)\right)dy\right|\\
&\leq \int_G |K_k(y)|\left|K_j^*(y^{-1}x) - K_j^*(x)\right|dy\\
&\leq \int_{2^{-k}\leq|y|\leq 2^{-k+1}} C_o|y|^{-Q}\left|K_j^*(y^{-1}x) - K_j^*(x)\right|dy,
\end{aligned}$$

where C_o is the maximum of $|\kappa_o|$ on the unit sphere $\{|x| = 1\}$. Thus after the change of variable $z = 2^k y$,

$$\left|K_k * K_j^*(x)\right| \leq \int_{1\leq|z|\leq 2} C_o|z|^{-Q} \left|K_j^*((2^{-k}z)^{-1}x) - K_j^*(x)\right| dz.$$

We want to estimate the L^1-norm with respect to x of the last expression. Hence we now look at

$$\int_G \left|K_j^*((2^{-k}z)^{-1}x) - K_j^*(x)\right| dx = \int_G \left|K_j\left(x_1\, 2^{-k}z\right) - K_j(x_1)\right| dx_1,$$

after the change of variable $x = x_1^{-1}$. Using $K_j = 2^{j\nu} K_0 \circ D_j$ and the change of variable $x_2 = 2^j x_1$, we obtain

$$\int_G \left|K_j\left(x_1\, 2^{-k}z\right) - K_j(x_1)\right| dx_1 = \int_G \left|K_0\left(x_2\, 2^{-k+j}z\right) - K_0(x_2)\right| dx_2.$$

Let $A_0 = \{1 \leq |x| \leq 2\}$ be the annulus with radii 1 and 2 around 0 and write momentarily $y^{-1} = 2^{-k+j}z$ with $z \in A_0$. We can write the last integral as

$$\int_G \left|K_0(xy^{-1}) - K_0(x)\right| dx = \int_{A_0\cap(A_0 y)} + \int_{A_0\setminus(A_0 y)} + \int_{(A_0 y)\setminus A_0}.$$

For the last two integrals, we see with a change of variable $x = x'y^{-1}$ that

$$\int_{A_0\setminus(A_0 y)} = \int_{A_0\setminus(A_0 y)} |K_0(x)|\, dx = \int_{(A_0 y)\setminus A_0} \left|K_0(x'y^{-1})\right| dx' = \int_{(A_0 y)\setminus A_0},$$

and

$$\int_{A_0\setminus(A_0 y)} |K_0| \leq \int_{\substack{|xy^{-1}|>2\\ 1\leq|x|\leq 2}} C_o|x|^{-Q} dx + \int_{\substack{|xy^{-1}|<1\\ 1\leq|x|\leq 2}} C_o|x|^{-Q} dx.$$

Thus

$$\begin{aligned}\int_G \left|K_0(xy^{-1}) - K_0(x)\right| dx &= \int_{A_0\cap(A_0 y)} |K_0(xy^{-1}) - K_0(x)| dx \\ &+2C_o\left(\int_{\substack{|xy^{-1}|>2\\ 1\leq|x|\leq 2}} |x|^{-Q} dx + \int_{\substack{|xy^{-1}|<1\\ 1\leq|x|\leq 2}} |x|^{-Q} dx\right).\end{aligned} \tag{3.55}$$

Since y^{-1} is relatively small, by (3.26) we get for the two integrals above

$$\begin{aligned}\int_{\substack{|xy^{-1}|>2\\ 1\leq|x|\leq 2}} + \int_{\substack{|xy^{-1}|<1\\ 1\leq|x|\leq 2}} &\leq \int_{2-C_1|y|<|x|\leq 2} + \int_{1\leq|x|<1+C_1|y|} \\ &= \ln\frac{2}{2 - C_1|y|} + \ln(1 + C_1|y|) \leq C|y|,\end{aligned}$$

(see Example 3.1.44), whereas by Proposition 3.1.40 we have for any $x \in A_0$,

$$\left|K_0(xy^{-1}) - K_0(x)\right| \le C|y|\ |x|^{-Q-1},$$

and so

$$\int_{A_0\cap(A_0y)} \left|K_0(xy^{-1}) - K_0(x)\right| dx \le C|y| \int_{1\le|x|\le 2} |x|^{-Q-1}dx \le C|y|.$$

We have obtained that the expression (3.55) is up to a constant less than 2^{-k+j} when $C_1 2^{j-k+1} < 1/2$ (and $y^{-1} = 2^{-k+j}z$, $z \in A_0$). This estimate gives

$$\begin{aligned} \|K_k * K_j^*\|_1 &\le C_o \int_{z\in A_0} |z|^{-Q} \int_G \left|K_0(x\,2^{-k+j}z) - K_0(x)\right| dx\ dz \\ &\le C_o \int_{z\in A_0} |z|^{-Q+1} C2^{-k+j}\ dz \le C2^{-k+j}. \end{aligned}$$

With a very minor modification, we can show in the same way that $\|K_k^* * K_j\|_1 \le C2^{-k+j}$.

This shows Claim (3.54) and concludes the proof of Theorem 3.2.30. □

Remark 3.2.31. In view of the proof, we can relax the smoothness condition in the hypotheses of Theorem 3.2.30: it suffices to assume that $\kappa_o \in C^1(G\backslash\{0\})$.

This ensures that we can apply Propositions 3.2.27 and 3.1.40 during the proof.

3.2.6 Properties of kernels of type ν, $\operatorname{Re}\nu \in [0, Q)$

The kernels and operators of type ν have been defined in Definition 3.2.9. Summarising results of the previous section, namely Corollary 3.2.11 for $\operatorname{Re}\nu \in (0, Q)$, and Theorem 3.2.30 for $\operatorname{Re}\nu = 0$, we can unite them as

Corollary 3.2.32. *Let G be a homogeneous Lie group and let $\nu \in \mathbb{C}$ with*

$$\operatorname{Re}\nu \in [0, Q).$$

Any operator of type ν on G is $(-\nu)$-homogeneous and extends to a bounded operator from $L^p(G)$ to $L^q(G)$ provided that

$$\frac{1}{p} - \frac{1}{q} = \frac{\operatorname{Re}\nu}{Q}, \qquad 1 < p \le q < \infty.$$

When considering kernels of type ν, we have regularly used the following property: if κ is a kernel of type ν then, fixing a homogeneous quasi-norm $|\cdot|$ on G, κ admits a maximum C_κ on the unit sphere $\{|x| = 1\}$, and by homogeneity we have

$$\forall x \in G\backslash\{0\} \qquad |\kappa(x)| \le C_\kappa |x|^{\operatorname{Re}\nu - Q}. \tag{3.56}$$

In particular, it is locally integrable if $\operatorname{Re}\nu > 0$ and defines a distribution on the whole group G in this case. In the case when $\operatorname{Re}\nu = 0$, by Proposition 3.2.27, κ also defines a distribution on G of the form

$$\kappa = p.v.\, \kappa_1 + c\delta_0,$$

where κ_1 is of type ν with vanishing average value and $c \in \mathbb{C}$ is a constant.

We can also deduce the type of a kernel from the following lemma:

Lemma 3.2.33. *Let κ be a kernel of type ν_κ with $\operatorname{Re}\nu_\kappa \in (0, Q)$. Let T be a homogeneous differential operator of homogeneous degree ν_T. If $\operatorname{Re}\nu_\kappa - \nu_T \in [0, Q)$ then $T\kappa$ defines a kernel of type $\nu_\kappa - \nu_T$.*

Proof. Clearly $T\kappa$ is a $(Q - \nu_\kappa + \nu_T)$-homogeneous distribution which coincides with a smooth function away from 0. □

Remark 3.2.34. We have obtained certain properties of convolution operators with kernels of type ν in Corollary 3.2.11 for $\operatorname{Re}\nu \in (0, Q)$, and in Theorem 3.2.30 for $\operatorname{Re}\nu = 0$. When composing two such types of operators, we have to deal with the convolution of two kernels and this is a problematic question in general. Indeed, the problems about convolving distributions on a non-compact Lie group are essentially the same as in the case of the abelian convolution on $\mathbb{R}^n$. The convolution $\tau_1 * \tau_2$ of two distributions $\tau_1, \tau_2 \in \mathcal{D}'(G)$ is well defined as a distribution provided that at most one of them has compact support, see Section 1.5. However, additional assumptions must be imposed in order to define convolutions of distributions with non-compact supports. Furthermore, the associative law

$$(\tau_1 * \tau_2) * \tau_3 = \tau_1 * (\tau_2 * \tau_3), \tag{3.57}$$

holds when at most one of the τ_j's has non-compact support, but not necessarily when only one of the τ_j's has compact support even if each convolution in (3.57) could have a meaning.

The following proposition establishes that there is no such pathology appearing when considering convolution with kernel of type ν with $\operatorname{Re}\nu \in [0, Q)$. This will be useful in the sequel.

Proposition 3.2.35. *Let G be a homogeneous Lie group.*

(i) Suppose $\nu \in \mathbb{C}$ with $0 \le \operatorname{Re}\nu < Q$, $p \ge 1$, $q > 1$, and $r \ge 1$ given by

$$\frac{1}{r} = \frac{1}{p} + \frac{1}{q} - \frac{\operatorname{Re}\nu}{Q} - 1.$$

If κ is a kernel of type ν, $f \in L^p(G)$, and $g \in L^q(G)$, then $f * (g * \kappa)$ and $(f * g) * \kappa$ are well defined as elements of $L^r(G)$, and they are equal.

(ii) *Suppose κ_1 is a kernel of type $\nu_1 \in \mathbb{C}$ with $\mathrm{Re}\,\nu_1 > 0$ and κ_2 is a kernel of type $\nu_2 \in \mathbb{C}$ with $\mathrm{Re}\,\nu_2 \geq 0$. We assume $\mathrm{Re}\,(\nu_1+\nu_2) < Q$. Then $\kappa_1 * \kappa_2$ is well defined as a kernel of type $\nu_1+\nu_2$. Moreover, if $f \in L^p(G)$ where*

$$1 < p < Q/(\mathrm{Re}\,(\nu_1+\nu_2))$$

then $(f*\kappa_1)*\kappa_2$ and $f*(\kappa_1*\kappa_2)$ belong to $L^q(G)$,

$$\frac{1}{q} = \frac{1}{p} - \frac{\mathrm{Re}\,(\nu_1+\nu_2)}{Q},$$

and they are equal.

Proof. Let us prove Part (i). By Corollary 3.2.11, Theorem 3.2.30 and Young's inequality (see Proposition 1.5.2), the mappings $(f,g) \mapsto f*(g*\kappa)$ and $(f,g) \mapsto (f*g)*\kappa$ are continuous from $L^p(G)\times L^q(G)$ to $L^r(G)$. They coincide when they have compact support, and hence in general.

Let us prove Part (ii). We fix a homogeneous quasi-norm $|\cdot|$ smooth away from the origin. We will use the general properties of kernels of type ν explained at the beginning of this section, especially estimate (3.56).

Let $x \neq 0$ be given. We can find $\epsilon > 0$ such that the balls

$$B(0,\epsilon) := \{y : |y| < \epsilon\} \quad \text{and} \quad B(x,\epsilon) := \{y : |xy^{-1}| < \epsilon\},$$

do not intersect. We note that these balls are different from those in Definition 3.1.33 (that are used throughout this book) but in this proof only, it will be more convenient for us to work with the balls defined as above.

If $\mathrm{Re}\,\nu_1$, $\mathrm{Re}\,\nu_2 > 0$, then both κ_1 and κ_2 are locally integrable and

$$\left|\kappa_1(xy^{-1})\kappa_2(y)\right| \leq C_{x,\epsilon} \begin{cases} |y|^{\mathrm{Re}\,\nu_2 - Q} & \text{for } y \in B(0,\epsilon), \\ |xy^{-1}|^{\mathrm{Re}\,\nu_1 - Q} & \text{for } y \in B(x,\epsilon), \\ O(|y|^{\mathrm{Re}\,(\nu_1+\nu_2)-2Q}) & y \notin B(0,\epsilon)\cup B(x,\epsilon). \end{cases}$$

Thus we can integrate $\kappa_1(xy^{-1})\kappa_2(y)$ against dy on $B(0,\epsilon)$, $B(x,\epsilon)$ and outside of $B(0,\epsilon)\cup B(x,\epsilon)$ to obtain the sum of three integrals absolutely convergent:

$$\left[\int_{y\in B(0,\epsilon)} + \int_{y\in B(x,\epsilon)} + \int_{\substack{|y|>\epsilon \\ |xy^{-1}|>\epsilon}}\right] \kappa_1(xy^{-1})\kappa_2(y)dy := \kappa(x).$$

This defines $\kappa(x)$ which is independent of ϵ small enough.

If $\mathrm{Re}\,\nu_2 = 0$, by Proposition 3.2.27, we may assume that κ_2 is the principal value of a homogeneous distribution with mean average 0 (see also Definition 3.2.26 and (3.53)). In this case, by smoothness of κ_1 away from 0 and Proposition 3.1.40,

$$\left|\left(\kappa_1(xy^{-1}) - \kappa_1(x)\right)\kappa_2(y)\right| \leq C_{x,\epsilon}|y|^{1-Q} \quad \text{for } y \in B(0,\epsilon),$$

and we obtain again the sum of three integrals absolutely convergent:

$$\int_{y\in B(0,\epsilon)} \left(\kappa_1(xy^{-1}) - \kappa_1(x)\right)\kappa_2(y)dy +$$
$$+ \left[\int_{y\in B(x,\epsilon)} + \int_{\substack{|y|>\epsilon \\ |xy^{-1}|>\epsilon}}\right] \kappa_1(xy^{-1})\kappa_2(y)dy =: \kappa(x).$$

This defines $\kappa(x)$ which is independent of ϵ small enough.

In both cases, we have defined a function κ on $G\backslash\{0\}$. A simple change of variables shows that κ is homogeneous of degree $\nu_1 + \nu_2 - Q$ (this is left to the reader interested in checking this fact).

Let us fix $\phi_1 \in \mathcal{D}(G)$ with $\phi_1 \equiv 1$ on $B(0, \epsilon/2)$ and $\phi_1 \equiv 0$ on the complement of $B(0, \epsilon)$. We fix again $x \neq 0$ and we set $\phi_2(y) = \phi_1(xy^{-1})$. Then ϕ_1 and ϕ_2 have disjoint supports and for $\text{Re}\,\nu_2 > 0$ it is easy to check that for $z \in B(x, \epsilon/2)$ we have $\kappa(z) = I_1 + I_2 + I_3$, where

$$\begin{aligned} I_1 &= \int_G \phi_1(y)\kappa_1(zy^{-1})\kappa_2(y)dy, \\ I_2 &= \int_G \phi_2(y)\kappa_1(zy^{-1})\kappa_2(y)dy = \int_G \phi_2(y^{-1}z)\kappa_1(y)\kappa_2(y^{-1}z)dy, \\ I_3 &= \int_G (1 - \phi_1(y) - \phi_2(y))\kappa_1(zy^{-1})\kappa_2(y)dy, \end{aligned}$$

with a similar formula for $\text{Re}\,\nu_2 = 0$. The integrands of I_1, I_2, and I_3 depend smoothly on z. Furthermore, one checks easily that their derivatives in z remains integrable. This shows that κ is smooth near each point $x \neq 0$. Since $\text{Re}\,(\nu_1+\nu_2) > 0$, κ is locally integrable on the whole group G. Hence the distribution $\kappa \in \mathcal{D}'(G)$ is a kernel of type $\nu_1 + \nu_2$.

We can check easily for $\phi \in \mathcal{D}(G)$,

$$\langle \kappa, \phi\rangle = \langle \kappa_1, \phi * \tilde{\kappa}_2\rangle = \langle \kappa_2, \tilde{\kappa}_1 * \phi\rangle.$$

So having (1.14) and (1.15) we define $\kappa_1 * \kappa_2 := \kappa$.

Let $f \in L^p(G)$ where $p > 1$ and

$$\frac{1}{q} = \frac{1}{p} - \frac{\text{Re}\,(\nu_1 + \nu_2)}{Q} > 0.$$

We observe that $(f * \kappa_1) * \kappa_2$ and $f * \kappa$ are in $L^q(G)$ by Corollary 3.2.11, Theorem 3.2.30, and Young's inequality (see Proposition 1.5.2). To complete the proof, it suffices to show that the distributions $(f * \kappa_1) * \kappa_2$ and $f * (\kappa_1 * \kappa_2)$ are equal. For this purpose, we write $\kappa_1 = \kappa_1^0 + \kappa_1^\infty$ with

$$\kappa_1^0 := \kappa_1\ 1_{|x|\leq 1} \quad \text{and} \quad \kappa_1^\infty := \kappa_1\ 1_{|x|>1}.$$

If $r = Q/(Q - \operatorname{Re}\nu_1)$ then $\kappa_1^0 \in L^{r-\epsilon}(G)$ and $\kappa_1^\infty \in L^{r+\epsilon}(G)$ for any $\epsilon > 0$. We take ϵ so small that $r - \epsilon > 1$ and

$$p^{-1} + (r+\epsilon)^{-1} - \operatorname{Re}\nu_2/Q - 1 > 0.$$

By Part (i), $(f * \kappa_1^0) * \kappa_2$ and $f * (\kappa_1^0 * \kappa_2)$ coincide as elements of $L^s(G)$ where

$$s^{-1} = p^{-1} + (r-\epsilon)^{-1} - \operatorname{Re}\nu_2/Q - 1.$$

And $(f * \kappa_1^\infty) * \kappa_2$ and $f * (\kappa_1^\infty * \kappa_2)$ coincide as elements of $L^t(G)$ where

$$t^{-1} = p^{-1} + (r+\epsilon)^{-1} - \operatorname{Re}\nu_2/Q - 1.$$

Thus $(f*\kappa_1)*\kappa_2$ and $f*\kappa$ coincide as elements of $L^s(G)$ and $L^t(G)$. This concludes the proof of Part (ii) and of Proposition 3.2.35. □

3.2.7 Fundamental solutions of homogeneous differential operators

On open sets or manifolds, general results about the existence of fundamental kernels of operators hold, see e.g. [Tre67, Theorems 52.1 and 52.2]. On a Lie group, we can study the case when the fundamental kernels are of the form $\kappa(x-y)$ in the abelian case and $\kappa(y^{-1}x)$ on a general Lie group, where κ is a distribution, often called a fundamental solution. It is sometimes possible and desirable to obtain the existence of such fundamental solutions for left or right invariant differential operators.

In this section, we first give a definition and two general statements valid on any connected Lie group, and then analyse in more detail the situation on homogeneous Lie groups.

Definition 3.2.36. Let L be a left-invariant differential operator on a connected Lie group G. A distribution κ in $\mathcal{D}'(G)$ is called a *(global) fundamental solution* of L if

$$L\kappa = \delta_0.$$

A distribution $\tilde{\kappa}$ on a neighbourhood Ω of 0 is called a *local fundamental solution* of L (at 0) if $L\tilde{\kappa} = \delta_0$ on Ω.

On $(\mathbb{R}^n, +)$, global fundamental solutions are often called *Green functions.*

Example 3.2.37. Fundamental solutions for the Laplacian $\Delta = \sum_j \partial_j^2$ on $\mathbb{R}^n$ are well-known

$$G(x) = \begin{cases} \frac{c_n}{|x|^{n-2}} + p(x) & \text{if} \quad n \geq 3 \\ c_2 \ln|x| + p(x) & \text{if} \quad n = 2 \\ x 1_{[0,\infty)}(x) + p(x) & \text{if} \quad n = 1 \end{cases}$$

where c_n is a (known) constant of n, p is any polynomial of degree ≤ 1, and $|\cdot|$ the Euclidean norm on $\mathbb{R}^n$.

Example 3.2.37 shows that fundamental solutions are not unique, unless some hypotheses, e.g. homogeneity (besides existence), are added.

Although, in practice, 'computing' fundamental solutions is usually difficult, they are useful and important objects.

Lemma 3.2.38. *Let L be a left-invariant differential operator with smooth coefficients on a connected Lie group G.*

1. *If L admits a fundamental solution κ, then for every distribution $u \in \mathcal{D}'(G)$ with compact support, the convolution $f = u * \kappa \in \mathcal{D}'(G)$ satisfies*
$$Lf = u$$
on G.
2. *An operator L admits a local fundamental solution if and only if it is locally solvable at every point.*

For the definition of locally solvability, see Definition A.1.4.

Proof. For the first statement,

$$L(u * \kappa) = u * L\kappa = u * \delta_0 = u.$$

For the second statement, if L is locally solvable, then at least at the origin, one can solve $L\tilde{\kappa} = \delta_0$ and this shows that L admits a local fundamental solution.

Conversely, let us assume that L admits a local fundamental solution $\tilde{\kappa}$ on the open neighbourhood Ω of 0. We can always find a function $\chi \in \mathcal{D}(\Omega)$ such that $\chi = 1$ on an open neighbourhood $\Omega_1 \subsetneq \Omega$ of 0; we define $\kappa_1 \in \mathcal{D}'(\Omega)$ by $\kappa_1 := \chi\tilde{\kappa}$ and view κ_1 also as a distribution with compact support. Then it is easy to check that $L\kappa_1 = \delta_0$ on Ω_1 but that

$$L\kappa_1 = \delta_0 + \Phi,$$

where Φ is a distribution whose support does not intersect Ω_1.

Let Ω_0 be an open neighbourhood of 0 such that

$$\Omega_0^{-1}\Omega_0 = \{x^{-1}y : x, y \in \Omega_0\} \subsetneq \Omega_1.$$

We can always find a function $\chi_1 \in \mathcal{D}(\Omega_0)$ which is equal to 1 on a neighbourhood $\Omega_0' \subsetneq \Omega_0$ of 0.

If now $u \in \mathcal{D}'(G)$, then the convolution $f = (\chi_1 u) * \kappa_1$ is well defined and

$$Lf = \chi_1 u + \chi_1 u * \Phi,$$

showing that $Lf = \chi_1 u$ on Ω_0 and hence $Lf = u$ on Ω_0'. Hence L is locally solvable at 0. By left-invariance, it is locally solvable at any point. □

Because of the duality between hypoellipticity and solvability, local fundamental solutions exist under the following condition:

Proposition 3.2.39. *Let L be a left-invariant hypoelliptic operator on a connected Lie group G. Then L^t is also left-invariant and it has a local fundamental solution.*

Proof. The first statement follows easily from the definition of L^t, and the second from the duality between solvability and hypoellipticity (cf. Theorem A.1.3) and Lemma 3.2.38. □

The next theorem describes some property of existence and uniqueness of global fundamental solutions in the context of homogeneous Lie groups.

Theorem 3.2.40. *Let L be a ν-homogeneous left-invariant differential operator on a homogeneous Lie group G. We assume that the operators L and L^t are hypoelliptic on a neighbourhood of 0. Then L admits a fundamental solution $\kappa \in \mathcal{S}'(G)$ satisfying:*

(a) if $\nu < Q$, the distribution κ is homogeneous of degree $\nu - Q$ and unique,

(b) if $\nu \geq Q$, $\kappa = \kappa_o + p(x)\ln|x|$ where

(i) $\kappa_o \in \mathcal{S}'(G)$ is a homogeneous distribution of degree $\nu - Q$, which is smooth away from 0,

(ii) p is a polynomial of degree $\nu - Q$ and,

(iii) $|\cdot|$ is any homogeneous quasi-norm, smooth away from the origin.

Necessarily κ is smooth on $G\backslash\{0\}$.

Remark 3.2.41. In case (a), the unique homogeneous fundamental solution is a kernel of type ν, with the uniqueness understood in the class of homogeneous distributions of degree $\nu - Q$. For case (b), Example 3.2.37 shows that one can not hope to always have a homogeneous fundamental solution.

The rest of this section is devoted to the proof of Theorem 3.2.40.

The proofs of Parts (a) and (b) as presented here mainly follow the original proofs of these results due to Folland in [Fol75] and Geller in [Gel83], respectively.

Proof of Theorem 3.2.40 Part (a). Let L be as in the statement and let $\nu < Q$. By Proposition 3.2.39, L admits a local fundamental solution at 0: there exist a neighbourhood Ω of 0 and a distribution $\tilde{\kappa} \in \mathcal{D}'(\Omega)$ such that $L\tilde{\kappa} = \delta_0$ on Ω. Note that by the hypoellipticity of L, $\tilde{\kappa}$ as well as any fundamental solution coincide with a smooth function away form 0. By shrinking Ω if necessary, we may assume that after having fixed a homogeneous quasi-norm, Ω is a ball around 0. So if $x \in \Omega$ and $r \in (0,1]$ then $rx \in \Omega$.

Folland observed that if κ exists then the distribution $h := \tilde{\kappa} - \kappa$ annihilates L on Ω, so it must be smooth on Ω, while

$$\kappa(x) = r^{Q-\nu}\tilde{\kappa}(rx) - r^{Q-\nu}h(rx)$$

yields
$$\kappa(x) = \lim_{r\to 0} r^{Q-\nu}\tilde{\kappa}(rx)$$
and
$$h(x) = \tilde{\kappa}(x) - \lim_{r\to 0} r^{Q-\nu}\tilde{\kappa}(rx).$$

Going back to our proof, Folland's idea was to define $h_r \in \mathcal{D}'(\Omega)$ by
$$h_r := \tilde{\kappa} - r^{Q-\nu}\tilde{\kappa}\circ D_r \quad \text{on } \Omega\backslash\{0\},\ r\in(0,1],$$
which makes sense in view of the smoothness of $\tilde{\kappa}$ on $\Omega\backslash\{0\}$. Since for any test function $\phi \in \mathcal{D}(\Omega)$,
$$\langle L(r^{Q-\nu}\tilde{\kappa}(r\,\cdot)),\phi\rangle = \langle r^Q(L\tilde{\kappa})(r\,\cdot)),\phi\rangle = \langle L\tilde{\kappa},\phi(r^{-1}\cdot)\rangle = \phi(r^{-1}0) = \phi(0),$$
we have $Lh_r = \delta_0 - \delta_0 = 0$. So h_r is in $N_L(\Omega) \subset C^\infty(\Omega)$ where the $\mathcal{D}'(\Omega)$ and $C^\infty(\Omega)$ topologies agree, see Theorem A.1.6. Let us show that
$$\exists \lim_{r\to 0} h_r \in h \in C^\infty(\Omega); \tag{3.58}$$
for this it suffices to show that $\{h_r\}$ is a Cauchy family in $\mathcal{D}'(\Omega)$.

We observe that if $s \le r$, we have
$$\begin{aligned} h_s(x) - h_r(x) &= r^{Q-\nu}\tilde{\kappa}(rx) - s^{Q-\nu}\tilde{\kappa}(sx) \\ &= r^{Q-\nu}\left(\tilde{\kappa}(rx) - \left(\frac{s}{r}\right)^{Q-\nu}\tilde{\kappa}\left(\frac{s}{r}rx\right)\right) \\ &= r^{Q-\nu}h_{\frac{s}{r}}(rx). \end{aligned} \tag{3.59}$$
In particular, setting $s = r^2$ in (3.59) we obtain
$$h_{r^2} = r^{Q-\nu}h_r\circ D_r + h_r.$$
This formula yields, first by substituting r by r^2,
$$\begin{aligned} h_{r^4} &= r^{2(Q-\nu)}h_{r^2}\circ D_{r^2} + h_{r^2} \\ &= r^{2(Q-\nu)}\left(r^{Q-\nu}h_r\circ D_r\circ D_{r^2} + h_r\circ D_{r^2}\right) + r^{Q-\nu}h_r\circ D_r + h_r \\ &= r^{3(Q-\nu)}h_r\circ D_{r^3} + r^{2(Q-\nu)}h_r\circ D_{r^2} + r^{Q-\nu}h_r\circ D_r + h_r. \end{aligned}$$
Continuing inductively, we obtain
$$h_{r^{2^\ell}} = \sum_{k=0}^{2^\ell-1} r^{k(Q-\nu)}h_r\circ D_{r^k}.$$
This implies
$$\forall n\in\mathbb{N}_0 \qquad \sup_{x\in(1-\epsilon)\Omega}|h_{r^{2^\ell}}(x)| \le (1-r^{Q-\nu})^{-1}\sup_{x\in(1-\epsilon)\Omega}|h_r(x)|,$$

and, since any $s \leq \frac{1}{2}$ can be expressed as $s = r^{2^\ell}$ for some $\ell \in \mathbb{N}_0$ and some $r \in [\frac{1}{4}, \frac{1}{2}]$,

$$\forall s \leq \frac{1}{2} \qquad \sup_{x \in (1-\epsilon)\Omega} |h_s(x)| \leq (1 - 2^{\nu - Q})^{-1} \sup_{\substack{x \in (1-\epsilon)\Omega \\ \frac{1}{4} \leq r \leq \frac{1}{2}}} |h_r(x)|.$$

Now the Schwartz-Treves lemma (see Theorem A.1.6) implies that the topologies of $\mathcal{D}'(\Omega)$ and $C^\infty(\Omega)$ on

$$N_L(\Omega) = \{f \in \mathcal{D}'(\Omega) \ : \ Tf = 0\} \subset C^\infty(\Omega)$$

coincide. Since $r \mapsto h_r$ is clearly continuous from $(0, 1]$ to $\mathcal{D}'(\Omega) \cap N_L(\Omega)$, $\{h_r,\, r \in [\frac{1}{4}, \frac{1}{2}]\}$ and $\{h_r,\, r \in [\frac{1}{2}, 1]\}$ are compact in $\mathcal{D}(\Omega)$. Therefore, we have

$$\begin{aligned} \sup_{\substack{x \in (1-\epsilon)\Omega \\ 0 < s \leq 1}} |h_s(x)| &\leq \sup_{\substack{x \in (1-\epsilon)\Omega \\ 0 < s \leq \frac{1}{2}}} |h_s(x)| + \sup_{\substack{x \in (1-\epsilon)\Omega \\ \frac{1}{2} \leq s \leq 1}} |h_s(x)| \\ &\leq (1 - 2^{\nu - Q})^{-1} \sup_{\substack{x \in (1-\epsilon)\Omega \\ \frac{1}{4} \leq r \leq \frac{1}{2}}} |h_r(x)| + \sup_{\substack{x \in (1-\epsilon)\Omega \\ \frac{1}{2} \leq s \leq 1}} |h_s(x)| = C_\epsilon < \infty, \end{aligned}$$

that is, the h_r's are uniformly bounded on $(1 - \epsilon)\Omega$. But if $s < r$, (3.59) implies

$$\sup_{x \in (1-\epsilon)\Omega} |h_s(x) - h_r(x)| \leq r^{Q-\nu} \sup_{x \in (1-\epsilon)\Omega} \left|h_{\frac{s}{r}}(rx)\right| \leq C_\epsilon r^{Q-\nu} \underset{r \to 0}{\longrightarrow} 0.$$

This shows that $\{h_r\}_{r \in (0,1]}$ is a Cauchy family of $C(K)$ for any compact subset K of Ω. Therefore, $\{h_r\}_{r \in (0,1]}$ is a Cauchy family of $\mathcal{D}'(\Omega)$ and Claim (3.58) is proved. Let $h \in C^\infty(\Omega)$ be the limit of $\{h_r\}$. Necessarily $Lh = 0$. We set

$$\kappa := \tilde{\kappa} - h \in \mathcal{D}'(\Omega).$$

Now, on one hand

$$L\kappa = L\tilde{\kappa} - Lh = \delta_0$$

and κ is smooth on $\Omega \backslash \{0\}$, and on the other,

$$\kappa(x) = \lim_{r \to 0} r^{Q-\nu} \tilde{\kappa}(rx),$$

so if $s \in (0, 1]$, then

$$\kappa(sx) = \lim_{r \to 0} r^{Q-\nu} \tilde{\kappa}(srx) = \lim_{r' = rs \to 0} \left(\frac{r'}{s}\right)^{Q-\nu} \tilde{\kappa}(r'x) = s^{\nu - Q} \kappa(x).$$

By requiring that the formula $\kappa(sx) = s^{\nu-Q}\kappa(x)$ holds for all $s > 0$ and $x \neq 0$, we can extend κ into a distribution defined on the whole space. The homogeneity of L guarantees that the equation $L\kappa = \delta_0$ holds globally.

Finally, if κ_1 were another fundamental solution of L satisfying (a), then $\kappa - \kappa_1$ would be $(\nu - Q)$-homogeneous with $\nu - Q < 0$; $\kappa - \kappa_1$ would also be smooth even at 0 since it annihilates L on G. Thus $\kappa - \kappa_1 = 0$. $\square$

Proof of Theorem 3.2.40 Part (b). Let L be as in the statement and let $\nu \geq Q$. Let also $\tilde{\kappa}$, Ω and h_r be defined as in the proof of part (a).

Geller noticed that Folland's idea could be adapted by taking higher order derivatives. Indeed from (3.59), we have

$$X^\alpha h_s(x) - X^\alpha h_r(x) = r^{Q-\nu+[\alpha]} X^\alpha h_{\frac{s}{r}}(rx);$$

if $\alpha \in \mathbb{N}_0^n$ is large so that $Q - \nu + [\alpha] > 0$, we can proceed as for h_r in the proof of Part (a) and obtain that $\{X^\alpha h_r\}_{r\in(0,1]}$ is a Cauchy family of $C^\infty(\Omega)$.

If $[\alpha] \leq \nu - Q$, the $C^\infty(\Omega)$-family $\{X^\alpha h_r\}_{r\in(0,1]}$ may not be Cauchy but by Taylor's theorem at the origin for homogeneous Lie groups, cf. Theorem 3.1.51,

$$\left| h_r(x) - P_{0,M}^{(h_r)}(x) \right| \leq C_M \sum_{\substack{|\alpha| \leq \lceil M \rfloor + 1 \\ [\alpha] > M}} |x|^{[\alpha]} \sup_{|z| \leq \eta^{\lceil M \rfloor + 1}|x|} |(X^\alpha h_r)(z)|,$$

for any x such that x and $\eta^{\lceil M \rfloor+1}x$ are in the ball Ω. Choosing $M = \nu - Q$ and setting the polynomial $p_r(x) := P_{0,M}^{(h_r)}(x)$ and the ball $\Omega' := \eta^{-(\lceil M \rfloor+1)}\Omega$, this shows that the $C^\infty(\Omega')$-family $\{h_r - p_r\}_{r\in(0,1]}$ is Cauchy. We set

$$C^\infty(\Omega') \ni h := \lim_{r\to 0}(h_r - p_r), \quad \kappa_o := \tilde{\kappa} - h \in \mathcal{D}(\Omega').$$

Note that $Lp_r = 0$, since the polynomial p_r is of degree $\nu - Q$ and the differential operator L is ν-homogeneous. Therefore, $L\kappa_o = \delta_0$ in Ω' and $\kappa_o \in C^\infty(\Omega'\backslash\{0\})$. Furthermore, if $[\alpha] > \nu - Q$ and $x \in \Omega'\backslash\{0\}$ then

$$\left(\frac{\partial}{\partial x}\right)^\alpha \kappa_o(x) = \lim_{r\to 0} r^{Q-\nu+[\alpha]} \left(\frac{\partial}{\partial x}\right)^\alpha \tilde{\kappa}(rx),$$

so if $s \in (0,1]$,

$$\begin{aligned} \left(\frac{\partial}{\partial x}\right)^\alpha \kappa_o(sx) &= \lim_{r\to 0} r^{Q-\nu+[\alpha]} \left(\frac{\partial}{\partial x}\right)^\alpha \tilde{\kappa}(rsx) \\ &= \lim_{r'=rs\to 0} \left(\frac{r'}{s}\right)^{Q-\nu+[\alpha]} \left(\frac{\partial}{\partial x}\right)^\alpha \tilde{\kappa}(r'x) = s^{\nu-Q-[\alpha]} \left(\frac{\partial}{\partial x}\right)^\alpha \kappa(x). \end{aligned}$$

One could describe this property as $\left(\frac{\partial}{\partial x}\right)^\alpha \kappa_o$ being homogeneous on $\Omega'\backslash\{0\}$. We conclude the proof by applying Lemma 3.2.42 below. □

In order to state Lemma 3.2.42, we first define the set $\mathcal{W}$ of all the possible homogeneous degrees $[\alpha]$, $\alpha \in \mathbb{N}_0^n$,

$$\mathcal{W} := \{\upsilon_1\alpha_1 + \ldots + \upsilon_n\alpha_n \ : \ \alpha_1, \ldots, \alpha_n \in \mathbb{N}_0\}. \tag{3.60}$$

In other words, $\mathcal{W}$ is the additive semi-group of $\mathbb{R}$ generated by 0 and $\mathcal{W}_A$.

For instance, in the abelian case $(\mathbb{R}^n, +)$ or on the Heisenberg group $\mathbb{H}_{n_o}$, with our conventions, $\mathcal{W} = \mathbb{N}_0$. This is also the case for a stratified Lie group or for a graded Lie group with $\mathfrak{g}_1$ non-trivial.

Lemma 3.2.42. *Let B be an open ball around the origin of a homogeneous Lie group G equipped with a smooth homogeneous quasi-norm $|\cdot|$. We consider the sets of functions $\mathcal{K}^\nu$ defined by*

$$\begin{aligned} &\text{if } \nu \in \mathbb{R}\backslash\mathcal{W} \quad \mathcal{K}^\nu := \{f \in C^\infty(B\backslash\{0\}) \ : \ f \text{ is } \nu\text{-homogeneous}\}, \\ &\text{if } \nu \in \mathcal{W} \quad \mathcal{K}^\nu := \{f \in C^\infty(B\backslash\{0\}) \ : \ f = f_1 + p(x)\ln|x|, \\ &\qquad \text{where } f_1 \text{ is } \nu\text{-homogeneous and } p \text{ is a } \nu\text{-homogeneous polynomial}\}, \end{aligned}$$

where $\mathcal{W}$ was defined in (3.60), and we say that a function f on B or $B\backslash\{0\}$ is ν-homogeneous when $f \circ D_s = s^\nu f$ on B for all $s \in (0,1)$.

For any $\nu \in \mathbb{R}$ and $f \in C^\infty(B\backslash\{0\})$, if $\left(\frac{\partial}{\partial x}\right)^\alpha f \in \mathcal{K}^{\nu-[\alpha]}$ with $[\alpha] > \nu$, then there exists $p \in \mathcal{P}_{<\nu}$ such that $f - p \in \mathcal{K}^\nu$.

Recall (see Definition 3.1.26) that $\mathcal{P}_{<M}$ denotes the set of polynomials P on G such that $D^\circ P < M$. It is empty if $M < 0$.

Proof of Lemma 3.2.42. By induction it suffices to prove that for any $\nu \in \mathbb{R}$ and $f \in C^\infty(B\backslash\{0\})$,

$$\begin{aligned} \frac{\partial(f - p_j)}{\partial x_j} \in \mathcal{K}^{\nu-\upsilon_j} \text{ with } p_j \in \mathcal{P}_{<\nu-\upsilon_j} \text{ for all } j = 1,\ldots,n \\ \Longrightarrow f - p \in \mathcal{K}^\nu \text{ for some } p \in \mathcal{P}_{<\nu}. \end{aligned} \tag{3.61}$$

To prove (3.61), we start by showing that for any $f \in C^\infty(B\backslash\{0\})$,

$$\frac{\partial f}{\partial x_j} \in \mathcal{K}^{\nu-\upsilon_j} \text{ for all } j = 1,\ldots,n \Longrightarrow f - c \in \mathcal{K}^\nu \text{ for some } c \in \mathbb{C}. \tag{3.62}$$

By convention (see Definition 3.1.26), a homogeneous polynomial of homogeneous degree which is not in $\mathcal{W}$ is 0. With this in mind we continue the proof of (3.62) in a unified way. We consider $f \in C^\infty(B\backslash\{0\})$ satisfying the hypothesis of (3.62): for each $j = 1,\ldots,n$, $\frac{\partial f}{\partial x_j} \in \mathcal{K}^{\nu-\upsilon_j}$ and there exists $p_j \in \mathcal{P}_{=\nu-\upsilon_j}$ such that $f - p_j \ln|\cdot|$ is a ν-homogeneous function on $\backslash\{0\}$. We define

$$A(r,x) := f(rx) - r^\nu f(x), \quad x \in B, \ r \in (0,1].$$

We see that

$$\begin{aligned} \frac{\partial A(r,x)}{\partial x_j} &= r^{\upsilon_j}\frac{\partial f}{\partial x_j}(rx) - r^\nu \frac{\partial f}{\partial x_j}(x) \\ &= r^{\upsilon_j} p_j(rx)\ln|rx| - r^\nu p_j(x)\ln|x| = r^\nu p_j(x)\ln r. \end{aligned}$$

Note that for any j,k we have

$$\frac{\partial p_j}{\partial x_k} = \frac{\partial p_k}{\partial x_j} \quad \text{since} \quad \frac{\partial}{\partial x_k}\frac{\partial}{\partial x_j}A(r,x) = \frac{\partial}{\partial x_j}\frac{\partial}{\partial x_k}A(r,x).$$

Because of this observation we can adapt the proof of the Poincaré Lemma to construct the polynomial

$$q(x) := c \sum_{k=1}^{n} \upsilon_k x_k p_k(x), \tag{3.63}$$

which is ν-homogeneous and satisfies

$$\begin{aligned}
\frac{\partial q}{\partial x_j} &= c \sum_{k=1}^{n} \upsilon_k x_k \frac{\partial p_k(x)}{\partial x_j} + c\upsilon_j p_j(x) = c \sum_{k=1}^{n} \upsilon_k x_k \frac{\partial p_j(x)}{\partial x_k} + c\upsilon_j p_j(x) \\
&= c\partial_{t=1} (p_j(tx)) + c\upsilon_j p_j(x) = c(\nu - \upsilon_j) p_j(x) + c\upsilon_j p_j(x) \\
&= p_j(x),
\end{aligned}$$

by choosing $c = \nu^{-1}$ if $\nu \neq 0$; if $\nu = 0$, the polynomials p_j and q are zero. So we have

$$\frac{\partial}{\partial x_j} (A(r,x) - q(x) r^\nu \ln r) = 0 \quad \text{for all } j = 1, \ldots, n.$$

Therefore,

$$A(r,x) = q(x) r^\nu \ln r + a(r) \quad \text{for some } a \in C^\infty((0,1]).$$

Replacing f by $f - (r^\nu \ln r)\, q$ we may assume that $q = 0$ in all the cases, so that

$$\forall r \in (0,1], \ x \in B \quad f(rx) - r^\nu f(x) = a(r). \tag{3.64}$$

Now if $0 < r, s < 1$, then using the formula just above twice, we get

$$\begin{aligned}
a(rs) &= f(rsx) - (rs)^\nu f(x) = a(r) + r^\nu f(sx) - (rs)^\nu f(x) \\
&= a(r) + r^\nu (a(s) + s^\nu f(x)) - (rs)^\nu f(x) \\
&= a(r) + r^\nu a(s).
\end{aligned}$$

Solving this functional equation and setting

$$f_o(x) := f(x) - a(|x|) \qquad (x \in G \backslash \{0\}),$$

for a particular solution a, we check easily that f_o is ν-homogeneous:

- If $\nu = 0$, then a satisfies the functional equation

  $$a(rs) = a(r) + a(s)$$

  and must, therefore, be of the form $a(r) = C \ln(r)$ for some constant $C \in \mathbb{C}$. Using (3.64) we obtain

  $$f_o(rx) = f(rx) - a(|rx|) = f(x) + a(r) - a(|rx|) = f(x) - C \ln |x| = f_o(x).$$

- If $\nu \neq 0$, then a satisfies the functional equation

$$a(r) + r^\nu a(s) = a(s) + s^\nu a(r)$$

and must therefore be of the form $a(r) = C(1-r^\nu)$ for some constant $C \in \mathbb{C}$. Using (3.64) we obtain

$$\begin{aligned} f_o(rx) &= f(rx) - C(1-|rx|^\nu) = r^\nu f(x) + C(1-r^\nu) - C(1-|rx|^\nu) \\ &= r^\nu \left(f(x) - C(1-|x|^\nu)\right) = r^\nu f_o(x). \end{aligned}$$

Hence (3.62) is proved and we can now go back to showing the main claim, that is, the one given in (3.61). Let f and p_j be as in the hypotheses of (3.61).

First we see that if $\nu < 0$, then all the polynomials p_j are zero and, inspired by the construction of q above, we check easily that

$$\frac{\partial}{\partial x_j}\left(\nu^{-1}\sum_{k=1}^n \upsilon_k x_k \frac{\partial f}{\partial x_k}\right) = \frac{\partial f}{\partial x_j},$$

thus f and $\nu^{-1}\sum_{k=1}^n \upsilon_k x_k \frac{\partial f}{\partial x_k}$ must coincide so (3.61) is proved in this case.

Let us assume $\nu \geq 0$. We claim that

$$\forall j,k = 1,\dots,n \qquad \frac{\partial p_k}{\partial x_j} = \frac{\partial p_j}{\partial x_k}. \tag{3.65}$$

This is certainly true if $\nu - \upsilon_j - \upsilon_k < 0$ since both are zero in this case. If instead $\nu - \upsilon_j - \upsilon_k \geq 0$ then the polynomial

$$\frac{\partial p_k}{\partial x_j} - \frac{\partial p_j}{\partial x_k} = \frac{\partial}{\partial x_j}\left(p_k - \frac{\partial f}{\partial x_k}\right) - \frac{\partial}{\partial x_k}\left(p_j - \frac{\partial f}{\partial x_j}\right),$$

is in $\mathcal{K}^{\nu-\upsilon_j-\upsilon_k}$ and thus must be zero. Indeed if a polynomial p is in some $\mathcal{K}^a$ then either $a \notin \mathcal{W}$ and then $p = 0$, or $a \in \mathcal{W}$ and $p(rx)$ is a polynomial in r of degree $\leq a$ with $r^{-a}p(rx)$ unbounded unless $p = 0$; in both cases, $p = 0$.

Therefore, we can construct q as above by (3.63) so that $\frac{\partial q}{\partial x_j} = p_j$. Then

$$\frac{\partial(f-q)}{\partial x_j} = \frac{\partial f}{\partial x_j} - p_j \in \mathcal{K}^{\nu-\upsilon_j} \text{ for all } j = 1,\dots,n,$$

so $f - q \in \mathcal{K}^\nu$ by (3.62).

This concludes the proof of Claim (3.61) and of Lemma 3.2.42. □

Remark 3.2.43. The class of functions $\mathcal{K}^\nu$ defined in Lemma 3.2.42 is also used in the definition of the calculus by Christ et al. [CGGP92].

As an application of Theorem 3.2.40, let us extend Liouville's Theorem to homogeneous Lie groups.

3.2.8 Liouville's theorem on homogeneous Lie groups

Let us consider the following statement and proof of Liouville's Theorem in $\mathbb{R}^n$:

Theorem 3.2.44 (Liouville). *Every harmonic tempered distribution is a polynomial.*

This means that if $f \in \mathcal{S}'(\mathbb{R}^n)$ and $\Delta f = 0$ in the sense of distributions where Δ is the canonical Laplacian, then f is a polynomial on $\mathbb{R}^n$.

Proof. Let $f \in \mathcal{S}'(\mathbb{R}^n)$ with $\Delta f = 0$. Then $|\xi|^2 \widehat{f} = 0$ where $\widehat{f}$ is the Euclidean Fourier transform of $f \in \mathcal{S}'(\mathbb{R}^n)$ on $\mathbb{R}^n$. Hence the distribution $\widehat{f}$ is supported at the origin and must be a linear combination of derivatives of the Dirac distribution at 0, see Proposition 1.4.2. Consequently f is a polynomial. □

Liouville's Theorem and its proof given above are also valid for any homogeneous elliptic constant-coefficient differential operator on $\mathbb{R}^n$. We now show the following generalisation for homogeneous Lie groups:

Theorem 3.2.45 (Liouville theorem on homogeneous Lie groups). *Let L be a homogeneous left-invariant differential operator on a homogeneous Lie group G. We assume that L and L^t are hypoelliptic on G. If the distribution $f \in \mathcal{S}'(G)$ satisfies $Lf = 0$ then f is a polynomial.*

The rest of this section is devoted to the proof of Theorem 3.2.45. We follow the proof given by Geller in [Gel83].

Let $\widehat{\cdot}$ denote the Euclidean Fourier transform on $\mathbb{R}^n$ (cf. (2.25)). In view of the proof of Theorem 3.2.44, we want to show that the distribution $\widehat{f}$ is supported at 0. For this purpose, it suffices to show that any test function $\phi \in \mathcal{S}(G)$ whose Euclidean Fourier transform is supported away from 0, that is, $\operatorname{supp} \widehat{\phi} \not\ni 0$, can be written as $L^t \psi$ for some $\psi \in \mathcal{S}(G)$. Indeed, denoting momentarily $\iota(x) = -x$ for $x \in G$ identified with $\mathbb{R}^n$, and by $\check{\cdot}$ the inverse Fourier transform on $\mathbb{R}^n$, we have $\check{\phi} = \widehat{\phi} \circ \iota$, so that $\operatorname{supp} \check{\phi} = \operatorname{supp} \widehat{\phi}$, and

$$\langle \widehat{f}, \check{\phi} \rangle = \langle f, \phi \rangle = \langle f, L^t \psi \rangle = \langle Lf, \psi \rangle = 0.$$

The set of functions ϕ with $0 \notin \operatorname{supp} \widehat{\phi}$ is contained in

$$\mathcal{S}_o(\mathbb{R}^n) := \left\{ \phi \in \mathcal{S}(\mathbb{R}^n) \ : \ \left(\frac{\partial}{\partial \xi} \right)^\alpha \widehat{\phi}(0) = 0, \ \forall \alpha \in \mathbb{N}_0^n \right\}.$$

We observe that the space $\mathcal{S}_o(\mathbb{R}^n)$ can be also described in terms of the group structure using the identification of G with $\mathbb{R}^n$, as

$$\mathcal{S}_o(\mathbb{R}^n) = \mathcal{S}_o(G) = \{ \phi \in \mathcal{S}(G) \ : \ \int_G x^\alpha \phi(x) dx = 0, \ \forall \alpha \in \mathbb{N}_0^n \}.$$

Indeed $\int_{\mathbb{R}^n} x^\alpha \phi(x) dx = c_\alpha (\frac{\partial}{\partial \xi})^\alpha \widehat{\phi}(0)$ with c_α a known non-zero constant. Here dx denotes the Lebesgue measure on $\mathbb{R}^n$ and the Haar measure on G since these two measures coincide via the identification of G with $\mathbb{R}^n$.

By Theorem 3.2.40, the operator L^t has a fundamental solution $\kappa \in \mathcal{S}'(G)$ satisfying Part (a) or (b) of the statement. Thus we need only showing that for any $\phi \in \mathcal{S}_o(G)$, the function $\psi := \phi * \kappa$ is not only smooth (cf. Lemma 3.1.55) but also Schwartz. This is done in the following lemma:

Lemma 3.2.46. *If $\phi \in \mathcal{S}_o(G)$ is a Schwartz function and $\kappa \in \mathcal{S}'(G)$ is a homogeneous distribution smooth away from the origin or a distribution of the form $\kappa = p(x)\ln|x|$ where p is a polynomial and $|\cdot|$ a homogeneous quasi-norm smooth away from the origin, then $\phi * \kappa \in \mathcal{S}(G)$.*

The end of this section is devoted to the proof of Lemma 3.2.46; this relies on consequences of the following versions of Hadamard's Lemma for $\mathcal{S}(\mathbb{R}^n)$ and $\mathcal{S}_o(\mathbb{R}^n)$:

Lemma 3.2.47 (Hadamard). *Let $f \in \mathcal{S}(\mathbb{R}^n)$ with $\int f = 0$. Then f can be written as*

$$f = \sum_{j=1}^n \frac{\partial f_j}{\partial x_j} \qquad \text{with} \quad f_j \in \mathcal{S}(\mathbb{R}^n)$$

In addition, if $f \in \mathcal{S}_o(\mathbb{R}^n)$, each function f_j can be also taken in $\mathcal{S}_o(\mathbb{R}^n)$.

Proof of Lemma 3.2.47. We fix $\chi_o \in \mathcal{D}(\mathbb{R}^n)$ such that $\chi_o(\xi) = 1$ if $|\xi| \leq 1$ and $\chi_o(\xi) = 0$ if $|\xi| > 2$. Since $\int f = 0$ we have $\widehat{f}(0) = 0$ and

$$\widehat{f}(\xi) = \chi_o\widehat{f} + (1-\chi_o)\widehat{f} = (\chi_o\widehat{f}) - (\chi_o\widehat{f})(0) + (1-\chi_o)\widehat{f}.$$

We can write

$$(\chi_o\widehat{f})(\xi) - (\chi_o\widehat{f})(0) = \int_0^1 \partial_t\left(\left(\chi_o\widehat{f}\right)(t\xi)\right)dt = \sum_{j=1}^n \xi_j \int_0^1 \frac{\partial(\chi_o\widehat{f})}{\partial \xi_j}(t\xi)dt,$$

and

$$(1-\chi_o)\widehat{f}(\xi) = \sum_{j=1}^n \xi_j^2 \frac{1-\chi_o(\xi)}{|\xi|^2}\widehat{f}(\xi) \quad (\text{here } |\xi|^2 = \sum_{j=1}^n \xi_j^2).$$

We set

$$h_j(\xi) := \int_0^1 \frac{\partial(\chi_o\widehat{f})}{\partial \xi_j}(t\xi)dt + \xi_j \frac{1-\chi_o(\xi)}{|\xi|^2}\widehat{f}(\xi).$$

The first term is compactly supported (in the ball of radius 2), whereas the second one is well defined and is identically 0 on the unit ball. Since both terms are smooth, $h_j \in \mathcal{S}(\mathbb{R}^n)$. We have obtained $\widehat{f} = \sum_j \xi_j h_j$. We define $f_j \in \mathcal{S}(\mathbb{R}^n)$ such that $\widehat{f_j} = c_j h_j$ where the constant c_j is such that $\widehat{\partial_j} = c_j\xi_j$. Hence $f = \sum_j \frac{\partial f_j}{\partial x_j}$.

Moreover, since

$$\left(\frac{\partial}{\partial x}\right)^\alpha h_j(0) = \left(\frac{\partial}{\partial x}\right)^\alpha \frac{\partial}{\partial \xi_j}\widehat{f}(0),$$

we see that if $f \in \mathcal{S}_o(\mathbb{R}^n)$ then $f_j \in \mathcal{S}_o(\mathbb{R}^n)$. □

We will use the following consequence of Lemma 3.2.47 (in fact only the second point):

Corollary 3.2.48. • *If $f \in \mathcal{S}_o(\mathbb{R}^n)$, then for any $M \in \mathbb{N}_0$,*

$$f = \sum_{|\alpha|=M} \left(\frac{\partial}{\partial x}\right)^\alpha f_\alpha \qquad \textit{with} \quad f_\alpha \in \mathcal{S}_o(\mathbb{R}^n).$$

- *If $f \in \mathcal{S}_o(G)$ where G is a homogeneous Lie groups, then for any $M \geq 1$, we can write f as a finite sum*

$$f = \sum_{[\alpha]>M} X^\alpha f_\alpha$$

with $f_\alpha \in \mathcal{S}_o(G)$.

Proof of Corollary 3.2.48. Both points are obtained recursively, the first one from Lemma 3.2.47 and the second from the following observation: if $f \in \mathcal{S}_o(G)$, there exists $g_j \in \mathcal{S}_o(G)$ such that $f = \sum_{j=1}^n X_j g_j$. Indeed writing f as in Lemma 3.2.47 and using (3.17) with Remark 3.1.29 (1), we set

$$g_j := f_j + \sum_{\substack{1\leq k\leq n\\ v_j<v_k}} (p_{j,k} f_j)$$

and we see that $g_j \in \mathcal{S}_o(G)$. □

We can now prove Lemma 3.2.46.

Proof of Lemma 3.2.46. Let κ be a distribution as in the statement. We can always decompose κ as the sum of $\kappa_0 + \kappa_\infty$, where κ_0 has compact support and κ_∞ is smooth. Indeed, let $\chi \in \mathcal{D}(G)$ be identically 1 on a neighbourhood of the origin and define κ_0 by

$$\langle \kappa_0, \phi\rangle := \langle \kappa, \chi\phi\rangle.$$

Then

$$\kappa_\infty := \kappa - \kappa_0$$

coincides with $(1-\chi)\kappa_o$, where κ_o is a smooth function on $G\backslash\{0\}$ either homogeneous or of the form $p(x)\ln|x|$; we denote by ν the homogeneous degree of the function κ_o or of the polynomial p.

Let $\phi \in \mathcal{S}_o(G)$. Since the distribution κ_0 is compactly supported, we get, by Lemma 3.1.55, that $\phi * \kappa_0 \in \mathcal{S}(G)$. Since, by Corollary 3.2.48, we can write ϕ as a (finite) linear combination of $X^\alpha f$ with $f \in \mathcal{S}_o(G)$ and $[\alpha]$ as large as we want. We observe that

$$(X^\alpha f) * \kappa_\infty = f * \tilde{X}^\alpha \kappa_\infty$$

and that for $[\alpha]$ larger that $|\nu| + N + 1$ for $N \in \mathbb{N}_0$ fixed, we have

$$|\tilde{X}^\alpha \kappa_\infty(x)| \le C_N(1 + |x|)^{-N}.$$

Thus

$$\begin{aligned} |(X^\alpha f) * \kappa_\infty(x)| &= |f * \tilde{X}^\alpha \kappa_\infty(x)| = \left| \int_G f(y) \tilde{X}^\alpha \kappa_\infty(y^{-1}x) dy \right| \\ &\le \int_G |f(y)| C_N (1 + |y^{-1}x|)^{-N} dy \\ &\le C_N C_o^N (1 + |x|)^{-N} \int_G |f(y)| (1 + |y|)^N dy, \end{aligned}$$

by (3.43). This shows that $\phi * \kappa_\infty \in \mathcal{S}(G)$. □

Hence Lemma 3.2.46 and Theorem 3.2.45 are proved.

Chapter 4

Rockland operators and Sobolev spaces

In this chapter, we study a special type of operators: the (homogeneous) Rockland operators. These operators can be viewed as a generalisation of sub-Laplacians to the non-stratified but still homogeneous (graded) setting. The terminology comes from a property conjectured by Rockland and eventually proved by Helffer and Nourrigat in [HN79], see Section 4.1.3.

First, we discuss these operators in general. Subsequently, we concentrate on positive Rockland operators and study the heat semi-group, the Bessel and Riesz potentials and the Sobolev spaces naturally associated with a positive Rockland operator. Most results concerning the heat semi-group are known [FS82, ch.3.B]. To the authors' knowledge, however, this chapter is the first systematic presentation of the fractional powers and the homogeneous and inhomogeneous Sobolev spaces associated with a positive Rockland operator on a graded Lie group.

In fact, this appears to be the greatest generality for such constructions, since the existence of a Rockland (differential) operator on a homogeneous Lie group implies that the group must admit a graded structure, see Proposition 4.1.3. In the case of stratified Lie groups, Sobolev spaces have been developed by Folland [Fol75] for $1 < p < \infty$, for the Rockland operator being a sub-Laplacian (see also [Sak79]). Since sub-Laplacians are not always available on graded Lie groups, our constructions are based on general positive Rockland operators. In particular, this allows one to still cover the case of stratified Lie groups, but permitting taking Rockland operators other than a canonical sub-Laplacian.

Although we define Sobolev spaces using a fixed Rockland operator, Theorem 4.4.20 shows that these spaces are actually independent of the choice of a homogeneous positive Rockland operator.

V. Fischer, M. Ruzhansky, *Quantization on Nilpotent Lie Groups*, Progress in Mathematics 314,
DOI 10.1007/978-3-319-29558-9_4

4.1 Rockland operators

We start with the discussion of general Rockland operators, giving definitions, examples, and then relating them to the hypoellipticity questions.

4.1.1 Definition of Rockland operators

The first definition of a Rockland operator uses the representations of the group. We use the notation which has become quite conventional nowadays in this part of the theory of group representations and which is explained in Section 1.7. In particular, $\widehat{G}$ denotes the unitary dual of G and $\mathcal{H}_\pi^\infty$ the smooth vectors of a representation $\pi \in \widehat{G}$, see Definition 1.7.2. For a left-invariant differential operator T we will denote $\pi(T) := d\pi(T)$, see Definition 1.7.4.

Definition 4.1.1. Let T be a left-invariant differential operator on a Lie group G. Then T satisfies the *Rockland condition* when

(R) for each representation $\pi \in \widehat{G}$, except for the trivial representation, the operator $\pi(T)$ is injective on $\mathcal{H}_\pi^\infty$, that is,

$$\forall v \in \mathcal{H}_\pi^\infty \qquad \pi(T)v = 0 \implies v = 0.$$

There is a similar definition of the Rockland condition for right-invariant differential operators, and also for left or right-invariant $L^2(G)$-bounded operators (for the latter, see Głowacki [Gło89, Gło91]). See also Section 4.4.8.

Definition 4.1.2. Let G be a homogeneous Lie group. A *Rockland operator* $\mathcal{R}$ on G is a left-invariant differential operator which is homogeneous of positive degree and satisfies the Rockland condition.

Some other authors may define non-homogeneous Rockland operators as operators of the form $\mathcal{R} = \sum_{[\alpha]\leq\nu} c_\alpha X^\alpha$ with the 'main' term $\sum_{[\alpha]=\nu} c_\alpha X^\alpha$ satisfying the Rockland property given in (R). Here we have chosen to assume that a Rockland operator is homogeneous to study directly the main term.

We will give examples of Rockland operators in Section 4.1.2. Before this, we show that their existence on a homogeneous Lie group implies that the group is graded and that the weights could be chosen in $\mathbb{N}$. This property influences the examples we can produce, and the subsequent development of the theory of pseudo-differential operators.

Proposition 4.1.3. *Let G be a homogeneous Lie group. If there exists a Rockland operator on G then the group G is graded.*

Furthermore, the dilations' weights $\upsilon_1, \ldots, \upsilon_n$ satisfy

$$a_1\upsilon_1 = \ldots = a_n\upsilon_n$$

for some integers $a_1, \ldots, a_n$.

This property was shown by Miller in [Mil80], with a small gap in the proof later corrected by ter Elst and Robinson (see [tER97]).

Proof of Proposition 4.1.3. Let G be a homogeneous Lie group. Its Lie algebra $\mathfrak{g}$ is endowed with the dilations $D_r = \text{Exp}(\ln r A)$. Let the number n' and $\{X_1, \ldots, X_n\}$ be the basis described in Lemma 3.1.14. We assume that there exists a ν-homogeneous Rockland operator $\mathcal{R}$ which we can write as

$$\mathcal{R} = \sum_{[\alpha]=\nu} c_\alpha X^\alpha.$$

We fix an integer $j \leq n'$. Let $\phi : \mathfrak{g} \to \mathbb{R}$ be the linear functional such that $\phi(X_k) = \delta_{j,k}$, that is, $\phi(X_j) = 1$ while $\phi(X_k) = 0$ for any $k \neq j$. Since $X_j \notin [\mathfrak{g}, \mathfrak{g}]$, ϕ is identically zero on $[\mathfrak{g}, \mathfrak{g}]$. We set for any $X \in \mathfrak{g}$:

$$\pi(\exp_G X) := \exp\left(i\phi(X)\right).$$

This defines a one-dimensional representation π of G. Indeed, if $x, y \in G$, we can write $x = \exp_G X$ and $y = \exp_G Y$ and we have

$$xy = \exp_G X \exp_G Y = \exp_G(X + Y + Z)$$

with $Z \in [\mathfrak{g}, \mathfrak{g}]$ by the Baker-Campbell-Hausdorff formula (see Theorem 1.3.2). Thus, $\phi(Z) = 0$ and we obtain

$$\begin{aligned} \pi(xy) &= \exp\left(i\phi(X + Y + Z)\right) = \exp\left(i\phi(X) + i\phi(Y)\right) \\ &= \exp\left(i\phi(X)\right)\exp\left(i\phi(Y)\right) = \pi(x)\pi(y). \end{aligned}$$

So π is a one-dimensional representation of G and we see that

$$\pi(X_k) = \partial_{t=0}\pi(e^{tX_k}) = \partial_{t=0}\exp\left(i\phi(tX_k)\right) = \partial_{t=0}\exp\left(it\phi(X_k)\right) = i\delta_{j,k}.$$

As π is a non-trivial one-dimensional representation of G and $\mathcal{R}$ satisfies the Rockland condition,

$$\pi(\mathcal{R}) = \sum_{[\alpha]=\nu} c_\alpha \pi(X^\alpha)$$

must be non-zero. We see that $\pi(X^\alpha)$ is always zero unless α is of the form ae_j for $a \in \mathbb{N}$ where e_j is the multi-index with 1 in the j-th place and zeros elsewhere; in this case $[\alpha] = \upsilon_j a$. So ν must be of the form $\nu = \upsilon_j a$ for some integer $a = a_j \in \mathbb{N}$ which may depend on j. And this is true for any $j = 1, \ldots, n'$.

Since $X_1, \ldots, X_{n'}$ generate the Lie algebra $\mathfrak{g}$, the other weights are linear combinations with coefficients in $\mathbb{N}_0$ of the υ_j's, $j \leq n'$. This shows that the operators $D'_r = \text{Exp}(\frac{\ln r}{\nu} A)$ are dilations over $\mathfrak{g}$ with rational weights. By Lemma 3.1.9, the group G is graded. $\square$

Remark 4.1.4. Proposition 4.1.3 and Remark 3.1.8 imply that the natural context for the study of Rockland operators is a graded Lie group endowed with a family of dilations with integer weights.

One may further assume that the weights have no common divisor other than 1 but we do not assume so unless we specify it.

From the proof of Proposition 4.1.3, we see:

Corollary 4.1.5. *Let G be a graded Lie group and let $\{X_1, \ldots, X_n\}$ be the basis described in Lemma 3.1.14. We keep the notation of the lemma.*

The homogeneous degree of any Rockland operator is a multiple of $\upsilon_1, \ldots, \upsilon_{n'}$.

If $\mathcal{R}$ is a Rockland operator satisfying $\mathcal{R}^t = \mathcal{R}$ then its homogeneous degree is even.

4.1.2 Examples of Rockland operators

On $(\mathbb{R}^n, +)$, it is easy to see that Rockland differential operators are exactly the operators $P(-i\partial_1, \ldots, -i\partial_n)$ where P is a polynomial which is homogeneous (for the standard dilations) and does not vanish except at zero. For instance homogeneous elliptic operators on $\mathbb{R}^n$ with constant coefficients are Rockland operators. More generally, let us prove that sub-Laplacians on a stratified Lie group are Rockland operators. First let us recall their definition.

Definition 4.1.6. If G is a stratified Lie group with a given basis $Z_1, \ldots, Z_p$ for the first stratum of its Lie algebra, then the left-invariant differential operator on G given by

$$Z_1^2 + \ldots + Z_p^2$$

is called a *sub-Laplacian.*

For example, the canonical sub-Laplacian of the Heisenberg group $\mathbb{H}_{n_o}$ is

$$X_1^2 + Y_1^2 + \ldots + X_{n_o}^2 + Y_{n_o}^2,$$

see Examples 1.6.4, 3.1.2 and 3.1.3 for our notation regarding the Heisenberg group.

Lemma 4.1.7. *Any sub-Laplacian on a stratified Lie group is a Rockland operator of homogeneous degree 2.*

This could be seen as a consequence of famous powerful theorems, namely from combining Hörmander's sums of squares and Helffer-Nourrigat (see Theorems A.1.2 and 4.1.12 in the sequel) but we prefer to give a direct and easy proof.

Proof. Let

$$\mathcal{R} = Z_1^2 + \ldots + Z_p^2$$

be a sub-Laplacian on the stratified Lie group G, where $Z_1, \ldots, Z_p$ is a given basis for the first stratum V_1 of the Lie algebra of G.

Clearly $\mathcal{R}$ is a homogeneous left-invariant differential operator of degree 2. Let $\pi \in \widehat{G}\backslash\{1\}$ and $v \in \mathcal{H}_\pi^\infty$ be such that $\pi(\mathcal{R})v = 0$. Then

$$\begin{aligned} 0 &= (\pi(\mathcal{R})v, v)_{\mathcal{H}_\pi} = (\pi(Z_1)^2 v, v)_{\mathcal{H}_\pi} + \ldots + (\pi(Z_p)^2 v, v)_{\mathcal{H}_\pi} \\ &= -(\pi(Z_1)v, \pi(Z_1)v)_{\mathcal{H}_\pi} - \ldots - (\pi(Z_p)v, \pi(Z_p)v)_{\mathcal{H}_\pi} \\ &= -\|\pi(Z_1)v\|^2_{\mathcal{H}_\pi} - \ldots - \|\pi(Z_p)v\|^2_{\mathcal{H}_\pi}, \end{aligned}$$

and hence

$$\pi(Z_1)v = \ldots = \pi(Z_p)v = 0.$$

Since $\{Z_1, \ldots, Z_p\}$ generates linearly the first stratum V_1 of $\mathfrak{g}$ and V_1 generates $\mathfrak{g}$ as a Lie algebra, we see that $\pi(X)v = 0$ for any vector $X \in \mathfrak{g}$. But since π is non-trivial and irreducible, this forces v to be zero. □

Looking at the proof of Lemma 4.1.7, it is not difficult to construct the 'classical' Rockland differential operators on graded Lie groups G:

Lemma 4.1.8. *Let G be a graded Lie group of dimension n, i.e. $G \sim \mathbb{R}^n$. We denote by $\{D_r\}_{r>0}$ the natural family of dilations on its Lie algebra $\mathfrak{g}$, and by $\upsilon_1, \ldots, \upsilon_n$ its weights. We fix a basis $\{X_1, \ldots, X_n\}$ of $\mathfrak{g}$ satisfying*

$$D_r X_j = r^{\upsilon_j} X_j, \quad j = 1, \ldots, n, \quad r > 0.$$

If ν_o is any common multiple of $\upsilon_1, \ldots, \upsilon_n$, the operator

$$\sum_{j=1}^n (-1)^{\frac{\nu_o}{\upsilon_j}} c_j X_j^{2\frac{\nu_o}{\upsilon_j}} \quad \textit{with} \quad c_j > 0, \tag{4.1}$$

is a Rockland operator of homogeneous degree $2\nu_o$.

Proof. The operator $\mathcal{R}$ given in (4.1) is clearly a homogeneous left-invariant differential operator of homogeneous degree $2\nu_o$. Let $\pi \in \widehat{G}\backslash\{1\}$ and $v \in \mathcal{H}_\pi^\infty$ be such that $\pi(\mathcal{R})v = 0$. Then

$$\begin{aligned} 0 &= (\pi(\mathcal{R})v, v)_{\mathcal{H}_\pi} = \sum_{j=1}^n (-1)^{\frac{\nu_o}{\upsilon_j}} c_j (\pi(X_j)^{2\frac{\nu_o}{\upsilon_j}} v, v)_{\mathcal{H}_\pi} \\ &= \sum_{j=1}^n c_j \|\pi(X_j)^{\frac{\nu_o}{\upsilon_j}} v\|_{\mathcal{H}_\pi}, \end{aligned}$$

and hence $\pi(X_j)^{\frac{\nu_o}{\upsilon_j}} v = 0$ for $j = 1, \ldots, n$.

Let us observe the following simple fact regarding any positive integer p and any $Z \in \mathfrak{U}(\mathfrak{g})$: the hypothesis $\pi(Z)^p v = 0$ implies that

- if p is odd then $\pi(Z)^{p+1} v = \pi(Z)\pi(Z)^p v = 0$,

- whereas if p is even then

$$0 = (\pi(Z)^p v, v)_{\mathcal{H}_\pi} = (-1)^{p/2}(\pi(Z)^{\frac{p}{2}} v, \pi(Z)^{\frac{p}{2}} v)_{\mathcal{H}_\pi} = (-1)^{p/2}\|\pi(Z)^{\frac{p}{2}} v\|^2_{\mathcal{H}_\pi},$$

and hence $\pi(Z)^{\frac{p}{2}} v = 0$.

Applying this argument inductively on $Z = X_j$ and $p = \nu_o/\upsilon_j$, $\nu_o/2\upsilon_j, \ldots$, we obtain that $\pi(X_j)v = 0$ for each j. Hence $v = 0$. □

Remark 4.1.9. By Proposition 4.1.3 and its proof, if a homogeneous Lie group G admits a Rockland operator, then, up to rescaling the dilations (cf. Remark 3.1.8), we may assume that the group G is graded and endowed with its natural family of dilations $\{D_r\}_{r>0}$. Lemma 4.1.8 gives the converse: on such a group, we can always find a Rockland operator.

The proof of Lemma 4.1.8 can easily be modified using an adapted basis constructed in Lemma 3.1.14 to obtain

Corollary 4.1.10. *Let G be a graded Lie group endowed with a family of dilations $\{D_r\}_{r>0}$. Let $\{X_1, \ldots, X_n\}$ be a basis of $\mathfrak{g}$ as in Lemma 3.1.14. In particular, the vectors $X_1, \ldots, X_{n'}$ generate the Lie algebra $\mathfrak{g}$.*

If ν_o is any common multiple of $\upsilon_1, \ldots, \upsilon_{n'}$, the operator

$$\sum_{j=1}^{n'} (-1)^{\frac{\nu_o}{\upsilon_j}} X_j^{2\frac{\nu_o}{\upsilon_j}}, \tag{4.2}$$

is a Rockland operator of homogeneous degree $2\nu_o$.

If the group G is stratified, the vectors $X_1, \ldots, X_{n'}$ span linearly the first stratum and we obtain the sub-Laplacian if we choose $\nu_o = \upsilon_1$.

From one Rockland operator, we can construct many since powers of a Rockland operator or its complex conjugate operator are Rockland:

Lemma 4.1.11. *Let $\mathcal{R}$ be a Rockland operator on a graded Lie group G endowed with a family of dilations with integer weights. Then the operators $\mathcal{R}^k$ for any $k \in \mathbb{N}$ and $\bar{\mathcal{R}}$ are also Rockland operators.*

The operator $\bar{\mathcal{R}}$ as an element of $\mathfrak{U}(\mathfrak{g})$ was defined in (1.8).

Proof. It is clear that $\bar{\mathcal{R}}$ and $\mathcal{R}^k$ are left-invariant homogeneous differential operators on G.

Let $\pi \in \widehat{G}\backslash\{1\}$. We have

$$\pi(\bar{\mathcal{R}}) = \overline{\pi(\mathcal{R})}.$$

This holds in fact for any left-invariant differential operator viewed as an element of $\mathfrak{U}(\mathfrak{g})$. Therefore, $\bar{\mathcal{R}}$ is Rockland. For the case of $\mathcal{R}^k$, let $v \in \mathcal{H}_\pi^\infty$ be such that $\pi(\mathcal{R}^k)v = 0$. Applying recursively the simple fact explained in the proof of Lemma 4.1.8, we obtain $\pi(\mathcal{R})v = 0$ and this implies $v = 0$ because $\mathcal{R}$ is Rockland. Therefore, $\mathcal{R}^k$ is also Rockland. □

4.1.3 Hypoellipticity and functional calculus

The analysis of left-invariant homogeneous operators on a nilpotent graded Lie group has played a very important role in the understanding of hypoellipticity. We refer the interested reader on this subject to the lecture notes by Helffer and Nier [HN05]. For the definition of hypoellipticity, see Section A.1.

In [Roc78], Rockland showed that if T is a homogeneous left-invariant differential operators on the Heisenberg group $\mathbb{H}_{n_o}$, then the hypoellipticity of T and T^t is equivalent to the Rockland condition (see Definition 4.1.1). He also asked whether this equivalence would be true for more general homogeneous Lie groups. Just afterwards, Beals showed [Bea77b] that the hypoellipticity of a homogeneous left-invariant differential operator on any homogeneous Lie group implies the Rockland condition. At the same time he also showed that the converse holds in some step-two cases. Eventually in [HN79], Helffer and Nourrigat settled what has become Rockland's conjecture by proving the following equivalence:

Theorem 4.1.12. *Let $\mathcal{R}$ be a left-invariant and homogeneous differential operator on a homogeneous Lie group G. The hypoellipticity of $\mathcal{R}$ is equivalent to $\mathcal{R}$ satisfying the Rockland condition.*

In this case, any operator of the form

$$\mathcal{R} + \sum_{[\alpha]<\nu} c_\alpha X^\alpha,$$

where ν is the degree of homogeneity of $\mathcal{R}$ and c_α any complex number, is also hypoelliptic.

The proof of Theorem 4.1.12 relies on the description of $\widehat{G}$ via Kirillov's orbit method.

Remark 4.1.13. 1. The hypotheses of Theorem 4.1.12 with the existence of a Rockland operator imply that the family of dilations of the group may be rescaled to have integer weights and consequently that the group may be viewed as graded, see Proposition 4.1.3. When describing properties of a Rockland operator $\mathcal{R}$ on a homogeneous Lie group G, unless stated otherwise, we will always assume that the group G is graded in such a way that the operator $\mathcal{R}$ is homogeneous for the natural family of dilations (with integer weights).

2. Combining the theorems of Hellfer-Nourrigat and of Hörmander (see Theorems 4.1.12 and A.1.2) gives another proof that the sub-Laplacians are Rockland operators, see Lemma 4.1.7.

3. If $\mathcal{R}$ is a Rockland operator formally self-adjoint, i.e. $\mathcal{R}^* = \mathcal{R}$ as elements of $\mathfrak{U}(\mathfrak{g})$, then $\mathcal{R}^t = \bar{\mathcal{R}}$ must also be Rockland by Lemma 4.1.11. Hence Theorem 4.1.12 implies that any formally self-adjoint Rockland operator satisfies the hypothesis of Theorem 3.2.40 and thus admits fundamental solutions. It also satisfies the hypothesis of the Liouville theorem as in Theorem 3.2.45.

4. Let us also mention an alternative reformulation of the Hellfer-Nourrigat theorem given by Rothschild [Rot83]: a left-invariant homogeneous operator $\mathcal{R}$ on a graded Lie group G is hypoelliptic if and only if there is no non-constant bounded function f on G such that $\mathcal{R}f = 0$ on G. The proof of this relies on the Liouville theorem from Section 3.2.8. Essentially, in one direction this is Beals' result as above, while in the other it will follow from Corollary 4.3.4.

Along the proof of Theorem 4.1.12 (see [HN79, Estimate (6.1)]), Helffer and Nourrigat also showed the following property which will be used in the sequel.

Corollary 4.1.14. *Let G be a graded Lie group endowed with a family of dilations with integer weights. Let $\mathcal{R}$ be a Rockland operator G of homogeneous degree ν. Then there exists $C > 0$ such that*

$$\forall \phi \in \mathcal{S}(G) \qquad \sum_{[\alpha]=\nu} \|X^\alpha \phi\|^2_{L^2(G)} \leq C \left(\|\mathcal{R}\phi\|^2_{L^2(G)} + \|\phi\|^2_{L^2(G)} \right).$$

After developing the Sobolev spaces on G, we will be actually able to prove its L^p-version, see Lemma 4.4.19.

The following property of Rockland differential operators is technically important and relies on hypoellipticity.

Proposition 4.1.15. *Let $\mathcal{R}$ be a Rockland operator on a graded Lie group G. We assume that $\mathcal{R}$ is formally self-adjoint. Let π be a strongly continuous unitary representation of G.*

Then the operators $\mathcal{R}$ and $\pi(\mathcal{R})$ densely defined on $\mathcal{D}(G) \subset L^2(G)$ and $\mathcal{H}^\infty_\pi \subset \mathcal{H}_\pi$, respectively, are essentially self-adjoint.

That $\mathcal{R}$ is formally self-adjoint means that $\mathcal{R}^* = \mathcal{R}$ as elements of the universal enveloping algebra $\mathfrak{U}(\mathfrak{g})$, see (1.9).

Before we prove it, let us point out its consequences:

Corollary 4.1.16 (Functional calculus of Rockland operators and their Fourier transform). *Let $\mathcal{R}$ be a Rockland operator on a graded Lie group G. We assume that $\mathcal{R}$ is formally self-adjoint as an element of $\mathfrak{U}(\mathfrak{g})$. Then $\mathcal{R}$ is essentially self-adjoint on $L^2(G)$ and we denote by $\mathcal{R}_2$ its self-adjoint extension on $L^2(G)$. Moreover, for each strongly continuous unitary representation π of G, $\pi(\mathcal{R})$ is essentially self-adjoint on $\mathcal{H}_\pi$ and we keep the same notation for its self-adjoint extension. Let E, E_π be the spectral measures of $\mathcal{R}_2$ and $\pi(\mathcal{R})$:*

$$\mathcal{R}_2 = \int_{\mathbb{R}} \lambda dE(\lambda) \quad \text{and} \quad \pi(\mathcal{R}) = \int_{\mathbb{R}} \lambda dE_\pi(\lambda).$$

For any Borel subset $B \subset \mathbb{R}$, the orthogonal projection $E(B)$ is left-invariant hence $E(B) \in \mathscr{L}_L(L^2(G))$. The group Fourier transform of its convolution kernel $E(B)\delta_0 \in \mathcal{K}(G)$ is

$$\mathcal{F}_G(E(B)\delta_0)(\pi) = E_\pi(B).$$

If ϕ is a measurable function on $\mathbb{R}$, the spectral multiplier operator $\phi(\mathcal{R}_2)$ is defined by

$$\phi(\mathcal{R}_2) := \int_{\mathbb{R}} \phi(\lambda) dE(\lambda),$$

and its domain $\text{Dom}(\phi(\mathcal{R}_2))$ *is the space of function* $f \in L^2(G)$ *such that the integral* $\int_{\mathbb{R}} |\phi(\lambda)|^2 d(E(\lambda)f, f)$ *is finite. It satisfies for all* $f \in \text{Dom}(\phi(\mathcal{R}_2))$ *and* $r > 0$*:*

$$f(r\,\cdot) \in \text{Dom}(\phi(r^{-\nu}\mathcal{R}_2)) \quad \text{and} \quad \phi(\mathcal{R}_2)f = \phi(r^{-\nu}\mathcal{R}_2)\,(f(r\,\cdot))\,(r^{-1}\,\cdot). \tag{4.3}$$

If π_1 is another strongly continuous representation such that $\pi_1 \sim_T \pi$, that is, T is a unitary operator satisfying $T\pi_1 = \pi T$, then $TE_{\pi_1} = E_\pi T$ and we have for any measurable function ϕ the equality

$$T\phi(\pi_1(\mathcal{R})) = \phi(\pi(\mathcal{R}))T. \tag{4.4}$$

Let $\phi \in L^\infty(\mathbb{R})$ be any measurable bounded function. Then the spectral multiplier operator $\phi(\mathcal{R}_2)$ is in $\mathscr{L}_L(L^2(G))$, that is, it is bounded on $L^2(G)$ and left-invariant. Its convolution kernel denoted by $\phi(\mathcal{R}_2)\delta_o$ is the unique tempered distribution $\phi(\mathcal{R}_2)\delta_o \in \mathcal{S}'(G)$ such that

$$\forall f \in \mathcal{S}(G) \quad \phi(\mathcal{R}_2)f = f * \phi(\mathcal{R}_2)\delta_o.$$

In fact $\phi(\mathcal{R}_2)\delta_o \in \mathcal{K}(G)$ and its group Fourier transform is

$$\mathcal{F}\{\phi(\mathcal{R}_2)\delta_o\}(\pi) = \phi(\pi(\mathcal{R})) = \int_{\mathbb{R}} \phi(\lambda) dE_\pi(\lambda). \tag{4.5}$$

Consequently, for any $f \in L^2(G)$,

$$\mathcal{F}\{\phi(\mathcal{R}_2)f\}(\pi) = \phi(\pi(\mathcal{R}))\widehat{f}(\pi). \tag{4.6}$$

We have for any $r > 0$ and $x \in G$:

$$\phi(r^\nu \mathcal{R}_2)\delta_o(x) = r^{-Q}\phi(\mathcal{R}_2)\delta_o(r^{-1}x). \tag{4.7}$$

For any $\phi \in L^\infty(\mathbb{R})$,

$$\{\phi(\mathcal{R}_2)\delta_0\}^* = \bar{\phi}(\mathcal{R})\delta_0, \quad \text{where} \quad \{\phi(\mathcal{R}_2)\delta_0\}^*(x) = \overline{\phi(\mathcal{R}_2)\delta_0}(x). \tag{4.8}$$

If ϕ is also real-valued, then $\phi(\mathcal{R}_2)$ is a self-adjoint operator and its kernel satisfies $\phi(\mathcal{R}_2)\delta_o = (\phi(\mathcal{R}_2)\delta_o)^*$, that is, in the sense of distributions,

$$\phi(\mathcal{R}_2)\delta_o(x) = \overline{\phi(\mathcal{R}_2)\delta_o}(x^{-1}).$$

If ϕ is real-valued and furthermore if $\mathcal{R}^t = \mathcal{R}$, then $\phi(\mathcal{R}_2)\delta_o$ is real-valued (as a distribution).

Remark 4.1.17. For any measurable function $\phi : \mathbb{R} \to \mathbb{C}$ such that for every $\pi_1 \in \operatorname{Rep} G$, the domain of $\phi(\pi_1(\mathcal{R}))$ contains $\mathcal{H}_{\pi_1}^\infty$, the corresponding $\widehat{G}$-field of operators $\{\phi(\pi(\mathcal{R})) : \mathcal{H}_\pi^\infty \to \mathcal{H}_\pi\}$ is well defined in the sense of Definition 1.8.13 because of (4.4). This is the case for instance if ϕ is bounded since in this case $\phi(\pi_1(\mathcal{R}))$ is a bounded and therefore defined on the whole space $\mathcal{H}_{\pi_1}$.

The rest of this section is devoted to the proof of Proposition 4.1.15 and Corollary 4.1.16; it may be skipped at first reading. Proposition 4.1.15 follows from a Theorem by Nelson and Stinespring [NS59, Theorem 2.2] regarding elliptic operators on Lie groups as well as the adaptation of its proof due to Folland and Stein [FS82, ch.3.B] to our case. Let us sketch briefly the ideas for the sake of completeness. Nelson and Stinespring's Theorem can be reformulated here as the following:

Proposition 4.1.18. *Let $\mathcal{R}$ be a Rockland operator on a graded Lie group G. We assume that $\mathcal{R}$ is formally self-adjoint as an element of $\mathfrak{U}(\mathfrak{g})$.*

If π is a strongly continuous unitary representation of G, then the closure of $\pi(\mathcal{R}^*)$ is the adjoint of $\pi(\mathcal{R})$.

Proof of Proposition 4.1.18. Let $v \in \mathcal{H}_\pi$ be orthogonal to the range of $\pi(\mathcal{R}) + \mathrm{I}$. Then for all $\phi \in \mathcal{D}(G)$,

$$0 = ((\pi(\mathcal{R}) + \mathrm{I})\pi(\phi)v, v)_{\mathcal{H}_\pi} = \int_G (\mathcal{R} + \mathrm{I})\phi(x)\ (\pi(x)^* v, v)_{\mathcal{H}_\pi}\, dx.$$

In other words, the continuous function f_π defined by

$$f_\pi(x) := (\pi(x)^* v, v)_{\mathcal{H}_\pi} = (v, \pi(x)v)_{\mathcal{H}_\pi}, \quad x \in G,$$

is a solution in the sense of distributions of the partial differential equation $(\mathcal{R} + \mathrm{I})f = 0$. By Theorem 4.1.12, the operator $\mathcal{R}+\mathrm{I}$ is hypoelliptic. Hence f_π is smooth on G and the equation $(\mathcal{R} + \mathrm{I})f_\pi = 0$ holds in the ordinary pointwise sense. We observe that for any $X \in \mathfrak{U}(\mathfrak{g})$ identified with a left-invariant vector field we have

$$Xf_\pi(x) = \partial_{t=0}\left\{(v, \pi(xe^{tX})v)_{\mathcal{H}_\pi}\right\} = (v, \pi(x)\pi(X)v)_{\mathcal{H}_\pi}.$$

Thus,

$$(\mathcal{R} + \mathrm{I})f_\pi(x) = (v, \pi(x)\pi(\mathcal{R})v)_{\mathcal{H}_\pi} + (v, \pi(x)v)_{\mathcal{H}_\pi}.$$

Therefore, $(\mathcal{R} + \mathrm{I})f_\pi(0) = 0$ implies

$$(v, \pi(\mathcal{R})v)_{\mathcal{H}_\pi} = -(v, v)_{\mathcal{H}_\pi} = -\|v\|_{\mathcal{H}_\pi}^2.$$

If $\mathcal{R}$ can be written as S^*S for some non-constant $S \in \mathfrak{U}(\mathfrak{g})$, then the left-hand side is equal to $\|\pi(S)v\|^2$ so $v = 0$. In the general case, we apply the argument above to $\mathcal{R}^*\mathcal{R} = \mathcal{R}^2$ which is also a Rockland operator by Lemma 4.1.11, and we obtain the desired conclusion thanks to the following lemma applied to $T = \pi(\mathcal{R})$, $T' = \pi(\mathcal{R}^*)$ and $\mathcal{D} = \mathcal{H}_\pi^\infty$. □

Lemma 4.1.19. *Let $\mathcal{D}$ be a dense vector subspace of a Hilbert space $\mathcal{H}$. Let T and T' be two linear operators on $\mathcal{H}$, whose domains are $\mathcal{D}$ and whose ranges are contained in $\mathcal{D}$ such that T' is contained in the adjoint of T. If $T'T$ is essentially self-adjoint then the closure of T' is the adjoint of T.*

Proof of Lemma 4.1.19. We denote by T_* the adjoint of T. Let (u, v) be an element of the graph of T_* which is orthogonal to the graph of T'. This means

$$v = T_* u \quad \text{and} \quad \forall w \in \mathcal{D} \qquad (u, w)_{\mathcal{H}} + (v, T'w)_{\mathcal{H}} = 0.$$

In particular, for $w = Tx$ with $x \in \mathcal{D}$, we obtain

$$0 = (u, Tx)_{\mathcal{H}} + (v, T'Tx)_{\mathcal{H}} = (v, x)_{\mathcal{H}} + (v, T'Tx)_{\mathcal{H}}, \quad x \in \mathcal{D}.$$

But it is not difficult to see that $\mathrm{I} + T'T$ has a dense range. Consequently $v = 0$. So $(u, w)_{\mathcal{H}} = 0$ for all $w \in \mathcal{D}$ and therefore $u = 0$. This shows that the graph of T_* contains no non-zero element orthogonal to the graph of T'; hence the closure of T' is T_*. □

Proof of Proposition 4.1.15. We apply Proposition 4.1.18 to the left regular action on $L^2(G)$ and the strongly continuous unitary representation π of G. □

Proof of Corollary 4.1.16. Applying the spectral theorem to the self-adjoint operators $\mathcal{R}_2$ and $\pi(\mathcal{R})$ (see, e.g., Rudin [Rud91, Part III]) we obtain the spectral measures E and E_π together with the definition of the spectral multipliers.

For each $x_o \in G$ and $r > 0$ we set for any Borel set $B \subset \mathbb{R}$ and any function $f \in L^2(G)$,

$$\begin{aligned} E^{(x_o)}(B)f &:= (E(B))\,(f(x_o\,\cdot))\,(x_o^{-1}\,\cdot), \\ E^{(r)}(B)f &:= \big(E(r^{-\nu}B)\big)\,(f(r\,\cdot))\,(r^{-1}\,\cdot), \end{aligned}$$

where the dilation of a subset of $\mathbb{R}$ is defined in the usual sense. It is not difficult to check that this defines new spectral measures $E^{(x_o)}$ and $E^{(r)}$ and, that for any function $f \in \mathcal{S}(G)$,

$$\begin{aligned} \int_{\mathbb{R}} \lambda dE^{(x_o)}(\lambda) f &= \int_{\mathbb{R}} \lambda d\,(E(\lambda))\,(f(x_o\,\cdot))\,(x_o^{-1}\,\cdot) = \mathcal{R}_2\,(f(x_o\,\cdot))\,(x_o^{-1}\,\cdot) \\ &= \mathcal{R}\,(f(x_o\,\cdot))\,(x_o^{-1}\,\cdot) = \mathcal{R}f = \mathcal{R}_2 f, \\ \int_{\mathbb{R}} \lambda dE^{(r)}(\lambda) f &= \int_{\mathbb{R}} (r^{-\nu}\lambda) d\,(E(\lambda))\,(f(r\,\cdot))\,(r^{-1}\,\cdot) = r^{-\nu}\mathcal{R}_2\,(f(r\,\cdot))\,(r^{-1}\,\cdot) \\ &= r^{-\nu}\mathcal{R}\,(f(r\,\cdot))\,(r^{-1}\,\cdot) = \mathcal{R}f = \mathcal{R}_2 f, \end{aligned}$$

since $\mathcal{R}$ is left-invariant and ν-homogeneous. By density of $\mathcal{S}(G)$ in $L^2(G)$, we have obtained for any $f \in L^2(G)$ that

$$\int_{\mathbb{R}} \lambda dE^{(x_o)}(\lambda) f = \mathcal{R}_2 f \quad \text{and} \quad \int_{\mathbb{R}} \lambda dE^{(r)}(\lambda) f = \mathcal{R}_2 f.$$

By uniqueness of the spectral measure of $\mathcal{R}_2$, the spectral measures $E^{(x_o)}$, $E^{(r)}$ and E coincide. For $E^{(r)}$ this implies (4.3).

For $E^{(x_o)}$ this means that for each Borel subset $B \subset \mathbb{R}$, the projection $E(B)$ is a left-invariant operator on $L^2(G)$. By the Plancherel theorem (see Section 1.8.2) the group Fourier transform of its convolution kernel $E(B)\delta_0 \in \mathcal{K}(G)$ satisfies

$$\forall f \in L^2(G) \qquad \pi(E(B)f) = \pi(E(B)\delta_0)\pi(f). \tag{4.9}$$

It is not difficult, using the uniqueness of the group Fourier transform, to check that

$$F : B \longmapsto \pi(E(B)\delta_0) =: F(B),$$

is a spectral measure on $\mathcal{H}_\pi$. Equality (4.9) can be rewritten for any $f \in L^2(G)$ as

$$\mathcal{F}_G\left(\int_{\mathbb{R}} \phi(\lambda)dE(\lambda)f\right)(\pi) = \left(\int_{\mathbb{R}} \phi(\lambda)dF(\lambda)\right)\widehat{f}(\pi), \tag{4.10}$$

with $\phi = 1_B$, that is, the characteristic function of a Borel subset $B \subset \mathbb{R}$. Hence Equality (4.10) also holds for a finite linear combination of characteristic functions, and then, passing through the limit carefully, for any $\phi \in L^\infty(\mathbb{R})$ with $f \in L^2(G)$ and $\phi(\lambda) = \lambda$ for $f \in \mathcal{S}(G)$. The latter yields

$$\begin{aligned}\left(\int_{\mathbb{R}} \lambda dF(\lambda)\right)\widehat{f}(\pi) &= \mathcal{F}_G\left(\int_{\mathbb{R}} \lambda dE(\lambda)f\right)(\pi)\\ &= \mathcal{F}_G(\mathcal{R}_2 f)(\pi) = \pi(\mathcal{R})\widehat{f}(\pi).\end{aligned}$$

Since the space $\mathcal{H}_\pi^\infty$ of smooth vectors is linearly spanned by elements of the form $\widehat{f}(\pi)v$, $f \in \mathcal{S}(G)$, $v \in \mathcal{H}_\pi$ (see Theorem 1.7.8), we have on $\mathcal{H}_\pi^\infty$

$$\int_{\mathbb{R}} \lambda dF(\lambda) = \pi(\mathcal{R}).$$

The uniqueness of the spectral measure E_π shows that

$$E_\pi(B) = F(B) = \pi(E(B)\delta_0).$$

Equality (4.5) follows from (4.10) for $\phi \in L^\infty(\mathbb{R})$.

If $\pi_1 \sim_T \pi$, then we set $E_\pi^{(T)} := TE_{\pi_1}T^{-1}$, where E_{π_1} denotes the spectral measure of $\pi_1(\mathcal{R})$. We check easily that $E_\pi^{(T)}$ is a spectral measure on $\mathcal{H}_\pi$ and that

$$\int_{\mathbb{R}} \lambda dE_\pi^{(T)} = T\int_{\mathbb{R}} \lambda dE_{\pi_1}T^{-1} = T\pi_1(\mathcal{R})T^{-1} = T\pi_1 T^{-1}(\mathcal{R}) = \pi(\mathcal{R}).$$

The property of the spectral measure E_π, that is, its uniqueness and the functional calculus, shows that $E_\pi^{(T)} = E_\pi$ and that (4.4) holds.

The rest of the statement follows from the Schwartz kernel theorem (see Corollary 3.2.1) and basic properties of the convolution. □

4.2 Positive Rockland operators

In this section we concentrate on positive Rockland operators, i.e. Rockland operators which are positive in the operator sense. Positive Rockland operators always exist on a graded Lie group, see Remark 4.2.4 below. Among Rockland operators, positive ones enjoy a number of additional useful properties. In particular, in this section, we analyse the heat semi-group associated to a positive Rockland operator and the corresponding heat kernel.

4.2.1 First properties

We shall be interested in Rockland differential operators which are positive in the sense of operators:

Definition 4.2.1. An operator T on a Hilbert space $\mathcal{H}$ is *positive* when for any vectors $v, v_1, v_2 \in \mathcal{H}$ in the domain of T, we have

$$(Tv_1, v_2)_{\mathcal{H}} = (v_1, Tv_2)_{\mathcal{H}} \quad \text{and} \quad (Tv, v)_{\mathcal{H}} \geq 0.$$

In the case of left-invariant differential operator, this is easily equivalent to

Proposition 4.2.2. *Let T be a left-invariant differential operator on a Lie group G. Then T is positive on $L^2(G)$ when T is formally self-adjoint, that is, $T^* = T$ in $\mathfrak{U}(\mathfrak{g})$, and satisfies*

$$\forall f \in \mathcal{D}(G) \qquad \int_G Tf(x)\overline{f(x)}\, dx \geq 0.$$

For the definition of T^*, see (1.9).

The following properties of positive operators are easy to prove:

Lemma 4.2.3. *1. A linear combination with non-negative coefficients of positive operators is a positive operator.*

2. If X is a left-invariant vector field and $p \in 2\mathbb{N}_0$, then the operator $(-1)^{\frac{p}{2}} X^p$ is positive on G.

3. If T is a positive differential operator on G then for any $k \in \mathbb{N}$ the differential operator T^k is also positive.

Proof. The first property is clear.

The second is true since each invariant vector field is essentially skew-symmetric, see Section 1.3.

Let us prove the third property. Let T be a positive differential operator and $k \in \mathbb{N}$. Clearly T^k is also formally self-adjoint and we obtain recursively if $k = 2\ell$:

$$\int_G T^k f(x)\overline{f(x)} dx = \int_G T^\ell f(x)\overline{T^\ell f(x)} dx = \int_G \left|T^\ell f(x)\right|^2 dx,$$

which is necessarily non-negative, whereas if $k = 2\ell + 1$,

$$\int_G T^k f(x)\overline{f(x)}dx = \int_G T(T^\ell f(x))\, \overline{T^\ell f(x)}dx,$$

which is non-negative since T is positive. $\square$

We observe that the signs of the coefficients of a positive differential operator can not be guessed, as the example $-(\partial_1 \pm \partial_2)^2$ on $\mathbb{R}^2$ shows.

Remark 4.2.4. By Lemma 4.2.3, Parts 1 and 2, we see that the examples in Section 4.1.2 yield positive Rockland operators. For instance, on stratified Lie groups, the sub-Laplacians give operators $-\mathcal{R}$ with $\mathcal{R}$ positive and Rockland. Also, the operators in (4.1) and (4.2) give positive Rockland operators. In particular, this shows that any graded Lie group admits a positive Rockland operator.

We may obtain other positive Rockland operators as powers of those since a direct consequence of Lemma 4.1.11 and Lemma 4.2.3, Part 3, is the following

Lemma 4.2.5. *Let $\mathcal{R}$ be a positive Rockland operator on a graded Lie group G. Then $\mathcal{R}^k$ for every $k \in \mathbb{N}$ and $\bar{\mathcal{R}} = \mathcal{R}^t$ are also positive Rockland operators.*

We fix a positive Rockland operator $\mathcal{R}$. By Proposition 4.2.2, $\mathcal{R}$ is essentially self-adjoint and we may adopt the same notation as in Corollary 4.1.16. Since $\mathcal{R}$ is positive, the spectrum of $\mathcal{R}_2$ is included in $[0, \infty)$ and we have

$$\mathcal{R}_2 = \int_0^\infty \lambda dE(\lambda).$$

Proposition 4.2.6. *Let $\mathcal{R}$ be a positive Rockland operator on a graded Lie group G. If $\pi \in \widehat{G}$, then the operator $\pi(\mathcal{R})$ is positive. Furthermore, if π is non-trivial and*

$$(\pi(\mathcal{R})v, v)_{\mathcal{H}_\pi} = 0$$

then $v = 0$.

Proof. By Proposition 4.1.15, $\pi(E(B)) = E_\pi(B)$. Since E is supported in $[0, \infty)$ then so is E_π and the operator $\pi(\mathcal{R})$ is positive:

$$\forall v \in \mathcal{H}_\pi^\infty \qquad (\pi(\mathcal{R})v, v)_{\mathcal{H}_\pi} = \int_0^\infty \lambda d(E_\pi(\lambda)v, v)_{\mathcal{H}_\pi} \geq 0.$$

If $(\pi(\mathcal{R})v, v)_{\mathcal{H}_\pi} = 0$ then the (real non-negative) measure $(E_\pi(\lambda)v, v)_{\mathcal{H}_\pi}$ is concentrated on $\{\lambda = 0\}$ and this means that $v = E_\pi(0)v$ is in the nullspace of $\pi(\mathcal{R})$. Thus $v = 0$ since $\mathcal{R}$ satisfies the Rockland condition and π is non-trivial. $\square$

4.2.2 The heat semi-group and the heat kernel

In this section, we fix a positive Rockland operator $\mathcal{R}$ which is homogeneous of degree $\nu \in \mathbb{N}$.

By the functional calculus (see Corollary 4.1.16), we define the multipliers

$$e^{-t\mathcal{R}_2} := \int_0^\infty e^{-t\lambda} dE(\lambda), \quad t > 0.$$

We then have

$$\|e^{-t\mathcal{R}_2}\|_{\mathscr{L}(L^2(G))} \leq \sup_{\lambda \geq 0} |e^{-t\lambda}| = 1 \quad \text{and} \quad e^{-t\mathcal{R}_2} e^{-s\mathcal{R}_2} = e^{-(t+s)\mathcal{R}_2},$$

since $e^{-s\lambda}e^{-t\lambda} = e^{-(t+s)\lambda}$. Thus $\{e^{-t\mathcal{R}_2}\}_{t>0}$ is a contraction semi-group of operators on $L^2(G)$ (see Section A.2). This semi-group is often called the *heat semi-group*. The corresponding convolution kernels $h_t \in \mathcal{S}'(G)$, $t > 0$, are called *heat kernels*. We summarise its main properties in the following theorem:

Theorem 4.2.7. *Let $\mathcal{R}$ be a positive Rockland operator on a graded Lie group G. Then the heat kernels h_t associated with $\mathcal{R}$ satisfy the following properties. Each function h_t is Schwartz and we have*

$$\begin{aligned}
\forall s, t > 0 \qquad h_t * h_s &= h_{t+s}, & (4.11)\\
\forall x \in G,\ t, r > 0 \qquad h_{r^\nu t}(rx) &= r^{-Q} h_t(x), & (4.12)\\
\forall x \in G \qquad h_t(x) &= \overline{h_t(x^{-1})}, & (4.13)\\
\int_G h_t(x)dx &= 1. & (4.14)
\end{aligned}$$

The function $h : G \times \mathbb{R} \to \mathbb{C}$ defined by

$$h(x,t) := \begin{cases} h_t(x) & \text{if } t > 0 \text{ and } x \in G, \\ 0 & \text{if } t \leq 0 \text{ and } x \in G, \end{cases}$$

is smooth on $(G \times \mathbb{R})\backslash\{(0,0)\}$ and satisfies

$$(\mathcal{R} + \partial_t)h = \delta_{0,0},$$

where $\delta_{0,0}$ is the delta-distribution at $(0,0) \in G \times \mathbb{R}$.

Having fixed a homogeneous norm $|\cdot|$ on G, we have for any $N \in \mathbb{N}_0$, $\alpha \in \mathbb{N}_0^n$ and $\ell \in \mathbb{N}_0$, that

$$\exists C = C_{\alpha,N,\ell} > 0 \quad \forall t \in (0,1] \quad \sup_{|x|=1} |\partial_t^\ell X^\alpha h_t(x)| \leq C_{\alpha,N} t^N. \tag{4.15}$$

The proof of Theorem 4.2.7 is given in the next section. We finish this section with some comments and some corollaries of this theorem.

Remark 4.2.8. 1. If the group is stratified and $\mathcal{R} = -\mathcal{L}$ where $\mathcal{L}$ is a sub-Laplacian, then $\mathcal{R}$ is of order two and the proof relies on Hunt's theorem [Hun56], cf. [FS82, ch1.G]. In this case, the heat kernel is real-valued and moreover non-negative. The heat semi-group is then a semi-group of contraction which preserves positivity.

2. The behaviour of the heat kernel in the general case is quite well understood. For instance, it can be extended to the complex right-half plane. Then the heat kernel h_z with $z \in \mathbb{C}$, $\operatorname{Re} z > 0$ decays exponentially. See [Dzi93, DHZ94, AtER94].

3. Since $\mathcal{R}_2$ is a positive operator, only the values of $\phi \in L^\infty(\mathbb{R})$ on $[0,\infty)$ are taken into account for the multipliers $\phi(\mathcal{R}_2)$. But in fact, the value at 0 can be neglected too, as a consequence of the property of the heat kernel. Indeed, from $h_t \in \mathcal{S}(G)$ and (4.12), it is not difficult to show

$$\|f * h_t\|_{L^2(G)} \underset{t\to\infty}{\longrightarrow} 0,$$

first for $f \in \mathcal{D}(G)$ and then by density for any $f \in L^2$. This shows

$$\|e^{-t\mathcal{R}_2} f\|_{L^2(G)} \underset{t\to\infty}{\longrightarrow} 0,$$

and therefore we have

$$\|\int_0^\epsilon dE(\lambda)\|_{L^2(G)} \underset{\epsilon\to 0}{\longrightarrow} 0.$$

4. Another consequence of the heat kernel being Schwartz, proved in [HJL85], is that the spectrum of $\pi(\mathcal{R})$ is discrete and lies in $(0,\infty)$ for any $\pi \in \widehat{G}\backslash\{1\}$. Indeed, it is easy to see that $\pi(\mathcal{R})$ is the infinitesimal generator of the semi-group $\{\pi(e^{-t\mathcal{R}})\}_{t>0}$ in $\mathcal{H}_\pi$ and that $\pi(e^{-t\mathcal{R}}) = \pi(h_t)$ is a compact operator since $h_t \in \mathcal{S}(G)$ (for this last property, see [CG90, Theorem 4.2.1]).

   Moreover, strong properties of the eigenvalue distributions of $\pi(\mathcal{R})$ are known, see [tER97].

Theorem 4.2.7 shows that the functions h_t provide a commutative approximation of the identity, see Remark 3.1.60. We already know that $\{e^{-t\mathcal{R}_2}\}_{t>0}$ is a strongly continuous contraction semi-group. Moreover, we have the following properties for any p:

Corollary 4.2.9. *The operators*

$$f \mapsto f * h_t, \quad t > 0,$$

form a strongly continuous semi-group on $L^p(G)$ for any $p \in [1,\infty)$ and on $C_o(G)$. Furthermore, for any $f \in \mathcal{D}(G)$ and any $p \in [1,\infty]$ (finite or infinite), we have the convergence

$$\left\| \frac{1}{t}(f * h_t - f) - \mathcal{R}f \right\|_p \longrightarrow_{t\to 0} 0. \tag{4.16}$$

Finally, we formulate a simple but useful corollary of Theorem 4.2.7.

Corollary 4.2.10. *Setting $r = t^{-\frac{1}{\nu}}$ in (4.12), we get*

$$\forall x \in G,\ t > 0 \qquad h_t(x) = t^{-\frac{Q}{\nu}} h_1(t^{-\frac{1}{\nu}} x) \tag{4.17}$$

and

$$\text{for } x \in G\backslash\{0\} \text{ fixed}, X_x^\alpha h(x,t) = \begin{cases} O(t^{-\frac{Q+[\alpha]}{\nu}}) \text{ as } t \to \infty, \\ O(t^N) \text{ for all } N \in \mathbb{N}_0 \text{ as } t \to 0. \end{cases} \tag{4.18}$$

Inequalities (4.18) are also valid for any x in a fixed compact subset of $G\backslash\{0\}$.

4.2.3 Proof of the heat kernel theorem and its corollaries

This section is entirely devoted to the proofs of Theorem 4.2.7 and Corollaries 4.2.9 and 4.2.10. This may be skipped at first reading. The proofs essentially follow the arguments of Folland and Stein [FS82, Ch. 4. B].

Since h_t is the convolution kernel of the $\mathcal{R}_2$-multiplier operator, Corollary 4.1.16 yield that $h_t \in \mathcal{S}'(G)$ is a distribution which satisfies Properties (4.12) and (4.13) for each $t > 0$ fixed. Note that (4.12) easily yields (4.17).

By the Schwartz kernel theorem (see Corollary 3.2.1), since $(0,\infty) \ni t \mapsto e^{-t\mathcal{R}_2} \in \mathscr{L}(L^2(G))$ is a strongly continuous mapping, the function $(0,\infty) \ni t \mapsto h_t \in \mathcal{S}'(G)$ is continuous. Consequently the mapping $(t,x) \mapsto h_t(x)$ is a distribution on $(0,\infty) \times G$.

By the properties of semi-groups (cf. Proposition A.2.3 (4)), we have

$$\forall \phi \in \mathcal{D}(G),\ t > 0, \qquad \partial_t(e^{-t\mathcal{R}_2}\phi) = -\mathcal{R}_2(e^{-t\mathcal{R}_2}\phi) = -\mathcal{R}(e^{-t\mathcal{R}_2}\phi).$$

Taking this equation at 0_G shows that $(t,x) \mapsto h_t(x)$ is a solution in the sense of distributions of the equation $(\partial_t + \mathcal{R})f = 0$ on $(0,\infty) \times G$.

The next lemma is independent of the rest of the proof and shows that $\partial_t + \mathcal{R}$ can be turned into a Rockland operator:

Lemma 4.2.11. *Let $\mathcal{R}$ be a positive Rockland operator on a graded Lie group G. We equip the group $H := G \times \mathbb{R}$ (which is the direct product of the groups G and $(\mathbb{R},+)$) with the dilations*

$$D_r(x,t) := (rx, r^\nu t), \quad x \in G,\ t \in \mathbb{R}.$$

The group H has become a homogeneous Lie group and the operators $\mathcal{R} + \partial_t$ and $\mathcal{R} - \partial_t$ are Rockland operators on H.

Proof of Lemma 4.2.11. The dual of H is easily seen to be isomorphic to $\widehat{G} \times \mathbb{R}$:

- if $\pi \in \widehat{G}$ and $\lambda \in \mathbb{R}$, we can construct the representation $\rho = \rho_{\pi,\lambda}$ of H on $\mathcal{H}_\rho = \mathcal{H}_\pi$ by $\rho(x,t) := e^{i\lambda t}\pi(x)$;

- conversely, any representation $\rho \in \widehat{H}$ can be realised into a representation of the form $\rho_{\pi,\lambda}$.

Let $\rho = \rho_{\pi,\lambda} \in \widehat{H}$. We observe that $\mathcal{H}_\rho^\infty = \mathcal{H}_\pi^\infty$, $\rho(\mathcal{R}) = \pi(\mathcal{R})$, and $\rho(\partial_t) = i\lambda$. If $v \in \mathcal{H}_\rho^\infty$ is such that $\rho(\mathcal{R} + \partial_t)v = 0$ then

$$0 = (\rho(\mathcal{R} \pm \partial_t)v, v)_{\mathcal{H}_\rho} = (\pi(\mathcal{R})v, v)_{\mathcal{H}_\pi} \pm i\lambda(v, v)_{\mathcal{H}_\pi} = (\pi(\mathcal{R})v, v)_{\mathcal{H}_\pi} \pm i\lambda\|v\|^2_{\mathcal{H}_\pi}.$$

Since, by Proposition 4.2.6, $(\pi(\mathcal{R})v, v)_{\mathcal{H}_\pi} \geq 0$, the real part of the previous equalities is $(\pi(\mathcal{R})v, v)_{\mathcal{H}_\pi} = 0$. Again by Proposition 4.2.6, necessarily $v = 0$. □

Remark 4.2.12. A similar proof implies that $\mathcal{R} \pm \partial_t^k$ for $k \in \mathbb{N}$ odd is a Rockland operator on the group $G \times \mathbb{R}$ endowed with the dilations $D_r(x, t) = (rx, r^{\nu/k}t)$.

Corollary 4.2.13. *The distribution $(t, x) \mapsto h_t(x)$ is smooth on $(0, \infty) \times G$ and satisfies the equation*

$$(\partial_t + \mathcal{R})f = 0.$$

Furthermore, for any $t > 0$, $h_t \in L^2(G)$ and

$$\int_G |h_t(x)|^2 dx = t^{-\frac{Q}{\nu}} \int_G |h_1(x)|^2 dx < \infty. \tag{4.19}$$

Proof. The operator $\partial_t + \mathcal{R}$ is Rockland on $G \times \mathbb{R}$ by Lemma 4.2.11, therefore hypoelliptic by the Hellfer-Nourrigat theorem (see Theorem 4.1.12). Since the distribution $(t, x) \mapsto h_t(x)$ is a solution of the equation $(\partial_t + \mathcal{R})f = 0$ on $(0, \infty) \times G$, it is in fact smooth.

Since $\mathcal{R}$ is a positive Rockland operator, $\mathcal{R}^t$ is also a positive Rockland operator (see Lemma 4.2.5) and we can apply Lemma 4.2.11 to both. Therefore, $\mathcal{R} + \partial_t$ and its transpose are Rockland and thus hypoelliptic on $G \times \mathbb{R}$. By the Schwartz-Treves theorem (see Theorem A.1.6), the distribution topology on $G \times (0, \infty)$ and the C^∞-topology agree on the the nullspace of $\mathcal{R} + \partial_t$

$$\mathcal{N} = \{f \in \mathcal{D}'(G \times (0, \infty)) \ : \ (\mathcal{R} + \partial_t)f = 0\}.$$

Since $(0, \infty) \ni t \mapsto h_t \in \mathcal{S}'(G)$ is continuous and $(t, x) \mapsto h_t(x)$ is smooth on $(0, \infty) \times G$, the mapping T defined via

$$T\phi(x, t) = (e^{-t\mathcal{R}_2}\phi)(x) = \int_G h_t(x)\phi(x)dx, \quad \phi \in L^2(G),\ x \in G,\ t > 0,$$

is continuous from $L^2(G)$ to $\mathcal{D}'(G \times (0, \infty))$. Furthermore, the semi-group properties imply that the range of T lies in $\mathcal{N}$. Therefore, the mapping

$$L^2(G) \ni \phi \longmapsto T\phi(0, 1) = \int_G \phi(x)h_1(x)dx,$$

is a continuous functional. Hence h_1 must be square integrable.

By homogeneity (see (4.17)), for any $t > 0$, we see that $h_t \in L^2(G)$ as a consequence of Corollary 4.2.10 and (4.19) must hold. □

We now define the function $h : G \times \mathbb{R} \to \mathbb{C}$ as in the statement of Theorem 4.2.7 by

$$h(x,t) := \begin{cases} h_t(x) & \text{if } t > 0 \text{ and } x \in G, \\ 0 & \text{if } t \le 0 \text{ and } x \in G. \end{cases}$$

By Corollary 4.2.13, the function h is smooth on $G \times (\mathbb{R}\backslash\{0\})$ and satisfies the equation $(\mathcal{R} + \partial_t)h = 0$ on $G \times (\mathbb{R}\backslash\{0\})$. However, it is not obvious that it is a distribution on $G \times \mathbb{R}$. Our next goal is to prove that it is indeed a distribution and that it satisfies the equation $(\mathcal{R} + \partial_t)h = 0$ on $G \times \mathbb{R}$.

It is easy to prove that h is a distribution under the assumption $\nu > Q/2$ since it is then locally integrable:

Lemma 4.2.14. *If $\nu > Q/2$, then h is locally integrable on $G \times \mathbb{R}$.*

Proof of Lemma 4.2.14. We assume $\nu > Q/2$. We see that for any $\epsilon > 0$ and $R > 0$, using the homogeneity property given in (4.19),

$$\begin{aligned} \int_0^\epsilon \int_{|x|<R} |h(x,t)| dx dt &\le \int_0^\epsilon \left(\int_{|x|<R} |h_t(x)|^2 dx \right)^{\frac12} \left(\int_{|x|<R} 1 dx \right)^{\frac12} dt \\ &\le |B(0,1)|^{\frac12} R^{Q/2} \left(\int_G |h_1(x)|^2 dx \right)^{\frac12} \int_0^\epsilon t^{-\frac{Q}{2\nu}} dt \\ &= C R^{Q/2} \epsilon^{1-\frac{Q}{2\nu}}, \end{aligned}$$

since we assumed $\nu > Q/2$. This shows that h is locally integrable on $G \times \mathbb{R}$ and hence defines a distribution. □

If we know that h is a distribution, being a solution of $(\mathcal{R} + \partial_t)h = \delta_{0,0}$ is almost granted:

Lemma 4.2.15. *Let us assume that $h \in \mathcal{D}'(G \times \mathbb{R})$ is a distribution and that*

- *either $h_1 \in L^2(G)$ and $\nu > Q/2$,*
- *or $h_1 \in L^1(G)$ (without restriction on $\nu > Q/2$).*

Then h satisfies the equation

$$(\mathcal{R} + \partial_t)h = \delta_{0,0}$$

as a distribution.

The proof of Lemma 4.2.15 will require the following technical property which is independent of the rest of the proof:

Lemma 4.2.16. *Let $\mathcal{R}$ be a positive Rockland operator on a graded Lie group $G \sim \mathbb{R}^n$ with homogeneous degree ν. If $m\nu \ge \lceil \frac{n}{2} \rceil$, the functions in the domain of $\mathcal{R}^m$ are continuous on Ω, i.e.*

$$\text{Dom}(\mathcal{R}^m) \subset C(\Omega),$$

where $C(\Omega)$ denotes the space of continuous functions on Ω. Furthermore, for any compact subset Ω of G, there exists a constant $C = C_{\Omega,\mathcal{R},G,m}$ such that

$$\forall \phi \in \mathrm{Dom}(\mathcal{R}^m) \qquad \sup_{x\in\Omega} |\phi(x)| \le C \left(\|\phi\|_{L^2} + \|\mathcal{R}^m\phi\|_{L^2}\right).$$

This is a (very) weak form of Sobolev embeddings. We will later on obtain stronger results in Theorem 4.4.25. The proof below uses Corollary 4.1.14 showed by Helffer and Nourrigat during their proof of Theorem 4.1.12.

Proof of Lemma 4.2.16. By the classical Sobolev embedding theorem on $\mathbb{R}^n$, see e.g. [Ste70a, p.124], if $\phi \in L^2(\mathbb{R}^n)$ together with $\partial_x^\alpha \phi \in L^2(\mathbb{R}^n)$ for any multi-index α satisfying $|\alpha| \le \lceil \frac{n}{2} \rceil$, then ϕ may be modified on a set of zero measure so that the resulting function, still denoted by ϕ, is continuous.

Furthermore, for any compact subset Ω of G, we may choose a closed ball $B(0,R)$ strictly containing Ω, and there exists a constant $C = C_{\Omega,R}$ independent of ϕ such that

$$\sup_\Omega |\phi| \le C \sum_{|\alpha|\le\lceil\frac{n}{2}\rceil} \|\partial_x^\alpha \phi\|_{L^2(B(0,R))}.$$

As the abelian derivatives may be expressed as linear combination of left-invariant ones, see Section 3.1.5, there exists another constant $C = C_R$ such that

$$\sum_{|\alpha|\le\lceil\frac{n}{2}\rceil} \|\partial_x^\alpha \psi\|_{L^2(B(0,R))} \le C \sum_{|\alpha|\le\lceil\frac{n}{2}\rceil} \|X^\alpha \psi\|_{L^2(B(0,R))}$$

for any ψ such that the right-hand side makes sense. By the corollary of the Helffer-Nourrigat theorem applied to $\mathcal{R}^m$ (see Corollary 4.1.14, see also Lemma 4.2.5), there exists $C = C_{\mathcal{R},m} > 0$ such that

$$\forall \psi \in \mathcal{S}(G) \qquad \sum_{[\alpha]\le m\nu} \|X^\alpha \psi\|_{L^2(G)} \le C \left(\|\mathcal{R}^m\psi\|_{L^2(G)} + \|\psi\|_{L^2(G)}\right).$$

The last two properties yield easily

$$\sum_{|\alpha|\le\lceil\frac{n}{2}\rceil} \|\partial_x^\alpha \psi\|_{L^2(B(0,R))} \le C \left(\|\mathcal{R}^m\psi\|_{L^2(G)} + \|\psi\|_{L^2(G)}\right),$$

for any function $\psi \in L^2(G)$ for which the right-hand side makes sense, for some constant $C = C_{R,\mathcal{R},m}$ independent of ψ, as long as $m\nu \ge \lceil \frac{n}{2} \rceil$. Together with the embedding property recalled at the beginning of the proof, this shows Lemma 4.2.16. □

We can now go back to the proof of the heat kernel theorem, and more precisely, the proof of Lemma 4.2.15.

Proof of Lemma 4.2.15. If we set for each $\epsilon > 0$ and $(x,t) \in G \times \mathbb{R}$,

$$h^{(\epsilon)}(x,t) := \begin{cases} h(x,t) & \text{if } t > \epsilon, \\ 0 & \text{if } t \le \epsilon, \end{cases}$$

it is clear that this defines a distribution $h^{(\epsilon)} \in \mathcal{D}'(G\times\mathbb{R})$ and that $\{h^{(\epsilon)}\}$ converges to h in $\mathcal{D}'(G \times \mathbb{R})$ as ϵ tends to 0. To prove that

$$(\mathcal{R} + \partial_t)h = \delta_{0,0},$$

it suffices to show that $(\mathcal{R} + \partial_t)h^{(\epsilon)}$ converges to $\delta_{0,0}$ in $\mathcal{D}'(G \times \mathbb{R})$ as ϵ tends to 0; this means:

$$\forall \phi \in \mathcal{D}(G \times \mathbb{R}) \qquad \langle h^{(\epsilon)}, (\mathcal{R}^t - \partial_t)\phi\rangle = \langle (\mathcal{R} + \partial_t)h^{(\epsilon)}, \phi\rangle \xrightarrow[\epsilon\to 0]{\mathcal{D}'} \phi(0).$$

Using the translation of the group $H = G \times \mathbb{R}$ which is the direct product of the groups G and $(\mathbb{R}, +)$, this is equivalent to the pointwise convergence in H:

$$\forall \phi \in \mathcal{D}(H),\ (x,t) \in H \qquad (\mathcal{R} + \partial_t)(\phi * h^{(\epsilon)})(x,t) \xrightarrow[\epsilon\to 0]{} \phi(x,t), \tag{4.20}$$

since

$$(\mathcal{R} + \partial_t)(\phi * h^{(\epsilon)})(x,t) = \phi * ((\mathcal{R} + \partial_t)h^{(\epsilon)})(x,t) = \langle (\mathcal{R} + \partial_t)h^{(\epsilon)}, \phi((x,t)\,\cdot^{-1})\rangle.$$

The above convolution is in H, given by

$$\begin{aligned} \left(\phi * h^{(\epsilon)}\right)(x,t) &= \int_G \int_{\mathbb{R}} \phi(y,u)\, h^{(\epsilon)}((y,u)^{-1}(x,t))dydu \\ &= \int_G \int_{u=-\infty}^{t-\epsilon} \phi(y,u)\, h(y^{-1}x, -u+t)dydu. \end{aligned}$$

We see that

$$\begin{aligned} (\mathcal{R} + \partial_t)(\phi * h^{(\epsilon)})(x,t) &= \int_G \int_{u=-\infty}^{t-\epsilon} \phi(y,u)\, (\mathcal{R}_x + \partial_t)h(y^{-1}x, -u+t)dydu \\ &\quad + \int_G \phi(y, t-\epsilon)\, h(y^{-1}x, \epsilon)dy, \end{aligned}$$

and the first term of the right hand side is zero since $(\mathcal{R} + \partial_t)h = 0$ on $G \times (0,\infty)$ and $\mathcal{R} + \partial_t$ is left-invariant on H. Hence

$$(\mathcal{R} + \partial_t)(\phi * h^{(\epsilon)})(x,t) = \phi(\cdot, t-\epsilon) * h_\epsilon(x), \tag{4.21}$$

using the convolution in H and G for the left and right hand sides respectively.

We now fix t and set $\phi_\epsilon(y) := \phi(y, t-\epsilon)$. Then

$$\phi(\cdot, t-\epsilon) * h_\epsilon = \phi_\epsilon * h_\epsilon,$$

and we can write

$$\phi_\epsilon * h_\epsilon - \phi_0 = (\phi_\epsilon - \phi_0) * h_\epsilon - (\phi_0 * h_\epsilon - \phi_0). \tag{4.22}$$

For the first term in the right-hand side of (4.22), we need to separate the case $h_1 \in L^2(G)$ with $\nu > Q/2$ from the case $h_1 \in L^1(G)$. Indeed if $h_1 \in L^2(G)$ with $\nu > Q/2$, then by (4.19),

$$\|h_\epsilon\|_2 = \epsilon^{-\frac{Q}{2\nu}} \|h_1\|_2$$

and the Cauchy-Schwartz inequality yields

$$\|(\phi_\epsilon - \phi_0) * h_\epsilon\|_\infty \leq \|\phi_\epsilon - \phi_0\|_2 \|h_\epsilon\|_2.$$

We easily obtain $\|\phi_\epsilon - \phi_0\|_2 \leq C\epsilon$ as $\phi \in \mathcal{D}(G \times \mathbb{R})$. Thus

$$\|(\phi_\epsilon - \phi_0) * h_\epsilon\|_\infty \leq C'\epsilon^{1-\frac{Q}{2\nu}} \longrightarrow_{\epsilon \to 0} 0,$$

since we assumed $\nu > Q/2$. If $h_1 \in L^1(G)$, then by (4.19), $\|h_\epsilon\|_1 = \|h_1\|_1$ and the Hölder inequality yields

$$\|(\phi_\epsilon - \phi_0) * h_\epsilon\|_\infty \leq \|\phi_\epsilon - \phi_0\|_\infty \|h_\epsilon\|_1 = \|h_1\|_1 \|\phi_\epsilon - \phi_0\|_\infty.$$

Again $\|\phi_\epsilon - \phi_0\|_2 \leq C\epsilon$ as $\phi \in \mathcal{D}(G \times \mathbb{R})$ thus

$$\|(\phi_\epsilon - \phi_0) * h_\epsilon\|_\infty \leq C'\epsilon \longrightarrow_{\epsilon \to 0} 0.$$

For the second term in the right-hand side of (4.22), the functional calculus of $\mathcal{R}_2$ yields the convergence in $L^2(G)$

$$\phi_0 * h_\epsilon = e^{-\epsilon \mathcal{R}_2} \phi_0 \longrightarrow_{\epsilon \to 0} \phi_0.$$

As $\mathcal{R}_2$ commutes with the $\mathcal{R}_2$-multiplier $e^{-\epsilon \mathcal{R}_2}$ and since $\phi_0 \in \mathcal{D}(G)$, $\mathcal{R}_2 \phi_0 = \mathcal{R}\phi_0$, we know that $\phi_0 * h_\epsilon = e^{-\epsilon \mathcal{R}_2} \phi_0 \in \text{Dom}(\mathcal{R}_2)$ and moreover

$$(\mathcal{R}\phi_0) * h_\epsilon = (\mathcal{R}_2 \phi_0) * h_\epsilon = e^{-\epsilon \mathcal{R}_2} \mathcal{R}_2 \phi_0 = \mathcal{R}_2 e^{-\epsilon \mathcal{R}_2} \phi_0 \overset{L^2(G)}{\longrightarrow}_{\epsilon \to 0} \mathcal{R}_2 \phi_0.$$

More generally, for any $m \in \mathbb{N}$, $\phi_0 * h_\epsilon = e^{-\epsilon \mathcal{R}_2} \phi_0 \in \text{Dom}(\mathcal{R}_2^m)$ and

$$\mathcal{R}_2^m e^{-\epsilon \mathcal{R}_2} \phi_0 \overset{L^2(G)}{\longrightarrow}_{\epsilon \to 0} \mathcal{R}_2^m \phi_0.$$

By Lemma 4.2.16, this implies that $\phi_0 * h_\epsilon - \phi_0$ is continuous on G. Furthermore, for any compact subset Ω of $G \sim \mathbb{R}^n$ and any $m \in \mathbb{N}$ with $m\nu > \lfloor \frac{n}{2} \rfloor$, we have

$$\sup_\Omega |\phi_0 * h_\epsilon - \phi_0| \leq C \left(\|\phi_0 * h_\epsilon - \phi_0\|_2 + \|\mathcal{R}^m (\phi_0 * h_\epsilon - \phi_0)\|_2 \right) \longrightarrow_{\epsilon \to 0} 0.$$

Hence we have obtained that both terms on the right-hand side of (4.22) go to zero for the supremum norm on any compact subset of G. Therefore, the expression in (4.21) tends to

$$(\mathcal{R} + \partial_t)(\phi * h^{(\epsilon)})(x,t) \longrightarrow_{\epsilon \to 0} \phi(\cdot, t - \epsilon) * h_\epsilon(x),$$

for t fixed, locally in x. This is even stronger than the pointwise convergence in H we wanted in (4.20) and concludes the proof of Lemma 4.2.15. □

Corollary 4.2.17. *Under the hypothesis of Lemma 4.2.15, h is smooth on $(G \times \mathbb{R})\backslash\{(0,0)\}$ and satisfies (4.15) and (4.18). Moreover, each function h_t is Schwartz on G and*

$$\int_G h_t(x)dx = 1.$$

Proof of Corollary 4.2.17. By Lemma 4.2.15, the distribution h annihilates the hypoelliptic operator $\mathcal{R}+\partial_t$ on $(G\times\mathbb{R})\backslash\{0\}$, and thus h is smooth on $(G\times\mathbb{R})\backslash\{0\}$. Since $h(x,t)=0$ for $t\leq 0$, this implies that $h(x,t)$ vanish to infinite order as $t\to 0$:

$$\forall x \in G\backslash\{0\},\ N\in\mathbb{N}_0 \quad \exists \epsilon>0,\ C>0 \quad \forall t\in(0,\epsilon) \quad |h(x,t)|\leq Ct^N.$$

We can choose $\epsilon=1$ since h is smooth on $G\times(0,\infty)$. In fact this estimate remains true for any x-derivatives $(\frac{\partial}{\partial x})^\alpha h(x,t)$. It is also uniform in x when x runs over a fixed compact set which does not contain 0. Choosing this compact set to be the unit sphere of a given quasi-norm $|\cdot|$, we have

$$\forall N\in\mathbb{N}_0 \quad \exists C>0 \quad \forall t\in(0,1] \quad \sup_{|x|=1}\left|\left(\frac{\partial}{\partial x}\right)^\alpha h(x,t)\right|\leq Ct^N.$$

We may replace the abelian derivatives $(\frac{\partial}{\partial x})^\alpha$ by the left-invariant ones, see Section 3.1.5. This implies (4.15).

Using the homogeneity of h (see Property (4.12) which was already proven and Proposition 3.1.23), we have

$$\forall x\in G,\ r>0 \qquad X^\alpha h(x,t)=r^{Q-[\alpha]}X^\alpha h_{r^\nu t}(rx),$$

and so, in particular, if $|x|\geq 1$ then we obtain, because of (4.15), that

$$|X^\alpha h_1(x)|=|x|^{-Q+[\alpha]}|X^\alpha h_{|x|^{-\nu}}(|x|^{-1}x)|\leq C_{\alpha,N}|x|^{-Q+[\alpha]-\nu N}.$$

Since h_1 is smooth on G, this shows that h_1 is Schwartz. This is also the case for h_t by homogeneity, see (4.17). Note that the same homogeneity property together with (4.15) implies (4.18).

Since each function h_t satisfies the homogeneity property given in (4.17) and is integrable, the functions h_t form a commutative approximation of the identity, see Remark 3.1.60. In particular,

$$\phi * h_t \longrightarrow_{t\to 0} c\phi \quad \text{in } L^2(G),$$

with $c=\int_G h_1(x)dx$. Since we know

$$\phi * h_t = e^{-t\mathcal{R}_2}\phi \longrightarrow_{\epsilon\to 0} \phi \quad \text{in } L^2(G),$$

this constant c must be equal to 1. By homogeneity,

$$\forall t>0 \qquad \int_G h_t(x)dx=\int_G h_1(x)dx=c=1.$$

□

Lemmata 4.2.14 and 4.2.15 imply Theorem 4.2.7 and Corollary 4.2.10 under the assumption $\nu > Q/2$. We now need to remove this assumption. For this, we will use the following formula which is a consequence of the principle of subordination:

Lemma 4.2.18. *For any $\gamma > 0$, we have*

$$e^{-\gamma} = \int_0^\infty \frac{e^{-s}}{\sqrt{\pi s}} e^{-\frac{\gamma^2}{4s}} ds. \tag{4.23}$$

Sketch of the proof of Lemma 4.2.18. We follow [Ste70a, p.61]. We start from the well known identity

$$\pi e^{-\gamma} = \int_{-\infty}^\infty \frac{e^{i\gamma x}}{1+x^2} dx, \tag{4.24}$$

which is an application of the Residue theorem to the function

$$z \mapsto \frac{e^{i\gamma z}}{z^2+1}.$$

In (4.24) we replace $1+x^2$ using

$$\frac{1}{1+x^2} = \int_0^\infty e^{-(1+x^2)u} du,$$

and we obtain the double integral

$$\pi e^{-\gamma} = \int_{-\infty}^\infty e^{i\gamma x} \int_0^\infty e^{-(1+x^2)u} du\, dx.$$

One can show that it is possible to invert the order of integration:

$$\pi e^{-\gamma} = \int_0^\infty e^{-u} \int_{-\infty}^\infty e^{i\gamma x} e^{-x^2 u} dx\, du.$$

It is well known that the inner integral in dx is equal to

$$\frac{e^{-\frac{\gamma^2}{4u}}}{\sqrt{\pi u}}.$$

And this shows (4.23). □

We can now finish the proofs of Theorem 4.2.7 and Corollary 4.2.10.

End of the proofs of Theorem 4.2.7 and Corollary 4.2.10. Since the case $\nu > Q/2$ is already proven, we may assume $\nu \le Q/2$.

For any $m \in \mathbb{N}_0$, $\mathcal{R}^{2^m}$ is a positive Rockland operator (see Lemma 4.2.5), with homogeneous degree $2^m\nu$. We denote by K_m the function on $G \times \mathbb{R}$ giving its heat kernel in the sense that if $t > 0$, $K_m(\cdot, t) \in \mathcal{S}'(G)$ is the kernel of $e^{-t\mathcal{R}_2^{2^m}}$

and if $t \leq 0$ then $K_m(x,t) = 0$ for any $x \in G$. This is possible since, by Corollary 4.2.13, K_m is smooth on $G \times (0,\infty)$. By homogeneity, it will always satisfy

$$\forall x \in G,\ t > 0 \qquad K_m(x,t) = t^{-\frac{Q}{\nu 2^m}} K_m(t^{-\frac{1}{\nu 2^m}} x, 1). \tag{4.25}$$

In (4.23), replacing γ by $t\lambda^{2^{m-1}}$, one finds that

$$e^{-t\lambda^{2^{m-1}}} = \int_0^\infty \frac{e^{-s}}{\sqrt{\pi s}} e^{-\frac{t^2\lambda^{2^m}}{4s}} ds.$$

Using the functional calculus on $\mathcal{R}$, that is, integrating against the spectral measure $dE(\lambda)$ of $\mathcal{R}_2$, we obtain formally that for any non-negative integer $m \in \mathbb{N}_0$ and $t > 0$,

$$e^{-t\mathcal{R}_2^{2^{m-1}}} = \int_0^\infty \frac{e^{-s}}{\sqrt{\pi s}} e^{-\frac{t^2}{4s}\mathcal{R}_2^{2^m}} ds, \tag{4.26}$$

and for the kernels of these operators,

$$K_{m-1}(x,t) = \int_0^\infty \frac{e^{-s}}{\sqrt{\pi s}} K_m(x, \frac{t^2}{4s}) ds. \tag{4.27}$$

It is not difficult to see that Formulae (4.26) and (4.27) hold as operators and continuous integrable functions respectively when, for instance, $K_m(\cdot,t)$ is integrable on G for each $t > 0$ and

$$\int_0^\infty \frac{e^{-s}}{\sqrt{\pi s}} \|K_m(\cdot, \frac{t^2}{4s})\|_{L^1(G)} ds < \infty.$$

Indeed under this hypothesis, $K_{m-1}(\cdot,t)$ is integrable on G for any fixed $t > 0$ and

$$\|K_{m-1}(\cdot,t)\|_{L^1(G)} \leq \int_0^\infty \frac{e^{-s}}{\sqrt{\pi s}} \|K_m(\cdot, \frac{t^2}{4s})\|_{L^1(G)} ds < \infty. \tag{4.28}$$

It is then a standard procedure to make sense of (4.26) by first integrating λ over $[0,N]$ and then letting N tend to infinity.

We first assume that $2^m \nu > Q/2$, so that the conclusion of Theorem 4.2.7 holds for K_m. In particular, $K_m(\cdot,1) \in \mathcal{S}(G)$ and by homogeneity, the L^1-norm of $K_m(\cdot,t)$ is

$$\int_G |K_m(x,t)| dx = \int_G |K_m(x,1)| dx,$$

is finite and independent of t. Therefore

$$\int_0^\infty \frac{e^{-s}}{\sqrt{\pi s}} \int_G |K_m(x, \frac{t^2}{4s})| dx ds = \int_G |K_m(x,1)| dx \int_0^\infty \frac{e^{-s}}{\sqrt{\pi s}} ds,$$

is finite. Consequently Formula (4.27) holds and by (4.28),

$$\|K_{m-1}(t,\cdot)\|_{L^1(G)} \leq \int_G |K_m(x,1)| dx \int_0^\infty \frac{e^{-s}}{\sqrt{\pi s}} ds < \infty.$$

By homogeneity, $\int_G |K_{m-1}(x,t)|dx$ must also be independent of $t>0$, while it is identically zero if $t\leq 0$. This implies that K_{m-1} is locally integrable on $G\times\mathbb{R}$ and that $K_{m-1}(\cdot,1)\in L^1(G)$. By Lemmata 4.2.14 and 4.2.15, K_{m-1} satisfy the properties of the heat kernel described in Theorem 4.2.7 and Corollary 4.2.10.

Now we can repeat the same reasoning with m replaced successively by $m-1, m-2,\ldots,2,1$. Since $K_0=h$, this concludes the proofs of Theorem 4.2.7 and Corollary 4.2.10. □

We still have to show Corollary 4.2.9.

Proof of Corollary 4.2.9. Since the heat kernels h_t, $t>0$, form a commutative approximation of the identity (see Theorem 4.2.7 and Remark 3.1.60 in Section 3.1.10), the operators $f\mapsto f*h_t$, $t>0$, form a strongly continuous semi-group on $L^p(G)$ for any $p\in[1,\infty)$ and on $C_o(G)$, see Lemma 3.1.58. It is naturally equibounded by $\|h_1\|$ since

$$\|f*h_t\|_p\leq\|f\|_p\|h_t\|_1 \quad\text{and}\quad \|h_t\|_1=\|h_1\|.$$

Let us prove the convergence in (4.16) for $p=\infty$. Let $f\in\mathcal{D}(G)$. By Lemma 4.2.16, for any compact subset $\Omega\subset G$,

$$\begin{aligned}&\sup_\Omega\left|\frac{1}{t}(f*h_t-f)-\mathcal{R}f\right|\\&\leq C\left(\left\|\frac{1}{t}(f*h_t-f)-\mathcal{R}f\right\|_2+\left\|\frac{1}{t}\mathcal{R}^m(f*h_t-f)-\mathcal{R}^{m+1}f\right\|_2\right),\end{aligned}$$

where m is an integer such that $m\nu\geq\lceil\frac{n}{2}\rceil$. Since $\mathcal{D}(G)\subset\text{Dom}(\mathcal{R})$ and

$$e^{-t\mathcal{R}_2}f=f*h_t,$$

we have for any integer $m'\in\mathbb{N}_0$ that

$$\begin{aligned}&\frac{1}{t}\mathcal{R}^{m'}(f*h_t-f)-\mathcal{R}^{m'+1}f=\frac{1}{t}\mathcal{R}_2^{m'}\left(e^{-t\mathcal{R}_2}f-f\right)-\mathcal{R}_2^{m'+1}f\\&=\frac{1}{t}\left(e^{-t\mathcal{R}_2}\mathcal{R}_2^{m'}f-\mathcal{R}_2^{m'}f\right)-\mathcal{R}_2^{m'+1}f=\frac{1}{t}\left((\mathcal{R}^{m'}f)*h_t-\mathcal{R}^{m'}f\right)-\mathcal{R}^{m'+1}f\\&\longrightarrow_{t\to0}0\quad\text{in } L^2(G).\end{aligned}$$

Therefore,

$$\sup_\Omega\left|\frac{1}{t}(f*h_t-f)-\mathcal{R}f\right|\longrightarrow_{t\to0}0.$$

We fix a quasi-norm $|\cdot|$. By Part 2 of Remark 3.2.16 and the existence of a homogeneous norm (Theorem 3.1.39), without loss of generality, we may assume $|\cdot|$ to be also a norm, that is, the triangular inequality is satisfied with constant 1; although we could give a proof without this hypothesis, it simplifies the constants below. Let $\bar{B}_R$ be a closed ball about 0 of radius R which contains the support of f. We choose $\Omega = \bar{B}_{2R}$ the closed ball about 0 and with radius $2R$. If $x \notin \Omega$, then since f is supported in $\bar{B}_R \subset \Omega$,

$$\left(\frac{1}{t}(f * h_t - f) - \mathcal{R}f\right)(x) = \frac{1}{t} f * h_t(x) = \frac{1}{t}\int_{|y|\leq R} f(y) h_t(y^{-1}x) dy,$$

hence

$$\left|\frac{1}{t} f * h_t(x)\right| \leq \frac{\|f\|_\infty}{t}\int_{|y|\leq R} |h_t(y^{-1}x)| dy = \frac{\|f\|_\infty}{t}\int_{|xt^{\frac{1}{\nu}} z^{-1}|\leq R} |h_1(z)| dz,$$

as h_t satisfies (4.17). Note that $\{z : |xt^{\frac{1}{\nu}} z^{-1}| \leq R\} \subset \{z : |t^{\frac{1}{\nu}} z| > R/2\}$ since

$$|t^{\frac{1}{\nu}} z| \leq R/2 \implies |xt^{\frac{1}{\nu}} z^{-1}| \geq |x| - |t^{\frac{1}{\nu}} z^{-1}| \geq \frac{3}{2} R.$$

Therefore

$$\int_{|xt^{\frac{1}{\nu}} z^{-1}|\leq R} |h_1(z)| dz \leq \int_{|z|>t^{-\frac{1}{\nu}} R/2} |h_1(z)| dz.$$

Since h_1 is Schwartz, we must have

$$\exists C \quad \forall z \in G\backslash\{0\} \quad |h_1(z)| \leq C|z|^{-a},$$

for $a = Q + 2\nu$ for instance. This together with the polar change of variable (cf. Proposition 3.1.42) yield

$$\int_{|z|>t^{-\frac{1}{\nu}} R/2} |h_1(z)| dz \leq C \int_{r=t^{-\frac{1}{\nu}} R/2}^{\infty} r^{-a-Q-1} dr = C' t^2.$$

Consequently, denoting by Ω^c the complement of Ω in G, we have

$$\sup_{\Omega^c}\left|\frac{1}{t}(f * h_t - f) - \mathcal{R}f\right| \leq C' t \longrightarrow_{t\to 0} 0.$$

This shows the convergence in (4.16) for $p = \infty$.

We proceed in a similar way to prove the convergence in (4.16) for p finite. As above we fix $f \in \mathcal{D}(G)$ supported in $\bar{B}_R$. We decompose

$$\begin{aligned}&\|\frac{1}{t}(f * h_t - f) - \mathcal{R}f\|_p \\ &\quad\leq \|\frac{1}{t}(f * h_t - f) - \mathcal{R}f\|_{L^p(\bar{B}_{2R})} + \|\frac{1}{t}(f * h_t - f) - \mathcal{R}f\|_{L^p(B_{2R}^c)}.\end{aligned}$$

For the first term,

$$\|\frac{1}{t}(f * h_t - f) - \mathcal{R}f\|_{L^p(\bar{B}_{2R})} \leq |\bar{B}_{2R}|^{\frac{1}{p}} \|\frac{1}{t}(f * h_t - f) - \mathcal{R}f\|_\infty \underset{t\to 0}{\longrightarrow} 0,$$

as we have already proved the convergence in (4.16) for $p=\infty$. For the second term, we obtain for the reasons explained in the case $p=\infty$ that

$$\begin{aligned}\|\frac{1}{t}(f*h_t-f)-\mathcal{R}f\|_{L^p(B_{2R}^c)} &= \frac{1}{t}\|f*h_t\|_{L^p(B_{2R}^c)}\\ &= \frac{1}{t}\left(\int_{|x|>2R}\left|\int_{|y|<R} f(y)\, h_t(y^{-1}x)dy\right|^p dx\right)^{\frac{1}{p}}\\ &\leq \frac{C}{t}\left(\int_{|x|>2R}\left(\int_{|y|<R} |f(y)|\, t^{-\frac{Q}{\nu}}|t^{-\frac{1}{\nu}}(y^{-1}x)|^{-a}dy\right)^p dx\right)^{\frac{1}{p}}\\ &\leq Ct^{-1+p(-\frac{Q}{\nu}+\frac{a}{\nu})}\|f\|_{L^1}\left(\int_{|x|>2R}(|x|-R)^{-ap}dx\right)^{\frac{1}{p}},\end{aligned}$$

where we have used that the reverse triangle inequality

$$|y^{-1}x|\geq |x|-|y|\geq |x|-R.$$

Consequently we obtain the convergence in (4.16) for p finite if we choose a large enough. □

4.3 Fractional powers of positive Rockland operators

In this section we aim at defining fractional powers of positive Rockland operators. We will carry out the construction on the scale of L^p-spaces for $1\leq p\leq\infty$, with $L^\infty(G)$ substituted by the space $C_o(G)$ of continuous functions vanishing at infinity. The extension of a positive Rockland operator $\mathcal{R}$ to $L^p(G)$ will be denoted by $\mathcal{R}_p$, and first we discuss the essential properties of such an extension. Then we define its complex powers. Before studying the corresponding Riesz and Bessel potentials, we will show that imaginary powers are continuous operators on L^p, $p\in(1,\infty)$.

4.3.1 Positive Rockland operators on L^p

We start by defining the analogue $\mathcal{R}_p$ of the operator $\mathcal{R}$ on $L^p(G)$.

Definition 4.3.1. Let $\mathcal{R}$ be a positive Rockland operator on a graded Lie group G.

For $p\in[1,\infty)$, we denote by $\mathcal{R}_p$ the operator such that $-\mathcal{R}_p$ is the infinitesimal generator of the semi-group of operators $f\mapsto f*h_t$, $t>0$, on $L^p(G)$.

We also denote by $\mathcal{R}_{\infty_o}$ the operator such that $-\mathcal{R}_{\infty_o}$ is the infinitesimal generator of the semi-group of operators $f\mapsto f*h_t$, $t>0$, on $C_o(G)$.

For the moment it seems that $\mathcal{R}_2$ denotes the self-adjoint extension of $\mathcal{R}$ on $L^2(G)$ and minus the generator of $f\mapsto f*h_t$, $t>0$, on $L^2(G)$. In the sequel, in fact in Theorem 4.3.3 below, we show that the two operators coincide and there is no conflict of notation.

The case $p = \infty$ is somewhat irrelevant and will be often replaced by $p = \infty_o$, especially when using duality. The next lemma aims at clarifying this point.

Lemma 4.3.2. • *If $p \in (1, \infty)$, any bounded linear functional on $L^p(G)$ can be realised by integration against a function in $L^{p'}(G)$, where p' is the conjugate exponent of p, that is, $\frac{1}{p} + \frac{1}{p'} = 1$. Consequently, the dual $L^p(G)'$ of $L^p(G)$ may be identified with $L^{p'}(G)$ and the corresponding norms coincide.*

- *If $p = 1$, any bounded linear functional on $L^1(G)$ can be realised by integration against a bounded function on G. Consequently, the dual $L^1(G)'$ of $L^1(G)$ may be identified with $L^\infty(G)$ and the corresponding norms coincide. In particular, $L^1(G)'$ contains $C_o(G)$.*
- *If $p = \infty_o$, any bounded linear functional on $C_o(G)$ can be realised by integration against a regular complex measure. Consequently, the dual $C_o(G)'$ of $C_o(G)$ may be identified with the Banach space $M(G)$ of regular complex measures endowed with the total mass $\|\cdot\|_{M(G)}$ as its norm, and the corresponding norms coincide. With this identification, $C_o(G)'$ contains $L^1(G)$ and the corresponding norms coincide.*

Proof. See, e.g., Rudin [Rud87, ch.6]. □

We can now describe the properties of $\mathcal{R}_p$.

Theorem 4.3.3. *Let $\mathcal{R}$ be a positive Rockland operator on a graded Lie group G. In this statement, $p \in [1, \infty) \cup \{\infty_o\}$.*

(i) The semi-group $\{f \mapsto f * h_t\}_{t>0}$ is strongly continuous and equicontinuous on $L^p(G)$ if $p \in [1, \infty)$ or on $C_o(G)$ if $p = \infty_o$:

$$\forall t > 0,\ \forall f \in L^p(G) \text{ or } C_o(G) \qquad \|f * h_t\|_p \le \|h_1\|_1 \|f\|_p.$$

Consequently, the operator $\mathcal{R}_p$ is closed. The domain of $\mathcal{R}_p$ contains $\mathcal{D}(G)$, and for $f \in \mathcal{D}(G)$ we have $\mathcal{R}_p f = \mathcal{R} f$.

(ii) The operator $\bar{\mathcal{R}}_p$ is the infinitesimal generator of the strongly continuous semi-group $\{f \mapsto f * \bar{h}_t\}_{t>0}$ on $L^p(G)$.

(iii) We use the identifications of Lemma 4.3.2. If $p \in (1, \infty)$ then the dual of $\mathcal{R}_p$ is $\bar{\mathcal{R}}_{p'}$. The dual of $\mathcal{R}_{\infty_o}$ restricted to $L^1(G)$ is $\bar{\mathcal{R}}_1$. The dual of $\mathcal{R}_1$ restricted to $C_o(G) \subset L^\infty(G)$ is $\bar{\mathcal{R}}_{\infty_o}$.

(iv) If $p \in [1, \infty)$, the operator $\mathcal{R}_p$ is the maximal restriction of $\mathcal{R}$ to $L^p(G)$, that is, the domain of $\mathcal{R}_p$ consists of all the functions $f \in L^p(G)$ such that the distributional derivative $\mathcal{R} f$ is in $L^p(G)$ and $\mathcal{R}_p f = \mathcal{R} f$.

The operator $\mathcal{R}_{\infty_o}$ is the maximal restriction of $\mathcal{R}$ to $C_o(G)$, that is, the domain of $\mathcal{R}_{\infty_o}$ consists of all the functions $f \in C_o(G)$ such that the distributional derivative $\mathcal{R} f$ is in $C_o(G)$ and $\mathcal{R}_p f = \mathcal{R} f$.

(v) *If* $p \in [1,\infty)$, *the operator* $\mathcal{R}_p$ *is the smallest closed extension of* $\mathcal{R}|_{\mathcal{D}(G)}$ *on* $L^p(G)$. *For* $p=2$, $\mathcal{R}_2$ *is the self-adjoint extension of* $\mathcal{R}$ *on* $L^2(G)$.

Proof. Part (i) is a consequence of Corollary 4.2.9, see also Section A.2.

Part (i) implies, intertwining with the complex conjugate, that $\{f \mapsto f * \bar{h}_t\}_{t>0}$ is also a strongly continuous semi-group on $L^p(G)$. On $\mathcal{D}(G)$, its infinitesimal operator coincide with $\bar{\mathcal{R}} = \mathcal{R}^t$ which is a positive Rockland operator (see Lemma 4.2.5) and it is easy to see that

$$\forall \phi \in \mathcal{D}(G),\ t>0 \qquad e^{-t\bar{\mathcal{R}}_2}\phi = \overline{e^{-t\mathcal{R}_2}\bar{\phi}} = \overline{\bar{\phi} * h_t} = \phi * \bar{h}_t.$$

This shows Part (ii).

For Part (iii), we observe that using (1.14) and (4.13), we have

$$\forall f_1, f_2 \in \mathcal{D}(G) \qquad \langle f_1 * h_t, f_2\rangle = \langle f_1, f_2 * \bar{h}_t\rangle. \tag{4.29}$$

Thus we have for any $f,g \in \mathcal{D}(G)$ and $p \in [1,\infty) \cup \{\infty_o\}$

$$\langle \frac{1}{t}(e^{-t\mathcal{R}_p}f - f), g\rangle = \frac{1}{t}\langle f * h_t - f, g\rangle = \frac{1}{t}\langle f, g * \bar{h}_t - g\rangle = \frac{1}{t}\langle f, e^{-t\bar{\mathcal{R}}_{p'}}g - g\rangle.$$

Here the brackets refer to the duality in the sense of distributions or, equivalently, to the duality explained in Lemma 4.3.2. Taking the limit as $t \to 0$ of the first and last expressions proves Part (iii).

We now prove Part (iv) for any $p \in [1,\infty) \cup \{\infty_o\}$. Let $f \in \text{Dom}(\mathcal{R}_p)$ and $\phi \in \mathcal{D}(G)$. Since $\mathcal{R}$ is formally self-adjoint, we know that $\mathcal{R}^t = \bar{\mathcal{R}}$, and by Part (i), we have $\mathcal{R}_q\phi = \mathcal{R}\phi$ for any $q \in [1,\infty) \cup \{\infty_o\}$. Thus by Part (iii) we have

$$\langle \mathcal{R}_p f, \phi\rangle = \langle f, \bar{\mathcal{R}}_{p'}\phi\rangle = \langle f, \mathcal{R}^t\phi\rangle = \langle \mathcal{R}f, \phi\rangle,$$

and $\mathcal{R}_p f = \mathcal{R}f$ in the sense of distributions. Thus

$$\text{Dom}(\mathcal{R}_p) \subset \{f \in L^p(G) \ :\ \mathcal{R}f \in L^p(G)\}.$$

We now prove the reverse inclusion. Let $f \in L^p(G)$ such that $\mathcal{R}f \in L^p(G)$. Let also $\phi \in \mathcal{D}(G)$. The following computations are justified by the properties of $\mathcal{R}$ and h_t (see Theorem 4.2.7), Fubini's Theorem, and (4.29):

$$\begin{aligned}
\langle f * h_t - f, \phi\rangle &= \langle f, \phi * \bar{h}_t - \phi\rangle = \langle f, \int_0^t \partial_s(\phi * \bar{h}_s)ds\rangle \\
&= \langle f, \int_0^t -\bar{\mathcal{R}}(\phi * \bar{h}_s)ds\rangle = -\langle f, \bar{\mathcal{R}}\int_0^t (\phi * \bar{h}_s)ds\rangle \\
&= -\langle \mathcal{R}f, \int_0^t \phi * \bar{h}_s ds\rangle = -\int_0^t \langle \mathcal{R}f, \phi * \bar{h}_s\rangle ds \\
&= -\int_0^t \langle (\mathcal{R}f) * h_s, \phi\rangle ds = -\langle \int_0^t (\mathcal{R}f) * h_s ds, \phi\rangle.
\end{aligned}$$

Therefore,

$$f * h_t - f = -\int_0^t (\mathcal{R}f) * h_s ds.$$

Let us recall the following general property: if $t \mapsto x_t$ is a continuous mapping from $[0,\infty)$ to a Banach space $\mathcal{X}$, then $\frac{1}{t}\int_0^t x_s ds$ converges to x_0 in the strong topology of $\mathcal{X}$ as $t \to 0$. We apply this property to $\mathcal{X} = L^p(G)$ and $t \mapsto (\mathcal{R}f) * h_t$; the hypotheses are indeed satisfied because of the properties of the heat kernel, see Theorem 4.2.7. Hence we have the following convergence in $L^p(G)$:

$$\frac{1}{t}(f * h_t - f) = -\frac{1}{t}\int_0^t (\mathcal{R}f) * h_s ds \underset{t\to 0}{\longrightarrow} -\mathcal{R}f.$$

This shows $f \in \mathrm{Dom}(\mathcal{R}_p)$ and concludes the proof of (iv).

Part (v) follows from (iv). This also shows that the self-adjoint extension of $\mathcal{R}$ coincides with $\mathcal{R}_2$ as defined in Definition 4.3.1 and concludes the proof of Theorem 4.3.3. □

Theorem 4.3.3 has the following couple of corollaries which will enable us to define the fractional powers of $\mathcal{R}_p$.

Corollary 4.3.4. *We keep the same setting and notation as in Theorem 4.3.3.*

(i) The operator $\mathcal{R}_p$ is injective on $L^p(G)$ for $p \in [1,\infty)$ and $\mathcal{R}_{\infty_o}$ is injective on $C_o(G)$, namely,

$$\text{for } p \in [1,\infty)\cup\{\infty_o\} : \qquad \forall f \in \mathrm{Dom}(\mathcal{R}_p) \qquad \mathcal{R}_p f = 0 \Longrightarrow f = 0.$$

(ii) If $p \in (1,\infty)$ then the operator $\mathcal{R}_p$ has dense range in $L^p(G)$. The operator $\mathcal{R}_{\infty_o}$ has dense range in $C_o(G)$. The closure of the range of $\mathcal{R}_1$ is the closed subspace $\{\phi \in L^1(G) : \int_G \phi = 0\}$ of $L^1(G)$.

Proof. Let $f \in \mathrm{Dom}(\mathcal{R}_p)$ be such that $\mathcal{R}_p f = 0$ for $p \in [1,\infty)\cup\{\infty_o\}$. By Theorem 4.3.3 (iv), $f \in \mathcal{S}'(G)$ and $\mathcal{R}f = 0$. In Remark 4.1.13 (3), we noticed that any positive Rockland operator satisfies the hypotheses of Liouville's Theorem for homogeneous Lie groups, that is, Theorem 3.2.45. Consequently f is a polynomial. Since f is also in $L^p(G)$ for $p \in [1,\infty)$ or in $C_o(G)$ for $p = \infty_o$, f must be identically zero. This proves (i).

For (ii), let Ψ be a bounded linear functional on $L^p(G)$ if $p \in [1,\infty)$ or on $C_o(G)$ if $p = \infty_o$ such that Ψ vanishes identically on Range$(\mathcal{R}_p)$. Then Ψ can be realised as the integration against a function $f \in L^{p'}(G)$ if $p \in [1,\infty)$ or a measure also denoted by $f \in M(G)$ if $p = \infty_o$, see Lemma 4.3.2. Using the distributional notation, we have

$$\Psi(\phi) = \langle f, \phi\rangle \qquad \forall \phi \in L^p(G) \quad \text{or} \quad \forall \phi \in C_o(G).$$

Then for any $\phi \in \mathcal{D}(G)$, we know that $\phi \in \text{Dom}(\mathcal{R}_p)$ and $\mathcal{R}_p\phi = \mathcal{R}\phi$ by Theorem 4.3.3 (i) thus

$$0 = \Psi(\mathcal{R}_p(\phi)) = \langle f, \mathcal{R}(\phi)\rangle = \langle \bar{\mathcal{R}} f, \phi\rangle,$$

since $\mathcal{R}^t = \bar{\mathcal{R}}$. This shows that $\bar{\mathcal{R}} f = 0$. Applying again Liouville's Theorem, this time to the positive Rockland operator $\bar{\mathcal{R}}$ (see Lemma 4.2.5), this shows that f is a polynomial. For $p \in (1,\infty)$, f being also a function in $L^{p'}(G)$, this implies that $f \equiv 0$. For $p = \infty_o$, $f \in M(G)$, this shows that f is an integrable polynomial on G hence $f \equiv 0$. For $p = 1$, f being a measurable bounded function and a polynomial, f must be constant, i.e. $f \equiv c$ for some $c \in \mathbb{C}$. This shows that if $p \in (1,\infty) \cup \{\infty_o\}$ then $\Psi = 0$ and $\text{Range}(\mathcal{R}_p)$ is dense in $L^p(G)$ or $C_o(G)$, whereas if $p = 1$ then $\Psi : L^1(G) \ni \phi \mapsto c\int_G \phi$. This shows (ii) for $p \in (1,\infty) \cup \{\infty_o\}$.

Let us study more precisely the case $p = 1$. It is easy to see that

$$\int_G X\phi(x)dx = -\int_G \phi(x)\ (X1)(x)dx = 0$$

holds for any $\phi \in L^1(G)$ such that $X\phi \in L^1(G)$. Consequently, for any $\phi \in \text{Dom}(\mathcal{R}_1)$, we know that ϕ and $\mathcal{R}\phi$ are in $L^1(G)$ thus $\int_G \mathcal{R}_1\phi = 0$. So the range of $\mathcal{R}_1$ is included in

$$S := \left\{\phi \in L^1(G)\ :\ \int_G \phi = 0\right\} \supset \text{Range}(\mathcal{R}_1).$$

Moreover, if Ψ_1 a bounded linear functional on S such that Ψ_1 is identically 0 on $\text{Range}(\mathcal{R}_1)$, by the Hahn-Banach Theorem (see, e.g. [Rud87, Theorem 5.16]), it can be extended into a bounded linear function Ψ on $L^1(G)$. As Ψ vanishes identically on $\text{Range}(\mathcal{R}_1) \subset S$, we have already proven that Ψ must be of the form

$$\Psi : L^1(G) \ni \phi \mapsto c\int_G \phi$$

for some constant $c \in \mathbb{C}$ and its restriction to S is $\Psi_1 \equiv 0$. This concludes the proof of Part (ii). □

Eventually, let us prove that the operator $\mathcal{R}_p$ is Komatsu-non-negative, see hypothesis (iii) in Section A.3:

Corollary 4.3.5. *For $p \in [1,\infty) \cup \{\infty_o\}$, and any $\mu > 0$, the operator $\mu\mathrm{I} + \mathcal{R}_p$ is invertible on $L^p(G)$, $p \in [1,\infty)$, and $C_o(G)$ for $p = \infty_o$, and the operator norm of $(\mu\mathrm{I} + \mathcal{R}_p)^{-1}$ is*

$$\|(\mu\mathrm{I} + \mathcal{R}_p)^{-1}\| \leq \|h_1\|\mu^{-1}.$$

Proof. Integrating the formula

$$(\mu + \lambda)^{-1} = \int_0^\infty e^{-t(\mu+\lambda)}dt,$$

against the spectral measure $dE(\lambda)$ of $\mathcal{R}_2$, we have formally

$$(\mu \mathrm{I} + \mathcal{R}_2)^{-1} = \int_0^\infty e^{-t(\mu\mathrm{I}+\mathcal{R}_2)} dt, \tag{4.30}$$

and the convolution kernel of the operator on the right-hand side is (still formally) given by

$$\kappa_\mu(x) := \int_0^\infty e^{-t\mu} h_t(x) dt.$$

From the properties of the heat kernel h_t (see Theorem 4.2.7 and Corollary 4.2.10), we see that the function κ_μ defined just above is continuous on G and that

$$\|\kappa_\mu\|_1 \leq \int_0^\infty e^{-t\mu} \|h_t\|_1 dt = \|h_1\| \int_0^\infty e^{-t\mu} dt = \frac{\|h_1\|}{\mu} < \infty.$$

As $\kappa_\mu \in L^1(G)$, it is a routine exercise to show that the operator

$$\int_0^\infty e^{-t(\mu\mathrm{I}+\mathcal{R}_2)} dt$$

is bounded on $L^2(G)$ with convolution kernel κ_μ (it suffices to consider integration over $[0, N]$ with $N \to \infty$). Moreover, Formula (4.30) holds in $\mathscr{L}(L^2(G))$.

For any $\phi \in \mathcal{D}(G)$ and $p \in [1, \infty) \cup \{\infty_o\}$, Theorem 4.3.3 (iv) implies

$$(\mu\mathrm{I} + \mathcal{R}_p)\phi = (\mu\mathrm{I} + \mathcal{R})\phi = (\mu\mathrm{I} + \mathcal{R}_2)\phi \in \mathcal{D}(G),$$

thus

$$((\mu\mathrm{I} + \mathcal{R}_p)\phi) * \kappa_\mu = ((\mu\mathrm{I} + \mathcal{R}_2)\phi) * \kappa_\mu = \phi.$$

This yields that the operator $(\mu\mathrm{I} + \mathcal{R}_p)^{-1} : \phi \mapsto \phi * \kappa_\mu$ is bounded on $L^p(G)$ if $p \in [1, \infty)$ and on $C_o(G)$ if $p = \infty_o$. Furthermore, its operator norm is

$$\|(\mu\mathrm{I} + \mathcal{R}_p)^{-1}\| \leq \|\kappa_\mu\|_1 \leq \|h_1\|\mu^{-1},$$

completing the proof. □

4.3.2 Fractional powers of operators $\mathcal{R}_p$

We now apply the general theory of fractional powers outlined in Section A.3 to the operators $\mathcal{R}_p$ and $\mathrm{I} + \mathcal{R}_p$.

Theorem 4.3.6. *Let $\mathcal{R}$ be a positive Rockland operator on a graded Lie group G. We consider the operators $\mathcal{R}_p$ defined in Definition 4.3.1. Let $p \in [1, \infty) \cup \{\infty_o\}$.*

1. *Let $\mathcal{A}$ denote either $\mathcal{R}$ or $\mathrm{I} + \mathcal{R}$.*

(a) *For every* $a \in \mathbb{C}$, *the operator* $\mathcal{A}_p^a$ *is closed and injective with* $(\mathcal{A}_p^a)^{-1} = \mathcal{A}_p^{-a}$. *We have* $\mathcal{A}_p^0 = \mathrm{I}$, *and for any* $N \in \mathbb{N}$, $\mathcal{A}_p^N$ *coincides with the usual powers of differential operators on* $\mathcal{S}(G)$ *and* $\mathrm{Dom}(\mathcal{A}^N) \cap \mathrm{Range}(\mathcal{A}^N)$ *is dense in* $\mathrm{Range}(\mathcal{A}_p)$.

(b) *For any* $a, b \in \mathbb{C}$, *in the sense of operator graph, we have* $\mathcal{A}_p^a \mathcal{A}_p^b \subset \mathcal{A}_p^{a+b}$. *If* $\mathrm{Range}(\mathcal{A}_p)$ *is dense then the closure of* $\mathcal{A}_p^a \mathcal{A}_p^b$ *is* $\mathcal{A}_p^{a+b}$.

(c) *Let* $a_o \in \mathbb{C}_+$.
- *If* $\phi \in \mathrm{Range}(\mathcal{A}_p^{a_o})$ *then* $\phi \in \mathrm{Dom}(\mathcal{A}_p^a)$ *for all* $a \in \mathbb{C}$ *with* $0 < -\mathrm{Re}\, a < \mathrm{Re}\, a_o$ *and the function* $a \mapsto \mathcal{A}_p^a \phi$ *is holomorphic in* $\{a \in \mathbb{C} : -\mathrm{Re}\, a_o < \mathrm{Re}\, a < 0\}$.
- *If* $\phi \in \mathrm{Dom}(\mathcal{A}_p^{a_o})$ *then* $\phi \in \mathrm{Dom}(\mathcal{A}_p^a)$ *for all* $a \in \mathbb{C}$ *with* $0 < \mathrm{Re}\, a < \mathrm{Re}\, a_o$ *and the function* $a \mapsto \mathcal{A}_p^a \phi$ *is holomorphic in* $\{a \in \mathbb{C} : 0 < \mathrm{Re}\, a < \mathrm{Re}\, a_o\}$.

(d) *For every* $a \in \mathbb{C}$, *the operator* $\mathcal{A}_p^a$ *is invariant under left translations.*

(e) *If* $p \in (1, \infty)$ *then the dual of* $\mathcal{A}_p$ *is* $\bar{\mathcal{A}}_{p'}$. *The dual of* $\mathcal{A}_{\infty_o}$ *restricted to* $L^1(G)$ *is* $\bar{\mathcal{A}}_1$. *The dual of* $\mathcal{A}_1$ *restricted to* $C_o(G) \subset L^\infty(G)$ *is* $\bar{\mathcal{A}}_{\infty_o}$.

(f) *If* $a, b \in \mathbb{C}_+$ *with* $\mathrm{Re}\, b > \mathrm{Re}\, a$, *then*

$$\exists C = C_{a,b} > 0 \quad \forall \phi \in \mathrm{Dom}(\mathcal{A}_p^b) \quad \|\mathcal{A}_p^a \phi\| \leq C \|\phi\|^{1 - \frac{\mathrm{Re}\, a}{\mathrm{Re}\, b}} \|\mathcal{A}_p^b \phi\|^{\frac{\mathrm{Re}\, a}{\mathrm{Re}\, b}}.$$

(g) *For any* $a \in \mathbb{C}_+$, $\mathrm{Dom}(\mathcal{A}_p^a)$ *contains* $\mathcal{S}(G)$.

(h) *If* $f \in \mathrm{Dom}(\mathcal{A}_p^a) \cap L^q(G)$ *for some* $q \in [1, \infty) \cup \{\infty_o\}$, *then* $f \in \mathrm{Dom}(\mathcal{A}_q^a)$ *if and only if* $\mathcal{A}_p^a f \in L^q(G)$, *in which case* $\mathcal{A}_p^a f = \mathcal{A}_q^a f$.

2. *For each* $a \in \mathbb{C}_+$, *the operators* $(\mathrm{I} + \mathcal{R}_p)^a$ *and* $\mathcal{R}_p^a$ *are unbounded and their domains satisfy for all* $\epsilon > 0$,

$$\mathrm{Dom}\left[(\mathrm{I} + \mathcal{R}_p)^a\right] = \mathrm{Dom}(\mathcal{R}_p^a) = \mathrm{Dom}\left[(\mathcal{R}_p + \epsilon \mathrm{I})^a\right].$$

3. *If* $0 < \mathrm{Re}\, a < 1$ *and* $\phi \in \mathrm{Range}(\mathcal{R}_p)$ *then*

$$\mathcal{R}_p^{-a} \phi = \frac{1}{\Gamma(a)} \int_0^\infty t^{a-1} e^{-t\mathcal{R}_p} \phi \, dt,$$

in the sense that $\lim_{N \to \infty} \int_0^N$ *converges in the norm of* $L^p(G)$ *or* $C_o(G)$.

4. *If* $a \in \mathbb{C}_+$, *then the operator* $(\mathrm{I} + \mathcal{R}_p)^{-a}$ *is bounded and for any* $\phi \in \mathcal{X}$ *with* $\mathcal{X} = L^p(G)$ *or* $C_o(G)$, *we have*

$$(\mathrm{I} + \mathcal{R}_p)^{-a} \phi = \frac{1}{\Gamma(a)} \int_0^\infty t^{a-1} e^{-t(\mathrm{I} + \mathcal{R}_p)} \phi \, dt,$$

in the sense of absolute convergence:

$$\int_0^\infty t^{a-1} \|e^{-t(\mathrm{I} + \mathcal{R}_p)} \phi\|_{\mathcal{X}} dt < \infty.$$

5. *For any $a, b \in \mathbb{C}$, the two (possibly unbounded) operators $\mathcal{R}_p^a$ and $(\mathrm{I}+\mathcal{R}_p)^b$ commute.*

6. *For any $a \in \mathbb{C}$, the operator $\mathcal{R}_p^a$ is homogeneous of degree νa.*

Recall (see Definition A.3.2) that the two (possibly unbounded) operators A and B commute when

$$x \in \mathrm{Dom}(AB) \cap \mathrm{Dom}(BA) \Longrightarrow ABx = BAx,$$

and that the domain of the product AB of two (possibly unbounded) operators A and B on the same Banach space $\mathcal{X}$ is formed by the elements $x \in \mathcal{X}$ such that $x \in \mathrm{Dom}(B)$ and $Bx \in \mathrm{Dom}(A)$.

Proof. The operator $\mathcal{R}_p$ is closed and densely defined by Theorem 4.3.3 (i), it is injective by Corollary 4.3.4 and Komatsu-non-negative in the sense of Section A.3 (iii) by Corollary 4.3.5. Therefore, $\mathcal{R}_p$ satisfies the hypotheses of Theorem A.3.4. Moreover, $\mathrm{I}+\mathcal{R}_p$ also satisfies these hypotheses by Remark A.3.3, and $-(\mathrm{I}+\mathcal{R}_p)$ generates an exponentially stable semi-group:

$$\|e^{-t(\mathrm{I}+\mathcal{R}_p)}\| \leq e^{-t}\|e^{-t\mathcal{R}_p}\| \leq \|h_1\|_1 e^{-t}.$$

Most of the statements then follow from the general properties of fractional powers constructed via the Balakrishnan formulae recalled in Section A.3. More precisely, from the Balakrishnan formula, for any $N \in \mathbb{N}$, $\mathcal{A}_p^N$ coincides with the usual powers of differential operators on $\mathcal{S}(G)$ and Part (1a) follows from Theorem A.3.4 (1) and (2) and Remark A.3.1.

The duality properties explained in Part (1e) for $p \in (1, \infty)$ hold for the Balakrishnan operators hence they hold for their maximal closure. The cases of $p = 1, \infty_o$ are similar and this proves Part (1e). The properties in Parts (1d), (5) and (6) hold for the Balakrishnan operators hence they hold for their maximal closure and these parts are proved.

Part (1b) follows from Theorem A.3.4 (4).

Part (1c) follows from Theorem A.3.4 (5).

Part (1f) follows from Theorem A.3.4 (6).

Part (1g) follows from Parts (1a) and (1c).

Part (1h) is certainly true for any $f \in \mathcal{S}(G)$ and $\mathrm{Re}\, a > 0$ via the Balakrishnan formulae. By analyticity (see Part (1c)) it is true for any $a \in \mathbb{C}$. The density of $\mathcal{D}(G)$ in $L^p(G)$ (or $C_o(G)$ if $p = \infty_o$) together with the maximality of $\mathcal{A}_p^a$ and the uniqueness of distributional convergence imply the result.

Part (2) follow from Theorem A.3.4 (8).

Parts (3) and (4) follows from Theorem A.3.4 (10).

This concludes the proof of Theorem 4.3.6. □

4.3.3 Imaginary powers of $\mathcal{R}_p$ and $\mathrm{I} + \mathcal{R}_p$

In this section, we show that imaginary powers of a positive Rockland operator $\mathcal{R}$ as well as $\mathrm{I} + \mathcal{R}$ are bounded operators on $L^p(G)$, $p \in (1, \infty)$. We prove this as a consequence of the theorem of singular integrals on homogeneous groups, see Section 3.2.3.

We start by showing that if $\mathcal{R}$ is a positive Rockland operator, then the imaginary powers of $\mathrm{I} + \mathcal{R}_p$ are bounded on $L^p(G)$:

Proposition 4.3.7. *Let $\mathcal{R}$ be a positive Rockland operator on a graded Lie group G. For any $\tau \in \mathbb{R}$ and $p \in (1, \infty)$, the operator $(\mathrm{I} + \mathcal{R}_p)^{i\tau}$ is bounded on $L^p(G)$. For any $p \in (1, \infty)$, there exists $C = C_{p,\mathcal{R}} > 0$ and $\theta > 0$ such that*

$$\forall \tau \in \mathbb{R} \qquad \|(\mathrm{I} + \mathcal{R}_p)^{i\tau}\|_{\mathscr{L}(L^p(G))} \leq C e^{\theta|\tau|}.$$

For any $p \in (1, \infty)$ and $a \in \mathbb{C}$, $\mathrm{Dom}((\mathrm{I} + \mathcal{R}_p)^a) = \mathrm{Dom}((\mathrm{I} + \mathcal{R}_p)^{\mathrm{Re}\, a})$.

The following technical result will be useful in the proof of Proposition 4.3.7 and in other proofs (see Sections 4.3.4 and 4.4.4).

Lemma 4.3.8. *Let $\mathcal{R}$ be a positive Rockland operator on a graded Lie group G. Let h_t be its heat kernel as in Section 4.2.2.*

1. *For any homogeneous quasi-norm $|\cdot|$, any multi-index $\alpha \in \mathbb{N}_0^n$, and any real number a with $0 < a < \frac{Q+[\alpha]}{\nu}$, there exists a constant $C > 0$ such that*

$$\int_0^\infty t^{a-1} |X^\alpha h_t(x)| dt \leq C |x|^{-Q-[\alpha]+\nu a}.$$

   For any homogeneous quasi-norm $|\cdot|$, any multi-index $\alpha \in \mathbb{N}_0^n$, there exists a constant $C > 0$ such that

$$\int_0^\infty |X^\alpha h_t(x)| e^{-t} dt \leq C |x|^{-Q-[\alpha]}.$$

2. *For any homogeneous quasi-norm $|\cdot|$, any multi-index $\alpha \in \mathbb{N}_0^n$, and any $t > 0$, we have*

$$\int_{|x| \geq 1/2} |X^\alpha h_t(x)| dx \leq t^{-\frac{[\alpha]}{\nu}} \|X^\alpha h_1\|_{L^1}.$$

3. *For any homogeneous quasi-norm $|\cdot|$, any multi-index $\alpha \in \mathbb{N}_0^n$, any $N \in \mathbb{N}$ and any $t \in (0, 1)$, there exists a constant $C > 0$ such that*

$$\int_{|x| \geq 1/2} |X^\alpha h_t(x)| dx \leq C t^N.$$

Proof of Lemma 4.3.8 . Let us prove Part 1. We write

$$\int_0^\infty t^{a-1}|X^\alpha h_t(x)|dt = \int_0^{|x|^\nu} + \int_{|x|^\nu}^\infty .$$

For the second integral, we use the property of homogeneity of h_t (see (4.12) or (4.17))

$$\begin{aligned}\int_{|x|^\nu}^\infty &= \int_{|x|^\nu}^\infty t^{a-1-\frac{Q+[\alpha]}{\nu}}|X^\alpha h_1(t^{-\frac{1}{\nu}}x)|dt \\ &\le (\frac{Q+[\alpha]}{\nu} - a)^{-1}\|X^\alpha h_1\|_\infty |x|^{\nu(a-\frac{Q+[\alpha]}{\nu})}.\end{aligned}$$

As $h_1 \in \mathcal{S}(G)$, $\|X^\alpha h_1\|_\infty$ is finite. For the first integral, we use again (4.12) to obtain

$$\begin{aligned}\int_0^{|x|^\nu} &= \int_0^{|x|^\nu} t^{a-1}|x|^{-(Q+[\alpha])}\left|X^\alpha h_{|x|^{-\nu}t}\left(\frac{x}{|x|}\right)\right| dt \\ &\le C_1 a^{-1}|x|^{\nu(a-\frac{Q+[\alpha]}{\nu})}.\end{aligned}$$

where $C_1 := \sup_{|y|=1, 0\le t_1\le 1}|X^\alpha h_{t_1}(y)|$ is finite by (4.15). Combining the two estimates above shows the estimates for the first integral in Part 1. We proceed in the same way for the second one:

$$\int_0^\infty |X^\alpha h_t(x)|e^{-t}dt = \int_0^{|x|^\nu} + \int_{|x|^\nu}^\infty .$$

We have (with C_1 as above)

$$\begin{aligned}\int_0^{|x|^\nu} &\le C_1|x|^{\nu(a-\frac{Q+[\alpha]}{\nu})}\int_0^{|x|^\nu} e^{-t}dt = C_1|x|^{\nu(a-\frac{Q+[\alpha]}{\nu})}(1-e^{-|x|^\nu}) \\ &\le C_1|x|^{\nu(a-\frac{Q+[\alpha]}{\nu})},\end{aligned}$$

whereas

$$\begin{aligned}\int_{|x|^\nu}^\infty &\le \|X^\alpha h_1\|_\infty (|x|^\nu)^{-\frac{Q+[\alpha]}{\nu}}\int_{|x|^\nu}^\infty e^{-t}dt = \|X^\alpha h_1\|_\infty |x|^{-(Q+[\alpha])}e^{-|x|^\nu} \\ &\le \|X^\alpha h_1\|_\infty |x|^{-(Q+[\alpha])}.\end{aligned}$$

We conclude in the same way as above and Part 1 is proved.

Let us prove Part 2. The property of homogeneity of h_t (see (4.17)) together with $h_1 \in \mathcal{S}(G)$ imply

$$\begin{aligned}\int_{|x|\ge 1/2}|X^\alpha h_t(x)|dx &= \int_{|x|\ge 1/2}|X^\alpha h_1(t^{-\frac{1}{\nu}}x)|t^{-\frac{[\alpha]+Q}{\nu}}dx \\ &= t^{-\frac{[\alpha]}{\nu}}\int_{t^{\frac{1}{\nu}}|x'|\ge 1/2}|X^\alpha h_1(x')|dx' \le t^{-\frac{[\alpha]}{\nu}}\int_G |X^\alpha h_1|,\end{aligned}$$

having used the change of variable $x' = t^{-\frac{1}{\nu}}x$. This shows Part 2.

Let us prove Part 3. The properties of the heat kernel, especially (4.12) and (4.15), imply

$$|X^\alpha h_t(x)| = |x|^{-[\alpha]-Q}|X^\alpha h_{|x|^{-\nu}t}(|x|^{-1}x)| \leq C|x|^{-[\alpha]-Q}(|x|^{-\nu}t)^N,$$

if $|x| \geq 1/2$ and $t \in (0,1)$ where $C = \sup_{|x'|=1, 0<t'<1} t'^{-N}|X^\alpha h_{t'}(x')|$ is finite. Hence

$$\int_{|x|\geq 1/2} |X^\alpha h_t(x)|dx \leq Ct^N \int_{|x|\geq 1/2} |x|^{-[\alpha]-Q-\nu N}dx.$$

This shows Part 3 and concludes the proof of Lemma 4.3.8. □

Proof of Proposition 4.3.7. By Theorem 4.3.6 (1), to show that $(\mathrm{I}+\mathcal{R}_p)^{i\tau}$ is bounded on $L^p(G)$ for some $p \in (1,\infty)$ and $\tau \in \mathbb{R}$, it suffices to show that $(\mathrm{I}+\mathcal{R}_2)^{i\tau}$ can be extended to an L^p-bounded operator. To do this, we will show that Corollary 3.2.21 can be applied to $(\mathrm{I}+\mathcal{R}_2)^{i\tau}$.

By functional calculus, $(\mathrm{I}+\mathcal{R}_2)^{i\tau}$ is bounded on $L^2(G)$. Part 1 of Lemma 4.3.8 together with the formula

$$\forall \lambda > 0 \qquad \lambda^{i\tau} = \frac{\lambda}{\Gamma(1-i\tau)} \int_0^\infty t^{-i\tau} e^{-\lambda t} dt,$$

and the functional calculus of $\mathcal{R}_2$ imply that the right convolution kernel of $(\mathrm{I}+\mathcal{R}_2)^{i\tau}$ is the tempered distribution κ which coincides with the smooth function away from 0 given via

$$\kappa(x) = \frac{1}{\Gamma(1-i\tau)} \int_0^\infty t^{-i\tau}(\mathrm{I}+\mathcal{R})h_t(x)e^{-t}dt, \qquad x \neq 0. \tag{4.31}$$

Using this formula, we have

$$\int_{|x|\geq 1/2} |\kappa(x)|dx \leq |\Gamma(1-i\tau)|^{-1} \int_{t=0}^\infty \int_{|x|\geq 1/2} (|h_t(x)| + |\mathcal{R}h_t(x)|)e^{-t}dxdt.$$

By Part 2 of Lemma 4.3.8, (and h_1 being Schwartz), the integrals

$$\int_{t=0}^\infty \int_{|x|\geq 1/2} |h_t(x)|e^{-t}dxdt \quad \text{and} \quad \int_{t=1}^\infty \int_{|x|\geq 1/2} |\mathcal{R}h_t(x)|e^{-t}dxdt,$$

are finite. By Part 3 of Lemma 4.3.8, the integral

$$\int_{t=0}^1 \int_{|x|\geq 1/2} |\mathcal{R}h_t(x)|e^{-t}dxdt \leq C \int_{t=0}^1 t^0 dt = C,$$

is finite. This shows that $\int_{|x|\geq 1/2} |\kappa(x)|dx$ is finite.

Using (4.31), we also obtain easily that

$$\sup_{0<|x|<1} |x|^{Q+[\alpha]}|X^\alpha \kappa(x)| \leq |\Gamma(1-i\tau)|^{-1} \sup_{0<|x|<1} |x|^{Q+[\alpha]} \int_0^\infty |X^\alpha h_t(x)| + |X^\alpha \mathcal{R} h_t(x)| dt,$$

and the right-hand side is finite by Lemma 4.3.8. Note that if we denote by $\kappa = \kappa_{\tau,\mathcal{R}}$ the kernel of $(\mathrm{I}+\mathcal{R}_2)^{i\tau}$, then we have

$$\kappa_{\tau,\mathcal{R}}(x^{-1}) = \overline{\kappa_{-\tau,\bar{\mathcal{R}}}(x)},$$

using the formula in (4.31) and

$$\begin{aligned} ((\mathrm{I}+\mathcal{R})h_t)(x^{-1}) &= ((\mathrm{I}-\partial_t)h_t)(x^{-1}) = \left((\mathrm{I}-\partial_t)\bar{h}_t\right)(x) \\ &= \overline{((\mathrm{I}-\partial_t)h_t)(x)} = \overline{((\mathrm{I}+\mathcal{R})h_t)(x)}, \end{aligned}$$

where we have used (4.13). Hence we also have that each quantity

$$\sup_{0<|x|<1} |x|^{Q+[\alpha]}|\tilde{X}^\alpha \kappa(x)| = \sup_{0<|x|<1} |x|^{Q+[\alpha]}|X^\alpha \kappa_{-\tau,\bar{\mathcal{R}}}(x)|$$

is finite.

The estimates above show that κ satisfies the hypotheses of Corollary 3.2.21 and therefore the operator $(\mathrm{I}+\mathcal{R}_2)^{i\tau}$ is bounded on $L^p(G)$, $p \in (1,\infty)$. The properties of the semi-group (see Theorem A.3.4 (3)) imply the rest of the statement in Proposition 4.3.7. □

Let us now prove the homogeneous case, that is, that the imaginary powers of a positive Rockland operator are bounded on $L^p(G)$:

Proposition 4.3.9. *Let $\mathcal{R}$ be a positive Rockland operator on a graded Lie group G. For any $\tau \in \mathbb{R}$ and $p \in (1,\infty)$, the operator $\mathcal{R}_p^{i\tau}$ is bounded on $L^p(G)$. For any $p \in (1,\infty)$, there exists $C = C_{p,\mathcal{R}} > 0$ and $\theta > 0$ such that*

$$\forall \tau \in \mathbb{R} \qquad \|\mathcal{R}_p^{i\tau}\|_{\mathscr{L}(L^p(G))} \leq Ce^{\theta|\tau|}.$$

For any $p \in (1,\infty)$ and $a \in \mathbb{C}$, $\mathrm{Dom}(\mathcal{R}_p^a) = \mathrm{Dom}(\mathcal{R}_p^{\mathrm{Re}\, a})$.

Proof of Proposition 4.3.9. Let $p \in (1,\infty)$ and $\tau \in \mathbb{R}$. Let us denote by $\mathcal{R}_{p,i\tau}$ the (possibly unbounded) operator given as the strong limit in $L^p(G)$ of $(\epsilon+\mathcal{R}_p)^{i\tau}\phi$ as $\epsilon \to 0$, for $\phi \in \mathrm{Dom}((\epsilon+\mathcal{R}_p)^{i\tau})$ for any $\epsilon \in (0,\epsilon_0)$ for some small $\epsilon_0 > 0$ and such that this strong limit exists. The domain of $\mathcal{R}_{p,i\tau}$ is naturally the space of all those functions ϕ. Note that the homogeneity of $\mathcal{R}$ implies

$$(\epsilon+\mathcal{R}_p)^{i\tau}\phi = \epsilon^{i\tau}(\mathrm{I}+\epsilon^{-1}\mathcal{R}_p)^{i\tau}\phi = \epsilon^{i\tau}(\mathrm{I}+\mathcal{R}_p)^{i\tau}\{\phi(\epsilon^{-1/\nu}\cdot)\}(\epsilon^{1/\nu}\cdot),$$

for any $\epsilon > 0$ and any $\phi \in L^p(G)$ such that

$$\phi(\epsilon^{-1/\nu}\cdot) \in \mathrm{Dom}((\mathrm{I}+\mathcal{R}_p)^{i\tau}).$$

By Proposition 4.3.7, $\text{Dom}((\text{I} + \mathcal{R}_p)^{i\tau}) = L^p(G)$ and the operator $(\text{I} + \mathcal{R}_p)^{i\tau}$ is bounded. Therefore for all $\phi \in L^p(G)$ and $\epsilon > 0$, ϕ is in $\text{Dom}((\epsilon + \mathcal{R}_p)^{i\tau})$ and we have

$$\begin{aligned} \|(\epsilon + \mathcal{R}_p)^{i\tau}\phi\|_{L^p(G)} &= \|(\text{I} + \mathcal{R}_p)^{i\tau}\{\phi(\epsilon^{-1/\nu}\cdot)\}(\epsilon^{1/\nu}\cdot)\|_{L^p(G)} \\ &= \epsilon^{-\frac{Q}{p\nu}}\|(\text{I} + \mathcal{R}_p)^{i\tau}\{\phi(\epsilon^{-1/\nu}\cdot)\}\|_{L^p(G)} \\ &\leq \epsilon^{-\frac{Q}{p\nu}}\|(\text{I} + \mathcal{R}_p)^{i\tau}\|_{\mathscr{L}(L^p(G))}\|\phi(\epsilon^{-1/\nu}\cdot)\|_{L^p(G)} \\ &= \|(\text{I} + \mathcal{R}_p)^{i\tau}\|_{\mathscr{L}(L^p(G))}\|\phi\|_{L^p(G)}. \end{aligned}$$

Consequently, $\mathcal{R}_{p,i\tau}$ extends to a bounded operator on $L^p(G)$. By Theorem A.3.4 (9), this implies that $\mathcal{R}_p^{i\tau}$ is also a bounded operator on $L^p(G)$ as $\mathcal{R}_p$ has dense range and domain by Corollary 4.3.4. As in the inhomogeneous case, the properties of the semi-group (see Theorem A.3.4 (3)) imply the rest of the statement in Proposition 4.3.9. □

Given the proof of Proposition 4.3.7, one would be tempted to study the convolution kernel of the operator $\mathcal{R}_2^{i\tau}$ in order to show the L^p-boundedness in the proof of Proposition 4.3.9. Indeed, following the same arguments as in the proof of Proposition 4.3.7, one shows that the kernel of $\mathcal{R}_2^{i\tau}$ coincides away from the origin with the smooth function

$$G\backslash\{0\} \ni x \mapsto \frac{1}{\Gamma(1 - i\tau)} \int_0^\infty t^{-i\tau}\mathcal{R}h_t(x)dt.$$

However, this function can not be in general a kernel of type $i\tau$: already for the usual Laplacian on $(\mathbb{R}^n, +)$ it is not the case. Indeed, in the Euclidean case, this function is radial and non-zero and its average on the sphere can therefore not vanish.

In the stratified case, Folland proved the L^p-boundedness of imaginary powers of the sub-Laplacian $-\mathcal{L}$ and $\text{I} + (-\mathcal{L})$ using general properties of semigroups preserving positivity together with the Laplace transform see [Fol75, Proposition 3.14 and Lemma 3.13]. More precisely, the boundedness follows from the Littlewood-Paley theory and the study of square functions associated with the semi-group. Note that in the case of a sub-Laplacian, the proof in [Fol75] yields a bound of the operator norm by $|\Gamma(1 - i\tau)|^{-1}$ up to a constant of p.

In our case, we applied a consequence of the theorem of Singular Integrals via Corollary 3.2.20 to obtain the L^p-boundedness of the imaginary powers of $\text{I} + \mathcal{R}$ and we have shown

$$\|\mathcal{R}_p^{i\tau}\|_{\mathscr{L}(L^p(G))} \leq \|(\text{I} + \mathcal{R}_p)^{i\tau}\|_{\mathscr{L}(L^p(G))}, \quad p \in (1, \infty),$$

in the proof of Proposition 4.3.9. We can follow the constants in the proof of the theorem of Singular Integrals (see Remark A.4.5 (2)) as well as in our application to show that $\|(\text{I} + \mathcal{R}_p)^{i\tau}\|_{\mathscr{L}(L^p(G))}$ is bounded up to a constant of p, by

$$(1 + |\Gamma(1 - i\tau)|^{-1})^{2|\frac{1}{p} - \frac{1}{2}|}.$$

However, we do not need these precise bounds as the bounds obtained from the general theory of semigroups as stated in Propositions 4.3.7 and 4.3.9 will be sufficient for our purpose in the proofs of interpolation properties for Sobolev spaces in Theorem 4.4.9 and Proposition 4.4.15.

4.3.4 Riesz and Bessel potentials

We mimic the usual terminology in the Euclidean setting, to define the Riesz and Bessel potentials associated with a positive Rockland operator.

Definition 4.3.10. Let $\mathcal{R}$ be a positive Rockland operator of homogeneous degree ν. We call the operators $\mathcal{R}^{-a/\nu}$ for $\{a \in \mathbb{C},\ 0 < \operatorname{Re} a < Q\}$ and $(\mathrm{I} + \mathcal{R})^{-a/\nu}$ for $a \in \mathbb{C}_+$, the *Riesz potential* and the *Bessel potential*, respectively.

In the sequel we will denote their kernels by $\mathcal{I}_a$ and $\mathcal{B}_a$, respectively, as defined in the following:

Corollary 4.3.11. *We keep the setting and notation of Theorem 4.3.3.*

(i) *Let $a \in \mathbb{C}$ with $0 < \operatorname{Re} a < Q$. The integral*

$$\mathcal{I}_a(x) := \frac{1}{\Gamma(\frac{a}{\nu})} \int_0^\infty t^{\frac{a}{\nu}-1} h_t(x) dt$$

converges absolutely for every $x \neq 0$. This defines a distribution $\mathcal{I}_a$ which is smooth away from the origin and $(a - Q)$-homogeneous.

For any $p \in (1, \infty)$, if $\phi \in \mathcal{S}(G)$ or, more generally, if $\phi \in L^q(G) \cap L^p(G)$ where $q \in [1, \infty)$ is given by $\frac{1}{q} - \frac{1}{p} = \frac{\operatorname{Re} a}{Q}$, then

$$\phi \in \operatorname{Dom}(\mathcal{R}_p^{-\frac{a}{\nu}}) \quad \text{and} \quad \mathcal{R}_p^{-\frac{a}{\nu}} \phi = \phi * \mathcal{I}_a \in L^p(G).$$

Consequently,

$$\forall \phi \in \mathcal{S}(G) \qquad \mathcal{R}_p^{\frac{a}{\nu}} \phi \in L^p(G) \quad \text{and} \quad \phi = (\mathcal{R}_p^{\frac{a}{\nu}} \phi) * \mathcal{I}_a.$$

(ii) *Let $a \in \mathbb{C}_+$. The integral*

$$\mathcal{B}_a(x) := \frac{1}{\Gamma(\frac{a}{\nu})} \int_0^\infty t^{\frac{a}{\nu}-1} e^{-t} h_t(x) dt$$

converges absolutely for every $x \neq 0$ and defines an integrable function $\mathcal{B}_a$ on G. The function $\mathcal{B}_a$ is always smooth away from 0.

If $\operatorname{Re} a > Q$, $\mathcal{B}_a$ is also smooth at 0.

If $\operatorname{Re} a > Q/2$, then $\mathcal{B}_a$ is square integrable: $\mathcal{B}_a \in L^2(G)$.

All the operators $(\mathrm{I}+\mathcal{R}_p)^{-a/\nu}$*,* $p \in [1,\infty) \cup \{\infty_o\}$*, are bounded convolution operators with the same (right convolution) kernel* $\mathcal{B}_a$*.*

If $a, b \in \mathbb{C}_+$*, then as integrable functions, we have*

$$\mathcal{B}_a * \mathcal{B}_b = \mathcal{B}_{a+b}.$$

Remark 4.3.12. In other words for Part (i), $\mathcal{I}_a$ is a kernel of type a and

$$\mathcal{R}_p^{-a/\nu}\delta_0 = \mathcal{I}_a.$$

This shows that if $\nu < Q$, $\mathcal{I}_1$ is a fundamental solution of $\mathcal{R}$, in fact, the unique homogeneous fundamental solution (cf. Theorem 3.2.40).

Note that we will show in Lemma 4.5.9 that more generally $X^\alpha \mathcal{B}_a \in L^2(G)$ whenever $\operatorname{Re} a > [\alpha] + Q/2$, as well as other L^1-estimates.

Proof of Corollary 4.3.11. The absolute convergence and the smoothness of $\mathcal{I}_a$ and $\mathcal{B}_a$ follow from Lemma 4.3.8.

For the homogeneity of $\mathcal{I}_a$, we use (4.12) and the change of variable $s = r^{-\nu}t$, to get

$$\begin{aligned}
\mathcal{I}_a(rx) &= \frac{1}{\Gamma(a/\nu)} \int_0^\infty t^{\frac{a}{\nu}-1} h_t(rx)\,dt \\
&= \frac{1}{\Gamma(a/\nu)} \int_0^\infty (r^\nu s)^{\frac{a}{\nu}-1} r^{-Q} h_s(x) r^\nu\, ds = r^{a-Q}\mathcal{I}_a(x).
\end{aligned}$$

Hence $\mathcal{I}_a$ is a kernel of type a with $0 < \operatorname{Re} a < Q$ (see Definition 3.2.9).

By Lemma 3.2.7, the operator $\mathcal{S}(G) \ni \phi \mapsto \phi * \mathcal{I}_a$ is homogeneous of degree $-a$, and by Proposition 3.2.8, it admits a bounded extension $L^q(G) \to L^p(G)$ when $\frac{1}{p} - \frac{1}{q} = \frac{\operatorname{Re}(a)}{Q}$.

Let $\phi \in \mathcal{R}^Q(\mathcal{S}(G))$. By Theorem 4.3.6, the function $a \mapsto \mathcal{R}_p^{-\frac{a}{\nu}}\phi$ is analytic on the strip $\{z \in \mathbb{C}, 0 < \operatorname{Re} z < Q\}$ and coincides there with

$$a \mapsto \frac{1}{\Gamma(\frac{a}{\nu})} \int_0^\infty t^{\frac{a}{\nu}-1} \phi * h_t\, dt.$$

But since the integral defining $\mathcal{I}_a(x)$ is absolutely convergent for all $x \in G\backslash\{0\}$, we have

$$\forall a \in \mathbb{C},\ \operatorname{Re} a \in (0,Q), \qquad \frac{1}{\Gamma(\frac{a}{\nu})} \int_0^\infty t^{\frac{a}{\nu}-1} \phi * h_t\, dt = \phi * \mathcal{I}_a,$$

and $a \mapsto \phi * \mathcal{I}_a$ is analytic on the strip $\{0 < \operatorname{Re} a < Q\}$.

Hence we have obtained that

$$\mathcal{R}_p^{-\frac{a}{\nu}}\phi = \phi * \mathcal{I}_a$$

holds for $\operatorname{Re} a \in (0, Q)$ and for any $\phi \in \mathcal{R}^Q(\mathcal{S}(G))$. Note that $\mathcal{R}^Q(\mathcal{S}(G))$ is dense in any $L^r(G)$, $r \in (1, \infty)$ as it suffices to apply Corollary 4.3.4 (ii) to the positive Rockland operator $\mathcal{R}^Q$. Then Corollary 3.2.32 concludes the proof of Part (i).

By Theorem 4.2.7,

$$\int_G |h_t| = \int_G |h_1| < \infty$$

for all $t > 0$, so

$$\int_G |\mathcal{B}_a(x)| dx \le \frac{1}{|\Gamma(\frac{a}{\nu})|} \int_0^\infty t^{\frac{\operatorname{Re} a}{\nu} - 1} e^{-t} \int_G |h_t(x)| dx\, dt = \frac{\Gamma(\frac{\operatorname{Re} a}{\nu})}{|\Gamma(\frac{a}{\nu})|} \|h_1\|_{L^1}, \quad (4.32)$$

and $\mathcal{B}_a$ is integrable.

By Theorem 4.3.6 Part (4), the integrable function $\mathcal{B}_a$ is the convolution kernel of $(\mathrm{I} + \mathcal{R}_p)^{-a/\nu}$.

Let us show the square integrability of $\mathcal{B}_a$. We compute for any $R > 0$:

$$\begin{aligned} |\Gamma(a/\nu)|^2 \int_{|x|<R} |\mathcal{B}_a(x)|^2 dx &= \int_{|x|<R} \Gamma(a/\nu)\mathcal{B}_a(x)\overline{\Gamma(a/\nu)\mathcal{B}_a(x)} dx \\ &= \int_{|x|<R} \int_0^\infty t^{\frac{a}{\nu}-1} e^{-t} h_t(x) dt \int_0^\infty s^{\frac{\bar{a}}{\nu}-1} e^{-s} \bar{h}_s(x) ds\, dx \\ &= \int_0^\infty \int_0^\infty s^{\frac{a}{\nu}-1} t^{\frac{\bar{a}}{\nu}-1} e^{-(t+s)} \int_{|x|<R} h_t(x)\bar{h}_s(x) dx\, dt ds. \end{aligned}$$

From the properties of the heat kernel (see (4.13) and (4.11)) we see that

$$\int_{|x|<R} h_t(x)\bar{h}_s(x) dx = \int_{|x|<R} h_t(x) h_s(x^{-1}) dx \underset{R\to\infty}{\longrightarrow} h_t * h_s(0),$$

$$\text{and } h_t * h_s(0) = h_{t+s}(0) = (t+s)^{-\frac{Q}{\nu}} h_1(0).$$

Therefore,

$$\begin{aligned} \int_G |\mathcal{B}_a(x)|^2 dx &= \frac{h_1(0)}{|\Gamma(a/\nu)|^2} \int_0^\infty \int_0^\infty s^{\frac{a}{\nu}-1} t^{\frac{\bar{a}}{\nu}-1} e^{-(t+s)} (t+s)^{-\frac{Q}{\nu}} dt ds \\ &= \frac{h_1(0)}{|\Gamma(a/\nu)|^2} \int_{s'=0}^1 s'^{\frac{a}{\nu}-1} (1-s')^{\frac{\bar{a}}{\nu}-1} ds' \int_{u=0}^\infty e^{-u} u^{2(\frac{\operatorname{Re} a}{\nu}-1)-\frac{Q}{\nu}+1} du, \end{aligned} \quad (4.33)$$

after the change of variables $u = s + t$ and $s' = s/u$. The integrals over s' and u converge when $\operatorname{Re} a > Q/2$. Thus $\mathcal{B}_a$ is square integrable under this condition.

The rest of the proof of Corollary 4.3.11 follows easily from the properties of the fractional powers of $\mathrm{I} + \mathcal{R}$. □

The proof of Corollary 4.3.11 implies:

Corollary 4.3.13. *We keep the notation of Corollary 4.3.11 and h_1 denotes the heat kernel at time $t = 1$ of $\mathcal{R}$.*

1. *For any $a \in \mathbb{C}_+$, the operator norm of $(\mathrm{I}+\mathcal{R}_p)^{-\frac{a}{\nu}}$ on $L^p(G)$ if $p \in [1,\infty)$ or on $C_o(G)$ if $p=\infty_o$ is bounded by $\|\mathcal{B}_a\|_1$ and we have*

$$\|\mathcal{B}_a\|_{L^1(G)} \leq \frac{\Gamma(\frac{\operatorname{Re} a}{\nu})}{|\Gamma\left(\frac{a}{\nu}\right)|}\|h_1\|_{L^1(G)}.$$

2. *If* $\operatorname{Re} a > Q/2$,

$$\|\mathcal{B}_a\|_{L^2(G)} = \left(h_1(0)\frac{\Gamma(\frac{2\operatorname{Re} a - Q}{\nu})}{\Gamma(\frac{2\operatorname{Re} a}{\nu})}\right)^{1/2}.$$

3. *If $p \in (1,2)$ and $a > Q(1-\frac{1}{p})$ then $\mathcal{B}_a \in L^p(G)$.*

Proof. The first statement follows from (4.32).

For the second part, Estimate (4.33) yields

$$\|\mathcal{B}_a\|_2^2 = h_1(0)C_a,$$

where

$$\begin{aligned} C_a &= |\Gamma(a/\nu)|^{-2}\int_{s'=0}^{1} s'^{\frac{a}{\nu}-1}(1-s')^{\frac{\bar{a}}{\nu}-1}ds'\int_{u=0}^{\infty} e^{-u}u^{2\frac{\operatorname{Re} a}{\nu}-\frac{Q}{\nu}-1}du \\ &= |\Gamma(\frac{a}{\nu})|^{-2}\frac{\Gamma(\frac{a}{\nu})\Gamma(\frac{\bar{a}}{\nu})}{\Gamma(\frac{a}{\nu}+\frac{\bar{a}}{\nu})}\Gamma(\frac{2\operatorname{Re} a - Q}{\nu}), \end{aligned}$$

thanks to the properties of the Gamma function (see equality (A.4)). We notice that

$$\Gamma(\frac{a}{\nu})\Gamma(\frac{\bar{a}}{\nu}) = \Gamma(\frac{a}{\nu})\overline{\Gamma(\frac{a}{\nu})} = |\Gamma(\frac{a}{\nu})|^2.$$

Thus the constant C_a simplifies into

$$C_a = \frac{\Gamma(\frac{2\operatorname{Re} a - Q}{\nu})}{\Gamma(\frac{a}{\nu}+\frac{\bar{a}}{\nu})}.$$

This shows the second part.

The third part is obtained by complex interpolation between Parts 1 and 2. More precisely, we fix $a>0$ and $b>Q/2$ and we consider the linear functional defined on simple functions in $L^1(G)$ via

$$T_z\phi = \int_G \mathcal{B}_{az+b(1-z)}(x)\phi(x)$$

for any $z \in \mathbb{C}$, $\operatorname{Re} z \in [0,1]$. We have

$$|T_z\phi| \leq \|\mathcal{B}_{az+b(1-z)}\|_1\|\phi\|_\infty.$$

Before applying Part 1 to $\|\mathcal{B}_{az+b(1-z)}\|_1$, let us mention that the Stirling formula (A.3) implies that for any $w \in \mathbb{C}_+$,

$$\begin{aligned}\frac{\Gamma(\operatorname{Re} w)}{|\Gamma(w)|} &\lesssim \sqrt{\frac{|w|}{\operatorname{Re} w}}\,\frac{(\frac{\operatorname{Re} w}{e})^{\operatorname{Re} w}}{|(\frac{w}{e})^{w}|}\\ &\lesssim \left(\frac{\operatorname{Re} w}{|w|}\right)^{\operatorname{Re} w-\frac{1}{2}} |w^{w-\operatorname{Re} w}|\\ &\lesssim \left(\frac{\operatorname{Re} w}{|w|}\right)^{\operatorname{Re} w-\frac{1}{2}} \exp\left(|\operatorname{Im} w| \ln |w|\right).\end{aligned}$$

This together with Part 1 then yield

$$\ln |T_z\phi| \leq \ln(\|\mathcal{B}_{az+b(1-z)}\|_1 \|\phi\|_\infty) \lesssim (1+|\operatorname{Im} z|)\ln(1+|\operatorname{Im} z|),$$

thus $\{T_z\}$ is an admissible family of operator (in the sense of Section A.6). The same arguments also show that

$$|T_{1+iy}\phi| \lesssim (1+|y|)^{-\frac{a}{\nu}+\frac{1}{2}} \exp\left(c|y|\ln(1+|y|)\right) \|\phi\|_\infty,$$

where c is a constant of a, b, ν.

The Cauchy-Schwartz estimate and Part 2 yield

$$|T_{iy}\phi| \leq \|\mathcal{B}_{aiy+b(1-iy)}\|_2 \|\phi\|_2,$$

and Part 2 implies that the quantity

$$\|\mathcal{B}_{aiy+b(1-iy)}\|_2 = \left(h_1(0)\frac{\Gamma(\frac{2b-Q}{\nu})}{\Gamma(\frac{2b}{\nu})}\right)^{1/2},$$

is independent of y. Hence we can apply Theorem A.6.1 to $\{T_z\}$: T_t extends to an L^{q_t}-bounded operator where $t \in (0,1)$ and $\frac{1}{q_t} = \frac{1-t}{2}$. Therefore $\mathcal{B}_{at+b(1-t)} \in L^{q_t'}$ where q_t' is the dual exponent to q_t, i.e. $\frac{1}{q_t} + \frac{1}{q_t'} = 1$. This shows Part 3 and concludes the proof of Corollary 4.3.13. □

We finish this section with some technical properties which will be useful in the sequel. The first one is easy to check.

Lemma 4.3.14. *If $\mathcal{R}$ is a positive Rockland operator with $\mathcal{B}_a$ being the kernel of the Bessel potential as given in Corollary 4.3.11, then $\bar{\mathcal{R}}$ is also a positive Rockland operator and $\bar{\mathcal{B}}_a$ is the kernel of the Bessel potential associated to $\bar{\mathcal{R}}$.*

Lemma 4.3.15. *We keep the notation of Corollary 4.3.11. If $a \in \mathbb{C}_+$, then the function*

$$x \mapsto |x|^N \mathcal{B}_a(x)$$

is integrable on G, where $|\cdot|$ denotes any homogeneous quasi-norm on G and N is any positive integer. Consequently, for any $\phi \in \mathcal{S}(G)$, the function $\phi * \mathcal{B}_a$ is Schwartz and

$$\phi \mapsto \phi * \mathcal{B}_a$$

acts continuously from $\mathcal{S}(G)$ to itself.

Note that we will show in Lemma 4.5.9 that, more generally,

$$|x|^b X^\alpha \mathcal{B}_a \in L^1(G) \quad \text{for } \operatorname{Re} a + b > [\alpha],$$

and that

$$X^\alpha \mathcal{B}_a \in L^2(G) \quad \text{for } \operatorname{Re} a > [\alpha] + Q/2.$$

Proof of Lemma 4.3.15. Let $|\cdot|$ be a homogeneous quasi-norm on G and $N \in \mathbb{N}$. We see that

$$\int_G |x|^N |\mathcal{B}_a(x)| dx \leq \frac{1}{|\Gamma(\frac{a}{\nu})|} \int_0^\infty t^{\frac{\operatorname{Re} a}{\nu}-1} e^{-t} \int_G |x|^N |h_t(x)| dx\, dt,$$

and using the homogeneity of the heat kernel (see (4.17)) and the change of variables $y = t^{-\frac{1}{\nu}} x$, we get

$$\int_G |x|^N |h_t(x)| dx = \int_G |t^{\frac{1}{\nu}} y|^N |h_1(y)| dy = c_N t^{\frac{N}{\nu}},$$

where $c_N = \| |y|^N h_1(y) \|_{L^1(dy)}$ is a finite constant since $h_1 \in \mathcal{S}(G)$. Thus,

$$\int_G |x|^N |\mathcal{B}_a(x)| dx \leq \frac{c_N}{|\Gamma(\frac{a}{\nu})|} \int_0^\infty t^{\frac{\operatorname{Re} a}{\nu}-1+\frac{N}{\nu}} e^{-t} dt < \infty,$$

and $x \mapsto |x|^N \mathcal{B}_a(x)$ is integrable.

Let $C_o \geq 1$ denote the constant in the triangle inequality for $|\cdot|$ (see Proposition 3.1.38 and also Inequality (3.43)). Let also $\phi \in \mathcal{S}(G)$. We have for any $N \in \mathbb{N}$ and $\alpha \in \mathbb{N}_0^n$:

$$\begin{aligned}
(1+|x|)^N \left| \tilde{X}^\alpha [\phi * \mathcal{B}_a](x) \right| &= (1+|x|)^N \left| \tilde{X}^\alpha \phi * \mathcal{B}_a(x) \right| \\
&\leq (1+|x|)^N \left| \tilde{X}^\alpha \phi \right| * |\mathcal{B}_a|(x) \\
&\leq C_o^N \left| (1+|\cdot|)^N \tilde{X}^\alpha \phi \right| * \left| (1+|\cdot|)^N \mathcal{B}_a(x) \right| (x) \\
&\leq C_o^N \left\| (1+|\cdot|)^N \tilde{X}^\alpha \phi \right\|_\infty \left\| (1+|\cdot|)^N \mathcal{B}_a \right\|_{L^1(G)}.
\end{aligned}$$

This shows that that $\phi * \mathcal{B}_a \in \mathcal{S}(G)$ and that $\phi \mapsto \phi * \mathcal{B}_a$ is continuous as a map of $\mathcal{S}(G)$ to itself (for a description of the Schwartz class, see Section 3.1.9). □

Corollary 4.3.16. *We keep the notation of Corollary 4.3.11.*

For any $a \in \mathbb{C}$ and $p \in [1,\infty) \cup \{\infty_o\}$, $\text{Dom}(\text{I}+\mathcal{R}_p)^a \supset \mathcal{S}(G)$ and, moreover,

$$(\text{I}+\mathcal{R}_p)^a(\mathcal{S}(G)) = \mathcal{S}(G). \tag{4.34}$$

Furthermore on $\mathcal{S}(G)$, $(\text{I}+\mathcal{R}_p)^a$ does not depend on $p \in [1,\infty) \cup \{\infty_o\}$ and acts continuously on $\mathcal{S}(G)$.

If $a \in \mathbb{C}_+$, we have

$$(\text{I}+\mathcal{R}_p)^a \left(\phi * \mathcal{B}_{a\nu}\right) = ((\text{I}+\mathcal{R}_p)^a \phi) * \mathcal{B}_{a\nu} = \phi \qquad (p \in [1,\infty) \cup \{\infty_o\}). \tag{4.35}$$

Proof. Formula (4.35) holds for each $p \in [1,\infty) \cup \{\infty_o\}$ by Theorem 4.3.6 and Corollary 4.3.11.

Let us show (4.34) in the case of $a = N \in \mathbb{N}$. By Theorem 4.3.6 (1a), we have the equality $(\text{I}+\mathcal{R}_p)^N\phi = (\text{I}+\mathcal{R})^N\phi$ for any $\phi \in \mathcal{S}(G)$ and $p \in (1,\infty)$. Hence $(\text{I}+\mathcal{R}_p)^N(\mathcal{S}(G)) = (\text{I}+\mathcal{R})^N(\mathcal{S}(G))$. The inclusion $(\text{I}+\mathcal{R})^N(\mathcal{S}(G)) \subset \mathcal{S}(G)$ is immediate. The converse follows easily from Lemma 4.3.15 together with (4.35). This proves (4.34) for $a = N \in \mathbb{N}$. This implies that for any $N \in \mathbb{N}$, $\mathcal{S}(G)$ is included in

$$\text{Dom}\left[(\text{I}+\mathcal{R}_p)^N\right] \cap \text{Range}\left[(\text{I}+\mathcal{R}_p)^N\right]$$

and we can apply the analyticity results (Part (1c)) of Theorem 4.3.6: fixing $\phi \in \mathcal{S}(G)$, the function $a \mapsto (\text{I}+\mathcal{R}_p)^a\phi$ is holomorphic in $\{a \in \mathbb{C} : -N < \text{Re}\, a < N\}$. We observe that by Corollary 4.3.11 (ii), if $-N < \text{Re}\, a < 0$, all the functions $(\text{I}+\mathcal{R}_p)^a\phi$ coincide with $\phi * \mathcal{B}_{a\nu}$ for any $p \in [1,\infty) \cup \{\infty_o\}$. This shows that for each $a \in \mathbb{C}$ fixed, $(\text{I}+\mathcal{R}_p)^a\phi$ is independent of p. Furthermore, it is Schwartz. Indeed if $\text{Re}\, a < 0$ this follow from Lemma 4.3.15. If $\text{Re}\, a \geq 0$, we write $a = a_o + a'$ with $a_o \in \mathbb{N}$ and $\text{Re}\, a' < 0$ and we have in the sense of operators

$$(\text{I}+\mathcal{R})^{a'}(\text{I}+\mathcal{R})^{a_o} \subset (\text{I}+\mathcal{R})^a.$$

The operator $(\text{I}+\mathcal{R})^{a_o}$ is a differential operator, hence maps $\mathcal{S}(G)$ to itself, and the operator $(\text{I}+\mathcal{R})^{a'}$ maps $\mathcal{S}(G)$ to itself by Lemma 4.3.15. Thus in any case $(\text{I}+\mathcal{R}_p)^a\phi \in \mathcal{S}(G)$ and is independent of p.

We have obtained that $(\text{I}+\mathcal{R}_p)^a(\mathcal{S}(G)) \subset \mathcal{S}(G)$ for any $p \in (1,\infty)$, $a \in \mathbb{C}$. As $\{(\text{I}+\mathcal{R}_p)^a\}^{-1} = (\text{I}+\mathcal{R}_p)^{-a}$ by Theorem 4.3.6 (1a), this proves the equality in (4.34) for any $a \in \mathbb{C}$. Lemma 4.3.15 says that this action is continuous if $\text{Re}\, a < 0$. This is also the case for $\text{Re}\, a \geq 0$ since we can proceed as above and write $a = a_o + a'$ with $a_o \in \mathbb{N}$ and $\text{Re}\, a' < 0$, the action of $(\text{I}+\mathcal{R})^{a_o}$ being continuous on $\mathcal{S}(G)$. This concludes the proof of Corollary 4.3.16. □

Corollary 4.3.16 implies that the following definition makes sense.

Definition 4.3.17. Let $\mathcal{R}$ be a positive Rockland operator of homogeneous degree ν and let $s \in \mathbb{R}$. For any tempered distribution $f \in \mathcal{S}'(G)$, we denote by $(\text{I}+\mathcal{R})^{s/\nu}f$ the tempered distribution defined by

$$\langle (\text{I}+\mathcal{R})^{s/\nu}f, \phi\rangle = \langle f, (\text{I}+\bar{\mathcal{R}})^{s/\nu}\phi\rangle, \quad \phi \in \mathcal{S}(G).$$

4.4 Sobolev spaces on graded Lie groups

In this section we define the (homogeneous and inhomogeneous) Sobolev spaces associated to a positive Rockland operator $\mathcal{R}$ and show that they satisfy similar properties to the Euclidean Sobolev spaces and to the Sobolev spaces defined and studied by Folland [Fol75] on stratified Lie groups. In Section 4.4.5, we show that the constructed spaces are actually independent of the choice of a positive Rockland operator $\mathcal{R}$ on a graded Lie group with which we start our construction. In Section 4.4.7, we list the main properties of our Sobolev spaces.

4.4.1 (Inhomogeneous) Sobolev spaces

We first need the following lemma:

Lemma 4.4.1. *We keep the notation of Theorem 4.3.6. For any $s \in \mathbb{R}$ and $p \in [1,\infty) \cup \{\infty_o\}$, the domain of the operator $(\mathrm{I}+\mathcal{R}_p)^{\frac{s}{\nu}}$ contains $\mathcal{S}(G)$, and the map*

$$f \longmapsto \|(\mathrm{I}+\mathcal{R}_p)^{\frac{s}{\nu}} f\|_{L^p(G)}$$

defines a norm on $\mathcal{S}(G)$. We denote it by

$$\|f\|_{L^p_s(G)} := \|(\mathrm{I}+\mathcal{R}_p)^{\frac{s}{\nu}} f\|_{L^p(G)}.$$

Moreover, any sequence in $\mathcal{S}(G)$ which is Cauchy for $\|\cdot\|_{L^p_s(G)}$ is convergent in $\mathcal{S}'(G)$.

We have allowed ourselves to write $\|\cdot\|_{L^\infty(G)} = \|\cdot\|_{L^{\infty_o}(G)}$ for the supremum norm. We may also write $\|\cdot\|_\infty$ or $\|\cdot\|_{\infty_o}$.

Proof. By Corollary 4.3.16, the domain of $(\mathrm{I}+\mathcal{R}_p)^{\frac{s}{\nu}}$ contains $\mathcal{S}(G)$. Since the operator $(\mathrm{I}+\mathcal{R}_p)^{\frac{s}{\nu}}$ is linear, it is easy to check that the map $f \mapsto \|(\mathrm{I}+\mathcal{R}_p)^{\frac{s}{\nu}} f\|_p$ is non-negative and satisfies the triangle inequality. Since $(\mathrm{I}+\mathcal{R}_p)^{s/\nu}$ is injective by Theorem 4.3.6, Part (1), we have that $\|f\|_{L^p_s(G)} = 0$ implies $f = 0$.

Clearly $\|\cdot\|_{L^p_0(G)} = \|\cdot\|_p$, so in the case of $s = 0$ a Cauchy sequence of Schwartz functions converges in L^p-norm, thus also in $\mathcal{S}'(G)$.

Let us assume $s > 0$. By Corollary 4.3.11 (ii), the operator $(\mathrm{I}+\mathcal{R}_p)^{-\frac{s}{\nu}}$ is bounded on $L^p(G)$. Hence we have

$$\|\cdot\|_{L^p(G)} \leq C\|\cdot\|_{L^p_s(G)}$$

on $\mathcal{S}(G)$. Consequently a $\|\cdot\|_{L^p_s(G)}$-Cauchy sequence of Schwartz functions converge in L^p-norm thus in $\mathcal{S}'(G)$.

Now let us assume $s < 0$. Let $\{f_\ell\}_{\ell\in\mathbb{N}}$ be a sequence of Schwartz functions which is Cauchy for the norm $\|\cdot\|_{L^p_s(G)}$. By (4.35) we have

$$f_\ell = \left((\mathrm{I}+\mathcal{R}_p)^{\frac{s}{\nu}} f_\ell\right) * \mathcal{B}_s.$$

Furthermore, if $\phi \in \mathcal{S}(G)$ then using (1.14) and (4.13), we have

$$\int_G f_\ell(x)\phi(x)dx = \int_G \left((\mathrm{I}+\mathcal{R}_p)^{\frac{s}{\nu}} f_\ell\right)(x) \ \ (\phi * \mathcal{B}_s)(x)\, dx. \tag{4.36}$$

By assumption the sequence $\{(\mathrm{I}+\mathcal{R}_p)^{\frac{s}{\nu}} f_\ell\}_{\ell\in\mathbb{N}}$ is $\|\cdot\|_{L^p(G)}$-Cauchy thus convergent in $L^p(G)$. By Lemma 4.3.15, $\phi * \mathcal{B}_s \in \mathcal{S}(G)$. Therefore, the right-hand side of (4.36) is convergent as $\ell \to \infty$. Hence the scalar sequence $\langle f_\ell, \phi\rangle$ converges for any $\phi \in \mathcal{S}(G)$. This shows that the sequence $\{f_\ell\}$ converges in $\mathcal{S}'(G)$. □

Lemma 4.4.1 allows us to define the (inhomogeneous) Sobolev spaces:

Definition 4.4.2. Let $\mathcal{R}$ be a positive Rockland operator on a graded Lie group G. We consider its L^p-analogue $\mathcal{R}_p$ and the powers of $(\mathrm{I}+\mathcal{R}_p)^a$ as defined in Theorems 4.3.3 and 4.3.6. Let $s \in \mathbb{R}$.

If $p \in [1,\infty)$, the *Sobolev space* $L^p_{s,\mathcal{R}}(G)$ is the subspace of $\mathcal{S}'(G)$ obtained by completion of $\mathcal{S}(G)$ with respect to the *Sobolev norm*

$$\|f\|_{L^p_{s,\mathcal{R}}(G)} := \|(\mathrm{I}+\mathcal{R}_p)^{\frac{s}{\nu}} f\|_{L^p(G)}, \quad f \in \mathcal{S}(G).$$

If $p = \infty_o$, the *Sobolev space* $L^{\infty_o}_{s,\mathcal{R}}(G)$ is the subspace of $\mathcal{S}'(G)$ obtained by completion of $\mathcal{S}(G)$ with respect to the *Sobolev norm*

$$\|f\|_{L^{\infty_o}_{s,\mathcal{R}}(G)} := \|(\mathrm{I}+\mathcal{R}_{\infty_o})^{\frac{s}{\nu}} f\|_{L^\infty(G)}, \quad f \in \mathcal{S}(G).$$

When the Rockland operator $\mathcal{R}$ is fixed, we may allow ourselves to drop the index $\mathcal{R}$ in $L^p_{s,\mathcal{R}}(G) = L^p_s(G)$ to simplify the notation.

We will see later that the Sobolev spaces actually do not depend on the Rockland operator $\mathcal{R}$, see Theorem 4.4.20.

By construction the Sobolev space $L^p_s(G)$ endowed with the Sobolev norm is a Banach space which contains $\mathcal{S}(G)$ as a dense subspace and is included in $\mathcal{S}'(G)$. The Sobolev spaces share many properties with their Euclidean counterparts.

Theorem 4.4.3. *Let $\mathcal{R}$ be a positive Rockland operator of homogeneous degree ν on a graded Lie group G. We consider the associated Sobolev spaces $L^p_s(G)$ for $p \in [1,\infty)\cup\{\infty_o\}$ and $s \in \mathbb{R}$.*

1. *If $s = 0$, then $L^p_0(G) = L^p(G)$ for $p \in [1,\infty)$ with $\|\cdot\|_{L^p_0(G)} = \|\cdot\|_{L^p(G)}$, and $L^{\infty_o}_0(G) = C_o(G)$ with $\|\cdot\|_{L^{\infty_o}_0(G)} = \|\cdot\|_{L^\infty(G)}$.*
2. *If $s > 0$, then for any $a \in \mathbb{C}$ with $\operatorname{Re} a = s$, we have*

$$L^p_s(G) = \operatorname{Dom}\left[(\mathrm{I}+\mathcal{R}_p)^{\frac{a}{\nu}}\right] = \operatorname{Dom}(\mathcal{R}_p^{\frac{a}{\nu}}) \subsetneq L^p(G),$$

and the following norms are equivalent to $\|\cdot\|_{L^p_s(G)}$:

$$f \longmapsto \|f\|_{L^p(G)} + \|(\mathrm{I}+\mathcal{R}_p)^{\frac{s}{\nu}} f\|_{L^p(G)}, \ f \longmapsto \|f\|_{L^p(G)} + \|\mathcal{R}_p^{\frac{s}{\nu}} f\|_{L^p(G)}.$$

3. *Let* $s \in \mathbb{R}$ *and* $f \in \mathcal{S}'(G)$.
   - *Given* $p \in (1,\infty)$, *we have* $f \in L^p_s(G)$ *if and only if the tempered distribution* $(\mathrm{I}+\mathcal{R}_p)^{s/\nu} f$ *defined in Definition 4.3.17 is in* $L^p(G)$, *in the sense that the linear mapping*
   $$\mathcal{S}(G) \ni \phi \mapsto \langle (\mathrm{I}+\mathcal{R})^{s/\nu} f, \phi\rangle = \langle f, (\mathrm{I}+\bar{\mathcal{R}}_{p'})^{s/\nu}\phi\rangle$$
   extends to a bounded functional on $L^{p'}(G)$ *where* p' *is the conjugate exponent of* p.
   - $f \in L^1_s(G)$ *if and only if* $(\mathrm{I}+\mathcal{R}_1)^{s/\nu} f \in L^1(G)$ *in the sense that the linear mapping*
   $$\mathcal{S}(G) \ni \phi \mapsto \langle (\mathrm{I}+\mathcal{R})^{s/\nu} f, \phi\rangle = \langle f, (\mathrm{I}+\bar{\mathcal{R}}_{\infty_o})^{s/\nu}\phi\rangle$$
   extends to a bounded functional on $C_o(G)$ *and is realised as a measure given by an integrable function.*
   - $f \in L^{\infty_o}_s(G)$ *if and only if* $(\mathrm{I}+\mathcal{R}_{\infty_o})^{s/\nu} f \in C_o(G)$ *in the sense that the linear mapping*
   $$\mathcal{S}(G) \ni \phi \mapsto \langle (\mathrm{I}+\mathcal{R})^{s/\nu} f, \phi\rangle = \langle f, (\mathrm{I}+\bar{\mathcal{R}}_1)^{s/\nu}\phi\rangle$$
   extends to a bounded functional on $L^1(G)$ *and is realised as integration against functions in* $C_o(G)$.
4. *If* $a, b \in \mathbb{R}$ *with* $a < b$ *and* $p \in [1,\infty) \cup \{\infty_o\}$, *then the following continuous strict inclusions hold*
   $$\mathcal{S}(G) \subsetneq L^p_b(G) \subsetneq L^p_a(G) \subsetneq \mathcal{S}'(G),$$
   and an equivalent norm for $L^p_b(G)$ *is*
   $$L^p_b(G) \ni f \longmapsto \|f\|_{L^p_a(G)} + \|\mathcal{R}_p^{\frac{b-a}{\nu}} f\|_{L^p_a(G)}.$$
5. *For* $p \in [1,\infty) \cup \{\infty_o\}$ *and any* $a, b, c \in \mathbb{R}$ *with* $a < c < b$, *there exists a positive constant* $C = C_{a,b,c}$ *such that for any* $f \in L^p_b$, *we have* $f \in L^p_c \cap L^p_a$ *and*
   $$\|f\|_{L^p_c} \leq C \|f\|^{1-\theta}_{L^p_a} \|f\|^{\theta}_{L^p_b},$$
   where $\theta := (c-a)/(b-a)$.

In Theorem 4.4.20, we will see that the definition of the Sobolev spaces and their properties given in Theorem 4.4.3 hold independently of the chosen Rockland operator $\mathcal{R}$.

From now on, we will often use the notation $L^p_0(G)$ since this allows us not to distinguish between the cases $L^p_0(G) = L^p(G)$ when $p \in [1,\infty)$ and $L^p_0(G) = C_o(G)$ when $p = \infty_o$.

In the proof of Part (2) of Theorem 4.4.3, we will need the following exercise in functional analysis:

Lemma 4.4.4. *Let T_1 and T_2 be two linear operators between two Banach spaces $\mathcal{X} \to \mathcal{Y}$. We assume that T_1 and T_2 are densely defined and share the same domain. We also assume that they are both closed injective operators and that T_2 is bijective with a bounded inverse. Then the graph norms of T_1 and T_2 are equivalent, that is,*

$$\begin{gathered}\exists C>0 \quad \forall x \in \operatorname{Dom}(T_1)=\operatorname{Dom}(T_2) \\ C^{-1}(\|x\|+\|T_2 x\|) \leq \|x\|+\|T_1 x\| \leq C(\|x\|+\|T_2 x\|).\end{gathered}$$

Sketch of the proof of Lemma 4.4.4. One can check easily that $T := T_1 T_2^{-1}$ defines a closed linear operator $T : \mathcal{Y} \to \mathcal{Y}$ defined on the whole space $\mathcal{Y}$. By the closed graph theorem (see, e.g., [Rud91, Theorem 2.15] or [RS80, Thm III. 12]), T is bounded. Furthermore, T is injective as the composition of two injective operators. It may not have a closed range in $\mathcal{Y}$ but one checks easily that the operator

$$(T_2^{-1}, T) : \left\{ \begin{array}{ccc} \mathcal{Y} & \longrightarrow & \mathcal{X} \times \mathcal{Y} \\ y & \longmapsto & (T_2^{-1} y, Ty) \end{array} \right. ,$$

has a closed range in $\mathcal{X} \times \mathcal{Y}$. Hence the restriction of (T_2^{-1}, T) onto its image is bounded with a bounded inverse (see e.g. [RS80, Thm III. 11]). Consequently,

$$\|T_2^{-1} y\| + \|Ty\| \asymp \|y\|$$

for any element $y \in \mathcal{Y}$, in particular of the form $y = T_2 x$, $x \in \operatorname{Dom}(T_2)$. $\square$

We can now prove Theorem 4.4.3.

Proof of Theorem 4.4.3. Part (1) is true since $(\mathrm{I}+\mathcal{R}_p)^{\frac{0}{\nu}} = \mathrm{I}$. Let us prove Part (2). So let $s > 0$. Clearly $L^p_s(G)$ coincides with the domain of the unbounded operator $(\mathrm{I}+\mathcal{R}_p)^{\frac{s}{\nu}}$ (see Theorem 4.3.6 (2)) hence it is a proper subspace of $L^p(G)$. As the operator $(\mathrm{I}+\mathcal{R}_p)^{-\frac{s}{\nu}}$ is bounded on $L^p(G)$, we have $\|\cdot\|_{L^p(G)} \leq C\|\cdot\|_{L^p_s(G)}$ on $L^p_s(G)$. So $\|\cdot\|_{L^p(G)} + \|\cdot\|_{L^p_s(G)}$ is a norm on $L^p_s(G)$ which is equivalent to the Sobolev norm. Theorem 4.3.6 implies that $\mathcal{R}_p^{\frac{s}{\nu}}$ and $(\mathrm{I}+\mathcal{R}_p)^{\frac{s}{\nu}}$ satisfy the hypotheses of Lemma 4.4.4. This shows part (2).

Part (3) follows from Part (2) and the duality properties of the spaces $L^p(G)$ and $C_o(G)$ in the case $s \geq 0$. We now consider the case $s < 0$. By Lemma 4.3.15 and Corollary 4.3.11 (and also Lemma 4.3.14), the mapping

$$T_{s,p',f} : \mathcal{S}(G) \ni \phi \longmapsto \langle f, (\mathrm{I}+\bar{\mathcal{R}}_{p'})^{s/\nu} \phi \rangle = \langle f, \phi * \bar{\mathcal{B}}_{-s} \rangle$$

is well defined for any $f \in \mathcal{S}'(G)$. If $T_{s,p',f}$ admits a bounded extension to a functional on $L_0^{p'}(G)$, then we denote this extension $\tilde{T}_{s,p',f}$ and we have

$$\|\tilde{T}_{s,p',f}\|_{\mathscr{L}(L_0^{p'},\mathbb{C})} = \|f\|_{L^p_s(G)}. \tag{4.37}$$

This is certainly so if $f \in \mathcal{S}(G)$. Furthermore a sequence $\{f_\ell\}_{\ell\in\mathbb{N}}$ of Schwartz functions is convergent for the Sobolev norm $\|\cdot\|_{L^p_s(G)}$ if and only if $\{\tilde{T}_{s,p',f_\ell}\}$ is convergent in $L^{p'}_0(G)$ (see Lemma 4.3.2). In the case of convergence, by Lemma 4.4.1, $\{f_\ell\}_{\ell\in\mathbb{N}}$ converges in the sense of distributions. Denoting this limit by $f \in \mathcal{S}'(G)$, we have

$$\left[\lim_{\ell\to\infty} \tilde{T}_{s,p',f_\ell}\right]\Big|_{\mathcal{S}(G)} = T_{s,p',f}.$$

It is easy to see, by linearity of $f_1 \mapsto T_{s,p',f_1}$ and (4.37), that $T_{s,p',f}$ extends to a continuous functional on $L^{p'}_0(G)$.

Conversely, let us consider a distribution $f \in \mathcal{S}'(G)$ such that $T_{s,p',f}$ extends to a bounded functional $\tilde{T}_{s,p',f}$ on $L^{p'}_0(G)$. If $\{f_\ell\}_{\ell\in\mathbb{N}}$ is a sequence of Schwartz functions converging to f in $\mathcal{S}'(G)$, then

$$\lim_{\ell\to\infty} T_{s,p',f_\ell}(\phi) = T_{s,p',f}(\phi)$$

for every $\phi \in \mathcal{S}(G)$, and using the density of $\mathcal{S}(G)$ in $L^{p'}_0(G)$ and the Banach-Steinhaus Theorem, this shows that $\{\tilde{T}_{s,p',f_\ell}\}$ converges to $\tilde{T}_{s,p',f}$ in the norm of the dual of $L^{p'}_0(G)$. This shows the case $s < 0$ and concludes the proof of Part (3).

Let us show Part (4). Let $a \le b$ and $p \in [1,\infty) \cup \{\infty_o\}$. By Theorem 4.3.6 (1), we have in the sense of operators

$$(\mathrm{I} + \mathcal{R}_p)^{\frac{a}{\nu}} \supset (\mathrm{I} + \mathcal{R}_p)^{\frac{a-b}{\nu}} (\mathrm{I} + \mathcal{R}_p)^{\frac{b}{\nu}}.$$

Since the operator $(\mathrm{I} + \mathcal{R}_p)^{\frac{a-b}{\nu}}$ is bounded, we have for any $f \in \mathcal{S}(G)$

$$\begin{aligned}\|f\|_{L^p_a(G)} &= \|(\mathrm{I} + \mathcal{R}_p)^{\frac{a}{\nu}} f\|_p = \|(\mathrm{I} + \mathcal{R}_p)^{\frac{a-b}{\nu}} (\mathrm{I} + \mathcal{R}_p)^{\frac{b}{\nu}} f\|_p \\ &\le \|(\mathrm{I} + \mathcal{R}_p)^{\frac{a-b}{\nu}}\|_{\mathscr{L}(L^p_0)} \|(\mathrm{I} + \mathcal{R}_p)^{\frac{b}{\nu}} f\|_p = \|(\mathrm{I} + \mathcal{R}_p)^{\frac{a-b}{\nu}}\|_{\mathscr{L}(L^p_0)} \|f\|_{L^p_b}.\end{aligned}$$

By density of $\mathcal{S}(G)$, this implies the continuous inclusion $L^p_b \subset L^p_a$. Note that we also have if $a < b$

$$\begin{aligned}\|f\|_{L^p_b(G)} &= \|(\mathrm{I} + \mathcal{R}_p)^{\frac{b-a}{\nu}} (\mathrm{I} + \mathcal{R}_p)^{\frac{a}{\nu}} f\|_p = \|(\mathrm{I} + \mathcal{R}_p)^{\frac{a}{\nu}} f\|_{L^p_{b-a}(G)} \\ &\asymp \|(\mathrm{I} + \mathcal{R}_p)^{\frac{a}{\nu}} f\|_{L^p(G)} + \|\mathcal{R}_p^{\frac{b-a}{\nu}} (\mathrm{I} + \mathcal{R}_p)^{\frac{a}{\nu}} f\|_{L^p(G)},\end{aligned}$$

by Part (2) above for any $f \in \mathcal{S}(G)$. By Theorem 4.3.6 (5), we can commute the operators $\mathcal{R}_p^{\frac{b-a}{\nu}}$ and $(\mathrm{I} + \mathcal{R}_p)^{\frac{a}{\nu}}$ in this last expression. Consequently, we have obtained for any $f \in \mathcal{S}(G)$,

$$\|f\|_{L^p_b(G)} \asymp \|f\|_{L^p_a(G)} + \|\mathcal{R}_p^{\frac{b-a}{\nu}} f\|_{L^p_a(G)}.$$

By density of $\mathcal{S}(G)$, this holds for any $f \in L^p_b(G)$. Since the operator $\mathcal{R}_p^{\frac{b-a}{\nu}}$ is unbounded, this also implies the strict inclusions given in Part (4).

Part (5) follows from Theorem 4.3.6 (1f) for the case of $a = 0$. For $f \in L^p_b$, we then apply this to $b-a, c-a$ instead of b and c and $\phi := (\mathrm{I}+\mathcal{R}_p)^{\frac{a}{\nu}} f \in L^p_{b-a}$ instead of f.

This concludes the proof of this part and of the whole theorem. □

Theorem 4.4.3 has the two following corollaries. The first one is an easy consequence of Part (3).

Corollary 4.4.5. *We keep the setting and notation of Theorem 4.4.3. Let $s < 0$ and $p \in [1,\infty) \cup \{\infty_o\}$. Let $f \in \mathcal{S}'(G)$.*

The tempered distribution f is in $L^p_s(G)$ if and only if the mapping

$$\mathcal{S}(G) \ni \phi \mapsto \langle f, \phi * \bar{\mathcal{B}}_{-s} \rangle$$

extends to a bounded linear functional on $L^{p'}_0(G)$ with the additional property that

- *for $p = 1$, this functional on $C_o(G)$ is realised as a measure given by an integrable function,*
- *if $p = \infty_o$, this functional on $L^1(G)$ is realised by integration against a function in $C_o(G)$.*

Corollary 4.4.6. *We keep the setting and notation of Theorem 4.4.3. Let $s \in \mathbb{R}$ and $p \in [1,\infty) \cup \{\infty_o\}$. Then $\mathcal{D}(G)$ is dense in $L^p_s(G)$.*

Proof of Corollary 4.4.6. This is certainly true for $s \geq 0$ (see the proof of Parts (1) and (2) of Theorem 4.4.3). For $s < 0$, it suffices to proceed as in the last part of the proof of Part (3) with a sequence of functions $f_\ell \in \mathcal{D}(G)$. □

Theorem 4.4.3, especially Part (3), implies the following property regarding duality of Sobolev spaces. This will be improved in Proposition 4.4.22 once we show in Theorem 4.4.20 that the Sobolev spaces are indeed independent of the considered Rockland operator.

Lemma 4.4.7. *Let $\mathcal{R}$ be a positive Rockland operator on a graded Lie group G. We consider the associated Sobolev spaces $L^p_{s,\mathcal{R}}(G)$. If $s \in \mathbb{R}$ and $p \in (1,\infty)$, the dual space of $L^p_{s,\mathcal{R}}(G)$ is isomorphic to $L^{p'}_{-s,\bar{\mathcal{R}}}(G)$ via the distributional duality, where p' is the conjugate exponent of p, $\frac{1}{p} + \frac{1}{p'} = 1$.*

Proof of Lemma 4.4.7. Clearly if $f \in L^p_{s,\mathcal{R}}(G)$ then for any $\phi \in \mathcal{S}(G)$,

$$\langle f, \phi \rangle = \langle f, (\mathrm{I}+\bar{\mathcal{R}}_{p'})^{\frac{s}{\nu}} (\mathrm{I}+\bar{\mathcal{R}}_{p'})^{-\frac{s}{\nu}} \phi \rangle = \langle (\mathrm{I}+\mathcal{R}_p)^{\frac{s}{\nu}} f, (\mathrm{I}+\bar{\mathcal{R}}_{p'})^{-\frac{s}{\nu}} \phi \rangle$$

by Theorem 4.3.6. Hence by Theorem 4.4.3 Part (3),

$$|\langle f, \phi \rangle| \leq \|(\mathrm{I}+\mathcal{R}_p)^{\frac{s}{\nu}} f\|_p \|(\mathrm{I}+\bar{\mathcal{R}}_{p'})^{-\frac{s}{\nu}} \phi\|_{p'}$$

and the linear function $\mathcal{S}(G) \ni \phi \mapsto \langle f, \phi \rangle$ extends to a bounded linear functional on $L^{p'}_{-s,\bar{\mathcal{R}}}(G)$. Conversely, let Ψ be a bounded linear functional on $L^{p'}_{-s,\bar{\mathcal{R}}}(G)$. Then since

$$(\mathrm{I} + \bar{\mathcal{R}}_{p'})^{s/\nu} \mathcal{S}(G) = \mathcal{S}(G) \subset L^{p'}_{-s,\bar{\mathcal{R}}}(G),$$

see Corollary 4.3.16 and Definition 4.4.2, the linear functional $\Psi \circ (\mathrm{I} + \bar{\mathcal{R}}_{p'})^{s/\nu}$ is well defined on $\mathcal{S}(G)$ and satisfies for any $\phi \in \mathcal{S}(G)$,

$$\begin{aligned} |\Psi \circ (\mathrm{I} + \bar{\mathcal{R}}_{p'})^{s/\nu}(\phi)| &= |\Psi \left((\mathrm{I} + \bar{\mathcal{R}}_{p'})^{s/\nu} \phi \right) | \\ &\leq C \|(\mathrm{I} + \bar{\mathcal{R}}_{p'})^{s/\nu} \phi\|_{L^{p'}_{-s,\bar{\mathcal{R}}}} = C \|\phi\|_{L^{p'}_0}. \end{aligned}$$

Therefore, $\Psi \circ (\mathrm{I} + \bar{\mathcal{R}}_{p'})^{s/\nu}$ extends into a bounded linear functional on $L^p_0(G)$. □

In the next statement, we show how to produce functions and converging sequences of Sobolev spaces using the convolution:

Proposition 4.4.8. *We keep the setting and notation of Theorem 4.4.3. Here* $a \in \mathbb{R}$ *and* $p \in [1, \infty) \cup \{\infty_o\}$.

(i) If $f \in L^p_0(G)$ *and* $\phi \in \mathcal{S}(G)$, *then* $f * \phi \in L^p_a$ *for any* a *and* p.

(ii) If $f \in L^p_a(G)$ *and* $\psi \in \mathcal{S}(G)$, *then*

$$(\mathrm{I} + \mathcal{R}_p)^{\frac{a}{\nu}} (\psi * f) = \psi * \left((\mathrm{I} + \mathcal{R}_p)^{\frac{a}{\nu}} f \right), \tag{4.38}$$

and $\psi * f \in L^p_a(G)$ *with*

$$\|\psi * f\|_{L^p_a(G)} \leq \|\psi\|_{L^1(G)} \|f\|_{L^p_a(G)}. \tag{4.39}$$

Furthermore, if $\int \psi = 1$, *writing*

$$\psi_\epsilon(x) := \epsilon^{-Q} \psi(\epsilon^{-1} x)$$

for each $\epsilon > 0$, *then* $\{\psi_\epsilon * f\}$ *converges to* f *in* $L^p_a(G)$ *as* $\epsilon \to 0$.

Proof of Proposition 4.4.8. Let us prove Part (i). Here $f \in L^p_0(G)$. By density of $\mathcal{S}(G)$ in $L^p_0(G)$, we can find a sequence of Schwartz functions $\{f_\ell\}$ converging to f in L^p_0-norm. Then $f_\ell * \phi \in \mathcal{S}(G)$ and for any $N \in \mathbb{N}$,

$$\mathcal{R}^N (f_\ell * \phi) = f_\ell * \mathcal{R}^N \phi \underset{\ell \to \infty}{\longrightarrow} f * \mathcal{R}^N \phi \quad \text{in } L^p_0(G),$$

thus $\mathcal{R}^N_p (f * \phi) = f * \mathcal{R}^N \phi \in L^p(G)$ and

$$\|f * \phi\|_{L^p_0(G)} + \|\mathcal{R}^N_p (f * \phi)\|_{L^p_0(G)} < \infty.$$

By Theorem 4.4.3 (4), this shows that $f * \phi$ is in $L^p_{\nu N}$ for any $N \in \mathbb{N}$, hence in any p-Sobolev spaces. This proves (i).

Let us prove Part (ii). We observe that both sides of Formula (4.38) always make sense as convolutions of a Schwartz function with a tempered distribution.

Let us first assume that $f \in \mathcal{S}(G)$. Formula (4.38) is true if $a < 0$ by Corollary 4.3.11 (ii) since then the $(\mathrm{I} + \mathcal{R}_p)^{\frac{a}{\nu}}$ is a convolution operator with an integrable convolution kernel. Formula (4.38) is also true if $a \in \nu\mathbb{N}_0$ as in this case $(\mathrm{I}+\mathcal{R}_p)^{\frac{a}{\nu}}$ is a left-invariant differential operator by Theorem 4.3.6 (1a). Hence Formula (4.38) holds for any $a > 0$ by writing $a = a_0 + a'$, $a_0 \in \nu\mathbb{N}_0$, $a' < 0$, and

$$(\mathrm{I} + \mathcal{R}_p)^{\frac{a}{\nu}} f = (\mathrm{I} + \mathcal{R}_p)^{\frac{a_0}{\nu}} (\mathrm{I} + \mathcal{R}_p)^{\frac{a'}{\nu}} f.$$

Together with Corollary 4.3.16, this shows that Formulae (4.38) and consequently (4.39) hold for any $a \in \mathbb{R}$ and $f \in \mathcal{S}(G)$.

By density of $\mathcal{S}(G)$ in $L^p_s(G)$ and (4.39), this shows that Formulae (4.38) and (4.39) hold for any $f \in L^p_s(G)$.

Hence $\psi * f \in L^p_a(G)$ with L^p_a-norm $\leq \|\psi\|_1 \|f\|_{L^p_a(G)}$.

If $\int_G \psi = 1$, by Lemma 3.1.58 (i),

$$\begin{aligned}\|\psi_\epsilon * f - f\|_{L^p_a(G)} &= \|(\mathrm{I} + \mathcal{R}_p)^{\frac{a}{\nu}} (\psi_\epsilon * f - f)\|_p \\ &= \|\psi_\epsilon * \left((\mathrm{I} + \mathcal{R}_p)^{\frac{a}{\nu}} f\right) - (\mathrm{I} + \mathcal{R}_p)^{\frac{a}{\nu}} f\|_p \longrightarrow_{\epsilon\to 0} 0,\end{aligned}$$

that is, $\{\psi_\epsilon * f\}$ converges to f in $L^p_a(G)$ as $\epsilon \to 0$. This proves (ii). □

4.4.2 Interpolation between inhomogeneous Sobolev spaces

In this section, we prove that interpolation between Sobolev spaces $L^p_a(G)$ works in the same way as its Euclidean counterpart.

Theorem 4.4.9. *Let $\mathcal{R}$ and $\mathcal{Q}$ be two positive Rockland operators on two graded Lie groups G and F. We consider their associated Sobolev spaces $L^p_a(G)$ and $L^q_b(F)$. Let $p_0, p_1, q_0, q_1 \in (1, \infty)$ and let a_0, a_1, b_0, b_1 be real numbers.*

We also consider a linear mapping T from $L^{p_0}_{a_0}(G) + L^{p_1}_{a_1}(G)$ to locally integrable functions on F. We assume that T maps $L^{p_0}_{a_0}(G)$ and $L^{p_1}_{a_1}(G)$ boundedly into $L^{q_0}_{b_0}(F)$ and $L^{q_1}_{b_1}(F)$, respectively.

Then T extends uniquely to a bounded mapping from $L^p_{a_t}(G)$ to $L^q_{b_t}(F)$ for $t \in [0, 1]$ where a_t, b_t, p_t, q_t are defined by

$$\left(a_t, b_t, \frac{1}{p_t}, \frac{1}{q_t}\right) = (1-t)\left(a_0, b_0, \frac{1}{p_0}, \frac{1}{q_0}\right) + t\left(a_1, b_1, \frac{1}{p_1}, \frac{1}{q_1}\right).$$

The idea of the proof is similar to the one of the Euclidean or stratified cases, see [Fol75, Theorem 4.7]. Some arguments will be modified since our estimates for $\|(\mathrm{I} + \mathcal{R})^{i\tau}\|_{\mathscr{L}(L^p)}$ are different from the ones obtained by Folland in [Fol75]. For this, compare Corollary 4.3.13 and Proposition 4.3.7 in this monograph with [Fol75, Proposition 4.3].

Proof of Theorem 4.4.9. By duality (see Lemma 4.4.7) and up to a change of notation, it suffices to prove the case

$$a_1 \geq a_0 \quad \text{and} \quad b_1 \leq b_0. \tag{4.40}$$

This fact is left to the reader to check. The idea is to interpolate between the operators formally given by

$$T_z = (\mathrm{I} + \mathcal{Q})^{\frac{b_z}{\nu_{\mathcal{Q}}}} T (\mathrm{I} + \mathcal{R})^{-\frac{a_z}{\nu_{\mathcal{R}}}}, \tag{4.41}$$

where $\nu_{\mathcal{R}}$ and $\nu_{\mathcal{Q}}$ denote the degrees of homogeneity of $\mathcal{R}$ and $\mathcal{Q}$, respectively, and the complex numbers a_z and b_z are defined by

$$(a_z, b_z) := z\,(a_1, b_1) + (1-z)\,(a_0, b_0),$$

for z in the strip

$$S := \{z \in \mathbb{C} \ : \ \mathrm{Re}\, z \in [0,1]\}.$$

In (4.41), we have abused the notation regarding the fractional powers of $\mathrm{I} + \mathcal{R}_p$ and $\mathrm{I} + \mathcal{Q}_q$ and removed p and q. This is possible by Corollary 4.3.16 and density of the Schwartz space in each Sobolev space. Hence (4.41) makes sense. We will use complex interpolation given by Theorem A.6.1, which requires to start with the space $\mathscr{B}$ of compactly supported simple functions on G (see Remark A.6.2). To solve this technical problem we proceed as in the proof of [Fol75, Theorem 4.7]: we will use the convolution of a function in $\mathscr{B}$ with a bump function χ_ϵ depending on ϵ at the end of the proof.

The hypotheses on T give that the operator norms

$$\|T\|_{\mathscr{L}(L^{p_j}_{a_j}, L^{q_j}_{b_j})} = \|(\mathrm{I} + \mathcal{Q})^{\frac{b_j}{\nu_{\mathcal{Q}}}} T (\mathrm{I} + \mathcal{R})^{-\frac{a_j}{\nu_{\mathcal{R}}}}\|_{\mathscr{L}(L^{p_j}, L^{q_j})}, \qquad j = 0, 1,$$

are finite.

By Corollary 4.3.16, for any $\phi \in \mathcal{S}(G)$ and $\psi \in \mathcal{S}(F)$, we have

$$\langle T_z \phi, \psi \rangle = \langle T (\mathrm{I} + \mathcal{R})^{-N - \frac{a_z}{\nu_{\mathcal{R}}}} (\mathrm{I} + \mathcal{R})^N \phi, (\mathrm{I} + \bar{\mathcal{Q}})^{-M + \frac{b_z}{\nu_{\mathcal{Q}}}} (\mathrm{I} + \bar{\mathcal{Q}})^M \psi \rangle$$

for any $M, N \in \mathbb{Z}$. In particular, for M and N large enough, Theorem 4.3.6 implies that

$$S \ni z \mapsto \langle T_z \phi, \psi \rangle$$

is analytic. With $M = N \in \mathbb{N}$ large enough, for instance the smallest integer with $N > a_1, a_0, b_1, b_0$, we get

$$|\langle T_z \phi, \psi \rangle| \leq \ A(z)\ B(z)\ \|T\|_{\mathscr{L}(L^{p_1}_{a_1}, L^{q_1}_{b_1})} \|\phi\|_{L^{p_1}_N} \|\psi\|_{L^{q_1}_N},$$

where $A(z)$ and $B(z)$ denote the operator norms

$$A(z) := \|(\mathrm{I} + \mathcal{R})^{-N + \frac{-a_z + a_1}{\nu_{\mathcal{R}}}}\|_{\mathscr{L}(L^{p_1})} \quad \text{and} \quad B(z) := \|(\mathrm{I} + \bar{\mathcal{Q}})^{-M + \frac{b_z - b_1}{\nu_{\mathcal{Q}}}}\|_{\mathscr{L}(L^{q_1})}.$$

We can write

$$A(z) = \|(\mathrm{I}+\mathcal{R})^{-(\alpha+\beta z)}\|_{\mathscr{L}(L^{p_1})} \quad \text{with } \alpha = N - \frac{a_1-a_0}{\nu_{\mathcal{R}}} > 0,\ \beta = \frac{a_1-a_0}{\nu_{\mathcal{R}}} \geq 0.$$

Thus

$$\begin{aligned} A(z) &\leq \|(\mathrm{I}+\mathcal{R})^{-(\alpha+\beta \operatorname{Re} z)}\|_{\mathscr{L}(L^{p_1})}\|(\mathrm{I}+\mathcal{R})^{-\beta \operatorname{Im} z}\|_{\mathscr{L}(L^{p_1})} \\ &\lesssim \|h_1\|_{L^1} e^{\theta\beta|\operatorname{Im} z|}, \end{aligned}$$

by Corollary 4.3.13 and Proposition 4.3.7 using the notation of their statements. We have a similar property for $B(z)$. This implies easily that there exists a constant C depending on $\phi, \psi, a_1, a_0, b_1, b_0$ and $F, G, \mathcal{R}, \mathcal{Q}$ such that we have

$$\forall z \in S \qquad \ln|\langle T_z\phi, \psi\rangle| \leq C(1+|\operatorname{Im} z|).$$

We now estimate operator norms of T_z for z on the boundary of the strip, that is, $z = j+iy$, $j=0,1$, $y \in \mathbb{R}$:

$$\begin{aligned} &\|T_z\|_{\mathscr{L}(L^{p_j},L^{q_j})} \\ &= \|(\mathrm{I}+\mathcal{Q})^{\frac{b_z}{\nu_{\mathcal{Q}}}} T(\mathrm{I}+\mathcal{R})^{-\frac{a_z}{\nu_{\mathcal{R}}}}\|_{\mathscr{L}(L^{p_j},L^{q_j})} \\ &= \|(\mathrm{I}+\mathcal{Q})^{\frac{b_z-b_j}{\nu_{\mathcal{Q}}}}(\mathrm{I}+\mathcal{Q})^{\frac{b_j}{\nu_{\mathcal{Q}}}} T(\mathrm{I}+\mathcal{R})^{\frac{-a_j}{\nu_{\mathcal{R}}}}(\mathrm{I}+\mathcal{R})^{\frac{a_j-a_z}{\nu_{\mathcal{R}}}}\|_{\mathscr{L}(L^{p_j},L^{q_j})} \\ &\leq \|(\mathrm{I}+\mathcal{Q}_{q_j})^{\frac{b_z-b_j}{\nu_{\mathcal{Q}}}}\|_{\mathscr{L}(L^{q_j})}\|T\|_{\mathscr{L}(L^{p_j}_{a_j},L^{q_j}_{b_j})}\|(\mathrm{I}+\mathcal{R}_{p_j})^{\frac{a_j-a_z}{\nu_{\mathcal{R}}}}\|_{\mathscr{L}(L^{p_j})} \\ &\qquad = \|(\mathrm{I}+\mathcal{Q}_{q_j})^{iy\frac{b_1-b_0}{\nu_{\mathcal{Q}}}}\|_{\mathscr{L}(L^{q_j})}\|T\|_{\mathscr{L}(L^{p_j}_{a_j},L^{q_j}_{b_j})}\|(\mathrm{I}+\mathcal{R}_{p_j})^{iy\frac{a_0-a_1}{\nu_{\mathcal{R}}}}\|_{\mathscr{L}(L^{p_j})}. \end{aligned}$$

Proposition 4.3.7 then implies

$$\|T_{j+iy}\|_{\mathscr{L}(L^{p_j},L^{q_j})} \leq C\|T\|_{\mathscr{L}(L^{p_j}_{a_j},L^{q_j}_{b_j})} e^{\theta_{\mathcal{R}}\frac{a_1-a_0}{\nu_{\mathcal{R}}}|y|} e^{\theta_{\mathcal{Q}}\frac{b_0-b_1}{\nu_{\mathcal{R}}}|y|},$$

where C, $\theta_{\mathcal{R}}$ and $\theta_{\mathcal{Q}}$ are positive constants obtained from the applications of Proposition 4.3.7 to $\mathcal{R}$ and $\mathcal{Q}$.

The end of the proof is now classical. We fix a non-negative function $\chi \in \mathcal{S}(G)$ with $\int_G \chi = 1$ and write

$$\chi_\epsilon(x) := \epsilon^{-Q}\chi(\epsilon^{-1}x)$$

for $\epsilon > 0$. If $f \in \mathscr{B}$, then $f * \chi_\epsilon \in \mathcal{S}(G)$ (see Lemma 3.1.59) and we can set for any $\epsilon > 0$, $z \in S$,

$$T_{z,\epsilon} f := T_z\left(f * \chi_\epsilon\right).$$

Clearly $T_{z,\epsilon}$ satisfy the hypotheses of Theorem A.6.1 (see also Remark A.6.2). Thus for any $t \in [0,1]$, there exists a constant $M_t > 0$ independent of ϵ such that

$$\forall f \in \mathscr{B} \qquad \|T_{t,\epsilon} f\|_{q_t} \leq M_t\|f\|_{p_t}.$$

For $p \in (1,\infty)$, we consider the space $\mathcal{V}_p$ of functions ϕ of the form $\phi = f * \chi_\epsilon$, with $f \in \mathscr{B}$ and $\epsilon > 0$, satisfying $\|f\|_p \leq 2\|f * \chi_\epsilon\|_p$. By Lemma 3.1.59, the space $\mathcal{V}_p$ contains $\mathcal{S}(G)$ and is dense in $L^p(G)$ for $p \in (1,\infty)$. Going back to the proof of Theorem 4.4.9, we have obtained for any $t \in [0,1]$ and $\phi = f * \chi_\epsilon \in \mathcal{V}_{p_t}$, that

$$\|T_t \phi\|_{q_t} = \|T_{t,\epsilon} f\|_{q_t} \leq M_t \|f\|_{p_t} \leq 2M_t \|\phi\|_{p_t}.$$

This shows that T_t extends to a bounded operator from $L^{p_t}(G)$ to $L^{q_t}(G)$. □

As a consequence of the interpolation properties, we have

Corollary 4.4.10. *Let $\kappa \in \mathcal{S}'(G)$ and let T_κ be its associated convolution operator*

$$T_\kappa : \mathcal{S}(G) \ni \phi \mapsto \phi * \kappa.$$

Let also $a \in \mathbb{R}$, $p \in (1,\infty)$ and let $\{\gamma_\ell, \ell \in \mathbb{Z}\}$ be a sequence of real numbers which tends to $\pm\infty$ as $\ell \to \pm\infty$. Assume that for any $\ell \in \mathbb{Z}$, the operator T_κ extends continuously to a bounded operator $L^p_{\gamma_\ell}(G) \to L^p_{a+\gamma_\ell}(G)$. Then the operator T_κ extends continuously to a bounded operator $L^p_\gamma(G) \to L^p_{a+\gamma}(G)$ for any $\gamma \in \mathbb{R}$. Furthermore, for any $c \geq 0$, we have

$$\sup_{|\gamma| \leq c} \|T_\kappa\|_{\mathscr{L}(L^p_\gamma, L^p_{a+\gamma})} \leq C_c \max\left(\|T_\kappa\|_{\mathscr{L}(L^p_{\gamma_\ell}, L^p_{a+\gamma_\ell})}, \|T_\kappa\|_{\mathscr{L}(L^p_{\gamma_{-\ell}}, L^p_{a+\gamma_{-\ell}})} \right)$$

where $\ell \in \mathbb{N}_0$ is the smallest integer such that $\gamma_\ell \geq c$ and $-\gamma_{-\ell} \geq c$.

4.4.3 Homogeneous Sobolev spaces

Here we define the homogeneous version of our Sobolev spaces and obtain their first properties. Many proofs are obtained by adapting the corresponding inhomogeneous cases and we may therefore allow ourselves to present them more succinctly. For technical reasons explained below, the definition of homogeneous Sobolev spaces is restricted to the case $p \in (1,\infty)$.

As in the inhomogeneous case, we first need the following lemma:

Lemma 4.4.11. *We keep the notation of Theorem 4.3.6.*

1. *For any $s \in \mathbb{R}$ and $p \in [1,\infty) \cup \{\infty_o\}$, the map $f \mapsto \|\mathcal{R}_p^{\frac{s}{\nu}} f\|_{L^p(G)}$ defines a norm on $\mathcal{S}(G) \cap \mathrm{Dom}(\mathcal{R}_p^{\frac{s}{\nu}})$. We denote it by*

$$\|f\|_{\dot{L}^p_s(G)} := \|\mathcal{R}_p^{\frac{s}{\nu}} f\|_{L^p(G)}.$$

2. *For any $s \leq 0$ and $p \in [1,\infty) \cup \{\infty_o\}$, $\mathcal{S}(G) \cap \mathrm{Dom}(\mathcal{R}_p^{\frac{s}{\nu}})$ contains $\mathcal{R}^{\lceil |s|\nu \rceil}(\mathcal{S}(G))$ which is dense in $\mathrm{Range}(\mathcal{R}_p)$ for $\|\cdot\|_{L^p(G)}$, and any sequence in $\mathcal{S}(G) \cap \mathrm{Dom}(\mathcal{R}_p^{\frac{s}{\nu}})$ which is Cauchy for $\|\cdot\|_{\dot{L}^p_s(G)}$ is convergent in $\mathcal{S}'(G)$.*

3. *If $s > 0$ and $p \in (1, \infty)$, then $\mathcal{S}(G) \subset \text{Dom}(\mathcal{R}_p^{\frac{s}{\nu}})$ and any sequence in $\mathcal{S}(G)$ which is Cauchy for $\|\cdot\|_{\dot{L}^p_s(G)}$ is convergent in $\mathcal{S}'(G)$.*

Proof of Lemma 4.4.11. The fact that the map $f \mapsto \|\mathcal{R}_p^{\frac{s}{\nu}} f\|_{L^p(G)}$ defines a norm on $\mathcal{S}(G)$ follows easily from Theorem 4.3.6 Part (1).

In the case $s = 0$, $\|\cdot\|_{\dot{L}^p_0(G)} = \|\cdot\|_{L^p(G)}$ and Part 2 is proved in this case.

Let $s < 0$ and $p \in [1, \infty) \cup \{\infty_o\}$. By Theorem 4.3.6 (especially Parts (1a) and (1c)), for any $N \in \mathbb{N}$ with $N > |s|/\nu$, $\text{Dom}(\mathcal{R}^{\frac{s}{\nu}})$ contains $\mathcal{R}^N(\mathcal{S}(G))$ and $\mathcal{R}^N(\mathcal{S}(G))$ is dense in Range$(\mathcal{R}_p)$. Consequently $\mathcal{S}(G) \cap \text{Dom}(\mathcal{R}_p^{\frac{s}{\nu}})$ contains $\mathcal{R}^N(\mathcal{S}(G))$ and is dense in Range$(\mathcal{R}_p)$. Let p' be the dual exponent of p, i.e. $\frac{1}{p} + \frac{1}{p'} = 1$ with the usual extension. Theorem 4.3.6 (1), and the duality properties of L^p as well as $\mathcal{R}^t = \bar{\mathcal{R}}$ imply

$$|\langle f, \phi\rangle| \le \|\mathcal{R}_p^{\frac{s}{\nu}} f\|_{L^p(G)} \|\bar{\mathcal{R}}_{p'}^{-\frac{s}{\nu}} \phi\|_{L^{p'}(G)},$$

for any $f \in \mathcal{S}(G) \cap \text{Dom}(\mathcal{R}_p^{\frac{s}{\nu}})$ and $\phi \in \mathcal{S}(G)$. Furthermore, as $\phi \in \mathcal{S}(G) \subset \text{Dom}(\mathcal{R}_{p'}^{-\frac{s}{\nu}})$, Theorem 4.3.6 (1) also yields for any $\phi \in \mathcal{S}(G)$

$$\begin{aligned} \|\bar{\mathcal{R}}_{p'}^{-\frac{s}{\nu}} \phi\|_{L^{p'}(G)} &\le \max\left(\|\bar{\mathcal{R}}_{p'}^{\lfloor \frac{|s|}{\nu} \rfloor} \phi\|_{L^{p'}(G)}, \|\bar{\mathcal{R}}_{p'}^{\lceil \frac{|s|}{\nu} \rceil} \phi\|_{L^{p'}(G)} \right) \\ &\le C \max_{[\alpha] = \lfloor \frac{|s|}{\nu} \rfloor, \lceil \frac{|s|}{\nu} \rceil} \|X^\alpha \phi\|_{L^{p'}(G)} \end{aligned}$$

for some constant $C = C_{N,\mathcal{R}}$. We have obtained that

$$|\langle f, \phi\rangle| \le C \|\mathcal{R}_p^{\frac{s}{\nu}} f\|_{L^p(G)} \max_{[\alpha]=N,N+1} \|X^\alpha \phi\|_{L^{p'}(G)}$$

for any $f \in \mathcal{S}(G) \cap \text{Dom}(\mathcal{R}_p^{\frac{s}{\nu}})$ and $\phi \in \mathcal{S}(G)$. This together with the properties of the Schwartz space (see Section 3.1.9) easily implies Part 2.

Let $s > 0$. By Theorem 4.3.6 (1g), $\mathcal{S}(G) \subset \text{Dom}(\mathcal{R}_p^{\frac{s}{\nu}})$.

Let $p \in (1, \infty)$. By Corollary 4.3.11 Part (i), if $s \in (0, \frac{Q}{p})$, then there exists $C > 0$ such that

$$\forall f \in \mathcal{S}(G) \qquad \|f\|_{L^q(G)} \le C \|\mathcal{R}_p^{\frac{s}{\nu}} f\|_{L^p(G)} = C \|f\|_{\dot{L}^p_s(G)},$$

where $q \in (1, \infty)$ is such that

$$\frac{1}{p} - \frac{1}{q} = \frac{s}{Q}.$$

Note that q is indeed in $(1, \infty)$ as $s < \frac{Q}{p}$. Hence if $\{f_\ell\} \subset \mathcal{S}(G)$ is Cauchy for $\|\cdot\|_{\dot{L}^p_s(G)}$, then $\{f_\ell\} \subset \mathcal{S}(G)$ is Cauchy for $\|\cdot\|_{L^q(G)}$ thus in $\mathcal{S}'(G)$. This shows Part 3 for any $s > 0$, $p \in (1, \infty)$ satisfying $ps < Q$.

If $s \in [N\frac{Q}{p}, (N+1)\frac{Q}{p})$ for some $N \in \mathbb{N}_0$, we write $s = s_1 + s'$ with $s' \in (0, \frac{Q}{p})$ and

$$s_1 \in [(N-1)\frac{Q}{p}, N\frac{Q}{p})$$

and by Corollary 4.3.11 Part (i) with Theorem 4.3.6 (1), we have

$$\exists C = C_{s',p} \qquad \forall f \in \mathcal{S}(G) \qquad \|\mathcal{R}_q^{\frac{s_1}{\nu}} f\|_{L^q} \leq C \|\mathcal{R}_p^{\frac{s}{\nu}} f\|_{L^p(G)},$$

where $q \in (1, \infty)$ is such that

$$\frac{1}{q} - \frac{1}{p} = \frac{s'}{Q}.$$

Hence if $\{f_\ell\} \subset \mathcal{S}(G)$ is Cauchy for $\|\cdot\|_{\dot{L}^p_s(G)}$, then $\{f_\ell\} \subset \mathcal{S}(G)$ is Cauchy for $\|\cdot\|_{\dot{L}^q_{s_1}(G)}$. Note that

$$s_1 \leq \frac{NQ}{p} < \frac{NQ}{q}.$$

Recursively, this shows Part 3. □

The use of Corollary 4.3.11 in the proof above requires $p \in (1, \infty)$. Moreover, by Corollary 4.3.4 (ii), the range of $\mathcal{R}_p$ is dense in $L^p(G)$ for $p \in (1, \infty_o]$. As we want to have a unified presentation for all the homogeneous spaces of any exponent $s \in \mathbb{R}$, we restrict the parameter p to be in $(1, \infty)$ only.

Definition 4.4.12. Let $\mathcal{R}$ be a Rockland operator of homogeneous degree ν on a graded Lie group G, and let $p \in (1, \infty)$. We denote by $\dot{L}^p_{s,\mathcal{R}}(G)$ the space of tempered distribution obtained by the completion of $\mathcal{S}(G) \cap \text{Dom}(\mathcal{R}_p^{\frac{s}{\nu}})$ for the norm

$$\|f\|_{\dot{L}^p_s(G)} := \|\mathcal{R}_p^{\frac{s}{\nu}} f\|_p, \quad f \in \mathcal{S}(G) \cap \text{Dom}(\mathcal{R}_p^{s/\nu}).$$

As in the inhomogeneous case, we will write $\dot{L}^p_s(G)$ or $\dot{L}^p_{s,\mathcal{R}}$ but often omit the reference to the Rockland operator $\mathcal{R}$. We will see in Theorem 4.4.20 that the homogeneous Sobolev spaces do not depend on a specific $\mathcal{R}$. Adapting the inhomogeneous case, one obtains easily:

Proposition 4.4.13. *Let G be a graded Lie group of homogeneous dimension Q. Let $\mathcal{R}$ be a positive Rockland operator of homogeneous degree ν on G. Let $p \in (1, \infty)$ and $s \in \mathbb{R}$.*

1. *We have*

$$\left(\mathcal{S}(G) \cap \text{Dom}(\mathcal{R}_p^{s/\nu})\right) \subsetneq \dot{L}^p_s(G) \subsetneq \mathcal{S}'(G).$$

   Equipped with the homogeneous Sobolev norm $\|\cdot\|_{\dot{L}^p_s(G)}$, the space $\dot{L}^p_s(G)$ is a Banach space which contains $\mathcal{S}(G) \cap \text{Dom}(\mathcal{R}_p^{s/\nu})$ as dense subspace.

2. *If $s > -Q/p$ then $\mathcal{S}(G) \subset \text{Dom}(\mathcal{R}_p^{s/\nu}) \subset \dot{L}^p_s(G)$. If $s < 0$ then $\mathcal{S}(G) \cap \text{Dom}(\mathcal{R}_p^{\frac{s}{\nu}})$ contains $\mathcal{R}^{\lceil |s|\nu \rceil}(\mathcal{S}(G))$ which is dense in $L^p(G)$.*

3. *If* $s = 0$, *then* $\dot{L}^p_0(G) = L^p(G)$ *for* $p \in (1, \infty)$ *with* $\|\cdot\|_{\dot{L}^p_0(G)} = \|\cdot\|_{L^p(G)}$.
4. *Let* $s \in \mathbb{R}$, $p \in (1, \infty)$ *and* $f \in \mathcal{S}'(G)$. *If* $f \in \dot{L}^p_s(G)$ *then* $\mathcal{R}_p^{s/\nu} f \in L^p(G)$ *in the sense that the linear mapping*
$$\left(\mathcal{S}(G) \cap \mathrm{Dom}(\bar{\mathcal{R}}_{p'}^{s/\nu})\right) \ni \phi \mapsto \langle f, \bar{\mathcal{R}}_{p'}^{s/\nu} \phi \rangle$$
is densely defined on $L^{p'}(G)$ *and extends to a bounded functional on* $L^{p'}(G)$ *where* p' *is the conjugate exponent of* p. *The converse is also true.*
5. *If* $1 < p < q < \infty$ *and* $a, b \in \mathbb{R}$ *with*
$$b - a = Q(\frac{1}{p} - \frac{1}{q}),$$
then we have the continuous inclusion
$$\dot{L}^p_b \subset \dot{L}^q_a$$
that is, for every $f \in \dot{L}^p_b$, *we have* $f \in \dot{L}^q_a$ *and there exists a constant* $C = C_{a,b,p,q,G} > 0$ *independent of* f *such that*
$$\|f\|_{\dot{L}^q_a} \leq C \|f\|_{\dot{L}^p_b}.$$
6. *For* $p \in (1, \infty)$ *and any* $a, b, c \in \mathbb{R}$ *with* $a < c < b$, *there exists a positive constant* $C = C_{a,b,c}$ *such that we have for any* $f \in \dot{L}^p_b$
$$\|f\|_{\dot{L}^p_c} \leq C \|f\|_{\dot{L}^p_a}^{1-\theta} \|f\|_{\dot{L}^p_b}^{\theta} \quad \text{where } \theta := (c - a)/(b - a).$$

Proof of Proposition 4.4.13. Parts (1), (2), and (3) follow from Lemma 4.4.11 and its proof. Part (4) follows easily by duality and Lemma 4.4.11. Parts (5) and (6) are an easy consequence of the property of the fractional powers of $\mathcal{R}$ on the L^p-spaces (cf. Theorem 4.3.6) and the operator $\mathcal{R}_p^{-s/\nu}$, $s \in (0, Q)$, being of type s and independent of p (cf. Corollary 4.3.11 (i)). □

Note that Part (2) of Proposition 4.4.13 can not be improved in general as the inclusions $\mathcal{S}(G) \subset \mathrm{Dom}(\mathcal{R}_p^{\frac{s}{\nu}})$ or $\mathcal{S}(G) \subset \dot{L}^s_p(G)$ can not hold in general for any group G as they do not hold in the Euclidean case i.e. $G = (\mathbb{R}^n, +)$ with the usual dilations. Indeed in the case of $\mathbb{R}^n$, $p = 2$, one can construct Schwartz functions which can not be in $\dot{L}^2_s$ with $s < -n/2$. It suffices to consider a function $\phi \in \mathcal{S}(G)$ satisfying $\widehat{\phi}(\xi) \equiv 1$ on a neighbourhood of 0 since then $|\xi|^s \widehat{\phi}(\xi)$ is not square integrable about 0 for $s < -n/2$.

As in the homogeneous case (see Lemma 4.4.7), Part (4) of Proposition 4.4.13 above implies the following property regarding duality of Sobolev spaces. This will be improved in Proposition 4.4.22 once we know (see Theorem 4.4.20) that homogeneous Sobolev spaces are indeed independent of the considered Rockland operator.

Lemma 4.4.14. *Let $\mathcal{R}$ be a positive Rockland operator on a graded Lie group G. We consider the associated homogeneous Sobolev spaces $\dot{L}^p_{s,\mathcal{R}}(G)$. If $s \in \mathbb{R}$ and $p \in (1,\infty)$, the dual space of $\dot{L}^p_{s,\mathcal{R}}(G)$ is isomorphic to $\dot{L}^{p'}_{-s,\bar{\mathcal{R}}}(G)$ via the distributional duality, where p' is the conjugate exponent of p, i.e. $\frac{1}{p}+\frac{1}{p'}=1$.*

The following interpolation property can be proved after a careful modification of the inhomogeneous proof:

Proposition 4.4.15. *Let $\mathcal{R}$ and $\mathcal{Q}$ be two positive Rockland operators on two graded Lie groups G and F respectively. We consider their associated homogeneous Sobolev spaces $\dot{L}^p_a(G)$ and $\dot{L}^q_b(F)$. Let $p_0, p_1, q_0, q_1 \in (1,\infty)$ and $a_0, a_1, b_0, b_1 \in \mathbb{R}$.*

We also consider a linear mapping T from $\dot{L}^{p_0}_{a_0}(G) + \dot{L}^{p_1}_{a_1}(G)$ to locally integrable functions on F. We assume that T maps $\dot{L}^{p_0}_{a_0}(G)$ and $\dot{L}^{p_1}_{a_1}(G)$ boundedly into $\dot{L}^{q_0}_{b_0}(F)$ and $\dot{L}^{q_1}_{b_1}(F)$, respectively.

Then T extends uniquely to a bounded mapping from $\dot{L}^p_{a_t}(G)$ to $\dot{L}^q_{b_t}(F)$ for $t \in [0,1]$, where a_t, b_t, p_t, q_t are defined by

$$\left(a_t, b_t, \frac{1}{p_t}, \frac{1}{q_t}\right) = (1-t)\left(a_0, b_0, \frac{1}{p_0}, \frac{1}{q_0}\right) + t\left(a_1, b_1, \frac{1}{p_1}, \frac{1}{q_1}\right).$$

Sketch of the proof of Proposition 4.4.15. By duality (see Lemma 4.4.14) and up to a change of notation, it suffices to prove the case $a_1 \geq a_0$ and $b_1 \leq b_0$. The idea is to interpolate between the operators formally given by

$$T_z = \mathcal{Q}^{z\frac{b_1-b_0}{\nu_{\mathcal{Q}}}} \mathcal{Q}^{\frac{b_0}{\nu_{\mathcal{Q}}}} T \mathcal{R}^{-\frac{a_0}{\nu_{\mathcal{R}}}} \mathcal{R}^{z\frac{a_0-a_1}{\nu_{\mathcal{R}}}}, \quad z \in S, \tag{4.42}$$

with the same notation for $\nu_{\mathcal{R}}$, $\nu_{\mathcal{Q}}$, a_z, b_z and S as in the proof of Theorem 4.4.9. In (4.42), we have abused the notation regarding the fractional powers of $\mathcal{R}_p$ and $\mathcal{Q}_q$ and removed p and q thanks to by Theorem 4.3.6 (1). Moreover, Theorem 4.3.6 implies that on $\mathcal{S}(G)$, each operator T_z, $z \in S$, coincides with

$$T_z = \mathcal{Q}^{(1-z)\frac{b_0-b_1}{\nu_{\mathcal{Q}}}} \mathcal{Q}^{\frac{b_1}{\nu_{\mathcal{Q}}}} T \mathcal{R}^{-\frac{a_1}{\nu_{\mathcal{R}}}} \mathcal{R}^{(1-z)\frac{a_1-a_0}{\nu_{\mathcal{R}}}},$$

and that for any $\phi \in \mathcal{S}(G)$ and $\psi \in \mathcal{S}(F)$, $z \mapsto \langle T_z\phi, \psi\rangle$ is analytic on S. We also have

$$|\langle T_z\phi, \psi\rangle| \leq \|T\|_{\mathscr{L}(\dot{L}^{p_1}_{a_1}, \dot{L}^{q_1}_{b_1})} \|\mathcal{R}^{\frac{-a_z+a_1}{\nu_{\mathcal{R}}}}\phi\|_{L^{p_1}} \|\bar{\mathcal{Q}}^{\frac{b_z-b_1}{\nu_{\mathcal{Q}}}}\psi\|_{L^{q_1}}.$$

Note that $-\mathrm{Re}\, a_z + a_1 \geq 0$ thus we have

$$\begin{aligned} \|\mathcal{R}^{\frac{-a_z+a_1}{\nu_{\mathcal{R}}}}\phi\|_{L^{p_1}} &\leq \|\mathcal{R}^{\frac{-\mathrm{Re}\, a_z+a_1}{\nu_{\mathcal{R}}}}\phi\|_{L^{p_1}} \|\mathcal{R}^{\frac{-\mathrm{Im}\, a_z}{\nu_{\mathcal{R}}}}\phi\|_{L^{p_1}} \\ &\lesssim \|\phi\|^{1-\alpha}_{L^{p_1}} \|\mathcal{R}^N\phi\|^{\alpha}_{L^{p_1}} e^{\theta\frac{|\mathrm{Im}\, a_z|}{\nu_{\mathcal{R}}}}, \end{aligned}$$

by Theorem 4.3.6 (1f) with N the smallest integer strictly greater than $-\mathrm{Re}\, a_z + a_1$ and $\alpha = (-\mathrm{Re}\, a_z + a_1)/N$, and by Proposition 4.3.9 using the notation of its statement. We have similar bounds for $\|\bar{\mathcal{Q}}^{\frac{b_z-b_1}{\nu_{\mathcal{Q}}}}\psi\|_{q_1}$ and all these estimates imply easily that there exists a constant depending on $\phi, \psi, a_1, a_0, b_1, b_0$ such that

$$\forall z \in S \qquad \ln|\langle T_z\phi, \psi\rangle| \leq C(1+|\operatorname{Im} z|).$$

For the estimate on the boundary of the strip, that is, $z = j+iy$, $j = 0, 1$, $y \in \mathbb{R}$, we see as in the proof of Theorem 4.4.9:

$$\|T_z\|_{\mathscr{L}(L^{p_j},L^{q_j})} \leq \|\mathcal{Q}_{q_j}^{iy\frac{b_1-b_0}{\nu_{\mathcal{Q}}}}\|_{\mathscr{L}(L^{q_j})}\|T\|_{\mathscr{L}(\dot{L}^{p_j}_{a_j},\dot{L}^{q_j}_{b_j})}\|\mathcal{R}_{p_j}^{iy\frac{a_0-a_1}{\nu_{\mathcal{R}}}}\|_{\mathscr{L}(L^{p_j})}.$$

Proposition 4.3.9 then implies

$$\|T_{j+iy}\|_{\mathscr{L}(L^{p_j},L^{q_j})} \leq C\|T\|_{\mathscr{L}(\dot{L}^{p_j}_{a_j},\dot{L}^{q_j}_{b_j})}e^{\theta_{\mathcal{R}}\frac{a_1-a_0}{\nu_{\mathcal{R}}}|y|}e^{\theta_{\mathcal{Q}}\frac{b_0-b_1}{\nu_{\mathcal{R}}}|y|},$$

where C, $\theta_{\mathcal{R}}$ and $\theta_{\mathcal{Q}}$ are positive constants obtained from the applications of Proposition 4.3.9 to $\mathcal{R}$ and $\mathcal{Q}$. We conclude the proof in the same way as for Theorem 4.4.9. □

4.4.4 Operators acting on Sobolev spaces

In this section we show that left-invariant differential operators act continuously on homogeneous and inhomogeneous Sobolev spaces. We will also show a similar property for operators of type ν, $\operatorname{Re}\nu = 0$.

In the statements and in the proofs of this section, we keep the same notation for an operator defined on a dense subset of some L^p-space and its possible bounded extensions to some Sobolev spaces in order to ease the notation.

Theorem 4.4.16. *Let G be a graded Lie group.*

1. *Let T be a left-invariant differential operator of homogeneous degree ν_T. Then for every $p \in (1,\infty)$ and $s \in \mathbb{R}$, T maps continuously $L^p_{s+\nu_T}(G)$ to $L^p_s(G)$. Fixing a positive Rockland operator $\mathcal{R}$ in order to define the Sobolev norms, it means that*

$$\exists C = C_{s,p,T} > 0 \qquad \forall \phi \in \mathcal{S}(G) \qquad \|T\phi\|_{L^p_s(G)} \leq C\|\phi\|_{L^p_{s+\nu_T}(G)}.$$

2. *Let T be a ν_T-homogeneous left-invariant differential operator. Then for every $p \in (1,\infty)$ and $s \in \mathbb{R}$, T maps continuously $\dot{L}^p_{s+\nu_T}(G)$ to $\dot{L}^p_s(G)$. Fixing a positive Rockland operator $\mathcal{R}$ in order to define the Sobolev norms, it means that*

$$\exists C = C_{s,p,T} > 0 \qquad \forall \phi \in \dot{L}^p_{s+\nu_T}(G) \qquad \|T\phi\|_{\dot{L}^p_s(G)} \leq C\|\phi\|_{\dot{L}^p_{s+\nu_T}(G)}.$$

We start the proof of Theorem 4.4.16 with studying the case of $T = X_j$. This uses the definition and properties of kernel of type 0, see Section 3.2.5.

Lemma 4.4.17. *Let $\mathcal{R}$ be a positive Rockland operator on a graded Lie group G and $\mathcal{I}_a$ the kernel of its Riesz operator as in Corollary 4.3.11.*

1. *For any $j = 1, \ldots, n$, $X_j \mathcal{I}_{\upsilon_j}$ is a kernel of type 0.*
2. *If κ is a kernel of type 0, then, for any $j = 1, \ldots, n$, $X_j(\kappa * \mathcal{I}_{\upsilon_j})$ is a kernel of type 0 and, more generally, for any multi-index $\alpha \in \mathbb{N}_0^n$, the kernel*
$$X^\alpha\left(\kappa * \mathcal{I}_{[\upsilon_1]}^{(*)^{\alpha_1}} * \ldots * \mathcal{I}_{[\upsilon_n]}^{(*)^{\alpha_n}}\right)$$
is of type 0.
3. *If T is an operator of type 0, then, for any $N \in \mathbb{N}$, $\mathcal{R}^N T \mathcal{R}_2^{-N}$ is an operator of type 0 hence it is bounded on $L^p(G)$, $p \in (1, \infty)$.*
4. *For any $j = 1, \ldots, n$ and for any $N \in \mathbb{N}_0$, $\mathcal{R}^N X_j \mathcal{R}_2^{-\frac{\upsilon_j}{\nu} - N}$ is an operator of type 0.*

In Part 2, we have used the notation
$$f^{(*)^m} = \underbrace{f * \ldots * f}_{m \text{ times}}$$

Proof of Lemma 4.4.17. We adopt the notation of the statement. By Corollary 4.3.11 (i), $\mathcal{I}_{\upsilon_j}$ is a kernel of type $\upsilon_j \in (0, Q)$ hence, by Lemma 3.2.33, $X_j \mathcal{I}_{\upsilon_j}$ is a kernel of type 0. This shows Part 1.

More generally, if κ is a kernel of type 0, then $\kappa * \mathcal{I}_{\upsilon_j}$ is a kernel of type υ_j by Proposition 3.2.35 (ii) hence by Lemma 3.2.33, $X_j(\kappa * \mathcal{I}_{\upsilon_j})$ is a kernel of type 0. Iterating this procedure shows Part 2.

Let T be an operator of type 0. We denote by κ its kernel. Let $N \in \mathbb{N}$. The operator $\mathcal{R}^N$ can be written as a linear combination of X^α, $\alpha \in \mathbb{N}_0^n$ with $[\alpha] = \nu N$. Using the spectral calculus of $\mathcal{R}$ to define and decompose $\mathcal{R}_2^{-N}$, this shows that the operator $\mathcal{R}^N T \mathcal{R}_2^{-N}$ can be written as a linear combination over $[\alpha] = \nu N$ of the operators $X^\alpha T \mathcal{R}_2^{-\frac{\upsilon_1}{\nu}\alpha_1} \ldots \mathcal{R}_2^{-\frac{\upsilon_n}{\nu}\alpha_n}$ whose kernel can be written as $X^\alpha\left(\kappa * \mathcal{I}_{[\upsilon_1]}^{(*)^{\alpha_1}} * \ldots * \mathcal{I}_{[\upsilon_n]}^{(*)^{\alpha_n}}\right)$. Part 2 implies that the operator $\mathcal{R}^N T \mathcal{R}_2^{-N}$ is of type 0. By Theorem 3.2.30, it is a bounded operator on $L^p(G)$, $p \in (1, \infty)$. This shows Part 3.

Part 4 follows from combining Parts 1 and 3. □

We can now finish the proof of Theorem 4.4.16.

Proof of Theorem 4.4.16. By Lemma 4.4.17, Part 4, $\mathcal{R}^N X_j \mathcal{R}_2^{-\frac{\upsilon_j}{\nu} - N}$ is an operator of type 0, hence bounded on $L^p(G)$, $p \in (1, \infty)$. The transpose of this operator is
$$(\mathcal{R}^N X_j \mathcal{R}_2^{-\frac{\upsilon_j}{\nu} - N})^t = -\bar{\mathcal{R}}_2^{-\frac{\upsilon_j}{\nu} - N} X_j \bar{\mathcal{R}}^N,$$

since $X_j^t = -X_j$ and $\mathcal{R}^t = \bar{\mathcal{R}}$. By duality, this operator is $L^{p'}$-bounded where $\frac{1}{p'} + \frac{1}{p} = 1$. As $\bar{\mathcal{R}}$ is also a positive Rockland operator, see Lemma 4.1.11, we can exchange the rôle of $\mathcal{R}$ and $\bar{\mathcal{R}}$. Hence we have obtained that the operators $\mathcal{R}^N X_j \mathcal{R}_2^{-\frac{v_j}{\nu}-N}$ and $\mathcal{R}_2^{-\frac{v_j}{\nu}-N} X_j \mathcal{R}^N$ are bounded on $L^p(G)$ for any $p \in (1,\infty)$ and $N \in \mathbb{N}$. This shows that X_j maps $\dot{L}^p_{v_j+N\nu}$ to $\dot{L}^p_{N\nu}$ and $\dot{L}^p_{-N\nu}$ to $\dot{L}^p_{-v_j-N\nu}$ continuously. The properties of interpolation, cf. Proposition 4.4.15, imply that X_j maps $\dot{L}^p_{v_j+s}$ to $\dot{L}^p_s$ continuously for any $s \in \mathbb{R}$, $p \in (1,\infty)$ and $j = 1, \ldots, n$.

Interpreting any X^α as a composition of operators X_j shows Part (2) for any $T = X^\alpha$, $\alpha \in \mathbb{N}_0^n$, with $\nu_T = [\alpha]$. As any ν_T-homogeneous left-invariant differential operator is a linear combination of X^α, $\alpha \in \mathbb{N}_0^n$, with $\nu_T = [\alpha]$, this shows Part (2).

Let us show Part (1). Let $\alpha \in \mathbb{N}_0^n$. If $s > 0$, then by Theorem 4.4.3 (4) and Part (2), we have for any $\phi \in \mathcal{S}(G)$

$$\begin{aligned} \|X^\alpha \phi\|_{L^p_s} &\lesssim \|X^\alpha \phi\|_{L^p} + \|X^\alpha \phi\|_{\dot{L}^p_s} \\ &\lesssim \|\phi\|_{\dot{L}^p_{[\alpha]}} + \|\phi\|_{\dot{L}^p_{s+[\alpha]}} \\ &\lesssim \|\phi\|_{L^p_{[\alpha]}} + \|\phi\|_{L^p_{s+[\alpha]}} \\ &\lesssim \|\phi\|_{L^p_{s+[\alpha]}}. \end{aligned}$$

This shows that X^α maps $L^p_{s+[\alpha]}$ to L^p_s continuously for any $s > 0$, $p \in (1,\infty)$ and any $\alpha \in \mathbb{N}_0^n$. The transpose $(X^\alpha)^t$ of X^α is a linear combination of X^β, $[\beta] = [\alpha]$, and will also have the same properties. By duality, this shows that X^α maps L^p_{-s} to $L^p_{-(s+[\alpha])}$ continuously for any $s > 0$, $p \in (1,\infty)$ and any $\alpha \in \mathbb{N}_0^n$. Together with the properties of interpolation (cf. Theorem 4.4.9), this shows that X^α maps $L^p_{s+[\alpha]}$ to L^p_s continuously for any $s \in \mathbb{R}$, $p \in (1,\infty)$ and any $\alpha \in \mathbb{N}_0^n$.

As any left invariant differential operator can be written as a linear combination of monomials X^α, this implies Part (1) and concludes the proof of Theorem 4.4.16. □

The ideas of the proofs above can be adapted to the proof of the following properties for the operators of type 0:

Theorem 4.4.18. *Let T be an operator of type $\nu \in \mathbb{C}$ on a graded Lie group G with* $\operatorname{Re}\nu = 0$. *Then for every $p \in (1,\infty)$ and $s \in \mathbb{R}$, T maps continuously $L^p_s(G)$ to $L^p_s(G)$ and $\dot{L}^p_s(G)$ to $\dot{L}^p_s(G)$. Fixing a positive Rockland operator $\mathcal{R}$ in order to define the Sobolev norms, it means that there exists $C = C_{s,p,T} > 0$ satisfying*

$$\forall \phi \in \mathcal{S}(G) \qquad \|T\phi\|_{L^p_s(G)} \leq C\|\phi\|_{L^p_s(G)}$$

and

$$\forall \phi \in \dot{L}^p_s \qquad \|T\phi\|_{\dot{L}^p_s(G)} \leq C\|\phi\|_{\dot{L}^p_s(G)}.$$

Proof. Let T be a operator of type $\nu_T \in \mathbb{C}$ with $\operatorname{Re}\nu_T = 0$. Proceeding as in the proof of Lemma 4.4.17 Part 3 yields that for any $N \in \mathbb{N}$, the operator $\mathcal{R}^N T \mathcal{R}_2^{-N}$

is of type ν_T. We can apply this to the transpose T^t of T as well as the operator T^t is also of type ν. By Theorem 3.2.30, the operators $\mathcal{R}^N T \mathcal{R}_2^{-N}$ and $\mathcal{R}^N T^t \mathcal{R}_2^{-N}$ are bounded on $L^p(G)$. This shows that T maps $\dot{L}^p_s$ to $\dot{L}^p_s$ continuously for $s = N$ and $s = -N$, $N \in \mathbb{N}_0$. By interpolation, this holds for any $s \in \mathbb{R}$ and this shows the statement for the homogeneous Sobolev spaces. If $s > 0$, then by Theorem 4.4.3 (4), using the continuity on homogeneous Sobolev spaces which has just been proven, we have for any $\phi \in \mathcal{S}(G)$

$$\|T\phi\|_{L^p_s} \lesssim \|T\phi\|_{L^p} + \|T\phi\|_{\dot{L}^p_s} \lesssim \|\phi\|_{L^p} + \|\phi\|_{\dot{L}^p_s} \lesssim \|\phi\|_{L^p_s}.$$

This shows that T maps L^p_s to L^p_s continuously for any $s > 0$, $p \in (1, \infty)$. Applying this to T^t, by duality, we also obtain this property for $s < 0$. The case $s = 0$ follows from Theorem 3.2.30. This concludes the proof of Theorem 4.4.18. □

Theorem 4.4.18 extends the result of Theorem 3.2.30, that is, the boundedness on $L^p(G)$ of an operator of type ν_T, $\operatorname{Re} \nu_T = 0$, from L^p-spaces to Sobolev spaces. Let us comment on similar results in related contexts:

- In the case of $\mathbb{R}^n$ (and similarly for compact Lie groups), the continuity on Sobolev spaces would be easy since T_κ would commute with the Laplace operator but the homogeneous setting requires a more substantial argument.
- Theorem 4.4.18 was shown by Folland in [Fol75, Theorem 4.9] on any stratified Lie group and for $\nu = 0$. However, the proof in that context uses the existence of a positive Rockland operator with a unique homogeneous fundamental solution, namely 'the' (any) sublaplacian. If we wanted to follow closely the same line of arguments, we would have to assume that the group is equipped with a Rockland operator with homogeneous degree ν with $\nu < Q$, see Remark 4.3.12. This is not always the case for a graded Lie group as the example of the three dimensional Heisenberg group with gradation (3.1) shows.
- The proof above is valid under no restriction in the graded case. Somehow the use of the homogeneous fundamental solution in the stratified case is replaced by the kernel of the Riesz potentials together with the properties of the Sobolev spaces proved so far.

4.4.5 Independence in Rockland operators and integer orders

In this Section, we show that the homogeneous and inhomogeneous Sobolev spaces do not depend on a particular choice of a Rockland operator. Consequently Theorems 4.4.3, 4.4.9, 4.4.16, and 4.4.18, Corollaries 4.4.6 and 4.4.10, Propositions 4.4.8 and 4.4.13 and 4.4.15, hold independently of any chosen Rockland operator $\mathcal{R}$.

We will need the following property:

Lemma 4.4.19. *Let $\mathcal{R}$ be a Rockland operator on G of homogeneous degree ν and let $\ell \in \mathbb{N}_0$, $p \in (1, \infty)$. Then the space $L^p_{\nu\ell}(G)$ is the collection of functions $f \in L^p(G)$ such that $X^\alpha f \in L^p(G)$ for any $\alpha \in \mathbb{N}_0^n$ with $[\alpha] = \nu\ell$. Moreover, the map*

$$\phi \mapsto \sum_{[\alpha]=\nu\ell} \|X^\alpha \phi\|_p$$

is a norm on $\dot{L}^p_{\nu\ell}(G)$ which is equivalent to the homogeneous Sobolev norm and the map

$$\phi \mapsto \|\phi\|_p + \sum_{[\alpha]=\nu\ell} \|X^\alpha \phi\|_p$$

is a norm on $L^p_{\nu\ell}(G)$ which is equivalent to the Sobolev norm.

Proof of Lemma 4.4.19. Writing

$$\mathcal{R}^\ell = \sum_{[\alpha]=\ell\nu} c_{\alpha,\ell} X^\alpha$$

we have on one hand,

$$\forall \phi \in \mathcal{S}(G) \qquad \|\mathcal{R}^\ell \phi\|_p \leq \max |c_{\alpha,\ell}| \sum_{[\alpha]=\ell\nu} \|X^\alpha \phi\|_p. \tag{4.43}$$

On the other hand, by Theorem 4.4.16 (2), for any $\alpha \in \mathbb{N}_0^n$, the operator X^α maps continuously $\dot{L}^p_{[\alpha]}(G)$ to $\dot{L}^p(G)$, hence

$$\exists C > 0 \quad \forall \phi \in \mathcal{S}(G) \qquad \sum_{[\alpha]=\ell\nu} \|X^\alpha \phi\|_p \leq C \|\phi\|_{\dot{L}^p_{[\alpha]}}.$$

This shows the property of Lemma 4.4.19 for homogeneous Sobolev spaces.

Adding $\|\phi\|_{L^p}$ on both sides of (4.43) implies by Theorem 4.4.3, Part (2):

$$\exists C > 0 \quad \forall \phi \in \mathcal{S}(G) \qquad \|\phi\|_{L^p_{\ell\nu}} \leq C \left(\|\phi\|_{L^p} + \sum_{[\alpha]=\ell\nu} \|X^\alpha \phi\|_p \right).$$

On the other hand, by Theorem 4.4.16 (1), for any $\alpha \in \mathbb{N}_0^n$, the operator X^α maps continuously $L^p_{[\alpha]}(G)$ to $L^p(G)$, hence

$$\exists C > 0 \quad \forall \phi \in \mathcal{S}(G) \qquad \sum_{[\alpha]=\ell\nu} \|X^\alpha \phi\|_p \leq C \|\phi\|_{L^p_{[\alpha]}}.$$

This shows the property of Lemma 4.4.19 for inhomogeneous Sobolev spaces and concludes the proof of Lemma 4.4.19. $\square$

One may wonder whether Lemma 4.4.19 would be true not only for integer exponents of the form $s = \nu\ell$ but for any integer s. In fact other inhomogeneous Sobolev spaces on a graded Lie group were defined by Goodman in [Goo76, Section III. 5.4] following this idea. More precisely the L^p Goodman-Sobolev space of order $s \in \mathbb{N}_0$ is given via the norm

$$\phi \longmapsto \sum_{[\alpha]\leq s} \|X^\alpha \phi\|_p \tag{4.44}$$

Goodman's definition does not use Rockland operators but makes sense only for integer exponents.

The L^p Goodman-Sobolev space of integer order s certainly contains $L^p_s(G)$. Indeed, proceeding almost as in the proof of Lemma 4.4.19, using Theorem 4.4.16 and Theorem 4.4.3, we have

$$\forall s \in \mathbb{N}_0 \quad \exists C = C_s > 0 \quad \forall \phi \in \mathcal{S}(G) \quad \sum_{[\alpha]\leq s} \|X^\alpha \phi\|_p \leq C\|\phi\|_{L^p_s}.$$

In fact, adapting the rest of the proof of Lemma 4.4.19, one could show easily that the L^p Goodman-Sobolev space of order $s \in \mathbb{N}_0$ with s proportional to the homogeneous degree ν of a positive Rockland operator coincides with our Sobolev spaces $L^p_s(G)$. Moreover, on any stratified Lie group, for any non-negative integer s without further restriction, they would coincide as well, see [Fol75, Theorem 4.10].

However, this equality between Goodman-Sobolev spaces and our Sobolev spaces is not true on any general graded Lie group. For instance this does not hold on a graded Lie groups whose weights are all strictly greater than 1. Indeed the L^p Goodman-Sobolev space of order $s = 1$ is $L^p(G)$ which contains $L^p_1(G)$ strictly (see Theorem 4.4.3 (4)). An example of such a graded Lie group was given by the gradation of the three dimensional Heisenberg group via (3.1).

We can now show the main result of this section, that is, that the Sobolev spaces on graded Lie groups are independent of the chosen positive Rockland operators.

Theorem 4.4.20. *Let G be a graded Lie group and $p \in (1, \infty)$. The homogeneous L^p-Sobolev spaces on G associated with any positive Rockland operators coincide. The inhomogeneous L^p-Sobolev spaces on G associated with any positive Rockland operators coincide. Moreover, in the homogeneous and inhomogeneous cases, the Sobolev norms associated to two positive Rockland operators are equivalent.*

Proof of Theorem 4.4.20. Positive Rockland operators always exist, see Remark 4.2.4 Let $\mathcal{R}_1$ and $\mathcal{R}_2$ be two positive Rockland operators on G of homogeneous degrees ν_1 and ν_2, respectively. By Lemma 4.2.5, $\mathcal{R}_1^{\nu_2}$ and $\mathcal{R}_2^{\nu_1}$ are two positive Rockland operators with the same homogeneous degree $\nu = \nu_1\nu_2$. Their associated homogeneous (respectively inhomogeneous) Sobolev spaces of exponent $\nu\ell = \nu_1\nu_2\ell$

for any $\ell \in \mathbb{N}_0$ coincide and have equivalent norms by Lemma 4.4.19. By interpolation (see Proposition 4.4.15, respectively Theorem 4.4.9), this is true for any Sobolev spaces of exponent $s \geq 0$, and by duality (see Lemma 4.4.14, respectively Lemma 4.4.7) for any exponent $s \in \mathbb{R}$. □

Corollary 4.4.21. *Let $\mathcal{R}^{(1)}$ and $\mathcal{R}^{(2)}$ be two positive Rockland operators on a graded Lie group G with degrees of homogeneity ν_1 and ν_2, respectively. Then for any $s \in \mathbb{C}$ and $p \in (1, \infty)$, the operators $(\mathrm{I} + \mathcal{R}^{(1)})^{\frac{s}{\nu_1}}(\mathrm{I} + \mathcal{R}^{(2)})^{-\frac{s}{\nu_2}}$ and $(\mathcal{R}^{(1)})^{\frac{s}{\nu_1}}(\mathcal{R}^{(2)})^{-\frac{s}{\nu_2}}$ extend boundedly on $L^p(G)$.*

Proof of Corollary 4.4.21. Let us prove the inhomogeneous case first. For any $a \in \mathbb{R}$, we view the operator $(\mathrm{I}+\mathcal{R}_p^{(2)})^{-\frac{a}{\nu_2}}$ as a bounded operator from $L^p(G)$ to $L^p_a(G)$ and use the norm $f \mapsto \|(\mathrm{I} + \mathcal{R}_p^{(1)})^{\frac{a}{\nu_1}} f\|_p$ on $L^p_a(G)$. This shows that the operator $(\mathrm{I} + \mathcal{R}^{(1)})^{\frac{s}{\nu_1}}(\mathrm{I} + \mathcal{R}^{(2)})^{-\frac{s}{\nu_2}}$ is bounded on $L^p(G)$, $p \in (1, \infty)$ for $s = a \in \mathbb{R}$. The case of $s \in \mathbb{C}$ follows from Proposition 4.3.7.

Let us prove the homogeneous case. For any $a \in \mathbb{R}$, we view the operator $(\mathcal{R}_p^{(2)})^{-\frac{a}{\nu_2}}$ as a bounded operator from $L^p(G)$ to $\dot{L}^p_a(G)$ and use the norm $f \mapsto \|(\mathcal{R}_p^{(1)})^{\frac{a}{\nu_1}} f\|_p$ on $\dot{L}^p_a(G)$. This shows that the operator $(\mathcal{R}^{(1)})^{\frac{s}{\nu_1}}(\mathcal{R}^{(2)})^{-\frac{s}{\nu_2}}$ is bounded on $L^p(G)$, $p \in (1, \infty)$ for $s = a \in \mathbb{R}$. The case of $s \in \mathbb{C}$ follows from Proposition 4.3.9. □

Thanks to Theorem 4.4.20, we can now improve our duality result given in Lemmata 4.4.7 and 4.4.14:

Proposition 4.4.22. *Let $L^p_s(G)$ and $\dot{L}^p_s(G)$, $p \in (1, \infty)$ and $s \in \mathbb{R}$, be the inhomogeneous and homogeneous Sobolev spaces on a graded Lie group G, respectively.*

For any $s \in \mathbb{R}$ and $p \in (1, \infty)$, the dual space of $L^p_s(G)$ is isomorphic to $L^{p'}_{-s}(G)$ via the distributional duality, and the dual space of $\dot{L}^p_s(G)$ is isomorphic to $\dot{L}^{p'}_{-s}(G)$ via the distributional duality. Here p' is the conjugate exponent of p if $p \in (1, \infty)$, i.e. $\frac{1}{p} + \frac{1}{p'} = 1$. Consequently the Banach spaces $L^p_s(G)$ and $\dot{L}^p_s(G)$ are reflexive.

4.4.6 Sobolev embeddings

In this section, we show local embeddings between the (inhomogeneous) Sobolev spaces and their Euclidean counterparts, and global embeddings in the form of an analogue of the classical fractional integration theorems of Hardy-Littlewood and Sobolev.

Local results

Recalling that G has a local topological structure of $\mathbb{R}^n$, one can wonder what is the relation between our Sobolev spaces $L^p_s(G)$ and their Euclidean counterparts $L^p_s(\mathbb{R}^n)$. The latter can also be seen as Sobolev spaces associated by the described construction to the abelian group $(\mathbb{R}^n, +)$, with Rockland operator being the Laplacian on $\mathbb{R}^n$.

By Proposition 3.1.28 the coefficients of vector fields X_j with respect to the abelian derivatives ∂_{x_k} are polynomials in the coordinate functions x_ℓ, and conversely the coefficients of ∂_{x_j}'s with respect to derivatives X_k are polynomials in the coordinate functions x_ℓ's. Hence, we can not expect any global embeddings between $L^p_s(G)$ and $L^p_s(\mathbb{R}^n)$.

It is convenient to define the local Sobolev spaces for $s \in \mathbb{R}$ and $p \in (1, \infty)$ as

$$L^p_{s,loc}(G) := \{f \in \mathcal{D}'(G) : \phi f \in L^p_s(G) \text{ for all } \phi \in \mathcal{D}(G)\}. \tag{4.45}$$

The following proposition shows that $L^p_{s,loc}(G)$ contains $L^p_s(G)$.

Proposition 4.4.23. *For any $\phi \in \mathcal{D}(G)$, $p \in (1, \infty)$ and $s \in \mathbb{R}$, the operator $f \mapsto f\phi$ defined for $f \in \mathcal{S}(G)$ extends continuously into a bounded map from $L^p_s(G)$ to itself. Consequently, we have*

$$L^p_s(G) \subset L^p_{s,loc}(G).$$

Proof. The Leibniz' rule for the X_j's and the continuous inclusions in Theorem 4.4.3 (4) imply easily that for any fixed $\alpha \in \mathbb{N}_0^n$ there exist a constant $C = C_{\alpha,\phi} > 0$ and a constant $C' = C'_{\alpha,\phi} > 0$ such that

$$\forall f \in \mathcal{D}(G) \quad \|X^\alpha(f\phi)\|_p \leq C \sum_{[\beta] \leq [\alpha]} \|X^\beta f\|_p \leq C' \|f\|_{L^p_{[\alpha]}(G)}.$$

Lemma 4.4.19 yields the existence of a constant $C'' = C''_{\alpha,\phi} > 0$ such that

$$\forall f \in \mathcal{D}(G) \quad \|f\phi\|_{L^p_{\ell\nu}(G)} \leq C'' \|f\|_{L^p_{\ell\nu}(G)}$$

for any integer $\ell \in \mathbb{N}_0$ and any degree of homogeneity ν of a Rockland operator.

This shows the statement for the case $s = \nu\ell$. The case $s > 0$ follows by interpolation (see Theorem 4.4.9), and the case $s < 0$ by duality (see Proposition 4.4.22). □

We can now compare locally the Sobolev spaces on graded Lie groups and on their abelian counterpart:

Theorem 4.4.24 (Local Sobolev embeddings). *For any $p \in (1, \infty)$ and $s \in \mathbb{R}$,*

$$L^p_{s/v_1,loc}(\mathbb{R}^n) \subset L^p_{s,loc}(G) \subset L^p_{s/v_n,loc}(\mathbb{R}^n).$$

Above, $L^p_{s,loc}(\mathbb{R}^n)$ denotes the usual local Sobolev spaces, or equivalently the spaces defined by (4.45) in the case of the abelian (graded) Lie group $(\mathbb{R}^n, +)$. Recall that v_1 and v_n are respectively the smallest and the largest weights of the dilations. In particular, in the stratified case, $v_1 = 1$ and v_n coincides with the number of steps in the stratification, and with the step of the nilpotent Lie group G. Hence in the stratified case we recover Theorem 4.16 in [Fol75].

Proof of Theorem 4.4.24. It suffices to show that the mapping $f \mapsto f\phi$ defined on $\mathcal{D}(G)$ extends boundedly from $L^p_{s/\upsilon_1}(\mathbb{R}^n)$ to $L^p_s(G)$ and from $L^p_s(G)$ to $L^p_{s/\upsilon_n,loc}(\mathbb{R}^n)$. By duality and interpolation (see Theorem 4.4.9 and Proposition 4.4.22), it suffices to show this for a sequence of increasing positive integers s.

For the $L^p_{s/\upsilon_1}(\mathbb{R}^n) \to L^p_s(G)$ case, we assume that s is divisible by the homogeneous degree of a positive Rockland operator. Then we use Lemma 4.4.19, the fact that the X^α may be written as a combination of the ∂_x^β with polynomial coefficients in the x_ℓ's and that $\max_{[\beta]\leq s}|\beta| = s/\upsilon_1$.

For the case of $L^p_s(G) \to L^p_{s/\upsilon_n,loc}(\mathbb{R}^n)$, we use the fact that the abelian derivative ∂_x^α, $|\alpha|\leq s$, may be written as a combination over the X^β, $|\beta|\leq s$, with polynomial coefficients in the x_ℓ's, that X^β maps $L^p \to L^p_{[\beta]}$ boundedly together with $\max_{|\beta|\leq s}[\beta] = s\upsilon_n$. □

Proceeding as in [Fol75, p.192], one can convince oneself that Theorem 4.4.24 can not be improved.

Global results

In this section, we show the analogue of the classical fractional integration theorems of Hardy-Littlewood and Sobolev. The stratified case was proved by Folland in [Fol75] (mainly Theorem 4.17 therein).

Theorem 4.4.25 (Sobolev embeddings). *Let G be a graded Lie group with homogeneous dimension Q.*

(i) If $1 < p < q < \infty$ and $a, b \in \mathbb{R}$ with

$$b - a = Q(\frac{1}{p} - \frac{1}{q})$$

then we have the continuous inclusion

$$L^p_b \subset L^q_a,$$

that is, for every $f \in L^p_b$, we have $f \in L^q_a$ and there exists a constant $C = C_{a,b,p,q,G} > 0$ independent of f such that

$$\|f\|_{L^q_a} \leq C\|f\|_{L^p_b}.$$

(ii) If $p \in (1,\infty)$ and

$$s > Q/p$$

then we have the inclusion

$$L^p_s \subset (C(G) \cap L^\infty(G)),$$

in the sense that any function $f \in L^p_s(G)$ admits a bounded continuous representative on G (still denoted by f). Furthermore, there exists a constant $C = C_{s,p,G} > 0$ independent of f such that

$$\|f\|_\infty \le C\|f\|_{L^p_s(G)}.$$

Proof. Let us first prove Part (i). We fix a positive Rockland operator $\mathcal{R}$ of homogeneous degree ν and we assume that $b > a$ and $p, q \in (1, \infty)$ satisfy $b - a = Q(\frac{1}{p} - \frac{1}{q})$. By Proposition 4.4.13 (5),

$$\|\mathcal{R}_q^{\frac{a}{\nu}}\phi\|_{L^q} \le C\|\mathcal{R}_p^{\frac{b}{\nu}}\phi\|_{L^p}.$$

We can apply this to (a, b) and to $(0, b - a)$. Adding the two corresponding estimates, we obtain

$$\|\phi\|_{L^q} + \|\mathcal{R}_q^{\frac{a}{\nu}}\phi\|_{L^q} \le C\left(\|\mathcal{R}_p^{\frac{b-a}{\nu}}\phi\|_{L^p} + \|\mathcal{R}_p^{\frac{b}{\nu}}\phi\|_{L^p}\right).$$

Since b, a, and $b - a$ are positive, by Theorem 4.4.3 (4), the left-hand side is equivalent to $\|\phi\|_{L^q_a}$ and both terms in the right-hand side are $\le C\|\phi\|_{L^p_b}$. Therefore, we have obtained that

$$\exists C = C_{a,b,p,q,\mathcal{R}} \quad \forall \phi \in \mathcal{S}(G) \qquad \|\phi\|_{L^q_a} \le C\|\phi\|_{L^p_b}.$$

By density of $\mathcal{S}(G)$ in the Sobolev spaces, this shows Part (i).

Let us prove Part (ii). Let $p \in (1, \infty)$ and $s > Q/p$. By Corollary 4.3.13, we know that

$$\mathcal{B}_s \in L^1(G) \cap L^{p'}(G),$$

where p' is the conjugate exponent of p. For any $f \in L^p_s(G)$, we have

$$f_s := (\mathrm{I} + \mathcal{R}_p)^{\frac{s}{\nu}} f \in L^p$$

and

$$f = (\mathrm{I} + \mathcal{R}_p)^{-\frac{s}{\nu}} f_s = f_s * \mathcal{B}_s.$$

Therefore, by Hölder's inequality,

$$\|f\|_\infty \le \|f_s\|_p \|\mathcal{B}_s\|_{p'} = \|\mathcal{B}_s\|_{p'} \|f\|_{L^p_s}.$$

Moreover, for almost every x, we have

$$f(x) = \int_G f_s(y)\mathcal{B}_s(y^{-1}x)dy = \int_G f_s(xz^{-1})\mathcal{B}_s(z)dz.$$

Thus for almost every x, x', we have

$$\begin{aligned} |f(x) - f(x')| &= \left|\int_G \left(f_s(xz^{-1}) - f_s(x'z^{-1})\right)\mathcal{B}_s(z)dz\right| \\ &\le \|\mathcal{B}_s\|_{p'} \|f_s(x\,\cdot) - f_s(x'\,\cdot)\|_p. \end{aligned}$$

As the left regular representation is continuous (see Example 1.1.2) we have

$$\|f_s(x\,\cdot) - f_s(x'\,\cdot)\|_{L^p(G)} \longrightarrow_{x'\to x} 0,$$

thus almost surely

$$|f(x) - f(x')| \longrightarrow_{x'\to x} 0.$$

Hence we can modify f so that it becomes a continuous function. This concludes the proof. □

From the Sobolev embeddings (Theorem 4.4.25 (ii)) and the description of Sobolev spaces with integer exponent (Lemma 4.4.19) the following property follows easily:

Corollary 4.4.26. *Let G be a graded Lie group, $p \in (1,\infty)$ and $s \in \mathbb{N}$. We assume that s is proportional to the homogeneous degree ν of a positive Rockland operator, that is, $\frac{s}{\nu} \in \mathbb{N}$, and that $s > Q/p$.*

Then if f is a distribution on G such that $f \in L^p(G)$ and $X^\alpha f \in L^p(G)$ when $\alpha \in \mathbb{N}_0^n$ satisfies $[\alpha] = s$, then f admits a bounded continuous representative (still denoted by f). Furthermore, there exists a constant $C = C_{s,p,G} > 0$ independent of f such that

$$\|f\|_\infty \le C\left(\|f\|_p + \sum_{[\alpha]=s} \|X^\alpha f\|_p\right).$$

The Sobolev embeddings, especially Corollary 4.4.26, enables us to define Schwartz seminorms not only in terms of the supremum norm, but also in terms of any L^p-norms:

Proposition 4.4.27. *Let $|\cdot|$ be a homogeneous norm on a graded Lie group G. For any $p \in [1,\infty]$, $a > 0$ and $k \in \mathbb{N}_0$, the mapping*

$$\mathcal{S}(G) \ni \phi \mapsto \|\phi\|_{\mathcal{S},a,k,p} := \sum_{[\alpha]\le k} \|(1+|\cdot|)^a X^\alpha \phi\|_p$$

is a continuous seminorm on the Fréchet space $\mathcal{S}(G)$.

Moreover, let us fix $p \in [1,\infty]$ and two sequences $\{k_j\}_{j\in\mathbb{N}}$, $\{a_j\}_{j\in\mathbb{N}}$, of non-negative integers and positive numbers, respectively, which go to infinity. Then the family of seminorms $\|\cdot\|_{\mathcal{S},a_j,k_j,p}$, $j \in \mathbb{N}$, yields the usual topology on $\mathcal{S}(G)$.

Proof of Proposition 4.4.27. One can check easily that the property

$$\begin{aligned} \forall 1 \le p,q \le \infty,\ a > 0,\ k \in \mathbb{N}_0, \quad \exists a' > 0,\ k' \in \mathbb{N}_0,\ C > 0,\\ \|\cdot\|_{\mathcal{S},a,k,p} \le \|\cdot\|_{\mathcal{S},a',k',q}, \end{aligned} \tag{4.46}$$

is a consequence of the following observations (applied to $X^\alpha\phi$ instead of ϕ):

1. If p and q are finite, by Hölder's inequality, we have

$$\|(1+|\cdot|)^a\phi\|_p \leq C\|(1+|\cdot|)^{a'}\phi\|_q$$

where C is a finite constant of the group G, p and q. In fact C is explicitly given by

$$C = \|(1+|\cdot|)^{-\frac{Q+1}{r}}\|_r = \left(|B(0,1)|\int_0^\infty (1+\rho)^{-(Q+1)}\rho^{Q-1}d\rho\right)^{\frac{1}{r}},$$

with $r\in(1,\infty)$ such that $\frac{1}{p}=\frac{1}{q}+\frac{1}{r}$.

2. If p is finite and $q=\infty$, we also have

$$\|(1+|\cdot|)^a\phi\|_p \leq C\|(1+|\cdot|)^{a+Q+1}\phi\|_\infty$$

where $C=\|(1+|\cdot|)^{-Q-1}\|_p$ is a finite constant.

3. In the case q is finite and $p=\infty$, let us prove that

$$\|(1+|\cdot|)^a\phi\|_\infty \leq C_{s,p}\sum_{[\alpha]\leq s}\|(1+|\cdot|)^a X^\alpha\phi\|_p. \tag{4.47}$$

Indeed first we notice that, by equivalence of the homogeneous quasi-norms (see Proposition 3.1.35), we may assume that the quasi-norm is smooth away from 0. We fix a function $\psi\in\mathcal{D}(G)$ such that

$$\psi(x)=\begin{cases}1 & \text{if } |x|\leq 1,\\ 0 & \text{if } |x|\geq 2.\end{cases}$$

We have easily

$$\|(1+|\cdot|)^a\phi\|_\infty \leq C_\psi\left(\|\phi\psi\|_\infty+\|\phi(1-\psi)|\cdot|^a\|_\infty\right). \tag{4.48}$$

By Corollary 4.4.26, there exist an integer $s\in\mathbb{N}$ such that

$$\|\phi\psi\|_\infty \leq C_{s,p}\sum_{[\alpha]\leq s}\|X^\alpha(\phi\psi)\|_p.$$

By the Leibniz rule (which is valid for any vector field) and Hölder's inequality, we have

$$\begin{aligned}\|X^\alpha(\phi\psi)\|_p &\leq C_\alpha\sum_{[\alpha_1]+[\alpha_2]\leq[\alpha]}\|X^{\alpha_1}\phi\, X^{\alpha_2}\psi\|_p\\ &\leq C_{\alpha,p}\sum_{[\alpha_1]+[\alpha_2]\leq[\alpha]}\|X^{\alpha_1}\phi\|_p\|X^{\alpha_2}\psi\|_\infty.\end{aligned}$$

Hence

$$\|\phi\psi\|_\infty \le C_{s,p,\psi} \sum_{[\alpha]\le s} \|X^\alpha \phi\|_p. \tag{4.49}$$

Following the same line of arguments, we have

$$\begin{aligned}
\|\phi(1-\psi)|\cdot|^a\|_\infty &\le C_{s,p} \sum_{[\alpha]\le s} \|X^\alpha(\phi(1-\psi)|\cdot|^a)\|_p \\
&\le C_{s,p} \sum_{[\alpha_1]+[\alpha_2]\le s} \|X^{\alpha_1}\phi\; X^{\alpha_2}\{(1-\psi)|\cdot|^a\}\|_p \\
&\le C_{s,p} \sum_{[\alpha_1]+[\alpha_2]\le s} \|(1+|\cdot|)^a X^{\alpha_1}\phi\|_p \|(1+|\cdot|)^{-a} X^{\alpha_2}\{(1-\psi)|\cdot|^a\}\|_\infty.
\end{aligned}$$

All the $\|\cdot\|_\infty$-norms above are finite since $X^{\alpha_2}\{(1-\psi)|\cdot|^a\}(x)=0$ if $|x|\le 1$ and for $|x|\ge 1$,

$$\begin{aligned}
|X^{\alpha_2}\{(1-\psi)|\cdot|^a\}(x)| &\le C_{\alpha_2} \sum_{[\alpha_3]+[\alpha_4]=[\alpha_2]} |X^{\alpha_3}(1-\psi)(x)|\; |X^{\alpha_4}|\cdot|^a|(x) \\
&\le C_{\alpha_2} \sum_{[\alpha_3]+[\alpha_4]=[\alpha_2]} \|X^{\alpha_3}(1-\psi)\,|_\infty |x|^{a-[\alpha_4]},
\end{aligned}$$

since $X^{\alpha_4}|\cdot|^a$ is a homogeneous function of degree $a-[\alpha_4]$. Hence we have obtained

$$\|\phi(1-\psi)|\cdot|^a\|_\infty \le C_{s,p,\psi} \sum_{[\alpha]\le s} \|(1+|\cdot|)^a X^\alpha \phi\|_p.$$

Together with (4.48) and (4.49), this shows (4.47).

4. If $p=q$ is finite or infinite, (4.46) is trivial.

Hence Property (4.46) holds. We also have directly for $p=q\in[1,\infty]$ and any $0<a\le a'$, $k\le k'$,

$$\|\cdot\|_{\mathcal{S},a,k,p} \le \|\cdot\|_{\mathcal{S},a',k',p}.$$

Consequently we can assume a' to be an integer in (4.46). This clearly implies that any family of seminorms $\|\cdot\|_{\mathcal{S},a_j,k_j,p}$, $j\in\mathbb{N}$, yields the same topology as the family of seminorms $\|\cdot\|_{\mathcal{S},N,N,\infty}$, $N\in\mathbb{N}$. The latter is easily equivalent to the topology given by the family of seminorms $\|\cdot\|_{\mathcal{S}(G),N}$ defined in Section 3.1.9. This is the usual topology on $\mathcal{S}(G)$. □

4.4.7 List of properties for the Sobolev spaces

In this section, we list the important properties of Sobolev spaces we have already obtained and also give some easy consequences regarding the special case of $p=2$.

Theorem 4.4.28. *Let G be a graded Lie group with homogeneous dimension Q.*

1. *Let $p \in (1,\infty)$ and $s \in \mathbb{R}$. The inhomogeneous Sobolev space $L^p_s(G)$ is a Banach space satisfying*

$$\mathcal{S}(G) \subsetneq L^p_s(G) \subset \mathcal{S}'(G).$$

The homogeneous Sobolev space $\dot{L}^p_s(G)$ is a Banach space satisfying

$$(\mathcal{S}(G) \cap \mathrm{Dom}(\mathcal{R}_p^{s/\nu})) \subsetneq \dot{L}^p_s(G) \subsetneq \mathcal{S}'(G).$$

Norms on the Banach spaces $L^p_s(G)$ and $\dot{L}^p_s(G)$ are given respectively by

$$\phi \mapsto \|(\mathrm{I} + \mathcal{R}_p)^{\frac{s}{\nu}} \phi\|_{L^p(G)} \qquad \text{and} \qquad \phi \mapsto \|\mathcal{R}_p^{\frac{s}{\nu}} \phi\|_{L^p(G)},$$

for any positive Rockland operator $\mathcal{R}$ (whose homogeneous degree is denoted by ν). All these homogeneous norms are equivalent, all these inhomogeneous norms are equivalent.

The continuous inclusions $L^p_a(G) \subset L^p_b(G)$ holds for any $a \geq b$ and $p \in (1,\infty)$.

2. *If $s = 0$ and $p \in (1,\infty)$, then $\dot{L}^p_0(G) = L^p_0(G) = L^p(G)$ with $\|\cdot\|_{\dot{L}^p_0(G)} = \|\cdot\|_{L^p_0(G)} = \|\cdot\|_{L^p(G)}$.*

3. *If $s > 0$ and $p \in (1,\infty)$, then we have*

$$L^p_s(G) = \dot{L}^p_s(G) \cap L^p(G),$$

and the inhomogeneous Sobolev norm (associated with a positive Rockland operator) is equivalent to

$$\|\cdot\|_{L^p_s(G)} \asymp \|\cdot\|_{L^p(G)} + \|\cdot\|_{\dot{L}^p_s(G)}.$$

4. *If T is a left-invariant differential operator of homogeneous degree ν_T, then T maps continuously $L^p_{s+\nu_T}(G)$ to $L^p_s(G)$ for every $s \in \mathbb{R}$, $p \in (1,\infty)$.*

If T is a ν_T-homogeneous left-invariant differential operator, then T maps continuously $\dot{L}^p_{s+\nu_T}(G)$ to $\dot{L}^p_s(G)$ for every $s \in \mathbb{R}$, $p \in (1,\infty)$.

5. *If $1 < p < q < \infty$ and $a, b \in \mathbb{R}$ with $b - a = Q(\frac{1}{p} - \frac{1}{q})$, then we have the continuous inclusions*

$$\dot{L}^p_b \subset \dot{L}^q_a \qquad \text{and} \qquad L^p_b \subset L^q_a.$$

If $p \in (1,\infty)$ and $s > Q/p$ then we have the following inclusion:

$$L^p_s \subset (C(G) \cap L^\infty(G)),$$

in the sense that any function $f \in L^p_s(G)$ *admits a bounded continuous representative on* G *(still denoted by* f*). Furthermore, there exists a constant* $C = C_{s,p,G} > 0$ *independent of* f *such that*

$$\|f\|_\infty \leq C\|f\|_{L^p_s(G)}.$$

6. *For* $p \in (1, \infty)$ *and any* $a, b, c \in \mathbb{R}$ *with* $a < c < b$*, there exists a positive constant* $C = C_{a,b,c}$ *such that we have for any* $f \in \dot{L}^p_b$

$$\|f\|_{\dot{L}^p_c} \leq C\|f\|^{1-\theta}_{\dot{L}^p_a}\|f\|^{\theta}_{\dot{L}^p_b}$$

and for any $f \in L^p_b$

$$\|f\|_{L^p_c} \leq C\|f\|^{1-\theta}_{L^p_a}\|f\|^{\theta}_{L^p_b}$$

where $\theta := (c - a)/(b - a)$.

7. *(Gagliardo-Nirenberg inequality) If* $q, r \in (1, \infty)$ *and* $0 < \sigma < s$ *then there exists* $C > 0$ *such that we have*

$$\forall f \in L^q(G) \cap \dot{L}^r_s(G) \qquad \|f\|_{\dot{L}^p_\sigma} \leq C\|f\|^{\theta}_{L^q}\|f\|^{1-\theta}_{\dot{L}^r_s},$$

where $\theta := 1 - \frac{\sigma}{s}$ *and* $p \in (1, \infty)$ *is given via* $\frac{1}{p} = \frac{\theta}{q} + \frac{1-\theta}{r}$.

8. *Let* s *be an integer which is proportional to the homogeneous degree of a positive Rockland operator. Let* $p \in (1, \infty)$*. Let* $f \in \mathcal{S}'(G)$*.*

   The membership of f *in* $L^p_s(G)$ *is equivalent to* $f \in L^p(G)$ *and* $X^\alpha f \in L^p(G)$*,* $\alpha \in \mathbb{N}^n_0$*,* $[\alpha] = s$*. Furthermore*

$$\phi \mapsto \|\phi\|_p + \sum_{[\alpha]=\nu\ell} \|X^\alpha \phi\|_p$$

   is a norm on the Banach space $L^p_s(G)$*.*

   The membership of f *in* $\dot{L}^p_s(G)$ *is equivalent to* $X^\alpha f \in L^p(G)$*,* $\alpha \in \mathbb{N}^n_0$*,* $[\alpha] = s$*. Furthermore*

$$\phi \mapsto \sum_{[\alpha]=\nu\ell} \|X^\alpha \phi\|_p$$

   is a norm on the Banach space $\dot{L}^p_s(G)$*.*

9. *(Interpolation) The inhomogeneous and homogeneous Sobolev spaces satisfy the properties of interpolation in the sense of Theorem 4.4.9 and Proposition 4.4.15 respectively.*

10. *(Duality) Let* $s \in \mathbb{R}$*. Let* $p \in (1, \infty)$ *and* p' *its conjugate exponent. The dual space of* $\dot{L}^p_s(G)$ *is isomorphic to* $\dot{L}^{p'}_{-s}(G)$ *via the distributional duality, and the dual space of* $L^p_s(G)$ *is isomorphic to* $\dot{L}^{p'}_{-s}(G)$ *via the distributional duality, Consequently, the Banach spaces* $L^p_s(G)$ *and* $\dot{L}^p_s(G)$ *are reflexive.*

Proof. Parts (1), (2), (3), and (6) follow from Theorem 4.4.3, Proposition 4.4.13 and Theorem 4.4.20.

Part (4) follows from Theorem 4.4.16 and Proposition 4.4.13.

Part (5) follows from Theorem 4.4.25 and Proposition 4.4.13 (5).

Part (7) follows from Parts (5) and (6).

Part (8) follows from Theorem 4.4.20.

For Part (9), see Theorem 4.4.9 and Proposition 4.4.15.

Part (10) follows from Lemmata 4.4.7 and 4.4.14 together with Theorem 4.4.20. □

Properties of $L^2_s(G)$

Here we discuss some special feature of the case $L^p(G)$, $p = 2$. Indeed $L^2(G)$ is a Hilbert space where one can use the spectral analysis of a positive Rockland operator.

Many of the proofs in Chapter 4 could be simplified if we had restricted the study to the case L^p with $p = 2$. For instance, let us consider a positive Rockland operator $\mathcal{R}$ and its self-adjoint extension $\mathcal{R}_2$ on $L^2(G)$. One can define the fractional powers of $\mathcal{R}_2$ and $\mathrm{I} + \mathcal{R}_2$ by functional analysis. Then one can obtain the properties of the kernels of the Riesz and Bessel potentials with similar methods as in Corollary 4.3.11.

In this case, one would not need to use the general theory of fractional powers of an operator recalled in Section A.3. Even if it is not useful, let us mention that the proof that $\mathcal{R}_2$ satisfies the hypotheses of Theorem A.3.4 is easy in this case: it follows directly from the Lumer-Phillips Theorem (see Theorem A.2.5) together with the heat semi-group $\{e^{-t\mathcal{R}_2}\}_{t>0}$ being an $L^2(G)$-contraction semi-group by functional analysis.

The proof of the properties of the associated Sobolev spaces $L^2_s(G)$ would be the same in this particular case, maybe slightly helped occasionally by the Hölder inequality being replaced by the Cauchy-Schwartz inequality. A noticeable exception is that Lemma 4.4.19 can be obtained directly in the case L^p, $p = 2$, from the estimates due to Helffer and Nourrigat (see Corollary 4.1.14).

The main difference between L^2 and L^p Sobolev spaces is the structure of Hilbert spaces of $L^2_s(G)$ whereas the other Sobolev spaces $L^p_s(G)$ are 'only' Banach spaces:

Proposition 4.4.29 (Hilbert space L^2_s). *Let G be a graded Lie group.*

For any $s \in \mathbb{R}$, $L^2_s(G)$ is a Hilbert space with the inner product given by

$$(f, g)_{L^2_s(G)} := \int_G (\mathrm{I} + \mathcal{R}_2)^{\frac{s}{\nu}} f(x)\, \overline{(\mathrm{I} + \mathcal{R}_2)^{\frac{s}{\nu}} g(x)} dx,$$

and $\dot{L}^2_s(G)$ is a Hilbert space with the inner product given by

$$(f, g)_{\dot{L}^2_s(G)} := \int_G \mathcal{R}_2^{\frac{s}{\nu}} f(x)\, \overline{\mathcal{R}_2^{\frac{s}{\nu}} g(x)} dx,$$

where $\mathcal{R}$ is a positive Rockland operator of homogeneous degree ν.

If $s > 0$, an equivalent inner product on $L^2_s(G)$ is

$$(f,g)_{L^2_s(G)} := \int_G f(x)\,\overline{g(x)}dx \;+\; \int_G \mathcal{R}_2^{\frac{s}{\nu}} f(x)\,\overline{\mathcal{R}_2^{\frac{s}{\nu}} g(x)}dx.$$

If $s = \nu\ell$ with $\ell \in \mathbb{N}_0$, an equivalent inner product on $L^2_s(G)$ is

$$(f,g) = (f,g)_{L^2(G)} + \sum_{[\alpha]=\nu\ell} (X^\alpha f, X^\alpha g)_{L^2(G)},$$

and an equivalent inner product on $\dot{L}^2_s(G)$ is

$$(f,g) = \sum_{[\alpha]=\nu\ell} (X^\alpha f, X^\alpha g)_{L^2(G)}.$$

Proposition 4.4.29 is easily checked, using the structure of Hilbert space of $L^2(G)$ and, for the last property, simplifying the proof of Lemma 4.4.19.

4.4.8 Right invariant Rockland operators and Sobolev spaces

We could have started with right-invariant (homogeneous) Rockland operators $\tilde{\mathcal{R}}$ instead of $\mathcal{R}$. We discuss here some links between the two operators and their Sobolev spaces.

Since both left and right invariant Rockland operators are differential operators, we can relate them by Formulae (1.11) for the derivatives X^α and $\tilde{X}^\alpha$. Then, given our analysis of $\mathcal{R}$, we can give some immediate properties of the right-invariant operator $\tilde{\mathcal{R}}$:

Proposition 4.4.30. *Let $\mathcal{R}$ be a positive Rockland operator. For any $\phi \in \mathcal{S}(G)$,*

$$\tilde{\mathcal{R}}\phi(x) = (\mathcal{R}^t\{\phi(\cdot^{-1})\})(x^{-1}) = (\bar{\mathcal{R}}\{\phi(\cdot^{-1})\})(x^{-1}),$$

because $\mathcal{R}^t = \bar{\mathcal{R}}$. Therefore, the spectral measure $\tilde{E}$ of $\tilde{\mathcal{R}}$ is given by

$$\tilde{E}(\phi)(x) = (\bar{E}\{\phi(\cdot^{-1})\})(x^{-1}), \quad \phi \in L^2(G),\ x \in G.$$

Consequently, the multipliers of $\tilde{\mathcal{R}}$ and $\mathcal{R}$ are linked by

$$m(\tilde{\mathcal{R}})(\phi)(x) = (m(\bar{\mathcal{R}})\{\phi(\cdot^{-1})\})(x^{-1}). \tag{4.50}$$

The operators $\mathcal{R}$ and $\tilde{\mathcal{R}}$ commute strongly, that is, their spectral measures E and $\tilde{E}$ commute. Moreover, for functions $f, g \in \mathcal{S}'(G)$ and $a \in \mathbb{C}$, we have

$$\begin{aligned} \mathcal{R}^a(f*g) &= f*\mathcal{R}^a g, \\ \tilde{\mathcal{R}}^a(f*g) &= (\tilde{\mathcal{R}}^a f)*g, \\ (\mathcal{R}^a f)*g &= f*\tilde{\mathcal{R}}^a g. \end{aligned}$$

We can give a right-invariant version of Definition 4.3.17:

Definition 4.4.31. Let $\mathcal{R}$ be a positive Rockland operator of homogeneous degree ν and let $s \in \mathbb{R}$. For any tempered distribution $f \in \mathcal{S}'(G)$, we denote by $(\mathrm{I}+\tilde{\mathcal{R}})^{s/\nu} f$ the tempered distribution defined by

$$\langle (\mathrm{I}+\tilde{\mathcal{R}})^{s/\nu} f, \phi \rangle := \langle f, (\mathrm{I}+\tilde{\tilde{\mathcal{R}}})^{s/\nu} \phi \rangle, \quad \phi \in \mathcal{S}(G).$$

The Sobolev spaces that we have introduced are based on the Sobolev spaces corresponding to left-invariant vector fields and left-invariant positive Rockland operators. We could have considered the right Sobolev spaces $\tilde{L}^p_s(G)$ defined via the Sobolev norms

$$f \mapsto \|(\mathrm{I}+\tilde{\mathcal{R}})^{s/\nu} f\|_{L^p}.$$

The relations between left and right vector fields in (1.11) easily implies that if $f \in L^p(G)$ is such that $X^\alpha f \in L^p(G)$ then $\tilde{f} : x \mapsto f(x^{-1})$ is in $L^p(G)$ and satisfies $\tilde{X}^\alpha \tilde{f} \in L^p(G)$. By Lemma 4.4.19, we see that the map $f \mapsto \tilde{f}$ must map continuously $L^p_s \to \tilde{L}^p_s$ for any $p \in (1,\infty)$ and s a multiple of the homogeneous degrees of positive Rockland operators.

More generally, the spectral calculus, see (4.50), implies

$$(\mathrm{I}+\tilde{\mathcal{R}}_2)^{s/\nu} f(x) = (\mathrm{I}+\mathcal{R}_2)^{s/\nu} \tilde{f}(x^{-1}), \qquad f \in \mathcal{S}(G),$$

where, again, $\tilde{f}(x) = f(x^{-1})$, and thus for any $p \in (1, \infty_o)$,

$$\|(\mathrm{I}+\tilde{\mathcal{R}}_p)^{s/\nu} f\|_{L^p(G)} = \|(\mathrm{I}+\mathcal{R}_p)^{s/\nu} \tilde{f}\|_{L^p(G)}, \qquad f \in \mathcal{S}(G).$$

This easily implies that $f \mapsto \tilde{f}$ maps continuously $L^p_s \to \tilde{L}^p_s$ for any $p \in (1,\infty)$ and any real exponent $s \in \mathbb{R}$. This is also an involution: $\tilde{\tilde{f}} = f$. Hence the map

$$\left\{ \begin{array}{ccc} L^p_s(G) & \longrightarrow & \tilde{L}^p_s(G) \\ f & \longmapsto & \tilde{f} \end{array} \right.$$

is an isomorphism of vector spaces.

Even if the left and right Sobolev spaces are isomorphic, they are not equal in general. Note that in the commutative case of $G = \mathbb{R}^n$, both left and right Sobolev spaces coincide. It is also the case on compact Lie groups, where the Sobolev spaces are associated with the Laplace-Beltrami operator (which is central) and coincide with localisation of the Euclidean Sobolev spaces [RT10a]. This is no longer the case in the nilpotent setting. Indeed, below we give an example of functions f (necessarily not symmetric, that is, $\tilde{f} \neq f$), in some $L^p_s(G)$ but not in $\tilde{L}^p_s(G)$.

Example 4.4.32. Let us consider the three dimensional Heisenberg group $\mathbb{H}_1$ and the canonical basis X, Y, T of its Lie algebra (see Example 1.6.4). Then $X = \partial_x - \frac{y}{2}\partial_t$ whereas $\tilde{X} = \partial_x + \frac{y}{2}\partial_t$ thus $\tilde{X} - X = y\partial_t$.

The Sobolev spaces are then associated with the natural sub-Laplacian $X^2 + Y^2$, see Example 6.1.1. Hence it is covered by the work of Folland [Fol75] on Sobolev spaces associated with sub-Laplacian on stratified Lie groups and consequently, $L^2_1(G)$ is the space of functions $f \in L^2(\mathbb{H}_1)$ such that Xf and Yf are both in $L^2(\mathbb{H}_1)$ [Fol75, Corollary 4.13].

One can find a smooth function $\phi \in C^\infty(\mathbb{R})$ such that $\phi, \phi' \in L^2(\mathbb{R})$ but $\int_{\mathbb{R}} |z|^2 |\phi'(z)|^2 dz = \infty$. For instance, we consider $\phi = \phi_1 * \psi$ where ψ is a suitable smoothing function (i.e. $\psi \in \mathcal{D}(G)$ is valued in $[0,1]$ with a 'small' support around 0), and the graph of the function ϕ_1 is given by isosceles triangles parametrised by $\ell \in \mathbb{N}$, with vertex at points (ℓ, ℓ^β), and base on the horizontal axis and with length $2/\ell^\alpha$. We then choose $\alpha, \beta \in \mathbb{R}$ with $2\beta \in (-3,-1)$ and $2\alpha > 2\beta + 1$. We also fix a smooth function $\chi : \mathbb{R} \to [0,1]$ supported on $[1/2, 2]$ with $\chi(1) = 1$. We define $f \in C^\infty(\mathbb{R}^3)$ via

$$f(x,y,t) = \phi\left(\frac{yx}{2} + t\right) \chi(x)\chi(t).$$

One checks easily that f, Xf and Yf are square integrable hence $f \in L^2_1(\mathbb{H}_1)$. However $y\partial_t f$ is not square integrable. As $\tilde{X} - X = y\partial_t f$, this shows that $(-X + \tilde{X})f \notin L^2(\mathbb{H}_1)$ and $\tilde{X}f$ can not be in L^2 thus f is not in $\tilde{L}^2_1(\mathbb{H}_1)$.

4.5 Hulanicki's theorem

We now turn our attention to Hulanicki's theorem which will be useful in the next chapter when we deal with pseudo-differential operators on graded Lie groups. An important consequence of Hulanicki's theorem is the fact that a Schwartz multiplier in (the L^2-self-adjoint extension of) a positive Rockland operator has a Schwartz kernel. This section is devoted to the statement and the proof of Hulanicki's theorem and its consequence regarding Schwartz multiplier.

From now on, we will allow ourselves to keep the same notation $\mathcal{R}$ for a positive Rockland operator and its self-adjoint extension $\mathcal{R}_2$ on $L^2(G)$ when no confusion is possible. In particular, when we define functions of $\mathcal{R}_2$ (see Corollary 4.1.16), that is, a multiplier $m(\mathcal{R}_2)$ defined using the spectral measure of $\mathcal{R}_2$ where $m \in L^\infty(\mathbb{R}_+)$ is a function, we may often write

$$m(\mathcal{R}_2) = m(\mathcal{R}),$$

in order to ease the notation. Furthermore, we denote the corresponding right-convolution kernel of this operator by

$$m(\mathcal{R})\delta_o.$$

4.5.1 Statement

Hulanicki proved in [Hul84] that if multipliers m satisfy Marcinkiewicz properties, then the kernels of $m(\mathcal{R})$ satisfy certain estimates:

Theorem 4.5.1 (Hulanicki). *Let $\mathcal{R}$ be a positive Rockland operator on a graded Lie group G. Let $|\cdot|$ be a fixed homogeneous quasi-norm on G. For any $M_1 \in \mathbb{N}, M_2 \geq 0$ there exist $C = C_{M_1,M_2} > 0$ and $k = k_{M_1,M_2} \in \mathbb{N}_0$, $k' = k'_{M_1,M_2} \in \mathbb{N}_0$ such that for any $m \in C^k[0,\infty)$, we have*

$$\sum_{[\alpha]\leq M_1} \int_G |X^\alpha m(\mathcal{R})\delta_o(x)| \ (1+|x|)^{M_2} dx \leq C \sup_{\substack{\lambda>0 \\ \ell=0,\dots,k \\ \ell'=0,\dots,k'}} (1+\lambda)^{\ell'} |\partial_\lambda^\ell m(\lambda)|,$$

in the sense that if the right-hand side is finite then the left-hand side is also finite and the inequality holds.

The main consequence of Theorem 4.5.1 is the following:

Corollary 4.5.2. *Let $\mathcal{R}$ be a positive Rockland operator on a graded Lie group G. If $\phi \in \mathcal{S}(\mathbb{R})$ then the kernel $\phi(\mathcal{R})\delta_o$ of $\phi(\mathcal{R})$ is Schwartz. Furthermore, the map associating a multiplier function with its kernel*

$$\mathcal{S}(\mathbb{R}) \ni \phi \longmapsto \phi(\mathcal{R})\delta_o \in \mathcal{S}(G), \tag{4.51}$$

is continuous between the Schwartz spaces.

The continuity of (4.51) means that for any continuous seminorm $\|\cdot\|$ on $\mathcal{S}(G)$ there exist $C > 0$ and $N \in \mathbb{N}$ such that for any $m \in \mathcal{S}(\mathbb{R})$ we have

$$\|m(\mathcal{R})\delta_o\| \leq C \sup_{x\in\mathbb{R},\ell\leq N} |(1+|x|)^N \partial^\ell m(x)|.$$

Examples of such Schwartz seminorms are $\|\cdot\|_{\mathcal{S}(G),N}$, $N \in \mathbb{N}$, defined in Section 3.1.9, and $\|\cdot\|_{\mathcal{S},a,k,p}$, $a > 0$, $k \in \mathbb{N}_0$, $p \in [1,\infty]$, defined in Proposition 4.4.27.

For completeness' sake, we include the proofs of Theorem 4.5.1 and Corollary 4.5.2 below. Before this, let us notice that Corollary 4.5.2 implies that the heat kernel of any Rockland operator is Schwartz. However, we will see that the proofs of Theorem 4.5.1 and Corollary 4.5.2 rely on the properties of the Bessel potentials which have been shown, in turn, using the properties of the heat kernel. Beside the properties of the Bessel potentials, the proof uses the functional calculus of $\mathcal{R}$ and the structure of G.

4.5.2 Proof of Hulanicki's theorem

This section is devoted to the proof of Theorem 4.5.1 and can be skipped at first reading.

We follow the essence of [Hul84], but we modify the original proof to take into account our presentation of the properties of Rockland operators as well as to bring some (small) simplifications. We also do not present some results obtained in [Hul84] on groups of polynomial growth. One of these simplifications is the fact that we fix a quasi-norm $|\cdot|$ which we assume to be a norm. Indeed, it is clear from the equivalence of quasi-norms (see Proposition 3.1.35) that it suffices to prove Hulanicki's theorem for one quasi-norm for it to hold for any quasi-norm. As a homogeneous norm exists by Theorem 3.1.39, we may assume that $|\cdot|$ is a norm without loss of generality. We could do without this but it simplifies the constants in the next pages.

First step

The first step in the proof can be summarised with the following lemma:

Lemma 4.5.3. *Let $m : [0,+\infty) \to \mathbb{C}$ be a function and let $\ell_o \in \mathbb{N}$. We define the function $F : (-\infty, 1) \to \mathbb{C}$ by*

$$F(\xi) := \begin{cases} m\left(\xi^{-\frac{1}{\ell_o}} - 1\right) & \text{if } 0 < \xi < 1, \\ 0 & \text{if } \xi \leq 0, \end{cases}$$

and we have

$$\forall \lambda \in [0,\infty) \qquad m(\lambda) = F\left((1+\lambda)^{-\ell_o}\right).$$

Furthermore, the following holds.

1. *The function F extends to a continuous function on $\mathbb{R}$ if and only if m is continuous on $[0,\infty)$ and $\lim_{\lambda\to+\infty} m(\lambda) = 0$.*
2. *The function F extends to a C^1 function on $\mathbb{R}$ if and only if m is C^1 on $[0,\infty)$ with $\lim_{\lambda\to+\infty} m(\lambda) = 0$ and $\lim_{\lambda\to+\infty}(1+\lambda)^{1+\ell_o} m'(\lambda) = 0$.*

   Let $k \in \mathbb{N}$. If $m \in C^k[0,+\infty)$ and

   $$\lim_{\lambda\to+\infty} (1+\lambda)^{1+j+k\ell_o} |m^{(j)}(\lambda)| = 0 \text{ for } j = 1, \ldots, k',$$

   then the function F extends to a function in $C^k(\mathbb{R})$
3. *Let $k \in \mathbb{N}$ and $m \in C^k[0,\infty)$. We assume that the suprema*

   $$\sup_{\lambda\geq 0}(1+\lambda)^{2+j+k\ell_o} |m^{(j)}(\lambda)|, \qquad j = 0, \ldots, k.$$

   are finite. Then we can construct an extension to $\mathbb{R}$, still denoted by F, such that the function $F \in C^k(\mathbb{R})$ is supported in $[0,2]$ and satisfies $\widehat{F}(0) = 0$ and for every $\ell \in \mathbb{Z}$,

   $$\left|\widehat{F}(\ell)\right| \leq C(1+|\ell|)^{-k} \sup_{\substack{\lambda\geq 0 \\ j=0,\ldots,k}} (1+\lambda)^{1+j+k\ell_o} |m^{(j)}(\lambda)|,$$

   where $C = C_{k,\ell_o}$ is a positive constant independent of m. Here $\widehat{F}(\ell)$, $\ell \in \mathbb{Z}$, denotes the Fourier coefficients of F in the sense of

   $$\widehat{F}(\ell) := \int_{-\pi}^{\pi} F(\xi) e^{-i\xi\ell} \frac{d\xi}{2\pi}.$$

Proof. Part (1) is easy to prove. Part (2) in the case of $k = 1$ follows easily from the following observations.

- If $\xi = (1+\lambda)^{-\ell_o}$, $\lambda > 0$ then
$$\frac{F(\xi) - F(0)}{\xi} = (1+\lambda)^{\ell_o} m(\lambda).$$

- We can compute formally for $\xi \in (0,1)$:
$$F'(\xi) = \frac{-\frac{1}{\ell_o}}{\xi^{\frac{1}{\ell_o}+1}} m'\left(\xi^{-\frac{1}{\ell_o}} - 1\right),$$
and in particular if $\xi = (1+\lambda)^{-\ell_o}$, $\lambda > 0$, then
$$F'(\xi) = -\frac{1}{\ell_o}(1+\lambda)^{1+\ell_o} m'(\lambda).$$

The general case of Part (2) follows from the following observation: $F^{(k')}(\xi)$ is a linear combination over $j = 1, \ldots, k'$ of
$$\xi^{-\frac{1}{\ell_o} - (k'-j) - j(\frac{1}{\ell_o}+1)} m^{(j)}\left(\xi^{-\frac{1}{\ell_o}} - 1\right) = \xi^{-\frac{1+j}{\ell_o} - k'} m^{(j)}\left(\xi^{-\frac{1}{\ell_o}} - 1\right).$$
The details are left to the reader.

Let us prove Part (3). Let $m \in C^k[0,\infty)$. Let P_k be the Taylor expansion of m at 0, that is, P_k is the polynomial of degree k such that we have for $\lambda > 0$ small,
$$m(\lambda) = P_k(\lambda) + o(|\lambda|^k).$$
We fix an arbitrary smooth function χ supported in $[0,2]$ and satisfying $\chi \equiv 1$ on $[0,1]$. We construct an extension of F, still denoted F, by setting
$$F(\xi) := \begin{cases} 0 & \text{if } \xi \leq 0, \\ m\left(\xi^{-\frac{1}{\ell_o}} - 1\right) & \text{if } 0 < \xi < 1, \\ P_k\left(\xi^{-\frac{1}{\ell_o}} - 1\right)\chi(\xi) & \text{if } \xi \geq 1. \end{cases}$$
We assume that the suprema given in the statement of Part 3 are finite. Clearly $F \in C^k(\mathbb{R})$ is supported in $[0,2]$. The proof of Part 2 implies easily
$$\|F^{(k')}\|_\infty \leq C \sum_{j=1}^{k'} \sup_{\lambda \geq 0} (1+\lambda)^{1+j+\ell_o k'} |m^{(j)}(\lambda)|, \tag{4.52}$$
where the constant $C = C_{k',\ell_o,\chi} > 0$ is independent on m.

The Fourier coefficient of F at 0 is

$$\begin{aligned}\widehat{F}(0) &= \int_{-\pi}^{\pi} F(\xi)\frac{d\xi}{2\pi} \\ &= \int_0^1 m(\xi^{-\frac{1}{\ell_o}}-1)\frac{d\xi}{2\pi} + \int_1^2 P_k\left(\xi^{-\frac{1}{\ell_o}}-1\right)\chi(\xi)\frac{d\xi}{2\pi} \\ &= \int_0^\infty m(\lambda)\frac{-\ell_o}{2\pi}\frac{d\lambda}{(1+\lambda)^{\ell_o+1}} + \int_1^2 P_k\left(\xi^{-\frac{1}{\ell_o}}-1\right)\chi(\xi)\frac{d\xi}{2\pi}.\end{aligned}$$

We can always assume that the function χ was chosen so that

$$\int_1^2 P_k\left(\xi^{-\frac{1}{\ell_o}}-1\right)\chi(\xi)\frac{d\xi}{2\pi} = \int_0^\infty m(\lambda)\frac{\ell_o}{2\pi}\frac{d\lambda}{(1+\lambda)^{\ell_o+1}}.$$

Indeed, it suffices to replace χ by $\chi + c\chi_1$ where $\chi_1 \in \mathcal{D}(\mathbb{R})$ is supported in $(1,2)$ and c a well chosen constant.

It is a simple exercise using integration by parts to show that the Fourier coefficients may be estimated by

$$\forall k' = 0,\ldots,k \quad \exists C = C_{k'} > 0 \quad \forall \ell \in \mathbb{Z} \qquad |\widehat{F}(\ell)| \leq C(1+|\ell|)^{-k'}\|F^{(k')}\|_\infty.$$

This together with (4.52) concludes the proof of Part (3). □

Second step

The second step consists in noticing that, with the notation of Lemma 4.5.3, studying the multiplier $m(\mathcal{R})$ and using the Fourier series of F leads to consider the operator $e^{i\ell(\mathrm{I}+\mathcal{R})^{-\ell_o}}$ and, more precisely, the properties of its convolution kernel.

Lemma 4.5.4. *Let $\mathcal{R}$ be a positive Rockland operator on a graded Lie group G. Let $\ell_o \in \mathbb{N}$ and $F_o(\xi) := e^{i\xi} - 1$, $\xi \in \mathbb{R}$. Then, for any $\ell \in \mathbb{Z}$, the convolution kernel of $F_o(\ell(\mathrm{I}+\mathcal{R})^{-\ell_o})$ is an integrable function:*

$$F_o(\ell(\mathrm{I}+\mathcal{R})^{-\ell_o})\delta_o \in L^1(G).$$

Proof of Lemma 4.5.4. Since $F_o(\ell\xi) = \sum_{j=1}^\infty \frac{(i\ell\xi)^j}{j!}$, we have at least formally

$$\begin{aligned}\kappa_\ell &:= \left\{F_o(\ell(\mathrm{I}+\mathcal{R})^{-\ell_o})\right\}\delta_o \\ &= \sum_{j=1}^\infty \frac{(i\ell)^j}{j!}(\mathrm{I}+\mathcal{R})^{-j\ell_o}\delta_0 = \sum_{j=1}^\infty \frac{(i\ell)^j}{j!}\mathcal{B}_{\nu j\ell_o},\end{aligned}$$

where $\mathcal{B}_a$ is the convolution kernel of the Bessel potentials, see Section 4.3.4, and ν is the degree of homogeneity of $\mathcal{R}$. In fact, by Corollary 4.3.11, we know that

$$\forall a \in \mathbb{C}_+ \quad \mathcal{B}_a \in L^1(G) \quad \text{and} \quad \mathcal{B}_{\nu j\ell_o} = \mathcal{B}_{\nu\ell_o} * \ldots * \mathcal{B}_{\nu\ell_o} := \mathcal{B}_{\nu\ell_o}^{*j}.$$

Thus in the Banach algebra $L^1(G)$ endowed with the convolution product, the series

$$\sum_{j=1}^{\infty} \frac{(i\ell)^j}{j!} \mathcal{B}_{\nu\ell_o}^{*j} = \sum_{j=1}^{\infty} \frac{(i\ell)^j}{j!} \mathcal{B}_{\nu j\ell_o},$$

is convergent in the L^1-norm. It is then a routine exercise to justify that κ_ℓ is in $L^1(G)$ and is also the convolution kernel of the multiplier $F_o(\ell(\mathrm{I}+\mathcal{R})^{-\ell_o})$. □

Unfortunately the brute force estimate

$$\|F_o(\ell(\mathrm{I}+\mathcal{R})^{-\ell_o})\delta_o\|_{L^1(G)} \leq \sum_{j=1}^{\infty} \frac{|\ell|^j}{j!} \|\mathcal{B}_{\nu\ell_o}\|_{L^1(G)}^j = \exp(\|\mathcal{B}_{\nu\ell_o}\|_{L^1(G)}|\ell|) - 1,$$

is exponential in ℓ and would be of no use for us. However, we notice that we can already modify the proof above to show:

Lemma 4.5.5. *We keep the notation of Lemma 4.5.4 and its proof. If* $|\cdot|$ *is a homogeneous quasi-norm on* G, *then for each* $\ell \in \mathbb{Z}$ *and* $a_o > 0$, *the function* $F_o(\ell(\mathrm{I}+\mathcal{R})^{-\ell_o})\delta_o$ *is integrable against a weight of the form* $(1+|\cdot|)^{a_o}$. *Moreover, if* $\nu\ell_o > Q/2$, *where* Q *is the homogeneous dimension of* G, *then the function* $F_o(\ell(\mathrm{I}+\mathcal{R})^{-\ell_o})\delta_o$ *is in* $L^2(G)$.

Proof of Lemma 4.5.4. One checks easily that if $\omega : G \to [1,\infty)$ is a continuous function satisfying

$$\exists C = C_\omega \quad \forall x, y \in G \qquad \omega(xy) \leq C\omega(x)\omega(y),$$

the subspace $L^1(w)$ of $L^1(G)$ of functions f which are integrable against w, is a Banach algebra for the norm

$$\|f\|_{L^1(\omega)} = \int_G |f(x)|\omega(x)dx.$$

Examples of such ω's are precisely weights of the form $(1+|\cdot|)^a$ with $|\cdot|$ being a quasi-norm on G. By Lemma 4.3.15, for any $a \in \mathbb{C}_+$ and $a_o > 0$,

$$\mathcal{B}_a \in L^1((1+|\cdot|)^{a_o}).$$

We keep the notation of the proof of Lemma 4.5.4 and proceed in the similar way but using $L^1((1+|\cdot|)^{a_o})$ instead of $L^1(G)$:

$$\|\kappa_\ell\|_{L^1((1+|\cdot|)^{a_o})} \leq \sum_{j=1}^{\infty} \frac{|\ell|^j}{j!} \|\mathcal{B}_{\nu\ell_o}\|_{L^1((1+|\cdot|)^{a_o})}^j = \exp(\|\mathcal{B}_{\nu\ell_o}\|_{L^1((1+|\cdot|)^{a_o})}|\ell|) - 1,$$

which is finite.

Let us now show that $\kappa_\ell \in L^2(G)$. We have

$$\|\kappa_\ell\|_2 \leq \sum_{j=1}^{\infty} \frac{|\ell|^j}{j!} \|\mathcal{B}_{\nu\ell_o}^{*j}\|_2$$

and

$$\|\mathcal{B}_{\nu\ell_o}^{*j}\|_2 \leq \|\mathcal{B}_{\nu\ell_o}\|_1 \|\mathcal{B}_{\nu\ell_o}^{*(j-1)}\|_2 \leq \ldots \leq \|\mathcal{B}_{\nu\ell_o}\|_1^{j-1} \|\mathcal{B}_{\nu\ell_o}\|_2.$$

We also know by Corollary 4.3.11 that $\mathcal{B}_a \in L^2(G)$ whenever $\operatorname{Re} a > Q/2$. Thus in this case,

$$\|\kappa_\ell\|_2 \leq \sum_{j=1}^{\infty} \frac{|\ell|^j}{j!} \|\mathcal{B}_{\nu\ell_o}\|_1^{j-1} \|\mathcal{B}_{\nu\ell_o}\|_2 < \infty,$$

finishing the proof. □

Using only the properties of Banach algebras, we obtain again that

$$\int_G |F_o(\ell(\mathrm{I}+\mathcal{R})^{-\ell_o})\delta_o|(1+|\cdot|)^{a_o} \quad \text{and} \quad \int_G |F_o(\ell(\mathrm{I}+\mathcal{R})^{-\ell_o})\delta_o|^2,$$

explode exponentially with ℓ which is not good enough for our subsequent analysis. However, we are going to show that $\int_G |F_o(\ell(\mathrm{I}+\mathcal{R})^{-\ell_o})\delta_o|(1+|\cdot|)^{a_o}$ actually grows polynomially in ℓ (see Lemma 4.5.6) and this is the crucial technical point in the proof of Theorem 4.5.1.

Main technical lemma

Lemma 4.5.6. *We keep the notation of Lemmata 4.5.4 and 4.5.5. We fix a homogeneous quasi-norm $|\cdot|$ on G and assume that $\nu\ell_o > Q/2$. Then for $a_o \geq 0$,*

$$\int_G (1+|x|)^{a_o} |F_o(\ell(\mathrm{I}+\mathcal{R})^{-\ell_o})\delta_o(x)| dx \leq C|\ell|^{3(a_o+\frac{Q}{2}+1)}, \tag{4.53}$$

where $C = C_{a_o,\ell_o,\mathcal{R},G,|\cdot|}$ is a positive constant independent of $\ell \in \mathbb{Z}$.

In the proof of Lemma 4.5.6 we will need the following easy lemma:

Lemma 4.5.7. *If $\mu_1, \ldots, \mu_{2m}$ are $2m$ measures in $M(G)$, then for any continuous function ϕ,*

$$\int_G \phi\, d\mu_1 * \ldots * \mu_{2m}$$
$$= \int_{G^m} \int_G \phi\; d\left\{\mu_2(y_1^{-1}\cdot)\right\} * \ldots * \left\{\mu_{2m}(y_m^{-1}\cdot)\right\} d\mu_1(y_1) \ldots d\mu_{2m-1}(y_m),$$

whenever the right or the left hand side is finite.

Here we have denoted by $\mu(y\,\cdot)$ the y-left-translated measure of a positive or complex Borel measure μ, that is, the measure given by

$$\int_{x\in G}\phi(x)d\mu_1(yx)=\int_G\phi(y^{-1}x)d\mu_1(x),\qquad \phi\in C_c(G).$$

Proof of Lemma 4.5.7. First let us observe that if $\mu_1,\ldots,\mu_m$ are m measures in $M(G)$, then, recursively, one can show readily that

$$\int_G\phi\,d(\mu_1*\ldots*\mu_m)=\int_{G^m}\phi(x_1\ldots x_m)\,d\mu_1(x_1)\ldots d\mu_m(x_m). \tag{4.54}$$

If $\mu_1,\ldots,\mu_{2m}$ are $2m$ measures in $M(G)$, then applying (4.54) for these $2m$ measures yields

$$\begin{aligned}\int_G&\phi\,d(\mu_1*\ldots*\mu_{2m})\\ &=\int_{G^{2m}}\phi(x_1x_2\ldots x_{2m-1}x_{2m})\\ &\qquad\qquad d\mu_1(x_1)\,d\mu_2(x_2)\ldots d\mu_{2m-1}(x_{2m-1})\,d\mu_{2m}(x_{2m})\\ &=\int_G\Bigg(\int_{G^m}\phi(x_2'x_4'\ldots x_{2m}')\\ &\qquad\qquad d\mu_2(x_1^{-1}x_2')d\mu_4(x_3^{-1}x_4')\ldots d\mu_{2m}(x_{2m-1}'x_{2m}^{-1})\Bigg)\\ &\qquad\qquad\qquad d\mu_1(x_1)\ldots d\mu_{2m-1}(x_{2m-1}),\end{aligned}$$

after the change of variables $x_2'=x_1x_2,\ldots,x_{2m}'=x_{2m-1}x_{2m}$. Using (4.54), we recognise an iterated convolution in the quantity in parenthesis. □

We now turn our attention to showing Lemma 4.5.6.

Proof of Lemma 4.5.6. By Theorem 3.1.39, we may assume that the homogeneous quasi-norm $|\cdot|$ is also a norm, that is, we assume that the triangle inequality holds with a constant $=1$. Moreover, we notice that it suffices to prove (4.53) with $\ell\in\mathbb{N}$. Indeed, as

$$\kappa_\ell:=F_o(\ell(\mathrm{I}+\mathcal{R})^{-\ell_o})\delta_0=e^{i\ell(\mathrm{I}+\mathcal{R})^{-\ell_o}}\delta_0-\delta_0,$$

we have

$$\bar{\kappa}_\ell=e^{-i\ell(\mathrm{I}+\bar{\mathcal{R}})^{-\ell_o}}\delta_0-\delta_0=F_o(-\ell(\mathrm{I}+\bar{\mathcal{R}})^{-\ell_o})\delta_0,$$

and $\kappa_0=0$. Since

$$F_o(\ell\xi)=e^{i\ell\xi}-1=(e^{i\xi})^\ell-1=(F_o(\xi)+1)^\ell-1,$$

we have in the unital Banach algebra $\mathbb{C}\delta_0\oplus L^1(G)$,

$$\kappa_\ell=(\kappa_1+\delta_o)^{*\ell}-\delta_0.$$

Let us fix $\ell \in \mathbb{N}$. We can decompose

$$\kappa_1 = f_0 + f_\infty,$$

where f_0 and f_∞ are the integrable functions defined via

$$f_0(x) := \kappa_1(x) 1_{|x| \leq \ell^2} \quad \text{and} \quad f_\infty(x) := \kappa_1(x) 1_{|x| > \ell^2}. \tag{4.55}$$

We can also write

$$\kappa_1 + \delta_0 = \mu_0 + f_\infty \qquad \text{where} \qquad \mu_0 := f_0 + \delta_0.$$

We now develop the non-commutative convolution product in $\mathbb{C}\delta_0 \oplus L^1(G)$,

$$\begin{aligned}
(\kappa_1 + \delta_o)^{*\ell} &= (\mu_0 + f_\infty)^{*\ell} = (\mu_0 + f_\infty) * \ldots * (\mu_0 + f_\infty) \\
&= \sum_{\alpha,\beta} \mu_0^{*\alpha_1} * f_\infty^{*\beta_1} * \ldots * \mu_0^{*\alpha_\ell} * f_\infty^{*\beta_\ell},
\end{aligned}$$

where the summation is over all sequences $\alpha = (\alpha_1, \ldots, \alpha_\ell)$, $\beta = (\beta_1, \ldots, \beta_\ell)$, of 0 and 1 such that $\alpha_1 + \ldots + \alpha_\ell + \beta_1 + \ldots + \beta_\ell = \ell$.

Let us fix two such sequences α and β. By Lemma 4.5.7 applied to

$$\mu_{2j-1} = \mu_0^{*\alpha_j}, \quad \text{and} \quad \mu_{2j} = f_\infty^{*\beta_j},$$

and $\phi = \omega^{a_o}$ where the positive function $\omega \in C(G)$ is defined by

$$\omega(x) := 1 + |x|, \quad x \in G,$$

we have for any $a_o > 0$,

$$\begin{aligned}
&\int_G \omega^{a_o} d\mu_0^{*\alpha_1} * f_\infty^{*\beta_1} * \ldots * \mu_0^{*\alpha_\ell} * f_\infty^{*\beta_\ell} \\
&\quad = \int_{G^\ell} \int_G \omega^{a_o} \, d\left\{\mu_0^{*\alpha_1}(y_1^{-\beta_1} \cdot)\right\} * \ldots * \left\{\mu_0^{*\alpha_\ell}(y_\ell^{-\beta_\ell} \cdot)\right\} \\
&\qquad\qquad\qquad df_\infty^{*\beta_1}(y_1) \ldots df_\infty^{*\beta_\ell}(y_\ell).
\end{aligned}$$

Let us also fix $(y_1, \ldots, y_\ell) \in G^\ell$. We notice that for any $\mu \in M(G)$ we have

$$(\delta_0(y\,\cdot)) * \mu = \mu(y\,\cdot),$$

since for any $\phi \in C_c(G)$,

$$\int_G \phi \, d\{(\delta_0(y\,\cdot)) * \mu\} = \int_G \phi(xz) \, d\delta_0(yx) \, d\mu(z) = \int_G \phi(y^{-1}x) \, d\mu(x).$$

Therefore, if $\alpha_1 = 0$,

$$\begin{aligned}\left\{\mu_0^{*\alpha_1}(y_1^{-\beta_1}\cdot)\right\} * \left\{\mu_0^{*\alpha_2}(y_2^{-\beta_2}\cdot)\right\} &= \left\{\delta_0(y_1^{-\beta_1}\cdot)\right\} * \left\{\mu_0^{*\alpha_2}(y_2^{-\beta_2}\cdot)\right\} \\ &= \left\{\mu_0^{*\alpha_2}(y_2^{-\beta_1}\cdot)\right\}(y_1^{-\beta_1}\cdot) \\ &= \left\{\mu_0^{*\alpha_2}(y_2^{-\beta_1}y_1^{-\beta_1}\cdot)\right\},\end{aligned}$$

whereas if $\alpha_1 = 1$, we have

$$\begin{aligned}&\left\{\mu_0^{*\alpha_1}(y_1^{-\beta_1}\cdot)\right\} * \left\{\mu_0^{*\alpha_2}(y_2^{-\beta_2}\cdot)\right\} \\ &\quad = \left\{\delta_0(y_1^{-\beta_1}\cdot)\right\} * \left\{\mu_0^{*\alpha_2}(y_2^{-\beta_2}\cdot)\right\} + \left\{f_0(y_1^{-\beta_1}\cdot)\right\} * \left\{\mu_0^{*\alpha_2}(y_2^{-\beta_2}\cdot)\right\} \\ &\quad = \left\{\mu_0^{*\alpha_2}(y_2^{-\beta_1}y_1^{-\beta_1}\cdot)\right\} + \left\{f_0(y_1^{-\beta_1}\cdot)\right\} * \left\{\mu_0^{*\alpha_2}(y_2^{-\beta_2}\cdot)\right\}.\end{aligned}$$

Recursively, we find that

$$\begin{aligned}&\left\{\mu_0^{*\alpha_1}(y_1^{-\beta_1}\cdot)\right\} * \ldots * \left\{\mu_0^{*\alpha_\ell}(y_\ell^{-\beta_\ell}\cdot)\right\} \\ &\quad = \sum_{j:\alpha_j=1} \left\{f_0(y_1^{-\beta_1}\ldots y_j^{-\beta_j}\cdot)\right\} * \left\{\mu_0^{*\alpha_{j+1}}(y_{j+1}^{-\beta_{j+1}}\cdot)\right\} * \ldots * \left\{\mu_0^{*\alpha_\ell}(y_\ell^{-\beta_\ell}\cdot)\right\}\end{aligned}$$

when $\alpha \neq 0$. If $\alpha = 0$, we compute directly:

$$\begin{aligned}\left\{\mu_0^{*\alpha_1}(y_1^{-\beta_1}\cdot)\right\} * \ldots * \left\{\mu_0^{*\alpha_\ell}(y_\ell^{-\beta_\ell}\cdot)\right\} &= \left\{\delta_o(y_1^{-\beta_1}\cdot)\right\} * \ldots * \left\{\delta_o(y_\ell^{-\beta_\ell}\cdot)\right\} \\ &= \delta_o\left(y_\ell^{-\beta_\ell}\ldots y_1^{-\beta_1}\cdot\right),\end{aligned}$$

so that

$$\begin{aligned}&\int_G \omega^{a_o}\, d\left\{\mu_0^{*\alpha_1}(y_1^{-\beta_1}\cdot)\right\} * \ldots * \left\{\mu_0^{*\alpha_\ell}(y_\ell^{-\beta_\ell}\cdot)\right\} \\ &\quad = \omega^{a_o}(y_1^{\beta_1}\ldots y_\ell^{\beta_\ell}) = \left(1 + |y_1^{\beta_1}\ldots y_\ell^{\beta_\ell}|\right)^{a_o} \\ &\quad \leq \left(1 + |y_1^{\beta_1}| + \ldots + |y_\ell^{\beta_\ell}|\right)^{a_o} \leq \left(1 + \ell \max_{j'=1,\ldots,\ell} |y_{j'}^{\beta_{j'}}|\right)^{a_o},\end{aligned}$$

since the quasi-norm $|\cdot|$ is assumed to be a norm.

If $\alpha_j \neq 0$, we notice that the measure given by

$$\left\{f_0(y_1^{-\beta_1}\ldots y_j^{-\beta_j}\cdot)\right\} * \left\{\mu_0^{*\alpha_{j+1}}(y_{j+1}^{-\beta_{j+1}}\cdot)\right\} * \ldots * \left\{\mu_0^{*\alpha_\ell}(y_\ell^{-\beta_\ell}\cdot)\right\}, \tag{4.56}$$

is compactly supported. Indeed recall that f_0 is supported in $\bar{B}_{\ell^2}$ where we denote by $\bar{B}_R := \{x \in G : |x| \leq R\}$ a closed ball of radius R about 0 for the chosen norm. Therefore

$$\operatorname{supp} \mu_0 \subseteq \{0\} \cup \operatorname{supp} f_0 \subseteq \bar{B}_{\ell^2}.$$

The general properties

$$\forall \mu_1, \mu_2 \in M(G) \quad \text{supp}(\mu_1 * \mu_2) \subset (\text{supp}\mu_1)(\text{supp}\mu_2) = \{x_1 x_2 \ : \ x_1 \in \text{supp}\mu_1, \ x_2 \in \text{supp}\mu_2\},$$

and

$$\forall \mu \in M(G), \ y \in G \qquad \text{supp}\left(\mu(\cdot\, y^{-1})\right) = (\text{supp}\mu)y = \{xy \ : \ x \in \text{supp}\mu\},$$

imply that the measure in (4.56) is supported in

$$\bar{B}_{\alpha_1 \ell^2} y_1^{\beta_1} \dots \bar{B}_{\alpha_{j-1}\ell^2} y_{j-1}^{\beta_{j-1}} \ \bar{B}_{\alpha_\ell \ell^2} y_\ell^{\beta_\ell} \dots y_j^{\beta_j}.$$

Using the properties of the norm $|\cdot|$, it is easy to check that the above subset of G is included in the closed ball about 0 of radius

$$(\sum_{j'=1}^{\ell} \alpha_{j'})\ell^2 + (\sum_{j'=1}^{\ell} \beta_{j'} |y_{j'}|) \le \ell^3 + \ell \max_{j'=1,\dots,\ell} |y_{j'}|,$$

since $\sum_{j'=1}^{\ell} \alpha_{j'}$ and $\sum_{j'=1}^{\ell} \beta_{j'}$ are $\le \ell$.

Note that if f is a measurable function supported in $\bar{B}_R$ then

$$\begin{aligned} \left|\int_G \omega^{a_o}(x) f(x) dx\right| &\le \int_{|x|\le R} (1+|x|)^{a_o} |f(x)| dx \\ &\le (1+R)^{a_o} \int_{|x|\le R} f(x) dx \\ &\le (1+R)^{a_o} |\bar{B}_R|^{1/2} \|f\|_2. \end{aligned}$$

We apply this to the function f in (4.56) and $R = \ell^3 + \ell \max_{j'=1,\dots,\ell} |y_{j'}|$. This leads us to look at the L^2-norm of this function which can be written as

$$f = T_{\mu_0^{*\alpha_\ell}(y_\ell^{-\beta_\ell}\,\cdot)} \dots T_{\mu_0^{*\alpha_{j+1}}(y_{j+1}^{-\beta_{j+1}}\,\cdot)} \left(f_0(y_1^{-\beta_1} \dots y_j^{-\beta_j}\,\cdot) \right).$$

Here we have used our usual notation for the convolution operator $T_\kappa : \phi \mapsto \phi * \kappa$ with right-convolution (distributional) kernel κ. Since such convolution operators are left-invariant, we have

$$\|T_{\mu_0^{*\alpha_{j'}}(y_{j'}^{-\beta_{j'}}\,\cdot)}\|_{\mathscr{L}(L^2(G))} = \|T_{\mu_0^{*\alpha_{j'}}}\|_{\mathscr{L}(L^2(G))} = \|T_{\mu_0}\|_{\mathscr{L}(L^2(G))}^{\alpha_{j'}}.$$

Again by left-invariance,

$$\left\| f_0(y_1^{-\beta_1} \dots y_j^{-\beta_j}\,\cdot) \right\|_{L^2(G)} = \|f_0\|_{L^2(G)} \le \|\kappa_1\|_{L^2(G)},$$

which is finite by Lemma 4.5.5 since we assume $\nu \ell_o > Q/2$. Therefore,

$$\|f\|_2 \le \|\kappa_1\|_{L^2(G)} \|T_{\mu_0}\|_{\mathscr{L}(L^2(G))}^{\sum_{j<j'\le \ell} \alpha_{j'}}.$$

We have obtained that for any $(y_1,\ldots,y_\ell)\in G^\ell$, α,β with $\alpha\neq 0$,

$$\begin{aligned}
&\left|\int_G \omega^{a_o} d\left\{f_0(y_1^{-\beta_1}\ldots y_j^{-\beta_j}\cdot)\right\}*\left\{\mu_0^{*\alpha_{j+1}}(y_{j+1}^{-\beta_{j+1}}\cdot)\right\}*\ldots*\left\{\mu_0^{*\alpha_\ell}(y_\ell^{-\beta_\ell}\cdot)\right\}\right|\\
&\leq \sum_{j:\alpha_j=1}\left(1+\ell^3+\ell\max_{j'=1,\ldots,\ell}|y_{j'}|\right)^{a_o}|\bar{B}_{\ell^3+\ell\max_{j'=1,\ldots,\ell}|y_{j'}|}|^{1/2}\\
&\qquad\qquad \|T_{\mu_0}\|_{\mathscr{L}(L^2(G))}^{\sum_{j<j'\leq\ell}\alpha_{j'}}\|\kappa_1\|_{L^2(G)}\\
&\leq C\ell\left(1+\ell^3+\ell\max_{j'=1,\ldots,\ell}|y_{j'}|\right)^{a_o+Q/2}\|T_{\mu_0}\|_{\mathscr{L}(L^2(G))}^{\sum_{j<j'\leq\ell}\alpha_{j'}},
\end{aligned}$$

with $C=\|\kappa_1\|_{L^2(G)}|\bar{B}_1|^{1/2}$. We now integrate against $df_\infty^{*\beta_1}(y_1)\ldots df_\infty^{*\beta_\ell}(y_\ell)$ to obtain

$$\begin{aligned}
&\left|\int_G \omega^{a_o} d\mu_0^{*\alpha_1}*f_\infty^{*\beta_1}*\ldots*\mu_0^{*\alpha_\ell}*f_\infty^{*\beta_\ell}\right|\\
&\leq \int_{G^\ell}\left|\int_G \omega^{a_o}\, d\left\{\mu_0^{*\alpha_1}(y_1^{-\beta_1}\cdot)\right\}*\ldots*\left\{\mu_0^{*\alpha_\ell}(y_\ell^{-\beta_\ell}\cdot)\right\}\right|\\
&\qquad\qquad d|f_\infty^{*\beta_1}|(y_1)\ldots d|f_\infty^{*\beta_\ell}|(y_\ell)\\
&\leq C\|T_{\mu_0}\|_{\mathscr{L}(L^2(G))}^{\sum_{j<j'\leq\ell}\alpha_{j'}}\\
&\qquad \int_{G^\ell}\ell\left(1+\ell^3+\ell\max_{j'=1,\ldots,\ell}|y_{j'}|\right)^{a_o+Q/2} d|f_\infty^{*\beta_1}|(y_1)\ldots d|f_\infty^{*\beta_\ell}|(y_\ell).
\end{aligned}$$

Let us estimate this last integral:

$$\begin{aligned}
&\int_{G^\ell}\ell\left(1+\ell^3+\ell\max_{j'=1,\ldots,\ell}|y_{j'}|\right)^{a_o+Q/2} d|f_\infty^{*\beta_1}|(y_1)\ldots d|f_\infty^{*\beta_\ell}|(y_\ell)\\
&\leq \sum_{q=0}^{\infty}(1+\ell^3+\ell(q+1))^{a_o+Q/2}\int_{q\leq\max_{j'}|y_{j'}|<q+1} d|f_\infty^{*\beta_1}|(y_1)\ldots d|f_\infty^{*\beta_\ell}|(y_\ell).
\end{aligned}$$

For each of these integrals, we see that either

$$\int_{q\leq\max_{j'}|y_{j'}|<q+1} d|f_\infty^{*\beta_1}|(y_1)\ldots d|f_\infty^{*\beta_\ell}|(y_\ell)=\begin{cases}0 & \text{if } \beta=0,\ q\geq 1 \text{ or } \beta\neq 0,\ q<\ell^2,\\ 1 & \text{if } \beta=0,\ q=0,\end{cases}$$

or if $\beta\neq 0$ and $q\geq\ell^2$,

$$\begin{aligned}
&\int_{q\leq\max_{j'}|y_{j'}|<q+1} d|f_\infty^{*\beta_1}|(y_1)\ldots d|f_\infty^{*\beta_\ell}|(y_\ell)\leq \|f_\infty\|_1^{|\beta|-1}\int_{q\leq|y|\leq q+1}|\kappa_1(y)|dy\\
&\leq \|f_\infty\|_1^{|\beta|-1}(1+q)^{-a}\int_G|\kappa_1|\omega^a,
\end{aligned}$$

for any $a > 0$ that will be suitably chosen. Indeed, by Lemma 4.5.5, this last integral is finite. Hence if $\beta \neq 0$, then

$$\int_{G^\ell} \ell \left(1 + \ell^3 + \ell \max_{j'=1,\ldots,\ell} |y_{j'}|\right)^{a_o + Q/2} d|f_\infty^{*\beta_1}|(y_1) \ldots d|f_\infty^{*\beta_\ell}|(y_\ell)$$
$$\leq \|f_\infty\|_1^{|\beta|-1} \left(\int_G |\kappa_1|\omega^a\right) \sum_{q=\ell^2}^{\infty} \ell \left(1 + \ell^3 + \ell(q+1)\right)^{a_o+Q/2} (1+q)^{-a}.$$

We choose $a := 2(a_o + Q/2 + 2)$ so that the sum

$$\ell \sum_{q=\ell^2}^{\infty} \left(1 + \ell^3 + \ell(q+1)\right)^{a_o+Q/2} (1+q)^{-a} \leq \ell(1+2\ell)^{a_o+Q/2} \sum_{q=\ell^2}^{\infty} (q+1)^{a_o+Q/2-a}$$
$$\leq \ell(1+2\ell)^{a_o+Q/2} \int_{\ell^2}^{\infty} x^{a_o+Q/2-a} dx \leq \ell(1+2\ell)^{a_o+Q/2} \frac{\ell^{2(a_o+Q/2-a+1)}}{a_o + Q/2 - a + 1},$$

is finite and bounded independently of ℓ. We set

$$C_a := \max\left(1, \left(\int_G |\kappa_1|\omega^a\right) \max_{\ell\in\mathbb{N}} \sum_{q=\ell^2}^{\infty} \left(1 + \ell^3 + \ell(q+1)\right)^{a_o+Q/2} (1+q)^{-a}\right).$$

We have obtained in the case α and β both non-zero:

$$\left|\int_G \omega^{a_o} d\mu_0^{*\alpha_1} * f_\infty^{*\beta_1} * \ldots * \mu_0^{*\alpha_\ell} * f_\infty^{*\beta_\ell}\right|$$
$$\leq CC_a \sum_{j:\alpha_j=1} \|T_{\mu_0}\|_{\mathscr{L}(L^2(G))}^{\sum_{j<j'\leq\ell}\alpha_{j'}} \|f_\infty\|_1^{|\beta|-1}$$
$$\leq CC_a \ell \max(1, \|T_{\mu_0}\|_{\mathscr{L}(L^2(G))})^{|\alpha|} \|f_\infty|_1^{|\beta|-1},$$

whereas in the case $\alpha \neq 0$ and $\beta = 0$,

$$\left|\int_G \omega^{a_o} d\mu_0^{*\alpha_1} * f_\infty^{*\beta_1} * \ldots * \mu_0^{*\alpha_\ell} * f_\infty^{*\beta_\ell}\right| \leq C \sum_{j:\alpha_j=1} \|T_{\mu_0}\|_{\mathscr{L}(L^2(G))}^{\sum_{j<j'\leq\ell}\alpha_{j'}} (1+\ell^3)^{a_o+Q/2}$$
$$\leq C(1+\ell^3)^{a_o+Q/2} \ell \max(1, \|T_{\mu_0}\|_{\mathscr{L}(L^2(G))})^{|\alpha|},$$

and in the case $\beta \neq 0$ and $\alpha = 0$,

$$\left|\int_G \omega^{a_o} d\mu_0^{*\alpha_1} * f_\infty^{*\beta_1} * \ldots * \mu_0^{*\alpha_\ell} * f_\infty^{*\beta_\ell}\right|$$
$$\leq \int_{G^\ell} \left(1 + \ell \max_{j'=1,\ldots,\ell} |y_{j'}^{\beta_{j'}}|\right)^{a_o} d|f_\infty^{*\beta_1}|(y_1) \ldots d|f_\infty^{*\beta_\ell}|(y_\ell)$$
$$\leq C \sum_{q=\ell^2}^{\infty} (1+\ell q)^{a_o} \|f_\infty\|_1^{|\beta|-1} \left(\int_G |\kappa_1|\omega^a\right) (1+q)^{-a}$$
$$\leq CC_a \|f_\infty\|_1^{|\beta|-1}.$$

We can now sum over α and β to obtain

$$\begin{aligned}\int_G |\kappa_\ell|\omega^{a_o} \leq & \sum_{\beta\neq 0} CC_a \|f_\infty\|_1^{|\beta|-1} \\ & + \sum_{\alpha\neq 0} C(1+\ell^3)^{a_o+Q/2}\ell \max(1, \|T_{\mu_0}\|_{\mathscr{L}(L^2(G))})^{|\alpha|} \\ & + \sum_{\substack{\alpha\neq 0\\ \beta\neq 0}} CC_a\ell \max(1, \|T_{\mu_0}\|_{\mathscr{L}(L^2(G))})^{|\alpha|} \|f_\infty\|_1^{|\beta|-1}.\end{aligned}$$

We need to estimate the operator norm of

$$T_{\mu_0} = \mathrm{I} + T_{\kappa_1} - T_{f_\infty}.$$

By functional calculus, the operator

$$\mathrm{I} + T_{\kappa_1} = \mathrm{I} + F_o((\mathrm{I}+\mathcal{R})^{-\ell_o}) = \exp(i(\mathrm{I}+\mathcal{R})^{-\ell_o}),$$

has norm

$$\begin{aligned}\|\mathrm{I} + T_{\kappa_1}\|_{\mathscr{L}(L^2(G))} &= \|\exp(i(\mathrm{I}+\mathcal{R})^{-\ell_o})\|_{\mathscr{L}(L^2(G))} \\ &\leq \sup_{\lambda\geq 0} |\exp(i(1+\lambda)^{-\ell_o})| \leq 1.\end{aligned}$$

For T_{f_∞}, we have

$$\|T_{f_\infty}\|_{\mathscr{L}(L^2(G))} \leq \|f_\infty\|_{L^1(G)}.$$

Hence

$$\|T_{\mu_o}\|_{\mathscr{L}(L^2(G))} \leq \|\mathrm{I} + T_{\kappa_1}\|_{\mathscr{L}(L^2(G))} + \|T_{f_\infty}\|_{\mathscr{L}(L^2(G))} \leq 1 + \|f_\infty\|_{L^1(G)},$$

and

$$\|f_\infty\|_1 + \max(1, \|T_{\mu_0}\|_{\mathscr{L}(L^2(G))}) \leq 1 + 2\|f_\infty\|_1.$$

Let us also estimate

$$\begin{aligned}\|f_\infty\|_{L^1(G)} &= \int_{|x|\geq \ell^2} |\kappa_1(x)|\omega(x)\frac{1}{1+|x|}dx \\ &\leq \frac{1}{1+\ell^2}\int_G |\kappa_1|\omega. \qquad (4.57)\end{aligned}$$

By Lemma 4.5.5,

$$c' := \int_G |\kappa_1|\omega$$

is finite. Thus we have obtained so far that

$$\begin{aligned}
\int_G |\kappa_\ell|\omega^{a_o} &\leq CC_a\ell \sum_{\beta\neq 0,\alpha} (1+2\|f_\infty\|_1)^{|\alpha|}\|f_\infty\|_1^{|\beta|-1} \\
&\quad +CC_a(1+\ell^3)^{a_o+\frac{Q}{2}+1}\sum_{\alpha\neq 0}(1+2\|f_\infty\|_1)^{|\alpha|} \\
&\leq CC_a\ell \sum_{\beta\neq 0,\alpha}\left(1+\frac{2c'}{1+\ell^2}\right)^{|\alpha|}\left(\frac{c'}{1+\ell^2}\right)^{|\beta|-1} \\
&\quad +CC_a(1+\ell^3)^{a_o+\frac{Q}{2}+1}\sum_{\alpha\neq 0}\left(1+\frac{2c'}{1+\ell^2}\right)^{|\alpha|} \\
&\leq CC_a(1+\ell^3)^{a_o+\frac{Q}{2}+1}\sum_{\beta,\alpha}\left(1+\frac{2c'}{1+\ell^2}\right)^{|\alpha|}\left(\frac{c'}{1+\ell^2}\right)^{|\beta|} \\
&= CC_a(1+\ell^3)^{a_o+\frac{Q}{2}+1}\left(1+\frac{3c'}{1+\ell^2}\right)^{\ell}.
\end{aligned}$$

Since

$$\max_{\ell\in\mathbb{N}}\left(1+\frac{3c'}{1+\ell^2}\right)^{\ell}<\infty,$$

we have proved what we wanted, that is,

$$\int_G |\kappa_\ell|\omega^{a_o}\leq C'(1+\ell^3)^{a_o+\frac{Q}{2}+1},$$

completing the proof of Lemma 4.5.6. □

The proof of Lemma 4.5.6 can be modified to obtain a similar property for $X^{\alpha_o}F_o(\ell(\mathrm{I}+\mathcal{R})^{-\ell_o})\delta_o$:

Lemma 4.5.8. *We keep the notation of Lemmata 4.5.4, 4.5.5 and 4.5.6. Then for any $\alpha_o\in\mathbb{N}_0^n$ with $\nu\ell_o>[\alpha_o]+Q/2$ and $a_o\geq 0$, we have*

$$\int_G (1+|x|)^{a_o}|X^{\alpha_o}F_o(\ell(\mathrm{I}+\mathcal{R})^{-\ell_o})\delta_o(x)|dx\leq C|\ell|^{3(a_o+\frac{Q}{2}-1)}, \tag{4.58}$$

where $C=C_{\alpha_o,a_o,\ell_o,\mathcal{R},G,|\cdot|}$ is a positive constant independent of $\ell\in\mathbb{Z}$.

The proof of Lemma 4.5.8 follows the one of Lemma 4.5.6 but with the derivatives X^{α_o} now applied to the last term of the convolution products. These convolution products have to be understood as convolutions between compactly supported distributions and integrable functions since we replace the definition of f_0 and f_∞ given in (4.55) by

$$f_0:=\kappa_1\chi(\ell^{-1}|\cdot|')\quad\text{and}\quad f_\infty:=\kappa_1(1-\chi)(\ell^{-1}|\cdot|'),$$

where

- $\chi \in \mathcal{D}(\mathbb{R})$ is a non-negative smooth function satisfying $\chi(x) = 1$ if $x \in [-1, 1]$ and $\chi(x) = 0$ if $x > 2$,
- and $|\cdot|'$ is a homogeneous norm which is smooth away from 0 (see Proposition 3.1.35 (i) or Remark 3.1.36).

We also need to replace $\omega = 1 + |\cdot|^{a_o}$ with

$$\omega(x) = \begin{cases} 1 & \text{if } |x| \leq 1, \\ |x| & \text{if } |x| \geq 1, \end{cases}$$

in order to make sense of the distribution $X^{\alpha_o}\delta_0$ with support at 0 being applied to the function ω which is smooth around 0. The conditions on the parameters given in Lemma 4.5.8 and the following lemma ensure that the new quantities $\int_G |X^{\alpha_o}\kappa|^2$ and $\int_G |X^{\alpha_o}\kappa|\omega^a$ appearing in the proof are finite.

Lemma 4.5.9. *Let $\mathcal{R}$ be a Rockland operator of homogeneous degree ν on a graded Lie group G and let $\mathcal{B}_a$, $a \in \mathbb{C}_+$, be the kernels of its Bessel potentials.*

1. *If $|\cdot|$ is a homogeneous quasi-norm on G, $b \geq 0$ and $\alpha \in \mathbb{N}_0^n$ with $\text{Re}\, a + b > [\alpha]$, then*

$$\int_G |x|^b |X^\alpha \mathcal{B}_a(x)| dx < \infty.$$

2. *If $\alpha \in \mathbb{N}_0^n$ with $\text{Re}\, a > [\alpha] + Q/2$, then*

$$\int_G |X^\alpha \mathcal{B}_a(x)|^2 dx < \infty.$$

Proof of Lemma 4.5.9. For the first part, we generalise the first part of Lemma 4.3.15, that is,

$$\int_G |x|^b |X^\alpha \mathcal{B}_a(x)| dx \leq \frac{1}{|\Gamma(\frac{a}{\nu})|} \int_0^\infty t^{\frac{\text{Re}\, a}{\nu} - 1} e^{-t} \int_G |x|^b |X^\alpha h_t(x)| dx\, dt,$$

and using the homogeneity of the heat kernel (see (4.17)) and the change of variables $y = t^{-\frac{1}{\nu}} x$, we get

$$\int_G |x|^b |X^\alpha h_t(x)| dx = \int_G |t^{\frac{1}{\nu}} y|^b t^{-\frac{[\alpha]}{\nu}} |X^\alpha h_1(y)| dy = c_{b,\alpha} t^{\frac{b-[\alpha]}{\nu}},$$

where $c_{b,\alpha} = \| |y|^b X^\alpha h_1(y) \|_{L^1(dy)}$ is a finite constant since $h_1 \in \mathcal{S}(G)$. Thus,

$$\int_G |x|^b |X^\alpha \mathcal{B}_a(x)| dx \leq \frac{c_{b,\alpha}}{|\Gamma(\frac{a}{\nu})|} \int_0^\infty t^{\frac{\text{Re}\, a}{\nu} - 1 + \frac{b-[\alpha]}{\nu}} e^{-t} dt$$

is finite whenever $\text{Re}\, a + b - [\alpha] > 0$.

For the second part of Lemma 4.5.9, we can not adapt the proof of the square integrability of $\mathcal{B}_a$ since we can not relate $X^\alpha \bar{h}(x)$ with $(X^\alpha h)(x^{-1})$. We proceed differently. Let $\phi \in \mathcal{D}(G)$. From the properties of the heat kernel and the definition of $\mathcal{B}_a$ we have

$$\mathcal{B}_a(x^{-1}) = \bar{\mathcal{B}}_a(x) = \overline{\mathcal{B}_{\bar{a}}(x)}$$

and so

$$\begin{aligned} \left|\int_G (X^\alpha \mathcal{B}_a)\phi\right| &= \left|\int_G \mathcal{B}_a X^\alpha \phi\right| \\ &= \left|(X^\alpha \phi) * \bar{\mathcal{B}}_{\bar{a}}(0)\right| \\ &\leq \left\|(X^\alpha \phi) * \bar{\mathcal{B}}_{\bar{a}}\right\|_\infty \\ &\leq C \left\|\left(\mathrm{I} + \bar{\mathcal{R}}_2\right)^{\frac{s}{\nu}} (X^\alpha \phi) * \bar{\mathcal{B}}_{\bar{a}}\right\|_2, \end{aligned}$$

by the Sobolev embeddings, see Theorem 4.4.25 for $s > Q/2$. But, since $\bar{\mathcal{B}}_{\bar{a}}$ is the right-convolution kernel of the Bessel potential $(\mathrm{I} + \bar{\mathcal{R}}_2)^{\frac{-\bar{a}}{\nu}}$, we have

$$\begin{aligned} \left\|\left(\mathrm{I} + \bar{\mathcal{R}}_2\right)^{\frac{s}{\nu}} (X^\alpha \phi) * \bar{\mathcal{B}}_{\bar{a}}\right\|_2 &= \left\|\left(\mathrm{I} + \bar{\mathcal{R}}_2\right)^{\frac{s-\mathrm{Re}\, a}{\nu}} (X^\alpha \varphi)\right\|_2 \\ &\leq C' \left\|\left(\mathrm{I} + \bar{\mathcal{R}}_2\right)^{\frac{s-\mathrm{Re}\, a+[\alpha]}{\nu}} \phi\right\|_2, \end{aligned}$$

by Theorem 4.4.16. Hence we have obtained

$$\left|\int_G X^\alpha \mathcal{B}_a \phi\right| \leq CC' \left\|\left(\mathrm{I} + \bar{\mathcal{R}}_2\right)^{\frac{s-\mathrm{Re}\, a+[\alpha]}{\nu}} \phi\right\|_2 \leq C_1 \|\phi\|_2,$$

if $s - \mathrm{Re}\, a + [\alpha] \leq 0$, from the Sobolev inclusions, see Theorem 4.4.3 (4). Hence, under this condition, $X^\alpha \mathcal{B}_a \in L^2(G)$. □

Proceeding as in the proof of Lemma 4.5.5, Lemma 4.5.9 yields

- $X^{\alpha_o} \kappa_\ell \in L^1((1 + |\cdot|)^{a_o})$ if $\nu \ell_o + a_o > [\alpha_o]$,
- and $X^{\alpha_o} \kappa_\ell \in L^2(G)$ if $\nu \ell_o > [\alpha_o] + Q/2$.

The details of the proof of Lemma 4.5.8 are left to the interested reader.

Last step

We can now conclude the proof of Theorem 4.5.1.

End of the proof of Theorem 4.5.1. We fix $a_o \in \mathbb{N}$ and $\alpha_o \in \mathbb{N}_0^n$. We consider $k \in \mathbb{N}$ and ℓ_o to be chosen in terms of a_o and α_o. Let $m \in C^k[0, \infty)$ satisfying the hypotheses of Lemma 4.5.3 Part 3 for ℓ_o and k, and let F be the corresponding function in $C^k(\mathbb{R})$. Since $\operatorname{supp} F \subset [0,2]$, we may develop F in the Fourier series:

$$F(\xi) = \sum_{\ell\in\mathbb{Z}} \widehat{F}(\ell) e^{i\xi\ell} = \sum_{\ell\in\mathbb{Z}\backslash\{0\}} \widehat{F}(\ell) e^{i\xi\ell}, \quad \xi \in (-\pi,\pi),$$

as $\widehat{F}(0) = 0$. Since

$$0 = F(0) = \sum_{\ell\in\mathbb{Z}\backslash\{0\}} \widehat{F}(\ell),$$

we also have for any $\xi \in [0,2]$ that

$$F(\xi) = \sum_{\ell\in\mathbb{Z}\backslash\{0\}} \widehat{F}(\ell)(e^{i\xi\ell} - 1).$$

This yields

$$m(\lambda) = \sum_{\ell\in\mathbb{Z}\backslash\{0\}} \widehat{F}(\ell)(e^{i\ell(1+\lambda)^{-\ell_o}} - 1), \quad \lambda \geq 0,$$

and at least formally

$$m(\mathcal{R})\delta_o = \sum_{\ell\in\mathbb{Z}\backslash\{0\}} \widehat{F}(\ell)\kappa_\ell, \tag{4.59}$$

where $\kappa_\ell := e^{i\ell(\mathrm{I}+\mathcal{R})^{-\ell_o}}\delta_0 - \delta_0$ as before.

By Lemma 4.5.8, if $\nu\ell_o > [\alpha_o] + Q/2$, then

$$\begin{aligned}
&\sum_{\ell\in\mathbb{Z}\backslash\{0\}} |\widehat{F}(\ell)| \int_G (1+|x|)^{a_o} |X^{\alpha_o}\kappa_\ell(x)| dx \\
&\quad \leq C \sum_{\ell\in\mathbb{Z}\backslash\{0\}} |\widehat{F}(\ell)||\ell|^{3(a_o+\frac{Q}{2}+1)} \\
&\quad \leq CC_k \sup_{\substack{\lambda\in\mathbb{R}\\ j=0,\dots,k}} (1+|\lambda|)^{1+j+k\ell_o} |m^{(j)}(\lambda)| \sum_{\ell\in\mathbb{Z}\backslash\{0\}} (1+|\ell|)^{-k}|\ell|^{3(a_o+\frac{Q}{2}+1)},
\end{aligned}$$

see Lemma 4.5.3 for the estimates of $|\widehat{F}(\ell)|$. This last sum converges provided that we have chosen $k > 3(a_o + \frac{Q}{2} + 1) + 2$. We assume that we have chosen such ℓ_o and k. One can now show easily that $m(\mathcal{R})\delta_o \in L^1(G)$ and that (4.59) holds in $L^1(G)$. Furthermore, $X^{\alpha_o} m(\mathcal{R})\delta_o \in L^1(G)$ and

$$\begin{aligned}
\int_G (1+|x|)^{a_o} |X^{\alpha_o} m(\mathcal{R})\delta_0(x)| dx &\leq \sum_{\ell\in\mathbb{Z}\backslash\{0\}} |\widehat{F}(\ell)| \int_G (1+|x|)^{a_o} |X^{\alpha_o}\kappa_\ell(x)| \\
&\leq C_k' \sup_{\substack{\lambda\geq 0\\ j=0,\dots,k}} (1+|\lambda|)^{1+j+k\ell_o} |m^{(j)}(\lambda)|.
\end{aligned}$$

This concludes the proof of Theorem 4.5.1. □

4.5.3 Proof of Corollary 4.5.2

Let us show Corollary 4.5.2.

Proof of Corollary 4.5.2. Applying Theorem 4.5.1 to the restriction of $m \in \mathcal{S}(\mathbb{R})$ to $[0,\infty)$, we have for any α and $a > 0$ that

$$\|m(\mathcal{R})\delta_0\|_{\mathcal{S},a,[\alpha],1} \leq C \sup_{\substack{\lambda \geq 0 \\ j=0,\ldots,k}} (1+|\lambda|)^{1+j+k\ell_o}|m^{(j)}(\lambda)|,$$

with $k := 3(a+\frac{Q}{2}+1)+3$ and ℓ_o the smallest integer such that $\nu\ell_o > [\alpha]+Q/2$. Clearly the right-hand side of this inequality is less than a $\mathcal{S}(\mathbb{R})$-seminorm of m, up to a constant depending on this seminorm. By Proposition 4.4.27, this concludes the proof of Corollary 4.5.2. □

We may simplify the proof of Corollary 4.5.2 by modifying the proof of Theorem 4.5.1 and choosing F independently of k for $m \in \mathcal{S}(\mathbb{R})$. Indeed, it suffices to set

$$F(\xi) := \begin{cases} m\left(\xi^{-\frac{1}{\ell_o}} - 1\right)\chi_1(\xi) & \text{if } \xi > 0, \\ 0 & \text{if } \xi \leq 0, \end{cases}$$

where $\chi_1 \in \mathcal{D}(\mathbb{R})$ is supported in $[-1,2]$ and satisfies $\chi_1 \equiv 1$ on $[0,1]$ together with $\widehat{F}(0) = 0$.

Remark 4.5.10. Behind this technical point lays the fact that the spectrum of $\mathcal{R}_2$ is contained in $[0,\infty)$ thus we can modify any function m as we see fit on $(-\infty,0)$ without changing the operator $m(\mathcal{R})$.

We had already used this idea in the proof of Theorem 4.5.1 indirectly, since different extensions of the function F will lead to the same formula in (4.59).

Note that we may also modify the function m at 0, see Remark 4.2.8 (3), but we did not use this point in the proofs of Theorem 4.5.1 or Corollary 4.5.2.

Chapter 5

Quantization on graded Lie groups

In this chapter we develop the theory of pseudo-differential operators on graded Lie groups. Our approach relies on using positive Rockland operators, their fractional powers and their associated Sobolev spaces studied in Chapter 4. As we have pointed out in the introduction, the graded Lie groups then become the natural setting for such analysis in the context of general nilpotent Lie groups.

The introduced symbol classes $S^m_{\rho,\delta}$ and the corresponding operator classes

$$\Psi^m_{\rho,\delta} = \operatorname{Op} S^m_{\rho,\delta},$$

for (ρ,δ) with $1 \geq \rho \geq \delta \geq 0$ and $\delta \neq 1$, have an operator calculus, in the sense that the set $\bigcup_{m\in\mathbb{R}} \Psi^m_{\rho,\delta}$ forms an algebra of operators, stable under taking the adjoint, and acting on the Sobolev spaces in such a way that the loss of derivatives is controlled by the order of the operator. Moreover, the operators that are elliptic or hypoelliptic within these classes allow for a parametrix construction whose symbol can be obtained from the symbol of the original operator.

During the construction of the pseudo-differential calculus $\cup_{m\in\mathbb{R}}\Psi^m_{\rho,\delta}$ on graded Lie groups in this chapter, there are several difficulties one has to overcome and which do not appear in the case of compact Lie groups as described in Chapter 2. The immediate one is the need to find a natural framework for discussing the symbols to which we will be associating the operators (quantization) and we will do so in Section 5.1. In Section 5.2 we define symbol classes leading to algebras of symbols and operators and discuss their properties. The symbol classes that we introduce are based on a positive Rockland operator on the group and contain all the left-invariant differential operators. As with Sobolev spaces, the symbol classes can be shown to be actually independent of the choice of a positive Rockland operator used in their definition. In Section 5.3 we show that the multipliers of Rockland operators are in the introduced symbol classes. We

V. Fischer, M. Ruzhansky, *Quantization on Nilpotent Lie Groups*, Progress in Mathematics 314,
DOI 10.1007/978-3-319-29558-9_5

investigate the behaviour of the kernels of operators corresponding to these symbols in Section 5.4, both at 0 and at infinity and show, in particular, that they are Calderón-Zygmund (in the sense of Coifman and Weiss, see Sections 3.2.3 and A.4). The symbolic calculus is established in Section 5.5. In Section 5.7 we show that the operators satisfy an analogue of the Calderón-Vaillancourt theorem. The construction of parametrices for elliptic and hypoelliptic operators in the calculus is carried out in Section 5.8.

Conventions

Throughout Chapter 5, G is always a graded Lie group, endowed with a family of dilations with integer weights. Its homogeneous dimension is denoted by Q. Also throughout, $\mathcal{R}$ will be a homogeneous positive Rockland operator of homogeneous degree ν. If G is a stratified Lie group, we can choose $\mathcal{R} = -\mathcal{L}$ with $\mathcal{L}$ a sub-Laplacian, or another homogeneous positive Rockland operator. Since it is a left-invariant differential operator, we denote by $\pi(\mathcal{R})$ the operator described in Definition 1.7.4. Both $\mathcal{R}$ and $\pi(\mathcal{R})$ and their properties have been extensively discussed in Chapter 4, especially Section 4.1.

Finally, when we write

$$\sup_{\pi \in \widehat{G}}$$

we always understand it as the essential supremum with respect to the Plancherel measure on $\widehat{G}$.

5.1 Symbols and quantization

The global quantization naturally occurs on any unimodular Lie (or locally compact) group of type 1 thanks to the Plancherel formula, see Subsection 1.8.2 for the Plancherel formula. The quantization was first noticed by Michael Taylor in [Tay86, Section I.3]. The case of locally compact type 1 groups was studied recently in [MR15]. The case of the compact Lie groups was described in Section 2.2.1. Here we describe the particular case of graded nilpotent Lie groups, with an emphasis on the technical meaning of the objects involved. A very brief outline of the constructions of this chapter appeared in [FR14a].

Formally, for a family of operators $\sigma(x, \pi)$ on $\mathcal{H}_\pi$ parametrised by $x \in G$ and $\pi \in \widehat{G}$, we associate the operator $T = \mathrm{Op}(\sigma)$ given by

$$T\phi(x) := \int_{\widehat{G}} \mathrm{Tr}\left(\pi(x)\sigma(x, \pi)\widehat{\phi}(\pi)\right) d\mu(\pi). \tag{5.1}$$

Again formally, the Fourier inversion formula implies that if $\sigma(x, \pi)$ does not depend on x and is the group Fourier transform of some function κ, i.e. if $\sigma(x, \pi) = \widehat{\kappa}(\pi)$, then $\mathrm{Op}(\sigma)$ is the convolution operator with right-convolution kernel κ, i.e.

$\mathrm{Op}(\sigma)\phi = \phi * \kappa$. We would like this to be true not only for (say) integrable functions κ but also for quite a large class of distributions, in order

$$\text{to quantize } X^\alpha = \mathrm{Op}(\sigma) \text{ by } \sigma(x,\pi) = \pi(X)^\alpha,$$

with $\pi(X)$ as in Definition 1.7.4.

The first problem is to make sense of the objects above. The dependence of σ on x is not problematic for the interpretation in the formula (5.1), but we have identified a unitary irreducible representation π with its equivalence class and the families of operators may be measurable in $\pi \in \mathrm{Rep}\, G$ but not defined for all $\pi \in \widehat{G}$. More worryingly, we would like to consider collections of operators which are unbounded, for instance such as $\pi(X)^\alpha$, $\pi \in \widehat{G}$. For these reasons, it may be difficult to give a meaning to the formula (5.1) in general.

Thus, our first task is to define a large class of collections of operators $\sigma(x,\pi)$, $x \in G$, $\pi \in \widehat{G}$, for which we can make sense of the quantization procedure. We will use the realisations

$$\mathcal{K}(G), \quad L^\infty(\widehat{G}), \quad \text{and} \quad \mathscr{L}_L(L^2(G))$$

of the von Neumann algebra of the group G described in Section 1.8.2. We will also use their generalisations

$$\mathcal{K}_{a,b}(G), \; L^\infty_{a,b}(\widehat{G}), \quad \text{and} \quad \mathscr{L}_L(L^2_a(G), L^2_b(G))$$

which we define in Section 5.1.2. In order to do so we use a special feature of our setting, namely the existence of positive Rockland operators and the corresponding L^2-Sobolev spaces.

5.1.1 Fourier transform on Sobolev spaces

In Section 4.3, we have discussed in detail the fractional powers of a positive Rockland operator $\mathcal{R}$ and of the operator $\mathrm{I}+\mathcal{R}$. In the sequel, we will also need to understand powers of the operators $\pi_1(\mathrm{I}+\mathcal{R})$, $\pi_1 \in \mathrm{Rep}\, G$. We now address this, and use it to extend the group Fourier transform to the Sobolev spaces $L^2_a(G)$.

From now on we will keep the same notation for the operators $\mathcal{R}$ and $\pi_1(\mathcal{R})$ (where $\pi_1 \in \mathrm{Rep}\,(G)$) and their respective self-adjoint extensions, see Proposition 4.1.15. We note that by Proposition 4.2.6 the operator $\pi_1(\mathcal{R})$ is also positive. We can consider the powers of $\mathrm{I}+\mathcal{R}$ and $\pi_1(\mathrm{I}+\mathcal{R}) = \mathrm{I}+\pi_1(\mathcal{R})$ as defined by the functional calculus

$$(\mathrm{I}+\mathcal{R})^{\frac{a}{\nu}} = \int_0^\infty (1+\lambda)^{\frac{a}{\nu}} dE(\lambda), \quad \pi_1(\mathrm{I}+\mathcal{R})^{\frac{a}{\nu}} = \int_0^\infty (1+\lambda)^{\frac{a}{\nu}} dE_{\pi_1}(\lambda),$$

where E and E_{π_1} are the spectral measures of $\mathcal{R}$ and $\pi_1(\mathcal{R})$, respectively, and ν is the homogeneous degree of $\mathcal{R}$, see Corollary 4.1.16.

Remark 5.1.1. If a/ν is a positive integer, there is no conflict of notation between

- the powers of $\pi_1(\mathrm{I}+\mathcal{R})$ as the infinitesimal representation of π_1 (see Definition 1.7.4) at $\mathrm{I}+\mathcal{R}\in\mathfrak{U}(\mathfrak{g})$
- and the operator $\pi_1(\mathrm{I}+\mathcal{R})^{\frac{a}{\nu}}$ defined by functional calculus.

Indeed, if $a=\nu$, the two coincide. If $a=\ell\nu$, $\ell\in\mathbb{N}$, then the operator $\pi_1(\mathrm{I}+\mathcal{R})^{\frac{a}{\nu}}$ defined by functional calculus coincides with the ℓ-th power of $\pi_1(\mathrm{I}+\mathcal{R})$. The case $a=0$ is trivial.

We can describe more concretely the operators $\pi_1(\mathrm{I}+\mathcal{R})^{\frac{a}{\nu}}$, $\pi_1\in\mathrm{Rep}\,G$.

Lemma 5.1.2. *Let $\mathcal{R}$ be a positive Rockland operator of homogeneous degree ν. As in Corollary 4.3.11, we denote by $\mathcal{B}_a$ the right-convolution kernels of its Bessel potentials $(\mathrm{I}+\mathcal{R})^{-\frac{a}{\nu}}$, $\mathrm{Re}\,a>0$.*

If $a\in\mathbb{C}$ with $\mathrm{Re}\,a<0$, then $\mathcal{B}_{-a}$ is an integrable function and

$$\forall\pi_1\in\mathrm{Rep}\,G\qquad \pi_1(\mathrm{I}+\mathcal{R})^{\frac{a}{\nu}}=\widehat{\mathcal{B}}_{-a}(\pi_1).$$

For any $a\in\mathbb{C}$ and any $\pi_1\in\mathrm{Rep}\,G$, the operator $\pi_1(\mathrm{I}+\mathcal{R})^{\frac{a}{\nu}}$ maps $\mathcal{H}_{\pi_1}^\infty$ onto $\mathcal{H}_{\pi_1}^\infty$ bijectively. Furthermore, the inverse of $\pi_1(\mathrm{I}+\mathcal{R})^{\frac{a}{\nu}}$ is $\pi_1(\mathrm{I}+\mathcal{R})^{-\frac{a}{\nu}}$ as operators acting on $\mathcal{H}_{\pi_1}^\infty$.

Proof. Let $a\in\mathbb{C}$, $\mathrm{Re}\,a<0$. Then the Bessel potential $(\mathrm{I}+\mathcal{R})^{\frac{a}{\nu}}$ coincides with the bounded operator with right-convolution kernel $\mathcal{B}_{-a}\in L^1(G)$, see Corollary 4.3.11. Therefore, $(\mathrm{I}+\mathcal{R})^{\frac{a}{\nu}}\in\mathscr{L}_L(L^2(G))$ and

$$\mathcal{F}_G\{(\mathrm{I}+\mathcal{R})^{\frac{a}{\nu}}f\}=\mathcal{F}_G\{f*\mathcal{B}_{-a}\}=\widehat{\mathcal{B}}_{-a}\widehat{f},\quad f\in L^2(G).$$

Now we apply Corollary 4.1.16 with the bounded multiplier given by $\phi(\lambda)=(1+\lambda)^{\frac{a}{\nu}}$, $\lambda\geq 0$. By Equality (4.5) in Corollary 4.1.16, we obtain

$$\mathcal{F}_G\{(\mathrm{I}+\mathcal{R})^{\frac{a}{\nu}}f\}=\pi(\mathrm{I}+\mathcal{R})^{\frac{a}{\nu}}\widehat{f},\quad f\in L^2(G).$$

The injectivity of the group Fourier transform on $\mathcal{K}(G)$ yields that $\widehat{\mathcal{B}}_{-a}(\pi)=\pi(\mathrm{I}+\mathcal{R})^{\frac{a}{\nu}}$ for any $\pi\in\widehat{G}$, and the first part of the statement is proved.

Let $a\in\mathbb{C}$. We apply Corollary 4.1.16 with the multiplier given by $\phi(\lambda)=(1+\lambda)^{\frac{a}{\nu}}$, $\lambda\geq 0$. Although this multiplier is unbounded, simple modifications of the proof show that Equality (4.5) in Corollary 4.1.16 still holds for f in the domain of the operator. Recall that the domain of $(\mathrm{I}+\mathcal{R})^{\frac{a}{\nu}}$ contains $\mathcal{S}(G)$ by Corollary 4.3.16 and moreover $(\mathrm{I}+\mathcal{R})^{\frac{a}{\nu}}\mathcal{S}(G)=\mathcal{S}(G)$. Consequently, if $\pi_1\in\mathrm{Rep}\,G$, we have

$$\pi_1\{(\mathrm{I}+\mathcal{R})^{\frac{a}{\nu}}f\}v=\pi_1(\mathrm{I}+\mathcal{R})^{\frac{a}{\nu}}\pi_1(f)v,\quad f\in\mathcal{S}(G),\ v\in\mathcal{H}_{\pi_1},$$

with $\pi_1(\mathrm{I}+\mathcal{R})^{\frac{a}{\nu}}$ defined spectrally. Recall that $\pi_1(f)v\in\mathcal{H}_{\pi_1}^\infty$ when $f\in\mathcal{S}(G)$ by Proposition 1.7.6 (iv), hence here $\pi_1\{(\mathrm{I}+\mathcal{R})^{\frac{a}{\nu}}f\}v\in\mathcal{H}_\pi^\infty$ as well. By Lemma 1.8.19,

$\pi_1(\text{I}+\mathcal{R})^{\frac{a}{\nu}}$ maps $\mathcal{H}_{\pi_1}^\infty$ to $\mathcal{H}_{\pi_1}^\infty$. The spectral calculus implies that as operators acting on $\mathcal{H}_{\pi_1}^\infty$, we have

$$\pi_1(\text{I}+\mathcal{R})^{\frac{a}{\nu}}\pi_1(\text{I}+\mathcal{R})^{-\frac{a}{\nu}} = \text{I}_{\mathcal{H}_{\pi_1}^\infty} \quad \text{and} \quad \pi_1(\text{I}+\mathcal{R})^{-\frac{a}{\nu}}\pi_1(\text{I}+\mathcal{R})^{\frac{a}{\nu}} = \text{I}_{\mathcal{H}_{\pi_1}^\infty}.$$

Consequently, the inverse of $\pi_1(\text{I}+\mathcal{R})^{\frac{a}{\nu}}$ is $\pi_1(\text{I}+\mathcal{R})^{-\frac{a}{\nu}}$ as operators defined on $\mathcal{H}_{\pi_1}^\infty$ and $\pi_1(\text{I}+\mathcal{R})^{\frac{a}{\nu}}\mathcal{H}_{\pi_1}^\infty = \mathcal{H}_{\pi_1}^\infty$. □

Lemma 5.1.2 and Remark 4.1.17 now imply easily

Corollary 5.1.3. *Let $\mathcal{R}$ be a positive Rockland operator of homogeneous degree ν. For any $a \in \mathbb{C}$, $\{\pi(\text{I}+\mathcal{R})^{\frac{a}{\nu}} : \mathcal{H}_\pi^\infty \to \mathcal{H}_\pi^\infty, \pi \in \widehat{G}\}$ is a measurable $\widehat{G}$-field of operators acting on smooth vectors (in the sense of Definition 1.8.14).*

Lemma 5.1.2 together with the Plancherel formula (see Section 1.8.2) and Corollary 4.3.11 also imply

Corollary 5.1.4. *Let $\mathcal{R}$ be a positive Rockland operator of homogeneous degree ν. For any $a \in \mathbb{R}$, we have*

$$a > Q/2 \qquad \Longrightarrow \qquad \{\pi(\text{I}+\mathcal{R})^{-\frac{a}{\nu}}, \pi \in \widehat{G}\} \in L^2(\widehat{G}),$$

and also, for $a > Q/2$,

$$\|\pi(\text{I}+\mathcal{R})^{-\frac{a}{\nu}}\|_{L^2(\widehat{G})} = \|\widehat{\mathcal{B}}_a(\pi)\|_{L^2(\widehat{G})} = \|\mathcal{B}_a\|_{L^2(G)} < \infty.$$

Note that an analogue of Corollary 5.1.4 for compact Lie groups may be obtained by noticing that (2.15) yields

$$m > n/2 \qquad \Longrightarrow \qquad \sum_{\pi\in\widehat{G}} d_\pi \|\pi(\text{I}-\mathcal{L}_G)^{-\frac{m}{2}}\|_{\text{HS}}^2 = \sum_{\pi\in\widehat{G}} d_\pi^2 \langle\pi\rangle^{-2m} < \infty.$$

The following statement describes an important property of the field $\{\pi(\text{I}+\mathcal{R})^{\frac{a}{\nu}}, \pi \in \widehat{G}\}$, in relation with the right Sobolev spaces (see Section 4.4.8 for right Sobolev spaces):

Proposition 5.1.5. *Let $\mathcal{R}$ be a positive Rockland operator on G of homogeneous degree ν. Let also $a \in \mathbb{R}$.*

If $f \in \tilde{L}_a^2(G)$, then $(\text{I}+\tilde{\mathcal{R}})^{\frac{a}{\nu}} f \in L^2(G)$ and there exists a field of operators $\{\sigma_\pi : \mathcal{H}_\pi^\infty \to \mathcal{H}_\pi\,,\ \pi \in \widehat{G}\}$ such that

$$\{\sigma_\pi \pi(\text{I}+\mathcal{R})^{\frac{a}{\nu}} : \mathcal{H}_\pi^\infty \to \mathcal{H}_\pi\,,\ \pi \in \widehat{G}\} \in L^2(\widehat{G}), \tag{5.2}$$

and for almost all $\pi \in \widehat{G}$,

$$\mathcal{F}_G\{(\text{I}+\tilde{\mathcal{R}})^{\frac{a}{\nu}} f\}(\pi) = \sigma_\pi \pi(\text{I}+\mathcal{R})^{\frac{a}{\nu}}. \tag{5.3}$$

Conversely, if $\{\sigma_\pi : \mathcal{H}_\pi^\infty \to \mathcal{H}_\pi\,,\ \pi \in \widehat{G}\}$ satisfies (5.2) then there exists a unique function $f \in \tilde{L}_a^2(G)$ satisfying (5.3).

In Proposition 5.1.5, $\sigma_\pi \pi(\mathrm{I}+\mathcal{R})^{\frac{a}{\nu}}$ is not obtained as the composition of (possibly) unbounded operators as in Definition A.3.2. Instead, for $\sigma_\pi \pi(\mathrm{I}+\mathcal{R})^{\frac{a}{\nu}}$, it is viewed as the composition of a field of operators defined on smooth vectors with a field of operators acting on smooth vectors, see Section 1.8.3.

In Proposition 5.1.5, we use the right Sobolev spaces associated with the positive Rockland operator $\mathcal{R}$. These spaces are in fact independent of the choice of a positive Rockland operator used in their definition, see Sections 4.4.5 and 4.4.8. Consequently, if (5.2) holds for one positive Rockland operator then (5.2) and (5.3) hold for any positive Rockland operator and the Sobolev norm of $f \in L^2(G)$, using one particular positive Rockland operator $\mathcal{R}$, is equal to the $L^2(\widehat{G})$-norm of (5.2).

Proof of Proposition 5.1.5. If $f \in \tilde{L}^2_a(G)$, then by Theorem 4.4.3 (3) (see also Section 4.4.8), we have that $f_a := (\mathrm{I}+\tilde{\mathcal{R}})^{\frac{a}{\nu}} f$ is in $L^2(G)$ and its Fourier transform is a field of bounded operators (in fact in the Hilbert-Schmidt class). By Lemma 5.1.2, $\pi(\mathrm{I}+\mathcal{R})^{-\frac{a}{\nu}}$ maps $\mathcal{H}_\pi^\infty$ onto itself. Hence we can define

$$\sigma_\pi := \pi(f_a)\pi(\mathrm{I}+\mathcal{R})^{-\frac{a}{\nu}},$$

as an operator defined on $\mathcal{H}_\pi^\infty$. One readily checks that the operators σ_π, $\pi \in \widehat{G}$, satisfy (5.2) and (5.3).

For the converse, if $\{\sigma_\pi : \mathcal{H}_\pi^\infty \to \mathcal{H}_\pi \, : \, \pi \in \widehat{G}\}$ satisfies (5.2) then we define the function

$$L^2(G) \ni f_a := \mathcal{F}_G^{-1}\{\sigma_\pi \pi(\mathrm{I}+\mathcal{R})^{\frac{a}{\nu}}\},$$

which is square integrable by the Plancherel theorem (see Theorem 1.8.11), and the function

$$f := (\mathrm{I}+\tilde{\mathcal{R}})^{-\frac{a}{\nu}} f_a,$$

which will be in $\tilde{L}^2_a(G)$ by Theorem 4.4.3 (3). One readily checks that the function f satisfies the properties described in the statement. □

We now aim at stating and proving a property similar to Proposition 5.1.5 for the left Sobolev spaces. It will use the composition of a field with $\pi(\mathrm{I}+\mathcal{R})^{\frac{a}{\nu}}$ on the left and this is problematic when we consider any general field $\sigma = \{\sigma_\pi : \mathcal{H}_\pi^\infty \to \mathcal{H}_\pi\}$ without utilising the composition of unbounded operators as in Definition A.3.2. To overcome this problem, we introduce the following notion:

Definition 5.1.6. Let $\pi_1 \in \operatorname{Rep} G$ and $a \in \mathbb{R}$. We denote by $\mathcal{H}^a_{\pi_1}$ the Hilbert space obtained by completion of $\mathcal{H}^\infty_{\pi_1}$ for the norm

$$\|\cdot\|_{\mathcal{H}^a_{\pi_1}} : v \longmapsto \|\pi_1(\mathrm{I}+\mathcal{R})^{\frac{a}{\nu}} v\|_{\mathcal{H}_{\pi_1}} := \|v\|_{\mathcal{H}^a_{\pi_1}},$$

where $\mathcal{R}$ is a positive Rockland operator on G of homogeneous degree ν.

We may call them the $\mathcal{H}_{\pi_1}$-Sobolev spaces. Note that in the case of the Schrödinger representation for the Heisenberg group, they coincide with Shubin-Sobolev spaces, see Section 6.4.3. More generally, if we realise an element $\pi \in \widehat{G}$ as a representation π_1 acting on some $L^2(\mathbb{R}^m)$ via the orbit methods, see Section 1.8.1, then we view the corresponding Sobolev spaces as tempered distributions: $\mathcal{H}^a_{\pi_1} \subset \mathcal{S}'(\mathbb{R}^m)$.

The following lemma is a routine exercise.

Lemma 5.1.7. *Let $\pi_1 \in \operatorname{Rep} G$ and $a \in \mathbb{R}$.*

1. *If $a = 0$, then $\mathcal{H}^a_{\pi_1} = \mathcal{H}_{\pi_1}$. If $a > 0$, we realise $\mathcal{H}^a_{\pi_1}$ as a subspace of $\mathcal{H}_{\pi_1}$ and it is the domain of the operator $\pi_1(\mathrm{I}+\mathcal{R})^{\frac{a}{\nu}}$. If $a < 0$, we realise $\mathcal{H}^a_{\pi_1}$ as a Hilbert space containing $\mathcal{H}_{\pi_1}$ and the operator $\pi_1(\mathrm{I}+\mathcal{R})^{\frac{a}{\nu}}$ extends uniquely to a bounded operator $\mathcal{H}^a_{\pi_1} \to \mathcal{H}_{\pi_1}$.*
2. *For any $a \in \mathbb{R}$, realising $\mathcal{H}^a_{\pi_1}$ as in Part 1, this space is independent of the positive Rockland operator $\mathcal{R}$ and two positive Rockland operators yield equivalent norms.*
3. *We have the continuous inclusions*
$$a < b \implies \mathcal{H}^b_{\pi_1} \subset \mathcal{H}^a_{\pi_1}.$$
For any $a, b \in \mathbb{R}$, the operator $\pi_1(\mathrm{I}+\mathcal{R})^{\frac{a}{\nu}}$ maps $\mathcal{H}^b_{\pi_1}$ to $\mathcal{H}^{b-a}_{\pi_1}$ injectively and continuously. In this way, $\mathcal{H}^a_{\pi_1}$ and $\mathcal{H}^{-a}_{\pi_1}$ are in duality via
$$\langle u, v\rangle_{\mathcal{H}^a_{\pi_1}\times\mathcal{H}^{-a}_{\pi_1}} := (\pi_1(\mathrm{I}+\mathcal{R})^{\frac{a}{\nu}}u, \pi_1(\mathrm{I}+\bar{\mathcal{R}})^{-\frac{a}{\nu}}\bar{v})_{\mathcal{H}_{\pi_1}}.$$
This duality extends the $\mathcal{H}_{\pi_1}$ duality in the sense that
$$\forall u \in \mathcal{H}^a_{\pi_1} \cap \mathcal{H}_{\pi_1},\ v \in \mathcal{H}^{-a}_{\pi_1} \cap \mathcal{H}_{\pi_1} \qquad \langle u, v\rangle_{\mathcal{H}^a_{\pi_1}\times\mathcal{H}^{-a}_{\pi_1}} = (u, \bar{v})_{\mathcal{H}_{\pi_1}}.$$
4. *If π_2 is another strongly continuous representation such that $\pi_1 \sim_T \pi_2$, that is, T is a unitary operator satisfying $T\pi_1 = \pi T_2$, then T maps $\mathcal{H}^\infty_{\pi_1}$ to $\mathcal{H}^\infty_{\pi_2}$ bijectively by Lemma 1.8.12 and extends uniquely to an isometric operator $\mathcal{H}^a_{\pi_1} \to \mathcal{H}^a_{\pi_2}$.*

Lemma 5.1.7, especially Part 4, shows that $\widehat{G}$-fields with domain or range on these Sobolev spaces make sense:

Definition 5.1.8. Let $a \in \mathbb{R}$. A $\widehat{G}$-field of operators $\sigma = \{\sigma_\pi : \mathcal{H}^\infty_\pi \to \mathcal{H}_\pi, \pi \in \widehat{G}\}$ defined on smooth vectors is *defined on the Sobolev spaces $\mathcal{H}^a_\pi$* when for each $\pi_1 \in \operatorname{Rep} G$, the operator σ_{π_1} is bounded on $\mathcal{H}^a_{\pi_1}$ in the sense that
$$\exists C \quad \forall v \in \mathcal{H}^\infty_{\pi_1} \quad \|\sigma_{\pi_1} v\|_{\mathcal{H}_{\pi_1}} \le C\|v\|_{\mathcal{H}^a_{\pi_1}}.$$

Thus, by density of $\mathcal{H}_{\pi_1}^\infty$ in $\mathcal{H}_{\pi_1}^a$, σ_{π_1} extends uniquely to a bounded operator defined on $\mathcal{H}_{\pi_1}^a$ for which we keep the same notation $\sigma_{\pi_1} : \mathcal{H}_{\pi_1}^a \to \mathcal{H}_{\pi_1}$.

Example 5.1.9. For any positive Rockland operator of degree ν, the field $\{\pi(\text{I} + \mathcal{R})^{\frac{a}{\nu}}, \pi \in \widehat{G}\}$, is defined on the Sobolev spaces $\mathcal{H}_\pi^a$. This is an easy consequence of Lemma 5.1.7, especially Part 3.

We will allow ourselves the shorthand notation

$$\sigma = \{\sigma_\pi : \mathcal{H}_\pi^a \to \mathcal{H}_\pi, \pi \in \widehat{G}\},$$

to indicate that the $\widehat{G}$-field of operators is defined on the Sobolev spaces $\mathcal{H}_\pi^a$.

Instead of Definition 5.1.8, we could also have defined $\widehat{G}$-fields of operators defined on $\mathcal{H}_\pi^a$-Sobolev spaces in a way similar to Definition 1.8.13 (where $\widehat{G}$-fields of operators defined on smooth vectors were defined). Naturally, these two viewpoints are equivalent since $\mathcal{H}_{\pi_1}^\infty$ is dense in $\mathcal{H}_{\pi_1}^a$.

However, in order to define $\widehat{G}$-fields of operators with range in the $\mathcal{H}_\pi^a$-Sobolev spaces, we have to adopt the latter viewpoint in the sense that we modify Definitions 1.8.13 and 1.8.14 (in this way, we make no further assumptions on the fields or on the Sobolev spaces):

Definition 5.1.10. Let $a \in \mathbb{R}$.

- A $\widehat{G}$-*field of operators defined on smooth vectors* with *range in the Sobolev spaces* $\mathcal{H}_\pi^a$ is a family of classes of operators $\{\sigma_\pi, \pi \in \widehat{G}\}$ where

$$\sigma_\pi := \{\sigma_{\pi_1} : \mathcal{H}_{\pi_1}^\infty \to \mathcal{H}_{\pi_1}^a, \pi_1 \in \pi\}$$

  for each $\pi \in \widehat{G}$ viewed as a subset of $\operatorname{Rep} G$, satisfying for any two elements σ_{π_1} and σ_{π_2} in σ_π:

$$\pi_1 \sim_T \pi_2 \Longrightarrow \sigma_{\pi_2} T = T\sigma_{\pi_1} \text{ on } \mathcal{H}_\pi^\infty.$$

  (Here we have kept the same notation for the intertwining operator T and its unique extension between Sobolev spaces $\mathcal{H}_{\pi_1}^a \to \mathcal{H}_{\pi_2}^a$, see Lemma 5.1.7 Part 4.)

- It is measurable when for one (and then any) choice of realisation $\pi_1 \in \pi$ and any vector $v_{\pi_1} \in \mathcal{H}_{\pi_1}^a$, as π runs over $\widehat{G}$, the resulting field $\{\sigma_\pi v_\pi, \pi \in \widehat{G}\}$ is μ-measurable whenever $\int_{\widehat{G}} \|v_\pi\|_{\mathcal{H}_\pi^a}^2 d\mu(\pi) < \infty$. (Here we assume that all the $\mathcal{H}_\pi^a$-norms are realised via a fixed positive Rockland operator.)

Unless otherwise stated, a $\widehat{G}$-field of operators defined on smooth vectors with range in the Sobolev spaces $\mathcal{H}_\pi^a$ is always assumed measurable. We will allow ourselves the shorthand notation

$$\sigma = \{\sigma_\pi : \mathcal{H}_\pi^\infty \to \mathcal{H}_\pi^a, \pi \in \widehat{G}\}$$

to indicate that the $\widehat{G}$-field of operators has range in the Sobolev space $\mathcal{H}_\pi^a$.

Naturally, if a $\widehat{G}$-field of operators is defined on smooth vectors $\sigma = \{\sigma_\pi : \mathcal{H}_\pi^\infty \to \mathcal{H}_\pi, \pi \in \widehat{G}\}$ with the usual range $\mathcal{H}_\pi = \mathcal{H}_\pi^0$, then it has *range in the Sobolev spaces* $\mathcal{H}_\pi^a$ when for each $\pi_1 \in \operatorname{Rep} G$ and any $v \in \mathcal{H}_{\pi_1}^\infty$, we have $\sigma_{\pi_1} v \in \mathcal{H}_{\pi_1}^a$.

Moreover, the following property of composition is easy to check: if σ_1 has range in $\mathcal{H}_\pi^a$ and σ_2 is defined on $\mathcal{H}_\pi^a$,

$$\text{i.e.} \quad \sigma_1 = \{\sigma_{1,\pi} : \mathcal{H}_\pi^\infty \to \mathcal{H}_\pi^a, \pi \in \widehat{G}\} \quad \text{and} \quad \sigma_2 = \{\sigma_{2,\pi} : \mathcal{H}_\pi^a \to \mathcal{H}_\pi, \pi \in \widehat{G}\},$$

then the following field

$$\sigma_2\sigma_1 := \{\sigma_{2,\pi}\sigma_{1,\pi} : \mathcal{H}_\pi^\infty \to \mathcal{H}_\pi, \pi \in \widehat{G}\}$$

makes sense as a $\widehat{G}$-field of operators defined on smooth vectors. This coincides or extends the definition of composition of fields (the first one acting on smooth vectors) given in Section 1.8.3.

We can apply this property of composition to $\sigma = \{\sigma_\pi : \mathcal{H}_\pi^\infty \to \mathcal{H}_\pi^a, \pi \in \widehat{G}\}$ and $\{\pi(\mathrm{I} + \mathcal{R})^{\frac{a}{\nu}}, \pi \in \widehat{G}\}$, see Example 5.1.9 for the latter, to obtain the $\widehat{G}$-field defined on smooth vectors by

$$\pi(\mathrm{I} + \mathcal{R})^{\frac{a}{\nu}}\sigma = \{\pi(\mathrm{I} + \mathcal{R})^{\frac{a}{\nu}}\sigma_\pi : \mathcal{H}_\pi^\infty \to \mathcal{H}_\pi, \pi \in \widehat{G}\}. \tag{5.4}$$

We can now state the proposition which will enable us to define the group Fourier transform of a function in a left or right Sobolev space.

Proposition 5.1.11. *Let $a \in \mathbb{R}$.*

(L) *If $f \in L_a^2(G)$, then $(\mathrm{I} + \mathcal{R})^{\frac{a}{\nu}} f \in L^2(G)$ and there exists a field of operators $\{\sigma_\pi : \mathcal{H}_\pi^\infty \to \mathcal{H}_\pi^a,\ \pi \in \widehat{G}\}$ such that*

$$\{\pi(\mathrm{I} + \mathcal{R})^{\frac{a}{\nu}}\sigma_\pi : \mathcal{H}_\pi^\infty \to \mathcal{H}_\pi,\ \pi \in \widehat{G}\} \in L^2(\widehat{G}), \tag{5.5}$$

$$\mathcal{F}_G\{(\mathrm{I} + \mathcal{R})^{\frac{a}{\nu}} f\}(\pi) = \pi(\mathrm{I} + \mathcal{R})^{\frac{a}{\nu}}\sigma_\pi, \quad \text{for almost all } \pi \in \widehat{G}, \tag{5.6}$$

where $\mathcal{R}$ is a positive Rockland operator on G of homogeneous degree ν.

Conversely, if $\{\sigma_\pi : \mathcal{H}_\pi^\infty \to \mathcal{H}_\pi^a,\ \pi \in \widehat{G}\}$ satisfies (5.5) for one positive Rockland operator $\mathcal{R}$, then there exists a unique function $f \in L_a^2(G)$ satisfying (5.6).

(R) *If $f \in \tilde{L}_a^2(G)$, then the (unique) field σ obtained in Proposition 5.1.5 can be extended uniquely into a field $\{\sigma_\pi : \mathcal{H}_\pi^a \to \mathcal{H}_\pi,\ \pi \in \widehat{G}\}$ defined on $\mathcal{H}_\pi^a$.*

Properties (L) *and* (R) *are independent of the choice of $\mathcal{R}$.*

In Proposition 5.1.11, $\pi(\mathrm{I} + \mathcal{R})^{\frac{a}{\nu}}\sigma_\pi$ is not obtained as the composition of (possibly) unbounded operators as in Definition A.3.2 but is understood via (5.4).

In Proposition 5.1.11, we use the left and right Sobolev spaces associated with the positive Rockland operator $\mathcal{R}$. These spaces are in fact independent of the choice of a positive Rockland operator used in their definition, see Sections 4.4.5 and 4.4.8. Consequently, if (5.5) hold for one positive Rockland operator then (5.5) and (5.6) hold for any positive Rockland operator and the Sobolev norm of $f \in L^2(G)$, using one particular positive Rockland operator $\mathcal{R}$, is equal to the $L^2(\widehat{G})$-norm of (5.5).

Proof of Proposition 5.1.11. Property (L). If $f \in L^2_a(G)$, then by Theorem 4.4.3 (3), we have that $f_a := (\mathrm{I}+\mathcal{R})^{\frac{a}{\nu}} f$ is in $L^2(G)$ and its Fourier transform is a field of bounded operators (in fact in the Hilbert-Schmidt class). By (5.4) we can define $\sigma = \{\sigma_\pi : \mathcal{H}_\pi^\infty \to \mathcal{H}_\pi^a\}$ via $\sigma_\pi := \pi(\mathrm{I}+\mathcal{R})^{-\frac{a}{\nu}}\pi(f_a)$. One readily checks that the field σ satisfies (5.2) and (5.3).

For the converse, if $\{\sigma_\pi : \mathcal{H}_\pi^\infty \to \mathcal{H}_\pi^a : \pi \in \widehat{G}\}$ satisfies (5.2) then we define the function

$$L^2(G) \ni f_a := \mathcal{F}_G^{-1}\{\pi(\mathrm{I}+\mathcal{R})^{\frac{a}{\nu}}\sigma_\pi\},$$

which is square integrable by the Plancherel theorem (see Theorem 1.8.11), and the function

$$f := (\mathrm{I}+\mathcal{R})^{-\frac{a}{\nu}} f_a,$$

which will be in $L^2_a(G)$ by Theorem 4.4.3 (3). One readily checks that the function f satisfies the properties described in the statement. This shows the property (L).

Property (R) follows easily from (5.2). □

From the proof above, one can check easily that if $f \in L^2_a(G)$ or $\tilde{L}^2_a(G)$ is also in any of the spaces where the group Fourier transform has already been defined, namely, $L^2(G)$ or $\mathcal{K}(G)$, then $\sigma = \{\sigma_\pi : \mathcal{H}_\pi^\infty \to \mathcal{H}_\pi, \pi \in \widehat{G}\}$ will coincide with the group Fourier transform of f. Hence we can extend the definition of the group Fourier transform to Sobolev spaces:

Definition 5.1.12. Let $a \in \mathbb{R}$. The group Fourier transform of $f \in L^2_a(G)$ or $f \in \tilde{L}^2_a(G)$ is the field σ of operators defined on smooth vectors given in Proposition 5.1.11.

This leads us to define the following spaces of fields of operators:

Definition 5.1.13. (L) Let $L^2_a(\widehat{G})$ denote the space of fields of operators σ with range in $\mathcal{H}_\pi^a$ and satisfying (5.5), that is,

$$\sigma = \{\sigma_\pi : \mathcal{H}_\pi^\infty \to \mathcal{H}_\pi^a\,,\ \pi \in \widehat{G}\},$$
$$\{\pi(\mathrm{I}+\mathcal{R})^{\frac{a}{\nu}}\sigma_\pi : \mathcal{H}_\pi^\infty \to \mathcal{H}_\pi\,,\ \pi \in \widehat{G}\} \in L^2(\widehat{G}),$$

for one (and then any) positive Rockland operator of homogeneous degree ν. We also set

$$\|\sigma\|_{L^2_a(\widehat{G})} := \|\pi(\mathrm{I}+\mathcal{R})^{\frac{a}{\nu}}\sigma_\pi\|_{L^2(\widehat{G})}. \tag{5.7}$$

(R) Let $\tilde{L}^2_a(\widehat{G})$ denote the space of fields of operators σ defined on $\mathcal{H}_\pi^a$ and satisfying (5.2), that is,

$$\sigma = \{\sigma_\pi : \mathcal{H}_\pi^a \to \mathcal{H}_\pi \,,\, \pi \in \widehat{G}\},$$
$$\{\sigma_\pi \pi(\mathrm{I}+\mathcal{R})^{\frac{a}{\nu}} : \mathcal{H}_\pi^\infty \to \mathcal{H}_\pi \,,\, \pi \in \widehat{G}\} \in L^2(\widehat{G}),$$

for one (and then any) positive Rockland operator of homogeneous degree ν. We also set

$$\|\sigma\|_{\tilde{L}^2_a(\widehat{G})} := \|\sigma_\pi \pi(\mathrm{I}+\mathcal{R})^{\frac{a}{\nu}}\|_{L^2(\widehat{G})}.$$

It is a routine exercise, using Proposition 5.1.11 and the properties of the Sobolev spaces (see Section 4.4), to show that

Proposition 5.1.14. *Let $a \in \mathbb{R}$. If $\mathcal{R}$ is a positive Rockland operator of homogeneous degree ν, the map $\|\cdot\|_{L^2_a(\widehat{G})}$ given by (5.7) is a norm on the vector space $L^2_a(\widehat{G})$. Endowed with this norm, $L^2_a(\widehat{G})$ is a Banach space which is independent of $\mathcal{R}$. Two norms corresponding to any two choices of Rockland operators via (5.7) are equivalent.*

The Fourier transform $\mathcal{F}_G$ is an isomorphism between Banach spaces acting from $L^2_a(G)$ onto $L^2_a(\widehat{G})$. It coincides with the usual Fourier transform on $L^2(G)$ for $a = 0$.

Let $\sigma = \{\sigma_\pi, \pi \in \widehat{G}\}$ be in $L^2_a(\widehat{G})$. Then

$$\{\pi(X)^\alpha \sigma_\pi, \pi \in \widehat{G}\}$$

is in $L^2_{a-[\alpha]}(\widehat{G})$ for any $\alpha \in \mathbb{N}_0^n$, and

$$\{\pi(\mathrm{I}+\mathcal{R})^{s/\nu} \sigma_\pi, \pi \in \widehat{G}\}$$

is in $L^2_{a-s}(\widehat{G})$ for any $s \in \mathbb{R}$. Furthermore, if $f = \mathcal{F}_G^{-1}\sigma \in L^2_a(G)$ then

$$\mathcal{F}_G(X^\alpha f)(\pi) = \pi(X)^\alpha \widehat{f}(\pi) \quad \text{and} \quad \mathcal{F}_G((\mathrm{I}+\mathcal{R})^{s/\nu} f)(\pi) = \pi(\mathrm{I}+\mathcal{R})^{s/\nu} \widehat{f}(\pi).$$

We have similar results for the right Sobolev spaces. Furthermore the adjoint map $\sigma \mapsto \sigma^*$ maps $L^2_a(\widehat{G}) \to \tilde{L}^2_a(\widehat{G})$ and $\tilde{L}^2_a(\widehat{G}) \to L^2_a(\widehat{G})$ isomorphically as Banach spaces.

Recall that the tempered distributions $X^\alpha f$ and $(\mathrm{I}+\mathcal{R})^{s/\nu} f$ used in the statement just above are respectively defined via

$$\langle X^\alpha f, \phi\rangle = \langle f, \{X^\alpha\}^t \phi\rangle, \quad \phi \in \mathcal{S}(G), \tag{5.8}$$

and

$$\langle (\mathrm{I}+\mathcal{R})^{s/\nu} f, \phi\rangle = \langle f, (\mathrm{I}+\bar{\mathcal{R}})^{s/\nu} \phi\rangle, \quad \phi \in \mathcal{S}(G). \tag{5.9}$$

For (5.9), see Definition 4.3.17. For (5.8), this is the composition of the formula obtained for one vector field (with polynomial coefficients) by integration by parts. See also (1.10) for the definition of $\{X^\alpha\}^t$.

In Corollary 1.8.3, we stated the inversion formula valid for any Schwartz function on any connected simply connected Lie group. Here we weaken the hypothesis using the Sobolev spaces in the context of a graded Lie group G:

Proposition 5.1.15 (Fourier inversion formula). *Let f be in the left Sobolev space $L^2_s(G)$ or in the right Sobolev space $\tilde{L}^2_s(G)$ with $s > Q/2$. Then for almost every $\pi \in \operatorname{Rep} G$, the operator $\widehat{f}(\pi)$ is trace class with*

$$\int_{\widehat{G}} \operatorname{Tr}|\widehat{f}(\pi)| d\mu(\pi) < \infty. \tag{5.10}$$

Furthermore, f is continuous on G, and for every $x \in G$ we have

$$f(x) = \int_{\widehat{G}} \operatorname{Tr}\left(\pi(x)\widehat{f}(\pi)\right) d\mu(\pi) = \int_{\widehat{G}} \operatorname{Tr}\left(\widehat{f}(\pi)\pi(x)\right) d\mu(\pi). \tag{5.11}$$

In the statement above, as $s > Q/2 > 0$, the field $\widehat{f}$ is in $L^2(\widehat{G})$, it is then a field of bounded operators (even in Hilbert-Schmidt classes) and so can be composed on the left and the right with $\pi(x)$. The (possibly infinite) traces

$$\operatorname{Tr}\left|\pi_1(x)\widehat{f}(\pi_1)\right|, \quad \operatorname{Tr}\left|\widehat{f}(\pi_1)\pi_1(x)\right| \quad \text{and} \quad \operatorname{Tr}\left|\widehat{f}(\pi_1)\right|$$

are equal for $\pi_1 \in \operatorname{Rep} G$ as π_1 is unitary. They are constant on the class of $\pi_1 \in \operatorname{Rep} G$ in $\widehat{G}$ and are, therefore, treated as depending on $\pi \in \widehat{G}$. They are finite for μ-almost all $\pi \in \widehat{G}$ in view of (5.10).

Note that (5.10) implies not only that the two expressions

$$\int_{\widehat{G}} \operatorname{Tr}\left(\pi(x)\widehat{f}(\pi)\right) d\mu(\pi) \quad \text{and} \quad \int_{\widehat{G}} \operatorname{Tr}\left(\widehat{f}(\pi)\pi(x)\right) d\mu(\pi)$$

make sense but that they are also equal by the properties of the trace since $\pi(x)$ is bounded.

Proof of Proposition 5.1.15. Let $\mathcal{R}$ be a positive Rockland operator of homogeneous degree ν. Let $f \in L^2_s(G)$ with $s > Q/2$. We set

$$f_s := (\mathrm{I} + \mathcal{R})^{\frac{s}{\nu}} f \in L^2(G).$$

The properties of the trace imply

$$\operatorname{Tr}|\widehat{f}(\pi)| = \operatorname{Tr}\left|\pi(\mathrm{I} + \mathcal{R})^{-\frac{s}{\nu}} \widehat{f_s}(\pi)\right| \leq \|\pi(\mathrm{I} + \mathcal{R})^{-\frac{s}{\nu}}\|_{\mathtt{HS}} \|\widehat{f_s}(\pi)\|_{\mathtt{HS}}.$$

Integrating against the Plancherel measure, we obtain by the Cauchy-Schwartz inequality

$$\int_{\widehat{G}} \mathrm{Tr}|\widehat{f}(\pi)|d\mu(\pi) \leq \|\pi(\mathrm{I}+\mathcal{R})^{-\frac{s}{\nu}}\|_{L^2(\widehat{G})}\|\widehat{f}_s\|_{L^2(\widehat{G})}.$$

By Corollary 5.1.4, $C_s := \|\pi(\mathrm{I}+\mathcal{R})^{-\frac{s}{\nu}}\|_{L^2(\widehat{G})}$ is a positive finite constant. Since $\|\widehat{f}_s(\pi)\|_{L^2(\widehat{G})}$ is equal to $\|f\|_{L^2_s(G)}$ which is finite, we have obtained (5.10).

Let $\phi \in \mathcal{S}(G)$. By the Plancherel formula, especially (1.30), we have

$$\begin{aligned}
(f,\phi)_{L^2(G)} &= (f_s, (\mathrm{I}+\mathcal{R})^{-\frac{s}{\nu}}\phi)_{L^2(G)} \\
&= \int_{\widehat{G}} \mathrm{Tr}\left(\mathcal{F}_G\{f_s\}(\pi)\ \left(\mathcal{F}_G\{(\mathrm{I}+\mathcal{R})^{-\frac{s}{\nu}}\phi\}(\pi)\right)^*\right) d\mu(\pi) \\
&= \int_{\widehat{G}} \mathrm{Tr}\left(\pi(\mathrm{I}+\mathcal{R})^{\frac{s}{\nu}}\widehat{f}(\pi)\ \widehat{\phi}(\pi)^*\pi(\mathrm{I}+\mathcal{R})^{-\frac{s}{\nu}}\right) d\mu(\pi) \\
&= \int_{\widehat{G}} \mathrm{Tr}\left(\widehat{f}(\pi)\ \widehat{\phi}(\pi)^*\right) d\mu(\pi).
\end{aligned}$$

Note that the two functions f_s and $(\mathrm{I}+\mathcal{R})^{\frac{s}{\nu}}\phi$ are both square integrable so all the traces above are finite.

We now fix a non-negative function $\chi \in \mathcal{D}(G)$ with compact support containing 0 and satisfying $\int_G \chi = 1$. We apply what precedes to $\phi := \chi_\epsilon$ given by

$$\chi_\epsilon(y) := \epsilon^{-Q}\chi(\epsilon^{-1}y), \quad \epsilon > 0,\ y \in G,$$

and obtain

$$(f,\chi_\epsilon)_{L^2(G)} = \int_{\widehat{G}} \mathrm{Tr}\left(\widehat{f}(\pi)\ \widehat{\chi}_\epsilon(\pi)^*\right) d\mu(\pi). \tag{5.12}$$

Let us show that the right hand-side of (5.12) converges to

$$\int_{\widehat{G}} \mathrm{Tr}\left(\widehat{f}(\pi)\ \widehat{\chi}_\epsilon(\pi)^*\right) d\mu(\pi) \longrightarrow_{\epsilon\to 0} \int_{\widehat{G}} \mathrm{Tr}\left(\widehat{f}(\pi)\right) d\mu(\pi). \tag{5.13}$$

Note that the right hand-side of (5.13) is finite by (5.10).

The integrand on the left-hand side is bounded by

$$\left|\mathrm{Tr}\left(\widehat{f}(\pi)\ \widehat{\chi}_\epsilon(\pi)^*\right)\right| \leq \|\widehat{\chi}_\epsilon(\pi)\|_{\mathscr{L}(\mathcal{H}_\pi)}\mathrm{Tr}|\widehat{f}(\pi)|,$$

and

$$\|\widehat{\chi}_\epsilon(\pi)\|_{\mathscr{L}(\mathcal{H}_\pi)} \leq \|\chi_\epsilon\|_{L^1(G)} = \|\chi\|_{L^1(G)}.$$

Hence

$$\left|\mathrm{Tr}\left(\widehat{f}(\pi)\ \widehat{\chi}_\epsilon(\pi)^*\right)\right| \leq \|\chi\|_{L^1(G)}\mathrm{Tr}|\widehat{f}(\pi)|,$$

and the right-hand side is μ-integrable by (5.10).

Let us show the convergence for every $\pi \in \widehat{G}$

$$\operatorname{Tr}\left(\widehat{f}(\pi)\ \widehat{\chi}_\epsilon(\pi)^*\right) \longrightarrow_{\epsilon\to 0} \operatorname{Tr}\left(\widehat{f}(\pi)\right). \tag{5.14}$$

In order to do this, we want to estimate the difference

$$\begin{aligned}\left|\operatorname{Tr}\left(\widehat{f}(\pi)\ \widehat{\chi}_\epsilon(\pi)^*\right) - \operatorname{Tr}\left(\widehat{f}(\pi)\right)\right| &= \left|\operatorname{Tr}\left(\widehat{f}(\pi)\ \left(\widehat{\chi}_\epsilon(\pi)^* - \mathrm{I}\right)\right)\right| \\ &\leq \|\widehat{\chi}_\epsilon(\pi)^* - \mathrm{I}\|_{L^\infty(\widehat{G})} \operatorname{Tr}\left|\widehat{f}(\pi)\right|.\end{aligned}$$

Since

$$\widehat{\chi}_\epsilon(\pi)^* = \int_G \chi_\epsilon(y)\pi(y)dy = \int_G \epsilon^{-Q}\chi(\epsilon^{-1}y)\pi(y)dy = \int_G \chi(z)\pi(\epsilon z)dz,$$

and as $\int_G \chi = 1$, we have

$$\begin{aligned}\|\widehat{\chi}_\epsilon(\pi)^* - \mathrm{I}\|_{\mathscr{L}(\mathcal{H}_\pi)} &= \|\int_G \chi(z)\left(\pi(\epsilon z) - \mathrm{I}\right) dz\|_{\mathscr{L}(\mathcal{H}_\pi)} \\ &\leq \int_G |\chi(z)|\ \|\pi(\epsilon z) - \mathrm{I}\|_{\mathscr{L}(\mathcal{H}_\pi)} dz \\ &\leq \sup_{z\in\operatorname{supp}\chi} \|\pi(\epsilon z) - \mathrm{I}\|_{\mathscr{L}(\mathcal{H}_\pi)} \int_G |\chi(z)|dz.\end{aligned}$$

As π is strongly continuous and $\operatorname{supp}\chi$ compact, we know that

$$\sup_{z\in\operatorname{supp}\chi} \|\pi(\epsilon z) - \mathrm{I}\|_{\mathscr{L}(\mathcal{H}_\pi)} \longrightarrow_{\epsilon\to 0} 0.$$

This implies the convergence in (5.14) for each $\pi \in \widehat{G}$.

We can now apply Lebesgue's dominated convergence theorem to obtain the convergence in (5.13).

By the Sobolev embeddings (see Theorem 4.4.25), f is continuous on G and it is a simple exercise to show that the left hand-side of (5.12) converges to

$$(f, \chi_\epsilon)_{L^2(G)} \longrightarrow_{\epsilon\to 0} f(0).$$

Hence we have obtained the inversion formula given in (5.11) at $x = 0$. Replacing f by its left translation $f(x\,\cdot)$ which is still in $L^2_s(G)$ with the same Sobolev norm, it is then easy to obtain (5.11) for every $x \in G$.

For the case of $f \in \tilde{L}^2_s(G)$ with $s > Q/2$, we set $f_s := (\mathrm{I} + \tilde{\mathcal{R}})^{\frac{s}{\nu}} f \in L^2(G)$ and we obtain similar properties as above, ending by using right translations to obtain (5.11). $\square$

5.1.2 The spaces $\mathcal{K}_{a,b}(G)$, $\mathscr{L}_L(L^2_a(G), L^2_b(G))$, and $L^\infty_{a,b}(\widehat{G})$

In this section we describe the spaces $\mathcal{K}_{a,b}(G)$, $\mathscr{L}_L(L^2_a(G), L^2_b(G))$ and $L^\infty_{a,b}(\widehat{G})$, extending the notion of the group von Neumann algebras discussed in Section 1.8.2, to the setting of Sobolev spaces.

Definition 5.1.16 (Spaces $\mathscr{L}_L(L^2_a(G), L^2_b(G))$ and $\mathcal{K}_{a,b}(G)$)**.** Let $a, b \in \mathbb{R}$. We denote by

$$\mathscr{L}_L(L^2_a(G), L^2_b(G))$$

the subspace of operators $T \in \mathscr{L}(L^2_a(G), L^2_b(G))$ which are left-invariant.

We denote by

$$\mathcal{K}_{a,b}(G)$$

the subspace of tempered distributions $f \in \mathcal{S}'(G)$ such that the operator $\mathcal{S}(G) \ni \phi \mapsto \phi * f$ extends to a bounded operator from $L^2_a(G)$ to $L^2_b(G)$.

If a positive Rockland operator $\mathcal{R}$ of homogeneous degree ν is fixed, then the $\mathcal{K}_{a,b}(G)$-norm is defined for any $f \in \mathcal{K}_{a,b}(G)$, as the operator norm of $\phi \mapsto \phi * f$ viewed as an operator from $L^2_a(G)$ to $L^2_b(G)$, i.e.

$$\|f\|_{\mathcal{K}_{a,b}} := \|\phi \mapsto \phi * f\|_{\mathscr{L}(L^2_a(G), L^2_b(G))}. \tag{5.15}$$

Here we have considered the Sobolev norms $\phi \mapsto \|(\mathrm{I} + \mathcal{R})^{\frac{c}{\nu}}\phi\|_2$ for $c = a, b$ for $L^2_a(G)$ and $L^2_b(G)$, respectively.

The vector space $\mathscr{L}_L(L^2_a, L^2_b)$ is a Banach subspace of $\mathscr{L}(L^2_a, L^2_b)$. Since the Sobolev spaces $L^2_a(G)$ are independent of the choice of a positive Rockland operator $\mathcal{R}$ (see Section 4.4.5), so are $\mathscr{L}_L(L^2_a(G), L^2_b(G))$ and also $\mathcal{K}_{a,b}(G)$. However, the norms on these spaces do depend on a choice of a positive Rockland operator $\mathcal{R}$.

We may often write $\mathcal{K}_{a,b}$ instead of $\mathcal{K}_{a,b}(G)$ to ease the notation when no confusion is possible.

We have the immediate properties:

Proposition 5.1.17. *1. If $a = b = 0$ then*

$$\mathcal{K}_{0,0} = \mathcal{K} \quad \textit{and} \quad \mathscr{L}_L(L^2_a, L^2_b) = \mathscr{L}_L(L^2).$$

The norms $\|\cdot\|_{\mathcal{K}_{0,0}}$ and $\|\cdot\|_{\mathcal{K}}$ (defined in (5.15) and in (1.37) respectively) coincide. For any $f \in \mathcal{K}$ we have

$$\|f^*\|_{\mathcal{K}} = \|f\|_{\mathcal{K}} \quad \textit{where} \quad f^*(x) = \bar{f}(x^{-1}),$$

and

$$\forall r > 0 \qquad \|f \circ D_r\|_{\mathcal{K}} = r^{-Q}\|f\|_{\mathcal{K}}.$$

2. *Fixing a positive Rockland operator $\mathcal{R}$, the mapping $f \mapsto \|f\|_{\mathcal{K}_{a,b}}$ defines a norm on the vector space $\mathcal{K}_{a,b}$ which becomes a Banach space. Any two positive Rockland operators produce equivalent norms on $\mathcal{K}_{a,b}$.*

3. *Let $a, b \in \mathbb{R}$. We have the continuous inclusion*
$$\mathcal{K}_{a,b}(G) \subset \mathcal{S}'(G).$$
Moreover if T_f denotes the convolution operator $\phi \mapsto \phi * f$ for $f \in \mathcal{S}'(G)$, then the following are equivalent:
$$\begin{aligned} f \in \mathcal{K}_{a,b} \quad &\Longleftrightarrow \quad T_f \in \mathscr{L}_L(L^2_a(G), L^2_b(G)) \\ &\Longleftrightarrow \quad (\mathrm{I}+\mathcal{R})^{\frac{b}{\nu}} T_f (\mathrm{I}+\mathcal{R})^{-\frac{a}{\nu}} \in \mathscr{L}_L(L^2(G)) \\ &\Longleftrightarrow \quad (\mathrm{I}+\mathcal{R})^{\frac{b}{\nu}} (\mathrm{I}+\tilde{\mathcal{R}})^{-\frac{a}{\nu}} f \in \mathcal{K}(G), \end{aligned}$$
where $\mathcal{R}$ is any positive Rockland operator of homogeneous degree ν.

4. *For any $c_1, c_2 \geq 0$ we have the inclusions*
$$\mathscr{L}_L(L^2_a, L^2_b) \subset \mathscr{L}_L(L^2_{a+c_1}, L^2_{b-c_2})$$
and
$$\mathcal{K}_{a,b} \subset \mathcal{K}_{a+c_1, b-c_2}.$$

5. *If $f \in \mathcal{K}_{a,b}$ then $X^\alpha f \in \mathcal{K}_{a,b-[\alpha]}$ for any $\alpha \in \mathbb{N}_0^n$ and $(\mathrm{I}+\mathcal{R})^{s/\nu} f \in \mathcal{K}_{a,b-s}$ for any $s \in \mathbb{R}$. Furthermore, X^α and $(\mathrm{I}+\mathcal{R})^{s/\nu}$ are bounded on $\mathcal{K}_{a,b}$:*
$$\|X^\alpha f\|_{\mathcal{K}_{a,b-[\alpha]}} \leq C_{a,b,[\alpha]} \|f\|_{\mathcal{K}_{a,b}}$$
and
$$\|(\mathrm{I}+\mathcal{R})^{s/\nu} f\|_{\mathcal{K}_{a,b-s}} \leq C'_{a,b,s} \|f\|_{\mathcal{K}_{a,b}}$$
for some positive finite constants $C_{a,b,[\alpha]}$ and $C'_{a,b,s}$ independent of f.

If $-a$ and b are in $\nu\mathbb{N}_0$, a norm equivalent to the $\mathcal{K}_{a,b}$-norm is
$$f \longmapsto \sum_{[\alpha]\leq -a,\ [\beta]\leq b} \|\tilde{X}^\alpha X^\beta f\|_{\mathcal{K}},$$
and if $a' \in [a, 0]$ and $b' \in [0, b]$ then
$$\|f\|_{\mathcal{K}_{a',b'}} \leq C_{a,b,a',b,\mathcal{R}} \sum_{[\alpha]\leq -a,\ [\beta]\leq b} \|\tilde{X}^\alpha X^\beta f\|_{\mathcal{K}}.$$

The definitions of the tempered distributions $X^\alpha f$ and $(\mathrm{I}+\mathcal{R})^{s/\nu} f$ were recalled in (5.8) and (5.9) respectively. For the proper definition of the operators $(\mathrm{I}+\mathcal{R})^{\frac{b}{\nu}}$, $(\mathrm{I}+\tilde{\mathcal{R}})^{-\frac{a}{\nu}}$, see Definitions 4.3.17 and 4.4.31.

Proof of Proposition 5.1.17. Part (1) follows from the properties of the von Neumann-algebras $\mathcal{K}(G)$ and $\mathscr{L}_L(L^2(G))$ as well as from the following two easy observations:

$$\forall \psi \in L^2(G) \qquad \|\psi \circ D_r\|_2 = r^{-\frac{Q}{2}} \|\psi\|_2,$$

and for any $f \in \mathcal{K}$, $\phi \in \mathcal{S}(G)$ and $r > 0$,

$$\phi * (f \circ D_r)(x) = r^{-Q} \left(\left(\phi \circ D_{\frac{1}{r}} \right) * f \right) (rx).$$

Part (2) is easy to check. Part (3) follows from the Schwartz kernel theorem, see Corollary 3.2.1. Parts (4) and (5), follow easily from the properties of the Sobolev spaces and Part (3). □

We now show that we can make sense of convolution of distributions in some $\mathcal{K}_{a,b}(G)$-spaces. The following lemma is almost immediate to check.

Lemma 5.1.18. *Let $f \in \mathcal{K}_{a,b}(G)$ and $g \in \mathcal{K}_{b,c}(G)$ for $a, b, c \in \mathbb{R}$, and let $T_f : \phi \mapsto \phi * f$ and $T_g : \phi \mapsto \phi * g$ be the associated operators. Then the operator $T_g T_f$ is continuous from $L^2_a(G)$ to $L^2_c(G)$ and its right-convolution kernel (as a continuous linear operator from $\mathcal{S}(G)$ to $\mathcal{S}'(G)$) is denoted by $h \in \mathcal{K}_{a,c}(G)$.*

If (f_n) and (g_n) are sequences of Schwartz functions converging to f in $\mathcal{K}_{a,b}(G)$ and g in $\mathcal{K}_{b,c}(G)$, respectively, then h is the limit of $f_n * g_n$ in $\mathcal{K}_{a,c}(G)$.

Consequently, with the notation of the lemma above, h coincides with the convolution of f with g whenever the convolution of f with g makes any technical sense, for instance, if the tempered distributions f and g (which are already assumed to be in $\mathcal{K}_{a,b}(G)$ and $\mathcal{K}_{b,c}(G)$ respectively) satisfy

- f and g are locally integrable functions with $|f| * |g| \in L^1(G)$,
- or at least one of the distributions f or g has compact support,
- or at least one of the distributions f or g is Schwartz.

Hence we may extend the notation and define:

Definition 5.1.19. If $f \in \mathcal{K}_{a,b}(G)$ and $g \in \mathcal{K}_{b,c}(G)$ for $a, b, c \in \mathbb{R}$, and $T_f : \phi \mapsto \phi * f$, $T_g : \phi \mapsto \phi * g$ are the associated operators, we denote by $f * g$ the distribution in $\mathcal{K}_{a,c}(G)$ which is the right convolution kernel of $T_g T_f$.

We obtain easily the following properties:

Corollary 5.1.20. *Let $f \in \mathcal{K}_{a,b}(G)$ and $g \in \mathcal{K}_{b,c}(G)$ for $a, b, c \in \mathbb{R}$. Then we have the following property of associativity for any $\phi \in \mathcal{S}(G)$*

$$\phi * (f * g) = (\phi * f) * g,$$

and more generally for any $h \in \mathcal{K}_{c,d}(G)$ (where $d \in \mathbb{R}$)

$$f * (g * h) = (f * g) * h,$$

as convolutions of an element of $\mathcal{K}_{a,b}(G)$ with an element of $\mathcal{K}_{b,d}(G)$ for the left-hand side, and of an element of $\mathcal{K}_{a,c}(G)$ with an element of $\mathcal{K}_{c,d}(G)$ for the right-hand side.

The rest of this section is devoted to the definition of the group Fourier transform of a distribution in $\mathcal{K}_{a,b}(G)$. We start by defining what will turn out to be the image of the group Fourier transform on $\mathcal{K}_{a,b}(G)$. We recall that $L^\infty(\widehat{G})$ is the space of measurable fields of operators on $\widehat{G}$ which are uniformly bounded, see Definition 1.8.8.

Definition 5.1.21. Let $a, b \in \mathbb{R}$. We denote by $L^\infty_{a,b}(\widehat{G})$ the space of fields of operators $\sigma = \{\sigma_\pi : \mathcal{H}^\infty_\pi \to \mathcal{H}^b_\pi\,,\ \pi \in \widehat{G}\}$ satisfying

$$\exists C > 0 \quad \forall \phi \in \mathcal{S}(G) \qquad \|\sigma\widehat{\phi}\|_{L^2_b(\widehat{G})} \leq C\|\phi\|_{L^2_a(G)}. \tag{5.16}$$

Here we assume that a positive Rockland operator has been fixed to define the norms on $L^2_b(\widehat{G})$ and $L^2_a(G)$.

For such a field σ, $\|\sigma\|_{L^\infty_{a,b}(\widehat{G})}$ denotes the infimum of the constant $C > 0$ satisfying (5.16).

We may sometimes abuse the notation and write $\|\sigma_\pi\|_{L^\infty_{a,b}(\widehat{G})}$ when no confusion is possible.

Note that as $\phi \in \mathcal{S}(G)$, its group Fourier transform acts on smooth vectors, see Example 1.8.18. Hence the composition $\sigma\widehat{\phi}$ above makes sense, see Section 1.8.3.

Naturally, the space $L^\infty_{a,b}(\widehat{G})$ introduced in Definition 5.1.21 is independent of the choice of a Rockland operator used to define the norms on $L^2_b(\widehat{G})$ and $L^2_a(G)$:

Lemma 5.1.22. *If $\{\sigma_\pi : \mathcal{H}^\infty_\pi \to \mathcal{H}^b_\pi, \pi \in \widehat{G}\}$ satisfies the condition in Definition 5.1.21 for one positive Rockland operator, then it satisfies the same property for any positive Rockland operator. Moreover, if $\mathcal{R}_1$ and $\mathcal{R}_2$ are two positive Rockland operators, and if $\|\sigma\|_{L^\infty_{a,b,\mathcal{R}_1}(\widehat{G})}$ and $\|\sigma\|_{L^\infty_{a,b,\mathcal{R}_2}(\widehat{G})}$ denote the corresponding infima, then there exists $C > 0$ independent of σ such that*

$$C^{-1}\|\sigma\|_{L^\infty_{a,b,\mathcal{R}_2}(\widehat{G})} \leq \|\sigma\|_{L^\infty_{a,b,\mathcal{R}_1}(\widehat{G})} \leq C\|\sigma\|_{L^\infty_{a,b,\mathcal{R}_2}(\widehat{G})}.$$

Proof. This follows easily from the independence of the Sobolev spaces on G and $\widehat{G}$ of the positive Rockland operators, see Section 4.4.5 and Proposition 5.1.14. □

If the field acts on smooth vectors, we can simplify Definition 5.1.21:

Lemma 5.1.23. *Let $\sigma = \{\sigma_\pi : \mathcal{H}^\infty_\pi \to \mathcal{H}^\infty_\pi\,,\ \pi \in \widehat{G}\}$ be a field acting on smooth vectors. Then $\sigma \in L^\infty_{a,b}(\widehat{G})$ if and only if*

$$\{\pi(\mathrm{I}+\mathcal{R})^{\frac{b}{\nu}}\sigma_\pi\,\pi(\mathrm{I}+\mathcal{R})^{-\frac{a}{\nu}} : \mathcal{H}^\infty_\pi \to \mathcal{H}^\infty_\pi, \pi \in \widehat{G}\} \in L^\infty(\widehat{G}), \tag{5.17}$$

where $\mathcal{R}$ is a positive Rockland operator of degree ν, and in this case,

$$\|\sigma\|_{L^\infty_{a,b}(\widehat{G})} = \|\pi(\mathrm{I}+\mathcal{R})^{\frac{b}{\nu}}\sigma_\pi\, \pi(\mathrm{I}+\mathcal{R})^{-\frac{a}{\nu}}\|_{L^\infty(\widehat{G})}.$$

Proof. This follows easily from the density of $\mathcal{S}(G)$ in $L^2_b(G)$. □

Note that the composition in (5.17) makes sense as all the fields involved act on smooth vectors. In Corollary 5.1.30, we will see a sufficient condition (which will be useful later) for a field to be acting on smooth vectors.

We can now characterise the elements of $\mathcal{K}_{a,b}(G)$ in terms of $L^\infty_{a,b}(\widehat{G})$:

Proposition 5.1.24. *Let $a, b \in \mathbb{R}$.*

(i) If $\sigma \in L^\infty_{a,b}(\widehat{G})$, then the operator $T_\sigma : \mathcal{S}(G) \to \mathcal{S}'(G)$ defined via

$$\widehat{T_\sigma\phi}(\pi) := \sigma_\pi \widehat{\phi}(\pi), \quad \phi \in \mathcal{S}(G),\ \pi \in \widehat{G}, \tag{5.18}$$

extends uniquely to an operator in $\mathscr{L}(L^2_a, L^2_b)$. Moreover,

$$\|T_\sigma\|_{\mathscr{L}(L^2_a, L^2_b)} = \|\sigma\|_{L^\infty_{a,b}(\widehat{G})}, \tag{5.19}$$

where the Sobolev norms are defined using a chosen positive Rockland operator $\mathcal{R}$ with homogeneous degree ν. The right convolution kernel $f \in \mathcal{S}'(G)$ of T_σ is in $\mathcal{K}_{a,b}(G)$.

(ii) Conversely, if $f \in \mathcal{K}_{a,b}(G)$ then there exists a unique $\sigma \in L^\infty_{a,b}(\widehat{G})$ such that

$$\widehat{\phi * f}(\pi) = \sigma_\pi \widehat{\phi}(\pi), \quad \phi \in \mathcal{S}(G),\ \pi \in \widehat{G}. \tag{5.20}$$

Furthermore, if f is also in any of the spaces where the group Fourier transform has already been defined, namely any Sobolev space $L^2_a(G)$ or $\mathcal{K}(G)$, then $\sigma = \{\sigma_\pi, \pi \in \widehat{G}\}$ will coincide with the group Fourier transform of f.

Proof. The properties of T_σ in Part (i) follow from the Plancherel theorem (Theorem 1.8.11) and the density of $\mathcal{S}(G)$ in $L^2(G)$. The right convolution kernel $f \in \mathcal{S}'(G)$ of T_σ is in $\mathcal{K}_{a,b}(G)$ by Proposition 5.1.17.

Conversely, let $f \in \mathcal{K}_{a,b}(G)$. By assumption the operator $T_f : \mathcal{S}(G) \ni \phi \mapsto \phi * f$ admits a bounded extension from $L^2_a(G)$ to $L^2_b(G)$. Thus the operator $(\mathrm{I}+\mathcal{R})^{\frac{b}{\nu}} T_f (\mathrm{I}+\mathcal{R})^{-\frac{a}{\nu}}$ is bounded on $L^2(G)$ and we denote by $f_{a,b} \in \mathcal{K}(G)$ its right convolution kernel. For any $\phi \in \mathcal{S}(G)$, we have $\phi_a := (\mathrm{I}+\mathcal{R})^{\frac{a}{\nu}}\phi \in \mathcal{S}(G)$ by Corollary 4.3.16 thus $\phi_a * f_{a,b} \in L^2(G)$ and we have

$$T_f\phi \in L^2_b(G) \quad \text{with} \quad T_f\phi = (\mathrm{I}+\mathcal{R})^{-\frac{b}{\nu}}(\phi_a * f_{a,b}).$$

Consequently $\mathcal{F}_G(T_f\phi) \in L^2_b(\widehat{G})$ and

$$\mathcal{F}_G(T_f\phi) = \pi(\mathrm{I}+\mathcal{R})^{-\frac{b}{\nu}} \widehat{f}_{a,b}\widehat{\phi}_a = \pi(\mathrm{I}+\mathcal{R})^{-\frac{b}{\nu}} \widehat{f}_{a,b}\pi(\mathrm{I}+\mathcal{R})^{\frac{a}{\nu}}\widehat{\phi}.$$

One checks easily that $\{\sigma_\pi : \mathcal{H}_\pi^\infty \to \mathcal{H}_\pi^b, \pi \in \widehat{G}\}$ defined via

$$\sigma_\pi := \pi(\mathrm{I}+\mathcal{R})^{-\frac{b}{\nu}} \widehat{f}_{a,b}(\pi)\, \pi(\mathrm{I}+\mathcal{R})^{\frac{a}{\nu}}$$

is in $L^\infty_{a,b}(\widehat{G})$ and satisfies (5.20). The rest of the proof of Part (ii) follows easily from the computations above and the uniqueness of the group Fourier transforms already defined. □

Thanks to Proposition 5.1.24, we can extend the definition of the group Fourier transform to $\mathcal{K}_{a,b}(G)$:

Definition 5.1.25 (The group Fourier transform on $\mathcal{K}_{a,b}(G)$)**.** The group Fourier transform of $f \in \mathcal{K}_{a,b}(G)$ is the field of operators $\{\sigma_\pi : \mathcal{H}_\pi^\infty \to \mathcal{H}_\pi^b, \pi \in \widehat{G}\}$ in $L^\infty_{a,b}(\widehat{G})$ associated to f by Proposition 5.1.24, and we write

$$\widehat{f}(\pi) := \pi(f) := \sigma_\pi, \quad \pi \in \widehat{G}.$$

As the next example implies, any left-invariant vector field is in some $\mathcal{K}_{a,b}(G)$ and their Fourier transform can be defined via Definition 5.1.25. As is shown in the proof below, this coincides with the infinitesimal representation of the corresponding element of $\mathfrak{U}(\mathfrak{g})$ defined in Section 1.7.

Example 5.1.26. Let $\alpha \in \mathbb{N}_0^n$. The operator X^α is in $\mathscr{L}(L^2_{[\alpha]}(G), L^2(G))$ and more generally in $\mathscr{L}(L^2_{[\alpha]+s}(G), L^2_s(G))$ for any $s \in \mathbb{R}$. Its right convolution kernel is the distribution $X^\alpha\delta_0$ defined via (see (5.8))

$$\langle X^\alpha\delta_0, \phi\rangle = \langle \delta_0, \{X^\alpha\}^t\phi\rangle = \{X^\alpha\}^t\phi(0),$$

which is in $\mathcal{K}_{[\alpha],0}$, and more generally in $\mathcal{K}_{s+[\alpha],s}$ for any $s \in \mathbb{R}$. Its group Fourier transform is

$$\mathcal{F}_G(X^\alpha\delta_0)(\pi) = \pi(X^\alpha) = \pi(X)^\alpha$$

and coincides with the infinitesimal representation on $\mathfrak{U}(\mathfrak{g})$. It is in $L^\infty_{s+[\alpha],s}(\widehat{G})$ for any $s \in \mathbb{R}$.

Proof. By Theorem 4.4.16, X^α maps $L^2_{[\alpha]}(G)$ continuously to $L^2(G)$ and, more generally, $L^2_{s+[\alpha]}(G)$ continuously to $L^2_s(G)$.

By Proposition 1.7.6, we have for any $\phi \in \mathcal{S}(G)$

$$\mathcal{F}_G(X^\alpha\phi)(\pi) = \pi(X^\alpha)\widehat{\phi}(\pi) = \pi(X)^\alpha\widehat{\phi}(\pi).$$

This shows that $\mathcal{F}_G(X^\alpha\delta_0)$ coincides with $\{\pi(X^\alpha), \pi \in \widehat{G}\}$. □

As our next example shows, when multipliers in a positive Rockland operator are in $\mathscr{L}_L(L^2_s(G), L^2_{s-b}(G))$, the group Fourier transform of their right convolution kernels can also be given via the functional calculus of the Rockland operators:

Example 5.1.27. Let $\mathcal{R}$ be a positive Rockland operator of homogeneous degree ν. Let m be a measurable function on $[0, \infty)$ satisfying

$$\exists C > 0 \quad \forall \lambda \geq 0 \quad |m(\lambda)| \leq C(1+\lambda)^{\frac{b}{\nu}}.$$

Then the operator $m(\mathcal{R})$ defined by the functional calculus of $\mathcal{R}$ extends uniquely to an operator in $\mathscr{L}_L(L^2_{s+b}(G), L^2_s(G))$ for any $s \in \mathbb{R}$. Its right convolution kernel $m(\mathcal{R})\delta_0$ is in $\mathcal{K}_{s+b,s}$ for any $s \in \mathbb{R}$. Its group Fourier transform is

$$\mathcal{F}_G(m(\mathcal{R})\delta_0)(\pi) = m(\pi(\mathcal{R}))$$

defined by the functional calculus of $\pi(\mathcal{R})$. It is in $L^\infty_{s+b,s}(\widehat{G})$ for any $s \in \mathbb{R}$. For a fixed $s \in \mathbb{R}$, we have

$$\begin{aligned} \|m(\mathcal{R})\|_{\mathscr{L}_L(L^2_{s+b}(G), L^2_s(G))} = \|m(\mathcal{R})\delta_0\|_{\mathcal{K}_{s+b,s}} &= \|m(\pi(\mathcal{R}))\|_{L^\infty_{s+b,s}(\widehat{G})} \\ &\leq \sup_{\lambda>0} (1+\lambda)^{-\frac{b}{\nu}} |m(\lambda)|, \end{aligned}$$

if we realise the Sobolev norms with $\mathcal{R}$.

We refer to Section 4.1.3 and Corollary 4.1.16 for the properties of the functional calculus of $\mathcal{R}_2$ and $\pi(\mathcal{R})$.

Proof. The function m_1 given by

$$m_1(\lambda) := m(\lambda)(1+\lambda)^{-\frac{b}{\nu}}, \quad \lambda \geq 0,$$

is measurable and bounded on $[0, \infty)$. The operator $m_1(\mathcal{R})$ defined by the functional calculus of $\mathcal{R}$ is therefore bounded on $L^2(G)$ with

$$\|m_1(\mathcal{R})\|_{\mathscr{L}(L^2(G))} \leq \sup_{\lambda \geq 0} |m_1(\lambda)|.$$

Again from the properties of the functional calculus of $\mathcal{R}$, we also have

$$m(\mathcal{R}) \supset m_1(\mathcal{R})(\mathrm{I}+\mathcal{R})^{\frac{b}{\nu}},$$

in the sense of operators. Since $\mathrm{Dom}(\mathrm{I}+\mathcal{R})^{b/\nu} \supset \mathcal{S}(G)$ (see Corollary 4.3.16), this shows that the domain of $m(\mathcal{R})$ contains $\mathcal{S}(G)$ and that

$$m_1(\mathcal{R}) = m(\mathcal{R})(\mathrm{I}+\mathcal{R})^{-\frac{b}{\nu}} \quad \text{on } \mathcal{S}(G).$$

The properties of the functional calculus of $\mathcal{R}$ yield for any $s \in \mathbb{R}$,

$$\begin{aligned} \|m_1(\mathcal{R})\|_{\mathscr{L}(L^2(G))} &= \|m_1(\mathcal{R})\|_{\mathscr{L}(L^2_s(G))} \\ &= \|m(\mathcal{R})(\mathrm{I}+\mathcal{R})^{-\frac{b}{\nu}}\|_{\mathscr{L}(L^2_s(G))} \\ &= \|m(\mathcal{R})\|_{\mathscr{L}(L^2_{s+b}(G), L^2_s(G))}. \end{aligned}$$

By Corollary 4.1.16, the kernel of $m_1(\mathcal{R})$ is the tempered distribution $m_1(\mathcal{R})\delta_0$ with Fourier transform $\{m_1(\pi(\mathcal{R})), \pi \in \widehat{G}\}$. Adapting the proof of Corollary 4.1.16, we see that

$$m_1(\pi(\mathcal{R})) = m(\pi(\mathcal{R}))(\mathrm{I} + \pi(\mathcal{R}))^{-\frac{b}{\nu}} \quad \text{on } \mathcal{H}_\pi^\infty, \quad \pi \in \widehat{G}.$$

It is now straightforward to check that the kernel of the operator $m(\mathcal{R})$ is in $\mathcal{K}_{s+b,s}$ and its Fourier transform is $\{m(\pi(\mathcal{R})), \pi \in \widehat{G}\}$. □

Naturally, any Schwartz function is in any $\mathcal{K}_{a,b}$ and one can readily estimate the associated norm:

Example 5.1.28. If $\phi \in \mathcal{S}(G)$, then for any $a, b \in \mathbb{R}$, the operator $T_\phi : \psi \mapsto \psi * \phi$ is in $\mathscr{L}(L_a^2(G), L_b^2(G))$, $\phi \in \mathcal{K}_{a,b}$ and $\widehat{\phi} \in L_{a,b}^\infty$. If we fix a positive Rockland operator $\mathcal{R}$ of homogeneous degree ν, then we have

$$\|T_\phi\|_{\mathscr{L}(L_a^2(G), L_b^2(G))} = \|\phi\|_{\mathcal{K}_{a,b}} = \|\widehat{\phi}\|_{L_{a,b}^\infty} \leq \|(\mathrm{I} + \mathcal{R})^{\frac{b}{\nu}}(\mathrm{I} - \tilde{\mathcal{R}})^{-\frac{a}{\nu}}\phi\|_{L^1(G)} < \infty,$$

where the norms on $\mathscr{L}(L_a^2(G), L_b^2(G))$, $\mathcal{K}_{a,b}$ and $L_{a,b}^\infty$ are defined with $\mathcal{R}$.

With Definition 5.1.25, we can reformulate Proposition 5.1.24 and parts of Proposition 5.1.17 and Corollary 5.1.20 as the following proposition.

Proposition 5.1.29. *1. Let $a, b \in \mathbb{R}$. The Fourier transform $\mathcal{F}_G$ maps $\mathcal{K}_{a,b}(G)$ onto $L_{a,b}^\infty(\widehat{G})$. Furthermore, $\mathcal{F}_G : \mathcal{K}_{a,b}(G) \to L_{a,b}^\infty(\widehat{G})$ is an isomorphism between Banach spaces. In particular, for $f \in \mathcal{K}_{a,b}(G)$,*

$$\|f\|_{\mathcal{K}_{a,b}} = \|\widehat{f}\|_{L_{a,b}^\infty(\widehat{G})}.$$

It coincides with the Fourier transform on $\mathcal{K}(G)$ for $a = b = 0$.

2. If $\sigma_1 \in L_{a_1,b_1}^\infty(\widehat{G})$ and $\sigma_2 \in L_{a_2,b_2}^\infty(\widehat{G})$ with $b_2 = a_1$, then their product $\sigma_1\sigma_2$ makes sense as the element of $L_{a_2,b_1}^\infty(\widehat{G})$ given by the Fourier transform of $(\mathcal{F}_G^{-1}\sigma_2) * (\mathcal{F}_G^{-1}\sigma_1)$.

In other words, if $f_1 \in \mathcal{K}_{a_1,b_1}(\widehat{G})$ and $f_2 \in \mathcal{K}_{a_2,b_2}(\widehat{G})$ with $b_2 = a_1$, then the Fourier transform of $f_2 * f_1 \in \mathcal{K}_{a_2,b_1}(\widehat{G})$ is

$$\mathcal{F}_G(f_2 * f_1) = \mathcal{F}_G(f_1)\mathcal{F}_G(f_2).$$

3. Let $\sigma = \{\sigma_\pi : \mathcal{H}_\pi^\infty \to \mathcal{H}_\pi, \pi \in \widehat{G}\} \in L_{a,b}^\infty(\widehat{G})$. Then we have for any $\alpha \in \mathbb{N}_0^n$,

$$\{\pi(X)^\alpha\sigma_\pi : \mathcal{H}_\pi^\infty \to \mathcal{H}_\pi, \pi \in \widehat{G}\} \in L_{a,b-[\alpha]}^\infty(\widehat{G}), \tag{5.21}$$

and for any $s \in \mathbb{R}$,

$$\{\pi(\mathrm{I} + \mathcal{R})^{s/\nu}\sigma_\pi : \mathcal{H}_\pi^\infty \to \mathcal{H}_\pi, \pi \in \widehat{G}\} \in L_{a,b-s}^2(\widehat{G}). \tag{5.22}$$

Furthermore, if $f = \mathcal{F}_G^{-1}\sigma \in \mathcal{K}_{a,b}(G)$ *then*

$$\mathcal{F}_G(X^\alpha f)(\pi) = \pi(X)^\alpha \widehat{f}(\pi) \quad \text{and} \quad \mathcal{F}_G((\mathrm{I}+\mathcal{R})^{s/\nu} f)(\pi) = \pi(\mathrm{I}+\mathcal{R})^{s/\nu}\widehat{f}(\pi).$$

The fields of operators in (5.21) and (5.22) are understood as compositions of fields of operators in $L^\infty_{a_2,b_2}$ and $L^\infty_{a_1,b_1}$ with $b_2 = a_1$, see Part 2 and Examples 5.1.26 and 5.1.27.

With the help of Proposition 5.1.29, we can now give a usefull sufficient condition for a field to act on smooth vectors and reformulate Corollary 4.4.10 into

Corollary 5.1.30. *Let* $a, b \in \mathbb{R}$ *and let* $\{\gamma_\ell, \ell \in \mathbb{Z}\}$ *be a sequence of real numbers which tends to* $\pm\infty$ *as* $\ell \to \pm\infty$. *Let* $\sigma \in L^\infty_{a+\gamma_\ell, b+\gamma_\ell}(\widehat{G})$ *for every* $\ell \in \mathbb{Z}$. *Then* σ *is a field of operators acting on smooth vectors:*

$$\sigma = \{\sigma_\pi : \mathcal{H}_\pi^\infty \to \mathcal{H}_\pi^\infty, \pi \in \widehat{G}\}.$$

Furthermore $\sigma \in L^\infty_{a+\gamma,b+\gamma}(\widehat{G})$ *for every* $\gamma \in \mathbb{R}$ *and for any* $c \geq 0$, *we have*

$$\sup_{|\gamma|\leq c} \|\sigma\|_{L^\infty_{a+\gamma,b+\gamma}(\widehat{G})} \leq C_c \max\left(\|\sigma\|_{L^\infty_{a+\gamma_\ell,b+\gamma_\ell}(\widehat{G})}, \|\sigma\|_{L^\infty_{a+\gamma_{-\ell},b+\gamma_{-\ell}}(\widehat{G})} \right),$$

where $\ell \in \mathbb{N}_0$ *is the smallest integer such that* $\gamma_\ell \geq c$ *and* $-\gamma_{-\ell} \geq c$.

Proof. By Proposition 5.1.29, $\pi(X)^\alpha \sigma \in L^\infty_{a+\gamma_\ell, b+\gamma_\ell-[\alpha]}$ for every $\alpha \in \mathbb{N}_0^n$ and every $\ell \in \mathbb{Z}$. Thus choosing $\gamma_\ell \geq [\alpha] - b$, we have $\pi(X)^\alpha \sigma\widehat{\phi} \in L^2(\widehat{G})$ for every $\phi \in \mathcal{S}(G)$. Realising $\pi \in \widehat{G}$ as a representation of G and fixing $v \in \mathcal{H}_\pi^\infty$, this implies that the mapping $x \mapsto \pi(x)\sigma_\pi\widehat{\phi}(\pi)v$ is smooth. Hence $\sigma_\pi\widehat{\phi}(\pi)v$ is smooth and $\sigma\widehat{\phi}$ acts on smooth vectors. As this holds for every $\phi \in \mathcal{D}(G)$, so does σ by Lemma 1.8.19. We conclude with Corollary 4.4.10. □

We end this section with one more technical property:

Lemma 5.1.31. *Let* $\sigma \in L^\infty_{a,b}(\widehat{G})$ *where* $a, b \in \mathbb{R}$. *Let* $\phi \in \mathcal{S}(G)$. *Then we have* $\sigma\widehat{\phi} \in \tilde{L}^2_s(\widehat{G})$ *for any* $s \in \mathbb{R}$ *and*

$$\int_{\widehat{G}} \mathrm{Tr} \left|\sigma_\pi \widehat{\phi}(\pi)\right| d\mu(\pi) < \infty. \tag{5.23}$$

Setting $f := \mathcal{F}_G^{-1}\sigma \in \mathcal{K}_{a,b}$, *the function* $\phi * f$ *is smooth and we have for any* $x \in G$ *the equality*

$$\phi * f(x) = \int_{\widehat{G}} \mathrm{Tr}\left(\pi(x)\sigma_\pi\widehat{\phi}(\pi)\right) d\mu(\pi).$$

Remark 5.1.32. The composition $\sigma\widehat{\phi}$ makes sense since σ is defined on smooth vectors and $\widehat{\phi}$ acts on smooth vectors. The composition $\pi(x)\sigma_\pi\pi(\phi)$ makes sense since $\pi(x)$ is bounded and $\sigma\widehat{\phi}$ is bounded (even in Hilbert Schmidt classes) since it is stated first that $\sigma\widehat{\phi} \in \tilde{L}^2_s(\widehat{G})$ for any s, hence in particular in $L^2(\widehat{G})$.

Proof. Let T_σ be the operator with right convolution kernel $f := \mathcal{F}_G^{-1}\sigma$. Then $T_\sigma \in \mathscr{L}(L^2_a(G), L^2_b(G))$ and $T_\sigma^* T_\sigma$ extends to an operator in $\mathscr{L}(L^2_a(G))$. For any $\phi \in \mathcal{S}(G)$, the definition of the adjoint and the duality between Sobolev spaces yield

$$\begin{aligned} \|T_\sigma \phi\|^2_{L^2(G)} &= \langle T_\sigma^* T_\sigma \phi, \bar{\phi}\rangle_{L^2_a(G)\times L^2_{-a}(G)} \\ &\leq \|T_\sigma^* T_\sigma\|_{\mathscr{L}(L^2_a(G))} \|\phi\|_{L^2_a(G)} \|\phi\|_{L^2_{-a}(G)}. \end{aligned}$$

This last expression is finite since $T_\sigma^* T_\sigma \in \mathscr{L}(L^2_a(G))$ and $\mathcal{S}(G) \subset L^2_{s'}(G)$ for any $s' \in \mathbb{R}$. Thus $T_\sigma \phi \in L^2(G)$ and its Fourier transform is $\sigma\widehat{\phi} \in L^2(\widehat{G})$. For any $s \in \mathbb{R}$, we may replace ϕ with $\phi_s = (\mathrm{I} + \mathcal{R})^{s/\nu}\phi \in \mathcal{S}(G)$ and $\sigma\widehat{\phi}_s \in L^2(\widehat{G})$ yields $\sigma\widehat{\phi} \in L^2_s(\widehat{G})$.

Applying Proposition 5.1.15 to $\sigma\widehat{\phi} \in \tilde{L}^2_s(\widehat{G})$ for some $s > Q/2$, we obtain (5.23). Note that $f := \mathcal{F}_G^{-1}\sigma$ is a tempered distribution so $\phi * f$ is smooth (see Lemma 3.1.55). The group Fourier transform of $\phi * f$ is $\sigma\widehat{\phi}$ by Proposition 5.1.29 Part 2 and Example 5.1.28. We now conclude with the inversion formula given in Proposition 5.1.15. □

5.1.3 Symbols and associated kernels

In this section we aim at establishing a one-to-one correspondence between a collection σ of operators parametrised by $G \times \widehat{G}$ and a function κ; this function will turn out to be the kernel of the operator naturally associated to σ. For the abstract setting behind measurable fields of operators and some of their properties we refer to Section B.1.6, especially to Proposition B.1.17, as well as Section 1.8.3.

Definition 5.1.33 (Symbols). A *symbol* is a field of operators $\{\sigma(x,\pi) : \mathcal{H}_\pi^\infty \to \mathcal{H}_\pi, \pi \in \widehat{G}\}$ depending on $x \in G$, satisfying for each $x \in G$

$$\exists a, b \in \mathbb{R} \qquad \sigma(x,\cdot) := \{\sigma(x,\pi) : \mathcal{H}_\pi^\infty \to \mathcal{H}_\pi, \pi \in \widehat{G}\} \in L^\infty_{a,b}(\widehat{G}).$$

Here we use the usual identifications of a strongly continuous irreducible unitary representation from $\mathrm{Rep}\, G$ with its equivalence class in $\widehat{G}$, and of a field of operators acting on the smooth vectors parametrised by $\widehat{G}$ with its equivalence class with respect to the Plancherel measure μ.

We will usually assume that the symbols are uniformly regular in x:

Definition 5.1.34 (Continuous and smooth symbols).

- A symbol $\{\sigma(x,\pi) : \mathcal{H}_\pi^\infty \to \mathcal{H}_\pi, \pi \in \widehat{G}\}$ is said to be *continuous* in $x \in G$ whenever there exists $a, b \in \mathbb{R}$ such that

$$\forall x \in G \quad \sigma(x,\cdot) := \{\sigma(x,\pi) : \mathcal{H}_\pi^\infty \to \mathcal{H}_\pi, \pi \in \widehat{G}\} \in L^\infty_{a,b}(\widehat{G}),$$

  and the map $x \mapsto \sigma(x,\cdot)$ is continuous from $G \sim \mathbb{R}^n$ to the Banach space $L^\infty_{a,b}(\widehat{G})$.

- A symbol $\sigma = \{\sigma(x,\pi) : \mathcal{H}_\pi^\infty \to \mathcal{H}_\pi, \pi \in \widehat{G}\}$ is said to be *smooth* in $x \in G$ whenever it is a field of operators depending smoothly in $x \in G$ (see Remark 1.8.16) and, for every $\beta \in \mathbb{N}_0^n$, the field $\{\partial_x^\beta \sigma(x,\pi) : \mathcal{H}_\pi^\infty \to \mathcal{H}_\pi, \pi \in \widehat{G}\}$ is continuous.

Important note: In the sequel, whenever we talk about symbols (on graded Lie groups), we always mean the symbols which are smooth in $x \in G$ in the sense of Definition 5.1.34 unless stated otherwise.

For a symbol as in Definition 5.1.34, we will usually write

$$\sigma = \{\sigma(x,\pi), (x,\pi) \in G \times \widehat{G}\},$$

but we may sometimes abuse the notation and refer to the symbol simply as $\sigma(x,\pi)$.

Lemma 5.1.35. *If* $\sigma = \{\sigma(x,\pi), (x,\pi) \in G \times \widehat{G}\}$ *is a symbol, then*

$$\kappa_x := \mathcal{F}_G^{-1}\{\sigma(x,\cdot)\}$$

is a tempered distribution and the map

$$G \ni x \longmapsto \kappa_x \in \mathcal{S}'(G)$$

is smooth.

In other words,

$$\kappa \in C^\infty(G, \mathcal{S}'(G)).$$

Here $C^\infty(G,\mathcal{S}'(G))$ denotes the set of smooth functions from G to $\mathcal{S}'(G)$.

Proof. As σ is a smooth symbol, for every $\beta \in \mathbb{N}_0^n$, there exists $a_\beta, b_\beta \in \mathbb{R}$ such that $G \ni x \mapsto \partial_x^\beta \sigma(x,\cdot) \in L^\infty_{a_\beta,b_\beta}(\widehat{G})$ is continuous. By Proposition 5.1.29, composing this with $\mathcal{F}_G^{-1}$ implies that $G \ni x \mapsto \partial_x^\beta \kappa_x \in \mathcal{K}_{a_\beta,b_\beta}$ is continuous. Since the inclusion $\mathcal{K}_{a_\beta,b_\beta} \subset \mathcal{S}'(G)$ is continuous, this implies that each map $G \ni x \mapsto \partial_x^\beta \kappa_x \in \mathcal{S}'(G)$ is continuous. Hence $G \ni x \mapsto \kappa_x \in \mathcal{S}'(G)$ is smooth. □

Definition 5.1.36 (Associated kernels). If σ is a symbol, then the tempered distribution

$$\kappa_x := \mathcal{F}_G^{-1}\{\sigma(x,\cdot)\} \in \mathcal{S}'(G)$$

is called its *associated kernel*, sometimes its right convolution kernel, or just a *kernel*. We may also call the smooth map $G \ni x \mapsto \kappa_x \in \mathcal{S}'(G)$ or the map $(x,y) \mapsto \kappa_x(y) = \kappa(x,y)$ the *kernel* associated with σ.

The smoothness of the map $x \mapsto \sigma(x,\cdot)$ implies easily:

Lemma 5.1.37. *If $\sigma = \{\sigma(x,\pi)\}$ is a symbol with kernel κ_x then for any $\beta \in \mathbb{N}_0^n$,*

$$X^\beta\sigma := \{X_x^\beta\sigma(x,\pi)\}, \quad \tilde{X}^\beta\sigma := \{\tilde{X}_x^\beta\sigma(x,\pi)\}, \text{ and } \partial_x^\beta\sigma := \{\partial_x^\beta\sigma(x,\pi)\},$$

are symbols with respective kernels

$$X_x^\beta\kappa_x, \quad \tilde{X}_x^\beta\kappa_x, \quad \text{and} \quad \partial_x^\beta\kappa_x.$$

Examples of symbols are the symbols in the classes $S^m_{\rho,\delta}(G)$ defined later on. Here are more specific examples of symbols which do not depend on $x \in G$.

Example 5.1.38. If $f \in \mathcal{K}_{a,b}(G)$, then $\widehat{f} = \{\widehat{f}(\pi) : \mathcal{H}_\pi^\infty \to \mathcal{H}_\pi, \pi \in \widehat{G}\}$ is a symbol with kernel f.

The following are particular instances of this case:

- $\widehat{\delta}_0 = \mathrm{I} = \{\mathrm{I} : \mathcal{H}_\pi^\infty \to \mathcal{H}_\pi^\infty, \pi \in \widehat{G}\}$ is a symbol and its kernel is the Dirac measure δ_0.
- For any $\alpha \in \mathbb{N}_0^n$, $\{\pi(X)^\alpha : \mathcal{H}_\pi^\infty \to \mathcal{H}_\pi^\infty, \pi \in \widehat{G}\}$ is a symbol with kernel $X^\alpha\delta_0$, see Example 5.1.26. It acts on smooth vectors, see Example 1.8.17, or alternatively Example 5.1.26 together with Corollary 5.1.30.
- If $\mathcal{R}$ is a positive Rockland operator of homogeneous degree ν and if m is a measurable function on $[0,\infty)$ satisfying

$$\exists C > 0 \quad \forall \lambda \geq 0 \quad |m(\lambda)| \leq C(1+\lambda)^{b/\nu},$$

then $\{m(\pi(\mathcal{R})) : \mathcal{H}_\pi^\infty \to \mathcal{H}_\pi, \pi \in \widehat{G}\}$ is a symbol with kernel $m(\mathcal{R})\delta_0$, see Example 5.1.27. By Corollary 5.1.30, this symbol also acts on smooth vectors

$$\{m(\pi(\mathcal{R})) : \mathcal{H}_\pi^\infty \to \mathcal{H}_\pi^\infty, \pi \in \widehat{G}\}.$$

5.1.4 Quantization formula

With the notion of symbol explained in Section 5.1.3, our quantization makes sense:

Theorem 5.1.39 (Quantization). *The quantization defined by formula (5.1) makes sense for any symbol $\sigma = \{\sigma(x,\pi)\}$. More precisely, for any $\phi \in \mathcal{S}(G)$ and $x \in G$, we have*

$$\mathrm{Op}(\sigma)\phi(x) = \int_{\widehat{G}} \mathrm{Tr}\left(\pi(x)\sigma(x,\pi)\widehat{\phi}(\pi)\right) d\mu(\pi) = \phi * \kappa_x(x), \tag{5.24}$$

where κ_x denotes the kernel of σ. The integral over $\widehat{G}$ in (5.24) is well-defined and absolutely convergent. We also have $\mathrm{Op}(\sigma)\phi \in C^\infty(G)$. Furthermore, the quantization mapping $\sigma \mapsto \mathrm{Op}(\sigma)$ is one-to-one and linear.

Proof. Lemma 5.1.31 (see also Remark 5.1.32) implies that the integral in (5.24) is well defined, absolutely convergent and is equal to $\phi * \kappa_x(x)$.

By Lemma 3.1.55, for each $x \in G$, the function $\phi * \kappa_x$ is smooth. By Lemma 5.1.35, $x \mapsto \kappa_x \in \mathcal{S}'(G)$ is smooth. Hence by composition, $x \mapsto \phi * \kappa_x(x)$ is smooth.

The quantization is clearly linear. Since the kernel is in one-to-one linear correspondence with the operator, and by Lemma 5.1.35 also with the symbol, the quantization $\sigma \mapsto \mathrm{Op}(\sigma)$ is one-to-one. □

Definition 5.1.40 (Notation). If an operator T is given by the formula (5.24) with symbol $\sigma(x,\pi)$, so that $T = \mathrm{Op}(\sigma)$, we will also write

$$\sigma = \sigma_T \quad \text{or} \quad \sigma(x,\pi) = \sigma_T(x,\pi) \quad \text{or even} \quad \sigma = \mathrm{Op}^{-1}(T).$$

This notation is justified since the quantization given by (5.24) is one-to-one by Theorem 5.1.39.

The operators associated with the symbols given in Example 5.1.38 are the ones alluded to in the introduction of this Section:

Continued Example 5.1.38: If $f \in \mathcal{K}_{a,b}(G)$, then $\mathrm{Op}(\widehat{f})$ is the convolution operator $\phi \mapsto \phi * f$ with the right convolution kernel f.

The following are particular instances of this case:

- $\mathrm{Op}(\mathrm{I}) = \mathrm{I}$ and, more generally, for any $\alpha \in \mathbb{N}_0^n$, $\mathrm{Op}(\pi(X)^\alpha) = X^\alpha$.

  These relations can also be expressed as

  $$\sigma_{\mathrm{I}}(x,\pi) = \mathrm{I}_{\mathcal{H}_\pi} \text{ and } \sigma_{X^\alpha}(x,\pi) = \pi(X)^\alpha.$$

- If $\mathcal{R}$ is a positive Rockland operator of homogeneous degree ν and if m is a measurable function on $[0,\infty)$ satisfying

  $$\exists C > 0 \quad \forall \lambda \geq 0 \quad |m(\lambda)| \leq C(1+\lambda)^{b/\nu},$$

  then $\mathrm{Op}(m(\pi(\mathcal{R}))) = m(\mathcal{R})$.

In these examples, the symbols are independent of x. However it is easy to produce x-dependent symbols out of them using the following two observations.

- If $\sigma = \{\sigma(x,\pi),(x,\pi)\in G\times\widehat{G}\}$ is a symbol and $c: G\to\mathbb{C}$ is a smooth function, then $c\sigma := \{c(x)\sigma(x,\pi),(x,\pi)\in G\times\widehat{G}\}$ is a symbol.
- If $\sigma = \{\sigma(x,\pi),(x,\pi)\in G\times\widehat{G}\}$ and $\tau = \{\tau(x,\pi),(x,\pi)\in G\times\widehat{G}\}$ are two symbols, then so is their sum $\sigma+\tau = \{\sigma(x,\pi)+\tau(x,\pi),(x,\pi)\in G\times\widehat{G}\}$.

Remark 5.1.41. 1. The observations just above together with Example 5.1.38 and its continuation above imply that any differential operator of the form

$$\sum_{[\alpha]\le M} c_\alpha(x)X^\alpha \quad \text{with smooth coefficients } c_\alpha \tag{5.25}$$

may be quantized, in the sense that $\sum_{[\alpha]\le M} c_\alpha(x)\pi(X)^\alpha$ is a (smooth) symbol and we have

$$\sum_{[\alpha]\le M} c_\alpha(x)X^\alpha = \mathrm{Op}\left(\sum_{[\alpha]\le M} c_\alpha(x)\pi(X)^\alpha\right).$$

The differential calculus is, by definition, the space of differential operators of the form

$$\sum_{|\alpha|\le d} b_\alpha(x)\partial_x^\alpha \quad \text{with smooth coefficients } b_\alpha,$$

or, equivalently, of the form (5.25), see (3.1.5). Hence, we have obtained that the differential calculus may be quantized. This could be viewed as 'the minimum requirement' for a notion of symbol and quantization on a manifold.

2. In order to achieve this, we had to consider and use fields of operators defined on smooth vectors in our definition of symbol. Indeed, for instance, the symbol associated to a left-invariant vector field X is $\{\pi(X)\}$ while $\pi(X)$ are defined on $\mathcal{H}_\pi^\infty$ but is not bounded on $\mathcal{H}_\pi$.

   This technicality has also the following advantage when we apply our theory in the setting of the Heisenberg group $\mathbb{H}_{n_o}$ in Chapter 6. Realising (almost all of) its dual group $\widehat{\mathbb{H}}_{n_o}$ via Schrödinger representations, the spaces of smooth vectors will coincide with the Schwartz space $\mathcal{S}(\mathbb{R}^{n_o})$. In this context, the symbols will be operators acting on $\mathcal{S}(\mathbb{R}^{n_o})$ (which are smoothly parametrised by points in $\mathbb{H}_{n_o}$).

3. With our notion of symbols and quantization, we also obtain part of the functional calculus of any Rockland operators. More precisely, if $\mathcal{R}$ is a positive Rockland operator, we obtain all the operators of the form $m(\mathcal{R})$ with $m:[0,\infty)\to\mathbb{C}$ a measurable function of (at most) polynomial growth at infinity.

4. The symbol classes that we have introduced are based on the quantization relying on writing the operators as operators with right-convolution kernels. There is an obvious parallel theory of quantization and of the corresponding symbols and their classes suited for problems based on the right-invariant operators. With natural modifications we could have considered at the same time right-invariant vector fields in Part (1) above and a quantization involving left-convolution kernels of operators, i.e. writing the same operators but now in the form $\phi \mapsto \kappa_x * \phi$. As an outcome, with natural modifications we would obtain a parallel theory with the same parallel collection of results to those presented here.

$\text{Op}(\sigma)$ as a limit of nice operators

The operators we have obtained as $\text{Op}(\sigma)$ for symbols σ are limits of 'nice operators' in the following sense:

Lemma 5.1.42. *If $\sigma = \{\sigma(x,\pi)\}$ is a symbol, we can construct explicitly a family of symbols $\sigma_\epsilon = \{\sigma_\epsilon(x,\pi)\}$, $\epsilon > 0$, in such a way that*

1. *the kernel $\kappa_\epsilon(x,y)$ of σ_ϵ is smooth in both x and y, and compactly supported in x,*
2. *if $\phi \in \mathcal{S}(G)$ then $\text{Op}(\sigma_\epsilon)\phi \in \mathcal{D}(G)$, and*
3. *$\text{Op}(\sigma_\epsilon)\phi \underset{\epsilon\to 0}{\longrightarrow} \text{Op}(\sigma)\phi$ uniformly on any compact subset of G.*

Proof of Lemma 5.1.42. We fix a number p such that $p/2$ is a positive integer divisible by all the weights $\upsilon_1, \ldots, \upsilon_n$. Therefore, if $|\cdot|_p$ is the quasi-norm given by (3.21), then the mapping $x \mapsto |x|_p^p$ is a p-homogeneous polynomial. We also fix $\chi_o \in C_c^\infty(\mathbb{R})$ with $\chi_o \geq 0$, $\chi_o = 1$ on $[1/2, 2]$ and $\chi_o = 0$ outside of $[1/4, 4]$. For any $\epsilon > 0$, we write

$$\chi_\epsilon(x) := \chi_o(\epsilon |x|_p^p).$$

Clearly $\chi_\epsilon \in \mathcal{D}(G)$.

If $\pi \in \widehat{G}$, we denote by $|\pi|$ the distance between the co-adjoint orbits corresponding to π and 1.

Applying the orbit method, one can construct explicitly for each $\pi \in \widehat{G}$ a basis $(v_{\ell,\pi})_{\ell=1}^\infty$ formed by smooth vectors and such that the field of vectors $\widehat{G} \ni \pi \mapsto v_{\ell,\pi}$ is measurable. We denote by $\text{proj}_{\epsilon,\pi}$ the orthogonal projection on the subspaces spanned by $v_{1,\pi}, \ldots, v_{\ell,\pi}$ where ℓ is the smallest integer such that $\ell > \epsilon^{-1}$.

We consider for any $\epsilon \in (0,1)$ the mapping

$$\sigma_\epsilon(x,\pi) := \chi_\epsilon(x) 1_{|\pi| \leq \epsilon^{-1}} \sigma(x,\pi) \circ \text{proj}_{\epsilon,\pi}.$$

By Definition 5.1.36, the symbol and the kernel are related by

$$\mathcal{F}_G(\kappa_{\epsilon,x})(\pi) = \sigma_\epsilon(x,\pi).$$

By the Fourier inversion formula (1.26), the corresponding kernel is

$$\kappa_{\epsilon,x}(y)=\kappa_\epsilon(x,y)=\chi_\epsilon(x)\int_{|\pi|\leq\epsilon^{-1}}\mathrm{Tr}\left(\sigma(x,\pi)\ \mathrm{proj}_{\epsilon,\pi}\pi(y)\right)d\mu(\pi),$$

which is smooth in x and y and compactly supported in x.

The corresponding operator is $\mathrm{Op}(\sigma_\epsilon)$, given for any $\phi\in\mathcal{S}(G)$ and $x\in G$ by

$$\begin{aligned}\mathrm{Op}(\sigma_\epsilon)\phi(x) &= \int_{\widehat{G}}\mathrm{Tr}\left(\pi(x)\sigma_\epsilon(x,\pi)\widehat{\phi}(\pi)\right)d\mu(\pi)\\ &= \chi_\epsilon(x)\int_{|\pi|\leq\epsilon^{-1}}\mathrm{Tr}\left(\pi(x)\sigma(x,\pi)\ \mathrm{proj}_{\epsilon,\pi}\widehat{\phi}(\pi)\right)d\mu(\pi).\end{aligned}$$

It is also given by

$$\mathrm{Op}(\sigma_\epsilon)\phi(x)=\phi*\kappa_{\epsilon,x}(x).$$

Clearly $\mathrm{Op}(\sigma_\epsilon)\phi$ is smooth and compactly supported.

Since

$$\widehat{G}\ni\pi\mapsto\mathrm{Tr}\left|\sigma(x,\pi)\widehat{\phi}(\pi)\right|$$

is integrable against μ, using the dominated convergence theorem, we obtain easily the uniform convergence of $\mathrm{Op}(\sigma_\epsilon)\phi$ to $\mathrm{Op}(\sigma)\phi$ on any compact set. □

5.2 Symbol classes $S^m_{\rho,\delta}$ and operator classes $\Psi^m_{\rho,\delta}$

In Section 5.2, we will define and study classes of symbols $S^m_{\rho,\delta}=S^m_{\rho,\delta}(G)$. By applying the quantization procedure described in Section 5.1, we will then obtain the corresponding classes of operators

$$\Psi^m_{\rho,\delta}=\mathrm{Op}(S^m_{\rho,\delta}).$$

In Section 5.5, we will show that this collection of operators $\cup_{m\in\mathbb{R}}\Psi^m_{\rho,\delta}$ forms an algebra and satisfies the usual properties expected from a symbolic calculus.

Before defining symbol classes, we need to define difference operators.

5.2.1 Difference operators

On compact Lie groups the difference operators were defined as acting on Fourier coefficients, see Definition 2.2.6. Its adaptation to our setting leads us to (densely) defined difference operators on $\mathcal{K}_{a,b}(G)$ viewed as fields.

Definition 5.2.1. For any $q \in C^\infty(G)$, we set

$$\Delta_q \widehat{f}(\pi) := \widehat{qf}(\pi) \equiv \pi(qf),$$

for any distribution $f \in \mathcal{D}'(G)$ such that $f \in \mathcal{K}_{a,b}$ and $qf \in \mathcal{K}_{a',b'}$ for some $a, b, a', b' \in \mathbb{R}$.

Recall that if $f \in \mathcal{D}'(G)$ and $q \in C^\infty(G)$, then the distribution $qf \in \mathcal{D}'(G)$ is defined via

$$\langle qf, \phi\rangle := \langle f, q\phi\rangle, \quad \phi \in \mathcal{D}(G), \tag{5.26}$$

which makes sense since $q\phi \in \mathcal{D}(G)$. In Definition 5.2.1, we assume that the two distributions f and qf are in $\cup_{a'',b''\in\mathbb{R}}\mathcal{K}_{a'',b''}$. Note that, as all the definitions of group Fourier transform coincide, different values for the parameters a, b, a', b' in Definition 5.2.1 yield the same fields of operators $\{\widehat{f}(\pi) : \mathcal{H}^\infty_\pi \to \mathcal{H}_\pi, \pi \in \widehat{G}\}$ and $\{\widehat{qf}(\pi) : \mathcal{H}^\infty_\pi \to \mathcal{H}_\pi, \pi \in \widehat{G}\}$. This justifies our use of the notation Δ_q without reference to the parameters a, b, a', b'.

Remark 5.2.2. In general, it is not possible to define an operator Δ_q on a single π, and it has to be viewed as acting on the 'whole' fields parametrised by $\widehat{G}$. For example, already on the commutative group $(\mathbb{R}^n, +)$, the difference operators corresponding to coordinate functions will satisfy

$$\Delta^\alpha \widehat{\phi}(\xi) = \left(\frac{1}{i}\frac{\partial}{\partial \xi}\right)^\alpha \widehat{\phi}(\xi), \quad \xi \in \mathbb{R}^n,$$

with appropriately chosen functions q, thus involving derivatives in the dual variable, see Example 5.2.6. Furthermore if q is not a coordinate function but for instance a (non-zero) smooth function with compact support, the corresponding difference operator is not local.

Also, on the Heisenberg group $\mathbb{H}_{n_o}$ (see Example 1.6.4), taking $q = t$ the central variable, and π_λ the Schrödinger representations (see Section 6.3.2), then Δ_t is expressed using derivatives in λ, see Lemma 6.3.6 and Remark 6.3.7.

Let us fix a basis of $\mathfrak{g}$. For the notation of the following proposition we refer to Section 3.1.3 where the spaces of polynomials on homogeneous Lie groups have been discussed, with the set $\mathcal{W}$ defined in (3.60). We will define the difference operators associated with the polynomials appearing in the Taylor expansions:

Proposition 5.2.3. *1. For each $\alpha \in \mathbb{N}_0^n$, there exists a unique homogeneous polynomial q_α of degree $[\alpha]$ satisfying*

$$\forall \beta \in \mathbb{N}_0^n \qquad X^\beta q_\alpha(0) = \delta_{\alpha,\beta} = \begin{cases} 1 & \text{if } \beta = \alpha, \\ 0 & \text{otherwise.} \end{cases}$$

2. The polynomials q_α, $\alpha \in \mathbb{N}_0^n$, form a basis of $\mathcal{P}$. Furthermore, for each $M \in \mathcal{W}$, the polynomials q_α, $[\alpha] = M$, form a basis of $\mathcal{P}_{[\alpha]=M}$.

3. *The Taylor polynomial of a suitable function f at a point $x \in G$ of homogeneous degree $M \in \mathcal{W}$ is*

$$P^{(f)}_{x,M}(y) = \sum_{[\alpha]\leq M} q_\alpha(y) X^\alpha f(x). \tag{5.27}$$

4. *For any $\alpha \in \mathbb{N}_0^n$, we have for any $x, y \in G$,*

$$q_\alpha(xy) = \sum_{[\alpha_1]+[\alpha_2]=[\alpha]} c_{\alpha_1,\alpha_2} q_{\alpha_1}(x) q_{\alpha_2}(y)$$

for some coefficients $c_{\alpha_1,\alpha_2} \in \mathbb{R}$ independent of x and y. Moreover, we have

$$c_{\alpha_1,0} = \begin{cases} 1 & \text{if } \alpha_1 = \alpha \\ 0 & \text{otherwise} \end{cases}, \quad c_{0,\alpha_2} = \begin{cases} 1 & \text{if } \alpha_2 = \alpha \\ 0 & \text{otherwise} \end{cases}.$$

Proof. For each $M \in \mathcal{W}$, by Corollary 3.1.31, there exists a unique polynomial $q_\alpha \in \mathcal{P}_{=M}$ satisfying $X^\beta q_\alpha(0) = \delta_{\alpha,\beta}$ for every $\beta \in \mathbb{N}_0^n$ with $[\beta] = M$, therefore for every $\beta \in \mathbb{N}_0^n$. This shows parts (1) and (2). Part (3) follows from the definition of a Taylor polynomial.

It remains to prove Part (4). For this it suffices to consider $q_\alpha(xy)$ as a polynomial in x and in y, using the bases $(q_{\alpha_1}(x))$ and $(q_{\alpha_2}(y))$. Therefore, $q_\alpha(xy)$ can be written as a finite linear combination of $q_{\alpha_1}(x)q_{\alpha_2}(y)$. Since

$$q_\alpha((rx)(ry)) = r^{[\alpha]} q_\alpha(xy),$$

this forces this linear combination to be over $\alpha_1, \alpha_2 \in \mathbb{N}_0^n$ satisfying $[\alpha_1] + [\alpha_2] = [\alpha]$. The conclusions about the coefficients follow by setting $y = 0$ and then $x = 0$, see also (3.14). □

In the case of $(\mathbb{R}^n, +)$ the polynomials q_α are the usual normalised monomials $(\alpha_1! \dots \alpha_n!)^{-1} x^\alpha$. But it is not usually the case on other groups:

Example 5.2.4. On the three dimensional Heisenberg group $\mathbb{H}_1$ where a point is described as $(x, y, t) \in \mathbb{R}^3$ (see Example 1.6.4), we compute directly that for degree 1 we have

$$q_{(1,0,0)} = x, \ q_{(0,1,0)} = y,$$

and for degree 2,

$$q_{(2,0,0)} = x^2, \ q_{(0,2,0)} = y^2, \ q_{(1,1,0)} = xy, \ q_{(0,0,1)} = t - \frac{1}{2}xy.$$

Definition 5.2.5. For each $\alpha \in \mathbb{N}_0^n$, the *difference operators* are

$$\Delta^\alpha := \Delta_{\tilde{q}_\alpha}, \quad \alpha \in \mathbb{N}_0^n,$$

where

$$\tilde{q}_\alpha(x) := q_\alpha(x^{-1})$$

and $q_\alpha \in \mathcal{P}_{=[\alpha]}$ is defined in Proposition 5.2.3.

The difference operators generalise the Euclidean derivatives with respect to the Fourier variable on $(\mathbb{R}^n, +)$ in the following sense:

Example 5.2.6. Let us consider the abelian group $G = (\mathbb{R}^n, +)$. We identify $\widehat{\mathbb{R}}^n$ with $\mathbb{R}^n$. If the Fourier transform of a function $\phi \in \mathcal{S}(\mathbb{R}^n)$ is given by

$$\mathcal{F}_G\phi(\xi) = (2\pi)^{-\frac{n}{2}} \int_{\mathbb{R}^n} e^{-ix\cdot\xi}\phi(x)dx, \quad \xi \in \mathbb{R}^n,$$

then

$$\Delta^\alpha \mathcal{F}_G\phi(\xi) = \int_{\mathbb{R}^n} e^{-ix\cdot\xi}(-x)^\alpha\phi(x)dx = \left(\frac{1}{i}\frac{\partial}{\partial\xi}\right)^\alpha \mathcal{F}_G\phi(\xi).$$

Thus, Δ^α coincide with the operators $D^\alpha = \left(\frac{1}{i}\frac{\partial}{\partial\xi}\right)^\alpha$ usually appearing in the Fourier analysis on $\mathbb{R}^n$.

Example 5.2.7. Δ^0 is the identity operator on each $\mathcal{K}_{a,b}(G)$.

Example 5.2.8. For $\mathrm{I} = \widehat{\delta}_o = \{\mathrm{I} : \mathcal{H}^\infty_\pi \to \mathcal{H}^\infty_\pi, \pi \in \widehat{G}\}$ and any $\alpha \in \mathbb{N}^n_0\backslash\{0\}$, we have $\Delta^\alpha \mathrm{I} = 0$.

Proof. We know that $\mathrm{I} = \widehat{\delta}_0$ (see Example 5.1.38). The distribution $\tilde{q}_\alpha\delta_0$ is defined by

$$\langle \tilde{q}_\alpha\delta_0, \phi\rangle = \langle \delta_0, \tilde{q}_\alpha\phi\rangle, \quad \phi \in \mathcal{D}(G),$$

see (5.26). Since

$$\langle \delta_0, \tilde{q}_\alpha\phi\rangle = (\tilde{q}_\alpha\phi)(0) = \tilde{q}_\alpha(0)\ \phi(0) = 0$$

we must have $q\delta_0 = 0$. Therefore, $\Delta^\alpha \mathrm{I} = \widehat{q\delta_0} = 0$. □

More generally, we have

Lemma 5.2.9. *Let $\alpha, \beta \in \mathbb{N}^n_0$. Then the symbol $\{\pi(X)^\beta : \mathcal{H}^\infty_\pi \to \mathcal{H}^\infty_\pi, \pi \in \widehat{G}\}$ (see Example 5.1.38) satisfies*

$$\Delta^\alpha \pi(X)^\beta = 0 \quad \text{if } [\alpha] > [\beta].$$

If $[\alpha] \le [\beta]$, then $\Delta^\alpha\pi(X)^\beta$ is a linear combination depending only on α, β, of the terms $\pi(X)^{\beta_2}$ with $[\beta_2] = [\beta] - [\alpha]$, that is,

$$\Delta^\alpha \pi(X)^\beta = \overline{\sum_{[\alpha]+[\beta_2]=[\beta]}} \pi(X)^{\beta_2}.$$

Proof of Lemma 5.2.9. We see that $\Delta^\alpha\pi(X)^\beta$ is the group Fourier transform of the distribution $\tilde{q}_\alpha X^\beta \delta_0$ defined via

$$\langle \tilde{q}_\alpha X^\beta\delta_0, \phi\rangle = \langle X^\beta\delta_0, \tilde{q}_\alpha\phi\rangle = \{X^\beta\}^t\{\tilde{q}_\alpha\phi\}(0)$$

for any $\phi \in \mathcal{D}(G)$, see Example 5.1.38. This is so as long as we prove that $\tilde{q}_\alpha X^\beta\delta_0$ is in some $\mathcal{K}_{a,b}$. Let us find another expression for this distribution. As $\{X^\beta\}^t$ is a $[\beta]$-homogeneous left-invariant differential operators, by the Leibniz formula for vector fields, we have

$$\{X^\beta\}^t\{\tilde{q}_\alpha\phi\} = \overline{\sum_{[\beta_1]+[\beta_2]=[\beta]}} X^{\beta_1}\tilde{q}_\alpha\, X^{\beta_2}\phi.$$

We easily see that $X^{\beta_1}\tilde{q}_\alpha \in \mathcal{P}_{=[\alpha]-[\beta_1]}$ and, therefore, by Part (2) of Proposition 5.2.3 we have

$$X^{\beta_1}\tilde{q}_\alpha = \overline{\sum_{[\alpha']=[\alpha]-[\beta_1]}} \tilde{q}_{\alpha'}.$$

Hence we have obtained

$$\{X^\beta\}^t\{\tilde{q}_\alpha\phi\} = \overline{\sum_{\substack{[\beta_1]+[\beta_2]=[\beta]\\ [\alpha']=[\alpha]-[\beta_1]}}} \tilde{q}_{\alpha'}\, X^{\beta_2}\phi,$$

and

$$\langle \tilde{q}_\alpha X^\beta\delta_0, \phi\rangle = \overline{\sum_{\substack{[\beta_1]+[\beta_2]=[\beta]\\ [\alpha']=[\alpha]-[\beta_1]}}} (\tilde{q}_{\alpha'}X^{\beta_2}\phi)(0) = \overline{\sum_{\substack{[\beta_1]+[\beta_2]=[\beta]\\ 0=[\alpha]-[\beta_1]}}} X^{\beta_2}\phi(0),$$

with the convention that the sum is zero if there are no such β_1, β_2. Thus

$$\tilde{q}_\alpha X^\beta\delta_0 = \overline{\sum_{\substack{[\beta_1]+[\beta_2]=[\beta]\\ [\alpha]=[\beta_1]}}} X^{\beta_2}\delta_0.$$

Since $X^{\beta_2}\delta_0 \in \mathcal{K}_{[\beta_2],0}$ (see Example 5.1.26), we see that $\tilde{q}_\alpha X^\beta\delta_0 \in \mathcal{K}_{[\beta],0}$. Furthermore, taking the group Fourier transform we obtain

$$\Delta^\alpha \pi(X)^\beta = \overline{\sum_{\substack{[\beta_1]+[\beta_2]=[\beta]\\ [\alpha]=[\beta_1]}}} \pi(X)^{\beta_2}.$$

This sum is zero if there are no such β_1, β_2, for instance if $[\beta] < [\alpha]$. □

Let us collect some properties of the difference operators.

Proposition 5.2.10. *(i) For any $\alpha \in \mathbb{N}_0^n$, the operator Δ^α is linear, its domain of definition contains $\mathcal{F}_G(\mathcal{S}(G))$ and $\Delta^\alpha\mathcal{F}_G(\mathcal{S}(G)) \subset \mathcal{F}_G(\mathcal{S}(G))$.*

(ii) For any $\alpha_1, \alpha_2 \in \mathbb{N}_0^n$, there exist constants $c_{\alpha_1,\alpha_2,\alpha} \in \mathbb{R}$, with $\alpha \in \mathbb{N}_0^n$ such that $[\alpha] = [\alpha_1] + [\alpha_2]$, so that for any $\phi \in \mathcal{S}(G)$, we have

$$\Delta^{\alpha_1}\left(\Delta^{\alpha_2}\widehat{\phi}\right) = \Delta^{\alpha_2}\left(\Delta^{\alpha_1}\widehat{\phi}\right) = \sum_{[\alpha]=[\alpha_1]+[\alpha_2]} c_{\alpha_1,\alpha_2,\alpha}\Delta^\alpha\widehat{\phi},$$

where the sum is taken over all $\alpha \in \mathbb{N}_0^n$ satisfying $[\alpha] = [\alpha_1] + [\alpha_2]$.

(iii) *For any* $\alpha \in \mathbb{N}_0^n$, *there exist constants* $c_{\alpha,\alpha_1,\alpha_2} \in \mathbb{R}$, $\alpha_1, \alpha_2 \in \mathbb{N}_0^n$, *with* $[\alpha_1] + [\alpha_2] = [\alpha]$, *such that for any* $\phi_1, \phi_2 \in \mathcal{S}(G)$, *we have*

$$\Delta^\alpha \left(\widehat{\phi_1}\ \widehat{\phi_2} \right) = \sum_{[\alpha_1]+[\alpha_2]=[\alpha]} c_{\alpha,\alpha_1,\alpha_2}\ \Delta^{\alpha_1}\widehat{\phi_1}\ \Delta^{\alpha_2}\widehat{\phi_2}, \qquad (5.28)$$

where the sum is taken over all $\alpha_1, \alpha_2 \in \mathbb{N}_0^n$ *satisfying* $[\alpha_1] + [\alpha_2] = [\alpha]$. *Moreover,*

$$c_{\alpha,\alpha_1,0} = \begin{cases} 1 & \text{if } \alpha_1 = \alpha \\ 0 & \text{otherwise} \end{cases}, \quad c_{\alpha,0,\alpha_2} = \begin{cases} 1 & \text{if } \alpha_2 = \alpha \\ 0 & \text{otherwise} \end{cases}.$$

The coefficients $c_{\alpha_1,\alpha_2,\alpha}$ in (ii) and $c_{\alpha,\alpha_1,\alpha_2}$ in (iii) are different in general. We interpret Formula (5.28) as the *Leibniz formula.*

Proof. Since the Schwartz space is stable under multiplication by polynomials, $\tilde{q}_\alpha\phi$ is Schwartz for any $\phi \in \mathcal{S}(G)$, and $\Delta^\alpha\widehat{\phi}(\pi) = \pi(\tilde{q}_\alpha\phi)$. This shows (i).

For Part (ii), we see that the polynomial $q_{\alpha_1}q_{\alpha_2}$ is homogeneous of degree $[\alpha_1]+[\alpha_2]$. Since $\{q_\alpha, [\alpha] = M\}$ is a basis of $\mathcal{P}_{=M}$ by Proposition 5.2.3, there exist constants $c_{\alpha_1,\alpha_2,\alpha} \in \mathbb{R}$, $\alpha_1, \alpha_2 \in \mathbb{N}_0^n$ with $[\alpha_1] + [\alpha_2] = [\alpha]$, satisfying

$$q_{\alpha_1}q_{\alpha_2} = \sum_{[\alpha_1]+[\alpha_2]=[\alpha]} c_{\alpha_1,\alpha_2,\alpha}\, q_\alpha.$$

Therefore

$$\begin{aligned} \Delta^{\alpha_1}\left(\Delta^{\alpha_2}\widehat{\phi}(\pi)\right) = \pi(\tilde{q}_{\alpha_1}\tilde{q}_{\alpha_2}\phi) &= \sum_{[\alpha_1]+[\alpha_2]=[\alpha]} c_{\alpha_1,\alpha_2,\alpha}\pi(\tilde{q}_\alpha\phi) \\ &= \sum_{[\alpha_1]+[\alpha_2]=[\alpha]} c_{\alpha_1,\alpha_2,\alpha}\Delta^\alpha\widehat{\phi}(\pi). \end{aligned}$$

This and the equality $\tilde{q}_{\alpha_1}\tilde{q}_{\alpha_2} = \tilde{q}_{\alpha_2}\tilde{q}_{\alpha_1}$ show (ii).

Let us prove (iii). By Proposition 5.2.3 (4),

$$\begin{aligned} \tilde{q}_\alpha(x)\ (\phi_2 * \phi_1)(x) &= \int_G q_\alpha(x^{-1}y\ y^{-1})\ \phi_2(y)\ \phi_1(y^{-1}x)\ dy \\ &= \sum_{[\alpha_1]+[\alpha_2]=[\alpha]} c_{\alpha_1,\alpha_2} \int_G q_{\alpha_2}(y^{-1})\phi_2(y)\ q_{\alpha_1}(x^{-1}y)\phi_1(y^{-1}x)\ dy \\ &= \sum_{[\alpha_1]+[\alpha_2]=[\alpha]} c_{\alpha_1,\alpha_2}\ (\tilde{q}_{\alpha_2}\phi_2) * (\tilde{q}_{\alpha_1}\phi_1), \end{aligned}$$

with constants depending on $\alpha, \alpha_1, \alpha_2$. Taking the Fourier transform implies the formula (5.28), with conclusions on coefficients following from Proposition 5.2.3. □

We will see that the difference operators Δ^α defined in Definition 5.2.5 appear in the general asymptotic formulae for adjoint and product of pseudo-differential operators in our context, see Sections 5.5.3 and 5.5.2.

5.2.2 Symbol classes $S^m_{\rho,\delta}$

In this section we define the symbol classes $S^m_{\rho,\delta} = S^m_{\rho,\delta}(G)$ of symbols on a graded Lie group G and discuss their properties. We use the notation for the symbol classes similar to the familiar ones on the Euclidean space and also on compact Lie groups.

Let us give the formal definition of our symbol classes.

Definition 5.2.11. Let $m, \rho, \delta \in \mathbb{R}$ with $0 \leq \rho \leq \delta \leq 1$. Let $\mathcal{R}$ be a positive Rockland operator of homogeneous degree ν. A symbol

$$\sigma = \{\sigma(x,\pi) : \mathcal{H}^\infty_\pi \to \mathcal{H}_\pi, (x,\pi) \in G \times \widehat{G}\}$$

is called a *symbol of order m and of type (ρ,δ)* whenever, for each $\alpha, \beta \in \mathbb{N}^n_0$ and $\gamma \in \mathbb{R}$, we have

$$\sup_{x\in G} \|X^\beta_x \Delta^\alpha \sigma(x,\cdot)\|_{L^\infty_{\gamma,\rho[\alpha]-m-\delta[\beta]+\gamma}(\widehat{G})} < \infty. \tag{5.29}$$

The *symbol class* $S^m_{\rho,\delta} = S^m_{\rho,\delta}(G)$ is the set of symbols of order m and of type (ρ,δ).

By Corollary 5.1.30, the symbols $X^\beta_x \Delta^\alpha \sigma$ are fields acting on smooth vectors. By Lemma 5.1.23, we can reformulate (5.29) as

$$\sup_{x\in G, \pi\in\widehat{G}} \|\pi(\mathrm{I}+\mathcal{R})^{\frac{\rho[\alpha]-m-\delta[\beta]+\gamma}{\nu}} X^\beta_x \Delta^\alpha \sigma(x,\pi)\pi(\mathrm{I}+\mathcal{R})^{-\frac{\gamma}{\nu}}\|_{\mathscr{L}(\mathcal{H}_\pi)} < \infty. \tag{5.30}$$

Recall that, as usual, the supremum in π in (5.30) has to be understood as the essential supremum with respect to the Plancherel measure.

Clearly, the converse holds: if σ is a symbol such that $X^\beta_x \Delta^\alpha \sigma$ are fields acting on smooth vectors for which (5.30) holds, then σ is in $S^m_{\rho,\delta}$.

We note that condition (5.30) requires one to fix a positive Rockland operator $\mathcal{R}$ in order to fix the norms of $L^\infty_{a',b'}(\widehat{G})$. However, the resulting class $S^m_{\rho,\delta}$ does not depend on the choice of $\mathcal{R}$, see Lemma 5.1.22.

If a positive Rockland operator $\mathcal{R}$ of homogeneous degree ν is fixed, then we set for $\sigma \in S^m_{\rho,\delta}$ and $a,b,c \in \mathbb{N}_0$,

$$\|\sigma\|_{S^m_{\rho,\delta},a,b,c} := \sup_{\substack{|\gamma|\leq c\\ [\alpha]\leq a,\, [\beta]\leq b}} \sup_{x\in G} \|X^\beta_x \Delta^\alpha \sigma(x,\cdot)\|_{L^\infty_{\gamma,\rho[\alpha]-m-\delta[\beta]+\gamma}(\widehat{G})}.$$

This quantity is also equal to

$$\|\sigma\|_{S^m_{\rho,\delta},a,b,c} = \sup_{x\in G,\, \pi\in\widehat{G}} \|\sigma(x,\pi)\|_{S^m_{\rho,\delta},a,b,c},$$

where we define for any symbol σ, $a,b,c \in \mathbb{N}_0$, and $(x,\pi) \in G \times \widehat{G}$ (fixed)

$$\|\sigma(x,\pi)\|_{S^m_{\rho,\delta},a,b,c} := \sup_{\substack{|\gamma|\leq c\\ [\alpha]\leq a,\, [\beta]\leq b}} \|\pi(\mathrm{I}+\mathcal{R})^{\frac{\rho[\alpha]-m-\delta[\beta]+\gamma}{\nu}} X^\beta_x \Delta^\alpha \sigma(x,\pi)\pi(\mathrm{I}+\mathcal{R})^{-\frac{\gamma}{\nu}}\|_{\mathscr{L}(\mathcal{H}_\pi)}.$$

Here, as always, the supremum has to be understood as the essential supremum with respect to the Plancherel measure.

Before making some comments, let us say that the classes of symbols we have just defined have the usual structures of symbol classes.

Proposition 5.2.12. *The symbol class $S^m_{\rho,\delta}$ is a vector space independent of any Rockland operator $\mathcal{R}$ used in (5.29) to consider the $L^\infty_{\gamma,\rho[\alpha]-m-\delta[\beta]+\gamma}(\widehat{G})$-norms. We have the continuous inclusions*

$$m_1 \le m_2, \quad \delta_1 \le \delta_2, \quad \rho_1 \ge \rho_2 \quad \Longrightarrow \quad S^{m_1}_{\rho_1,\delta_1} \subset S^{m_2}_{\rho_2,\delta_2}. \tag{5.31}$$

We fix a positive Rockland operator $\mathcal{R}$. For any $m \in \mathbb{R}$, $\rho, \delta \ge 0$, the resulting maps $\|\cdot\|_{S^m_{\rho,\delta},a,b,c}$, $a,b,c \in \mathbb{N}_0$, are seminorms over the vector space $S^m_{\rho,\delta}$ which endow $S^m_{\rho,\delta}$ with the structure of a Fréchet space.

We may replace the family of seminorms $\|\cdot\|_{S^m_{\rho,\delta},a,b,c}$, $a,b,c \in \mathbb{N}_0$, by

$$\sigma \longmapsto \sup_{\substack{[\alpha]\le a,\ x\in G \\ [\beta]\le b}} \|X^\beta_x \Delta^\alpha \sigma(x,\cdot)\|_{L^\infty_{\gamma_\ell,\rho[\alpha]-m-\delta[\beta]+\gamma_\ell}(\widehat{G})}, \qquad a,b \in \mathbb{N}_0, \ \ell \in \mathbb{Z},$$

where the sequence $\{\gamma_\ell, \ell \in \mathbb{Z}\}$ of real numbers satisfies $\gamma_\ell \underset{\ell\to\pm\infty}{\longrightarrow} \pm\infty$.

Two different positive Rockland operators give equivalent families of seminorms. The topology on $S^m_{\rho,\delta}$ is independent of the choice of the Rockland operator $\mathcal{R}$.

Proof. Using Corollary 5.1.30 and Lemma 5.1.22, this is a routine exercise. □

Remark 5.2.13. Let us make some comments about Definition 5.2.11:

1. In the abelian case, that is, $\mathbb{R}^n$ endowed with the addition law, and $\mathcal{R} = -\mathcal{L}$ with $\mathcal{L}$ being the Laplace operator, $S^m_{\rho,\delta}$ boils down to the usual Hörmander class, in view of the difference operators corresponding to the derivatives, see Example 5.2.6.

2. In the case of compact Lie groups with $\mathcal{R}$ being the (positive) Laplacian, a similar definition leads to the one considered in (2.26) since the operator $\pi(\mathrm{I}+\mathcal{R})$ is scalar. However, here, in the case of non-abelian graded Lie groups, the operator $\mathcal{R}$ can not have a scalar Fourier transform.

3. The presence of the parameter γ is included to facilitate proving that the space of symbols $\cup_{m\in\mathbb{R}} S^m_{\rho,\delta}$, with suitable restrictions on ρ, δ, forms an algebra of operators later on. It already has enabled us to see that the symbols are fields of operators acting on smooth vectors and therefore can be composed without using the composition of unbounded operators (in Definition A.3.2).

   We will see in Theorem 5.5.20 that in fact we can remove this γ. By this we mean that a symbol σ is in $S^m_{\rho,\delta}$ if and only if the condition in (5.29) holds for any $\alpha, \beta \in \mathbb{N}_0^n$ and $\gamma = 0$. Furthermore, the seminorms $\|\cdot\|_{S^m_{\rho,\delta},a,b,0}$, $a, b \in \mathbb{N}_0$, yield the topology of $S^m_{\rho,\delta}$.

4. We could have used other families of difference operators instead of the Δ^α's to define the symbol classes $S^m_{\rho,\delta}$. For instance, we could have used any family of difference operators associated with a family $\{p_\alpha\}_{\alpha\in\mathbb{N}_0^n}$ of homogeneous polynomials on G which satisfy

   - for each $\alpha \in \mathbb{N}_0^n$, p_α is of homogeneous degree $[\alpha]$,
   - and $\{p_\alpha\}_{\alpha\in\mathbb{N}_0^n}$ is a basis of $\mathcal{P}(G)$.

   Indeed, in this case, the following properties hold.

   - Any $\tilde{q}_\alpha$ is a linear combination of p_β, $[\beta] = [\alpha]$.
   - Conversely, any p_α is a linear combination of $\tilde{q}_\beta$, $[\beta] = [\alpha]$.

   Thus,

   - any Δ^α is a linear combination of Δ_{p_β}, $[\beta] = [\alpha]$.
   - Conversely, any Δ_{p_α} is a linear combination of Δ^β, $[\beta] = [\alpha]$.

   It is then easy to see that a symbol σ is in $S^m_{\rho,\delta}$ if and only if for each $\alpha, \beta \in \mathbb{N}_0^n$ and $\gamma \in \mathbb{R}$,

$$\sup_{x\in G} \|X_x^\beta \Delta_{p_\alpha} \sigma(x,\cdot)\|_{L^\infty_{\gamma,\rho[\alpha]-m-\delta[\beta]+\gamma}(\widehat{G})} < \infty.$$

   Note that this implies that the symbol class $S^m_{\rho,\delta}$ does not depend on a particular choice of realisation of G through a basis of $\mathfrak{g}$ (of eigenvectors for the dilations) but only on the graded Lie group G and its homogeneous structure.

   For such a family Δ_{p_α}, the same proof as for Proposition 5.2.10 shows a Leibniz formula in the sense of (5.28).

   Although we could use 'easier' difference operators to define our symbol classes, for instance Δ_{x^α}, $\alpha \in \mathbb{N}_0^n$, we choose to present our analysis with the difference operators Δ^α given in Definition 5.2.5. Note that the asymptotic formulae for composition and adjoint in (5.57) and (5.60) will be expressed in terms of the difference operators Δ^α and derivatives X_x^α.

   Note that the change of difference operators explained just above is linear, whereas in the compact case, one can use many more difference operators to define the symbol classes $S^m_{\rho,\delta}$, see Section 2.2.2.

The type $(1,0)$ can be thought of as the basic class of symbols and the types (ρ,δ) as its generalisations. There are certain limitations on the parameters (ρ,δ) coming from reasons similar to the ones in the Euclidean settings. For type $(1,0)$, we set

$$S^m := S^m_{1,0},$$

and

$$\|\sigma(x,\pi)\|_{S^m_{1,0},a,b,c} = \|\sigma(x,\pi)\|_{a,b,c}, \ \|\sigma\|_{S^m_{1,0},a,b,c} = \|\sigma\|_{a,b,c}, \text{ etc.} \ldots$$

We also define the class of smoothing symbols

Definition 5.2.14. We set

$$S^{-\infty} := \bigcap_{m\in\mathbb{R}} S^m.$$

One checks easily that

$$S^{-\infty} = \bigcap_{m\in\mathbb{R}} S^m_{\rho,\delta},$$

independently of ρ and δ as long as $0 \le \delta \le \rho \le 1$ and $\rho \neq 0$. Moreover, $S^{-\infty}$ is equipped with the topology of projective limit induced by $\cap_{m\in\mathbb{R}} S^m_{\rho,\delta}$, again independently of ρ and δ.

We will see in Corollary 5.4.10 that the symbols in $S^{-\infty}$ really deserve to be called smoothing.

5.2.3 Operator classes $\Psi^m_{\rho,\delta}$

The pseudo-differential operators of order $m \in \mathbb{R} \cup \{-\infty\}$ and type (ρ,δ) are obtained by the quantization

$$\text{Op}(\sigma)\phi(x) = \int_{\widehat{G}} \text{Tr}\left(\pi(x)\sigma(x,\pi)\widehat{\phi}(\pi)\right) d\mu(\pi),$$

justified in Theorem 5.1.39, from the symbols of the same order and type, that is,

$$\Psi^m_{\rho,\delta} := \text{Op}(S^m_{\rho,\delta}).$$

They inherit a structure of topological vector spaces from the classes of symbols,

$$\|\text{Op}(\sigma)\|_{\Psi^m_{\rho,\delta},a,b,c} := \|\sigma\|_{S^m_{\rho,\delta},a,b,c}.$$

For type $(1,0)$, we set as for the corresponding symbol classes:

$$\Psi^m := \Psi^m_{1,0}.$$

Continuity on $\mathcal{S}(G)$

By Theorem 5.1.39, any operator in the operator classes defined above maps Schwartz functions to smooth functions. Let us show that in fact it acts continuously on the Schwartz space:

Theorem 5.2.15. *Let $T \in \Psi^m_{\rho,\delta}$ where $m \in \mathbb{R}$, $1 \geq \rho \geq \delta \geq 0$. Then for any $\phi \in \mathcal{S}(G)$, $T\phi \in \mathcal{S}(G)$. Moreover the operator T act continuously on $\mathcal{S}(G)$: for any seminorm $\|\cdot\|_{\mathcal{S}(G),N}$ there exist a constant $C>0$ and a seminorm $\|\cdot\|_{\mathcal{S}(G),N'}$ such that for every $\phi \in \mathcal{S}(G)$,*

$$\|T\phi\|_{\mathcal{S}(G),N} \leq C\|\phi\|_{\mathcal{S}(G),N'}.$$

The constant C can be chosen as $C_1\|T\|_{\Psi^m_{\rho,\delta},a,b,c}$ where C_1 is a constant of and the seminorm $\|\cdot\|_{\Psi^m_{\rho,\delta},a,b,c}$ depend on G, m, ρ,δ, and on the seminorm $\|\cdot\|_{\mathcal{S}(G),N}$.

In other words, the mapping $T \mapsto T$ from $\Psi^m_{\rho,\delta}$ to the space $\mathscr{L}(\mathcal{S}(G))$ of continuous operators on $\mathcal{S}(G)$ is continuous (it is clearly linear).

Our proof of Theorem 5.2.15 will require the following preliminary result on the right convolution kernels:

Proposition 5.2.16. *Let $\sigma = \{\sigma(x,\pi)\}$ be in $S^m_{\rho,\delta}$ with $1 \geq \rho \geq \delta \geq 0$. Let κ_x denote its associated kernel. If $m < -Q/2$ then for any $x \in G$, the distribution κ_x is square integrable and*

$$\begin{aligned} \|\kappa_x\|_{L^2(G)} &\leq C \sup_{\pi\in\widehat{G}} \|\pi(\mathrm{I}+\mathcal{R})^{\frac{-m}{\nu}}\sigma(x,\pi)\|_{\mathscr{L}(\mathcal{H}_\pi)}, \\ \|\kappa_x\|_{L^2(G)} &\leq C \sup_{\pi\in\widehat{G}} \|\sigma(x,\pi)\pi(\mathrm{I}+\mathcal{R})^{\frac{-m}{\nu}}\|_{\mathscr{L}(\mathcal{H}_\pi)}, \end{aligned}$$

with $C = C_m > 0$ a finite constant independent of σ and x.

The proof below will show that we can choose $C_m = \|\mathcal{B}_{-m}\|_{L^2(G)}$ the L^2-norm of the right-convolution kernel of the Bessel potential of the positive Rockland operator $\mathcal{R}$.

Proof of Proposition 5.2.16. We write

$$\begin{aligned} \|\sigma(x,\pi)\|_{\mathtt{HS}} &= \|\pi(\mathrm{I}+\mathcal{R})^{\frac{m}{\nu}}\pi(\mathrm{I}+\mathcal{R})^{\frac{-m}{\nu}}\sigma(x,\pi)\|_{\mathtt{HS}} \\ &\leq \|\pi(\mathrm{I}+\mathcal{R})^{\frac{m}{\nu}}\|_{\mathtt{HS}}\|\pi(\mathrm{I}+\mathcal{R})^{\frac{-m}{\nu}}\sigma(x,\pi)\|_{\mathscr{L}(\mathcal{H}_\pi)}, \end{aligned}$$

which shows

$$\|\sigma(x,\pi)\|_{\mathtt{HS}} \leq \sup_{\pi_1\in\widehat{G}} \|\pi_1(\mathrm{I}+\mathcal{R})^{\frac{-m}{\nu}}\sigma(x,\pi_1)\|_{\mathscr{L}(\mathcal{H}_{\pi_1})}\|\pi(\mathrm{I}+\mathcal{R})^{\frac{m}{\nu}}\|_{\mathtt{HS}}.$$

Squaring and integrating against the Plancherel measure, we obtain

$$\int_{\widehat{G}} \|\sigma(x,\pi)\|^2_{\mathtt{HS}}d\mu(\pi) \leq \sup_{\pi_1\in\widehat{G}} \|\pi_1(\mathrm{I}+\mathcal{R})^{\frac{-m}{\nu}}\sigma(x,\pi_1)\|^2_{\mathscr{L}(\mathcal{H}_{\pi_1})} \int_{\widehat{G}} \|\pi(\mathrm{I}+\mathcal{R})^{\frac{m}{\nu}}\|^2_{\mathtt{HS}}d\mu(\pi).$$

By the Plancherel formula and Corollary 5.1.4, if $m < -Q/2$, we have

$$C_m^2 := \int_{\widehat{G}} \|\pi(\mathrm{I}+\mathcal{R})^{\frac{m}{\nu}}\|_{\mathtt{HS}}^2 d\mu(\pi) = \|\mathcal{B}_{-m}\|^2_{L^2(G)} < \infty.$$

This gives the first estimate in the statement. For the second estimate, we write

$$\sigma(x,\pi) = \sigma(x,\pi)\pi(\mathrm{I}+\mathcal{R})^{\frac{-m}{\nu}}\pi(\mathrm{I}+\mathcal{R})^{\frac{m}{\nu}},$$

and adapt the ideas above. □

We can now prove Theorem 5.2.15.

Proof of Theorem 5.2.15. Let $T \in \Psi^m_{\rho,\delta}$ where $m \in \mathbb{R}$, $1 \geq \rho \geq \delta \geq 0$. Then for any $\phi \in \mathcal{S}(G)$, $T\phi$ is smooth by Theorem 5.1.39.

Let $\kappa : (x,y) \mapsto \kappa_x(y)$ be the kernel associated with T. Let $\mathcal{R}$ be a positive Rockland operator of homogeneous degree ν. The properties of $\mathcal{R}$ (see Sections 4.3 and 4.4.8) yield for any $\phi \in \mathcal{S}(G)$ and $x \in G$ that

$$\begin{aligned} T\phi(x) &= \int_G \phi(y)\kappa_x(y^{-1}x)dy \\ &= \int_G \left[(\mathrm{I}+\mathcal{R})^{-N}\{(\mathrm{I}+\mathcal{R})^N\phi\}(y)\right] \ \kappa_x(y^{-1}x)dy \\ &= \int_G \{(\mathrm{I}+\mathcal{R})^N\phi\}(y)\ \{(\mathrm{I}+\tilde{\mathcal{R}})^{-N}\kappa_x\}(y^{-1}x)dy, \end{aligned}$$

thus, by the Cauchy-Schwartz inequality,

$$|T\phi(x)| \leq \|(\mathrm{I}+\mathcal{R})^N\phi\|_{L^2(G)}\|(\mathrm{I}+\tilde{\mathcal{R}})^{-N}\kappa_x\|_{L^2(G)}.$$

Since $\mathcal{F}_G\{(\mathrm{I}+\tilde{\mathcal{R}})^{-N}\kappa_x\}(\pi) = \sigma(x,\pi)\pi(\mathrm{I}+\mathcal{R})^{-N}$ yields a symbol in $S^{m-N\nu}_{\rho,\delta}$, by Proposition 5.2.16, we have

$$\|(\mathrm{I}+\tilde{\mathcal{R}})^{-N}\kappa_x\|_{L^2(G)} \leq C \sup_{\pi\in\widehat{G}} \|\sigma(x,\pi)\pi(\mathrm{I}+\mathcal{R})^{-N}\|_{\mathscr{L}(\mathcal{H}_\pi)},$$

whenever $m - N\nu < -Q/2$. Note that in this case,

$$\sup_{\pi\in\widehat{G}} \|\sigma(x,\pi)\pi(\mathrm{I}+\mathcal{R})^{-N}\|_{\mathscr{L}(\mathcal{H}_\pi)} \leq \|\sigma\|_{S^m_{\rho,\delta},0,0,|m|}\|\pi(\mathrm{I}+\mathcal{R})^{-N+\frac{m}{\nu}}\|_{\mathscr{L}(\mathcal{H}_\pi)},$$

and by functional calculus

$$\|\pi(\mathrm{I}+\mathcal{R})^{-N+\frac{m}{\nu}}\|_{\mathscr{L}(\mathcal{H}_\pi)} \leq \sup_{\lambda\geq 0}(1+\lambda)^{-N+\frac{m}{\nu}} \leq 1.$$

Thus if we choose $N \in \mathbb{N}_0$ such that $N > (m+\frac{Q}{2})/\nu$, then

$$|T\phi(x)| \leq C\|\sigma\|_{S^m_{\rho,\delta},0,0,|m|}\|(\mathrm{I}+\mathcal{R})^N\phi\|_{L^2(G)}.$$

This shows that $T\phi$ is bounded.

Let $\beta\in\mathbb{N}_0^n$. Using the Leibniz property of vector fields, we easily obtain

$$X^\beta T\phi(x)=\sum_{[\beta_1]+[\beta_2]=[\beta]} c_{\beta_1,\beta_2,\beta}\int_G \phi(y)X^{\beta_1}_{x_1=x}X^{\beta_2}_{x_2=y^{-1}x}\kappa_{x_1}(x_2)dy.$$

As above, we can insert powers of $\mathrm{I}+\mathcal{R}$. Noticing that the symbol

$$\mathcal{F}_G\{(\mathrm{I}+\tilde{\mathcal{R}})_{x_1}^{-N}X^{\beta_1}_{x_1=x}X^{\beta_2}\kappa_{x_1}\}=\pi(X)^{\beta_2}X_x^{\beta_1}\sigma(x,\pi)\pi(\mathrm{I}+\mathcal{R})^{-N}$$

is in $S^{m+\delta[\beta_1]+[\beta_2]-N\nu}_{\rho,\delta}$, we proceed as above to obtain

$$\begin{aligned}\left|X^\beta T\phi(x)\right| &\leq C_1\sum_{[\beta_1]+[\beta_2]=[\beta]}\|(\mathrm{I}+\mathcal{R})^N\phi\|_{L^2(G)}\|\pi(X)^{\beta_2}X_x^{\beta_1}\sigma(x,\pi)\pi(\mathrm{I}+\mathcal{R})^{-N}\|_{L^2(\widehat{G})}\\ &\leq C_2\|\sigma\|_{S^m_{\rho,\delta},0,[\beta],|m|+[\beta]}\|(\mathrm{I}+\mathcal{R})^N\phi\|_{L^2(G)}.\end{aligned}$$

as long as $N>(m+[\beta]+\frac{Q}{2})/\nu$.

Let $\alpha\in\mathbb{N}_0^n$. Proceeding as in the proof of Proposition 5.2.3 (4), we can write

$$(xy)^\alpha=\sum_{[\alpha_1]+[\alpha_2]=[\alpha]} c'_{\alpha,\alpha_1,\alpha_2}\, q_{\alpha_1}(x)\, q_{\alpha_2}(y).$$

Using this, we easily obtain

$$\begin{aligned}x^\alpha T\phi(x) &= \int_G (y\ \ y^{-1}x)^\alpha\phi(y)\kappa_x(y^{-1}x)dy\\ &= \sum_{[\alpha_1]+[\alpha_2]=[\alpha]} c'_{\alpha,\alpha_1,\alpha_2}\int_G q_{\alpha_1}(y)\phi(y)q_{\alpha_2}(y^{-1}x)\kappa_x(y^{-1}x)dy.\end{aligned}$$

Noticing that

$$\mathcal{F}_G\{(\mathrm{I}+\tilde{\mathcal{R}})^{-N}\{q_{\alpha_2}\kappa_x\}=\{\Delta^{\alpha_2}\sigma(x,\cdot)\}\ \pi(\mathrm{I}+\mathcal{R})^{-N}\in S^{m-N\nu-\rho[\alpha_2]}_{\rho,\delta},$$

we can now proceed as in the first paragraph above to obtain

$$\begin{aligned}|x^\alpha T\phi(x)| &\leq C_1\sum_{[\alpha_1]+[\alpha_2]=[\alpha]}\|(\mathrm{I}+\mathcal{R})_y^N\{q_{\alpha_1}\phi\}\|_2\|(\mathrm{I}+\tilde{\mathcal{R}})^{-N}\{q_{\alpha_2}\kappa_x\}\|_2\\ &\leq C_2\|\sigma(x,\pi)\|_{S^m_{\rho,\delta},[\alpha],0,|m|+\rho[\alpha]}\sum_{[\alpha_1]\leq[\alpha]}\|(\mathrm{I}+\mathcal{R})_y^N\{q_{\alpha_1}\phi\}\|_2\end{aligned}$$

as long as $N>(m+Q/2)/\nu$.

We can combine the two paragraphs above to show that for any $\alpha,\beta\in\mathbb{N}_0^n$, we have

$$\left|x^\alpha X^\beta T\phi(x)\right|\leq C\|\sigma(x,\pi)\|_{S^m_{\rho,\delta},[\alpha],[\beta],|m|+[\beta]+\rho[\alpha]}\sum_{[\alpha_1]\leq[\alpha]}\|(\mathrm{I}+\mathcal{R})_y^N\{q_{\alpha_1}\phi\}\|_2,$$

as long as $N > (m + [\beta] + Q/2)/\nu$. By Lemma 3.1.56, we have

$$\sum_{[\alpha_1]\leq[\alpha]} \|(\mathrm{I} + \mathcal{R})_y^N \{q_{\alpha_1}\phi\}\|_2 \leq C'\|\phi\|_{\mathcal{S}(G),N'}$$

for some $N' \in \mathbb{N}$ depending on N and α, and $T\phi$ is a Schwartz function. Furthermore, these estimates also imply the rest of Theorem 5.2.15. □

Theorem 5.2.15 shows that composing two operators in (possibly different) $\Psi^m_{\rho,\delta}$ makes sense as the composition of operators acting on the Schwartz space. We will see that in fact, the composition of $T_1 \in \Psi^{m_1}_{\rho,\delta}$ with $T_2 \in \Psi^{m_2}_{\rho,\delta}$ is T_1T_2 in $\Psi^{m_1+m_2}_{\rho,\delta}$, see Theorem 5.5.3.

We will see that our classes of pseudo-differential operators are stable under taking the formal L^2-adjoint, see Theorem 5.5.12. This together with Theorem 5.2.15 will imply the continuity of our operators on the space $\mathcal{S}'(G)$ of tempered distributions, see Corollary 5.5.13.

Returning to our exposition, before proving that the introduced classes of symbols $\cup_{m\in\mathbb{R}} S^m_{\rho,\delta}$ and of the corresponding operators $\cup_{m\in\mathbb{R}}\Psi^m_{\rho,\delta}$ are stable under composition and taking the adjoint, let us give some examples.

5.2.4 First examples

As it should be, $\cup_{m\in\mathbb{R}}\Psi^m$ contains the left-invariant differential operators. More precisely, the following lemma implies that $\sum_{[\beta]\leq m} c_\beta X^\beta \in \Psi^m$. The coefficients c_α here are constant and it is easy to relax this condition with each function c_α being smooth and bounded together with all of its left derivatives.

Lemma 5.2.17. *For any $\beta_o \in \mathbb{N}_0^n$, the operator $X^{\beta_o} = \mathrm{Op}(\pi(X)^{\beta_o})$ is in $\Psi^{[\beta_o]}$.*

Proof. By Lemma 5.2.9, we have

$$\Delta^\alpha \pi(X)^{\beta_o} = \begin{cases} 0 & \text{if } [\alpha] > [\beta_o], \\ \sum\limits_{[\alpha]+[\beta_2]=[\beta_o]} \pi(X)^{\beta_2} & \text{if } [\alpha] \leq [\beta_o]. \end{cases}$$

Recall that, by Example 5.1.26, $\{\pi(X)^\beta, \pi \in \widehat{G}\} \in L^\infty_{\gamma+[\beta],\gamma}(\widehat{G})$ for any $\gamma \in \mathbb{R}, \beta \in \mathbb{N}_0^n$. So $\{\Delta^\alpha\pi(X)^{\beta_o}, \pi \in \widehat{G}\}$ is zero if $[\alpha] > [\beta_o]$ whereas it is in $L^\infty_{\gamma+[\beta_o]-[\alpha],\gamma}(\widehat{G})$ for any $\gamma \in \mathbb{R}$ if $[\alpha] \leq [\beta_o]$. □

Remark 5.2.18. Lemma 5.2.17 implies that $\cup_{m\in\mathbb{R}}\Psi^m$ contains the left-invariant differential calculus, that is, the space of left-invariant differential operators.

One could wonder whether it also contains the right-invariant differential calculus, since we can quantize any differential operator, see Remark 5.1.41 (1). This is false in general, see Example 5.2.19 below. Thus, if one is interested in dealing with problems based on the setting of right-invariant operators one can use the corresponding version of the theory based on the right-invariant Rockland operator, see Remark 5.1.41 (4).

Example 5.2.19. Let us consider the three dimensional Heisenberg group $\mathbb{H}_1$ and the canonical basis X, Y, T of its Lie algebra (see Example 1.6.4). Then the right invariant vector field $\tilde{X}$ can not be in $\cup_{m\in\mathbb{R}}\Psi^m$.

Proof of the statement in Example 5.2.19 . We have already seen that any operator $A \in \Psi^m$ acts continuously on the Schwartz space, cf. Theorem 5.2.15. We will see later (see Corollary 5.7.2) that it also acts on Sobolev spaces with a loss of derivative controlled by its order m. By this, we mean that, if an operator A in Ψ^m is homogeneous of degree ν_A, then we must have

$$\forall s \in \mathbb{R} \quad \exists C > 0 \quad \forall f \in \mathcal{S}(G) \qquad \|Af\|_{L^2_{s-m}} \leq C\|f\|_{L^2_s},$$

and when $s+m$ and s are non-negative, we realise the Sobolev norm as $\|f\|_{L^2_s} = \|f\|_{L^2} + \|\mathcal{R}^{\frac{s}{\nu}} f\|_{L^2}$ for some positive Rockland operator of degree ν, cf. Theorem 4.4.3 Part (2). Applying the inequality to dilated functions $f \circ D_r$ and letting $r \to \infty$ yield that $m \geq \nu_A$.

Applying this to the case of $\tilde{X}$ shows that if $\tilde{X}$ were in some Ψ^m then $m \geq 1$ and $\tilde{X}$ would map L^2_1 to L^2_{1-m} hence to L^2 continuously. We have already shown in the proof of Example 4.4.32 that this is not possible. □

An example of a smoothing operator is given via convolution with a Schwartz function:

Lemma 5.2.20. *Let $\kappa \in \mathcal{S}(G)$. We denote by $T_\kappa : \phi \mapsto \phi * \kappa$ the corresponding convolution operator. Its symbol σ_{T_κ} is independent of x and is given by*

$$\sigma_{T_\kappa}(\pi) = \widehat{\kappa}(\pi).$$

Furthermore, the mapping

$$\mathcal{S}(G) \ni \kappa \mapsto T_\kappa \in \Psi^{-\infty}$$

is continuous.

Proof. For the first part, see Example 5.1.38 and its continuation.

For any $\kappa \in \mathcal{S}(G)$, we have $\tilde{q}_\alpha\kappa \in \mathcal{S}(G)$ for any $\alpha \in \mathbb{N}_0^n$, and

$$(\mathrm{I}+\mathcal{R})^a(\mathrm{I}+\tilde{\mathcal{R}})^b\kappa \in \mathcal{S}(G)$$

for any $a, b \in \mathbb{N}$ (see also (4.34) and Proposition 4.4.30). For any $m \in \mathbb{R}$, $\gamma \in \mathbb{R}$ and $\alpha \in \mathbb{N}_0^n$, we have by (1.38)

$$\begin{aligned}
\|\Delta^\alpha\widehat{\kappa}\|_{L^\infty_{\gamma,[\alpha]-m+\gamma}(\widehat{G})} &= \|\pi(\mathrm{I}+\mathcal{R})^{\frac{[\alpha]-m+\gamma}{\nu}}\Delta^\alpha\pi(\kappa)\pi(\mathrm{I}+\mathcal{R})^{-\frac{\gamma}{\nu}}\|_{L^\infty(\widehat{G})} \\
&\leq \|(\mathrm{I}+\mathcal{R})^{\frac{[\alpha]-m+\gamma}{\nu}}(\mathrm{I}+\tilde{\mathcal{R}})^{-\frac{\gamma}{\nu}}\{\tilde{q}_\alpha\kappa\}\|_{L^1(G)}.
\end{aligned}$$

As $\kappa \in \mathcal{S}(G)$, this L^1-norm is finite and this shows that $\sigma_{T_\kappa} \in \Psi^{-\infty}$. More precisely, this L^1-norm is less or equal to

$$\begin{cases} \|\mathcal{B}_\gamma\|_1 \|(\mathrm{I}+\mathcal{R})^a\{\tilde{q}_\alpha \kappa\}\|_1 & \text{if } \gamma \text{ and } \frac{[\alpha]-m+\gamma}{\nu} > 0 \text{ and } a = \lceil \frac{[\alpha]-m+\gamma}{\nu} \rceil, \\ \|\mathcal{B}_{-\frac{[\alpha]-m+\gamma}{\nu}}\|_1 \|(\mathrm{I}+\tilde{\mathcal{R}})^b\{\tilde{q}_\alpha \kappa\}\|_1 & \text{if } \gamma \text{ and } \frac{[\alpha]-m+\gamma}{\nu} < 0 \text{ and } b = \lceil -\frac{\gamma}{\nu} \rceil, \end{cases}$$

where $\lceil x \rceil$ denotes the smallest integer $> x$ and $\mathcal{B}_\gamma$ is the right-convolution kernel of the Bessel potential of $\mathcal{R}$, see Corollary 4.3.11. By Proposition 4.4.27, these quantities can be estimated by Schwartz seminorms. □

More generally, the operators and symbols with kernels 'depending on x' but satisfying the following property are smoothing:

Lemma 5.2.21. *Let $\kappa : (x, y) \mapsto \kappa_x(y)$ be a smooth function on $G \times G$ such that, for each multi-index $\beta \in \mathbb{N}_0^n$ and each Schwartz seminorm $\|\cdot\|_{\mathcal{S}(G),N}$, the following quantity*

$$\sup_{x\in G} \|X_x^\beta \kappa_x\|_{\mathcal{S}(G),N} < \infty,$$

is finite.

Then the symbol σ given via $\sigma(x, \pi) = \widehat{\kappa}_x(\pi)$ is smoothing. Furthermore for any seminorm $\|\cdot\|_{S^m,a,b,c}$, there exists $C > 0$ and $\beta \in \mathbb{N}_0^n$, $N \in \mathbb{N}_0$ such that

$$\|\sigma\|_{S^m,a,b,c} \leq C \sup_{x\in G} \|X_x^\beta \kappa_x\|_{\mathcal{S}(G),N}.$$

Proof of Lemma 5.2.21. By (1.38), we have

$$\sup_{\pi\in\widehat{G}} \|\sigma(x,\pi)\|_{\mathscr{L}(\mathcal{H}_\pi)} = \sup_{\pi\in\widehat{G}} \|\widehat{\kappa}_x(\pi)\|_{\mathscr{L}(\mathcal{H}_\pi)} \leq \|\kappa_x\|_{L^1(G)}.$$

More generally, for any $\gamma_1, \gamma_2 \in \mathbb{R}$, denoting by $N_1, N_2 \in \mathbb{N}_0$ integers such that $\gamma_1 \leq N_1$ $\gamma_2 \leq N_2$, we have

$$\begin{aligned} &\sup_{\pi\in\widehat{G}} \|\pi(\mathrm{I}+\mathcal{R})^{\gamma_1} X_x^\beta \Delta^\alpha \sigma(x,\pi)\ \pi(\mathrm{I}+\mathcal{R})^{\gamma_2}\|_{\mathscr{L}(\mathcal{H}_\pi)} \\ &\quad \leq \sup_{\pi\in\widehat{G}} \|\pi(\mathrm{I}+\mathcal{R})^{N_1} X_x^\beta \Delta^\alpha \sigma(x,\pi)\ \pi(\mathrm{I}+\mathcal{R})^{N_2}\|_{\mathscr{L}(\mathcal{H}_\pi)} \\ &\qquad = \sup_{\pi\in\widehat{G}} \|\mathcal{F}_G\{(\mathrm{I}+\mathcal{R})^{N_1}(\mathrm{I}+\tilde{\mathcal{R}})^{N_2} X_x^\beta q_\alpha \kappa_x\}(\pi)\|_{\mathscr{L}(\mathcal{H}_\pi)} \\ &\qquad \leq \|(\mathrm{I}+\mathcal{R})^{N_1}(\mathrm{I}+\tilde{\mathcal{R}})^{N_2} q_\alpha X_x^\beta \kappa_x\|_{L^1(G)}. \end{aligned}$$

This last L^1-norm is, up to a constant, less or equal than a Schwartz seminorm of $X_x^\beta \kappa_x$, see Section 3.1.9. This implies the statement. □

In Theorem 5.4.9, we will see that the converse holds, that is, that any smoothing operator has an associated kernel as in Lemma 5.2.21.

5.2.5 First properties of symbol classes

We summarise in the next theorem some properties of the symbol classes which follow from their definition.

Theorem 5.2.22. *Let $1 \geq \rho \geq \delta \geq 0$.*

(i) Let $\sigma \in S^m_{\rho,\delta}$ have kernel κ_x and order $m \in \mathbb{R}$.

1. *For every $x \in G$ and $\gamma \in \mathbb{R}$,*

$$\tilde{q}_\alpha X^\beta \kappa_x \in \mathcal{K}_{\gamma,\rho[\alpha]-m-\delta[\beta]+\gamma}.$$

2. *If $\beta_o \in \mathbb{N}_0^n$ then the symbol $\{X_x^{\beta_o}\sigma(x,\pi), (x,\pi) \in G \times \widehat{G}\}$ is in $S^{m+\delta[\beta_o]}_{\rho,\delta}$ with kernel $X_x^{\beta_o}\kappa_x$, and*

$$\|X_x^{\beta_o}\sigma(x,\pi)\|_{S^{m+\delta[\beta_o]}_{\rho,\delta},a,b,c} \leq C_{b,\beta_o}\|\sigma(x,\pi)\|_{S^m_{\rho,\delta},a,b+[\beta_o],c}.$$

3. *If $\alpha_o \in \mathbb{N}_0^n$ then the symbol $\{\Delta^{\alpha_o}\sigma(x,\pi), (x,\pi) \in G \times \widehat{G}\}$ is in $S^{m-\rho[\alpha_o]}_{\rho,\delta}$ with kernel $\tilde{q}_{\alpha_o}\kappa_x$, and*

$$\|\Delta^{\alpha_o}\sigma(x,\pi)\|_{S^{m-\rho[\alpha_o]}_{\rho,\delta},a,b,c} \leq C_{a,\alpha_o}\|\sigma(x,\pi)\|_{S^m_{\rho,\delta},a+[\alpha_o],b,c}.$$

4. *The symbol*

$$\sigma^* := \{\sigma(x,\pi)^*, (x,\pi) \in G \times \widehat{G}\}$$

is in $S^m_{\rho,\delta}$ with kernel κ_x^* given by

$$\kappa_x^*(y) = \bar{\kappa}_x(y^{-1}),$$

and

$$\|\sigma(x,\pi)^*\|_{S^m_{\rho,\delta},a,b,c} = \sup_{\substack{|\gamma|\leq c \\ [\alpha]\leq a,\ [\beta]\leq b}} \|\pi(\mathrm{I}+\mathcal{R})^{-\frac{\gamma}{\nu}} X_x^\beta \Delta^\alpha \sigma(x,\pi)\pi(\mathrm{I}+\mathcal{R})^{\frac{\rho[\alpha]-m-\delta[\beta]+\gamma}{\nu}}\|_{\mathscr{L}(\mathcal{H}_\pi)}.$$

(ii) Let $\sigma_1 \in S^{m_1}_{\rho,\delta}$ and $\sigma_2 \in S^{m_2}_{\rho,\delta}$ have kernels κ_{1x} and κ_{2x}, respectively. Then

$$\sigma(x,\pi) := \sigma_1(x,\pi)\sigma_2(x,\pi)$$

defines the symbol σ in $S^m_{\rho,\delta}$, $m = m_1 + m_2$, with kernel $\kappa_{2x} * \kappa_{1x}$ with the convolution in the sense of Definition 5.1.19. Furthermore,

$$\|\sigma(x,\pi)\|_{S^m_{\rho,\delta},a,b,c} \leq C\|\sigma_1(x,\pi)\|_{S^{m_1}_{\rho,\delta},a,b,c+\rho a+|m_2|+\delta b}\|\sigma_2(x,\pi)\|_{S^{m_2}_{\rho,\delta},a,b,c},$$

where the constant $C = C_{a,b,c,m_1,m_2} > 0$ does not depend on σ_1, σ_2.

Note that, in Part (ii), the composition $\sigma(x,\pi) := \sigma_1(x,\pi)\sigma_2(x,\pi)$ may be understood as the composition of two fields of operators acting on smooth vectors as well as the composition of $\sigma_1(x,\cdot) \in L^\infty_{\gamma_1,\gamma_1-m_1}(\widehat{G})$ with $\sigma_2(x,\cdot) \in L^\infty_{\gamma_2,\gamma_2-m_2}(\widehat{G})$ for any choice of $\gamma_1,\gamma_2 \in \mathbb{R}$ such that $\gamma_1 - m_1 = \gamma_2$.

Proof. Properties (1), (2), (3), and (4) of (i) are straightforward to check.

Let us prove Part (ii). By Property (1) of (i), or by the definition of symbol classes,

$$\kappa_{jx} \in \mathcal{K}_{\gamma_j,-m_j+\gamma_j} \quad \text{for any} \quad \gamma_j \in \mathbb{R},\ j=1,2,$$

thus choosing $\gamma = \gamma_2$ and $\gamma_1 = -m_2 + \gamma_2$, we have by Corollary 5.1.20

$$\kappa_{2x} * \kappa_{1x} \in \mathcal{K}_{\gamma,-m+\gamma} \quad \text{for any} \quad \gamma \in \mathbb{R}.$$

Its group Fourier transform is

$$\pi(\kappa_{1x})\pi(\kappa_{2x}) = \sigma_1(x,\pi)\sigma_2(x,\pi) = \sigma(x,\pi).$$

Therefore, σ is a symbol with kernel $\kappa_{2x} * \kappa_{1x}$.

Let $\alpha,\beta \in \mathbb{N}_0^n$ and $\gamma \in \mathbb{R}$. From the Leibniz rules for Δ^α (see Proposition 5.2.10) and X^β, the operator

$$\pi(\mathrm{I}+\mathcal{R})^{\frac{\rho[\alpha]-m-\delta[\beta]+\gamma}{\nu}} X_x^\beta \Delta^\alpha \sigma(x,\pi)\pi(\mathrm{I}+\mathcal{R})^{-\frac{\gamma}{\nu}},$$

is a linear combination over $\beta_1,\beta_2,\alpha_1,\alpha_2 \in \mathbb{N}^n$ satisfying $[\beta_1]+[\beta_2]=[\beta]$, $[\alpha_1]+[\alpha_2]=[\alpha]$, of terms

$$\pi(\mathrm{I}+\mathcal{R})^{\frac{\rho[\alpha]-m-\delta[\beta]+\gamma}{\nu}} X_x^{\beta_1}\Delta^{\alpha_1}\sigma_1(x,\pi) X_x^{\beta_2}\Delta^{\alpha_2}\sigma_2(x,\pi)\pi(\mathrm{I}+\mathcal{R})^{-\frac{\gamma}{\nu}},$$

whose operator norm is bounded by

$$\begin{aligned}\|\pi(\mathrm{I}+\mathcal{R})^{\frac{\rho[\alpha]-m-\delta[\beta]+\gamma}{\nu}} X_x^{\beta_1}\Delta^{\alpha_1}\sigma_1(x,\pi)\pi(\mathrm{I}+\mathcal{R})^{-\frac{\rho[\alpha_2]-m_2-\delta[\beta_2]+\gamma}{\nu}}\|_{\mathscr{L}(\mathcal{H}_\pi)} \\ \|\pi(\mathrm{I}+\mathcal{R})^{\frac{\rho[\alpha_2]-m_2-\delta[\beta_2]+\gamma}{\nu}} X_x^{\beta_2}\Delta^{\alpha_2}\sigma_2(x,\pi)\pi(\mathrm{I}+\mathcal{R})^{-\frac{\gamma}{\nu}}\|_{\mathscr{L}(\mathcal{H}_\pi)}.\end{aligned}$$

This shows that the inequality between the seminorms of σ, σ_1 and σ_2 given in (ii) holds. Consequently σ is a symbol of order $m = m_1 + m_2$ and of type (ρ,δ), and (ii) is proved. □

A direct consequence of Part (ii) of Theorem 5.2.22 is that the symbols in the introduced symbol classes form an algebra:

Corollary 5.2.23. *Let $1 \ge \rho \ge \delta \ge 0$. The collection of symbols $\bigcup_{m\in\mathbb{R}} S^m_{\rho,\delta}$ forms an algebra.*

Furthermore, if $\sigma_0 \in S^{-\infty}$ and $\sigma \in S^m_{\rho,\delta}$ is of order $m \in \mathbb{R}$, then $\sigma_0\sigma$ and $\sigma\sigma_0$ are also in $S^{-\infty}$.

The fact that the symbol classes $\bigcup_{m\in\mathbb{R}} S^m_{\rho,\delta}$ form an algebra does not imply directly the same property for the operator classes $\bigcup_{m\in\mathbb{R}} \Psi^m_{\rho,\delta}$ since our quantization is not an algebra morphism, that is, $\text{Op}(\sigma_1\sigma_2)$ is not equal in general to $\text{Op}(\sigma_1)\text{Op}(\sigma_2)$. However, we will show that indeed $\bigcup_{m\in\mathbb{R}} \Psi^m_{\rho,\delta}$ is an algebra of operators, cf. Theorem 5.5.3, and we will often use the following property:

Lemma 5.2.24. *Let σ_1 and σ_2 be as in Theorem 5.2.22, (ii). We assume that σ_2 does not depend on x: $\sigma_2 = \{\sigma_2(\pi) : \pi \in \widehat{G}\}$. Then*

$$\sigma(x,\pi) := \sigma_1(x,\pi)\sigma_2(\pi)$$

defines the symbol σ in $S^m_{\rho,\delta}$, $m = m_1 + m_2$ and

$$\text{Op}(\sigma) = \text{Op}(\sigma_1)\text{Op}(\sigma_2)$$

Proof. We keep the notation of the statement. Let κ_{1x} and κ_2 be the convolution kernels of σ_1 and σ_2 respectively. Hence κ_2 is a function on G independent of x. By Theorem 5.2.22(ii), $\kappa_2 * \kappa_{1x}$ is the convolution kernel of σ, thus

$$\forall \phi \in \mathcal{S}(G) \qquad \text{Op}(\sigma)(\phi)(x) = \phi * (\kappa_2 * \kappa_{1x}).$$

As $\phi * \kappa_2 = \text{Op}(\sigma_2)\phi$, this implies easily that $\text{Op}(\sigma)$ is the composition of $\text{Op}(\sigma_1)$ with $\text{Op}(\sigma_2)$. □

The following will also be useful, for instance in the estimates for the kernels in Section 5.4.1.

Corollary 5.2.25. *Let $1 \geq \rho \geq \delta \geq 0$. Let $\sigma \in S^m_{\rho,\delta}$ have kernel κ_x. If β_1 and β_2 are in $\mathbb{N}_0^n$, then*

$$\{\pi(X)^{\beta_1}\sigma(x,\pi)\,\pi(X)^{\beta_2}, (x,\pi) \in G \times \widehat{G}\} \in S^{m+[\beta_1]+[\beta_2]}_{\rho,\delta}$$

with kernel $X_y^{\beta_1}\tilde{X}_y^{\beta_2}\kappa_x(y)$. Furthermore, for any a,b,c there exists $C = C_{a,b,c,\beta_1,\beta_2}$ independent of σ such that

$$\|\pi(X)^{\beta_1}\sigma(x,\pi)\pi(X)^{\beta_2}\|_{S^m_{\rho,\delta},a,b,c} \leq C\|\sigma\|_{S^m_{\rho,\delta},a,b,c+\rho a+[\beta_1]+[\beta_2]+\delta b}.$$

If $\beta_2 = 0$, for any a,b,c there exists $C = C_{a,b,c,\beta_1}$ independent of σ such that

$$\|\pi(X)^{\beta_1}\sigma\|_{S^m_{\rho,\delta},a,b,c} \leq C\|\sigma\|_{S^m_{\rho,\delta},a,b,c}.$$

Proof. The first part follows directly from Theorem 5.2.22 Part (ii) together with Lemma 5.2.17.

We need to show a better estimate for $\beta_2 = 0$. Let $\alpha, \beta_o \in \mathbb{N}_0^n$. By the Leibniz formula (see (5.28)), we have

$$\begin{aligned} &X_x^{\beta_o}\Delta^\alpha\{\pi(X)^{\beta_1}\sigma(x,\pi)\} \\ &\quad = \sum_{[\alpha_1]+[\alpha_2]=[\alpha]} c_{\alpha,\alpha_1,\alpha_2}\{\Delta^{\alpha_1}\pi(X)^{\beta_1}\}\ \{X_x^{\beta_o}\Delta^{\alpha_2}\sigma(x,\pi)\}. \end{aligned}$$

Hence, denoting $m_o := m + \delta[\beta_o]$, we have

$$\begin{aligned}
&\|\pi(\mathrm{I}+\mathcal{R})^{\frac{\rho[\alpha]-m_o-[\beta_1]+\gamma}{\nu}} X_x^{\beta_o}\Delta^\alpha\{\pi(X)^{\beta_1}\sigma(x,\pi)\}\pi(\mathrm{I}+\mathcal{R})^{-\frac{\gamma}{\nu}}\|_{\mathscr{L}(\mathcal{H}_\pi)} \\
&\quad\le C\sum_{[\alpha_1]+[\alpha_2]=[\alpha]} \|\pi(\mathrm{I}+\mathcal{R})^{\frac{\rho[\alpha]-m_o-[\beta_1]+\gamma}{\nu}}\Delta^{\alpha_1}\pi(X)^{\beta_1}\pi(\mathrm{I}+\mathcal{R})^{-\frac{\rho[\alpha_2]-m_o+\gamma}{\nu}}\|_{\mathscr{L}(\mathcal{H}_\pi)} \\
&\qquad\qquad \|\pi(\mathrm{I}+\mathcal{R})^{\frac{\rho[\alpha_2]-m_o+\gamma}{\nu}} X_x^{\beta_o}\Delta^{\alpha_2}\sigma(x,\pi)\pi(\mathrm{I}+\mathcal{R})^{-\frac{\gamma}{\nu}}\|_{\mathscr{L}(\mathcal{H}_\pi)}.
\end{aligned}$$

As $\{\pi(X)^{\beta_1}\}\in S_{1,0}^{[\beta_1]}$ by Lemma 5.2.17, each quantity

$$\sup_{|\gamma|\le c,\pi\in\widehat{G}} \|\pi(\mathrm{I}+\mathcal{R})^{\frac{\rho[\alpha]-m_o-[\beta_1]+\gamma}{\nu}}\Delta^{\alpha_1}\pi(X)^{\beta_1}\pi(\mathrm{I}+\mathcal{R})^{-\frac{\rho[\alpha_2]-m_o+\gamma}{\nu}}\|_{\mathscr{L}(\mathcal{H}_\pi)}<\infty$$

is finite for any $c>0$ and $\alpha_1,\alpha_2\in\mathbb{N}_0^n$ such that $[\alpha_1]+[\alpha_2]=[\alpha]$. This implies

$$\begin{aligned}
&\sup_{|\gamma|\le c,\pi\in\widehat{G}} \|\pi(\mathrm{I}+\mathcal{R})^{\frac{\rho[\alpha]-m_o-[\beta_1]+\gamma}{\nu}} X_x^{\beta_o}\Delta^\alpha\{\pi(X)^{\beta_1}\sigma(x,\pi)\}\pi(\mathrm{I}+\mathcal{R})^{-\frac{\gamma}{\nu}}\|_{\mathscr{L}(\mathcal{H}_\pi)} \\
&\quad\le C'\sum_{[\alpha_2]\le[\alpha]} \sup_{\substack{|\gamma|\le c\\ \pi\in\widehat{G}}} \|\pi(\mathrm{I}+\mathcal{R})^{\frac{\rho[\alpha_2]-m_o+\gamma}{\nu}} X_x^{\beta_o}\Delta^{\alpha_2}\sigma(x,\pi)\pi(\mathrm{I}+\mathcal{R})^{-\frac{\gamma}{\nu}}\|_{\mathscr{L}(\mathcal{H}_\pi)}.
\end{aligned}$$

Taking the supremum over $[\alpha]\le a$ and $[\beta]\le b$ yields the stated estimate. □

5.3 Spectral multipliers in positive Rockland operators

In this section we show that multipliers in positive Rockland operators belong to the introduced symbol classes Ψ^m.

The main result is stated in Proposition 5.3.4. This will allow us to use the Littlewood-Paley decompositions associated with a positive Rockland operator, and therefore will enter most of the subsequent proofs.

5.3.1 Multipliers in one positive Rockland operator

The precise class of multiplier functions that we consider is the following:

Definition 5.3.1. Let $\mathcal{M}_m$ be the space of functions $f\in C^\infty(\mathbb{R}_+)$ such that the following quantities for all $\ell\in\mathbb{N}_0$ are finite:

$$\|f\|_{\mathcal{M}_m,\ell} := \sup_{\lambda>0,\,\ell'=0,\dots,\ell} (1+\lambda)^{-m+\ell'}|\partial_\lambda^{\ell'} f(\lambda)|.$$

In other words, the class of functions f that appears in the definition above are the functions which are smooth on $\mathbb{R}_+=(0,\infty)$ and have the symbolic behaviour at infinity of the Hörmander class $S^m_{1,0}(\mathbb{R})$ on the real line. However, we rather prefer the notation $\mathcal{M}_m$ in order not to create any confusion between these classes and the classes $S^m_{\rho,\delta}(G)$ defined on the group G.

Example 5.3.2. For any $m \in \mathbb{R}$, the function $\lambda \mapsto (1+\lambda)^m$ is in $\mathcal{M}_m$.

It is a routine exercise to check that $\mathcal{M}_m$ endowed with the family of maps $\|\cdot\|_{\mathcal{M}_{m,\ell}}$, $\ell \in \mathbb{N}_0$, is a Fréchet space. Furthermore, it satisfies the following property.

Lemma 5.3.3. *If $f_1 \in \mathcal{M}_{m_1}$ and $f_2 \in \mathcal{M}_{m_2}$ then $f_1 f_2 \in \mathcal{M}_{m_1+m_2}$ with*

$$\|f_1 f_2\|_{\mathcal{M}_{m_1+m_2,\ell}} \leq C_\ell \|f_1\|_{\mathcal{M}_{m_1,\ell}} \|f_2\|_{\mathcal{M}_{m_2,\ell}}.$$

Proof. This follows from the Leibniz formula for $|\partial^{\ell'}(f_1 f_2)|$ and from the following inequality which holds for $\lambda > 0$ and $\ell'_1, \ell'_2 \leq \ell$:

$$(1+\lambda)^{-m_1-m_2+\ell'_1+\ell'_2} |\partial_\lambda^{\ell'_1} f_1(\lambda)|\, |\partial_\lambda^{\ell'_2} f_2(\lambda)| \leq \|f_1\|_{\mathcal{M}_{m_1,\ell}} \|f_2\|_{\mathcal{M}_{m_2,\ell}},$$

which implies the claim. □

The main property of this section is

Proposition 5.3.4. *Let $m \in \mathbb{R}$ and let $\mathcal{R}$ be a positive Rockland operator of homogeneous degree ν. If $f \in \mathcal{M}_{\frac{m}{\nu}}$, then $f(\mathcal{R})$ is in Ψ^m and its symbol $\{f(\pi(\mathcal{R})), \pi \in \widehat{G}\}$ satisfies*

$$\forall a,b,c \in \mathbb{N}_0 \qquad \exists \ell \in \mathbb{N},\ C>0 \ : \qquad \|f(\pi(\mathcal{R}))\|_{a,b,c} \leq C \|f\|_{\mathcal{M}_{\frac{m}{\nu}},\ell},$$

with ℓ and C independent of f.

Proof. First let us show that it suffices to show Proposition 5.3.4 for $m < -\nu$. If $f \in \mathcal{M}_{\frac{m}{\nu}}$ with $m \geq -\nu$, then we define

- $m_2 \geq \nu$ such that $\frac{m_2}{\nu}$ is the smallest integer strictly larger than $\frac{m}{\nu}$,
- $f_1(\lambda) := (1+\lambda)^{-\frac{m_2}{\nu}} f(\lambda)$ and $f_2(\lambda) := (1+\lambda)^{\frac{m_2}{\nu}}$.

By Example 5.3.2 and Lemma 5.3.3, we see that $f_1 \in \mathcal{M}_{\frac{m_1}{\nu}}$ with $m_1 = m - m_2$. By Lemma 5.2.17, we see that $f_2(\pi(\mathcal{R})) \in S^{m_2}$. If Proposition 5.3.4 holds for $m_1 < -\nu$, then we can apply it to f_1 and hence $f_1(\pi(\mathcal{R})) \in S^{m_1}$. Thus the product

$$f(\pi(\mathcal{R})) = f_1(\pi(\mathcal{R})) f_2(\pi(\mathcal{R}))$$

is in $S^{m_1+m_2} = S^m$.

Therefore, as claimed above, it suffices to show Proposition 5.3.4 for $m < -\nu$.

Now we show that we may assume that f is supported away from 0. Indeed, if $f \in \mathcal{M}_{\frac{m}{\nu}}$, we extend it smoothly to $\mathbb{R}$ and we write

$$f = f\chi_o + f(1-\chi_o),$$

where $\chi_o \in \mathcal{D}(\mathbb{R})$ is identically 1 on $[-1,1]$. Since $f\chi_o \in \mathcal{D}(\mathbb{R})$, by Hulanicki's theorem (cf. Corollary 4.5.2), the kernel of $(f\chi_o)(\mathcal{R})$ is Schwartz and by Lemma 5.2.20, we have $(f\chi_o)(\mathcal{R}) \in \Psi^{-\infty}$ with suitable inequalities for the seminorms. Thus we just have to prove the result for $f(1-\chi_o)$ which is supported in $[1,\infty)$ where $\lambda \asymp 1+\lambda$. The statement then follows from the following lemma. □

Showing Proposition 5.3.4 boils then down to

Lemma 5.3.5. *Let $m < -\nu$. If $f \in C^\infty(\mathbb{R})$ is supported in $[1,\infty)$ and satisfies*

$$\forall \ell \in \mathbb{N}_0 \quad \exists C_\ell \quad \forall \lambda \geq 1 \qquad |\partial_\lambda^\ell f(\lambda)| \leq C_\ell |\lambda|^{\frac{m}{\nu}-\ell},$$

then $f(\mathcal{R}) \in \Psi^m$, and for any $a,b,c \in \mathbb{N}_0$ we have

$$\|f(\mathcal{R})\|_{\Psi^m,a,b,c} \leq C \sup_{\lambda \geq 1, \ell'=0,\ldots,\ell} |\lambda|^{-\frac{m}{\nu}+\ell'} |\partial_\lambda^{\ell'} f(\lambda)|,$$

with $\ell = \ell_{m,a,b,c} \in \mathbb{N}$ and $C = C_{m,a,b,c} > 0$ independent of f.

The proof of Lemma 5.3.5 relies on the following consequence of Hulanicki's theorem (see Theorem 4.5.1).

Lemma 5.3.6. *Let $\mathcal{R}$ be a positive Rockland operator on a graded Lie group G.*

Let $m \in \mathcal{D}(\mathbb{R})$ and $\alpha_o \in \mathbb{N}_0^n$. We denote by $m(\mathcal{R})\delta_0$ the kernel of the multiplier $m(\mathcal{R})$ and we set

$$\kappa(x) := x^{\alpha_o} m(\mathcal{R})\delta_0(x).$$

The function κ is Schwartz.

For any $p \in (1,\infty)$, $N \in \mathbb{N}$ and $a \in \mathbb{R}$ with $0 \leq a \leq N\nu$, there exist $C > 0$ and $k \in \mathbb{N}$ such that for any $\phi \in \mathcal{S}(G)$,

$$\|\mathcal{R}^N(\phi * \kappa)\|_p \leq C \sup_{\substack{\lambda>0 \\ \ell=0,\ldots,k}} (1+\lambda)^k |\partial_\lambda^\ell m(\lambda)| \; \|\mathcal{R}_p^{\frac{a}{\nu}} \phi\|_{L^p(G)}.$$

Proof of Lemma 5.3.6. By Hulanicki's Theorem 4.5.1 or Corollary 4.5.2, $\kappa \in \mathcal{S}(G)$.

It suffices to prove the result with X^α, $[\alpha] = N\nu$, instead of $\mathcal{R}^N$. By Corollary 3.1.30, we can write X^α as a finite sum of $\tilde{X}^\beta p_{\alpha,\beta}$ with $p_{\alpha,\beta}$ a homogeneous polynomial of homogeneous degree $[\beta]-[\alpha] \geq 0$. We then have

$$X^\alpha(\phi * \kappa) = \phi * X^\alpha \kappa = \sum \phi * (\tilde{X}^\beta p_{\alpha,\beta}\kappa) = \sum (X^\beta \phi) * (p_{\alpha,\beta}\kappa).$$

Therefore, by Proposition 4.4.30,

$$\begin{aligned} \|X^\alpha(\phi * \kappa)\|_p &\leq \sum \|(\mathcal{R}^{\frac{-[\beta]+a}{\nu}} X^\beta \phi) * (\tilde{\mathcal{R}}^{\frac{[\beta]-a}{\nu}} p_{\alpha,\beta}\kappa)\|_p \\ &\leq \sum \|\mathcal{R}^{\frac{-[\beta]+a}{\nu}} X^\beta \phi\|_p \|\tilde{\mathcal{R}}^{\frac{[\beta]-a}{\nu}} p_{\alpha,\beta}\kappa\|_1. \end{aligned}$$

By Theorem 4.4.16, Part 2,

$$\|\mathcal{R}^{\frac{-[\beta]+a}{\nu}} X^\beta \phi\|_p \leq C \|\mathcal{R}^{\frac{a}{\nu}} \phi\|_p.$$

And we have

$$\|\tilde{\mathcal{R}}^{\frac{[\beta]-a}{\nu}} p_{\alpha,\beta}\kappa\|_1 = \|\mathcal{R}^{\frac{[\beta]-a}{\nu}} \tilde{p}_{\alpha,\beta}\tilde{\kappa}\|_1,$$

see Section 4.4.8. By Theorem 4.3.6, since $[\beta]\geq[\alpha]=N\nu\geq a$, we obtain

$$\|\mathcal{R}^{\frac{[\beta]-a}{\nu}}\tilde{p}_{\alpha,\beta}\tilde{\kappa}\|_1\leq C\|\tilde{p}_{\alpha,\beta}\tilde{\kappa}\|_1^{1-\frac{[\beta]-a}{\nu N}}\|\mathcal{R}^N\tilde{p}_{\alpha,\beta}\tilde{\kappa}\|_1^{\frac{[\beta]-a}{\nu N}}.$$

Note that because of (4.8), we have

$$\tilde{\kappa}(x):=(-1)^{|\alpha_o|}x^{\alpha_o}\bar{m}(\mathcal{R})\delta_0(x).$$

By Hulanicki's theorem (see Theorem 4.5.1), $\|\tilde{p}_{\alpha,\beta}\tilde{\kappa}\|_1$ and $\|\mathcal{R}^{\frac{b}{\nu}}\tilde{p}_{\alpha,\beta}\tilde{\kappa}\|_1$ are

$$\lesssim\sup_{\substack{\lambda>0\\ \ell=0,\ldots,k}}(1+\lambda)^k|\partial_\lambda^\ell m(\lambda)|,$$

for a suitable k, therefore this is also the case for $\|\tilde{\mathcal{R}}^{\frac{[\beta]-a}{\nu}}p_{\alpha,\beta}\kappa\|_1$.

Combining all these inequalities shows the desired result. □

Proof of Lemma 5.3.5. Let f be as in the statement. We need to show for any $\alpha\in\mathbb{N}_0^n$ that the convolution operator with right convolution kernel $\tilde{q}_\alpha f(\mathcal{R})\delta_0$ maps $L^2_\gamma(G)$ boundedly to $L^2_{[\alpha]-m+\gamma}(G)$ for any $\gamma\in\mathbb{R}$. It is sufficient to prove this for γ in a sequence going to $+\infty$ and $-\infty$ (see Proposition 5.2.12) and, in fact, only for a sequence of positive γ since

$$(\tilde{q}_\alpha f(\mathcal{R})\delta_0)^*=(-1)^{|\alpha|}\tilde{q}_\alpha\bar{f}(\mathcal{R})\delta_0.$$

At the end of the proof, we will see that, because of the equivalence between the Sobolev norms, it actually suffices to prove that for a fixed γ in this sequence, the operators given by

$$\phi\longmapsto\phi*(\tilde{q}_\alpha f(\mathcal{R})\delta_0)\text{ and }\phi\longmapsto\mathcal{R}^{\frac{[\alpha]-m+\gamma}{\nu}}\left(\left\{\mathcal{R}^{-\frac{\gamma}{\nu}}\phi\right\}*(\tilde{q}_\alpha f(\mathcal{R})\delta_0)\right),\quad(5.32)$$

are bounded on $L^2(G)$. So, we first prove this by decomposing f and applying the Cotlar-Stein lemma.

We fix a dyadic decomposition: there exists a non-negative function $\eta\in\mathcal{D}(\mathbb{R})$ supported in $[1/2,2]$ and satisfying

$$\forall\lambda\geq1\qquad1=\sum_{j\in\mathbb{N}_0}\eta_j(\lambda)\quad\text{where}\quad\eta_j(\lambda):=\eta(2^{-j}\lambda).$$

We set for $j\in\mathbb{N}_0$ and $\lambda\geq1$,

$$\begin{aligned}f_j(\lambda)&:=\lambda^{-\frac{m}{\nu}}f(\lambda)\eta_j(\lambda),\\ f^{(j)}(\lambda)&:=f_j(2^j\lambda),\\ g_j(\lambda)&:=\lambda^{\frac{m}{\nu}}f^{(j)}(\lambda).\end{aligned}$$

One obtains easily that for any $j \in \mathbb{N}_0$ and $\ell \in \mathbb{N}_0$, we have

$$\begin{aligned}
\partial^\ell f_j(\lambda) &= \overline{\sum_{\ell_1+\ell_2+\ell_3=\ell}} \lambda^{-\frac{m}{\nu}-\ell_1} (\partial^{\ell_2} f)(\lambda)\, 2^{-j\ell_3} (\partial^{\ell_3}\eta)(2^{-j}\lambda), \\
|\partial^\ell f_j(\lambda)| &\leq C_\ell \sup_{\substack{\lambda\geq 1\\ \ell'\leq \ell}} \lambda^{-\frac{m}{\nu}+\ell'}\, |\partial_\lambda^{\ell'} f(\lambda)| \overline{\sum_{\ell_1+\ell_2+\ell_3=\ell}} \lambda^{-\ell_1}\lambda^{-\ell_2} 2^{-j\ell_3} |(\partial^{\ell_3}\eta)(2^{-j}\lambda)|,
\end{aligned}$$

where $\overline{\sum}$ stands for a linear combination of its terms with some constants. As η is supported in $[1/2,2]$ and since $\lambda \asymp 2^j$, we have

$$\lambda^{-\ell_1}\lambda^{-\ell_2} 2^{-j\ell_3} \asymp 2^{-j\ell_1+\ell_2+\ell_3},$$

so that

$$\overline{\sum_{\ell_1+\ell_2+\ell_3=\ell}} \lambda^{-\ell_1}\lambda^{-\ell_2} 2^{-j\ell_3} |(\partial^{\ell_3}\eta)(2^{-j}\lambda)| \leq C_{\ell,\eta} 2^{-j\ell}.$$

Therefore, we have obtained

$$|\partial^\ell f_j(\lambda)| \leq C_\ell \sup_{\substack{\lambda\geq 1\\ \ell'\leq \ell}} \lambda^{-\frac{m}{\nu}+\ell'}\, |\partial_\lambda^{\ell'} f(\lambda)|\, 2^{-j\ell}.$$

Hence, for each $j \in \mathbb{N}_0$, $f^{(j)}$ is smooth and supported in $[1/2,2]$, and satisfies for any $\ell \in \mathbb{N}_0$ the estimate

$$|\partial^\ell f^{(j)}(\lambda)| = |2^{j\ell}\partial^\ell f_j(\lambda)| \leq C_\ell \sup_{\substack{\lambda\geq 1\\ \ell'\leq \ell}} \lambda^{-\frac{m}{\nu}+\ell'}\, |\partial_\lambda^{\ell'} f(\lambda)|.$$

Consequently, each g_j is smooth and supported in $[1/2,2]$, and satisfies

$$\forall \ell \in \mathbb{N}_0 \qquad \sup_{\substack{\lambda\in[\frac{1}{2},2]\\ \ell'=0,\ldots,\ell}} |\partial^{\ell'} g_j(\lambda)| \leq C_\ell \sup_{\substack{\lambda\geq 1\\ \ell'\leq \ell}} \lambda^{-\frac{m}{\nu}+\ell'}\, |\partial_\lambda^{\ell'} f(\lambda)|. \tag{5.33}$$

Clearly $f(\lambda)$ is the sum of the terms

$$2^{j\frac{m}{\nu}} g_j(2^{-j}\lambda) = f(\lambda)\eta_j(\lambda)$$

over $j \in \mathbb{N}_0$ and this sum is uniformly locally finite with respect to λ. Furthermore, since the functions f and g_j are continuous and bounded, the operators $f(\mathcal{R})$ and $g_j(2^{-j}\mathcal{R})$ defined by the functional calculus are bounded on $L^2(G)$ by Corollary 4.1.16. Therefore, we have in the strong operator topology of $\mathscr{L}(L^2(G))$ that

$$f(\mathcal{R}) = \sum_{j=0}^{\infty} 2^{j\frac{m}{\nu}} g_j(2^{-j}\mathcal{R}),$$

and in $\mathcal{K}(G)$ or $\mathcal{S}'(G)$ that

$$f(\mathcal{R})\delta_o = \sum_{j=0}^{\infty} 2^{j\frac{m}{\nu}} g_j(2^{-j}\mathcal{R})\delta_o.$$

We fix $\alpha \in \mathbb{N}_0^n$. For each $j \in \mathbb{N}_0$, by Hulanicki's theorem (see Corollary 4.5.2), $g_j(2^{-j}\mathcal{R})\delta_o$ is Schwartz, thus so is

$$K_j := 2^{j\frac{m}{\nu}} \tilde{q}_\alpha g_j(2^{-j}\mathcal{R})\delta_o$$

and also (see (4.8))

$$K_j^* =: K_j^*(x) = \bar{K}_j(x^{-1}) = (-1)^{|\alpha|} 2^{j\frac{m}{\nu}} \tilde{q}_\alpha \bar{g}_j(2^{-j}\mathcal{R})\delta_o(x^{-1}).$$

We claim that for any $a, b \in \mathbb{R}$ satisfying

- either $b \in \nu\mathbb{N}_0$ and $a \in [0, b)$
- or $b \geq 0$ and $a < \lfloor b/\nu \rfloor$

there exist $\ell \in \mathbb{N}$ and $C > 0$ such that for all $j \in \mathbb{N}_0$, we have

$$\|\tilde{\mathcal{R}}^{-\frac{a}{\nu}} \mathcal{R}^{\frac{b}{\nu}} K_j\|_{\mathcal{K}} \leq C(2^{\frac{j}{\nu}})^{m-[\alpha]-a+b} \sup_{\substack{\lambda \geq 1 \\ \ell' \leq \ell}} \lambda^{-\frac{m}{\nu}+\ell'} |\partial_\lambda^{\ell'} f(\lambda)|, \tag{5.34}$$

and the same is true for $\mathcal{R}^{-\frac{a}{\nu}} \tilde{\mathcal{R}}^{\frac{b}{\nu}} K_j^*$.

Let us prove this claim. By homogeneity (see (4.3)), we see that

$$g_j(2^{-j}\mathcal{R})\delta_o(x) = (2^{-\frac{j}{\nu}})^{-Q} g_j(\mathcal{R})\delta_o(2^{\frac{j}{\nu}} x),$$

thus

$$\begin{aligned} K_j(x) &= 2^{j\frac{m}{\nu}} (2^{\frac{j}{\nu}})^{-[\alpha]} \tilde{q}_\alpha(2^{\frac{j}{\nu}} x)\ (2^{-\frac{j}{\nu}})^{-Q} g_j(\mathcal{R})\delta_o(2^{\frac{j}{\nu}} x) \\ &= (2^{\frac{j}{\nu}})^{m-[\alpha]+Q} \left(\tilde{q}_\alpha g_j(\mathcal{R})\delta_o\right)(2^{\frac{j}{\nu}} x). \end{aligned}$$

More generally, by Part (7) of Theorem 4.3.6 for $\mathcal{R}$ and consequently for $\tilde{\mathcal{R}}$ (see (4.50)) we have

$$\tilde{\mathcal{R}}^{-\frac{a}{\nu}} \mathcal{R}^{\frac{b}{\nu}} K_j = (2^{\frac{j}{\nu}})^{m-[\alpha]+Q-a+b} \left(\tilde{\mathcal{R}}^{-\frac{a}{\nu}} \mathcal{R}^{\frac{b}{\nu}} \{\tilde{q}_\alpha g_j(\mathcal{R})\delta_o\}\right) \circ D_{2^{\frac{j}{\nu}}},$$

whenever it makes sense (that is, K_j is in the L^2-domain of $\mathcal{R}^{\frac{b}{\nu}}$ such that $\mathcal{R}^{\frac{b}{\nu}} K_j$ is in the L^2-domain of $\tilde{\mathcal{R}}^{-\frac{a}{\nu}}$). Consequently, by Proposition 5.1.17 (1), with norms possibly infinite, we have

$$\|\tilde{\mathcal{R}}^{-\frac{a}{\nu}} \mathcal{R}^{\frac{b}{\nu}} K_j\|_{\mathcal{K}} = (2^{\frac{j}{\nu}})^{m-[\alpha]-a+b} \left\|\tilde{\mathcal{R}}^{-\frac{a}{\nu}} \mathcal{R}^{\frac{b}{\nu}} \{\tilde{q}_\alpha g_j(\mathcal{R})\delta_o\}\right\|_{\mathcal{K}}.$$

Since $(\tilde{\mathcal{R}}^{-\frac{a}{\nu}}\mathcal{R}^{\frac{b}{\nu}}K_j)^* = \tilde{\mathcal{R}}^{\frac{b}{\nu}}\mathcal{R}^{-\frac{a}{\nu}}K_j^*$ for any a, b whenever it makes sense, or by the same argument as above, we also have

$$\|\tilde{\mathcal{R}}^{-\frac{a}{\nu}}\mathcal{R}^{\frac{b}{\nu}}K_j^*\|_{\mathcal{K}} = (2^{\frac{j}{\nu}})^{m-[\alpha]-a+b}\left\|\tilde{\mathcal{R}}^{-\frac{a}{\nu}}\mathcal{R}^{\frac{b}{\nu}}\{\tilde{q}_\alpha \bar{g}_j(\mathcal{R})\delta_o\}\right\|_{\mathcal{K}}.$$

Therefore, if $b \in \nu\mathbb{N}_0$ and $a \in [0, b)$, by Lemma 5.3.6, there exist $\ell = \ell_{a,b} \in \mathbb{N}$ such that

$$\begin{aligned}
\left\|\tilde{\mathcal{R}}^{-\frac{a}{\nu}}\mathcal{R}^{\frac{b}{\nu}}\{\tilde{q}_\alpha g_j(\mathcal{R})\delta_o\}\right\|_{\mathcal{K}} &\leq C_{a,b} \sup_{\substack{\lambda>0\\ \ell'=0,\ldots,\ell}} (1+\lambda)^\ell |\partial_\lambda^{\ell'} g_j(\lambda)| \\
&\leq C_{a,b} \sup_{\substack{\lambda>0\\ \ell'=0,\ldots,\ell}} |\partial_\lambda^{\ell'} g_j(\lambda)|,
\end{aligned}$$

since each g_j is supported in $[1/2, 2]$. As g_j satisfies (5.33), we have shown Claim (5.34) in the case $b \in \nu\mathbb{N}_0$ and $a \in [0, b)$.

If $a < \lfloor b/\nu \rfloor$ then we can apply the result we have just obtained to $\nu(\lfloor b/\nu \rfloor)$ and $\nu\lceil b/\nu \rceil$. Using Theorem 4.3.6 we then have for any $\phi \in \mathcal{S}(G)$, with $\theta := \lfloor\frac{b}{\nu}\rfloor\lceil\frac{b}{\nu}\rceil^{-1}$, that

$$\begin{aligned}
\|\mathcal{R}^{\frac{b}{\nu}}\phi\|_2 &\leq C\|\mathcal{R}^{\lfloor\frac{b}{\nu}\rfloor}\phi\|_2^{1-\theta}\|\mathcal{R}^{\lceil\frac{b}{\nu}\rceil}\phi\|_2^\theta \\
&\leq C\left(\sup_{\substack{\lambda>0\\ \ell'=0,\ldots,\ell}} |\partial_\lambda^{\ell'} g_j(\lambda)|\ \|\mathcal{R}^{\frac{a}{\nu}}\phi\|_2\right)^{1-\theta+\theta},
\end{aligned}$$

for some ℓ. This shows Claim (5.34) in the case $a < \lfloor b/\nu \rfloor$.

We set $T_j : \mathcal{S}(G) \ni \phi \mapsto \phi * K_j$. We want to apply the Cotlar-Stein lemma (Theorem A.5.2) to two families of $L^2(G)$-bounded operators: first to T_j, $j \in \mathbb{N}_0$, and then to

$$T_{j,\beta,\gamma} : \phi \longmapsto \phi * \mathcal{R}^{\frac{\beta}{\nu}}\tilde{\mathcal{R}}^{-\frac{\gamma}{\nu}}K_j,\ j \in \mathbb{N}_0.$$

where $\gamma \in \nu\mathbb{N}$ is such that $\beta := [\alpha] - m + \gamma > 0$.

Let us check the hypothesis of the Cotlar-Stein lemma for T_j. By Claim (5.34) for $a = b = 0$, there exists $\ell \in \mathbb{N}_0$ such that for any $j, k \in \mathbb{N}_0$,

$$\begin{aligned}
&\max\left(\|T_j^*T_k\|_{\mathscr{L}(L^2(G))}, \|T_jT_k^*\|_{\mathscr{L}(L^2(G))}\right) \\
&\quad\leq C\max\left(\|T_j^*\|_{\mathscr{L}(L^2(G))}\|T_k\|_{\mathscr{L}(L^2(G))}, \|T_j\|_{\mathscr{L}(L^2(G))}\|T_k^*\|_{\mathscr{L}(L^2(G))}\right) \\
&\quad\leq C2^{\frac{j+k}{\nu}(m-[\alpha])}(\sup_{\substack{\lambda\geq1\\ \ell'\leq\ell}} \lambda^{-\frac{m}{\nu}+\ell'}\ |\partial_\lambda^{\ell'} f(\lambda)|)^2 \\
&\quad\leq C2^{\frac{|j-k|}{\nu}(m-[\alpha])}(\sup_{\substack{\lambda\geq1\\ \ell'\leq\ell}} \lambda^{-\frac{m}{\nu}+\ell'}\ |\partial_\lambda^{\ell'} f(\lambda)|)^2,
\end{aligned}$$

since $m-[\alpha]<0$.

Let us check the hypothesis of the Cotlar-Stein lemma for $T_{j,\beta,\gamma}$. By Proposition 4.4.30 the right convolution kernel of the operator $T^*_{j,\beta,\gamma}T_{k,\beta,\gamma}$ is given by

$$(\mathcal{R}^{\frac{\beta}{\nu}}\tilde{\mathcal{R}}^{-\frac{\gamma}{\nu}}K_k)*(\tilde{\mathcal{R}}^{\frac{\beta}{\nu}}\mathcal{R}^{-\frac{\gamma}{\nu}}K_j^*)=(\tilde{\mathcal{R}}^{-\frac{\gamma}{\nu}}\mathcal{R}^{\frac{\gamma}{\nu}}K_k)*(\tilde{\mathcal{R}}^{\frac{2\beta-\gamma}{\nu}}\mathcal{R}^{-\frac{\gamma}{\nu}}K_j^*).$$

Therefore, its operator norm is

$$\begin{aligned}\|T^*_{j,\beta,\gamma}T_{k,\beta,\gamma}\|_{\mathscr{L}(L^2(G))} &\leq \|\tilde{\mathcal{R}}^{-\frac{\gamma}{\nu}}\mathcal{R}^{\frac{\gamma}{\nu}}K_k\|_{\mathcal{K}}\|\tilde{\mathcal{R}}^{\frac{2\beta-\gamma}{\nu}}\mathcal{R}^{-\frac{\gamma}{\nu}}K_j^*\|_{\mathcal{K}}.\\ &\leq 2^{\frac{k}{\nu}(m-[\alpha]-\gamma+\gamma)}2^{\frac{j}{\nu}(m-[\alpha]-\gamma+2\beta-\gamma)}\left(\sup_{\substack{\lambda\geq 1\\ \ell'\leq\ell}}\lambda^{-\frac{m}{\nu}+\ell'}\ |\partial_\lambda^{\ell'}f(\lambda)|\right)^2,\end{aligned}$$

for some ℓ, thanks to Claim (5.34) with $a=b=\gamma\in\nu\mathbb{N}$ and with $b=2\beta-\gamma=2[\alpha]-2m+\gamma$ and $a=\gamma$. So we have obtained

$$\|T^*_{j,\beta,\gamma}T_{k,\beta,\gamma}\|_{\mathscr{L}(L^2(G))}\leq 2^{\frac{k-j}{\nu}(m-[\alpha])}\left(\sup_{\substack{\lambda\geq 1\\ \ell'\leq\ell}}\lambda^{-\frac{m}{\nu}+\ell'}\ |\partial_\lambda^{\ell'}f(\lambda)|\right)^2.$$

Since the adjoint of $T^*_{j,\beta,\gamma}T_{k,\beta,\gamma}$ is $T^*_{k,\beta,\gamma}T_{j,\beta,\gamma}$, we may replace $k-j$ above by $|k-j|$.

We proceed in a similar way for the operator norm of $T_{j,\beta,\gamma}T^*_{k,\beta,\gamma}$ whose right convolution kernel is

$$(\mathcal{R}^{\frac{\beta}{\nu}}\tilde{\mathcal{R}}^{-\frac{\gamma}{\nu}}K_k^*)*(\tilde{\mathcal{R}}^{\frac{\beta}{\nu}}\mathcal{R}^{-\frac{\gamma}{\nu}}K_j)=(\mathcal{R}^{\frac{2\beta-\gamma}{\nu}}\tilde{\mathcal{R}}^{-\frac{\gamma}{\nu}}K_k^*)*(\tilde{\mathcal{R}}^{\frac{\gamma}{\nu}}\mathcal{R}^{\frac{-\gamma}{\nu}}K_j).$$

Therefore, we obtain

$$\begin{aligned}&\max\left(\|T^*_{j,\beta,\gamma}T_{k,\beta,\gamma}\|_{\mathscr{L}(L^2(G))},\|T_{j,\beta,\gamma}T^*_{k,\beta,\gamma}\|_{\mathscr{L}(L^2(G))}\right)\\ &\qquad\leq C2^{\frac{|k-j|}{\nu}(m-[\alpha])}(\sup_{\substack{\lambda\geq 1\\ \ell'\leq\ell}}\lambda^{-\frac{m}{\nu}+\ell'}\ |\partial_\lambda^{\ell'}f(\lambda)|)^2.\end{aligned}$$

By the Cotlar-Stein lemma (see Theorem A.5.2), $\sum T_j$ and $\sum_j T_{j,\beta,\gamma}$ converge in the strong operator topology of $\mathscr{L}(L^2(G))$ and the resulting operators have operator norms, up to a constant, less or equal than

$$\sup_{\lambda\geq 1,\ell\leq k}\lambda^{-\frac{m}{\nu}+\ell}\ |\partial_\lambda^\ell f(\lambda)|.$$

Clearly $\sum T_j$ and $\sum_j T_{j,\beta,\gamma}$ coincide on $\mathcal{S}(G)$ with the operators in (5.32), respectively. Using the equivalence between the two Sobolev norms (Theorem 4.4.3, Part

4), this implies

$$\begin{aligned}\|\phi * (\tilde{q}_\alpha f(\mathcal{R})\delta_0)\|_{L^2_\beta(G)} &\leq C\left(\|\phi * (\tilde{q}_\alpha f(\mathcal{R})\delta_0)\|_2 + \|\mathcal{R}^{\frac{\beta}{\nu}}\left(\phi * (\tilde{q}_\alpha f(\mathcal{R})\delta_0)\right\|_2\right)\\ &\leq C\sup_{\substack{\lambda\geq 1\\ \ell'\leq\ell}} \lambda^{-\frac{m}{\nu}+\ell'}\,|\partial_\lambda^{\ell'} f(\lambda)|\left(\|\phi\|_2 + \|\mathcal{R}^{\frac{\gamma}{\nu}}\phi\|_2\right)\\ &\leq C\sup_{\substack{\lambda\geq 1\\ \ell'\leq\ell}} \lambda^{-\frac{m}{\nu}+\ell'}\,|\partial_\lambda^{\ell'} f(\lambda)|\|\phi\|_{L^2_\gamma(G)}.\end{aligned}$$

We have obtained that the convolution operator with the right convolution kernel $\tilde{q}_\alpha f(\mathcal{R})\delta_0$ maps $L^2_\gamma(G)$ boundedly to $L^2_{m-[\alpha]+\gamma}(G)$ for any $\gamma \in \nu\mathbb{N}$ such that $m - [\alpha] + \gamma > 0$, with operator norm less or equal than

$$\sup_{\lambda\geq 1,\ell'\leq\ell} \lambda^{-\frac{m}{\nu}+\ell'}\,|\partial_\lambda^{\ell'} f(\lambda)|,$$

up to a constant, with ℓ depending on γ. This concludes the proof of Lemma 5.3.5. □

Hence the proof of Proposition 5.3.4 is now complete.

Looking back at the proof of Proposition 5.3.4, we see that we can assume that f depends on $x \in G$ in the following way:

Corollary 5.3.7. *Let $\mathcal{R}$ be a positive Rockland operator of homogeneous degree ν. Let $m \in \mathbb{R}$ and $0 \leq \delta \leq 1$. Let*

$$f : G \times \mathbb{R}_+ \ni (x, \lambda) \mapsto f_x(\lambda) \in \mathbb{C}$$

be a smooth function. We assume that for every $\beta \in \mathbb{N}_0^n$, $X_x^\beta f_x \in \mathcal{M}_{\frac{m+\delta[\beta]}{\nu}}$. Then $\sigma(x,\pi) = f_x(\pi(\mathcal{R}))$ defines a symbol σ in $S^m_{1,\delta}$ which satisfies

$$\forall a, b, c \in \mathbb{N}_0 \qquad \exists \ell \in \mathbb{N},\ C > 0\ : \qquad \|\sigma\|_{S^m_{1,\delta},a,b,c} \leq C \sup_{[\beta]\leq b} \|X_x^\beta f_x\|_{\mathcal{M}_{\frac{m+\delta[\beta]}{\nu}},\ell},$$

with ℓ and C independent of f.

5.3.2 Joint multipliers

To a certain extent, we can tensorise the property in Proposition 5.3.4. But we need to define the tensorisation of the space $\mathcal{M}_m$ and the multipliers of two Rockland operators.

First, we define the space $\mathcal{M}_{m_1} \otimes \mathcal{M}_{m_2}$ of functions $f \in C^\infty(\mathbb{R}_+ \times \mathbb{R}_+)$ such that

$$\|f\|_{\mathcal{M}_{m_1}\otimes\mathcal{M}_{m_2},\ell} := \sup_{\substack{\lambda_1,\lambda_2>0\\ \ell_1',\ell_2'=0,\dots,\ell}} (1+\lambda_1)^{-m_1+\ell_1'}(1+\lambda_2)^{-m_2+\ell_2'}|\partial_{\lambda_1}^{\ell_1'}\partial_{\lambda_2}^{\ell_2'} f(\lambda_1,\lambda_2)|,$$

is finite for every $\ell \in \mathbb{N}_0$. It is a routine exercise to check that $\mathcal{M}_{m_1} \otimes \mathcal{M}_{m_2}$ is a Fréchet space.

Secondly, we observe that if $\mathcal{L}$ and $\mathcal{R}$ are two Rockland operators on G which commute strongly, meaning that their spectral measures $E_{\mathcal{L}}$ and $E_{\mathcal{R}}$ commute, then we can define their common spectral measure $E_{\mathcal{L},\mathcal{R}}$ via

$$E_{\mathcal{L},\mathcal{R}}(B_1 \times B_2) := E_{\mathcal{L}}(B_1)E_{\mathcal{R}}(B_2), \qquad \text{for } B_1, B_2 \text{ Borel subsets of } \mathbb{R},$$

and we can also define the multipliers in $\mathcal{L}$ and $\mathcal{R}$ by

$$f(\mathcal{L},\mathcal{R}) := \int_{\mathbb{R}_+ \times \mathbb{R}_+} f(\lambda_1, \lambda_2) dE_{\mathcal{L},\mathcal{R}}(\lambda_1, \lambda_2),$$

for any $f \in L^\infty(\mathbb{R}_+ \times \mathbb{R}_+)$.

Corollary 5.3.8. *Let $\mathcal{L}$ and $\mathcal{R}$ be two positive Rockland operators on G of respective degrees $\nu_{\mathcal{L}}$ and $\nu_{\mathcal{R}}$. We assume that $\mathcal{L}$ and $\mathcal{R}$ commute strongly, that is, their spectral measures $E_{\mathcal{L}}$ and $E_{\mathcal{R}}$ commute. If $f \in \mathcal{M}_{\frac{m_1}{\nu_{\mathcal{L}}}} \otimes \mathcal{M}_{\frac{m_2}{\nu_{\mathcal{R}}}}$ then $f(\mathcal{L},\mathcal{R})$ is in $\Psi^{m_1+m_2}$. Furthermore, we have for any $a, b, c \in \mathbb{N}_0$,*

$$\|f(\mathcal{L},\mathcal{R})\|_{\Psi^{m_1+m_2},a,b,c} \leq C\|f\|_{\mathcal{M}_{\frac{m_1}{\nu_{\mathcal{L}}}} \otimes \mathcal{M}_{\frac{m_2}{\nu_{\mathcal{R}}}},\ell},$$

where ℓ and $C > 0$ are independent of f.

Proof. By uniqueness, the spectral measure $E_{\mathcal{L},\mathcal{R}}$ is invariant under left translations. Denoting by $\pi(E_{\mathcal{L},\mathcal{R}})$ for $\pi \in \widehat{G}$ its group Fourier transform, we see that the group Fourier transform of a multiplier $f(\mathcal{L},\mathcal{R})$ for $f \in L^\infty(\mathbb{R}_+ \times \mathbb{R}_+)$ is

$$\pi(f(\mathcal{L},\mathcal{R})) = \int_{\mathbb{R}_+ \times \mathbb{R}_+} f(\lambda_1, \lambda_2) d\pi(E_{\mathcal{L},\mathcal{R}})(\lambda_1, \lambda_2),$$

since it is true for a function f of the form $f(\lambda_1, \lambda_2) = f_1(\lambda_1) f_2(\lambda_2)$ with $f_1, f_2 \in L^\infty(\mathbb{R}_+)$, by Corollary 5.3.7.

We fix $\eta \in C^\infty(\mathbb{R})$ supported in $[-\frac{1}{2}, \frac{1}{2}]$ such that

$$\forall \lambda' \in \mathbb{R} \qquad \sum_{j' \in \mathbb{Z}} \eta(\lambda' + j') = 1.$$

We also fix another function $\tilde{\eta} \in C^\infty(\mathbb{R})$ supported in $[-1, 1]$ such that $\tilde{\eta} = 1$ on $[-\frac{1}{2}, \frac{1}{2}]$. For any $j', k' \in \mathbb{Z}$, we define $\psi_{j',k'} \in C^\infty(\mathbb{R})$ by

$$\psi_{j',k'}(\lambda') := e^{-ik'(\lambda'-j')}\tilde{\eta}(\lambda' - j').$$

It is easy to show that for any $\ell' \in \mathbb{N}_0$ there exists $C = C_{\ell'} > 0$ such that

$$\forall j', k' \in \mathbb{Z} \qquad \|\psi_{j',k'}\|_{\mathcal{M}_m,\ell'} \leq C(1 + |k'|)^{\ell'}(1 + |j'|)^{-m+\ell'}.$$

Since the symbols form an algebra (see Section 5.2.5), and by Proposition 5.3.4, writing $m = m_1 + m_2$, we have for any $j_1, j_2, k_1, k_2 \in \mathbb{Z}$:

$$\begin{aligned}
&\|\psi_{j_1,k_1}(\pi(\mathcal{L}))\psi_{j_2,k_2}(\pi(\mathcal{R}))\|_{S^m,a,b,c} \\
&\quad \leq C\|\psi_{j_1,k_1}(\pi(\mathcal{L}))\|_{S^{m_1},a_1,b_1,c_1}\|\psi_{j_2,k_2}(\pi(\mathcal{R}))\|_{S^{m_2},a_2,b_2,c_2} \\
&\quad \leq C(1+|k_1|)^{\ell_1}(1+|j_1|)^{-\frac{m_1}{\nu_\mathcal{L}}+\ell_1}(1+|k_2|)^{\ell_2}(1+|j_2|)^{-\frac{m_2}{\nu_\mathcal{R}}+\ell_2}
\end{aligned} \tag{5.35}$$

for some $\ell_1, \ell_2 \in \mathbb{N}_0$.

Let f be as in the statement. We extend f to a smooth function supported in $(-1,\infty)^2$ and decompose it as a locally finite sum:

$$f = \sum_{j\in\mathbb{Z}^2} f_j \qquad \text{where} \quad f_j(\lambda) = f(\lambda)\eta(\lambda_1 - j_1')\eta(\lambda_2 - j_2'), \quad \lambda = (\lambda_1, \lambda_2).$$

For each $j \in \mathbb{Z}$, we view $f_j(\cdot + j)$ as a smooth function supported in $[-1,1]\times[-1,1]$ and we expand it in the Fourier series

$$f_j(\lambda + j) = \sum_{k\in\mathbb{Z}^2} c_{j,k} e^{-ik\cdot\lambda}.$$

The hypothesis on f implies that for any $\ell_1, \ell_2 \in \mathbb{N}_0$, we have

$$\begin{aligned}
|c_{j,k}| \quad \leq \quad & C_{\ell_1,\ell_2}\|f\|_{\mathcal{M}_{\frac{m_1}{\nu_\mathcal{L}}}\otimes\mathcal{M}_{\frac{m_2}{\nu_\mathcal{R}}},\ell_1+\ell_2}(1+|k_1|)^{-\ell_1}(1+|k_2|)^{-\ell_2} \times \\
& \times(1+|j_1|)^{\frac{m_1}{\nu_\mathcal{L}}-\ell_1}(1+|j_2|)^{\frac{m_2}{\nu_\mathcal{R}}-\ell_2}.
\end{aligned} \tag{5.36}$$

We have obtained that (taking different ℓ's)

$$\sum_{j,k\in\mathbb{Z}^2} |c_{j,k}|\|\psi_{j_1,k_1}\|_{\mathcal{M}_{\frac{m_1}{\nu_\mathcal{L}}},\ell_1}\|\psi_{j_2,k_2}\|_{\mathcal{M}_{\frac{m_2}{\nu_\mathcal{R}}},\ell_2} < \infty.$$

We have therefore obtained the following decomposition of f in the Fréchet space $\mathcal{M}_{\frac{m_1}{\nu_\mathcal{L}}} \otimes \mathcal{M}_{\frac{m_2}{\nu_\mathcal{R}}}$,

$$f(\lambda_1, \lambda_2) = \sum_{j,k\in\mathbb{Z}^2} c_{j,k}\psi_{j_1,k_1}(\lambda_1)\psi_{j_2,k_2}(\lambda_2).$$

And so for any a, b, c with ℓ_1, ℓ_2 as in (5.35),

$$\begin{aligned}
&\|f(\pi(\mathcal{L}),\pi(\mathcal{R}))\|_{S^m,a,b,c} \leq \sum_{j,k\in\mathbb{Z}^2} |c_{j,k}|\|\psi_{j_1,k_1}(\pi(\mathcal{L}))\psi_{j_2,k_2}(\pi(\mathcal{R}))\|_{S^m,a,b,c} \\
&\leq \sum_{j,k\in\mathbb{Z}^2} |c_{j,k}|C(1+|k_1|)^{\ell_1}(1+|j_1|)^{-\frac{m_1}{\nu_\mathcal{L}}+\ell_1}(1+|k_2|)^{\ell_2}(1+|j_2|)^{-\frac{m_2}{\nu_\mathcal{R}}+\ell_2} \\
&\leq C\|f\|_{\mathcal{M}_{\frac{m_1}{\nu_\mathcal{L}}}\otimes\mathcal{M}_{\frac{m_2}{\nu_\mathcal{R}}},\ell_1+\ell_2+4},
\end{aligned}$$

by (5.37) with $\ell_1 + 2$ and $\ell_2 + 2$. This shows that $f(\pi(\mathcal{L}),\pi(\mathcal{R})) \in S^m$ and the desired inequalities for the seminorms. □

Corollary 5.3.8 could be generalised by considering a finite family of positive Rockland operators which commute strongly between themselves (i.e. with commuting spectral measures), with symbols possibly depending on x in a similar way to Corollary 5.3.7.

5.4 Kernels of pseudo-differential operators

In this section we obtain estimates for the kernels of operators in the classes $\Psi^m_{\rho,\delta}$ (cf. Section 5.4.1) and some consequences for smoothing operators (cf. Section 5.4.2) and for operators of Calderón-Zygmund type in the calculus (cf. Section 5.4.4). We will also show the L^p boundedness of Ψ^0 in Section 5.4.4.

For technical reasons which will become apparent in Section 5.5.2, we will also consider the seminorms:

$$\|\sigma\|_{S^{m,R}_{\rho,\delta},a,b} := \sup_{\substack{(x,\pi)\in G\times\widehat{G}\\ [\alpha]\leq a,[\beta]\leq b}} \|\Delta^\alpha X^\beta_x \sigma(x,\pi)\pi(\mathrm{I}+\mathcal{R})^{-\frac{m-\rho[\alpha]+\delta[\beta]}{\nu}}\|_{\mathscr{L}(\mathcal{H}_\pi)}, \tag{5.37}$$

where $\mathcal{R}$ is a positive Rockland operator of homogeneous degree ν. The superscript R indicates that the powers of $\mathrm{I}+\mathcal{R}$ are 'on the right'. As for the $S^m_{\rho,\delta}$-seminorms, this is a seminorm which is equivalent to a similar seminorm for another positive Rockland operator.

5.4.1 Estimates of the kernels

This section is devoted to describing the behaviour of the kernel of an operator with symbol in the class $S^m_{\rho,\delta}$. As usual in this chapter, G is a graded Lie group of homogeneous dimension Q. Our results in this section may be summarised in the following theorem.

Theorem 5.4.1. *Let $\sigma = \{\sigma(x,\pi)\}$ be in $S^m_{\rho,\delta}$ with $1\geq\rho\geq\delta\geq 0$, $\rho\neq 0$. Then its associated kernel $\kappa : (x,y)\mapsto\kappa_x(y)$ is smooth on $G\times(G\backslash\{0\})$. We also fix a homogeneous quasi-norm $|\cdot|$ on G.*

(i) Away from 0, κ_x has a Schwartz decay:

$$\forall M\in\mathbb{N}\quad \exists C>0,\ a,b,c\in\mathbb{N}:\quad \forall(x,y)\in G\times G$$
$$|y|>1 \Longrightarrow |\kappa_x(y)|\leq C\sup_{\pi\in\widehat{G}}\|\sigma(x,\pi)\|_{S^m_{\rho,\delta},a,b,c}|y|^{-M}.$$

(ii) Near 0, we have

– *if $Q+m>0$, κ_x behaves like $|y|^{-\frac{Q+m}{\rho}}$: there exists $C>0$ and $a,b,c\in\mathbb{N}$ such that*

$$\forall(x,y)\in G\times(G\backslash\{0\})\quad |\kappa_x(y)|\leq C\sup_{\pi\in\widehat{G}}\|\sigma(x,\pi)\|_{S^m_{\rho,\delta},a,b,c}|y|^{-\frac{Q+m}{\rho}};$$

– *if* $Q + m = 0$, κ_x *behaves like* $\ln|y|$*: there exists* $C > 0$ *and* $a, b, c \in \mathbb{N}$ *such that*

$$\forall(x, y) \in G \times (G\backslash\{0\}) \quad |\kappa_x(y)| \leq C \sup_{\pi \in \widehat{G}} \|\sigma(x, \pi)\|_{S^m_{\rho,\delta}, a,b,c} \ln|y|;$$

– *if* $Q + m < 0$, κ_x *is continuous on* G *and bounded:*

$$\sup_{z \in G} |\kappa_x(z)| \leq C \sup_{\pi \in \widehat{G}} \|\sigma(x, \pi)\|_{S^m_{\rho,\delta}, 0,0,0}.$$

Moreover, it is possible to replace the seminorm $\|\cdot\|_{S^m_{\rho,\delta}, a,b,c}$ *in (i) and (ii) with a seminorm* $\|\cdot\|_{S^{m,R}_{\rho,\delta}, a,b}$ *given in (5.37).*

Remark 5.4.2. Using Theorem 5.2.22 (i) Parts (3) and (2), and Corollary 5.2.25, we obtain similar properties for $X_y^{\beta_1} \tilde{X}_y^{\beta_2} (X_x^{\beta_o} \tilde{q}_\alpha(y) \kappa_x(y))$.

We start the proof of Theorem 5.4.1 with consequences of Proposition 5.2.16 as preliminary results on the right convolution kernels and then proceed to analysing the behaviour of these kernels both at zero and at infinity.

Proposition 5.2.16 has the following consequences:

Corollary 5.4.3. *Let* $\sigma = \{\sigma(x, \pi)\}$ *be in* $S^m_{\rho,\delta}$ *with* $1 \geq \rho \geq \delta \geq 0$*. Let* κ_x *denote its associated kernel.*

1. *If* $\alpha, \beta_1, \beta_2, \beta_o \in \mathbb{N}_0^n$ *are such that*

$$m - \rho[\alpha] + [\beta_1] + [\beta_2] + \delta[\beta_o] < -Q/2,$$

then the distribution $X_z^{\beta_1} \tilde{X}_z^{\beta_2} (X_x^{\beta_o} \tilde{q}_\alpha(z) \kappa_x(z))$ *is square integrable and for every* $x \in G$ *we have*

$$\int_G \left| X_z^{\beta_1} \tilde{X}_z^{\beta_2} (X_x^{\beta_o} \tilde{q}_\alpha(z) \kappa_x(z)) \right|^2 dz \leq C \sup_{\pi \in \widehat{G}} \|\sigma(x, \pi)\|^2_{S^m_{\rho,\delta}, a,b,c}$$

where $a = [\alpha]$, $b = [\beta_o]$, $c = \rho[\alpha] + [\beta_1] + [\beta_2] + \delta[\beta_o]$ *and* $C = C_{m,\alpha,\beta_1,\beta_2,\beta_o} > 0$ *is a constant independent of* σ *and* x*. If* $\beta_1 = 0$ *then we may replace the seminorm* $\|\cdot\|_{S^m_{\rho,\delta}, a,b,c}$ *with a seminorm* $\|\cdot\|_{S^{m,R}_{\rho,\delta}, a,b}$ *given in (5.37).*

2. *For any* $\alpha, \beta_1, \beta_2, \beta_o \in \mathbb{N}_0^n$ *satisfying*

$$m - \rho[\alpha] + [\beta_1] + [\beta_2] + \delta[\beta_o] < -Q,$$

the distribution $z \mapsto X_z^{\beta_1} \tilde{X}_z^{\beta_2} X_x^{\beta_o} \tilde{q}_\alpha(z) \kappa_x(z)$ *is continuous on* G *for every* $x \in G$ *and we have*

$$\sup_{z \in G} \left| X_z^{\beta_1} \tilde{X}_z^{\beta_2} \left\{ X_x^{\beta_o} \tilde{q}_\alpha(z) \kappa_x(z) \right\} \right| \leq C \sup_{\pi \in \widehat{G}} \|\sigma(x, \pi)\|_{S^m_{\rho,\delta}, [\alpha], [\beta_o], [\beta_2]},$$

where $C = C_{m,\alpha,\beta_1,\beta_2,\beta_o} > 0$ is a constant independent of σ and x. If $\beta_1 = 0$ then we may replace the seminorm $\|\cdot\|_{S^m_{\rho,\delta},[\alpha],[\beta_o],[\beta_2]}$ with the seminorm $\|\cdot\|_{S^{m,R}_{\rho,\delta},[\alpha],[\beta_o]}$, see (5.37).

Consequently, if $\rho > 0$ then the map $\kappa : (x,y) \mapsto \kappa_x(y)$ is smooth on $G \times (G \setminus\{0\})$.

Proof. Part (1) follows from Proposition 5.2.16 together with Theorem 5.2.22 (i) Parts (3) and (2), and Corollary 5.2.25 . Now by the Sobolev inequality in Theorem 4.4.25 (ii), if the right-hand side of the following inequality is finite:

$$\sup_{z\in G}\left|X_z^{\beta_1}\tilde{X}_z^{\beta_2}\left\{X_x^{\beta_o}\tilde{q}_\alpha(z)\kappa_x(z)\right\}\right| \leq C\left\|(\mathrm{I}+\mathcal{R}_z)^{\frac{s}{\nu}}X_z^{\beta_1}\tilde{X}_z^{\beta_2}\left\{X_x^{\beta_o}\tilde{q}_\alpha(z)\kappa_x(z)\right\}\right\|_{L^2(dz)},$$

for $s > Q/2$, then the distribution

$$z \mapsto X_z^{\beta_1}\tilde{X}_z^{\beta_2}\left\{X_x^{\beta_o}\tilde{q}_\alpha(z)\kappa_x(z)\right\}$$

is continuous and the inequality of Part (2) holds. By Theorem 4.4.16,

$$\begin{aligned}
&\left\|(\mathrm{I}+\mathcal{R}_z)^{\frac{s}{\nu}}X_z^{\beta_1}\tilde{X}_z^{\beta_2}\left\{X_x^{\beta_o}\tilde{q}_\alpha(z)\kappa_x(z)\right\}\right\|_{L^2(dz)}\\
&\leq C\left\|(\mathrm{I}+\mathcal{R})^{\frac{s+[\beta_1]}{\nu}}(\mathrm{I}+\tilde{\mathcal{R}})^{\frac{[\beta_2]}{\nu}}\left\{X_x^{\beta_o}\tilde{q}_\alpha(z)\kappa_x(z)\right\}\right\|_{L^2(dz)}\\
&\leq C\left\|\pi(\mathrm{I}+\mathcal{R})^{\frac{s+[\beta_1]}{\nu}}X_x^{\beta_o}\Delta^\alpha\sigma(x,\pi)\pi(\mathrm{I}+\mathcal{R})^{\frac{[\beta_2]}{\nu}}\right\|_{L^2(\widehat{G})},
\end{aligned}$$

by the Plancherel formula (1.28). By Proposition 5.2.16 (together with Theorem 5.2.22 (ii)) as long as

$$m + s + [\beta_1] - \rho[\alpha] + \delta[\beta_o] + [\beta_2] < -Q/2,$$

since

$$(\mathrm{I}+\mathcal{R})^{\frac{s+[\beta_1]}{\nu}}(\mathrm{I}+\tilde{\mathcal{R}})^{\frac{[\beta_2]}{\nu}}\left\{X_x^{\beta_o}\tilde{q}_\alpha(z)\kappa_x(z)\right\}$$

is the kernel of the symbol

$$\pi(\mathrm{I}+\mathcal{R})^{\frac{s+[\beta_1]}{\nu}}X_x^{\beta_o}\Delta^\alpha\sigma(x,\pi)\pi(\mathrm{I}+\mathcal{R})^{\frac{[\beta_2]}{\nu}},$$

we have

$$\left\|\pi(\mathrm{I}+\mathcal{R})^{\frac{s+[\beta_1]}{\nu}}X_x^{\beta_o}\Delta^\alpha\sigma(x,\pi)\pi(\mathrm{I}+\mathcal{R})^{\frac{[\beta_2]}{\nu}}\right\|_{L^2(\widehat{G})} \leq C\|\sigma(x,\pi)\|_{S^m_{\rho,\delta},[\alpha],[\beta_o],[\beta_2]},$$

if $s + [\beta_1] \leq \rho[\alpha] - m - \delta[\beta_o] - [\beta_2]$. This shows Part (2). □

Estimates at infinity

We will now prove better estimates for the kernel than the ones stated in Corollary 5.4.3. First let us show that the kernel has a Schwartz decay away from the origin.

Proposition 5.4.4. *Let $\sigma = \{\sigma(x,\pi)\}$ be in $S^m_{\rho,\delta}$ with $1 \geq \rho \geq \delta \geq 0$. Let κ_x denote its associated kernel.*

We assume that $\rho > 0$ and we fix a homogeneous quasi-norm $|\cdot|$ on G. Then for any $M \in \mathbb{R}$ and any $\alpha, \beta_1, \beta_2, \beta_o \in \mathbb{N}_0^n$ there exist $C > 0$ and $a, b, c \in \mathbb{N}$ independent of σ such that for all $x \in G$ and $z \in G$ satisfying $|z| \geq 1$, we have

$$\left| X_z^{\beta_1} \tilde{X}_z^{\beta_2} (X_x^{\beta_o} \tilde{q}_\alpha(z) \kappa_x(z)) \right| \leq C \sup_{\pi \in \widehat{G}} \|\sigma(x,\pi)\|_{S^m_{\rho,\delta},a,b,c} |z|^{-M}.$$

Furthermore, if $\beta_1 = 0$ then we may replace the seminorm $\|\cdot\|_{S^m_{\rho,\delta},a,b,c}$ with a seminorm $\|\cdot\|_{S^{m,R}_{\rho,\delta},a,b}$ given in (5.37).

Proof. We start by proving the stated result for $\alpha = \beta_1 = \beta_2 = \beta_o = 0$ and for the homogeneous quasi-norm $|\cdot|_p$ given by (3.21). Here $p > 0$ is a positive number to be chosen suitably. We also fix a number $b_o > 0$ and a function $\eta_o \in C^\infty(\mathbb{R})$ valued in $[0,1]$ with $\eta_o \equiv 0$ on $(-\infty, \frac{1}{2}]$ and $\eta_o \equiv 1$ on $[1, \infty)$. We set

$$\eta(x) := \eta_o(b_o^{-p}|x|_p^p).$$

Therefore, η is a smooth function on G such that $\eta(z) = 1$ if $|z|_p \geq b_o$. Consequently,

$$\begin{aligned} \sup_{|z|_p \geq b_o} \left| |z|_p^M \kappa_x(z) \right| &\leq \sup_{z \in G} \left| |z|_p^M \kappa_x(z) \eta(z) \right| \\ &\leq C \sum_{[\beta'] \leq \lceil Q/2 \rceil} \left\| X_z^{\beta'} \left\{ |z|_p^M \kappa_x(z) \eta(z) \right\} \right\|_{L^2(G,dz)} \end{aligned} \tag{5.38}$$

by the Sobolev inequality in Theorem 4.4.25.

We study each term separately. We assume that $p/2$ is a positive integer divisible by all the weights $\upsilon_1, \ldots, \upsilon_n$ and we introduce the polynomial

$$|z|_p^p = \sum_{j=1}^n |z_j|^{\frac{p}{\upsilon_j}}$$

and its inverse, so that

$$\begin{aligned} X_z^{\beta'} \left\{ |z|_p^M \kappa_x(z) \eta(z) \right\} &= X_z^{\beta'} \left\{ |z|_p^M |z|_p^{-p} |z|_p^p \kappa_x(z)\, \eta(z) \right\} \\ &= \overline{\sum_{[\beta_1'] + [\beta_2'] = [\beta']}} X_z^{\beta_1'} \left\{ |z|_p^M |z|_p^{-p} \eta(z) \right\} X_z^{\beta_2'} \left\{ |z|_p^p \kappa_x(z) \right\}, \end{aligned}$$

where $\overline{\sum}$ means taking a linear combination, that is, a sum involving some constants. We observe that, using a polar change of coordinates,

$$\|X_z^{\beta_1'}\left\{|z|_p^M |z|_p^{-p}\eta(z)\right\}\|_{L^2(G,dz)} < \infty$$

as long as $2(M-p-[\beta_1'])+Q-1 < -1$. We assume that p has been chosen so that $2(M-p)+Q<0$. Therefore, all these L^2-norms can be viewed as constants. By the Cauchy-Schwartz inequality and the properties of Sobolev spaces, we obtain

$$\begin{aligned}\|X_z^{\beta'}\left\{|z|_p^M \kappa_x(z)\eta(z)\right\}\|_{L^2(G,dz)} &\leq C\sum_{[\beta_2']\leq[\beta']} \|X_z^{\beta_2'}\left\{|z|_p^p\kappa_x(z)\right\}\|_{L^2(G,dz)} \\ &\leq C\sum_{[\beta_2']\leq[\beta']}\sum_{[\alpha]\leq p} \|X_z^{\beta_2'}\left\{\tilde{q}_\alpha\kappa_x\right\}\|_2,\end{aligned}$$

since $|z|_p^p = \sum_{j=1}^n z_j^{\frac{p}{\upsilon_j}}$ is a polynomial of homogeneous degree p. Therefore, by Corollary 5.4.3 Part (1), we get

$$\|X_z^{\beta'}\left\{|z|_p^M \kappa_x(z)\eta(z)\right\}\|_{L^2(G,dz)} \leq C\sup_{\pi\in\widehat{G}}\|\sigma(x,\pi)\|_{S^m_{\rho,\delta},p,0,\rho p+[\beta']}$$

if $\rho p - m > Q/2+[\beta']$. We choose p accordingly. Combining this with (5.38) yields

$$\sup_{|z|_p\geq b_o}\left||z|_p^M\kappa_x(z)\right| \leq C\sup_{\pi\in\widehat{G}}\|\sigma(x,\pi)\|_{S^m_{\rho,\delta},p,0,\rho p+\lceil Q/2\rceil}.$$

Therefore, we have obtained the result for the homogeneous norm $|\cdot|_p$ and $\alpha=\beta_1=\beta_2=\beta_o=0$.

The full result follows for any homogeneous norm and indices $\alpha,\beta_1,\beta_2,\beta_o$ from the equivalence of any two homogeneous norms and by Theorem 5.2.22 (i) Parts (3) and (2), and Corollary 5.2.25. □

Remark 5.4.5.
1. During the proof of Proposition 5.4.4, we have obtained the following statement which is quantitatively more precise. We keep the setting of Proposition 5.4.4. Then for any $M\in\mathbb{R}$ and $b_o>0$, there exists $C=C_{M,b_o,m}>0$ such that

$$\sup_{|z|_p\geq b_o}\left||z|_p^M\kappa_x(z)\right| \leq C\sup_{\pi\in\widehat{G}}\|\sigma(x,\pi)\|_{S^m_{\rho,\delta},p,0,\rho p+\lceil Q/2\rceil},$$

   where $p\in\mathbb{N}$ is the smallest positive integer such that $p/2$ is divisible by all the weights $\upsilon_1,\ldots,\upsilon_n$ and $p>\max(Q/2+M,\frac{1}{\rho}(m+Q+1))$.

2. Combining Part (1) above, Theorem 5.2.22 (i) Parts (3) and (2), and Corollary 5.2.25, it is possible (but not necessarily useful) to obtain a concrete expression for the numbers a,b,c appearing in Proposition 5.4.4, in terms of $m,\rho,\delta,\alpha,\beta_1,\beta_2,\beta_o$ and of Q.

   Furthermore, the same statement is true for $|z|\geq b_o$ for an arbitrary lower bound $b_o>0$. However, the constant C may depend on b_o.

Estimates at the origin

We now prove a singular estimate for the kernel near the origin which is (therefore) not covered by Corollary 5.4.3 (2).

Proposition 5.4.6. *Let $\sigma = \{\sigma(x,\pi)\}$ be in $S^m_{\rho,\delta}$ with $1 \geq \rho \geq \delta \geq 0$. Let κ_x denote its associated kernel.*

We assume that $\rho > 0$ and we fix a homogeneous quasi-norm $|\cdot|$ on G. Then for any $\alpha, \beta_1, \beta_2, \beta_o \in \mathbb{N}_0^n$ with $Q + m + \delta[\beta_o] - \rho[\alpha] + [\beta_1] + [\beta_2] \geq 0$ there exist a constant $C > 0$ and computable integers $a, b, c \in \mathbb{N}_0$ independent of σ such that for all $x \in G$ and $z \in G\backslash\{0\}$, we have that if

$$Q + m + \delta[\beta_o] - \rho[\alpha] + [\beta_1] + [\beta_2] > 0,$$

then

$$\left| X_z^{\beta_1} \tilde{X}_z^{\beta_2} (X_x^{\beta_o} \tilde{q}_\alpha(z) \kappa_x(z)) \right| \leq C \sup_{\pi \in \widehat{G}} \|\sigma(x,\pi)\|_{S^m_{\rho,\delta},a,b,c} |z|^{-\frac{Q+m+\delta[\beta_o]-\rho[\alpha]+[\beta_1]+[\beta_2]}{\rho}},$$

and if

$$Q + m + \delta[\beta_o] - \rho[\alpha] + [\beta_1] + [\beta_2] = 0,$$

then

$$\left| X_z^{\beta_1} \tilde{X}_z^{\beta_2} (X_x^{\beta_o} \tilde{q}_\alpha(z) \kappa_x(z)) \right| \leq C \sup_{\pi \in \widehat{G}} \|\sigma(x,\pi)\|_{S^m_{\rho,\delta},a,b,c} \ln|z|.$$

In both estimates, if $\beta_1 = 0$ then we may replace the seminorm $\|\cdot\|_{S^m_{\rho,\delta},a,b,c}$ with a seminorm $\|\cdot\|_{S^{m,R}_{\rho,\delta},a,b}$ given in (5.37).

During the proof of Proposition 5.4.6, we will need the following technical lemma which is of interest on its own.

Lemma 5.4.7. *Let $\sigma = \{\sigma(x,\pi)\}$ be in $S^m_{\rho,\delta}$ with $1 \geq \rho \geq \delta \geq 0$. Let $\eta \in \mathcal{D}(\mathbb{R})$ and $c_o > 0$. We also fix a positive Rockland operator $\mathcal{R}$ of homogeneous degree ν with corresponding seminorms for the symbol classes $S^m_{\rho,\delta}$.*

Then for any $\ell \in \mathbb{N}_0$, the symbols given by

$$\sigma_{L,\ell}(x,\pi) := \eta(2^{-\ell c_o}\pi(\mathcal{R}))\sigma(x,\pi) \quad \text{and} \quad \sigma_{R,\ell}(x,\pi) := \sigma(x,\pi)\eta(2^{-\ell c_o}\pi(\mathcal{R})),$$

are in $S^{-\infty}$. Moreover, for any $m_1 \in \mathbb{R}$ and $a, b, c \in \mathbb{N}_0$, there exists a constant $C = C_{m,m_1,\rho,\delta,a,b,c,\eta,c_o} > 0$ such that for any $\ell \in \mathbb{N}_0$ we have

$$\|\sigma_{L,\ell}(x,\pi)\|_{S^{m_1}_{\rho,\delta},a,b,c} \leq C \sup_{\pi \in \widehat{G}} \|\sigma(x,\pi)\|_{S^m_{\rho,\delta},a,b,c} 2^{\ell \frac{c_o}{\nu}(m-m_1)}.$$

The same holds for $\sigma_{R,\ell}(x,\pi)$, but with a possibly different seminorm on the right hand side.

Only for $\sigma_{R,\ell}(x,\pi)$, we also have for the seminorm $\|\cdot\|_{S^{m,R}_{\rho,\delta},a,b}$ given in (5.37), the estimate

$$\|\sigma_{R,\ell}(x,\pi)\|_{S^{m_1,R}_{\rho,\delta},a,b} \leq C \sup_{\pi \in \widehat{G}} \|\sigma(x,\pi)\|_{S^{m,R}_{\rho,\delta},a,b} 2^{\ell \frac{c_o}{\nu}(m-m_1)}.$$

Proof of Lemma 5.4.7. For each $\ell \in \mathbb{N}_0$, the symbol $\eta(2^{-\ell c_o}\pi(\mathcal{R}))$ is in $S^{-\infty}$ by Proposition 5.3.4. Therefore, by Theorem 5.2.22 (ii) and the inclusions (5.31), $\sigma_{L,\ell}$ and $\sigma_{R,\ell}$ are in $S^{-\infty}$.

Let us fix $\alpha_o, \beta_o \in \mathbb{N}_0^n$ and $\gamma \in \mathbb{R}$. By the Leibniz formula (see (5.28)),

$$\begin{aligned}
&\pi(\mathrm{I}+\mathcal{R})^{\frac{\rho[\alpha_o]-m_1-\delta[\beta_o]+\gamma}{\nu}} X_x^{\beta_o}\Delta^{\alpha_o}\sigma_{L,\ell}\pi(\mathrm{I}+\mathcal{R})^{-\frac{\gamma}{\nu}} \\
&= \pi(\mathrm{I}+\mathcal{R})^{\frac{\rho[\alpha_o]-m_1-\delta[\beta_o]+\gamma}{\nu}} X_x^{\beta_o}\Delta^{\alpha_o}\left\{\eta(2^{-\ell c_o}\pi(\mathcal{R}))\sigma(x,\pi)\right\}\pi(\mathrm{I}+\mathcal{R})^{-\frac{\gamma}{\nu}} \\
&= \sum_{[\alpha_1]+[\alpha_2]=[\alpha_o]} c_{\alpha_1,\alpha_2}\pi(\mathrm{I}+\mathcal{R})^{\frac{\rho[\alpha_o]-m_1-\delta[\beta_o]+\gamma}{\nu}} \Delta^{\alpha_1}\eta(2^{-\ell c_o}\pi(\mathcal{R})) \\
&\qquad\qquad X_x^{\beta_o}\Delta^{\alpha_2}\sigma(x,\pi)\pi(\mathrm{I}+\mathcal{R})^{-\frac{\gamma}{\nu}}.
\end{aligned}$$

Therefore, taking the operator norm, we obtain

$$\begin{aligned}
&\|\pi(\mathrm{I}+\mathcal{R})^{\frac{\rho[\alpha_o]-m_1-\delta[\beta_o]+\gamma}{\nu}} X_x^{\beta_o}\Delta^{\alpha_o}\sigma_{L,\ell}\pi(\mathrm{I}+\mathcal{R})^{-\frac{\gamma}{\nu}}\| \\
&\leq C\sum_{[\alpha_1]+[\alpha_2]=[\alpha_o]} \|\pi(\mathrm{I}+\mathcal{R})^{\frac{\rho[\alpha_o]-m_1-\delta[\beta_o]+\gamma}{\nu}} \Delta^{\alpha_1}\eta(2^{-\ell c_o}\pi(\mathcal{R}))\pi(\mathrm{I}+\mathcal{R})^{-\frac{\rho[\alpha_2]-m-\delta[\beta_o]+\gamma}{\nu}}\| \\
&\qquad\qquad \|\pi(\mathrm{I}+\mathcal{R})^{\frac{\rho[\alpha_2]-m-\delta[\beta_o]+\gamma}{\nu}} X_x^{\beta_o}\Delta^{\alpha_2}\sigma(x,\pi)\pi(\mathrm{I}+\mathcal{R})^{-\frac{\gamma}{\nu}}\| \\
&\leq C\|\sigma(x,\pi)\|_{S^m_{\rho,\delta},[\alpha_o],[\beta_o],|\gamma|} \\
&\quad \sum_{[\alpha_1]+[\alpha_2]=[\alpha_o]} \|\pi(\mathrm{I}+\mathcal{R})^{\frac{\rho[\alpha_o]-m_1-\delta[\beta_o]+\gamma}{\nu}} \Delta^{\alpha_1}\eta(2^{-\ell c_o}\pi(\mathcal{R}))\pi(\mathrm{I}-\mathcal{R})^{-\frac{\rho[\alpha_2]-m-\delta[\beta_o]+\gamma}{\nu}}\|.
\end{aligned}$$

By Proposition 5.3.4,

$$\begin{aligned}
&\|\pi(\mathrm{I}+\mathcal{R})^{\frac{\rho[\alpha_o]-m_1-\delta[\beta_o]+\gamma}{\nu}} \Delta^{\alpha_1}\eta(2^{-\ell c_o}\pi(\mathcal{R}))\pi(\mathrm{I}+\mathcal{R})^{-\frac{\rho[\alpha_2]-m-\delta[\beta_o]+\gamma}{\nu}}\| \\
&\qquad\qquad\qquad \leq C\|\eta(2^{-\ell c_o}\cdot)\|_{\mathcal{M}_{\frac{m_2}{\nu}},k},
\end{aligned}$$

for some k, where m_2 is such that

$$[\alpha_1] - m_2 = \rho[\alpha_o] - m_1 - \delta[\beta_o] + \gamma - (\rho[\alpha_2] - m - \delta[\beta_o] + \gamma),$$

that is,

$$m_2 = m_1 - m + [\alpha_1](1-\rho).$$

Now, we can estimate

$$\begin{aligned}
\|\eta(2^{-\ell c_o}\cdot)\|_{\mathcal{M}_{\frac{m_2}{\nu}},k} &= \sup_{\lambda>0,\,k'=0,\dots,k} (1+\lambda)^{k'-\frac{m_2}{\nu}}\partial_\lambda^{k'}(\eta(2^{-\ell c_o}\lambda)) \\
&= \sup_{\lambda>0,\,k'=0,\dots,k} (1+\lambda)^{k'-\frac{m_2}{\nu}}2^{-\ell c_o k'}(\partial^{k'}\eta)(2^{-\ell c_o}\lambda) \\
&\leq C2^{-\ell c_o\frac{m_2}{\nu}}.
\end{aligned}$$

Therefore,

$$\sum_{[\alpha_1]+[\alpha_2]=[\alpha_o]} \|\pi(\mathrm{I}+\mathcal{R})^{\frac{\rho[\alpha]-m_1+\gamma}{\nu}} \Delta^{\alpha_1}\eta(2^{-\ell c_o}\pi(\mathcal{L}))\pi(\mathrm{I}+\mathcal{R})^{-\frac{\rho[\alpha_2]-m-\delta[\beta_o]+\gamma}{\nu}}\|$$

$$\leq C \sum_{[\alpha_1]+[\alpha_2]=[\alpha_o]} 2^{-\ell c_o \frac{m_1-m+[\alpha_1](1-\rho)}{\nu}} \leq C 2^{-\ell c_o \frac{m_1-m}{\nu}},$$

and we have shown that

$$\begin{aligned} &\|\pi(\mathrm{I}+\mathcal{R})^{\frac{\rho[\alpha_o]-m_1-\delta[\beta_o]+\gamma}{\nu}} X_x^{\beta_o}\Delta^{\alpha_o}\sigma_{L,\ell}\pi(\mathrm{I}+\mathcal{R})^{-\frac{\gamma}{\nu}}\| \\ &\leq C_{\alpha_o}\|\sigma(x,\pi)\|_{S^m_{\rho,\delta},[\alpha_o],[\beta_o],|\gamma|} 2^{-\ell c_o \frac{m_1-m}{\nu}}. \end{aligned}$$

The desired property for $\sigma_{L,\ell}$ follows easily. The property for $\sigma_{R,\ell}$ may be obtained by similar methods and its proof is left to the reader. □

Proof of Proposition 5.4.6. By Theorem 5.2.22 (i) Parts (3) and (2), and Corollary 5.2.25, it suffices to show the statement for $\alpha = \beta_1 = \beta_2 = \beta_o = 0$. By equivalence of homogeneous quasi-norms (Proposition 3.1.35), we may assume that the homogeneous quasi-norm is $|\cdot|_p$ given by (3.21) where $p > 0$ is such that $p/2$ is the smallest positive integer divisible by all the weights $\upsilon_1, \ldots, \upsilon_n$. Since κ_x decays faster than any polynomial away from the origin (more precisely see Proposition 5.4.4), it suffices to prove the result for $|z|_p < 1$.

So let $\sigma \in S^m_{\rho,\delta}$ with $Q + m \geq 0$. By Lemma 5.4.11 (to be shown in Section 5.4.2) we may assume that the kernel $\kappa : (x, y) \mapsto \kappa_x(y)$ of σ is smooth on $G \times G$ and compactly supported in x. By Proposition 5.4.4 it is also Schwartz in y.

We fix a positive Rockland operator $\mathcal{R}$ of homogeneous degree ν and a dyadic decomposition of its spectrum: we choose two functions $\eta_0, \eta_1 \in \mathcal{D}(\mathbb{R})$ supported in $[-1, 1]$ and $[1/2, 2]$, respectively, both valued in $[0, 1]$ and satisfying

$$\forall \lambda > 0 \qquad \sum_{\ell=0}^{\infty} \eta_\ell(\lambda) = 1,$$

where for $\ell \in \mathbb{N}$ we set

$$\eta_\ell(\lambda) := \eta_1(2^{-(\ell-1)\nu}\lambda).$$

For each $\ell \in \mathbb{N}_0$, the symbol $\eta_\ell(\pi(\mathcal{R}))$ is in $S^{-\infty}$ by Proposition 5.3.4 and its kernel $\eta_\ell(\mathcal{R})\delta_0$ is Schwartz by Corollary 4.5.2. Furthermore, by the functional calculus, $\sum_{\ell=0}^N \eta_\ell(\mathcal{R})$ converges in the strong operator topology of $\mathscr{L}(L^2(G))$ to the identity operator I as $N \to \infty$, and thus $\sum_{\ell=0}^N \eta_\ell(\mathcal{R})\delta_0$ converges in $\mathcal{K}(G)$ and in $\mathcal{S}'(G)$ to the Dirac measure δ_0 at the origin as $N \to \infty$.

By Theorem 5.2.22 (ii), the symbol σ_ℓ given by

$$\sigma_\ell(x, \pi) := \sigma(x, \pi)\eta_\ell(\pi(\mathcal{R})), \quad (x, \pi) \in G \times \widehat{G},$$

is in $S^{-\infty}$. The kernel associated with σ_ℓ is κ_ℓ given by

$$\kappa_\ell(x,y) = \kappa_{\ell,x}(y) = (\eta_\ell(\mathcal{R})\delta_0) * \kappa_x(y).$$

For each x, we have $\kappa_{\ell,x} \in \mathcal{S}(G)$. The sum $\sum_{\ell=0}^N \kappa_{\ell,x}$ converges in $\mathcal{S}'(G)$ to κ_x as $N \to \infty$ since

$$\sum_{\ell=0}^{N} \mathrm{Op}(\sigma_\ell(x,\cdot)) = \mathrm{Op}(\sigma(x,\cdot)) \sum_{\ell=0}^{N} \eta_\ell(\mathcal{R})$$

converges to $\mathrm{Op}(\sigma(x,\cdot))$ in the strong operator topology of $\mathscr{L}(L^2(G), L^2_{-m}(G))$. This convergence is in fact stronger. Indeed, by Lemma 5.4.7,

$$\|\sigma_\ell\|_{S^{m_1}_{\rho,\delta},a,b,c} \leq C \sup_{\pi\in\widehat{G}} \|\sigma\|_{S^{m}_{\rho,\delta},a',b',c'} 2^{\ell(m-m_1)},$$

thus

$$\sum_{\ell\in\mathbb{N}} \|\sigma_\ell\|_{S^{m_1}_{\rho,\delta},a,b,c} < \infty$$

if $m_1 > m$. Consequently, the sum $\sum_\ell \sigma_\ell$ is convergent in $S^{m_1}_{\rho,\delta}$ and, fixing $x \in G$, the sum $\sum_\ell \sup_{z\in S} |\kappa_{\ell,x}(z)|$ is convergent where S is any compact subset of $G\backslash\{0\}$ by Proposition 5.4.4 or more precisely the first part in Remark 5.4.5. Necessarily, the limit of $\sum_\ell \sigma_\ell$ is σ and the limit of $\sum_\ell \kappa_{\ell,x}$ for the uniform convergence on any compact subset of $G\backslash\{0\}$ is κ_x with

$$|\kappa_x(z)| \leq \sum_{\ell=0}^{\infty} |\kappa_{\ell,x}(z)|, \qquad z \in G\backslash\{0\}.$$

By Corollary 5.4.3 (2), for any $m_1 < -Q$ and $r \in \mathbb{N}_0$, we have

$$\begin{aligned} \sup_{z\in G} |z|_p^{pr} |\kappa_{\ell,x}(z)| &\leq C \sum_{[\alpha]=pr} \sup_{\pi\in\widehat{G}} \|\Delta^\alpha \sigma_\ell(x,\pi)\|_{S^{m_1}_{\rho,\delta},0,0,0} \\ &\leq C c_{\sigma,r} 2^{\ell(m-m_1-\rho pr)} \end{aligned} \tag{5.39}$$

by Lemma 5.4.7 and its proof, with $c_{\sigma,r} := \sup_{\pi\in\widehat{G}} \|\sigma(x,\pi)\|_{S^{m}_{\rho,\delta},pr,0,0}$.

We write $|z|_p \sim 2^{-\ell_o}$ in the sense that $\ell_o \in \mathbb{N}_0$ is the only integer satisfying $|z|_p \in (2^{-(\ell_o+1)}, 2^{-\ell_o}]$.

Let us assume that $Q + m > 0$. We use (5.39) with $r = 0$ and m_1 such that $m - m_1 = (Q+m)/\rho$. In particular,

$$m_1 = m(1 - \frac{1}{\rho}) - \frac{Q}{\rho} < -Q.$$

The sum over $\ell = 0, \ldots, \ell_o - 1$, can be estimated as

$$\begin{aligned} \sum_{\ell=0}^{\ell_o-1} |\kappa_{\ell,x}(z)| &\leq \sum_{\ell=0}^{\ell_o-1} C c_{\sigma,0} 2^{\ell(m-m_1)} \leq c_{\sigma,0} 2^{\ell_o(m-m_1)} \\ &\leq C c_{\sigma,0} |z|_p^{-\frac{Q+m}{\rho}}. \end{aligned}$$

We now choose $r \in \mathbb{N}$ and $m_1 < -Q$ such that

$$m - m_1 - \rho p r < 0 \quad \text{and} \quad pr(1-\rho) + m - m_1 = \frac{Q+m}{\rho}.$$

More precisely, we set $r := \lceil (m+Q)/(\rho p) \rceil$, that is, r is the largest integer strictly greater than $(m+Q)/(\rho p)$, while m_1 is defined by the equality just above; in particular,

$$m - m_1 > \frac{Q+m}{\rho} - (1-\rho)\frac{Q+m}{\rho} \quad \text{thus} \quad m_1 < -Q.$$

We may use (5.39) and sum over $\ell = \ell_o, \ell_o + 1 \ldots$, to get

$$\sum_{\ell=\ell_o}^{\infty} |z|_p^{pr} |\kappa_{\ell,x}(z)| \leq C c_{\sigma,r} \sum_{\ell=\ell_o}^{\infty} 2^{\ell(m-m_1-\rho p r)} \leq C c_{\sigma,r} 2^{\ell_o(m-m_1-\rho p r)}.$$

Therefore, we obtain

$$\begin{aligned}\sum_{\ell=\ell_o}^{\infty} |\kappa_{\ell,x}(z)| &\leq C c_{\sigma,r} 2^{\ell_o(m-m_1-\rho p r)} |z|_p^{-pr} \\ &\leq C c_{\sigma,r} |z|_p^{-pr-(m-m_1-\rho p r)} = C c_{\sigma,r} |z|_p^{-\frac{Q+m}{\rho}}.\end{aligned}$$

This yields the desired estimate for κ_x when $Q + m < 0$.

Let us assume that $Q + m = 0$. Using (5.39) with $r = 0$ and $m_1 = -m$, we obtain

$$\begin{aligned}\sum_{\ell=0}^{\ell_o-1} |\kappa_{\ell,x}(z)| &\leq \sum_{\ell=0}^{\ell_o-1} C c_{\sigma,0} 2^{\ell(m-m_1)} \leq c_{\sigma,0} \ell_o \\ &\leq C c_{\sigma,0} \ln |z|_p.\end{aligned}$$

Proceeding as above for the sum over $\ell \geq \ell_o$, we obtain that $\sum_{\ell=\ell_o}^{\infty} |\kappa_{\ell,x}(z)|$ is bounded. This yields the desired estimate for κ_x in the case $Q + m = 0$. □

Remark 5.4.8. It is possible to obtain a concrete expression for the numbers a, b, c appearing in Proposition 5.4.6, in terms of $m, \rho, \delta, \alpha, \beta_1, \beta_2, \beta_o$ and of Q.

5.4.2 Smoothing operators and symbols

The kernel estimates obtained in Section 5.4.1 allow us to characterise smoothing operators in terms of their kernels. Moreover they also imply that the operators in $\Psi^{-\infty}$ map the tempered distribution to smooth functions and enable the construction of sequences of smoothing operators converging in $\Psi^m_{\rho,\delta}$

Theorem 5.4.9. *1. If $T \in \Psi^{-\infty}$, then its associated kernel $\kappa : (x,y) \mapsto \kappa_x(y)$ is a smooth function on $G \times G$ such that for each $x \in G$, $y \mapsto \kappa_x(y)$ is Schwartz. Moreover, for each multi-index $\beta \in \mathbb{N}_0^n$ and each Schwartz seminorm $\|\cdot\|_{\mathcal{S}(G),N}$, there exist a constant $C > 0$ and a seminorm $\|\cdot\|_{S^m,a,b,c}$ (both independent of T) such that*

$$\sup_{x\in G} \|X_x^\beta \kappa_x\|_{\mathcal{S}(G),N} \leq C\|\sigma\|_{S^m,a,b,c}.$$

The converse is true, see Lemma 5.2.21.

2. If $T \in \Psi^{-\infty}$, then T extends to a continuous mapping from $\mathcal{S}'(G)$ to $C^\infty(G)$ via

$$Tf(x) = f * \kappa_x(x)$$

where $f \in \mathcal{S}'(G)$, $x \in G$, and κ_x is the kernel associated with T.

Furthermore, for any compact subset $K \subset G$ and any multi-index $\beta \in \mathbb{N}_0^n$, there exists a constant $C > 0$ and a seminorm $\|\cdot\|_{\mathcal{S}'(G),N}$ such that

$$\sup_{x\in K} |\partial^\beta Tf(x)| \leq C|f\|_{\mathcal{S}'(G),N}.$$

Moreover C can be chosen as $C_1\|\sigma\|_{S^m,a,b,c}$, and $C_1 > 0$ and N can be chosen independently of f and T.

Part 1 may be rephrased as stating that the map between the smoothing operators and their associated kernels is a Fréchet isomorphism between $\Psi^{-\infty}$ and the space $C_b^\infty(G, \mathcal{S}(G))$ of functions $\kappa \in C^\infty(G \times G)$ satisfying

$$\sup_{x\in G} \|X_x^\beta \kappa_x\|_{\mathcal{S}(G),N} < \infty.$$

Here $C_b^\infty(G, \mathcal{S}(G))$ is endowed with the Fréchet structure given via the seminorms

$$\kappa \longmapsto \max_{[\beta]\leq N} \sup_{x\in G} \|X_x^\beta \kappa_x\|_{\mathcal{S}(G),N} < \infty, \qquad N \in \mathbb{N}_0.$$

Part 2 may be rephrased as stating that the mapping $T \mapsto T$ from $\Psi^{-\infty}$ to the space $\mathscr{L}(\mathcal{S}'(G), C^\infty(G))$ of linear continuous mappings from $\mathcal{S}'(G)$ to $C^\infty(G)$ is continuous (it is clearly linear).

Proof. Part 1 follows easily from Theorem 5.4.1 and Remark 5.4.2. By Lemma 3.1.55, for any tempered distribution $f \in \mathcal{S}'(G)$, the function $f * \kappa_x$ is smooth on G and the function $x \mapsto f * \kappa_x(x)$ is smooth on G. Hence T extends to $\mathcal{S}'(G)$ and $Tf \in C^\infty$ if $f \in \mathcal{S}'(G)$.

Note that Lemma 3.1.55 also implies the existence of a positive constant C and $N \in \mathbb{N}_0$ such that

$$|f * \kappa_x(z)| \leq C(1+|z|)^N \|f\|_{\mathcal{S}'(G),N} \|\kappa_x\|_{\mathcal{S}(G),N}.$$

Using the Leibniz property for vector fields, one checks easily that for any multi-index $\beta \in \mathbb{N}_0^n$, we have

$$X^\beta (Tf)(x) = \sum_{[\beta_1]+[\beta_2]=[\beta]} c_{\beta,\beta_1,\beta_2} X^{\beta_1}_{x_1=x} (f * X^{\beta_2}_{x_2=x} \kappa_{x_2})(x_1).$$

Thus, proceeding as above, passing from left derivatives to the right, and using Lemma 3.1.55, we get

$$\begin{aligned}
|X^\beta (Tf)(x)| &\leq C \sum_{[\beta_1]+[\beta_2]=[\beta]} (1+|x|)^{[\beta_1]} |(\tilde{X}^{\beta_1}_{x_1=x} (f * (X^{\beta_2}_{x_2=x} \kappa_{x_2}))(x_1)| \\
&\leq C \sum_{[\beta_1]+[\beta_2]=[\beta]} (1+|x|)^{[\beta_1]} |(\tilde{X}^{\beta_1}_{x_1=x} f) * (X^{\beta_2}_{x_2=x} \kappa_{x_2})(x_1)| \\
&\leq C \sum_{[\beta_1]+[\beta_2]=[\beta]} (1+|x|)^{[\beta_1]+N} \|\tilde{X}^{\beta_1} f\|_{\mathcal{S}'(G),N} \|X^{\beta_2}_{x_2=x} \kappa_{x_2}\|_{\mathcal{S}(G),N} \\
&\leq C(1+|x|)^{N_2} \|f\|_{\mathcal{S}'(G),N_1} \|X^{\beta_2}_{x_2=x} \kappa_{x_2}\|_{\mathcal{S}(G),N}
\end{aligned}$$

with a new constant $C > 0$ and integers $N_2, N_1, N \in \mathbb{N}_0$. This shows that $f \mapsto Tf$ is continuous from $\mathcal{S}'(G)$ to $C^\infty(G)$.

Using Part 1, the inequality above also shows the continuity of $T \mapsto T$ from $\Psi^{-\infty}$ to the space of continuous mappings from $\mathcal{S}'(G)$ to $C^\infty(G)$. This concludes the proof of Theorem 5.4.9. □

Using the stability of taking the adjoint, reasoning by duality from Part 2 of Theorem 5.4.9, will yield the fact that smoothing operators map distributions with compact support to Schwartz functions, see Corollary 5.5.13.

Note that the proof of Part 2 of Theorem 5.4.9 yields the more precise result:

Corollary 5.4.10. *If $T \in \Psi^{-\infty}$ and $f \in \mathcal{S}'(G)$, then Tf is smooth and all its left-derivatives $X^\beta Tf$, $\beta \in \mathbb{N}_0^n$, have polynomial growth. More precisely, for any multi-index $\beta \in \mathbb{N}_0^n$, there exist a constant $C > 0$, and integer $M \in \mathbb{N}_0$ and a seminorm $\|\cdot\|_{\mathcal{S}'(G),N}$ such that*

$$|X^\beta Tf(x)| \leq C(1+|x|)^M \|f\|_{\mathcal{S}'(G),N}.$$

Moreover C can be chosen as $C_1 \|\sigma\|_{S^m,a,b,c}$, and $C_1 > 0$ and N, M can be chosen independently of f and T.

5.4.3 Pseudo-differential operators as limits of smoothing operators

In the proof of Lemma 5.1.42, for a given symbol σ, we constructed a sequence of symbols σ_ϵ such that $\mathrm{Op}(\sigma_\epsilon)$ is a sequence of 'nice operators' converging towards $\mathrm{Op}(\sigma)$ in a certain sense. If we assume that $\sigma \in S^m_{\rho,\delta}$, then we can construct a sequence of smoothing operators with a convergence in $\Psi^m_{\rho,\delta}$ described in the next lemma and its corollary. These operators are therefore 'nice' since they have Schwartz associated kernels in the sense of Theorem 5.4.9.

Lemma 5.4.11. *Let $1 \geq \rho \geq \delta \geq 0$. If $\sigma = \{\sigma(x,\pi)\}$ is in $S^m_{\rho,\delta}$, then we can construct a family $\sigma_\epsilon = \{\sigma_\epsilon(x,\pi)\}$, $\epsilon > 0$, in $S^{-\infty}$, satisfying the following properties:*

1. *For each $\epsilon > 0$, the x-support of each σ_ϵ is compact, or in other words, the function $x \mapsto \sup_{\pi\in\widehat{G}} \|\sigma(x,\pi)\|_{\mathscr{L}(\mathcal{H}_\pi)}$ is zero outside a compact set in G. Hence the kernel $\kappa_\epsilon : (x,y) \mapsto \kappa_{\epsilon,x}(y)$ associated with each symbol σ_ϵ is Schwartz on $G \times G$ and compactly supported in x.*
2. *For any seminorm $\|\cdot\|_{S^{m_1}_{\rho,\delta},a,b,c}$, there exist a constant $C = C_{a,b,c,m,m_1\rho,\delta} > 0$ such that*
$$\forall \epsilon \in (0,1) \qquad \|\sigma_\epsilon\|_{S^{m_1}_{\rho,\delta},a,b,c} \leq C\|\sigma\|_{S^m_{\rho,\delta},a,b,c}\epsilon^{\frac{m_1-m}{\nu}},$$
and when $m \leq m_1$,
$$\forall \epsilon \in (0,1) \qquad \|\sigma_\epsilon - \sigma\|_{S^{m_1}_{\rho,\delta},a,b,c} \leq C\|\sigma\|_{S^m_{\rho,\delta},a,b,c+\rho a}\epsilon^{\frac{m_1-m}{\nu}}.$$
Here ν is the degree of homogeneity of the positive Rockland operator used to define the seminorms.

   Consequently, when $m < m_1$, the convergence $\sigma_\epsilon \to \sigma$ as $\epsilon \to 0$ holds in $S^{m_1}_{\rho,\delta}$.
3. *If $\phi \in \mathcal{S}(G)$ then $\mathrm{Op}(\sigma_\epsilon)\phi \in \mathcal{D}(G)$ and the convergence*
$$\mathrm{Op}(\sigma_\epsilon)\phi \underset{\epsilon\to 0}{\longrightarrow} \mathrm{Op}(\sigma)\phi$$
holds uniformly on any compact subset of G and also in $\mathcal{S}(G)$.

Remark 5.4.12. As the construction will show, the symbols σ_ϵ are constructed independently of the order $m \in \mathbb{R}$.

Proof of Lemma 5.4.11. We consider the function χ_ϵ on G constructed in Lemma 5.1.42. Let $\eta \in \mathcal{D}(\mathbb{R})$ be such that $\eta \equiv 1$ on $[0,1]$. Let $\mathcal{R}$ be a positive Rockland operator. Let $\sigma \in S^m_{\rho,\delta}$. We set
$$\sigma_\epsilon(x,\pi) = \chi_\epsilon(x)\sigma(x,\pi)\eta(\epsilon\,\pi(\mathcal{R})).$$

Arguing as in Lemma 5.4.7 and its proof yields that
$$\{\sigma(x,\pi)\eta(\epsilon\,\pi(\mathcal{R})), (x,\pi) \in G \times \widehat{G}\}$$
is in $S^{-\infty}$. Moreover, for any $m_1 \in \mathbb{R}$ and $a,b,c \in \mathbb{N}_0$, there exists a constant $C = C_{m,m_1,\rho,\delta,a,b,c,\eta} > 0$ such that for any $\ell \in \mathbb{N}_0$ we have
$$\|\sigma(x,\pi)\eta(\epsilon\,\pi(\mathcal{R}))\|_{S^{m_1}_{\rho,\delta},a,b,c} \leq C \sup_{\pi\in\widehat{G}} \|\sigma(x,\pi)\|_{S^m_{\rho,\delta},a,b,c}\epsilon^{\frac{m_1-m}{\nu}}.$$

From this, it is clear that Property (1) and the first estimate in Property (2) hold.

Let us prove the second estimate in Property (2). We notice that

$$\begin{aligned}
&\|\pi(\mathrm{I}+\mathcal{R})^{-\frac{m_1}{\nu}}\left(\sigma(x,\pi)\eta(\epsilon\,\pi(\mathcal{R}))-\sigma(x,\pi)\right)\|_{\mathscr{L}(\mathcal{H}_\pi)}\\
&\quad=\|\pi(\mathrm{I}+\mathcal{R})^{-\frac{m_1}{\nu}}\sigma(x,\pi)\left(\eta(\epsilon\,\pi(\mathcal{R}))-\mathrm{I}\right)\|_{\mathscr{L}(\mathcal{H}_\pi)}\\
&\quad\le\|\pi(\mathrm{I}+\mathcal{R})^{-\frac{m_1}{\nu}}\sigma(x,\pi)\pi(\mathrm{I}+\mathcal{R})^{\frac{m_1-m}{\nu}}\|_{\mathscr{L}(\mathcal{H}_\pi)}\\
&\qquad\qquad\|\pi(\mathrm{I}+\mathcal{R})^{\frac{m-m_1}{\nu}}\left(\eta(\epsilon\,\pi(\mathcal{R}))-\mathrm{I}\right)|_{\mathscr{L}(\mathcal{H}_\pi)},
\end{aligned}$$

and the spectral calculus properties (cf. Corollary 4.1.16) imply

$$\begin{aligned}
&\sup_{\pi\in\widehat{G}}\|\pi(\mathrm{I}+\mathcal{R})^{\frac{m-m_1}{\nu}}\left(\eta(\epsilon\,\pi(\mathcal{R}))-\mathrm{I}\right)\|_{\mathscr{L}(\mathcal{H}_\pi)}\\
&=\|(\mathrm{I}+\mathcal{R})^{\frac{m-m_1}{\nu}}\left(\eta(\epsilon\,\mathcal{R})-\mathrm{I}\right)\|_{\mathscr{L}(L^2(G))}\le\sup_{\lambda>0}(1+\lambda)^{\frac{m-m_1}{\nu}}|\eta(\epsilon\lambda)-1|.
\end{aligned}$$

One checks easily that

$$\begin{aligned}
\sup_{\lambda>0}(1+\lambda)^{\frac{m-m_1}{\nu}}|\eta(\epsilon\lambda)-1| \;&\le\; \|\eta-1\|_\infty\sup_{\lambda>\epsilon^{-1}}(1+\lambda)^{\frac{m-m_1}{\nu}}\\
&\le\; t(1+\epsilon^{-1})^{\frac{m-m_1}{\nu}}\le C\epsilon^{\frac{m_1-m}{\nu}},
\end{aligned}$$

provided that $m-m_1\le 0$. Hence

$$\begin{aligned}
\sup_{(x,\pi)\in G\times\widehat{G}}\;&\|\pi(\mathrm{I}+\mathcal{R})^{-\frac{m_1}{\nu}}\left(\sigma(x,\pi)\eta(\epsilon\,\pi(\mathcal{R}))-\sigma(x,\pi)\right)\|_{\mathscr{L}(\mathcal{H}_\pi)}\\
&\le C\|\sigma\|_{S^m_{\rho,\delta},0,0,|m_1-m|}\,\epsilon^{\frac{m_1-m}{\nu}}.
\end{aligned}$$

More generally, we can introduce derivatives in x and difference operators and use the Leibniz properties (cf. Proposition 5.2.10):

$$\begin{aligned}
&X_x^\beta\Delta^\alpha\left(\sigma(x,\pi)\eta(\epsilon\,\pi(\mathcal{R}))-\sigma(x,\pi)\right)\\
&\quad=\sum_{[\alpha_1]+[\alpha_2]=[\alpha]}c_{\alpha,\alpha_1,\alpha_2}X_x^\beta\Delta^{\alpha_1}\sigma(x,\pi)\;\;\Delta^{\alpha_2}\left(\eta(\epsilon\,\pi(\mathcal{R}))-\mathrm{I}\right),
\end{aligned}$$

so that the quantity

$$\|\pi(\mathrm{I}+\mathcal{R})^{\frac{-m_1+\rho[\alpha]-\delta[\beta]-\gamma}{\nu}}X_x^\beta\Delta^\alpha\left(\sigma(x,\pi)\eta(\epsilon\,\pi(\mathcal{R}))-\sigma(x,\pi)\right)\pi(\mathrm{I}+\mathcal{R})^{\frac{\gamma}{\nu}}\|_{\mathscr{L}(\mathcal{H}_\pi)}$$

is, up to a constant, less or equal to the sum over $[\alpha_1]+[\alpha_2]=[\alpha]$ of

$$\begin{aligned}
&\|\pi(\mathrm{I}+\mathcal{R})^{\frac{-m_1+\rho[\alpha]-\delta[\beta]-\gamma}{\nu}}X_x^\beta\Delta^{\alpha_1}\sigma(x,\pi)\pi(\mathrm{I}+\mathcal{R})^{\frac{m_1-m-\rho[\alpha_2]+\gamma}{\nu}}\|_{\mathscr{L}(\mathcal{H}_\pi)}\\
&\quad\times\|\pi(\mathrm{I}+\mathcal{R})^{-\frac{m_1-m-\rho[\alpha_2]+\gamma}{\nu}}\Delta^{\alpha_2}(\eta(\epsilon\,\pi(\mathcal{R}))-\mathrm{I})\pi(\mathrm{I}+\mathcal{R})^{\frac{\gamma}{\nu}}\|_{\mathscr{L}(\mathcal{H}_\pi)}.
\end{aligned}$$

Applying Proposition 5.3.4, we obtain

$$\|\pi(\mathrm{I}+\mathcal{R})^{-\frac{m_1-m-\rho[\alpha_2]+\gamma}{\nu}}\Delta^{\alpha_2}(\eta(\epsilon\,\pi(\mathcal{R}))-\mathrm{I})\pi(\mathrm{I}+\mathcal{R})^{\frac{\gamma}{\nu}}\|_{\mathscr{L}(\mathcal{H}_\pi)}\leq C\epsilon^{\frac{m-m_1}{\nu}}.$$

Collecting the estimates and taking the supremum over $[\alpha]\leq a$, $[\beta]\leq b$, $|\gamma|\leq c$ yield the second estimate in Property (2).

Property (3) follows from Property (2) and the continuity of $\sigma\mapsto \mathrm{Op}(\sigma)$ from $S^{m_1}_{\rho,\delta}$ to $\mathscr{L}(\mathcal{S}(G))$, see Theorem 5.2.15. □

Keeping the notation of Lemma 5.4.11, we can also show that the kernels κ_ϵ converge in some sense towards the kernel of σ. In order to make this more precise, let us define the space $C^\infty_b(G,\mathcal{S}'(G))$ as the space of functions $x\mapsto\kappa_x\in\mathcal{S}'(G)$ such that for each $x\in G$, $y\mapsto\kappa_x(y)$ is a tempered distribution and, for any $\beta\in\mathbb{N}_0^n$, the map $x\mapsto X_x^\beta\kappa_x$ is continuous and bounded on G. This definition is motivated by the following property:

Lemma 5.4.13. *If $\sigma\in S^m_{\rho,\delta}$ then its associated kernel $\kappa=\kappa^{(\sigma)}$ is in $C^\infty_b(G,\mathcal{S}'(G))$ defined above. Furthermore, the map*

$$\sigma\mapsto\kappa^{(\sigma)}$$

from $S^m_{\rho,\delta}$ to $C^\infty_b(G,\mathcal{S}'(G))$ is continuous.

Naturally, we have endowed $C^\infty_b(G,\mathcal{S}'(G))$ with the structure of Fréchet space given by the seminorms

$$\kappa\longmapsto\max_{[\beta]\leq N}\sup_{x\in G}\|X_x^\beta\kappa_x\|_{\mathcal{S}'(G),N},\qquad N\in\mathbb{N}_0.$$

Proof of Lemma 5.4.13. By Lemma 5.1.35, if σ is a symbol then its kernel is in $C^\infty(G,\mathcal{S}'(G))$. Adapting slightly its proof yields

$$\sup_{x\in G}\|X_x^\beta\kappa_x\|_{\mathcal{S}'(G)}\leq C\sup_{x\in G}\|X_x^\beta\sigma(x,\cdot)\|_{L^\infty_{0,-m-\delta[\beta]}(\widehat{G})}.$$

As the inverse Fourier transform is one-to-one and continuous from $L^\infty_{0,-m-\delta[\beta]}(\widehat{G})$ to $\mathcal{S}'(G)$, this shows the continuity of the map $\sigma\mapsto\kappa^{(\sigma)}$ from $S^m_{\rho,\delta}$ to $C^\infty_b(G,\mathcal{S}'(G))$. □

We can now express the convergence in distribution of the sequence of kernels κ_ϵ constructed in the proof of Lemma 5.4.11:

Corollary 5.4.14. *We keep the notation of Lemma 5.4.11. The sequence of kernels κ_ϵ converges towards the kernel κ associated with σ in $C^\infty_b(G,\mathcal{S}'(G))$. If $\rho>0$, the convergence is also uniform on any compact subset of $G\times(G\backslash\{0\})$.*

Proof. The statement follows from the convergence of σ_ϵ to σ in $S^{m_1}_{\rho,\delta}$ for $m_1<m$ by Part 2 of Lemma 5.4.11, together with Lemma 5.4.13 for the first part and Corollary 5.4.3 for the second part. □

5.4.4 Operators in Ψ^0 as singular integral operators

From the kernel estimates obtained in Section 5.4.1, one can show easily that the operators in Ψ^0 are Calderón-Zygmund, and generalise this to some classes $\Psi^m_{\rho,\delta}$, see Theorem 5.4.16. We are then led to study the L^2-boundedness.

First let us notice that thanks to the kernel estimates, our operators admit a representation as singular integrals in the following sense:

Lemma 5.4.15. *Let κ_x be the kernel associated with $T \in \Psi^m_{\rho,\delta}$ with $m \in \mathbb{R}$ and $1 \geq \rho \geq \delta \geq 0$ with $\rho \neq 0$. For any $f \in \mathcal{S}'(G)$ and any $x_0 \in G$ such that $f \equiv 0$ on a neighbourhood of x_0, the integral*

$$\int_G f(y)\kappa_{x_0}(y^{-1}x_0)dy$$

makes distributional sense and defines a smooth function at x_0.

This coincides with Tf if $f \in \mathcal{S}(G)$.

Proof. Let T and κ_x be as in the statement. Let $f \in \mathcal{S}'(G)$ and $x_0 \in G$. We assume that there exists a bounded open set Ω_2 containing x_0 and where $f \equiv 0$. Let $\Omega \subsetneq \Omega_1 \subsetneq \Omega_2$ be open subsets of Ω_2 such that $x_0 \in \Omega$, $\bar{\Omega} \subset \Omega_1$, and $\bar{\Omega}_1 \subset \Omega_2$. We can find $\chi_1, \chi \in \mathcal{D}(G)$ such that $\chi_1 \equiv 1$ on Ω_1 but $\chi_1 \equiv 0$ outside Ω_2, $\chi \equiv 1$ on Ω but $\chi \equiv 0$ outside Ω_1. At least formally, we have

$$\chi(x)\int_G f(y)\kappa_x(y^{-1}x)dy = \int_G f(y)\ \chi(x)(1-\chi_1)(y)\kappa_x(y^{-1}x)dy,$$

since $f \equiv 0$ on $\{\chi_1 = 1\}$. Clearly the function $(x, y) \mapsto \chi(x)(1-\chi_1)(y)$ is smooth on $G \times G$ and supported away from the diagonal $\{(x, y) \in G \times G : x = y\}$. By Theorem 5.4.1, the function

$$y \longmapsto \chi(x)(1-\chi_1)(y)\kappa_x(y^{-1}x),$$

is Schwartz and this yields a smooth mapping $G \to \mathcal{S}(G)$ (which is also compactly supported). The rest of the statement follows easily. $\square$

In Corollary 5.5.13, we will see that an operator in $\Psi^m_{\rho,\delta}$ extends naturally to $\mathcal{S}'(G)$. Lemma 5.4.15 and its proof above will then imply that the operator admits a singular representation for any tempered distribution in the sense that the following formula makes sense and holds

$$Tf(x) = \int_G f(y)\kappa_x(y^{-1}x)dy,$$

for any $f \in \mathcal{S}'(G)$ and any $x \in G$ such that $f \equiv 0$ on a neighbourhood of x. We will not use this.

We can now give sufficient condition for operator in some $\Psi^m_{\rho,\delta}$ to be Calderón-Zygmund.

Theorem 5.4.16. *1. If $T \in \Psi^0$ then the operator T is Calderón-Zygmund in the sense of Definition 3.2.15.*

2. If $T \in \Psi^m_{\rho,\delta}$ with

$$m \leq (\rho - 1)Q,$$

$1 \geq \rho \geq \delta \geq 0$ and $\rho \neq 0$, then the operator T is Calderón-Zygmund in the sense of Definition 3.2.15.

In Parts 1 and 2, the constants appearing in the Definition 3.2.15 are $\gamma = 1$ and, up to constants of the group, given by seminorms of $T \in \Psi^m_{\rho,\delta}$.

Proof. We fix a homogeneous quasi-norm $|\cdot|$ on G.

Let $T \in \Psi^0$. We denote by κ its associated kernel. Then its integral kernel κ_o is formally given via $\kappa_o(x, y) = \kappa_x(y^{-1}x)$. By Theorem 5.4.1, for any two distinct points $y, x \in G$, we have

$$|\kappa_o(x, y)| = |\kappa_x(y^{-1}x)| \leq C|y^{-1}x|^{-Q}.$$

Using Remark 5.4.2 as well and the Leibniz property for vector fields, we obtain

$$|(X_j)_x\kappa_o(x, y)| \leq |(X_j)_{x_1=x}\kappa_{x_1}(y^{-1}x)| + |(X_j)_{x_2=x}\kappa_x(y^{-1}x_2)| \leq C|y^{-1}x|^{-(Q+v_j)},$$

and

$$|(X_j)_y\kappa_o(x, y)| \leq |(\tilde{X}_j)_{z=y^{-1}x}\kappa_x(z)| \leq C|y^{-1}x|^{-(Q+v_j)}.$$

Hence κ_o satisfies the hypotheses of Lemma 3.2.19. This shows Part 1.

Let us now assume that $T \in \Psi^m_{\rho,\delta}$. Again, let κ be its associated kernel. Let $\chi \in C^\infty(G)$ be supported in the unit ball $\{x \in G : |x| \leq 1\}$ and such that $\chi \equiv 1$ on $\{x \in G : |x| \leq 1/2\}$. By Theorem 5.4.1 and Remark 5.4.2 together with Lemma 5.2.21, the operator given by $\phi \mapsto \phi * \{(1-\chi)\kappa\}$ is smoothing (as $\rho \neq 0$) hence it is a Calderón-Zygmund operator by Part 1. Thus we just have to study the operator $\phi \mapsto \phi * \{\chi\kappa\}$. Its integral kernel is κ_o given via

$$\kappa_o(x, y) = \chi(y^{-1}x)\kappa_x(y^{-1}x).$$

Proceeding as above, in particular by Theorem 5.4.1, we have

$$\begin{aligned} |\kappa_o(x, y)| &= |(\chi\kappa_x)(y^{-1}x)| \lesssim |y^{-1}x|^{-\frac{Q+m}{\rho}}, \\ |(X_j)_y\kappa_o(x, y)| &= |(\tilde{X}_j)_{z=y^{-1}x}\kappa_x(z)| \lesssim |y^{-1}x|^{-\frac{Q+m+v_j}{\rho}}, \end{aligned}$$

and κ_o is supported on $\{(x, y) \in G : |y^{-1}x| \leq 1\}$ where we have

$$\begin{aligned} |(X_j)_x\kappa_o(x, y)| &\leq |(X_j)_{x_1=x}\kappa_{x_1}(y^{-1}x)| + |(X_j)_{x_2=x}\kappa_x(y^{-1}x_2)| \\ &\lesssim |y^{-1}x|^{-\frac{Q+m+\delta v_j}{\rho}} + |y^{-1}x|^{-\frac{Q+m+v_j}{\rho}} \lesssim |y^{-1}x|^{-\frac{Q+m}{\rho} - \frac{\delta}{\rho}v_j} \\ &\lesssim |y^{-1}x|^{-\frac{Q+m+v_j}{\rho}}, \end{aligned}$$

since $|y^{-1}x| \leq 1$. Hence if $(Q + m)/\rho \leq Q$, we can apply Lemma 3.2.19. □

In order to apply the singular integrals theorem (Theorem A.4.4), we still need to show that the operators are L^2-bounded. In the case $(\rho, \delta) = (1, 0)$, it is not very difficult to adapt the Euclidean case to show that the operators in Ψ^0 are L^2-bounded.

Theorem 5.4.17. *If $T \in \Psi^0$ then T extends to a bounded operator on $L^2(G)$. Furthermore, there exist constants $C > 0$ and $a, b, c \in \mathbb{N}_0$ of the group such that*

$$\forall f \in \mathcal{S}(G) \qquad \|Tf\|_{L^2(G)} \leq C\|T\|_{\Psi^m,a,b,c}\|f\|_{L^2(G)}.$$

During the proof of Theorem 5.4.17, we will need the following observation:

Lemma 5.4.18. *The collection of operators Ψ^0 is invariant under left translations in the sense that*

$$T \in \Psi^0 \Longrightarrow \forall x_o \in G \quad \tau_{x_o} T \tau_{x_o}^{-1} \in \Psi^0, \qquad \text{where} \qquad \tau_{x_o} : f \mapsto f(x_o\,\cdot).$$

Furthermore, if κ_x is the kernel of T and $\sigma = \text{Op}^{-1}(T)$ is its symbol, then the operator $\tau_{x_o} T \tau_{x_o}^{-1}$ has $\kappa_{x_o x}$ as kernel and $\sigma(x_o x, \pi)$ as symbol, and

$$\|T\|_{\Psi^0,a,b,c} = \|\tau_{x_o} T \tau_{x_o}^{-1}\|_{\Psi^0,a,b,c}.$$

Proof of Lemma 5.4.18. Let $T \in \Psi^0$ and let κ_x be its kernel. Then

$$\begin{aligned}
\tau_{x_o} T \tau_{x_o}^{-1} f(x) &= T(\tau_{x_o}^{-1} f)(x_o x) = (\tau_{x_o}^{-1} f) * \kappa_{x_o x}(x_o x) \\
&= \int_G f(x_o^{-1} y) \kappa_{x_o x}(y^{-1} x_o x) dy \\
&= \int_G f(z) \kappa_{x_o x}(z^{-1} x) dz
\end{aligned}$$

after the change of variable $z = x_o^{-1} y$. Therefore

$$\tau_{x_o} T \tau_{x_o}^{-1} f(x) = f * \kappa_{x_o x}(x).$$

Since $\mathcal{F}_G(\kappa_{x_o x})(\pi) = \sigma(x_o x, \pi)$ if σ denotes the symbol of T, we see that $\kappa_{x_o x}$ is the kernel associated to the symbol $\{\sigma(x_o x, \pi), (x, \pi) \in G \times \widehat{G}\}$ and the corresponding operator is $\tau_{x_o} T \tau_{x_o}^{-1}$. The rest of the statement follows easily. □

Proof of Theorem 5.4.17. The proof follows the Euclidean case as given in [Ste93, ch. VI §2]. Let $T \in \Psi^0$ and let $\sigma = \text{Op}^{-1}(T)$ be its symbol. We claim that it suffices to show Theorem 5.4.17 under the additional assumption that the kernel κ associated with σ is smooth in x and Schwartz in y, and such that $G \ni x \mapsto \kappa_x \in \mathcal{S}(G)$ is smooth. Indeed, this would imply that Theorem 5.4.17 is proved for each operator $T_\epsilon = \text{Op}(\sigma_\epsilon)$ where σ_ϵ is as in Lemma 5.4.11. The properties (2) and (3) in Lemma 5.4.11 allow to pass through the limit as $\epsilon \to 0$ and imply then the theorem. This shows our earlier claim and hence we may assume that $G \ni x \mapsto \kappa_x \in \mathcal{S}(G)$ is smooth.

We fix $|\cdot|$ to be the homogeneous quasi-norm $|\cdot|_p$ given by (3.21), where $p>0$ is such that $p/2$ is the smallest positive integer divisible by all the weights $\upsilon_1,\ldots,\upsilon_n$. The balls are defined by $B(x_o,r):=\{x\in G:|x^{-1}x_o|<r\}$. We denote by $C_o\geq 1$ a constant such that for all $x,y\in G$, we have

$$|xy|\leq C_o(|x|+|y|) \quad \text{and} \quad |y|\leq \frac{|x|}{2}\Longrightarrow ||xy|-|x||\leq C_o|y|,$$

see the triangle inequality in Proposition 3.1.38 and its converse (3.26).

Let $f\in\mathcal{S}(G)$ and let us write it as

$$f=f_1+f_2,$$

where f_1 and f_2 are two smooth functions supported in $B(0,4C_o)$ and outside of $B(0,2C_o)$, respectively, and satisfying $|f_1|,|f_2|\leq|f|$.

First, we claim that there exists a constant $C>0$ of the group such that

$$\int_{B(0,1)}|Tf_1(x)|^2dx\leq C\|\sigma\|^2_{S^0,0,\lceil Q/2\rceil,0}\ \|f_1\|^2_{L^2(G)}. \tag{5.40}$$

Let us prove this. We fix a function $\chi\in\mathcal{D}(G)$ which is identically 1 on $B(0,1)$. Then

$$\begin{aligned}\int_{B(0,1)}|Tf_1(x)|^2dx &\leq \int_{B(0,1)}|\chi(x)\ f_1*\kappa_x(x)|^2dx\\ &\leq \int_{B(0,1)}\sup_{z\in G}|\chi(z)\ f_1*\kappa_z(x)|^2dx.\end{aligned}$$

We now use the Sobolev inequality in Theorem 4.4.25 to get

$$\sup_{z\in G}|\chi(z)\ f_1*\kappa_z(x)|^2\leq C\sum_{[\alpha]\leq\lceil Q/2\rceil}\int_G\left|X_z^\alpha\{\chi(z)\ f_1*\kappa_z(x)\}\right|^2dz.$$

Since

$$X_z^\alpha\{\chi(z)\ f_1*\kappa_z(x)\}=f_1*X_z^\alpha\{\chi(z)\kappa_z\}(x),$$

we have obtained

$$\begin{aligned}\int_{B(0,1)}|Tf_1(x)|^2dx &\leq \int_{B(0,1)}C\sum_{[\alpha]\leq\lceil Q/2\rceil}\int_G\left|f_1*X_z^\alpha\{\chi(z)\kappa_z\}(x)\right|^2dzdx\\ &=C\sum_{[\alpha]\leq\lceil Q/2\rceil}\int_G\int_{B(0,1)}\left|f_1*X_z^\alpha\{\chi(z)\kappa_z\}(x)\right|^2dxdz,\end{aligned}$$

by Fubini's property. But the integral over $B(0,1)$ can be estimated using Plancherel's Theorem (see Theorem 1.8.11) by

$$\begin{aligned}\int_{B(0,1)}\left|f_1*X_z^\alpha\{\chi(z)\kappa_z\}(x)\right|^2dx &\leq \|f_1*X_z^\alpha\{\chi(z)\kappa_z\}\|_2^2\\ &\leq \|\pi(X_z^\alpha\{\chi(z)\kappa_z\})\|^2_{L^\infty(\widehat{G})}\|f_1\|_2^2.\end{aligned}$$

Now the Leibniz formula for X_z^α gives

$$\begin{aligned}\|\pi(X_z^\alpha\{\chi(z)\kappa_z\})\|_{\mathscr{L}(L^2(G))} &\leq \sum_{[\alpha_1]+[\alpha_2]=[\alpha]} c_{\alpha_1,\alpha_2}\|\pi\left(X^{\alpha_1}\chi(z)X_z^{\alpha_2}\kappa_z\}\right)\|_{\mathscr{L}(L^2(G))} \\ &\leq C_\alpha \max_{[\beta]\leq[\alpha]} \|\pi\left(X_z^\beta\kappa_z\}\right)\|_{\mathscr{L}(L^2(G))} \sum_{[\alpha_1]\leq[\alpha]} |X^{\alpha_1}\chi(z)|.\end{aligned}$$

Since $\pi\left(X_z^\beta\kappa_z\right) = X_z^\beta\sigma(z,\pi)$, we have obtained

$$\begin{aligned}&\int_{B(0,1)} |f_1 * X_z^\alpha\{\chi(z)\kappa_z\}(x)|^2\, dx \\ &\qquad \leq C \max_{[\beta]\leq[\alpha]} \|X_z^\beta\sigma(z,\pi)\|^2_{L^\infty(\widehat{G})}\|f_1\|_2^2 \sum_{[\alpha_1]\leq[\alpha]} |X^{\alpha_1}\chi(z)|^2.\end{aligned}$$

Therefore,

$$\begin{aligned}\int_{B(0,1)} |Tf_1(x)|^2 dx &\leq C \sum_{[\alpha]\leq\lceil Q/2\rceil} \int_G\int_{B(0,1)} |f_1 * X_z^\alpha\{\chi(z)\kappa_z\}(x)|^2\, dxdz \\ &\leq C \max_{[\beta]\leq\lceil Q/2\rceil} \sup_{z\in G} \|X_z^\beta\sigma(z,\pi)\|^2_{L^\infty(\widehat{G})} \|f_1\|_2^2.\end{aligned}$$

This concludes the proof of Claim (5.40).

Secondly, we claim that for any $r \in \mathbb{N}$, there exists a constant $C = C_r > 0$ such that

$$\int_{B(0,1)} |Tf_2(x)|^2 dx \leq C\|\sigma\|^2_{S^0,pr,0,pr} \|(1+|\cdot|)^{-pr} f_2\|^2_{L^2(G)}. \tag{5.41}$$

Let us prove this. We write

$$Tf_2(x) = \int_{y\notin B(0,2C_o)} f_2(y)|y^{-1}x|^{-pr}(|\cdot|^{pr}\kappa_x)(y^{-1}x)dy.$$

If $x \in B(0,1)$ and $y \notin B(0,2C_o)$, then

$$|y^{-1}| - |y^{-1}x| \leq C_o|x| \leq C_o \quad \text{thus} \quad |y^{-1}x| \geq |y| - C_o \geq \frac{1}{2}|y| \geq \frac{1}{4}(1+|y|),$$

and

$$\begin{aligned}|Tf_2(x)| &\leq \int_{y\notin B(0,2C_o)} |f_2(y)| \left(\frac{1}{4}(1+|y|)\right)^{-pr} \left|(|\cdot|^{pr}\kappa_x)(y^{-1}x)\right| dy \\ &\leq 4^{pr}\|(1+|\cdot|)^{-pr} f_2\|_{L^2(G)} \|(|\cdot|^{pr}\kappa_x)\|_{L^2(G)},\end{aligned}$$

after having used the Cauchy-Schwartz inequality. Integrating the square of the left-hand side over $x \in B(0,1)$, and taking the supremum over $x \in B(0,1)$ of the right-hand side, we obtain

$$\int_{B(0,1)} |Tf_2(x)|^2 dx \leq 4^{2pr} \sup_{x\in B(0,1)} \| |\cdot|^{pr}\kappa_x\|^2_{L^2(G)} \ \|(1+|\cdot|)^{-pr} f_2\|^2_{L^2(G)}. \quad (5.42)$$

Now writing $|z|_p^{pr} = \sum_{[\alpha]=pr} c_\alpha \tilde{q}_\alpha(z)$, we have

$$\| |\cdot|^{pr}\kappa_x\|^2_{L^2(G)} \leq C_r \sum_{[\alpha]=pr} \|\tilde{q}_\alpha \kappa_x\|^2_{L^2(G)}$$

and since by Corollary 5.4.3 (1), if $[\alpha] > Q/2$,

$$\|\tilde{q}_\alpha \kappa_x\|^2_{L^2(G)} \leq C_\alpha \sup_{\pi\in\widehat{G}} \|\sigma(x,\pi)\|^2_{S^m_{\rho,\delta},[\alpha],0,[\alpha]},$$

we have obtained that if $pr > Q/2$, then

$$\sup_{x\in B(0,1)} \| |\cdot|^{pr}\kappa_x\|^2_{L^2(G)} \leq C_r \|\sigma\|^2_{S^0,pr,0,pr}.$$

This and (5.42) show Claim (5.41).

Now, combining together Claims (5.40) and (5.41), we obtain

$$\int_{B(0,1)} |Tf(x)|^2 dx \leq C_r \|T\|^2_{\Psi^0,pr,\lceil Q/2\rceil,pr} \ \|(1+|\cdot|)^{-pr} f\|^2_{L^2(G)},$$

and this is so for any $f \in \mathcal{S}(G)$. Therefore, by Lemma 5.4.18 (and its notation), we have for any $x_o \in G$, that

$$\begin{aligned}
\int_{B(x_o,1)} |Tf(x)|^2 dx &= \int_{|x_o^{-1}x|<1} |Tf(x)|^2 dx = \int_{B(0,1)} |Tf(x_o x')|^2 dx' \\
&= \int_{B(0,1)} |\tau_{x_o}(Tf)(x')|^2 dx' = \int_{B(0,1)} |(\tau_{x_o} T \tau_{x_o}^{-1})(\tau_{x_o} f)(x')|^2 dx' \\
&\leq C_r \|\tau_{x_o} T \tau_{x_o}^{-1}\|^2_{\Psi^0,pr,\lceil Q/2\rceil,pr} \ \|(1+|\cdot|)^{-pr} \tau_{x_o} f\|^2_{L^2(G)} \\
&= C_r \|T\|^2_{\Psi^0,pr,\lceil Q/2\rceil,pr} \ \|(1+|\cdot|)^{-pr} \tau_{x_o} f\|^2_{L^2(G)}.
\end{aligned}$$

Integrating over $x_o \in G$, we obtain for the left hand side,

$$\begin{aligned}
\int_G \int_{B(x_o,1)} |Tf(x)|^2 dx dx_o &= \int_G \int_G 1_{|x_o^{-1}x|<1} |Tf(x)|^2 dx dx_o \\
&= \int_G \int_G 1_{|y|<1} |Tf(x)|^2 dx dy = |B(0,1)| \|Tf\|_2^2,
\end{aligned}$$

and for the last term in the right hand side,

$$\begin{aligned}\int_G \|(1+|\cdot|)^{-pr}\tau_{x_o} f\|_{L^2(G)}^2 dx_o &= \int_G\int_G \left|(1+|x|)^{-pr} f(x_o x)\right|^2 dx dx_o \\ &= \|f\|_2^2 \int_G (1+|x|)^{-2pr} dx.\end{aligned}$$

Assuming $-2pr+Q<0$, this last integral is finite.

We have obtained that if $r > Q/2p$ (for instance $r=\lceil Q/2p\rceil$) then $pr > Q/2$ and

$$|B(0,1)|\|Tf\|_2^2 \le C\|T\|^2_{\Psi^0,pr,\lceil Q/2\rceil,pr}\|f\|_2^2.$$

This concludes the proof of Theorem 5.4.17. □

Remark 5.4.19. More precisely we have obtained that if $T\in\Psi^0$, then

$$\|Tf\|_2 \le C\|T\|_{\Psi^0,pr,\lceil Q/2\rceil,pr}\|f\|_2,$$

where $r := \lceil\frac{Q}{2p}\rceil$, and $p\in\mathbb{R}$ is such that $p/2$ is the smallest positive integer divisible by all the weights $\upsilon_1,\ldots,\upsilon_n$.

Theorem 5.4.16 and Theorem 5.4.17 show that any operator of order 0 and of type (1,0) satisfies the hypotheses of the singular integrals theorem, see Sections 3.2.3 and A.4. Therefore, we have the following corollary:

Corollary 5.4.20. *If $T\in\Psi^0$ then T extends to a bounded operator on $L^p(G)$ for any $p\in(1,\infty)$. Furthermore, there exist constants $a,b,c\in\mathbb{N}_0$ such that*

$$\forall p\in(1,\infty)\quad \exists C>0\quad \forall f\in\mathcal{S}(G)\qquad \|Tf\|_{L^p(G)} \le C\|T\|_{\Psi^0,a,b,c}\|f\|_{L^p(G)}.$$

5.5 Symbolic calculus

In this section we present elements of the symbolic calculus of operators with symbols in the classes $S^m_{\rho,\delta}$. In particular, we will discuss asymptotic sums of symbols, adjoints, and compositions.

5.5.1 Asymptotic sums of symbols

We now establish a nilpotent analogue of the asymptotic sum of symbols of decreasing orders going to $-\infty$.

Theorem 5.5.1. *We assume $1\ge\rho\ge\delta\ge 0$. Let $\{\sigma_j\}_{j\in\mathbb{N}_0}$ be a sequence of symbols such that $\sigma_j\in S^{m_j}_{\rho,\delta}$ with m_j strictly decreasing to $-\infty$. Then there exists $\sigma\in S^{m_0}_{\rho,\delta}$, unique modulo $S^{-\infty}$, such that*

$$\forall M\in\mathbb{N}\qquad \sigma-\sum_{j=0}^{M}\sigma_j \in S^{m_{M+1}}_{\rho,\delta}. \tag{5.43}$$

Definition 5.5.2. Under the hypotheses and conclusions of Theorem 5.5.1, we write

$$\sigma \sim \sum_j \sigma_j.$$

Proof. We keep the notation of the statement. We also fix a positive Rockland operator $\mathcal{R}$ of homogeneous degree ν on G. Let $\chi \in C^\infty(\mathbb{R})$ with $\chi_{|(-\infty,1/2)} = 0$ and $\chi_{|[1,\infty)} = 1$. We fix $t \in (0,1)$.

Let us check that for any seminorm $\|\cdot\|_{S^{m_0}_{\rho,\delta},a,b,c}$, there exists a constant $C = C_{a,b,c} > 0$ such that for any $t \in (0,1)$ and any $j \in \mathbb{N}$, we have

$$\|\sigma_j(x,\pi)\chi(t\pi(\mathcal{R}))\|_{S^{m_0}_{\rho,\delta},a,b,c} \leq C\|\sigma_j(x,\pi)\|_{S^{m_0}_{\rho,\delta},a,b,c+\rho a+m_0-m_j} t^{\frac{m_0-m_j}{\nu}}. \tag{5.44}$$

Indeed, from the Leibniz formula (see Formula (5.28)), we obtain easily

$$\begin{aligned}
&\|\pi(\mathrm{I}+\mathcal{R})^{\frac{\rho[\alpha_o]-m_0-\delta[\beta_o]+\gamma}{\nu}} X_x^{\beta_o}\Delta^{\alpha_o}\left(\sigma_j(x,\pi)\chi(t\pi(\mathcal{R}))\right)\pi(\mathrm{I}+\mathcal{R})^{-\frac{\gamma}{\nu}}\|_{\mathscr{L}(\mathcal{H}_\pi)} \\
&\lesssim \sum_{[\alpha_1]+[\alpha_2]=[\alpha_o]} \|\pi(\mathrm{I}+\mathcal{R})^{\frac{\rho[\alpha_o]-m_0-\delta[\beta_o]+\gamma}{\nu}} X_x^{\beta_o}\Delta^{\alpha_1}\sigma_j(x,\pi) \\
&\qquad\qquad \Delta^{\alpha_2}\chi(t\pi(\mathcal{R}))\ \pi(\mathrm{I}+\mathcal{R})^{-\frac{\gamma}{\nu}}\|_{\mathscr{L}(\mathcal{H}_\pi)} \\
&\lesssim \sum_{[\alpha_1]+[\alpha_2]=[\alpha_o]} \|\sigma_j(x,\pi)\|_{S^{m_0}_{\rho,\delta},[\alpha_1],[\beta_o],\rho([\alpha_o]-[\alpha_1])+m_0-m_j+|\gamma|} \\
&\qquad\qquad \|\pi(\mathrm{I}+\mathcal{R})^{\frac{\rho[\alpha_2]-m_0+m_j+\gamma}{\nu}}\Delta^{\alpha_2}\chi(t\pi(\mathcal{R}))\pi(\mathrm{I}+\mathcal{R})^{-\frac{\gamma}{\nu}}\|_{\mathscr{L}(\mathcal{H}_\pi)}.
\end{aligned}$$

By the functional calculus, we have

$$\begin{aligned}
&\|\pi(\mathrm{I}+\mathcal{R})^{\frac{\rho[\alpha_2]-m_0+m_j+\gamma}{\nu}}\Delta^{\alpha_2}\chi(t\pi(\mathcal{R}))\pi(\mathrm{I}+\mathcal{R})^{-\frac{\gamma}{\nu}}\|_{\mathscr{L}(\mathcal{H}_\pi)} \\
&\leq \|\pi(\mathrm{I}+\mathcal{R})^{\frac{[\alpha_2]-m_0+m_j+\gamma}{\nu}}\Delta^{\alpha_2}\chi(t\pi(\mathcal{R}))\pi(\mathrm{I}+\mathcal{R})^{-\frac{\gamma}{\nu}}\|_{\mathscr{L}(\mathcal{H}_\pi)} \\
&\lesssim \sup_{\substack{k'\leq k\\ \lambda>0}} (1+\lambda)^{\frac{-m_0+m_j}{\nu}+k'}|\partial_\lambda^{k'}\{\chi(t\lambda)\}| \lesssim t^{\frac{m_0-m_j}{\nu}},
\end{aligned}$$

by Proposition 5.3.4 for some $k \in \mathbb{N}_0$. This shows (5.44).

Let us choose strictly increasing sequences $\{a_\ell\}$, $\{b_\ell\}$ and $\{c_\ell\}$ of positive integers. For each ℓ there exists $C_\ell > 0$ such that for any $j \in \mathbb{N}$ and $t \in (0,1)$, we have

$$\|\sigma_j(x,\pi)\chi(t\pi(\mathcal{R}))\|_{S^{m_0}_{\rho,\delta},a_\ell,b_\ell,c_\ell} \leq C_\ell\|\sigma_j(x,\pi)\|_{S^{m_0}_{\rho,\delta},a_\ell,b_\ell,c_\ell+\rho a_\ell+m_0-m_j} t^{\frac{m_0-m_j}{\nu}}.$$

We may assume that the constants C_ℓ are increasing with ℓ.

We now choose a decreasing sequence of numbers $\{t_j\}$ such that for any $j \in \mathbb{N}$,

$$t_j \in (0, 2^{-j}) \quad \text{and} \quad C_j \sup_{\substack{x\in G\\ \pi\in\widehat{G}}} \|\sigma_j(x,\pi)\|_{S^{m_0}_{\rho,\delta},a_j,b_j,c_j+\rho a_j+m_0-m_j} t_j^{\frac{m_0-m_j}{\nu}} \leq 2^{-j}.$$

For any $j \in \mathbb{N}$, we define the symbols

$$\tilde{\sigma}_j(x,\pi) := \sigma_j(x,\pi)\chi(t_j\pi(\mathcal{R})).$$

For any $\ell \in \mathbb{N}$, the sum

$$\sum_{j=0}^{\infty} \|\tilde{\sigma}_j\|_{S^{m_0}_{\rho,\delta},a_\ell,b_\ell,c_\ell} \leq \sum_{j=0}^{\ell} \|\tilde{\sigma}_j\|_{S^{m_0}_{\rho,\delta},a_\ell,b_\ell,c_\ell} + \sum_{j=\ell+1}^{\infty} 2^{-j},$$

is finite. Since $S^{m_0}_{\rho,\delta}$ is a Fréchet space, we obtain that

$$\sigma := \sum_{j=0}^{\infty} \tilde{\sigma}_j,$$

is a symbol in $S^{m_0}_{\rho,\delta}$.

Starting the sequence at m_{M+1}, the same proof gives

$$\sum_{j=M+1}^{\infty} \tilde{\sigma}_j \in S^{m_{M+1}}_{\rho,\delta}.$$

By Proposition 5.3.4, each symbol given by $(1-\chi)(t_j\pi(\mathcal{R}))$ is in $S^{-\infty}$. Thus by Theorem 5.2.22 (ii) and the inclusions (5.31), each symbol given by $\sigma_j(x,\pi)(1-\chi)(t_j\pi(\mathcal{R}))$ is in $S^{-\infty}$. Therefore, the symbol given by

$$\sigma(x,\pi) - \sum_{j=0}^{M} \sigma_j(x,\pi) = \sum_{j=0}^{M} \sigma_j(x,\pi)(1-\chi)(t_j\pi(\mathcal{R})) + \sum_{j=M+1}^{\infty} \tilde{\sigma}_j(x,\pi),$$

is in $S^{m_{M+1}}_{\rho,\delta}$. This shows (5.43) for σ.

If τ is another symbol as in the statement of the theorem, then for any $M \in \mathbb{N}$,

$$\sigma - \tau = \left(\sigma - \sum_{j=0}^{M} \sigma_j\right) - \left(\tau - \sum_{j=0}^{M} \sigma_j\right)$$

is in $S^{m_{M+1}}$. Thus $\sigma - \tau \in S^{-\infty}$. □

We note that the proof above does not produce a symbol σ depending continuously on $\{\sigma_j\}$, the same as in the abelian case.

5.5.2 Composition of pseudo-differential operators

In this section, we show that the class of operators $\cup_{m\in\mathbb{R}}\Psi^m_{\rho,\delta}$ is an algebra:

Theorem 5.5.3. *Let $1 \geq \rho \geq \delta \geq 0$ with $\delta \neq 1$ and $m_1, m_2 \in \mathbb{R}$. If $T_1 \in \Psi^{m_1}_{\rho,\delta}$ and $T_2 \in \Psi^{m_2}_{\rho,\delta}$ are two pseudo-differential operators of type (ρ, δ), then their composition $T_1 T_2$ is in $\Psi^{m_1+m_2}_{\rho,\delta}$. Moreover, the mapping*

$$(T_1, T_2) \mapsto T_1 T_2$$

is continuous from $\Psi^{m_1}_{\rho,\delta} \times \Psi^{m_2}_{\rho,\delta}$ to $\Psi^{m_1+m_2}_{\rho,\delta}$.

Since any operator in $\Psi^{m}_{\rho,\delta}$ maps $\mathcal{S}(G)$ to itself continuously (see Theorem 5.2.15), the composition of any two operators in $\Psi^{m_1}_{\rho,\delta}$ and $\Psi^{m_2}_{\rho,\delta}$ defines an operator in $\mathscr{L}(\mathcal{S}(G))$.

Let us start the proof of Theorem 5.5.3 with observing that the symbol of $T_1 T_2$ is necessarily known and unique at least formally or under favourable conditions such as between smoothing operators:

Lemma 5.5.4. *Let σ_1 and σ_2 be two symbols in $S^{-\infty}$ and let κ_1 and κ_2 be their associated kernels. We set*

$$\kappa_x(y) := \int_G \kappa_{2,xz^{-1}}(yz^{-1}) \kappa_{1,x}(z) dz, \qquad x, y \in G.$$

Then $\sigma(x, \pi) = \pi(\kappa_x)$ defines a smooth symbol σ in the sense of Definition 5.1.34. Furthermore, it satisfies

$$\mathrm{Op}(\sigma_1)\mathrm{Op}(\sigma_2) = \mathrm{Op}(\sigma).$$

and

$$\sigma(x, \pi) = \int_G \kappa_{1,x}(z) \pi(z)^* \sigma_2(xz^{-1}, \pi)\, dz, \tag{5.45}$$

In particular, if $\sigma_2(x, \pi)$ is independent of x then $\sigma_1 \circ \sigma_2 = \sigma_1 \sigma_2$.

We will often write

$$\sigma := \sigma_1 \circ \sigma_2.$$

Proof of Lemma 5.5.4. We keep the notation of the statement. Clearly $\kappa : (x, y) \mapsto \kappa_x(y)$ is smooth on $G \times G$, compactly supported in x. Furthermore, κ_x is integrable in y since

$$\begin{aligned}
\int_G |\kappa_x(y)| dy &\leq \int_G \int_G |\kappa_2(xz^{-1}, yz^{-1}) \kappa_1(x, z)| dz dy \\
&\leq \int_G \int_G |\kappa_{2,xz^{-1}}(w)| dw\, |\kappa_1(x, z)| dz \\
&\leq \max_{x' \in G} \int_G |\kappa_{2,x'}(w)| dw \int_G |\kappa_{1,x}(z)| dz.
\end{aligned}$$

Therefore, $\sigma(x, \pi) = \pi(\kappa_x)$ defines a symbol σ in the sense of Definition 5.1.33.

Using the Leibniz formula iteratively, one obtains easily that for any $\beta_o \in \mathbb{N}_0^n$, $\tilde{X}_x^{\beta_o}\kappa_x(y)$ is a linear combination of

$$\int_G \tilde{X}_{x_2=xz^{-1}}^{\beta_2}\kappa_{2,x_2}(yz^{-1})\tilde{X}_{x_1=x}^{\beta_1}\kappa_{1,x_1}(z)dz, \qquad [\beta_1]+[\beta_2]=[\beta_o].$$

Hence proceeding as above

$$\int_G |\tilde{X}_x^{\beta_o}\kappa_x(y)|dy \lesssim \sum_{[\beta_1]+[\beta_2]=[\beta_o]} \max_{x_2\in G}\int_G |\tilde{X}_{x_2}^{\beta_2}\kappa_{2,x_2}(w)|dw \int_G |\tilde{X}_x^{\beta_1}\kappa_{1,x}(z)|dz.$$

This together with the link between abelian and right-invariant derivatives (see Section 3.1.5, especially 3.17) implies easily that σ is a smooth symbol in the sense of Definition 5.1.34.

The properties of κ_1 and κ_2 (see Theorem 5.4.9) justify the equalities

$$\begin{aligned}
\mathrm{Op}(\sigma_1)\mathrm{Op}(\sigma_2)\phi(x) &= \int_G T_2\phi(y)\kappa_{1,x}(y^{-1}x)dy \\
&= \int_G\int_G \phi(z)\kappa_{2,y}(z^{-1}y)\kappa_{1,x}(y^{-1}x)dzdy \\
&= \int_G\int_G \phi(z)\kappa_{2,xw^{-1}}(z^{-1}xw^{-1})\kappa_{1,x}(w)dzdw \\
&= \int_G \phi(z)\kappa_x(z^{-1}x)dz = \phi*\kappa_x(x),
\end{aligned}$$

with the change of variables $y^{-1}x=w$. This yields $T_1T_2=\mathrm{Op}(\sigma)$. We have then finally

$$\begin{aligned}
\sigma(x,\pi) &= \widehat{\kappa}_x(\pi) = \int_G \kappa_x(y)\pi(y)^*dy \\
&= \int_G\int_G \kappa_{2,xz^{-1}}(yz^{-1})\kappa_{1,x}(z)\pi(z)^*\pi(yz^{-1})^*dydz \\
&= \int_G \kappa_{1,x}(z)\pi(z)^*\sigma_2(xz^{-1},\pi)\ dz,
\end{aligned}$$

after an easy change of variable. □

From Lemma 5.5.4 and its proof, we see that if $T=\mathrm{Op}(\sigma_1)\mathrm{Op}(\sigma_2)$ then the symbol σ of T is not $\sigma_1\sigma_2$ in general, unless the symbol $\{\sigma_2(x,\pi)\}$ does not depend on $x\in G$ for instance. However, we can link formally σ with σ_1 and σ_2 in the following way: using the vector-valued Taylor expansion (see (5.27)) for $\sigma_2(x,\pi)$ in the variable x, we have

$$\sigma_2(xz^{-1},\pi) \approx \sum_\alpha q_\alpha(z^{-1})X_x^\alpha\sigma_2(x,\pi),$$

Thus, implementing this in the expression (5.45), we obtain informally

$$\begin{aligned}\sigma(x,\pi) \quad &\approx \quad \int_G \kappa_{1,x}(z)\pi(z)^* \sum_\alpha q_\alpha(z^{-1}) X_x^\alpha \sigma_2(x,\pi)\, dz \\ &= \sum_\alpha \int_G q_\alpha(z^{-1})\kappa_{1,x}(z)\pi(z)^* dz\ X_x^\alpha \sigma_2(x,\pi) \\ &= \sum_\alpha \Delta^\alpha \sigma_1(x,\pi)\ X_x^\alpha \sigma_2(x,\pi).\end{aligned}$$

We will show that in fact these formal manipulations effectively give the asympotitcs, see Corollary 5.5.8. From Theorem 5.2.22, we know that if $\sigma_1 \in S^{m_1}_{\rho,\delta}$, $\sigma_2 \in S^{m_2}_{\rho,\delta}$ then

$$\Delta^\alpha \sigma_1\ X_x^\alpha \sigma_2 \in S^{m_1+m_2-(\rho-\delta)[\alpha]}_{\rho,\delta}. \tag{5.46}$$

The main problem with the informal approach above is that one needs to estimate the remainder

$$\sigma_1 \circ \sigma_2 - \sum_{[\alpha]\leq M} \Delta^\alpha \sigma_1\ X_x^\alpha \sigma_2.$$

We will first show how to estimate this remainder in the case of $\rho > \delta$ using the following property.

Lemma 5.5.5. *We fix a positive Rockland operator of homogeneous degree ν. Let $m_1, m_2 \in \mathbb{R}$, $1 \geq \rho \geq \delta \leq 0$ with $\rho \neq 0$ and $\delta \neq 1$, $\beta_0 \in \mathbb{N}_0^n$, and $M, M_1 \in \mathbb{N}_0$. We assume that*

$$\begin{cases} \frac{m_2+\delta(c_{\beta_0}+\upsilon_n)}{1-\delta} \leq \nu M_1 < M - Q - m_1 - \delta[\beta_0] + \rho(Q+\upsilon_1), \\ m_2 + \delta(c_{\beta_0} + \upsilon_n + M) \leq \nu M_1 < -Q - m_1 - \delta[\beta_0] + \rho(Q+M), \end{cases} \tag{5.47}$$

where

$$c_{\beta_0} := \max_{\substack{[\beta_{02}]\leq[\beta_0] \\ [\beta']\geq[\beta_{02}],\ |\beta'|\geq|\beta_{02}|}} [\beta'].$$

If $M \geq \nu M_1$, only the second condition may be assumed.

Then there exist a constant $C > 0$, and two pseudo-norms $\|\cdot\|_{S^{m_1,R}_{\rho,\delta},a_1,b_1}$, $\|\cdot\|_{S^{m_2}_{\rho,\delta},0,b_2,0}$, such that for any $\sigma_1, \sigma_2 \in S^{-\infty}$ and any $(x,\pi) \in G \times \widehat{G}$ we have

$$\begin{aligned}&\|X_x^{\beta_0}\big(\sigma_1\circ\sigma_2(x,\pi) - \sum_{[\alpha]\leq M} \Delta^\alpha \sigma_1(x,\pi)\ X_x^\alpha \sigma_2(x,\pi)\big)\|_{\mathscr{L}(\mathcal{H}_\pi)} \\ &\leq C\|\sigma_1\|_{S^{m_1,R}_{\rho,\delta},a_1,b_1} \|\sigma_2\|_{S^{m_2}_{\rho,\delta},0,b_2,0}.\end{aligned}$$

In the proof of Lemma 5.5.5, we will use the following easy consequence of the estimates of the kernels given in Theorem 5.2.22.

Lemma 5.5.6. *Let $\sigma \in S^m_{\rho,\delta}$ with $1 \geq \rho \geq \delta \geq 0$ with $\rho \neq 0$. We denote by κ_x its associated kernel. For any $\gamma \in \mathbb{R}$, if $\gamma + Q > \max(\frac{m+Q}{\rho}, 0)$ then there exist a constant $C > 0$ and a seminorm $\|\cdot\|_{S^m_{\rho,\delta},a,b,c}$ such that*

$$\int_G |z|^\gamma |\kappa_x(z)| dz \leq C \|\sigma\|_{S^m_{\rho,\delta},a,b,c}.$$

We may replace $\|\cdot\|_{S^m_{\rho,\delta},a,b,c}$ with $\|\cdot\|_{S^{m,R}_{\rho,\delta},a,b}$.

Proof of Lemma 5.5.6. We keep the notation and the statement and write

$$\int_G |z|^\gamma |\kappa_x(z)| dz = \int_{|z|\geq 1} + \int_{|z|<1}.$$

The estimate for large $|z|$ given in Theorem 5.4.1 easily implies that the integral $\int_{|z|\geq 1}$ is bounded up to a constant of γ, m, ρ, δ, by a seminorm of σ. The estimate for small $|z|$ yield

$$\int_{|z|\leq 1} |z|^\gamma |\kappa_x(z)| dz \lesssim \begin{cases} \int_{|z|\leq 1} |z|^{\gamma - \frac{m+Q}{\rho}} dz & \text{if } m+Q > 0, \\ \int_{|z|\leq 1} |z|^\gamma |\ln|z|| dz & \text{if } m+Q = 0, \\ \int_{|z|\leq 1} |z|^\gamma dz & \text{if } m+Q < 0. \end{cases}$$

Using the polar change of coordinates yields the result. □

Proof of Lemma 5.5.5, case $\beta_0 = 0$. By Lemma 5.5.4 and the observations that follow, we have

$$\begin{aligned} &\sigma(x,\pi) - \sum_{[\alpha]\leq M} \Delta^\alpha \sigma_1(x,\pi)\ X^\alpha_x \sigma_2(x,\pi) \\ &= \int_G \kappa_{1,x}(z)\pi(z)^* \left(\sigma_2(xz^{-1},\pi) - \sum_{[\alpha]\leq M} q_\alpha(z^{-1}) X^\alpha_x \sigma_2(x,\pi) \right) dz \\ &= \int_G \kappa_{1,x}(z)\pi(z)^* R^{\sigma_2(\cdot,\pi)}_{x,M}(z^{-1}) dz, \end{aligned}$$

where $R^{\sigma_2(\cdot,\pi)}_{x,M}$ denotes the remainder of the (vector-valued) Taylor expansion of $v \mapsto \sigma_2(xv,\pi)$ of order M at 0. We now introduce powers of $\pi(\mathrm{I}+\mathcal{R})$ near $\pi(z)^*$

$$\pi(z)^* = \pi(z)^* \pi(\mathrm{I}+\mathcal{R})^{M_1} \pi(\mathrm{I}+\mathcal{R})^{-M_1} = \sum_{[\beta]\leq \nu M_1} \pi(z)^* \pi(X)^\beta \pi(\mathrm{I}+\mathcal{R})^{-M_1}$$

and we notice that

$$\pi(z)^* \pi(X)^\beta = (-1)^{|\beta|} \left(\pi(X)^\beta \pi(z)\right)^* = (-1)^{|\beta|} \left(\tilde{X}^\beta_z \pi(z)\right)^*. \tag{5.48}$$

We integrate by parts and obtain

$$\begin{aligned}\sigma(x,\pi) - \sum_{[\alpha]\le M} \Delta^\alpha\sigma_1(x,\pi)\ X_x^\alpha\sigma_2(x,\pi) &= \overline{\sum_{[\beta_1]+[\beta_2]\le\nu M_1}} \int_G \tilde{X}_{z_1=z}^{\beta_1}\kappa_{1,x}(z_1)\pi(z)^*\tilde{X}_{z_2=z}^{\beta_2} R_{x,M}^{\pi(\mathrm{I}+\mathcal{R})^{-M_1}\sigma_2(\cdot,\pi)}(z_2^{-1})dz\\ &= \overline{\sum_{[\beta_1]+[\beta_2]\le\nu M_1}} \int_G \tilde{X}_{z_1=z}^{\beta_1}\kappa_{1,x}(z_1)\pi(z)^* R_{x,M-[\beta_2]}^{\pi(\mathrm{I}+\mathcal{R})^{-M_1}X^{\beta_2}\sigma_2(\cdot,\pi)}(z^{-1})dz\end{aligned}$$

by Lemma 3.1.50. Taking the operator norm, we have

$$\begin{aligned}&\|\sigma(x,\pi) - \sum_{[\alpha]\le M} \Delta^\alpha\sigma_1(x,\pi)\ X_x^\alpha\sigma_2(x,\pi)\|_{\mathscr{L}(\mathcal{H}_\pi)}\\ &\lesssim \sum_{[\beta_1]+[\beta_2]\le\nu M_1} \int_G |\tilde{X}_{z_1=z}^{\beta_1}\kappa_{1,x}(z_1)|\ \|R_{x,M-[\beta_2]}^{\pi(\mathrm{I}+\mathcal{R})^{-M_1}X^{\beta_2}\sigma_2(\cdot,\pi)}(z^{-1})\|_{\mathscr{L}(\mathcal{H}_\pi)}dz.\end{aligned}$$

The adapted statement of Taylor's estimates remains valid for vector-valued function, see Theorem 3.1.51 and Remark 3.1.52 (3), so we have

$$\begin{aligned}&\|R_{x,M-[\beta_2]}^{\pi(\mathrm{I}+\mathcal{R})^{-M_1}X^{\beta_2}\sigma_2(\cdot,\pi)}(z^{-1})\|_{\mathscr{L}(\mathcal{H}_\pi)}\\ &\lesssim \sum_{\substack{|\gamma|\le\lceil(M-[\beta_2])_+\rfloor+1\\ [\gamma]>(M-[\beta_2])_+}} |z|^{[\gamma]} \sup_{x_1\in G} \|\pi(\mathrm{I}+\mathcal{R})^{-M_1}X_{x_1}^\gamma X_{x_1}^{\beta_2}\sigma_2(x_1,\pi)\|_{\mathscr{L}(\mathcal{H}_\pi)}.\end{aligned}$$

We have obtained that

$$\begin{aligned}&\|\sigma(x,\pi) - \sum_{[\alpha]\le M} \Delta^\alpha\sigma_1(x,\pi)\ X_x^\alpha\sigma_2(x,\pi)\|_{\mathscr{L}(\mathcal{H}_\pi)}\\ &\lesssim \sum_{\substack{[\gamma]>(M-[\beta_2])_+\\ |\gamma|\le\lceil(M-[\beta_2])_+\rfloor+1}} \int_G |z|^{[\gamma]}|\tilde{X}_{z_1=z}^{\beta_1}\kappa_{1,x}(z_1)|dz\\ &\qquad \sup_{x_1\in G} \|\pi(\mathrm{I}+\mathcal{R})^{-M_1}X_{x_1}^\gamma X_{x_1}^{\beta_2}\sigma_2(x_1,\pi)\ |_{\mathscr{L}(\mathcal{H}_\pi)}.\end{aligned}$$

If $M-[\beta_2]\le 0$, the integrals above are finite by Lemma 5.5.6 and the suprema are bounded by a $S_{\rho,\delta}^{m_2}$-seminorm in σ_2 when

$$\begin{cases} m_1+[\beta_1]+Q<\rho(Q+\upsilon_1)\\ -\nu M_1+m_2+\delta(\upsilon_n+[\beta_2])\le 0\end{cases},$$

and it suffices

$$\begin{cases} m_1+\nu M_1-M+Q<\rho(Q+\upsilon_1)\\ -\nu M_1+m_2+\delta(\upsilon_n+\nu M_1)\le 0\end{cases}.$$

If $M - [\beta_2] > 0$, the integrals above are finite by Lemma 5.5.6 and the suprema are bounded by a $S^{m_2}_{\rho,\delta}$-seminorm in σ_2 when

$$\begin{cases} m_1 + [\beta_1] + Q < \rho(Q + [\gamma]) \\ -\nu M_1 + m_2 + \delta([\gamma] + [\beta_2]) \le 0 \end{cases},$$

and it suffices

$$\begin{cases} m_1 + \nu M_1 + Q < \rho(Q + M) \\ -\nu M_1 + m_2 + \delta(\upsilon_n + M) \le 0 \end{cases}.$$

Our conditions on M and M_1 ensure that the sufficient conditions above are satisfied. Collecting the various estimates yields the statement in the case $\rho \neq 0$ and $\beta_0 = 0$. □

Proof of Lemma 5.5.5, general case. Using Formula (5.45), the Leibniz property for left invariant vector fields easily implies that

$$X_x^{\beta_0} \sigma_1 \circ \sigma_2(x,\pi) = \overline{\sum_{[\beta_{01}]+[\beta_{02}]=[\beta_0]}} \int_G X_x^{\beta_{01}} \kappa_{1,x}(z)\pi(z)^* X^{\beta_{02}}_{x_2=x} \sigma_2(x_2 z^{-1},\pi)\, dz.$$

Proceeding as in the case $\beta_0 = 0$, we have

$$X_x^{\beta_0}\left(\sigma_1 \circ \sigma_2(x,\pi) - \sum_{[\alpha]\le M} \Delta^\alpha \sigma_1(x,\pi)\ X_x^\alpha \sigma_2(x,\pi)\right)$$
$$= \overline{\sum_{[\beta_{01}]+[\beta_{02}]=[\beta_0]}} \int_G X_x^{\beta_{01}} \kappa_{1,x}(z)\pi(z)^* R_{0,M}^{X^{\beta_{02}}_{x_2=x}\sigma_2(x_2\,\cdot,\pi)}(z^{-1})\, dz.$$

Introducing the powers of $\pi(\mathrm{I}+\mathcal{R})$, each integral on the right-hand side above is equal to

$$\overline{\sum_{[\beta_1]+[\beta_2]\le \nu M_1}} \int_G \tilde{X}^{\beta_1}_{z_1=z} X_x^{\beta_{01}} \kappa_{1,x}(z_1)\pi(z)^*$$
$$R_{0,M-[\beta_2]}^{\pi(\mathrm{I}+\mathcal{R})^{-M_1} X^{\beta_{02}}_{x_2=x} X^{\beta_2}\sigma_2(x_2\,\cdot,\pi)}(z^{-1})\, dz, \tag{5.49}$$

by Corollary 3.1.53. We use a more precise version for the Taylor remainder than in the proof of the case $\beta_0 = 0$:

$$\|R_{0,M-[\beta_2]}^{\pi(\mathrm{I}+\mathcal{R})^{-M_1} X^{\beta_{02}}_{x_2=x} X^{\beta_2}\sigma_2(x_2\,\cdot,\pi)}(z^{-1})\|_{\mathscr{L}(\mathcal{H}_\pi)}$$
$$\le C_M \sum_{\substack{[\gamma]>(M-[\beta_2])_+ \\ |\gamma|\le \lceil (M-[\beta_2])_+ \rfloor +1}} |z|^{[\gamma]} S(z, M_1, \gamma, \beta_{02}, \beta_2),$$

where $S(z, M_1, \gamma, \beta_{02}, \beta_2)$ denotes the supremum

$$S(z, M_1, \gamma, \beta_{02}, \beta_2) := \sup_{|y|\le \eta^{\lceil M \rfloor +1}|z|} \|\pi(\mathrm{I}+\mathcal{R})^{-M_1} X_y^\beta X^{\beta_{02}}_{x_2=x} X_y^{\beta_2}\sigma_2(x_2 y,\pi)\|_{\mathscr{L}(\mathcal{H}_\pi)}.$$

For any reasonable function $f : G \to \mathbb{C}$, the definitions of left and right-invariant vector fields imply

$$X_x^\beta f(xy) = \tilde{X}_y^\beta f(xy) \tag{5.50}$$

and the properties of left or right-invariant vector fields (see Section 3.1.5) then yield

$$X_x^\beta f(xy) = \tilde{X}_y^\beta f(xy) = \sum_{\substack{|\beta'|\leq|\beta| \\ [\beta']\geq[\beta]}} Q_{\beta,\beta'}(y) X_y^{\beta'} f(xy), \tag{5.51}$$

where $Q_{\beta,\beta'}$ are $([\beta'] - [\beta])$-homogeneous polynomials. Therefore

$$S(z, M_1, \gamma, \beta_{02}, \beta_2) \lesssim \sum_{\substack{[\beta'_{02}]\geq[\beta_{02}] \\ |\beta'_{02}|\leq|\beta_{02}|}} |z|^{[\beta'_{02}]-[\beta_{02}]} \tilde{S}(M_1, [\gamma] + [\beta'_{02}] - [\beta_2]),$$

where $\tilde{S}(M_1, [\beta_0])$ denotes the supremum

$$\tilde{S}(M_1, [\beta_0]) := \sup_{[\gamma']=[\beta_0]} \sup_{x_1\in G} \|\pi(\mathrm{I} + \mathcal{R})^{-M_1} X_{x_1}^{\gamma'} \sigma_2(x_1, \pi)\|_{\mathscr{L}(\mathcal{H}_\pi)}.$$

We then obtain that (5.49) is bounded up to a constant by

$$\sum_{[\beta_1]+[\beta_2]\leq\nu M_1} \int_G |\tilde{X}_{z_1=z}^{\beta_1} X_x^{\beta_{01}} \kappa_{1,x}(z_1)| \sum_{\substack{[\gamma]>(M-[\beta_2])_+ \\ |\gamma|\leq\lceil(M-[\beta_2])_+\rceil+1}} |z|^{[\gamma]}$$
$$\sum_{\substack{[\beta'_{02}]\geq[\beta_{02}] \\ |\beta'_{02}|\leq|\beta_{02}|}} |z|^{[\beta'_{02}]-[\beta_{02}]} \tilde{S}(M_1, [\gamma] + [\beta'_{02}] + [\beta_2])\, dz.$$

We conclude in the same way as in the case $\beta_0 = 0$. □

To take into account the difference operator, we will use the following observation.

Lemma 5.5.7. *Let $\sigma_1, \sigma_2 \in S^{-\infty}$. For any $\alpha \in \mathbb{N}_0^n$, $\Delta^\alpha(\sigma_1 \circ \sigma_2)$ is a linear combination independent of σ_1, σ_2 of $(\Delta^{\alpha_1}\sigma_1) \circ (\Delta^{\alpha_2}\sigma_2)$, over $\alpha_1, \alpha_2 \in \mathbb{N}_0^n$ satisfying $[\alpha_1] + [\alpha_2] = [\alpha]$. It is the same linear combination as in the Leibniz rule (5.28).*

Proof of Lemma 5.5.7. We keep the notation of Lemma 5.5.4 and adapt the proof of the Leibniz rule for Δ^α given in Proposition 5.2.10. By Proposition 5.2.3 (4), we have

$$\begin{aligned} \tilde{q}_\alpha(y)\kappa_x(y) &= \int_G \tilde{q}_\alpha(yz^{-1}z)\kappa_{2,xz^{-1}}(yz^{-1})\kappa_{1,x}(z)dz \\ &= \overline{\sum_{[\alpha_1]+[\alpha_2]=[\alpha]}} \int_G \tilde{q}_{\alpha_2}(yz^{-1})\kappa_{2,xz^{-1}}(yz^{-1})\ \tilde{q}_{\alpha_1}(z)\kappa_{1,x}(z)dz, \end{aligned}$$

where $\overline{\sum}$ denotes a linear combination. Lemma 5.5.4 implies easily the statement. □

Proof of Theorem 5.5.3 with $\rho > \delta$. We assume $\rho > \delta$. We fix a positive Rockland operator $\mathcal{R}$ of homogeneous degree ν. Let us show that for any $\alpha_0, \beta_0 \in \mathbb{N}_0^n$, and $M_0 \in \mathbb{N}$, there exists $M \geq M_0$, a constant $C > 0$ and seminorms $\|\cdot\|_{S^{m_1,R}_{\rho,\delta},a_1,b_1}$, $\|\cdot\|_{S^{m_2}_{\rho,\delta},a_2,b_2,c_2}$ such that for any $\sigma_1, \sigma_2 \in S^{-\infty}$ we have

$$\begin{aligned}\left\|X_x^{\beta_0}\Delta^{\alpha_0}\tau_M(x,\pi)\ \pi(\mathrm{I}+\mathcal{R})^{-\frac{m-(\rho-\delta)M_0-\rho[\alpha_0]+\delta[\beta_0]}{\nu}}\right\|_{\mathscr{L}(\mathcal{H}_\pi)} \\ \leq C\|\sigma_1\|_{S^{m_1,R}_{\rho,\delta},a_1,b_1}\|\sigma_2\|_{S^{m_2}_{\rho,\delta},a_2,b_2,c_2},\end{aligned} \tag{5.52}$$

where we have denoted $m = m_1 + m_2$ and

$$\tau_M := \sigma_1 \circ \sigma_2 - \sum_{[\alpha]\leq M} \Delta^\alpha \sigma_1 X_x^\alpha \sigma_2.$$

By Lemma 5.5.7, it suffices to show (5.52) only for $\alpha_0 = 0$.

Let $\beta_0 \in \mathbb{N}_0$ and $M_0 \in \mathbb{N}$. We fix $m_2' := -m_1 + (\rho-\delta)M_0 - \delta[\beta_0]$. As $\rho > \delta$, we can find $M \geq \max(M_0, \upsilon_1)$ such that

$$(-Q - m_1 - \delta[\beta_0] + \rho(Q+M)) - (m_2' + \delta(c_{\beta_0} + \upsilon_n + M)) \geq \nu.$$

This shows that we can find M_1 satisfying the second condition in (5.47) for m_1, m_2' and therefore also the first. Hence we can apply Lemma 5.5.5 to M, M_1 and the symbols σ_1 and $\sigma_2\pi(\mathrm{I}+\mathcal{R})^{-\frac{m-(\rho-\delta)M_0+\delta[\beta_0]}{\nu}}$, with orders m_1 and m_2'. The left-hand side of (5.52) is then bounded up to a constant by

$$\begin{aligned}\|\sigma_1\|_{S^{m_1,R}_{\rho,\delta},a_1,b_1}\|\sigma_2\pi(\mathrm{I}+\mathcal{R})^{-\frac{m-(\rho-\delta)M_0+\delta[\beta_0]}{\nu}}\|_{S^{m_2'}_{\rho,\delta},0,b_2,0} \\ \lesssim \|\sigma_1\|_{S^{m_1,R}_{\rho,\delta},a_1,b_1}\|\sigma_2\|_{S^{m_2}_{\rho,\delta},0,b_2,c_2}.\end{aligned}$$

Hence (5.52) is proved.

Using (5.46), classical considerations imply that (5.52) yield that for any $M_0 \in \mathbb{N}_0$, and any seminorm $\|\cdot\|_{S^{m-M_0(\rho-\delta),R}_{\rho,\delta},a,b}$, there exist a constant $C > 0$ and two seminorms $\|\cdot\|_{S^{m_1,R}_{\rho,\delta},a_1,b_1}$, $\|\cdot\|_{S^{m_2}_{\rho,\delta},a_2,b_2,c_2}$ such that for any $\sigma_1, \sigma_2 \in S^{-\infty}$ we have

$$\|\tau_{M_0}\|_{S^{m-M_0(\rho-\delta),R}_{\rho,\delta},a,b} \leq C\|\sigma_1\|_{S^{m_1,R}_{\rho,\delta},a_1,b_1}\|\sigma_2\|_{S^{m_2}_{\rho,\delta},a_2,b_2,c_2}. \tag{5.53}$$

In Section 5.5.4, we will see that for any seminorm $\|\cdot\|_{S^{\tilde{m}}_{\rho,\delta},\tilde{a},\tilde{b},\tilde{c}}$ there exist a constant $C > 0$ and a seminorm $\|\cdot\|_{S^{m,R}_{\rho,\delta},a,b}$ such that

$$\forall \sigma \in S^{-\infty} \qquad \|\sigma\|_{S^{\tilde{m}}_{\rho,\delta},\tilde{a},\tilde{b},\tilde{c}} \leq C\|\sigma\|_{S^{\tilde{m},R}_{\rho,\delta},a,b}. \tag{5.54}$$

Inequalities (5.54) together with (5.53) and Lemma 5.4.11 (to pass from $S^{-\infty}$ to $S^{m_1}_{\rho,\delta}, S^{m_2}_{\rho,\delta}$) conclude the proof of Theorem 5.5.3 in the case $\rho > \delta$. □

Note that the proof of the case $\rho > \delta$ above also shows:

Corollary 5.5.8. *We assume $1 \geq \rho > \delta \geq 0$. If $\sigma_1 \in S^{m_1}_{\rho,\delta}$ and $\sigma_2 \in S^{m_2}_{\rho,\delta}$, then there exists a unique symbol σ in $S^m_{\rho,\delta}$, $m = m_1 + m_2$, such that*

$$\mathrm{Op}(\sigma) = \mathrm{Op}(\sigma_1)\mathrm{Op}(\sigma_2). \tag{5.55}$$

Moreover, for any $M \in \mathbb{N}_0$, we have

$$\{\sigma - \sum_{[\alpha]\leq M} \Delta^\alpha \sigma_1 \; X_x^\alpha \sigma_2\} \in S^{m-(\rho-\delta)M}_{\rho,\delta}. \tag{5.56}$$

Furthermore, the mapping

$$\left\{ \begin{array}{ccc} S^m_{\rho,\delta} & \longrightarrow & S^{m-(\rho-\delta)M}_{\rho,\delta} \\ \sigma & \longmapsto & \{\sigma - \sum_{[\alpha]\leq M} \Delta^\alpha \sigma_1 \; X_x^\alpha \sigma_2\} \end{array} \right. ,$$

is continuous.

Consequently, we can also write

$$\sigma \sim \sum_{j=0}^{\infty} \left(\sum_{[\alpha]=j} \Delta^\alpha \sigma_1 \; X_x^\alpha \sigma_2 \right), \tag{5.57}$$

in the sense of an asymptotic expansion as in Definition 5.5.2.

The case $\rho = \delta$ is more delicate to prove but relies on the same kind of arguments as above. If $\rho = \delta$, the asymptotic formula (5.56) does not bring any improvement and, in this sense, is not interesting.

We will need the following variation of the properties given in Lemma 5.5.6 obtained using Corollary 5.4.3 instead of Theorem 5.4.1.

Lemma 5.5.9. *Let $\sigma \in S^m_{\rho,\delta}$ with $1 \geq \rho \geq \delta \geq 0$. We denote by κ_x its associated kernel. Let $\gamma \geq 0$ and $m < -Q$. Then there exist a constant $C > 0$ and a seminorm $\|\cdot\|_{S^m_{\rho,\delta},a,b,c}$ such that*

$$\int_G |z|^\gamma |\kappa_x(z)| dz \leq C \|\sigma\|_{S^m_{\rho,\delta},a,b,c}.$$

We may replace $\|\cdot\|_{S^m_{\rho,\delta},a,b,c}$ with $\|\cdot\|_{S^{m,R}_{\rho,\delta},a,b}$

Proof of Lemma 5.5.9. By Part 2 of Corollary 5.4.3, $z \mapsto |\kappa_x(z)|$ is a continuous bounded function if $m - \rho\gamma < -Q$ hence the integral $\int_{|z|<1} |z|^\gamma |\kappa_x(z)| dz$ is finite. By the Cauchy-Schwartz inequality, we have

$$\begin{aligned} \int_{|z|>1} |z|^\gamma |\kappa_x(z)| dz &\leq \sqrt{\int_{|z|>1} |z|^{-Q-\frac{1}{2}}} \sqrt{\int_{|z|>1} |z|^{2\gamma+Q+\frac{1}{2}} |\kappa_x(z)|^2 dz} \\ &\lesssim \sum_{[\alpha]=M} \|\tilde{q}_\alpha \kappa_x\|_{L^2(G)}, \end{aligned}$$

where $M/2 \in \mathbb{N}$ is the smallest integer divisible by $\upsilon_1, \ldots, \upsilon_n$ satisfying $M \geq 2\gamma + Q + \frac{1}{2}$, having chosen (3.21) with $p = M$ for quasi-norm. By Part 1 of Corollary 5.4.3, the sum above is finite when $m - \rho M < -Q/2$, which holds true. □

Using Lemma 5.5.9 instead of Lemma 5.5.6 in the proof of Lemma 5.5.10 produces the following result.

Lemma 5.5.10. *We fix a positive Rockland operator of homogeneous degree ν. Let $m_1 \in \mathbb{R}$, $1 \geq \rho \geq \delta \leq 0$ with $\delta \neq 1$, $\beta_0 \in \mathbb{N}_0^n$, and $M, M_1 \in \mathbb{N}_0$. We assume that*

$$\begin{cases} m_1 + \nu M_1 < -Q \\ -\nu M_1 + m_2 + \delta(c_{\beta_0} + \upsilon_n + \max(\nu M_1, M)) \leq 0 \end{cases},$$

where

$$c_{\beta_0} := \max_{\substack{[\beta_{02}] \leq [\beta_0] \\ [\beta'] \geq [\beta_{02}],\ |\beta'| \geq |\beta_{02}|}} [\beta'].$$

Then there exist a constant $C > 0$, and two seminorms $\|\cdot\|_{S^{m_1,R}_{\rho,\delta}, a_1, b_1}$, $\|\cdot\|_{S^{m_2}_{\rho,\delta},0,b_2,0}$, such that for any $\sigma_1, \sigma_2 \in S^{-\infty}$ and any $(x,\pi) \in G \times \widehat{G}$ we have

$$\begin{aligned} &\|X_x^{\beta_0}(\sigma_1 \circ \sigma_2(x,\pi) - \sum_{[\alpha] \leq M} \Delta^\alpha \sigma_1(x,\pi)\ X_x^\alpha \sigma_2(x,\pi))\|_{\mathscr{L}(\mathcal{H}_\pi)} \\ &\leq C \|\sigma_1\|_{S^{m_1,R}_{\rho,\delta},a_1,b_1} \|\sigma_2\|_{S^{m_2}_{\rho,\delta},0,b_2,0}. \end{aligned}$$

The details of the proof of Lemma 5.5.10 are left to the reader. The first inequality in the statement just above shows that we will require the ability to choose m_1 as negative as one wants. We can do this thanks to the following remark:

Lemma 5.5.11. *Let $\sigma_1, \sigma_2 \in S^{-\infty}$. For any $X \in \mathfrak{g}$ and any $\sigma_1, \sigma_2 \in S^{-\infty}$, we have*

$$(\sigma_1 \pi(X)) \circ \sigma_2 = \sigma_1 \circ (X_x \sigma_2) + \sigma_1 \circ (\pi(X)\sigma_2).$$

More generally, for any $\beta \in \mathbb{N}_0^n$, we have

$$\{\sigma_1 \pi(X)^\beta\} \circ \sigma_2 = \overline{\sum_{[\beta_1]+[\beta_2]=[\beta]}} \sigma_1 \circ \{\pi(X)^{\beta_1} X_x^{\beta_2} \sigma_2\},$$

where $\overline{\sum}$ denotes a linear combination independent of σ_1, σ_2.

Note that in the expression above, $\pi(X)^{\beta_1}$ and $X_x^{\beta_2}$ commute.

Proof of Lemma 5.5.7. We keep the notation of Lemma 5.5.4. Using integration by parts and the Leibniz formula, we obtain

$$\begin{aligned} (\sigma_1 \pi(X)) \circ \sigma_2\ (x,\pi) &= \int_G \tilde{X}_{z_1=z} \kappa_{1,x}(z_1) \pi(z)^* \sigma_2(xz^{-1},\pi)\, dz \\ &= -\int_G \kappa_{1,x}(z) \left(\tilde{X}_{z_1=z} \pi(z_1)^* \sigma_2(xz^{-1},\pi) + \pi(z)^* \tilde{X}_{z_2=z} \sigma_2(xz_2^{-1},\pi) \right) dz \\ &= \int_G \kappa_{1,x}(z) \left(\pi(z)^* \pi(X) \sigma_2(xz^{-1},\pi) + \pi(z)^* X_{x_2=xz^{-1}} \sigma_2(x_2,\pi) \right) dz. \end{aligned}$$

This shows the first formula. The next formula is obtained recursively. □

We can now sketch the proof of Theorem 5.5.3 in the case $\rho = \delta$.

Sketch of the proof of Theorem 5.5.3 with $\rho = \delta$. We assume $\rho = \delta \in [0, 1)$. Writing $\sigma_1 = \sigma_1 \pi(\mathrm{I} + \mathcal{R})^{-N} \pi(\mathrm{I} + \mathcal{R})^N$ and using Lemma 5.5.11, it suffices to prove (5.52) for m_1 as negative as one wants. We proceed as in the proof of the case $\rho > \delta$ replacing Lemma 5.5.5 with Lemma 5.5.10. The details are left to the reader. □

5.5.3 Adjoint of a pseudo-differential operator

Here we prove that the classes $\Psi^m_{\rho,\delta}$ are stable under taking the formal adjoints of operators.

Theorem 5.5.12. *We assume $1 \geq \rho \geq \delta \geq 0$ with $\delta \neq 1$ and $m \in \mathbb{R}$. If $T \in \Psi^m_{\rho,\delta}$ then its formal adjoint T^* is also in $\Psi^m_{\rho,\delta}$. Moreover, the mapping $T \mapsto T^*$ is continuous on $\Psi^m_{\rho,\delta}$.*

Recall that the formal adjoint of an operator $T : \mathcal{S}(G) \to \mathcal{S}'(G)$ is the operator $T^* : \mathcal{S}(G) \to \mathcal{S}'(G)$ defined by

$$\forall \phi, \psi \in \mathcal{S}(G) \qquad \int_G T\phi(x)\, \overline{\psi(x)}\, dx = \int_G \phi(x)\, \overline{T^*\psi(x)}\, dx.$$

We observe that the operator $T = \mathrm{Op}(\sigma) \in \Psi^m_{\rho,\delta}$ maps $\mathcal{S}(G)$ to itself continuously (see Theorem 5.2.15) and therefore has a formal adjoint T^*.

Before beginning the proof of Theorem 5.5.12, let us point out some of its consequences.

Corollary 5.5.13. 1. *We assume $1 \geq \rho \geq \delta \geq 0$ with $\delta \neq 1$, and $m \in \mathbb{R}$.*

Any $T \in \Psi^m_{\rho,\delta}$ extends uniquely to a continuous operator on $\mathcal{S}'(G)$. Furthermore the mapping $T \mapsto T$ from $\Psi^m_{\rho,\delta}$ to the space $\mathscr{L}(\mathcal{S}'(G))$ of continuous operators on $\mathcal{S}'(G)$ is linear and continuous.

2. *Any smoothing operator $T \in \Psi^{-\infty}$ maps continuously the space $\mathcal{E}'(G)$ of compactly supported distributions to the Schwartz space $\mathcal{S}(G)$. Furthermore the mapping $T \mapsto T$ from $\Psi^{-\infty}$ to the space $\mathscr{L}(\mathcal{E}'(G), \mathcal{S}(G))$ of continuous mappings from $\mathcal{E}'(G)$ to $\mathcal{S}(G)$ is linear and continuous.*

Proof of Corollary 5.5.13. We admit Theorem 5.5.12 (whose proof is given below). The statement then follows by classical arguments of duality and Theorem 5.2.15 for Part 1, and Part 2 of Theorem 5.4.9 for Part 2. □

Let us start the proof of Theorem 5.5.12 by observing that the symbol $\sigma^{(*)}$ of the adjoint T^* of $T = \mathrm{Op}(\sigma)$ is necessarily known and unique at least formally or under favourable conditions such as in the case of a smoothing operator:

Lemma 5.5.14. *Let $\sigma \in S^{-\infty}$ and let $\kappa : (x, y) \mapsto \kappa_x(y)$ be its associated kernel. We set*

$$\kappa_x^{(*)}(y) := \bar{\kappa}_{xy^{-1}}(y^{-1}), \quad x, y \in G.$$

Then $\kappa^{(*)} : (x, y) \mapsto \kappa_x^{(*)}(y)$ is smooth on $G \times G$ and for every $\alpha \in \mathbb{N}_0^n$, $x \mapsto X^\alpha \kappa_x^{(*)}$ is continuous from G to $\mathcal{S}(G)$.

The symbol $\sigma^{(*)}$ defined via

$$\sigma^{(*)}(x, \pi) := \mathcal{F}_G(\kappa_x^{(*)})(\pi), \quad (x, \pi) \in G \times \widehat{G},$$

is a smooth symbol in the sense of Definition 5.1.34 and satisfies

$$(\mathrm{Op}(\sigma))^* = \mathrm{Op}(\sigma^{(*)}).$$

In particular, if σ does not depend on x, then $\sigma^{(*)} = \sigma^*$.

Note that this operation is an involution since

$$\kappa_x(y) = \bar{\kappa}^{(*)}_{xy^{-1}}(y^{-1}).$$

Recall that if $\sigma = \{\sigma(x, \pi), (x, \pi) \in G \times \widehat{G}\}$ then we have defined the adjoint symbol

$$\sigma^* = \{\sigma(x, \pi)^*, (x, \pi) \in G \times \widehat{G}\},$$

(see Theorem 5.2.22). Hence we may write

$$\sigma^*(x, \pi) := \sigma(x, \pi)^*.$$

Proof of Lemma 5.5.14. By Corollary 3.1.30, we have

$$\begin{aligned}
X_x^{\beta_o}\{\kappa_x^{(*)}(y)\} &= X_x^{\beta_o}\{\bar{\kappa}_{xy^{-1}}(y^{-1})\} = (-1)^{|\beta_o|}\tilde{X}^{\beta_o}_{y_1=y^{-1}}\{\bar{\kappa}_{xy_1}(y^{-1})\} \\
&= (-1)^{|\beta_o|} \sum_{|\beta|\leq|\beta_o|,\, [\beta]\geq[\beta_o]} Q_{\beta_o,\beta}(y^{-1}) X^{\beta}_{y_1=y^{-1}}\{\bar{\kappa}_{xy_1}(y^{-1})\} \\
&= (-1)^{|\beta_o|} \sum_{|\beta|\leq|\beta_o|,\, [\beta]\geq[\beta_o]} Q_{\beta_o,\beta}(y^{-1}) X^{\beta}_{x_1=xy^{-1}}\{\bar{\kappa}_{x_1}(y^{-1})\},
\end{aligned}$$

where the $Q_{\beta_o,\beta}$'s are $([\beta_o] - [\beta])$-homogeneous polynomials. The regularity of κ described in Theorem 5.4.9 implies that $\kappa^{(*)} : (x, y) \mapsto \kappa_x^{(*)}(y)$ is smooth in x and y (but maybe not compactly supported in x), and it is also Schwartz in y in such a way that all the mappings $G \ni x \mapsto X_x^\alpha \kappa_x^{(*)} \in \mathcal{S}(G)$ are continuous. Clearly $\sigma^{(*)}(x, \pi) = \pi(\kappa_x^{(*)})$ defines a smooth symbol $\sigma^{(*)}$.

Let $\phi, \psi \in \mathcal{S}(G)$ and let $x \in G$. The regularity of κ described in Theorem 5.4.9 justifies easily the following computations:

$$\begin{aligned}
\int_G (\mathrm{Op}(\sigma)\phi)(x)\overline{\psi(x)}dx &= \int_G \phi * \kappa_x(x)\bar{\psi}(x)dx = \int_G\int_G \phi(z)\kappa_x(z^{-1}x)\bar{\psi}(x)dzdx \\
&= \int_G\int_G \phi(z)\bar{\kappa}^{(*)}_{x(z^{-1}x)^{-1}}((z^{-1}x)^{-1})\bar{\psi}(x)dzdx \\
&= \int_G\int_G \phi(z)\overline{\kappa_z^{(*)}(x^{-1}z)\psi(x)}dzdx \\
&= \int_G \phi(z)\overline{\psi * \kappa_z^{(*)}}(z)dz.
\end{aligned}$$

This shows that $\mathrm{Op}(\sigma)^*\psi(z) = \psi * \kappa_z^{(*)}(z)$. □

In general, $\sigma^{(*)}$ is not the adjoint σ^* of the symbol σ, unless for instance it does not depend on $x \in G$. However, we can perform formal considerations to link $\sigma^{(*)}$ with σ^* in the following way: using the Taylor expansion for κ_x^* in x (see equality (5.27)), we obtain

$$\kappa_x^{(*)}(y) = \kappa^*_{xy^{-1}}(y) \approx \sum_\alpha q_\alpha(y^{-1})X_x^\alpha \kappa_x^*(y) = \sum_\alpha \tilde{q}_\alpha(y)X_x^\alpha\kappa_x^*(y).$$

Thus, taking the group Fourier transform at $\pi \in \widehat{G}$, we get

$$\sigma^{(*)}(x,\pi) = \pi(\kappa_x^{(*)}) \approx \sum_\alpha \pi(\tilde{q}_\alpha(y)X_x^\alpha\kappa_x^*(y)) = \sum_\alpha \Delta^\alpha X_x^\alpha \sigma(x,\pi)^*.$$

From Theorem 5.2.22 we know that if $\sigma \in S^m_{\rho,\delta}$ then

$$\Delta^\alpha X_x^\alpha \sigma(x,\pi)^* \in S^{m-(\rho-\delta)[\alpha]}_{\rho,\delta}. \tag{5.58}$$

From these formal computations we see that the main problem is to estimate the remainder coming from the use of the Taylor expansion. This is the purpose of the following technical lemma.

Lemma 5.5.15. *We fix a positive Rockland operator of homogeneous degree ν. Let $m \in \mathbb{R}$, $1 \geq \rho \geq \delta \geq 0$ with $\rho \neq 0$ and $\delta \neq 1$, $\beta_0 \in \mathbb{N}_0^n$, and $M, M_1 \in \mathbb{N}_0$. We assume that $M \geq \nu M_1$ and $(\rho-\delta)M + \rho Q > m + \delta[\beta_0] + \nu M_1 + Q$. Then there exist a constant $C > 0$, and a seminorm $\|\cdot\|_{S^m_{\rho,\delta},a,b,0}$, such that for any $\sigma \in S^{-\infty}$ and any $(x,\pi) \in G \times \widehat{G}$ we have*

$$\|X_x^{\beta_0}(\sigma^{(*)}(x,\pi) - \sum_{[\alpha]\leq M} \Delta^\alpha X_x^\alpha \sigma^*(x,\pi))\pi(\mathrm{I}+\mathcal{R})^{M_1}\|_{\mathscr{L}(\mathcal{H}_\pi)} \leq C\,\|\sigma\|_{S^m_{\rho,\delta},a,b,0}.$$

Proof of Lemma 5.5.15, case $\beta_0 = 0$. By Lemma 5.5.14 and the observations that follow, we have

$$\begin{aligned}
&\sigma^{(*)}(x,\pi) - \sum_{[\alpha]\leq M} \Delta^\alpha X_x^\alpha \sigma^*(x,\pi) \\
&\qquad = \int_G \left(\kappa^*_{xz^{-1}}(z) - \sum_{[\alpha]\leq M} q_\alpha(z^{-1}) X_x^\alpha \kappa_x^*(z) \right) \pi(z)^* dz \\
&\qquad = \int_G R_{x,M}^{\kappa_x^*(z)}(z^{-1})\pi(z)^* dz,
\end{aligned}$$

where $R_{x,M}^{\kappa_x^*(z)}$ denotes the remainder of the (vector-valued) Taylor expansion of $v \mapsto \kappa^*_{xv}(z)$ of order M at 0. Using (5.48), we can integrate by parts to obtain

$$\begin{aligned}
&\big(\sigma^{(*)}(x,\pi) - \sum_{[\alpha]\leq M} \Delta^\alpha X_x^\alpha \sigma^*(x,\pi)\big)\pi(\mathrm{I}+\mathcal{R})^{M_1} \\
&\qquad = \overline{\sum_{[\beta_1]+[\beta_2]\leq \nu M_1}} \int_G \tilde{X}^{\beta_1}_{z_1=z} R_{x,M}^{\tilde{X}^{\beta_2}_{z_2=z}\kappa_x^*(z_2)}(z_1^{-1})\pi(z)^* dz \\
&\qquad = \overline{\sum_{[\beta_1]+[\beta_2]\leq \nu M_1}} \int_G R_{x_1=x,M-[\beta_1]}^{\tilde{X}^{\beta_2}_{z_2=z}X^{\beta_1}_{x_1}\kappa_{x_1}^*(z_2)}(z^{-1})\pi(z)^* dz.
\end{aligned}$$

Taking the operator norm, we have

$$\begin{aligned}
&\|\big(\sigma^{(*)}(x,\pi) - \sum_{[\alpha]\leq M} \Delta^\alpha X_x^\alpha \sigma^*(x,\pi)\big)\pi(\mathrm{I}+\mathcal{R})^{M_1}\|_{\mathscr{L}(\mathcal{H}_\pi)} \\
&\qquad \lesssim \sum_{[\beta_1]+[\beta_2]\leq \nu M_1} \int_G |R_{x_1=x,M-[\beta_1]}^{\tilde{X}^{\beta_2}_{z_2=z}X^{\beta_1}_{x_1}\kappa_{x_1}^*(z_2)}(z^{-1})| dz.
\end{aligned}$$

For $|z| < 1$, we will use Taylor's theorem, see Theorem 3.1.51:

$$|R_{x_1=x,M-[\beta_1]}^{\tilde{X}^{\beta_2}_{z_2=z}X^{\beta_1}_{x_1}\kappa_{x_1}^*(z_2)}(z^{-1})| \lesssim \sum_{\substack{|\gamma|\leq \lceil (M-[\beta_1])_+\rfloor+1 \\ [\gamma]>(M-[\beta_1])_+}} |z|^{[\gamma]} \sup_{x_1\in G} |X_z^\gamma \tilde{X}^{\beta_2}_{z_2=z} X^{\beta_1}_{x_1}\kappa^*_{x_1}(z_2)|,$$

together with the estimate for z near the origin given in Theorem 5.4.1. The link between left and right derivatives, see (1.11), implies

$$\sup_{x_1\in G} |X_z^\gamma \tilde{X}^{\beta_2}_{z_2=z} X^{\beta_1}_{x_1}\kappa^*_{x_1}(z_2)| = \sup_{x_1\in G} |X_z^\gamma X^{\beta_2}_{z_2=z} X^{\beta_1}_{x_1}\kappa_{x_1}(z_2)|.$$

Proceeding as in the proof of Lemma 5.5.6, we obtain that the integral

$$\begin{aligned}
&\int_{|z|<1} |R_{x,M-[\beta_1]}^{\tilde{X}^{\beta_2}_{z_2=z}X^{\beta_1}_{x}\kappa_x^*(z_2)}(z^{-1})| dz \\
&\qquad \lesssim \sum_{\substack{|\gamma|\leq \lceil (M-[\beta_1])_+\rfloor+1 \\ [\gamma]>(M-[\beta_1])_+}} \int_{|z|<1} |z|^{[\gamma]} \sup_{x_1\in G} |X_{x_1}^\gamma X^{\beta_2}_{z_2=z} X^{\beta_1}_{x_1}\kappa_{x_1}(z_2)| dz
\end{aligned}$$

is finite whenever $[\gamma] + Q > (m + [\beta_2] + \delta([\gamma] + [\beta_1]) + Q)/\rho$ with the indices as above. These conditions are implied by the hypotheses of the statement. The estimates for z large given in Theorem 5.4.1 show directly that the integral

$$\int_{|z|>1} |R^{\tilde{X}^{\beta_2}_{z_2=z} X^{\beta_1}_{x_1} \kappa^*_{x_1}(z_2)}_{x_1=x, M-[\beta_1]}(z^{-1})| dz,$$

is finite. Collecting the various estimates yields the statement in the case $\rho \neq 0$ and $\beta_0 = 0$. □

Proof of Lemma 5.5.15, general case. We proceed as above and introduce the derivatives with respect to x. We obtain

$$X^{\beta_0}_x \big(\sigma^{(*)}(x,\pi) - \sum_{[\alpha]\leq M} \Delta^\alpha X^\alpha_x \sigma^*(x,\pi)\big) = \int_G R^{X^{\beta_0}_x \kappa^*_x \cdot(z)}_{0,M}(z^{-1})\pi(z)^* dz.$$

And adding $(\mathrm{I}+\mathcal{R})^{M_1}$, we have

$$\begin{aligned} &X^{\beta_0}_x \big(\sigma^{(*)}(x,\pi) - \sum_{[\alpha]\leq M} \Delta^\alpha X^\alpha_x \sigma^*(x,\pi)\big)(\mathrm{I}+\mathcal{R})^{M_1} \\ &\quad = \sum_{[\beta_1]+[\beta_2]\leq \nu M_1} \int_G R^{\tilde{X}^{\beta_2}_{z_2=z} X^{\beta_1}_{x_1} X^{\beta_0}_x \kappa^*_{xx_1}(z_2)}_{x_1=0, M-[\beta_1]}(z^{-1})\pi(z)^* dz. \end{aligned}$$

Taking the operator norm, we have

$$\begin{aligned} &\|X^{\beta_0}_x \big(\sigma^{(*)}(x,\pi) - \sum_{[\alpha]\leq M} \Delta^\alpha X^\alpha_x \sigma^*(x,\pi)\big)\pi(\mathrm{I}+\mathcal{R})^{M_1}\|_{\mathscr{L}(\mathcal{H}_\pi)} \\ &\quad \lesssim \sum_{[\beta_1]+[\beta_2]\leq \nu M_1} \int_G |R^{\tilde{X}^{\beta_2}_{z_2=z} X^{\beta_1}_{x_1} X^{\beta_0}_x \kappa^*_{xx_1}(z_2)}_{x_1=0, M-[\beta_1]}(z^{-1})| dz. \end{aligned}$$

For $|z| < 1$, we use the more precise version of Taylor's theorem than in the case $\beta_0 = 0$:

$$\begin{aligned} &|R^{\tilde{X}^{\beta_2}_{z_2=z} X^{\beta_1}_{x_1} X^{\beta_0}_x \kappa^*_{xx_1}(z_2)}_{x, M-[\beta_1]}(z^{-1})| \\ &\quad \lesssim \sum_{\substack{|\gamma|\leq \lceil (M-[\beta_1])_+ \rfloor + 1 \\ [\gamma] > (M-[\beta_1])_+}} |z|^{[\gamma]} \sup_{|y|\leq \eta^{\lceil (M-[\beta_1])_+ \rfloor+1}|z|} |X^\gamma_y \tilde{X}^{\beta_2}_{z_2=z} X^{\beta_1}_y X^{\beta_0}_x \kappa^*_{xy}(z_2)|. \end{aligned}$$

We proceed as in the proof of Lemma 5.5.5, that is, we use (5.51) to obtain

$$\begin{aligned} &\sup_{|y|\leq \eta^{\lceil (M-[\beta_1])_+ \rfloor+1}|z|} |X^\gamma_y \tilde{X}^{\beta_2}_{z_2=z} X^{\beta_1}_y X^{\beta_0}_x \kappa^*_{xy}(z_2)| \\ &\quad \lesssim \sum_{\substack{[\beta_0']\geq[\beta_0] \\ |\beta_0'|\leq|\beta_0|}} |z|^{[\beta_0']-[\beta_0]} \sup_{\substack{x_1\in G \\ [\gamma_0]=[\gamma]+[\beta_0']}} |X^{\gamma_0}_{x_1} \tilde{X}^{\beta_2}_{z_2=z} \kappa^*_{x_1}(z_2)|. \end{aligned}$$

We conclude by adapting the case $\beta_0 = 0$. □

To take into account the difference operator, we will use the following observation.

Lemma 5.5.16. *For any $\alpha \in \mathbb{N}_0^n$ and $\sigma \in S^{-\infty}$, $\Delta^\alpha \sigma^{(*)}$ can be written as a linear combination (independent of σ) of $\{\Delta^{\alpha'}\sigma\}^{(*)}$ over $\alpha' \in \mathbb{N}_0^n$, $[\alpha'] = [\alpha]$. This is the same linear combination as when writing $\Delta^\alpha \sigma^*$ as a linear combination of $\{\Delta^{\alpha'}\sigma\}^*$.*

Proof of Lemma 5.5.16. For $\sigma \in S^{-\infty}$, let κ_σ be the kernel associated with the symbol σ and similarly for any other symbol.

Let us prove Part 1. We have

$$\{\tilde{q}_\alpha \kappa_{\sigma^{(*)},x}\}(y) = \tilde{q}_\alpha(y)\bar{\kappa}_{\sigma,xy^{-1}}(y^{-1}).$$

As $\bar{q}_\alpha$ is a $[\alpha]$-homogeneous polynomial, by Proposition 5.2.3, $\overline{\tilde{q}_\alpha}$ is a linear combination of $\tilde{q}_{\alpha'}$ over multi-indices $\alpha' \in \mathbb{N}_0^n$ satisfying $[\alpha'] = [\alpha]$. Hence

$$\{\tilde{q}_\alpha \kappa_{\sigma^{(*)},x}\}(y) = \overline{\sum_{[\alpha']=[\alpha]}} \overline{\tilde{q}_{\alpha'} \kappa_{\sigma,xy^{-1}}}(y^{-1}) = \overline{\sum_{[\alpha']=[\alpha]}} \{\tilde{q}_{\alpha'}\kappa_\sigma\}^{(*)}(y),$$

where $\overline{\sum}$ means taking a linear combination. Taking the Fourier transform, we obtain

$$\mathcal{F}_G\{\tilde{q}_\alpha \kappa_{\sigma^{(*)},x}\}(\pi) = \Delta^\alpha \sigma^{(*)}(x,\pi) = \overline{\sum_{[\alpha']=[\alpha]}} \{\Delta^{\alpha'}\sigma\}^{(*)}.$$

□

We can now prove Theorem 5.5.12 in the case $\rho > \delta$.

Proof of Theorem 5.5.12 with $\rho > \delta$. We assume $\rho > \delta$. We fix a positive Rockland operator of homogeneous degree ν. Let us show that for any $\alpha_0, \beta_0 \in \mathbb{N}_0^n$, and $M_0 \in \mathbb{N}$, there exists $M \geq M_0$, a constant $C > 0$ and a seminorm $\|\cdot\|_{S^m_{\rho,\delta},a_1,b_1,0}$, such that for any $\sigma \in S^{-\infty}$ we have

$$\begin{aligned}\|X_x^{\beta_0}\Delta^{\alpha_0}\tau_M(x,\pi)\ \pi(\mathrm{I}+\mathcal{R})^{-\frac{m-(\rho-\delta)M_0-\rho[\alpha_0]+\delta[\beta_0]}{\nu}}&\|_{\mathscr{L}(\mathcal{H}_\pi)} \\ &\leq C\|\sigma\|_{S^m_{\rho,\delta},a_1,b_1,0},\end{aligned} \tag{5.59}$$

where we have denoted $\tau_M := \sigma^{(*)} - \sum_{[\alpha]\leq M} \Delta^\alpha X_x^\alpha \sigma^*$. By Lemma 5.5.16, it suffices to show (5.59) only for $\alpha_0 = 0$.

Let $\beta_0 \in \mathbb{N}_0$ and $M_0 \in \mathbb{N}$. Let $M_1 \in \mathbb{N}_0$ be the smallest non-negative integer such that

$$-\frac{m-(\rho-\delta)M_0+\delta[\beta_0]}{\nu} \leq M_1.$$

We choose $M \geq \max(M_0, \nu M_1)$ such that $(\rho-\delta)M + \rho Q > m + \delta[\beta_0] + \nu M_1 + Q$. This is possible as $\rho > \delta$. Then (5.59) follows from the application of Lemma 5.5.15 to M, M_1 and the symbol σ.

Using (5.58), classical considerations imply that (5.59) yields that for any $M_0 \in \mathbb{N}_0$, and any seminorm $\|\cdot\|_{S^{m-M_0(\rho-\delta),R}_{\rho,\delta},a,b}$, there exist a constant $C>0$ and a seminorm $\|\cdot\|_{S^m_{\rho,\delta},a_1,b_1,0}$, such that for any $\sigma_1,\sigma_2 \in S^{-\infty}$ we have

$$\|\tau_{M_0}\|_{S^{m-M_0(\rho-\delta),R}_{\rho,\delta},a,b} \leq C\|\sigma\|_{S^m_{\rho,\delta},a_1,b_1,0}.$$

We can then conclude as in the proof of Theorem 5.5.3 in the case $\rho>\delta$. □

In fact, we have obtained a much more precise result:

Corollary 5.5.17. *We assume $1\geq\rho>\delta\geq 0$. If $\sigma\in S^m_{\rho,\delta}$, then there exists a unique symbol $\sigma^{(*)}$ in $S^m_{\rho,\delta}$ such that*

$$(\mathrm{Op}(\sigma))^* = \mathrm{Op}(\sigma^{(*)}).$$

Furthermore, for any $M\in\mathbb{N}_0$,

$$\{\sigma^{(*)}(x,\pi) - \sum_{[\alpha]\leq M} X_x^\alpha \Delta^\alpha \sigma^*(x,\pi)\} \in S^{m-(\rho-\delta)M}_{\rho,\delta}.$$

Moreover, the mapping

$$\left\{\begin{array}{ccc} S^m_{\rho,\delta} & \longrightarrow & S^{m-(\rho-\delta)M}_{\rho,\delta} \\ \sigma & \longmapsto & \{\sigma^{(*)}(x,\pi) - \sum_{[\alpha]\leq M} X_x^\alpha \Delta^\alpha \sigma^*(x,\pi)\} \end{array}\right. ,$$

is continuous.

Consequently, we can also write

$$\sigma^{(*)} \sim \sum_{j=0}^{\infty}\left(\sum_{[\alpha]=j} X_x^\alpha\Delta^\alpha\sigma^*\right), \tag{5.60}$$

where the asymptotic was defined in Definition 5.5.2.

As for composition, in the case $\rho=\delta$, the asymptotic formula does not bring any improvement and, in this sense, is not interesting. The proof of this case is more delicate to prove but relies on the same kind of arguments as above. Using Lemma 5.5.9 instead of Lemma 5.5.6 in the proof of Lemma 5.5.15 produces the following result:

Lemma 5.5.18. *We fix a positive Rockland operator of homogeneous degree ν. Let $m\in\mathbb{R}$, $1\leq\rho\leq\delta\leq 0$ with $\delta\neq 1$, $\beta_0\in\mathbb{N}_0^n$, and $M,M_1\in\mathbb{N}_0$. We assume that*

$$M\geq \nu M_1 \quad \text{and} \quad m+\delta(M+c_{\beta_0})+\nu M_1 < -Q,$$

where

$$c_{\beta_0} := \max_{\substack{[\beta_0']\leq[\beta_0] \\ [\beta']\geq[\beta_0'],\ |\beta'|\geq|\beta_0'|}} [\beta'].$$

Then there exist a constant $C > 0$, and a seminorm $\|\cdot\|_{S^{m,R}_{\rho,\delta},a,b}$, such that for any $\sigma \in S^{-\infty}$ and any $(x,\pi) \in G \times \widehat{G}$ we have

$$\|X_x^{\beta_0}\big(\sigma^{(*)}(x,\pi) - \sum_{[\alpha]\leq M} \Delta^\alpha X_x^\alpha \sigma^*(x,\pi)\big)\pi(\mathrm{I}+\mathcal{R})^{M_1}\|_{\mathscr{L}(\mathcal{H}_\pi)} \leq C\|\sigma\|_{S^{m,R}_{\rho,\delta},a,b}.$$

The details of the proof of Lemma 5.5.18 are left to the reader. The conditions in the statement just above show that we will require the ability to choose m as negative as one wants. We can do this thanks to the following remark.

Lemma 5.5.19. *For any $\sigma \in S^{-\infty}$ and any $X \in \mathfrak{g}$, we have*

$$\{\pi(X)\sigma\}^{(*)} = -\sigma^{(*)}(x,\pi)\ \pi(X) - \{X_x\sigma\}^{(*)}(x,\pi).$$

More generally, for any $\beta \in \mathbb{N}_0^n$, we have

$$\{\pi(X)^\beta\sigma\}^{(*)} = \overline{\sum_{[\beta_1]+[\beta_2]=[\beta]}} \{X_x^{\beta_1}\sigma\}^{(*)}\pi(X)^{\beta_2},$$

where $\overline{\sum}$ denotes a linear combination independent of σ_1, σ_2.

Proof of Lemma 5.5.19. We keep the notation of Lemma 5.5.14. The kernel of $\sigma^{(*)}\pi(X)$ is given via

$$\tilde{X}_y\kappa_x^{(*)}(y) = \tilde{X}_y\{\bar{\kappa}_{xy^{-1}}(y^{-1})\} = -X_{x_1=xy^{-1}}\bar{\kappa}_{x_1}(y^{-1}) - X_{y_2=y^{-1}}\bar{\kappa}_{xy^{-1}}(y_2),$$

having used (5.50) and the Leibniz property for vector fields. Hence we recognise:

$$\tilde{X}_y\kappa_x^{(*)}(y) = -(X_x\kappa_x)^{(*)}(y) - (X\kappa_x)^{(*)}(y),$$

and

$$\sigma^{(*)}\pi(X) = -(X_x\sigma)^{(*)} - (\pi(X)\sigma)^{(*)}.$$

This shows the first formula. The second formula is obtained recursively. □

We can now show sketch the proof of Theorem 5.5.3 in the case $\rho = \delta$.

Sketch of the proof of Theorem 5.5.3 with $\rho = \delta$. We assume $\rho = \delta \in [0,1)$. Writing $\sigma = \pi(\mathrm{I}+\mathcal{R})^N\pi(\mathrm{I}+\mathcal{R})^{-N}\sigma$ and using Lemma 5.5.19, it suffices to prove (5.59) for m as negative as one wants. We proceed as in the proof of the case $\rho > \delta$ replacing Lemma 5.5.15 with Lemma 5.5.18. The details are left to the reader. □

5.5.4 Simplification of the definition of $S^m_{\rho,\delta}$

In this section, we show that it is possible to choose $\gamma = 0$ in the definition of symbols as it was pointed out in Remark 5.2.13 Part (3). This simplifies the definition of the symbol classes $S^m_{\rho,\delta}$ given in Definition 5.2.11. We will also show a pivotal argument in the proof of Theorems 5.5.3 and 5.5.12, namely Inequalities (5.54).

Theorem 5.5.20. *Let $m, \rho, \delta \in \mathbb{R}$ with $1 \geq \rho \geq \delta \geq 0$ and $\delta \neq 1$.*

(L) *A symbol $\sigma = \{\sigma(x,\pi), (x,\pi) \in G \times \widehat{G}\}$ is in $S^m_{\rho,\delta}$ if and only if for each $\alpha, \beta \in \mathbb{N}_0^n$, the field of operators*

$$X_x^\beta \Delta^\alpha \sigma = \{X_x^\beta \Delta^\alpha \sigma(x,\pi) : \mathcal{H}_\pi^\infty \to \mathcal{H}_\pi, (x,\pi) \in G \times \widehat{G}\}$$

is in $L^\infty_{0,\rho[\alpha]-m-\delta[\beta]}(\widehat{G})$ uniformly in $x \in G$, that is,

$$\sup_{x\in G} \|X_x^\beta \Delta^\alpha \sigma(x,\cdot)\|_{L^\infty_{0,\rho[\alpha]-m-\delta[\beta]}(\widehat{G})} < \infty. \tag{5.61}$$

Furthermore, the family of seminorms

$$\sigma \longmapsto \|\sigma\|_{S^m_{\rho,\delta},a,b,0} = \sup_{\substack{[\alpha]\leq a\\ [\beta]\leq b}} \sup_{x\in G} \|X_x^\beta \Delta^\alpha \sigma(x,\cdot)\|_{L^\infty_{0,\rho[\alpha]-m-\delta[\beta]}(\widehat{G})}, \quad a,b \in \mathbb{N}_0,$$

yields the topology of $S^m_{\rho,\delta}$.

(R) *A symbol $\sigma = \{\sigma(x,\pi), (x,\pi) \in G \times \widehat{G}\}$ is in $S^m_{\rho,\delta}$ if and only if for each $\alpha, \beta \in \mathbb{N}_0^n$, the field of operators*

$$X_x^\beta \Delta^\alpha \sigma = \{X_x^\beta \Delta^\alpha \sigma(x,\pi) : \mathcal{H}_\pi^\infty \to \mathcal{H}_\pi, (x,\pi) \in G \times \widehat{G}\}$$

is in $L^\infty_{m+\delta[\beta]-\rho[\alpha],0}(\widehat{G})$ uniformly in $x \in G$, that is,

$$\sup_{x\in G} \|X_x^\beta \Delta^\alpha \sigma(x,\cdot)\|_{L^\infty_{m+\delta[\beta]-\rho[\alpha],0}(\widehat{G})} < \infty. \tag{5.62}$$

Furthermore, the family of seminorms

$$\sigma \longmapsto \|\sigma\|_{S^{m,R}_{\rho,\delta},a,b} = \sup_{\substack{[\alpha]\leq a\\ [\beta]\leq b}} \sup_{x\in G} \|X_x^\beta \Delta^\alpha \sigma(x,\cdot)\|_{L^\infty_{m+\delta[\beta]-\rho[\alpha],0}(\widehat{G})}, \quad a,b \in \mathbb{N}_0,$$

yields the topology of $S^m_{\rho,\delta}$.

In other words,

(R) a symbol $\sigma = \{\sigma(x,\pi), (x,\pi) \in G \times \widehat{G}\}$ is in $S^m_{\rho,\delta}$ if and only if for each $\alpha, \beta \in \mathbb{N}_0^n$, the field of operators

$$X_x^\beta \Delta^\alpha \sigma = \{X_x^\beta \Delta^\alpha \sigma(x,\pi) : \mathcal{H}_\pi^\infty \to \mathcal{H}_\pi, (x,\pi) \in G \times \widehat{G}\}$$

is defined on smooth vectors and satisfy

$$\sup_{x\in G, \pi\in\widehat{G}} \|X_x^\beta \Delta^\alpha \sigma(x,\cdot)\pi(\mathrm{I}+\mathcal{R})^{\frac{\rho[\alpha]-m-\delta[\beta]}{\nu}}\|_{\mathscr{L}(\mathcal{H}_\pi)} < \infty$$

for one (and then any) positive Rockland operator $\mathcal{R}$ of homogeneous degree ν (as the symbol is given by a field of operators defined on smooth vectors, and since $\pi(\mathrm{I}+\mathcal{R})^{\frac{s}{\nu}}$ acts on smooth vectors, this condition makes sense);

(L) a symbol $\sigma = \{\sigma(x,\pi),(x,\pi) \in G \times \widehat{G}\}$ is in $S^m_{\rho,\delta}$ if and only if for each $\alpha, \beta \in \mathbb{N}_0^n$, the field of operators

$$X_x^\beta \Delta^\alpha \sigma = \{X_x^\beta \Delta^\alpha \sigma(x,\pi) : \mathcal{H}_\pi^\infty \to \mathcal{H}_\pi^{\rho[\alpha]-m-\delta[\beta]}, (x,\pi) \in G \times \widehat{G}\}$$

is defined on smooth vectors and has range in $\mathcal{H}_\pi^{\rho[\alpha]-m-\delta[\beta]}$, and satisfies

$$\sup_{x\in G, \pi \in \widehat{G}} \|\pi(\mathrm{I}+\mathcal{R})^{\frac{\rho[\alpha]-m-\delta[\beta]}{\nu}} X_x^\beta \Delta^\alpha \sigma(x,\cdot)\|_{\mathscr{L}(\mathcal{H}_\pi)} < \infty$$

for one (and then any) positive Rockland operator $\mathcal{R}$ of homogeneous degree ν. The notion of a field having range in a Sobolev space $\mathcal{H}_\pi^s$ is described in Definition 5.1.10 and allows us to compose on the left with $\pi(\mathrm{I}+\mathcal{R})^{\frac{s}{\nu}}$ with $s = \rho[\alpha] - m - \delta[\beta]$ here, see (5.4).

Naturally, the condition does not depend on the choice of the positive Rockland operator $\mathcal{R}$.

Theorem 5.5.20 makes it considerably easier to check whether a symbol is in one of our symbol classes. However using the definition 'with any γ' has the advantages

1. that we see easily that the symbols are fields of operators acting on smooth vectors,
2. that we see easily that the symbols in $S^m_{\rho,\delta}$, $m \in \mathbb{R}$, form an algebra (cf. Theorem 5.2.22),
3. and that the properties for the multipliers in $\mathcal{R}$ in Proposition 5.3.4 are for the definition 'with any γ'.

While showing Theorem 5.5.20, we will also finish the proofs of Theorems 5.5.3 and 5.5.12. Indeed, an important argument used in the proof of Theorems 5.5.3 and 5.5.12 (i.e. the properties of stability under composition and taking the adjoint) is Inequality (5.54) which can easily be seen as equivalent to Part 2 of Theorem 5.5.20.

Before showing Theorem 5.5.20, let us summarise what has been shown in the proofs of Theorems 5.5.3 and 5.5.12 up to before the use of Inequality (5.54):

$$\|\sigma_1 \circ \sigma_2\|_{S^{m_1+m_2,R}_{\rho,\delta},a,b} \lesssim \|\sigma_1\|_{S^{m_1,R}_{\rho,\delta},a_1,b_1} \|\sigma_2\|_{S^{m_2}_{\rho,\delta},a_2,b_2,c_2}, \tag{5.63}$$

$$\|\sigma^{(*)}\|_{S^{m,R}_{\rho,\delta},a,b} \lesssim \|\sigma\|_{S^m_{\rho,\delta},a',b',0}; \tag{5.64}$$

these estimates are valid for any $\sigma, \sigma_1, \sigma_2 \in S^{-\infty}$ in the sense that for any seminorm on the left hand side, one can find seminorms on the right.

Proof of Theorem 5.5.20. Using Estimate (5.64) together with the properties of taking the adjoint and of the difference operators together, one checks easily that the two families of seminorms $\{\|\cdot\|_{S^{m,R}_{\rho,\delta},a,b}, a,b\in\mathbb{N}\}$ and $\{\|\cdot\|_{S^m_{\rho,\delta},a,b,0}, a,b\in\mathbb{N}\}$ yield the same topology on $S^{-\infty}$ and that taking the adjoint of a symbol is continuous for this topology. Consequently, for any $\gamma\in\mathbb{R}$, any symbol $\sigma\in S^{-\infty}$ and any seminorm $\|\cdot\|_{S^{m,R}_{\rho,\delta},a,b}$, we have

$$\|\pi(\mathrm{I}+\mathcal{R})^{\frac{\gamma}{\nu}}\sigma\|_{S^{m+\gamma,R}_{\rho,\delta},a,b}\lesssim\|\sigma^*\pi(\mathrm{I}+\mathcal{R})^{\frac{\gamma}{\nu}}\|_{S^{m+\gamma,R}_{\rho,\delta},a_1,b_1}\lesssim\|\sigma^*\|_{S^{m,R}_{\rho,\delta},a_2,b_2},$$

having used (5.63) and the fact that $\pi(\mathrm{I}+\mathcal{R})^{\frac{\gamma}{\nu}}\in S^\gamma$. As taking the adjoint is a continuous operator for the $S^{m,R}$-topology, we have obtained

$$\|\pi(\mathrm{I}+\mathcal{R})^{\frac{\gamma}{\nu}}\sigma\|_{S^{m+\gamma,R}_{\rho,\delta},a,b}\lesssim\|\sigma\|_{S^{m,R}_{\rho,\delta},a_3,b_3}.$$

One checks easily that

$$\forall a,b,c\in\mathbb{N}_0\qquad\|\sigma\|_{S^m_{\rho,\delta},a,b,c}\leq\max_{|\gamma|\leq c}\|\pi(\mathrm{I}+\mathcal{R})^{\frac{\gamma}{\nu}}\sigma\|_{S^{m+\gamma,R}_{\rho,\delta},a,b},$$

whereas

$$\forall a,b\in\mathbb{N}_0\qquad\|\sigma\|_{S^{m,R}_{\rho,\delta},a,b}\leq\|\sigma\|_{S^m_{\rho,\delta},a,b,|m|+\rho a+\delta b}.$$

This easily implies that the topologies on $S^{-\infty}$ coming from the two families of seminorms $\{\|\cdot\|_{S^m_{\rho,\delta},a,b,c}, a,b,c\in\mathbb{N}_0\}$ and $\{\|\cdot\|_{S^{m,R}_{\rho,\delta},a,b}, a,b\in\mathbb{N}_0\}$ coincide. This together with Lemma 5.4.11 (to pass from $S^{-\infty}$ to $S^m_{\rho,\delta}$) concludes the proof of Theorem 5.5.20. □

5.6 Amplitudes and amplitude operators

In this section, we discuss the notion of an amplitude extending that of the symbol, to functions/operators depending on both space variables x and y. This allows for another way of writing pseudo-differential operators as amplitude operators, analogous to Formula (2.27) in the case of compact groups. However, as in the classical theory, or as in Theorem 2.2.15 in the case of compact groups, we can show that amplitude operators with symbols in suitable amplitude classes reduce to pseudo-differential operator with symbols in corresponding symbol classes, with asymptotic formulae relating amplitudes to symbols.

5.6.1 Definition and quantization

Following the Euclidean and compact cases, it is natural to define amplitudes in the following way, extending the notion of symbols from Definitions 5.1.33 and 5.1.34:

Definition 5.6.1. An *amplitude* is a field of operators

$$\{\mathcal{A}(x,y,\pi):\mathcal{H}^\infty_\pi\to\mathcal{H}_\pi,\pi\in\widehat{G}\}$$

depending on $x, y \in G$, satisfying for each $x, y \in G$

$$\exists a, b \in \mathbb{R} \qquad \mathcal{A}(x, y, \cdot) := \{\mathcal{A}(x, y, \pi) : \mathcal{H}_\pi^\infty \to \mathcal{H}_\pi, \pi \in \widehat{G}\} \in L_{a,b}^\infty(\widehat{G}).$$

- An amplitude $\{\mathcal{A}(x, y, \pi) : \mathcal{H}_\pi^\infty \to \mathcal{H}_\pi, \pi \in \widehat{G}\}$ is said to be *continuous* in $x, y \in G$ whenever there exists $a, b \in \mathbb{R}$ such that

$$\forall x, y \in G \quad \mathcal{A}(x, y, \cdot) := \{\mathcal{A}(x, y, \pi) : \mathcal{H}_\pi^\infty \to \mathcal{H}_\pi, \pi \in \widehat{G}\} \in L_{a,b}^\infty(\widehat{G}),$$

  and the map $(x, y) \mapsto \mathcal{A}(x, y, \cdot)$ is continuous from $G \times G \sim \mathbb{R}^n \times \mathbb{R}^n$ to the Banach space $L_{a,b}^\infty(\widehat{G})$.

- An amplitude $\mathcal{A} = \{\mathcal{A}(x, y, \pi) : \mathcal{H}_\pi^\infty \to \mathcal{H}_\pi, \pi \in \widehat{G}\}$ is said to be *smooth* in $x, y \in G$ whenever it is a field of operators depending smoothly on $(x, y) \in G \times G$ (see Remark 1.8.16) and, for every $\beta_1, \beta_2 \in \mathbb{N}_0^n$, the field $\{\partial_x^{\beta_1} \partial_y^{\beta_2} \mathcal{A}(x, y, \pi) : \mathcal{H}_\pi^\infty \to \mathcal{H}_\pi, \pi \in \widehat{G}\}$ is continuous.

Clearly if an amplitude $\mathcal{A} = \{\mathcal{A}(x, y, \pi)\}$ does not depend on y, that is, $\mathcal{A}(x, y, \pi) = \sigma(x, \pi)$, then it defines a symbol $\sigma = \{\sigma(x, \pi)\}$. More generally any amplitude $\mathcal{A} = \{\mathcal{A}(x, y, \pi)\}$ defines a symbol σ given by $\sigma(x, \pi) = \mathcal{A}(x, x, \pi)$. In Section 5.6.2, we will define amplitude classes and give other examples of amplitudes.

Similarly to the symbol case, one can associate a kernel with an amplitude:

Definition 5.6.2. Let $\mathcal{A}$ be an amplitude. For each $(x, y) \in G \times G$, let $\kappa_{x,y} \in \mathcal{S}'(G)$ be the unique distribution such that

$$\mathcal{F}_G(\kappa_{x,y})(\pi) = \mathcal{A}(x, y, \pi).$$

The map $G \times G \ni (x, y) \mapsto \kappa_{x,y} \in \mathcal{S}'(G)$ is called its *kernel*.

As in the symbol case, the map $G \times G \ni (x, y) \mapsto \kappa_{x,y} \in \mathcal{S}'(G)$ is smooth, see Lemma 5.1.35 for the proof of this as well as for the existence and uniqueness of $\kappa_{x,y}$ in the case of symbols.

Before defining the amplitude quantization, we need to open a (quick) parenthesis to describe the following property from distribution theory:

Lemma 5.6.3. *Let $G \times G \ni (x, y) \mapsto \kappa_{x,y} \in \mathcal{S}'(G)$ be a continuous mapping. For each x, we consider the distribution $\tilde{\kappa}_x$ defined by*

$$\int_G \tilde{\kappa}_x(y)\phi(y)dy = \lim_{\epsilon \to 0} \int_{G \times G} \kappa_{x,w}(y^{-1}x)\phi(y)\psi_\epsilon(wy^{-1})dydw,$$

where $\phi \in \mathcal{D}(G)$, $\psi_1 \in \mathcal{D}(G)$, $\int_G \psi_1 = 1$ and $\psi_\epsilon(z) = \epsilon^{-Q}\psi(\epsilon^{-1}z)$, $\epsilon > 0$.

Indeed this limit exists and is independent of the choice of ψ_1.

This defines a continuous map $G \ni x \longmapsto \tilde{\kappa}_x \in \mathcal{D}'(G)$.

Proof of Lemma 5.6.3. Since $\kappa_{x,y} \in \mathcal{S}'(G)$, there exists a seminorm $\|\cdot\|_{\mathcal{S}(G),N}$ such that

$$\forall\phi \in \mathcal{S}(G) \qquad |\langle \kappa_{x,y}, \phi\rangle| \leq C_{x,y,N}\|\phi\|_{\mathcal{S}(G),N}.$$

Furthermore, since the map $G \times G \ni (x,y) \mapsto \kappa_{x,y} \in \mathcal{S}'(G)$ is smooth, we obtain that the constant $C_{x,y,N} = \|\kappa_{x,y}\|_{\mathcal{S}'(G),N}$ can be chosen locally uniform with respect to x and y. Furthermore, fixing two compacts K_1 and K_2 of G, there exists a seminorm $\|\cdot\|_{\mathcal{S}(G),N}$ (depending on K_1 and K_2) such that the map

$$((x,y),(x',y')) \in (K_1 \times K_2) \times (K_1 \times K_2) \mapsto \|\kappa_{x,y} - \kappa_{x',y'}\|_{\mathcal{S}'(G),N},$$

is uniformly continuous. This is easily proved using a cover of the compacts $K_1 \times K_2$ by balls of sufficiently small radius, and the continuity at each centre of these balls.

For any $\psi_1 \in \mathcal{D}(G)$, $\epsilon > 0$ and $x \in G$, we define the distribution $T_{\psi_1,\epsilon,x}$ by

$$T_{\psi_1,\epsilon,x}(\phi) := \int_{G\times G} \kappa_{x,w}(y^{-1}x)\phi(y)\psi_\epsilon(wy^{-1})dydw,$$

where $\phi \in \mathcal{D}(G)$ is supported in a fixed compact $K \subset G$. Using the change of variable from w to z with $z = \epsilon^{-1}(wy^{-1})$, so that $w = (\epsilon z)y$, we obtain

$$T_{\psi_1,\epsilon,x}(\phi) = \int_{G\times G} \kappa_{x,(\epsilon z)y}(y^{-1}x)\phi(y)\psi_1(z)dydz.$$

Therefore, for any $\epsilon_1, \epsilon_2 \in (0,1)$, we get

$$\begin{aligned}
&|(T_{\psi_1,\epsilon_1,x} - T_{\psi_1,\epsilon_2,x})(\phi)| \\
&\quad = \left|\int_{G\times G} \left(\kappa_{x,(\epsilon_1 z)y}(y^{-1}x) - \kappa_{x,(\epsilon_2 z)y}(y^{-1}x)\right)\phi(y)\psi_1(z)dydz\right| \\
&\quad \leq \sup_{\substack{z\in\text{supp}\psi_1\\ y\in\text{supp}\phi}} \|\kappa_{x,(\epsilon_1 z)y} - \kappa_{x,(\epsilon_2 z)y}\|_{\mathcal{S}'(G),N}\|\phi\|_{\mathcal{S}(G),N}\|\psi_1\|_{L^1(G)},
\end{aligned}$$

where $\|\cdot\|_{\mathcal{S}(G),N}$ is chosen with respect to the compact sets

$$\{x\} \quad \text{and} \quad \{(\epsilon z)y, \epsilon \in [0,1], \ z \in \text{supp}\psi_1, \ y \in K_2\}.$$

This shows that the scalar sequence $(T_{\psi_1,\epsilon,x}(\phi))$ converges as $\epsilon \to 0$ and that the linear map

$$\psi_1 \in \mathcal{D}(G) \longmapsto \lim_{\epsilon\to 0} T_{\psi_1,\epsilon,x}(\phi), \tag{5.65}$$

extends continuously to $L^1(K_o) \to \mathbb{C}$ for any compact $K_o \subset G$. Thus the map given in (5.65) is given by integration against a locally bounded function on G.

Let us show that the map given in (5.65) is invariant under left or right translation. Indeed, modifying the argument above we obtain

$$\begin{aligned}&\left|T_{\psi_1,\epsilon,x}(\phi) - T_{\psi_1(\cdot y_o^{-1}),\epsilon,x}(\phi)\right|\\&= \left|\int_{G\times G} \left(\kappa_{x,(\epsilon z)y} - \kappa_{x,(\epsilon(zy_o))y}\right)(y^{-1}x)\phi(y)\psi_1(z)dydz\right|\\&\leq \sup_{\substack{z\in\mathrm{supp}\psi_1\\ y\in\mathrm{supp}\phi}} \|\kappa_{x,(\epsilon z)y} - \kappa_{x,(\epsilon(zy_o))y}\|_{\mathcal{S}'(G),N}\|\phi\|_{\mathcal{S}(G),N}\|\psi_1\|_{L^1(G)}\end{aligned}$$

for a suitable seminorm $\|\cdot\|_{\mathcal{S}(G),N}$, (depending locally on y_o). Since the two sequences $((\epsilon z)y)_{\epsilon>0}$ and $((\epsilon(zy_o))y)_{\epsilon>0}$ converge to y in G, we see that

$$\lim_{\epsilon\to 0} T_{\psi_1,\epsilon,x}(\phi) = \lim_{\epsilon\to 0} T_{\psi_1(\cdot y_o^{-1}),\epsilon,x}(\phi),$$

and the same is true for right translation. Therefore, the locally bounded function given by the mapping (5.65) is a constant which we denote by $T_{0,x}(\phi)$:

$$\lim_{\epsilon\to 0} T_{\psi_1,\epsilon,x}(\phi) = T_{0,x}(\phi)\int_G \psi_1.$$

One checks easily that $T_{0,x}(\phi)$, $\phi\in\mathcal{D}(G)$, $\mathrm{supp}\,\phi\subset K$, defines a distribution $\tilde{\kappa}_x\in\mathcal{D}'(G)$ which is therefore independent of ψ_1. Refining the argument given above shows that $\tilde{\kappa}_x\in\mathcal{D}'(G)$ depends continuously on $x\in G$. □

If $G\times G\ni(x,y)\mapsto\kappa_{x,y}\in\mathcal{S}'(G)$ is a continuous mapping, we will allow ourselves to denote the distribution defined in Lemma 5.6.3 by

$$\tilde{\kappa}_x(y) := \kappa_{x,y}(y^{-1}x).$$

This closes our parenthesis about distribution theory.

We can now define the operator

$$T = \mathrm{AOp}(\mathcal{A})$$

associated with an amplitude $\mathcal{A}=\{\mathcal{A}(x,y,\pi)\}$ with amplitude kernel $\kappa_{x,y}$, by

$$T\phi(x) := \int_G \phi(y)\kappa_{x,y}(y^{-1}x)dy, \quad \phi\in\mathcal{D}(G),\ x\in G. \tag{5.66}$$

The quantization defined by formula (5.66) makes sense for any amplitude $\mathcal{A}=\{\mathcal{A}(x,y,\pi)\}$. Clearly the quantization mapping $\mathcal{A}\mapsto\mathrm{AOp}(\mathcal{A})$ is linear. However, as in the Euclidean or compact cases, it is injective but not necessarily 1-1 since different amplitudes may lead to the same operator, in contrast to the situation for symbols, cf. Theorem 5.1.39.

Remark 5.6.4. If an amplitude $\mathcal{A} = \{\mathcal{A}(x, y, \pi)\}$ does not depend on y, that is, $\mathcal{A}(x, y, \pi) = \sigma(x, \pi)$, then the corresponding symbol $\sigma = \{\sigma(x, \pi)\}$ yield the same operator:

$$\mathrm{AOp}(\mathcal{A}) = \mathrm{Op}(\sigma)$$

since in this case the amplitude $\kappa_{x,y}$ is a function/distribution κ_x independent of y which coincides with the kernel of the symbol σ.

As in the symbol case in Lemma 5.1.42, we may see $\mathrm{AOp}(\mathcal{A})$ as a limit of nice operators in the following sense:

Lemma 5.6.5. *If $\mathcal{A} = \{\mathcal{A}(x, y, \pi)\}$ is an amplitude, we can construct explicitly a family of amplitudes $\mathcal{A}_\epsilon = \{\mathcal{A}_\epsilon(x, y, \pi)\}$, $\epsilon > 0$, in such a way that*

1. *the kernel $\kappa_{\epsilon,x,y}(z)$ of $\mathcal{A}_\epsilon$ is smooth in both x, y and z, and compactly supported in x and y,*
2. *the associated kernel $\tilde{\kappa}_{\epsilon,x}(y) = \kappa_{\epsilon,x,y}(y^{-1}x)$ is smooth and compactly supported in both x, y,*
3. *if $\phi \in \mathcal{S}(G)$ then $\mathrm{AOp}(\mathcal{A}_\epsilon)\phi \in \mathcal{D}(G)$, and*
4. *$\mathrm{AOp}(\mathcal{A}_\epsilon)\phi \underset{\epsilon \to 0}{\longrightarrow} \mathrm{AOp}(\mathcal{A})\phi$ uniformly on any compact subset of G.*

Proof of Lemma 5.6.5. We use the same notation $\chi_\epsilon \in \mathcal{D}(G)$, $|\pi|$ and $\mathrm{proj}_{\epsilon,\pi}$ as in the proof of Lemma 5.1.42. We consider for any $\epsilon \in (0, 1)$ the amplitude given by

$$\mathcal{A}_\epsilon(x, y, \pi) := \chi_\epsilon(x)\chi_\epsilon(y)1_{|\pi| \leq \epsilon^{-1}}\mathcal{A}(x, y, \pi) \circ \mathrm{proj}_{\epsilon,\pi}.$$

By Definition 5.6.2 and the Fourier inversion formula (1.26), the corresponding kernel is

$$\kappa_{\epsilon,x,y}(z) = \chi_\epsilon(x)\chi_\epsilon(y) \int_{|\pi| \leq \epsilon^{-1}} \mathrm{Tr}\left(\mathcal{A}(x, y, \pi)\, \mathrm{proj}_{\epsilon,\pi}\pi(z)\right) d\mu(\pi),$$

which is smooth in x, y and z and compactly supported in x and y. The rest follows easily. □

There is a simple relation between the amplitudes of an operator and its adjoint, much simpler than in the symbol case:

Proposition 5.6.6. *Let $\mathcal{A}$ be an amplitude. Then $\mathcal{B}$ given by*

$$\mathcal{B}(x, y, \pi) := \mathcal{A}(y, x, \pi)^*$$

is also an amplitude. Furthermore, the formal adjoint of the operator $T = \mathrm{AOp}(\mathcal{A})$ is $T^* = \mathrm{AOp}(\mathcal{B})$. If $\{\kappa_{x,y}(z)\}$ is the kernel of $\mathcal{A}$, then the kernel of $\mathcal{B}$ is given via $(x, y, z) \mapsto \bar{\kappa}_{y,x}(z^{-1})$.

Proof. On one hand, from the amplitude quantization in (5.66), we compute for $\phi, \psi \in \mathcal{D}(G)$, that

$$(T\phi, \psi) = \int_G \int_G \phi(y)\kappa_{x,y}(y^{-1}x)\bar{\psi}(x)dy\,dx = (\phi, T^*\psi),$$

therefore

$$T^*\psi(y) = \int_G \bar{\kappa}_{x,y}(y^{-1}x)\psi(x)dx$$

or, equivalently,

$$T^*\psi(x) = \int_G \bar{\kappa}_{y,x}(x^{-1}y)\psi(y)dy.$$

One the other hand, the amplitude kernel for $\mathcal{B}$ is $\kappa'_{x,y}$ satisfying

$$\pi(\kappa'_{x,y}) = \mathcal{B}(x, y, \pi) = \mathcal{A}(y, x, \pi)^* = \pi(\kappa_{y,x})^* = \pi(\kappa^*_{y,x}),$$

with $\kappa^*_{y,x}(z) = \bar{\kappa}_{y,x}(z^{-1})$, and therefore,

$$\kappa'_{x,y}(z) = \kappa^*_{y,x}(z) = \bar{\kappa}_{y,x}(z^{-1}).$$

By (5.66), this implies that $T^* = \mathrm{AOp}(\mathcal{B})$. □

5.6.2 Amplitude classes

Again similarly to the symbol case, we may define the amplitude classes $AS^m_{\rho,\delta}$. This is done in analogy to Definition 5.2.11 for symbols and its equivalent reformulation in (5.29).

Definition 5.6.7. Let $m, \rho, \delta \in \mathbb{R}$ with $1 \geq \rho \geq \delta \geq 1$. An amplitude $\mathcal{A}$ is called an *amplitude of order* m *and of type* (ρ, δ) whenever, for each $\alpha, \beta \in \mathbb{N}_0^n$ and $\gamma \in \mathbb{R}$, the field $\{X_x^{\beta_1} X_y^{\beta_2} \Delta^\alpha \mathcal{A}(x, y, \pi)\}$ is in $L^\infty_{\gamma,\rho[\alpha]-m-\delta([\beta_1]+[\beta_2])+\gamma}(\widehat{G})$ uniformly in $(x, y) \in G$, i.e. if

$$\sup_{x,y\in G} \|X_x^{\beta_1} X_y^{\beta_2} \Delta^\alpha \mathcal{A}(x, y, \cdot)\|_{L^\infty_{\gamma,\rho[\alpha]-m-\delta([\beta_1]+[\beta_2])+\gamma}(\widehat{G})} < \infty. \tag{5.67}$$

In this case, proceeding in a similar way to $S^m_{\rho,\delta}$ in Section 5.2.2, we see that the fields of operators $X_x^{\beta_1} X_y^{\beta_2} \Delta^\alpha \mathcal{A}(x, y, \cdot)$ act on smooth vectors and (5.67) implies

$$\sup_{\substack{x,y\in G\\ \pi\in\widehat{G}}} \|\pi(\mathrm{I} + \mathcal{R})^{\frac{\rho[\alpha]-m-\delta([\beta_1]+[\beta_2])+\gamma}{\nu}} X_x^{\beta_1} X_y^{\beta_2} \Delta^\alpha \mathcal{A}(x, y, \cdot)\pi(\mathrm{I} + \mathcal{R})^{-\frac{\gamma}{\nu}}\|_{\mathscr{L}(\mathcal{H}_\pi)} < \infty. \tag{5.68}$$

The converse also holds.

The *amplitude class* $AS^m_{\rho,\delta} = AS^m_{\rho,\delta}(G)$ is the set of amplitudes of order m and of type (ρ,δ). We also define

$$AS^{-\infty} := \bigcap_{m\in\mathbb{R}} AS^m_{\rho,\delta},$$

the class of smoothing amplitudes. As in the case of symbols, the class $AS^{-\infty}$ is independent of ρ and δ and can be denoted just by $AS^{-\infty}$.

It is a routine exercise to check that each amplitude class $AS^m_{\rho,\delta}$ is a vector space and that we have the inclusions

$$m_1 \le m_2, \quad \delta_1 \le \delta_2, \quad \rho_1 \ge \rho_2 \quad \Longrightarrow \quad AS^{m_1}_{\rho_1,\delta_1} \subset AS^{m_2}_{\rho_2,\delta_2}. \tag{5.69}$$

We assume that a positive Rockland operator $\mathcal{R}$ of degree ν is fixed. If $\mathcal{A}$ is an amplitude and $a,b,c\in[0,\infty)$, we set

$$\begin{aligned}
&\|\mathcal{A}(x,y,\pi)\|_{AS^m_{\rho,\delta},a,b,c}\\
&:= \sup_{\substack{|\gamma|\le c\\ [\alpha]\le a,\ [\beta_1],[\beta_2]\le b}} \|\pi(\mathrm{I}+\mathcal{R})^{\frac{\rho[\alpha]-m-\delta([\beta_1]+[\beta_2])+\gamma}{\nu}} X_x^{\beta_1}X_y^{\beta_2}\Delta^\alpha\mathcal{A}(x,y,\pi)\pi(\mathrm{I}+\mathcal{R})^{-\frac{\gamma}{\nu}}\|_{\mathscr{L}(\mathcal{H}_\pi)},
\end{aligned}$$

and

$$\|\mathcal{A}\|_{AS^m_{\rho,\delta},a,b,c} := \sup_{(x,y)\in G\times G,\ \pi\in\widehat{G}} \|\mathcal{A}(x,y,\pi)\|_{AS^m_{\rho,\delta},a,b,c}.$$

Again, one checks easily that the resulting maps $\|\cdot\|_{S^m_{\rho,\delta},a,b,c}$, $a,b,c\in[0,\infty)$, are seminorms over the vector space $AS^m_{\rho,\delta}$. Furthermore, taking a,b,c as non-negative integers, they endow $AS^m_{\rho,\delta}$ with the structure of a Fréchet space. The class of smoothing amplitudes $AS^{-\infty}$ is then equipped with the topology of projective limit. Similarly to the case of symbols in Proposition 5.2.12, two different positive Rockland operators give equivalent families of seminorms.

The inclusions given in (5.69) are continuous for these topologies.

Symbols in $S^m_{\rho,\delta}$ are examples of amplitudes in $AS^m_{\rho,\delta}$ which do not depend on y. Conversely, if an amplitude $\mathcal{A}=\{\mathcal{A}(x,y,\pi)\}$ in $AS^m_{\rho,\delta}$ does not depend on y, that is, $\mathcal{A}(x,y,\pi)=\sigma(x,\pi)$, then it defines a symbol $\sigma=\{\sigma(x,\pi)\}$ in $S^m_{\rho,\delta}$. More generally we check easily:

Lemma 5.6.8. *If $\mathcal{A}=\{\mathcal{A}(x,y,\pi)\}$ is in $AS^m_{\rho,\delta}$, then the symbol σ given by*

$$\sigma(x,\pi) := \mathcal{A}(x,x,\pi)$$

is in $S^m_{\rho,\delta}$.

A wider class of examples is given by the following property which can be shown by an easy adaption of Proposition 5.3.4 and Corollary 5.3.7:

Corollary 5.6.9. *Let $\mathcal{R}$ be a positive Rockland operator of degree ν. Let $m \in \mathbb{R}$ and $0 \leq \delta < 1$. Let $f : G \times G \times \mathbb{R}^+ \ni (x, y, \lambda) \mapsto f_{x,y}(\lambda) \in \mathbb{C}$ be a smooth function. We assume that for every $\beta_1, \beta_2 \in \mathbb{N}_0^n$, we have*

$$X_x^{\beta_1} X_y^{\beta_2} f_{x,y} \in \mathcal{M}_{\frac{m+\delta([\beta_1]+[\beta_2])}{\nu}},$$

where $\mathcal{M}$ is as in Definition 5.3.1. Then

$$\mathcal{A}(x, y, \pi) = f_{x,y}(\pi(\mathcal{R}))$$

defines an amplitude $\mathcal{A}$ in $AS_{1,\delta}^m$ which satisfies

$$\forall a, b, c \in \mathbb{N}_0 \qquad \exists \ell \in \mathbb{N},\ C > 0$$
$$\|\mathcal{A}\|_{AS_{1,\delta}^m, a,b,c} \leq C \sup_{[\beta_1],[\beta_2] \leq b} \|X_x^{\beta_1} X_y^{\beta_2} f_{x,y}\|_{\mathcal{M}_{\frac{m+\delta[\beta_1+\beta_2]}{\nu}}, \ell},$$

with ℓ and C independent of f.

This can also be generalised easily to multipliers in a finite family of strongly commuting positive Rockland operators.

5.6.3 Properties of amplitude classes and kernels

One can readily prove properties for the amplitudes similar to the ones already established for symbols. Here we note that although the subsequent properties would follow also from Theorem 5.6.14 in the sequel and from the corresponding properties of symbols in Section 5.2.5, we now indicate what can be shown concerning amplitudes and their classes by a simple adaptation of proofs of the corresponding properties for symbols.

Proceeding as in Section 5.2.5, we also have the following properties for the amplitude classes:

Proposition 5.6.10. *Let $1 \geq \rho \geq \delta \geq 0$ and $\delta \neq 1$.*

(i) Let $\mathcal{A} \in AS_{\rho,\delta}^m$ have kernel $\kappa_{x,y}$. Then we have the following properties.

1. *For every $x, y \in G$ and $\gamma \in \mathbb{R}$, $\tilde{q}_\alpha X_x^{\beta_1} X_y^{\beta_2} \kappa_{x,y} \in \mathcal{K}_{\gamma, \rho[\alpha]-m-\delta[\beta_1+\beta_2]+\gamma}$, where we recall the notation $\tilde{q}_\alpha(x) = q_\alpha(x^{-1})$.*
2. *If $\beta_1, \beta_2 \in \mathbb{N}_0^n$ then the amplitude $\{X_x^{\beta_1} X_y^{\beta_2} \mathcal{A}(x, y, \pi), (x, y, \pi) \in G \times G \times \widehat{G}\}$ is in $AS_{\rho,\delta}^{m+\delta[\beta_1+\beta_2]}$ with kernel $X_x^{\beta_1} X_y^{\beta_2} \kappa_{x,y}$, and*

$$\|X_x^{\beta_1} X_y^{\beta_2} \mathcal{A}(x, y, \pi)\|_{AS_{\rho,\delta}^{m+\delta[\beta_1+\beta_2]}, a,b,c} \leq C \|\mathcal{A}(x, y, \pi)\|_{AS_{\rho,\delta}^m, a, b+[\beta_1+\beta_2], c},$$

with $C = C_{b,\beta_1,\beta_2}$.

3. *If $\alpha_o \in \mathbb{N}_0^n$ then the amplitude $\{\Delta^{\alpha_o}\mathcal{A}(x,y,\pi),(x,y,\pi)\in G\times G\times\widehat{G}\}$ is in $AS^{m-\rho[\alpha_o]}_{\rho,\delta}$ with kernel $\tilde{q}_{\alpha_o}\kappa_{x,y}$, and*

$$\|\Delta^{\alpha_o}\mathcal{A}(x,\pi)\|_{S^{m-\rho[\alpha_o]}_{\rho,\delta},a,b,c} \leq C_{a,\alpha_o}\|\mathcal{A}(x,\pi)\|_{S^m_{\rho,\delta},a+[\alpha_o],b,c}.$$

4. *The symbol $\{\mathcal{A}(x,y,\pi)^*,(x,\pi)\in G\times G\times\widehat{G}\}$ is in $AS^m_{\rho,\delta}$ with kernel $\kappa^*_{x,y}$ given by $\kappa^*_{x,y}(z)=\bar{\kappa}_{y,x}(z^{-1})$, and*

$$\|\mathcal{A}(x,y,\pi)^*\|_{AS^m_{\rho,\delta},a,b,c} = \sup_{\substack{|\gamma|\leq c\\ [\alpha]\leq a,\,[\beta_1],[\beta_2]\leq b}} \|\pi(\mathrm{I}+\mathcal{R})^{-\frac{\gamma}{\nu}}X_x^{\beta_1}X_y^{\beta_2}\Delta^\alpha\mathcal{A}(x,y,\pi)\pi(\mathrm{I}+\mathcal{R})^{\frac{\rho[\alpha]-m-\delta([\beta_1]+[\beta_2])+\gamma}{\nu}}\|_{\mathscr{L}(\mathcal{H}_\pi)}.$$

(ii) Let $\mathcal{A}_1\in AS^{m_1}_{\rho,\delta}$ and $\mathcal{A}_2\in AS^{m_2}_{\rho,\delta}$ have kernels $\kappa_{1,x,y}$ and $\kappa_{2,x,y}$, respectively. Then

$$\mathcal{A}(x,y,\pi) := \mathcal{A}_1(x,y,\pi)\mathcal{A}_2(x,y,\pi)$$

defines the amplitude $\mathcal{A}$ in $S^m_{\rho,\delta}$, $m=m_1+m_2$, with kernel $\kappa_{2,x,y}*\kappa_{1,x,y}$ with the convolution in the sense of Definition 5.1.19. Furthermore,

$$\|\mathcal{A}(x,y,\pi)\|_{S^m_{\rho,\delta},a,b,c} \leq C\|\mathcal{A}_1(x,y,\pi)\|_{S^{m_1}_{\rho,\delta},a,b,c+\rho a+|m_2|+\delta b}\|\mathcal{A}_2(x,y,\pi)\|_{S^{m_2}_{\rho,\delta},a,b,c},$$

where the constant $C=C_{a,b,c}>0$ does not depend on $\mathcal{A}_1,\mathcal{A}_2$.

A direct consequence of Part (ii) of Proposition 5.6.10 is that the amplitudes in the introduced amplitude classes form an algebra:

Corollary 5.6.11. *Let $1\geq\rho\geq\delta\geq 0$ and $\delta\neq 1$. The collection of symbols $\bigcup_{m\in\mathbb{R}} AS^m_{\rho,\delta}$ forms an algebra.*

Furthermore, if $\mathcal{A}_0\in AS^{-\infty}$ is smoothing and $\mathcal{A}\in AS^m_{\rho,\delta}$ is of order $m\in\mathbb{R}$, then $\mathcal{A}_0\mathcal{A}$ and $\mathcal{A}\mathcal{A}_0$ are also in $AS^{-\infty}$.

Another consequence of Part (ii) together with Lemma 5.2.17 gives the following property:

Corollary 5.6.12. *Let $1\geq\rho\geq\delta\geq 0$ and $\delta\neq 1$. Let $\mathcal{A}\in AS^m_{\rho,\delta}$ have kernel $\kappa_{x,y}$. If β and $\tilde{\beta}$ are in $\mathbb{N}_0^n$, then*

$$\{\pi(X)^\beta\mathcal{A}\,\pi(X)^{\tilde{\beta}},(x,\pi)\in G\times\widehat{G}\}\in AS^{m+[\beta]+[\tilde{\beta}]}_{\rho,\delta}$$

with kernel $X_z^\beta\tilde{X}_z^{\tilde{\beta}}\kappa_{x,y}(z)$. Furthermore, for any a,b,c there exists $C=C_{a,b,c}$ independent of $\mathcal{A}$ such that

$$\|\pi(X)^\beta\mathcal{A}\pi(X)^{\tilde{\beta}}\|_{AS^m_{\rho,\delta},a,b,c} \leq C\|\mathcal{A}\|_{AS^m_{\rho,\delta},a,b,c+\rho a+[\beta]+[\tilde{\beta}]+\delta b}.$$

Proceeding as in Section 5.4.1, taking into account the dependence in x and y, we obtain

Proposition 5.6.13. *Let $\mathcal{A} = \{\mathcal{A}(x,y,\pi)\}$ be in $AS^m_{\rho,\delta}$ with $1 \geq \rho \geq \delta \geq 0$. Let $\kappa_{x,y}$ denote its associated kernel.*

1. *If $\alpha, \beta_1, \beta_2, \beta_o, \beta_o' \in \mathbb{N}_0^n$ are such that*
$$m - \rho[\alpha] + [\beta_1] + [\beta_2] + \delta([\beta_o] + [\beta_o']) < -Q/2,$$
then the distribution $X_z^{\beta_1}\tilde{X}_z^{\beta_2}(X_x^{\beta_o}X_y^{\beta_o'}\tilde{q}_\alpha(z)\kappa_{x,y}(z))$ is square integrable and for every $x \in G$ we have
$$\int_G \left|X_z^{\beta_1}\tilde{X}_z^{\beta_2}(X_x^{\beta_o}X_y^{\beta_o'}\tilde{q}_\alpha(z)\kappa_{x,y}(z))\right|^2 dz \leq C \sup_{\pi\in\widehat{G}} \|\mathcal{A}(x,\pi)\|^2_{AS^m_{\rho,\delta},a,b,c}$$
where $a = [\alpha]$, $b = [\beta_o] + [\beta_o']$, $c = \rho[\alpha] + [\beta_1] + [\beta_2] + \delta([\beta_o] + [\beta_o'])$ and $C = C_{m,\alpha,\beta_1,\beta_2,\beta_o,\beta_o'} > 0$ is a constant independent of $\mathcal{A}$ and x, y.
2. *For any $\alpha, \beta_1, \beta_2, \beta_o, \beta_o' \in \mathbb{N}_0^n$ satisfying*
$$m - \rho[\alpha] + [\beta_1] + [\beta_2] + \delta([\beta_o] + [\beta_o']) < -Q,$$
the distribution $z \mapsto X_z^{\beta_1}\tilde{X}_z^{\beta_2}X_x^{\beta_o}X_y^{\beta_o'}\tilde{q}_\alpha(z)\kappa_{x,y}(z)$ is continuous on G for every $(x,y) \in G \times G$ and we have
$$\sup_{z\in G}\left|X_z^{\beta_1}\tilde{X}_z^{\beta_2}\left\{X_x^{\beta_o}X_y^{\beta_o'}\tilde{q}_\alpha(z)\kappa_{x,y}(z)\right\}\right| \leq C \sup_{\pi\in\widehat{G}} \|\mathcal{A}(x,\pi)\|_{AS^m_{\rho,\delta},[\alpha],[\beta_o]+[\beta_o'],[\beta_2]},$$
where $C = C_{m,\alpha,\beta_1,\beta_2,\beta_o,\beta_o'} > 0$ is a constant independent of $\mathcal{A}$ and x, y.

We now assume $\rho > 0$. Then the map $\kappa : (x,y,z) \mapsto \kappa_{x,y}(z)$ is smooth on $G \times G \times (G \setminus \{0\})$. Fixing a homogeneous quasi-norm $|\cdot|$ on G, we have the following more precise estimates:

at infinity: *For any $M \in \mathbb{R}$ and any $\alpha, \beta_1, \beta_2, \beta_o, \beta_o' \in \mathbb{N}_0^n$ there exist $C > 0$ and $a, b, c \in \mathbb{N}$ independent of $\mathcal{A}$ such that for all $x \in G$ and $z \in G$ satisfying $|z| \geq 1$, we have*
$$\left|X_z^{\beta_1}\tilde{X}_z^{\beta_2}(X_x^{\beta_o}X_y^{\beta_o'}\tilde{q}_\alpha(z)\kappa_{x,y}(z))\right| \leq C \sup_{\pi\in\widehat{G}} \|\mathcal{A}(x,y,\pi)\|_{AS^m_{\rho,\delta},a,b,c}|z|^{-M}.$$

at the origin: *For any $\alpha, \beta_1, \beta_2, \beta_o, \beta_o' \in \mathbb{N}_0^n$ with $Q+m+\delta([\beta_o]+[\beta_o'])-\rho[\alpha]+[\beta_1]+[\beta_2] \geq 0$ there exist a constant $C > 0$ and computable integers $a, b, c \in \mathbb{N}_0$ independent of $\mathcal{A}$ such that for all $x \in G$ and $z \in G\backslash\{0\}$, we have, if*
$$Q + m + \delta([\beta_o] + [\beta_o']) - \rho[\alpha] + [\beta_1] + [\beta_2] > 0,$$

then

$$\left|X_z^{\beta_1}\tilde{X}_z^{\beta_2}(X_x^{\beta_o}X_y^{\beta'_o}\tilde{q}_\alpha(z)\kappa_{x,y}(z))\right|$$
$$\leq C\sup_{\pi\in\widehat{G}}\|\mathcal{A}(x,\pi)\|_{AS^m_{\rho,\delta},a,b,c}|z|^{-\frac{Q+m+\delta([\beta_o]+[\beta'_o])-\rho[\alpha]+[\beta_1]+[\beta_2]}{\rho}},$$

and if

$$Q+m+\delta([\beta_o]+[\beta'_o])-\rho[\alpha]+[\beta_1]+[\beta_2]=0,$$

then

$$\left|X_z^{\beta_1}\tilde{X}_z^{\beta_2}(X_x^{\beta_o}X_y^{\beta'_o}\tilde{q}_\alpha(z)\kappa_{x,y}(z))\right|\leq C\sup_{\pi\in\widehat{G}}\|\mathcal{A}(x,y,\pi)\|_{AS^m_{\rho,\delta},a,b,c}\ln|z|.$$

5.6.4 Link between symbols and amplitudes

Symbols can be viewed as amplitudes which do not depend on the second variable of the group. Then $S^m_{\rho,\delta}\subset AS^m_{\rho,\delta}$ and, by Remark 5.6.4, we have the inclusion

$$\Psi^m_{\rho,\delta}=\mathrm{Op}(S^m_{\rho,\delta})\subset\mathrm{AOp}(AS^m_{\rho,\delta}).$$

The next theorem shows the converse, namely, that the class of operators $\mathrm{AOp}(AS^m_{\rho,\delta})$ is included in $\Psi^m_{\rho,\delta}$. Therefore this will show that the amplitude quantization of $AS^m_{\rho,\delta}$ coincides with the symbol quantization of $S^m_{\rho,\delta}$.

Theorem 5.6.14. *Let $\mathcal{A}\in AS^m_{\rho,\delta}$ with $1\geq\rho\geq\delta\geq0$, $\delta\neq1$. Then $\mathrm{AOp}(\mathcal{A})$ is in $\Psi^m_{\rho,\delta}$, that is, there exists a (unique) symbol $\sigma\in S^m_{\rho,\delta}$ such that*

$$\mathrm{AOp}(\mathcal{A})=\mathrm{Op}(\sigma).$$

Furthermore, for any $M\in\mathbb{N}_0$, the map

$$\left\{\begin{array}{ccc} AS^m_{\rho,\delta} & \longrightarrow & S^{m-(\rho-\delta)(M+1)}_{\rho,\delta} \\ \mathcal{A} & \longmapsto & \sigma(x,\pi)-\sum_{[\alpha]\leq M}\Delta^\alpha X_y^\alpha\mathcal{A}(x,y,\pi)|_{y=x}\end{array}\right.,$$

is continuous. If $\rho>\delta$, we have the asymptotic expansion

$$\sigma(x,\pi)\sim\sum_\alpha\Delta^\alpha X_y^\alpha\mathcal{A}(x,y,\pi)|_{y=x}.$$

The proof of Theorem 5.6.14 is in essence close to the proofs of product and adjoint of operators in $\cup_{m\in\mathbb{R}}\Psi^m_{\rho,\delta}$, see Theorems 5.5.12 and 5.5.3. As for these theorems, it is helpful to understand formally the steps of the rigorous proof.

From the amplitude quantization in (5.66), we see that if $\mathrm{AOp}(\mathcal{A})$ can be written as $\mathrm{Op}(\sigma)$, then, denoting by $\kappa_{\sigma,x}$ the symbol kernel and by $\kappa_{\mathcal{A},x,y}$ the amplitude kernel, we have

$$\mathrm{AOp}(\mathcal{A})(\phi)(x)=\int_G\phi(y)\kappa_{\mathcal{A},x,y}(y^{-1}x)dy=\int_G\phi(xz^{-1})\kappa_{\mathcal{A},x,xz^{-1}}(z)dz$$

whereas

$$\mathrm{Op}(\sigma)(\phi)(x) = \int_G \phi(y)\kappa_{\sigma,x}(y^{-1}x)dy = \int_G \phi(xz^{-1})\kappa_{\sigma,x}(z)dz.$$

Therefore, formally we must have

$$\kappa_{\mathcal{A},x,xz^{-1}}(z) = \kappa_{\sigma,x}(z) \quad \left(\text{or equivalently } \kappa_{\mathcal{A},x,y}(y^{-1}x) = \kappa_{\sigma,x}(y^{-1}x)\right).$$

Using the Taylor expansion in $y = xz^{-1}$ for $\kappa_{\mathcal{A},x,y}$ at x, we have (again formally)

$$\kappa_{\sigma,x}(z) = \kappa_{\mathcal{A},x,xz^{-1}}(z) \approx \sum_\alpha \tilde{q}_\alpha(z) X_y^\alpha \kappa_{\mathcal{A},x,y}(z)|_{y=x}. \tag{5.70}$$

Note that the group Fourier transform in z of each term in the sum above is

$$\begin{aligned}\mathcal{F}_{z\in G}\{\tilde{q}_\alpha(z)X^\alpha_{y=x}\kappa_{\mathcal{A},x,y}(z)\}(\pi) &= \Delta^\alpha X^\alpha_{y=x}\mathcal{F}_{z\in G}\{\kappa_{\mathcal{A},x,y}(z)\}(\pi)\\ &= \Delta^\alpha X^\alpha_{y=x}\mathcal{A}(x,y,\pi).\end{aligned}$$

Taking the group Fourier transform in z on both sides of (5.70), we obtain still formally that

$$\sigma(x,\pi) \approx \sum_\alpha \Delta^\alpha X_y^\alpha \mathcal{A}(x,y,\pi)|_{y=x}.$$

As in the proofs of Theorems 5.5.12 and 5.5.3, the crucial point is to control the remainder while using Taylor's expansion. The method is similar as in the proof of Theorem 5.5.12 and the adaptation is easy and left to the reader.

Note that Theorem 5.6.14 together with Proposition 5.6.6 give another proof of Theorem 5.5.12. This is not surprising given the similarity between the proof of Theorems 5.6.14 and 5.5.12.

5.7 Calderón-Vaillancourt theorem

In this section, we prove the analogue of the Calderón-Vaillancourt theorem, now in the setting of graded Lie groups. This extends the L^2-boundedness of operators in the class $\Psi^0_{1,0}$ given in Theorem 5.4.17 to the classes $\Psi^0_{\rho,\delta}$.

Theorem 5.7.1. *Let $T \in \Psi^0_{\rho,\delta}$ with $1 \geq \rho \geq \delta \geq 0$ and $\delta \neq 1$. Then T extends to a bounded operator on $L^2(G)$.*

Moreover, there exist a constant $C > 0$ and a seminorm $\|\cdot\|_{\Psi^0_{\rho,\delta},a,b,c}$ with computable integers $a,b,c \in \mathbb{N}_0$ independent of T such that

$$\forall \phi \in \mathcal{S}(G) \qquad \|T\phi\|_{L^2(G)} \leq C\|T\|_{\Psi^0_{\rho,\delta},a,b,c}\|\phi\|_{L^2(G)}.$$

Before showing Theorem 5.7.1, let us mention that together with the pseudo-differential calculus, it implies the following boundedness on Sobolev spaces L^2_s.

Corollary 5.7.2. *Let $T \in \Psi^m_{\rho,\delta}$ with $1 \geq \rho \geq \delta \geq 0$ and $\delta \neq 1$. Then for any $s \in \mathbb{R}$, the operator T extends to a continuous operator from $L^2_s(G)$ to $L^2_{s-m}(G)$:*

$$\forall \phi \in \mathcal{S}(G) \qquad \|T\phi\|_{L^2_{s-m}(G)} \leq C_{s,m,\rho,\delta}\|T\|_{\Psi^m_{\rho,\delta},a,b,c}\|\phi\|_{L^2_s(G)},$$

with some (computable) integers a, b, c depending on s, m, ρ, δ.

Proof of Corollary 5.7.2. Let $\mathcal{R}$ be a positive Rockland operator. By the composition theorem (e.g. Theorem 5.5.3), we have

$$(\mathrm{I}+\mathcal{R})^{\frac{-m+s}{\nu}} T(\mathrm{I}+\mathcal{R})^{-\frac{s}{\nu}} \in \Psi^0_{\rho,\delta}.$$

Therefore, by Theorem 5.7.1, we have

$$\begin{aligned}\|(\mathrm{I}+\mathcal{R})^{\frac{-m+s}{\nu}} T(\mathrm{I}+\mathcal{R})^{-\frac{s}{\nu}}\phi\|_{\mathscr{L}(L^2(G)} &\lesssim \|(\mathrm{I}+\mathcal{R})^{\frac{-m+s}{\nu}} T(\mathrm{I}+\mathcal{R})^{-\frac{s}{\nu}}\|_{\Psi^0_{\rho,\delta},a_1,b_1,c_1} \\ &\lesssim \|T\|_{\Psi^m_{\rho,\delta},a_2,b_2,c_2},\end{aligned}$$

by Theorem 5.5.3. □

Remark 5.7.3. Combining the results obtained so far, for each (ρ, δ) with $1 \geq \rho \geq \delta \geq 0$ and $\delta \neq 1$, we have therefore obtained an operator calculus, in the sense that the set $\bigcup_{m\in\mathbb{R}} \Psi^m_{\rho,\delta}$ forms an algebra of operators, stable under taking the adjoint, and acting on the Sobolev spaces in such a way that the loss of derivatives in L^2 is controlled by the order of the operator.

Note that the L^2-boundedness in the case $(\rho, \delta) = (1, 0)$ was already proved by different methods, see Theorem 5.4.17 and its proof. With the same proof as in the corollary above, one obtains easily boundedness for L^p-Sobolev spaces in this case:

Corollary 5.7.4. *Let $T \in \Psi^m_{1,0}$. Then for any $s \in \mathbb{R}$ and $p \in (1, \infty)$ the operator T extends to a continuous operator from $L^p_s(G)$ to $L^p_{s-m}(G)$:*

$$\forall \phi \in \mathcal{S}(G) \qquad \|T\phi\|_{L^p_{s-m}(G)} \leq C_{s,m,\rho,\delta}\|T\|_{\Psi^m_{\rho,\delta},a,b,c}\|\phi\|_{L^p_s(G)},$$

with some (computable) integers a, b, c depending on s, m, ρ, δ.

Proof of Corollary 5.7.4. As above, $(\mathrm{I}+\mathcal{R})^{\frac{-m+s}{\nu}} T(\mathrm{I}+\mathcal{R})^{-\frac{s}{\nu}} \in \Psi^0$ therefore, by Corollary 5.4.20 we have

$$\begin{aligned}\|(\mathrm{I}+\mathcal{R})^{\frac{-m+s}{\nu}} T(\mathrm{I}+\mathcal{R})^{-\frac{s}{\nu}}\phi\|_{\mathscr{L}(L^p(G))} &\lesssim \|(\mathrm{I}+\mathcal{R})^{\frac{-m+s}{\nu}} T(\mathrm{I}+\mathcal{R})^{-\frac{s}{\nu}}\|_{\Psi^0,a_1,b_1,c_1} \\ &\lesssim \|T\|_{\Psi^0,a_2,b_2,c_2},\end{aligned}$$

by Theorem 5.5.3. □

The rest of this section is devoted to the proof of the Calderón-Vaillancourt Theorem, that is, Theorem 5.7.1. In Section 5.7.2, we prove the result for $\rho = \delta = 0$.

The proof will rely on an analogue on G of the familiar decomposition of $\mathbb{R}^n$ into unit cubes presented in Section 5.7.1. The case $\rho = \delta \in (0,1)$ will be proved in Section 5.7.4 and its proof relies on the case $\rho = \delta = 0$ and on a bilinear estimate proved in Section 5.7.3. The case of $\rho = \delta \in [0,1)$ will then be proved and this will imply Theorem 5.7.1 thanks to the continuous inclusions between symbol classes (see (5.31)).

5.7.1 Analogue of the decomposition into unit cubes

In this section, we present an analogue of the dyadic cubes, more precisely we construct a useful covering of the general homogeneous Lie group G by unit balls and the corresponding partition of unity with a number of advantageous properties. The proof is an adaptation of [FS82, Lemma 7.14].

Lemma 5.7.5. *Let $|\cdot|$ be a fixed homogeneous quasi-norm on the homogeneous Lie group G. We denote by $C_o \geq 1$ a constant for the triangle inequality*

$$\forall x, y \in G \qquad |xy| \leq C_o(|x| + |y|). \tag{5.71}$$

Denoting by $B(x,R)$ the $|\cdot|$-ball centred at point x with radius R,

$$B(x,R) := \{y \in G \ : \ |x^{-1}y| < R\},$$

there exists a maximal family $\{B(x_i, \frac{1}{2C_o})\}_{i=1}^{\infty}$ of disjoint balls of radius $\frac{1}{2C_o}$, and we choose one such family. Then the following properties hold:

1. *The balls $\{B(x_i, 1)\}_{i=1}^{\infty}$ cover G.*
2. *For any $C \geq 1$, no point of G belongs to more than $\lceil (4C_o^2 C)^Q \rceil$ of the balls $\{B(x_i, C)\}_{i=1}^{\infty}$.*
3. *There exists a sequence of functions $\chi_i \in \mathcal{D}(G)$, $i \in \mathbb{N}$, such that each χ_i is supported in $B(x_i, 2)$ and satisfies $0 \leq \chi_i \leq 1$ while we have $\sum_{i=1}^{\infty} \chi_i = 1$. Moreover, for any $\beta \in \mathbb{N}_0^n$, $X^\beta \chi_i$ is uniformly bounded in $i \in \mathbb{N}$.*
4. *For any $p_1 > Q + 1$, we have*

$$\exists C_{p_1} > 0 \quad \forall i_o \in \mathbb{N} \qquad \sum_{i=1}^{\infty} (1 + |x_{i_o}^{-1} x_i|)^{-p_1} \leq C_{p_1} < \infty.$$

Remark 5.7.6. The conclusion of Part (4) is rough but will be sufficient for our purposes. We note, however, that if the quasi-norm in Lemma 5.7.5 is actually a norm, i.e. if the constant C_o in (5.71) is equal to one, $C_o = 1$, then the conclusion of Part (4) of Lemma 5.7.5 holds true for all $p_1 > Q$. This will be proved together with the lemma.

Proof of Lemma 5.7.5 and of Remark 5.7.6. If $x \in G$ then by maximality there exists i such that the distance from x to $B(x_i, \frac{1}{2C_o})$ is $< 1/(2C_o)$. Denoting by y a point in $\bar{B}(x_i, \frac{1}{2C_o})$ which realises the distance, we have

$$|x_i^{-1}x| \le C_o(|x_i^{-1}y| + |y^{-1}x|) < C_o\left(\frac{1}{2C_o} + \frac{1}{2C_o}\right) = 1.$$

This proves Part (1).

If x is in all the balls $B(x_{i_\ell}, C)$, $\ell = 1, \ldots, \ell_o$, then

$$\forall y \in \cup_{\ell=1}^{\ell_o} B(x_{i_\ell}, C) \quad \exists \ell \in [1, \ell_o] \quad |x^{-1}y| \le C_o(|x^{-1}x_{i_\ell}| + |x_{i_\ell}^{-1}y|) \le C_o 2C.$$

This shows that $B(x, 2C_oC)$ contains $\cup_{\ell=1}^{\ell_o} B(x_{i_\ell}, C)$ and, therefore, it must contain the disjoint balls $\cup_{\ell=1}^{\ell_o} B(x_{i_\ell}, \frac{1}{2C_o})$. Taking the Haar measure and denoting $c_1 := |B(0,1)|$, we have

$$|\cup_{\ell=1}^{\ell_o} B(x_{i_\ell}, \frac{1}{2C_o})| = \ell_o c_1 \left(\frac{1}{2C_o}\right)^Q \le |B(x, 2C_oC)| = (2C_oC)^Q\, c_1.$$

This proves Part (2).

Let us fix $\chi \in \mathcal{D}(G)$ satisfying $0 \le \chi \le 1$ with $\chi = 1$ on $B(0,1)$ and $\chi = 0$ on $B(0,2)$. The sum $\sum_{i'=1}^{\infty} \chi(x_{i'}^{-1}\,\cdot)$ is locally finite by Part (2); it is a smooth function with values between 1 and $\lceil (4C_o^2 \times 2)^Q \rceil$. We define

$$\chi_i(x) := \frac{\chi(x_i^{-1}x)}{\sum_{i'=1}^{\infty} \chi(x_{i'}^{-1}x)}.$$

This gives Part (3).

To prove Part (4), we fix a point x_{i_o} and observe that if $x \in G$ is in one of the balls $B(x_i, \frac{1}{2C_o})$ with $|x_{i_o}^{-1}x_i| \in [\ell, \ell+1)$ for some $\ell \in \mathbb{N}$, let us say $B(x_{i_1}, \frac{1}{2C_o})$, then

$$|x_{i_o}^{-1}x| \le C_o(|x_{i_1}^{-1}x| + |x_{i_o}^{-1}x_{i_1}|) \le C_o(\frac{1}{2C_o} + \ell + 1).$$

This yields the inclusion

$$\sqcup_{|x_{i_o}^{-1}x_i| \in [\ell, \ell+1)} B(x_i, \frac{1}{2C_o}) \;\subset\; B(x_{i_o}, C_o(\frac{1}{2C_o} + \ell + 1)).$$

The measure of the left hand side is $c_1(2C_o)^{-Q}\text{card}\{i : |x_{i_o}^{-1}x_i| \in [\ell, \ell+1)\}$ and the measure of the right hand side is $c_1(C_o(\frac{1}{2C_o} + \ell + 1))^Q$. Therefore,

$$\text{card}\{i : |x_{i_o}^{-1}x_i| \in [\ell, \ell+1)\} \le c\ell^Q.$$

Now we decompose

$$\sum_{i=1}^{\infty}(1+|x_{i_o}^{-1}x_i|)^{-p_1} = \sum_{|x_i^{-1}x_{i_o}|<1}(1+|x_{i_o}^{-1}x_i|)^{-p_1} + \sum_{\ell=1}^{\infty}\sum_{|x_i^{-1}x_{i_o}|\in[\ell,\ell+1)}(1+|x_{i_o}^{-1}x_i|)^{-p_1}.$$

By Part (2) the first sum on the right hand side is $\leq \lceil (4C_o^2)^Q \rceil$ whereas from the observation just above, the second sum is $\leq \sum_{\ell=0}^{\infty}(1+\ell)^{-p_1}c'(1+\ell)^Q$. This last sum being convergent whenever $-p_1+Q<-1$, Part (4) is proved.

Let us finally prove Remark 5.7.6, that is, Part (4) of the lemma for $p_1>Q$ provided that $C_o=1$. This will follow by the same argument as above if we can show a refined estimate

$$\text{card}\{i : |x_{i_o}^{-1}x_i| \in [\ell,\ell+1)\} \leq c\ell^{Q-1}.$$

We claim that this estimate holds true. Since $C_o=1$, we can estimate

$$|x_{i_o}^{-1}x| \geq |x_{i_o}^{-1}x_{i_1}| - |x_{i_1}^{-1}x| > \ell - \frac{1}{2C_o} = \ell - \frac{1}{2}.$$

We also have $C_o(\frac{1}{2C_o}+\ell+1)=\ell+\frac{3}{2}$. Consequently, we have the inclusion

$$\sqcup_{|x_{i_o}^{-1}x_i|\in[\ell,\ell+1)} B(x_i,\frac{1}{2C_o}) \subset B(x_{i_o},\ell+\frac{3}{2})\backslash B(x_{i_o},\ell-\frac{1}{2}),$$

with the measure on the right hand side being $c_1(\ell+\frac{3}{2})^Q - c_1(\ell-\frac{1}{2})^Q$. Therefore,

$$\text{card}\{i : |x_{i_o}^{-1}x_i| \in [\ell,\ell+1)\} \leq c\ell^{Q-1},$$

so that the required claim is proved. □

5.7.2 Proof of the case $S^0_{0,0}$

This section is devoted to the proof of the following result which is a particular case of Theorem 5.7.1. We also give an explicit estimate on the number of derivatives and differences of the symbol needed for the L^2-boundedness.

Proposition 5.7.7. *Let $T\in\Psi^0_{0,0}$. Then T extends to a bounded operator on $L^2(G)$. Furthermore, if we fix a positive Rockland operator $\mathcal{R}$ (in order to define the semi-norms on $\Psi^m_{\rho,\delta}$) then*

$$\forall\phi\in\mathcal{S}(G) \qquad \|T\phi\|_{L^2(G)} \leq C\|T\|_{\Psi^0_{0,0},a,b,c}\|\phi\|_{L^2(G)},$$

where $C>0$ and $a,b,c\in\mathbb{N}_0$ are independent of T. In particular, this estimate holds with $a=rp_o$, $b=r\nu+\lceil\frac{Q}{2}\rceil$, $c=r\nu$, where ν is the degree of $\mathcal{R}$, $p_o/2$ is the smallest positive integer divisible by $v_1,\ldots,v_n$, and $r\in\mathbb{N}_0$ is the smallest integer such that $rp_o>Q+1$.

Throughout Section 5.7.2, we fix the homogeneous norm $|\cdot| = |\cdot|_{p_o}$ given by (3.21), where $p_o/2$ is the smallest positive integer divisible by $\upsilon_1, \ldots, \upsilon_n$. We fix a maximal family $\{B(x_i, \frac{1}{2C_o})\}_{i=1}^{\infty}$ of disjoint balls and a sequence of functions $(\chi_i)_{i=1}^{\infty}$ so that the properties of Lemma 5.7.5 hold. We also fix $\psi_0, \psi_1 \in \mathcal{D}(\mathbb{R})$ supported in $[-1,1]$ and $[1/2,2]$, respectively, such that $0 \le \psi_0, \psi_1 \le 1$ and

$$\forall \lambda \geq 0 \qquad \sum_{j=0}^{\infty} \psi_j(\lambda) = 1 \ \text{ with } \ \psi_j(\lambda) := \psi_1(2^{-(j-1)}\lambda), \ j \in \mathbb{N}.$$

Let us start the proof of Proposition 5.7.7. Let $\sigma \in S^0_{0,0}$.

For each $I = (i,j) \in \mathbb{N} \times \mathbb{N}_0$, we define

$$\sigma_I(x,\pi) := \chi_i(x)\sigma(x,\pi)\psi_j(\pi(\mathcal{R})).$$

We denote by T_I and κ_I the corresponding operator and kernel.

Roughly speaking, the parameters i and j correspond to localising in space and frequency, respectively. The localisation in space corresponds to the covering of G by the balls centred at the x_i's, while the localisation in frequency is determined by the spectral projection of $\mathcal{R}$ to the $L^2(G)$-eigenspaces corresponding to eigenvalues close to each 2^j.

It is not difficult to see that each T_I is bounded on $L^2(G)$:

Lemma 5.7.8. *Each operator T_I is bounded on $L^2(G)$.*

Since σ_I is localised both in space and in frequency, we may use one of the two localisations.

Proof of Lemma 5.7.8 using frequency localisation. Let $\alpha, \beta \in \mathbb{N}_0^n$. By the Leibniz formulae for difference operators (see Proposition 5.2.10) and for vector fields, we have

$$X_x^\beta \Delta^\alpha \sigma_I(x,\pi) = \overline{\sum_{\substack{[\beta_1]+[\beta_2]=[\beta] \\ [\alpha_1]+[\alpha_2]=[\alpha]}}} X_x^{\beta_1}\chi_i(x)\ X_x^{\beta_2}\Delta^{\alpha_1}\sigma(x,\pi)\ \Delta^{\alpha_2}\psi_j(\pi(\mathcal{R})).$$

Therefore,

$$\begin{aligned}
&\|\pi(\mathrm{I}+\mathcal{R})^{\frac{[\alpha]+\gamma}{\nu}} X_x^\beta \Delta^\alpha \sigma_I(x,\pi)\pi(\mathrm{I}+\mathcal{R})^{-\frac{\gamma}{\nu}}\|_{\mathscr{L}(\mathcal{H}_\pi)} \\
&\leq C \sum_{\substack{[\beta_2]\leq[\beta] \\ [\alpha_1]+[\alpha_2]=[\alpha]}} \|\pi(\mathrm{I}+\mathcal{R})^{\frac{[\alpha]+\gamma}{\nu}} X_x^{\beta_2}\Delta^{\alpha_1}\sigma(x,\pi)\ \Delta^{\alpha_2}\psi_j(\pi(\mathcal{R}))\pi(\mathrm{I}+\mathcal{R})^{-\frac{\gamma}{\nu}}\|_{\mathscr{L}(\mathcal{H}_\pi)} \\
&\leq C \sum_{\substack{[\beta_2]\leq[\beta] \\ [\alpha_1]+[\alpha_2]=[\alpha]}} \|\pi(\mathrm{I}+\mathcal{R})^{\frac{[\alpha]+\gamma}{\nu}} X_x^{\beta_2}\Delta^{\alpha_1}\sigma(x,\pi)\pi(\mathrm{I}+\mathcal{R})^{-\frac{[\alpha_2]+\gamma}{\nu}}\|_{\mathscr{L}(\mathcal{H}_\pi)} \\
&\qquad\qquad \|\pi(\mathrm{I}+\mathcal{R})^{\frac{[\alpha_2]+\gamma}{\nu}} \Delta^{\alpha_2}\psi_j(\pi(\mathcal{R}))\pi(\mathrm{I}+\mathcal{R})^{-\frac{\gamma}{\nu}}\|_{\mathscr{L}(\mathcal{H}_\pi)}.
\end{aligned}$$

Therefore, by Lemma 5.4.7, we obtain

$$\|\sigma_I\|_{S^0_{1,0},a,b,c} \leq \|\sigma\|_{S^0_{0,0},a,b,c+a} 2^{ja/\nu}.$$

This shows that the operator T_I is in Ψ^0 and is therefore bounded on $L^2(G)$ by Theorem 5.4.17. □

Proof of Lemma 5.7.8 using space localisation. Another proof is to apply the following lemma since the symbol $\sigma_I(x,\pi)$ has compact support in x. □

Lemma 5.7.9. *Let $\sigma(x,\pi)$ be a symbol (in the sense of Definition 5.1.33) supported in $x \in S$, and assume that S is compact. Then the operator norm of the associated operator on $L^2(G)$ is*

$$\|\mathrm{Op}(\sigma)\|_{\mathscr{L}(L^2(G))} \leq C|S|^{1/2} \sup_{\substack{x\in G\\ [\beta]\leq\lceil\frac{Q}{2}\rceil}} \|X_x^\beta \sigma(x,\pi)\|_{L^\infty(\widehat{G})}.$$

Proof of Lemma 5.7.9. Let $T=\mathrm{Op}(\sigma)$ and let κ_x be the associated kernel. We have by the Sobolev inequality in Theorem 4.4.25,

$$\begin{aligned} |T\phi(x)|^2 &= |\phi*\kappa_x(x)|^2 \leq \sup_{x_o\in G} |\phi*\kappa_{x_o}(x)|^2 \\ &\leq C \sum_{[\beta]\leq\lceil\frac{Q}{2}\rceil} \left\|\phi * X_{x_o}^\beta \kappa_{x_o}(x)\right\|^2_{L^2(dx_o)}. \end{aligned}$$

Hence

$$\begin{aligned} \|T\phi\|^2_{L^2(G)} &\leq C \sum_{[\beta]\leq\lceil\frac{Q}{2}\rceil} \int_G\int_G |\phi * X_{x_o}^\beta \kappa_{x_o}(x)|^2 dx_o dx \\ &\leq C \sum_{[\beta]\leq\lceil\frac{Q}{2}\rceil} \int_G \|\phi * X_{x_o}^\beta \kappa_{x_o}\|^2_{L^2(dx)} dx_o \\ &\leq C|S| \sup_{x_o\in G, [\beta]\leq\lceil\frac{Q}{2}\rceil} \left\|\phi * X_{x_o}^\beta \kappa_{x_o}(x)\right\|^2_{L^2(dx)}. \end{aligned}$$

Now by Plancherel's Theorem,

$$\left\|\phi * X_{x_o}^\beta \kappa_{x_o}(x)\right\|_{L^2(dx)} \leq \|\phi\|_{L^2(dx)} \|X_{x_o}^\beta \sigma(x_o,\pi)\|_{L^\infty(\widehat{G})}.$$

This implies that the L^2-operator norm of T is

$$\leq C|S|^{1/2} \sup_{x_o\in G, [\beta]\leq\lceil\frac{Q}{2}\rceil} \|X_{x_o}^\beta \sigma(x_o,\pi)\|_{L^\infty(\widehat{G})},$$

and concludes the proof of Lemma 5.7.9. □

Let us go back to the proof of Proposition 5.7.7. The approach is to apply the following version of Cotlar's lemma:

Lemma 5.7.10 (Cotlar's lemma here)**.** *Suppose that $r \in \mathbb{N}_0$ is such that $rp_o > Q+1$ and that there exists $A_r > 0$ satisfying for all $(I, I') \in \mathbb{N} \times \mathbb{N}_0$:*

$$\max\left(\|T_I T_{I'}^*\|_{\mathscr{L}(L^2(G))}, \|T_I^* T_{I'}\|_{\mathscr{L}(L^2(G))}\right) \leq A_r 2^{-|j-j'|r}(1+|x_{i'}^{-1}x_i|)^{-rp_o}.$$

Then $T = \text{Op}(\sigma)$ is L^2-bounded with operator norm $\leq C\sqrt{A_r}$.

Lemma 5.7.10 can be easily shown, adapting for instance the proof given in [Ste93, ch. VII §2] using Part (4) of Lemma 5.7.5. Indeed, the numbering of the sequence of operators to which the Cotlar-Stein lemma (see Theorem A.5.2) is applied is not important, and the condition $rp_o > Q+1$ is motivated by Lemma 5.7.5, Part (4). This is left to the reader.

Lemma 5.7.11 which follows gives the operator norm for $T_I T_{I'}^*$ and $T_I^* T_{I'}$. Combining Lemmata 5.7.10 and 5.7.11 gives the proof of Proposition 5.7.7.

Lemma 5.7.11. *1. For any $r \in \mathbb{N}_0$, the operator norm of $T_I T_{I'}^*$ on $L^2(G)$ is*

$$\|T_I T_{I'}^*\|_{\mathscr{L}(L^2(G))} \leq C_r 1_{|j-j'|\leq 1}(1+|x_{i'}^{-1}x_i|)^{-rp_o}\|\sigma\|^2_{S^0_{0,0}, rp_o, \lceil \frac{Q}{2}\rceil, 0}.$$

2. For any $r \in \mathbb{N}_0$, the operator norm of $T_I^* T_{I'}$ on $L^2(G)$ is

$$\|T_I^* T_{I'}\|_{\mathscr{L}(L^2(G))} \leq C_r 1_{|x_{i'}^{-1}x_i|\leq 4C_o} 2^{-|j-j'|r}\|\sigma\|^2_{S^0_{0,0}, 0\ r\nu+\lceil \frac{Q}{2}\rceil, r\nu}.$$

In the proof of Lemma 5.7.11, we will also use the symbols σ_i, $i \in \mathbb{N}$, given by

$$\sigma_i(x,\pi) := \chi_i(x)\ \sigma(x,\pi),$$

and the corresponding operators $T_i = \text{Op}(\sigma_i)$ and kernels κ_i. We observe that σ_i is compactly supported in x, therefore by Lemma 5.7.9, the operator T_i is bounded on $L^2(G)$.

Proof of Lemma 5.7.11 Part (1). We have (see the end of Lemma 5.5.4)

$$T_I = \text{Op}(\sigma_I) = T_i\ \psi_j(\mathcal{R}),$$

thus

$$T_I T_{I'}^* = T_i \psi_j(\mathcal{R})\psi_{j'}(\mathcal{R})T_{i'}^*.$$

Since $\psi_j(\mathcal{R})\psi_{j'}(\mathcal{R}) = (\psi_j\psi_{j'})(\mathcal{R})$, this is 0 if $|j-j'| > 1$. Let us assume $|j-j'| \leq 1$. We set

$$T_{i'j'j} := T_{i'} \circ (\psi_j\psi_{j'})(\mathcal{R}) = \text{Op}\left(\sigma_{i'} \circ (\psi_j\psi_{j'})\left(\pi(\mathcal{R})\right)\right),$$

see again the end of Lemma 5.5.4. Therefore $T_I T_{I'}^* = T_i T_{i'j'j}^*$, and we have by the Sobolev inequality in Theorem 4.4.25,

$$\begin{aligned} |T_I T_{I'}^* \phi(x)| &= \left| \int_G T_{i'j'j}^* \phi(z) \ \kappa_{ix}(z^{-1}x) dz \right| \\ &\leq \sup_{x_o} \left| \int_G T_{i'j'j}^* \phi(z) \ \kappa_{ix_o}(z^{-1}x) dz \right| 1_{x \in B(x_i,2)} \\ &\leq C \sum_{[\beta] \leq \lceil \frac{Q}{2} \rceil} \left\| X_{x_o}^\beta \int_G T_{i'j'j}^* \phi(z) \kappa_{ix_o}(z^{-1}x) dz \right\|_{L^2(dx_o)} 1_{x \in B(x_i,2)}. \end{aligned}$$

Hence,

$$\|T_I T_{I'}^* \phi\|_{L^2} \leq C \sum_{[\beta] \leq \lceil \frac{Q}{2} \rceil} \left\| \int_G T_{i'j'j}^* \phi(z) X_{x_o}^\beta \kappa_{ix_o}(z^{-1}x) dz \ 1_{x \in B(x_i,2)} \right\|_{L^2(dx_o dx)}.$$

The idea of the proof is to use a quantity which will help the space localisation; so we introduce this quantity $1 + |z^{-1}x|^{rp_o}$ and its inverse, where the integer $r \in \mathbb{N}$ is to be chosen suitably. Notice that for the inverse we have

$$(1 + |z^{-1}x|^{rp_o})^{-1} \leq C_r (1 + |z^{-1}x|)^{-rp_o} \leq C_r (1 + |x_{i'}^{-1} x_i|)^{-rp_o},$$

for any $z \in \text{supp}\chi_{i'}$ and $x \in B(x_i, 2)$. Therefore, we obtain

$$\begin{aligned} &\left\| \int_G T_{i'j'j}^* \phi(z) X_{x_o}^\beta \kappa_{ix_o}(z^{-1}x) dz \ 1_{x \in B(x_i,2)} \right\|_{L^2(dx_o dx)} \\ &= \left\| \int_G T_{i'j'j}^* \phi(z) \frac{1 + |z^{-1}x|^{rp_o}}{1 + |z^{-1}x|^{rp_o}} X_{x_o}^\beta \kappa_{ix_o}(z^{-1}x) dz \ 1_{x \in B(x_i,2)} \right\|_{L^2(dx_o, dx)} \\ &\leq C(1 + |x_{i'}^{-1} x_i|)^{-rp_o} \left\| T_{i'j'j}^* \phi(z_1) \right\|_{L^2(dz_1)} \\ &\qquad \left\| (1 + |z_2^{-1}x|^{rp_o}) X_{x_o}^\beta \kappa_{ix_o}(z_2^{-1}x) \ 1_{x \in B(x_i,2)} \right\|_{L^2(dz_2, dx_o, dx)} \end{aligned}$$

by the observation just above and the Cauchy-Schwartz inequality. The last term can be estimated as

$$\begin{aligned} &\left\| (1 + |z_2^{-1}x|^{rp_o}) X_{x_o}^\beta \kappa_{ix_o}(z_2^{-1}x) \ 1_{x \in B(x_i,2)} \right\|_{L^2(dz_2, dx_o, dx)} \\ &\quad \leq |B(x_i, 2)| \sup_{x_o \in G} \left\| (1 + |z'|^{rp_o}) X_{x_o}^\beta \kappa_{ix_o}(z') \right\|_{L^2(dz')} \\ &\quad \leq C \sup_{x_o \in G} \sum_{[\alpha]=0}^{rp_o} \left\| X_{x_o}^\beta \Delta^\alpha \sigma_i(x_o, \pi) \right\|_{L^\infty(\widehat{G})} \end{aligned}$$

by the Plancherel theorem and Theorem 5.2.22, since $|z'|^{rp_o}$ can be written as a linear combination of $\tilde{q}_\alpha(z)$, $[\alpha] = rp_o$. Combining the estimates above, we have obtained

$$\|T_I T_{I'}^* \phi\|_{L^2} \leq C(1+|x_{i'}^{-1}x_i|)^{-rp_o} \left\|T_{i'j'j}^*\phi\right\|_{L^2} \sup_{\substack{x_o\in G\\ [\beta']\leq \lceil \frac{Q}{2}\rceil, [\alpha]\leq rp_o}} \left\|\Delta^\alpha X_{x_o}^{\beta'}\sigma(x_o,\pi)\right\|_{L^\infty(\widehat{G})}.$$

The supremum is equal to $\|\sigma\|_{S^0_{0,0}, rp_o, \lceil \frac{Q}{2}\rceil, 0}$. So we now want to study the operator norm of $T_{i'j'j}^*$, which is equal to the operator norm of $T_{i'j'j}$. Since the symbol of $T_{i'j'j}$ is localised in space we may apply Lemma 5.7.9 and obtain

$$\begin{aligned}
&\|T_{i'j'j}^*\|_{\mathscr{L}(L^2(G))} = \|T_{i'j'j}\|_{\mathscr{L}(L^2(G))} = \|\mathrm{Op}\left(\sigma_i\left(\psi_j\psi_{j'}\right)(\pi(\mathcal{R}))\right)\|_{\mathscr{L}(L^2(G))}\\
&\leq C|B(x_i,2)|^{1/2} \sup_{\substack{x\in G\\ [\beta]\leq\lceil Q/2\rceil}} \|X_x^\beta\left\{\chi_i(x)\sigma(x,\pi)\left(\psi_j\psi_{j'}\right)(\pi(\mathcal{R}))\right\}\|_{L^\infty(\widehat{G})}\\
&\leq C \sup_{\substack{x\in G,\,\pi\in\widehat{G}\\ [\beta]\leq\lceil Q/2\rceil}} \sum_{[\beta_1]+[\beta_2]=[\beta]} |X^{\beta_1}\chi_i(x)|\ \|X_x^{\beta_2}\sigma(x,\pi)\|_{\mathscr{L}(\mathcal{H}_\pi)}\|\left(\psi_j\psi_{j'}\right)(\pi(\mathcal{R}))\|_{\mathscr{L}(\mathcal{H}_\pi)}\\
&\leq C \sup_{\substack{x\in G,\,\pi\in\widehat{G}\\ [\beta_2]\leq\lceil Q/2\rceil}} \|X_x^{\beta_2}\sigma(x,\pi)\|_{\mathscr{L}(\mathcal{H}_\pi)} = C\|\sigma\|_{S^0_{0,0}, 0, \lceil Q/2\rceil, 0},
\end{aligned}$$

since the $X^{\beta_2}\chi_i$'s are uniformly bounded on G and over i.

Thus, we have obtained

$$\|T_I T_{I'}^*\phi\|_{L^2} \leq C(1+|x_{i'}^{-1}x_i|)^{-rp_o}\|\sigma\|_{S^0_{0,0},0,\lceil Q/2\rceil,0}\,\|\phi\|_{L^2}\,\|\sigma\|_{S^0_{0,0},rp_o,\lceil\frac{Q}{2}\rceil,0},$$

and this concludes the proof of the first part of Lemma 5.7.11. □

Proof of Lemma 5.7.11 Part (2). Recall that each $\kappa_{Ix}(y)$ is supported, with respect to x, in the ball $B(x_i,2)$. We compute easily that the kernel of $T_I^*T_{I'}$ is

$$\kappa_{I*I'}(x,w) = \int_G \kappa_{I'xz^{-1}}(wz^{-1})\kappa^*_{Ixz^{-1}}(z)dz.$$

Therefore, $\kappa_{I*I'}$ is identically 0 if there is no z such that $xz^{-1}\in B(x_i,2)\cap B(x_{i'},2)$. So if $|x_{i'}^{-1}x_i| > 4C_o$ (which implies $B(x_i,2)\cap B(x_{i'},2)=\emptyset$) then $T_I^*T_{I'}=0$. So we may assume $|x_{i'}^{-1}x_i|\leq 4C_o$.

The idea of the proof is to use a quantity which will help the frequency localisation; so we introduce this quantity $(\mathrm{I}+\mathcal{R})^r$ and its inverse, where the integer $r\in\mathbb{N}$ is to be chosen suitably. We can write

$$T_I^*T_{I'} = T_I^*T_{i'}\psi_{j'}(\mathcal{R}) = T_I^*T_{i'}(\mathrm{I}+\mathcal{R})^r\ (\mathrm{I}+\mathcal{R})^{-r}\psi_{j'}(\mathcal{R}).$$

By the functional calculus (see Corollary 4.1.16),

$$\|(\mathrm{I}+\mathcal{R})^{-r}\psi_{j'}(\mathcal{R})\|_{\mathscr{L}(L^2(G))} = \sup_{\lambda\geq 0}(1+\lambda)^{-r}\psi_{j'}(\lambda)\leq C_r 2^{-j'r}.$$

Thus we need to study $T_I^* T_{i'}(\mathrm{I}+\mathcal{R})^r$. We see that its kernel is

$$\begin{aligned}\kappa_x(w) &= \int_G (\mathrm{I}+\tilde{\mathcal{R}})^r \kappa_{i' xz^{-1}}(wz^{-1})\kappa^*_{I xz^{-1}}(z)dz \\ &= \int_G (\mathrm{I}+\tilde{\mathcal{R}})^r \kappa_{i' xw^{-1}z}(z)\kappa^*_{I xw^{-1}z}(z^{-1}w)dz.\end{aligned}$$

We introduce $(\mathrm{I}+\mathcal{R})^r(\mathrm{I}+\mathcal{R})^{-r}$ on the first term of the integrand acting on the variable of $\kappa_{i'xw^{-1}z}$, and then integrate by parts to obtain

$$\begin{aligned}\kappa_x(w) &= \overline{\sum_{[\beta_1]+[\beta_2]+[\beta_3]=r\nu}} \int_G X^{\beta_1}_{z_1=z}(\mathrm{I}+\mathcal{R})^{-r}(\mathrm{I}+\tilde{\mathcal{R}})^r \kappa_{i' xw^{-1}z_1}(z) \\ &\qquad\qquad X^{\beta_2}_{z_2=z} X^{\beta_3}_{z_3=z} \kappa^*_{I xw^{-1}z_2}(z_3^{-1}w)dz \\ &= \overline{\sum_{[\beta_1]+[\beta_2]+[\beta_3]=r\nu}} \int_G X^{\beta_1}_{z_1=xw^{-1}z}(\mathrm{I}+\mathcal{R})^{-r}(\mathrm{I}+\tilde{\mathcal{R}})^r \kappa_{i' z_1}(z) \\ &\qquad\qquad X^{\beta_2}_{z_2=xw^{-1}z}(X^{\beta_3}\kappa_{I z_2})^*(z^{-1}w)dz.\end{aligned}$$

Re-interpreting this in terms of operators, we obtain

$$\begin{aligned}T_I^* T_{i'}(\mathrm{I}+\mathcal{R})^r = \overline{\sum_{[\beta_1]+[\beta_2]+[\beta_3]=r\nu}} & \mathrm{Op}\left(\pi(X^{\beta_3})X_x^{\beta_2}\sigma_I(x,\pi)\right)^* \\ & \mathrm{Op}\left(\pi(\mathrm{I}+\mathcal{R})^{-r}X_x^{\beta_1}\sigma_{i'}(x,\pi)\pi(\mathrm{I}+\mathcal{R})^r\right).\end{aligned}$$

By Lemma 5.7.9,

$$\begin{aligned}&\|\mathrm{Op}\left(\pi(\mathrm{I}+\mathcal{R})^{-r}X_x^{\beta_1}\sigma_{i'}(x,\pi)\pi(\mathrm{I}+\mathcal{R})^r\right)\|_{\mathscr{L}(L^2(G))} \\ &\quad \le C \sup_{\substack{x\in G \\ [\beta]\le\lceil\frac{Q}{2}\rceil}} \|\pi(\mathrm{I}+\mathcal{R})^{-r}X_x^{\beta}X_x^{\beta_1}\sigma_{i'}(x,\pi)\pi(\mathrm{I}+\mathcal{R})^r\|_{L^\infty(\widehat{G})} \\ &\quad \le \|\sigma\|_{S^0_{0,0},0,[\beta_1]+\lceil\frac{Q}{2}\rceil,r\nu},\end{aligned}$$

and

$$\begin{aligned}&\|\mathrm{Op}\left(\pi(X^{\beta_3})X_x^{\beta_2}\sigma_I(x,\pi)\right)\|_{\mathscr{L}(L^2(G))} \\ &\quad \le \sup_{[\beta]\le\lceil\frac{Q}{2}\rceil} \|\pi(X^{\beta_3})X_x^{\beta}X_x^{\beta_2}\sigma_i(x,\pi)\psi_j(\pi(\mathcal{R}))\|_{L^\infty(\widehat{G})} \\ &\quad \le \sup_{[\beta]\le\lceil\frac{Q}{2}\rceil} \|\pi(X^{\beta_3})\pi(\mathrm{I}+\mathcal{R})^{-\frac{[\beta_3]}{\nu}}\|_{L^\infty(\widehat{G})} \times \\ &\qquad\quad \times \|\pi(\mathrm{I}+\mathcal{R})^{\frac{[\beta_3]}{\nu}}X_x^{\beta+\beta_2}\sigma_i(x,\pi)\pi(\mathrm{I}+\mathcal{R})^{-\frac{[\beta_3]}{\nu}}\|_{L^\infty(\widehat{G})} \times \\ &\qquad\quad \times \|\pi(\mathrm{I}+\mathcal{R})^{\frac{[\beta_3]}{\nu}}\psi_j(\pi(\mathcal{R}))\|_{L^\infty(\widehat{G})} \\ &\quad \le C_\beta 2^{j\frac{[\beta_3]}{\nu}}\|\sigma\|_{S^0_{0,0},0,[\beta_2]+\lceil\frac{Q}{2}\rceil,[\beta_3]},\end{aligned}$$

by Lemma 5.4.7. Hence we have obtained

$$\begin{aligned}\|T_I^* T_{I'}\|_{\mathscr{L}(L^2(G))} &\leq C_r 2^{-j'r} \sum_{[\beta_1]+[\beta_2]+[\beta_3]=r\nu} \|\sigma\|^2_{S^0_{0,0},0,r\nu+\lceil\frac{Q}{2}\rceil,r\nu} 2^{j\frac{[\beta_3]}{\nu}} \\ &\leq C_r 2^{(j-j')r} \|\sigma\|^2_{S^0_{0,0},0,r\nu+\lceil\frac{Q}{2}\rceil,r\nu}.\end{aligned}$$

This shows Part 2 of Lemma 5.7.11 up to the fact that we should have $-|j-j'|$ instead of $(j-j')$ but this can be deduced easily by reversing the rôle of I and I', and using $\|T\|_{\mathscr{L}(L^2(G))} = \|T^*\|_{\mathscr{L}(L^2(G))}$. □

This concludes the proof of Lemma 5.7.11. Therefore, by Lemma 5.7.10, Proposition 5.7.7 is also proved.

5.7.3 A bilinear estimate

In this section, we prove a bilinear estimate which will be the major ingredient in the proof of the L^2-boundedness for operators of orders 0 in the case $\rho = \delta \in (0,1)$ in Section 5.7.4.

Note that if $f, g \in \mathcal{S}(G)$ and if $\gamma \in \mathbb{N}_0$ then the Leibniz properties together with the properties of the Sobolev spaces (cf. Theorem 4.4.28, especially the Sobolev embeddings in Part (5)) imply

$$\begin{aligned}\|(\mathrm{I}+\mathcal{R})^\gamma(fg)\|_{L^2(G)} &\lesssim \sum_{[\beta_1]+[\beta_2]\leq\nu\gamma} \|X^{\alpha_1} f\, X^{\alpha_2} g\|_{L^2(G)} \\ &\lesssim \sum_{[\beta_1]+[\beta_2]\leq\nu\gamma} \|X^{\alpha_1} f\|_{L^\infty(G)} \|X^{\alpha_2} g\|_{L^2(G)} \\ &\lesssim \sum_{[\beta_1]+[\beta_2]\leq\nu\gamma} \|X^{\alpha_1} f\|_{H^s(G)} \|X^{\alpha_2} g\|_{L^2(G)} \\ &\lesssim \|f\|_{H^{s+\nu\gamma}(G)} \|g\|_{H^{\nu\gamma}(G)},\end{aligned}$$

where $s > Q/2$. As usual, $\mathcal{R}$ is a positive Rockland operator of homogeneous degree ν; we denote by E its spectral decomposition, see Corollary 4.1.16. Consequently, if f, g are localised in the spectrum of $\mathcal{R}$ in the sense that $f = E(I_i)f$, $g = E(I_j)g$, where I_i, I_j are the dyadic intervals given via

$$I_j := (2^{j-2}, 2^j), \quad j \in \mathbb{N}, \quad \text{and} \quad I_0 := [0,1), \tag{5.72}$$

we obtain easily

$$\|(\mathrm{I}+\mathcal{R})^\gamma(fg)\|_{L^2(G)} \lesssim \|f\|_{L^2(G)} \|g\|_{L^2(G)} 2^{(\gamma+\frac{s}{\nu})\max(i,j)}. \tag{5.73}$$

Our aim in this section is to prove a similar result but for $\gamma \ll 0$:

Proposition 5.7.12. *Let $\mathcal{R}$ be a positive Rockland operator of homogeneous degree ν. As usual, we denote by E its spectral decomposition. There exists a constant $C > 0$ such that for any $\gamma \in \mathbb{R}$ with $\gamma + Q/(2\nu) < 0$, for any $i, j \in \mathbb{N}_0$ with $|i - j| > 3$, we have*

$$\begin{aligned} &\forall f, g \in L^2(G) \qquad f = E(I_i)f \quad \text{and} \quad g = E(I_j)g \\ &\Longrightarrow \|(\mathrm{I} + \mathcal{R})^\gamma (fg)\|_{L^2(G)} \leq C\|f\|_{L^2}\|g\|_{L^2} 2^{(\gamma + \frac{Q}{2\nu}) \max(i,j)}. \end{aligned}$$

The intervals I_i, I_j were defined via (5.72). The proof of Proposition 5.7.12 relies on the following lemma:

Lemma 5.7.13. *Let $\mathcal{R}$ be a positive Rockland operator. As in Corollary 4.1.16, for any strongly continuous unitary representation π_1 on G, E_{π_1} denotes the spectral decomposition of $\pi_1(\mathcal{R})$. There exists a 'gap' constant $a \in \mathbb{N}$ such that for any $i, j, k \in \mathbb{N}_0$ with $k < j - a$ and $i \leq j - 4$, we have*

$$\forall \tau, \pi \in \widehat{G} \qquad E_{\tau \otimes \pi}(I_i)\big(E_\tau(I_j) \otimes E_\pi(I_k)\big) = 0.$$

and

$$\forall \tau, \pi \in \widehat{G} \qquad \big(E_\tau(I_j) \otimes E_\pi(I_k)\big) E_{\tau \otimes \pi}(I_i) = 0.$$

Proof of Lemma 5.7.13. We keep the notation of the statement. We also set

$$\mathcal{H}_{\pi_1, j} := E_{\pi_1}(I_j), \qquad j \in \mathbb{N}_0,$$

for any strongly continuous unitary representation π_1 on G. We can write $\mathcal{R}$ as a linear combination

$$\mathcal{R} = \sum_{[\alpha]=\nu} c_\alpha X^\alpha,$$

for some complex coefficients c_α. For any strongly continuous unitary representation π_1, we have

$$\pi_1(\mathcal{R}) = \sum_{[\alpha]=\nu} c_\alpha \pi_1(X)^\alpha.$$

Let $\tau, \pi \in \widehat{G}$. We consider the strongly continuous unitary representation $\pi_1 = \tau \otimes \pi$. For any $X \in \mathfrak{g}$, its infinitesimal representation is given via $\pi_1(X) = X_{x=0}\{\pi_1(x)\}$, see Section 1.7. Consequently, we have for any $u \in \mathcal{H}_\tau, v \in \mathcal{H}_\pi$,

$$\begin{aligned} \pi_1(X)(u, v) &= X_{x=0}\pi_1(x)(u, v) \\ &= X_{x=0}\tau(x)u \otimes \pi(x)v \\ &= \tau(X)u \otimes v + u \otimes \pi(X)v. \end{aligned}$$

In other words,

$$(\tau \otimes \pi)(X) = \tau(X) \otimes \mathrm{I}_{\mathcal{H}_\pi} + \mathrm{I}_{\mathcal{H}_\tau} \otimes \pi(X).$$

We obtain iteratively

$$(\tau\otimes\pi)(X)^\alpha=\tau(X)^\alpha\otimes \mathrm{I}_{\mathcal{H}_\pi}+\mathrm{I}_{\mathcal{H}_\tau}\otimes\pi(X)^\alpha+\overline{\sum_{\substack{[\beta_1]+[\beta_2]=[\alpha]\\0<[\beta_1],[\beta_2]<[\alpha]}}}\tau(X)^{\beta_1}\otimes\pi(X)^{\beta_2},$$

where $\overline{\sum}$ denotes a linear combination which depends only on $\alpha\in\mathbb{N}_0^n$ and on the structure of G but not on $\tau,\pi\in\widehat{G}$. This easily implies

$$\begin{aligned}(\tau\otimes\pi)(\mathcal{R})&=\sum_{[\alpha]=\nu}c_\alpha(\tau\otimes\pi)(X)^\alpha\\&=\tau(\mathcal{R})\otimes \mathrm{I}_{\mathcal{H}_\pi}+\mathrm{I}_{\mathcal{H}_\tau}\otimes\pi(\mathcal{R})+\overline{\sum_{\substack{[\beta_1]+[\beta_2]=\nu\\0<[\beta_1],[\beta_2]<\nu}}}\tau(X)^{\beta_1}\otimes\pi(X)^{\beta_2},\end{aligned}$$

where $\overline{\sum}$ denotes a linear combination which depends only on $\mathcal{R}$ and on the structure of G but not on π,τ. Hence there exists a constant $C>0$ independent of π,τ such that for any $u\in\mathcal{H}_\tau$, $v\in\mathcal{H}_\pi$, we have

$$\begin{aligned}\|(\tau\otimes\pi)(\mathcal{R})(u\otimes v)\|_{\mathcal{H}_{\tau\otimes\pi}}\ \geq\ &\|\tau(\mathcal{R})u\|_{\mathcal{H}_\tau}\|v\|_{\mathcal{H}_\pi}-\|u\|_{\mathcal{H}_\tau}\|\pi(\mathcal{R})v\|_{\mathcal{H}_\pi}\\&-C\sum_{\substack{[\beta_1]+[\beta_2]=\nu\\0<[\beta_1],[\beta_2]<\nu}}\|\tau(X)^{\beta_1}u\|_{\mathcal{H}_\tau}\|\pi(X)^{\beta_2}v\|_{\mathcal{H}_\pi}.\end{aligned}$$

If $u\in\mathcal{H}_{\tau,j}$ then from the properties of the functional calculus of $\tau(\mathcal{R})$, we have

$$\|\tau(\mathcal{R})u\|_{\mathcal{H}_\tau}\in\|u\|_{\mathcal{H}_\tau}I_j.$$

Furthermore, the properties of the functional calculus of $\mathcal{R}$ and $\tau(\mathcal{R})$ yield

$$\|\tau(X)^{\beta_1}u\|_{\mathcal{H}_\tau}\leq\|\tau(X)^{\beta_1}E_\tau(I_j)\|_{\mathscr{L}(\mathcal{H}_\tau)}\|u\|_{\mathcal{H}_\tau},$$

and, as $X^{\beta_1}\mathcal{R}^{-\frac{[\beta_1]}{\nu}}$ is bounded on $L^2(G)$ by Theorem 4.4.16, we have

$$\begin{aligned}\|\tau(X)^{\beta_1}E_\tau(I_j)\|_{\mathscr{L}(\mathcal{H}_\tau)}\ &\leq\ \|X^{\beta_1}E(I_j)\|_{\mathscr{L}(L^2(G))}\\&\leq\ \|X^{\beta_1}\mathcal{R}^{-\frac{[\beta_1]}{\nu}}\|_{\mathscr{L}(L^2(G))}\|\mathcal{R}^{\frac{[\beta_1]}{\nu}}E(I_j)\|_{\mathscr{L}(L^2(G))}\\&\lesssim\ 2^{j\frac{[\beta_1]}{\nu}}.\end{aligned}$$

We have similar inequalities for $v\in\mathcal{H}_{\pi,k}$. For any unit vectors $u\in\mathcal{H}_{\tau,j}$ and $v\in\mathcal{H}_{\pi,k}$ with $j,k\in\mathbb{N}$, we then have

$$\|(\tau\otimes\pi)(\mathcal{R})(u\otimes v)\|_{\mathcal{H}_{\tau\otimes\pi}}\geq 2^{j-2}-2^k-C_1\sum_{\substack{[\beta_1]+[\beta_2]=\nu\\0<[\beta_1],[\beta_2]<\nu}}2^{\frac{j[\beta_1]+k[\beta_2]}{\nu}},$$

where the constant C_1 depends only on $\mathcal{R}$ and on the structure of G. We notice that

$$\sum_{\substack{[\beta_1]+[\beta_2]=\nu \\ 0<[\beta_1],[\beta_2]<\nu}} 2^{\frac{j[\beta_1]+k[\beta_2]}{\nu}} = 2^j \sum_{\substack{[\beta_1]+[\beta_2]=\nu \\ 0<[\beta_1],[\beta_2]<\nu}} 2^{\frac{[\beta_2]}{\nu}(k-j)} \leq 2^j C' 2^{-av_1},$$

if $k-j \leq -a$. Here C' is a constant which depends on the structure of G and on ν. We choose $a \in \mathbb{N}$ the smallest integer such that

$$CC'2^{-av_1+2} < 1/2 \quad \text{and} \quad 2^{-a+3} < 1/2.$$

Note that a depends only on the structure of G and on $\mathcal{R}$. When $k-j \leq -a$, we have obtained

$$\begin{aligned}\|(\tau\otimes\pi)(\mathcal{R})(u\otimes v)\|_{\mathcal{H}_{\tau\otimes\pi}} &\geq 2^{j-2} - 2^k - C2^jC'2^{-av_1} \\ &= 2^{j-2}(1-CC'2^{-av_1+2}) - 2^k \\ &> 2^{j-3} - 2^{j-a} > 2^{j-4}.\end{aligned}$$

This implies that $u\otimes v$ can not be in $\mathcal{H}_{\tau\otimes\mathcal{R},\pi}$ for $i \in \mathbb{N}_0$ such that $2^i \leq 2^{j-4}$. This shows the first equality of the statement when $i,j,k \in \mathbb{N}$. The case of $k=0$ or $i=0$ requires to modify slightly some constants above and is left to the reader. This shows the first equality of the statement and the second follows by taking the adjoint. This concludes the proof of Lemma 5.7.13. □

Proof of Proposition 5.7.12. We keep the notation of Proposition 5.7.12 and Lemma 5.7.13. We notice that it suffices to prove the statement for large enough $\max(i,j)$ and that the rôles of i and j are symmetric. Hence we may assume that $i \leq j-4$ and that $j \geq a$ where a is the 'gap' constant of Lemma 5.7.13

Let $f, g \in L^2(G)$ such that $f = E(I_i)f$ and $g = E(I_j)g$. The inverse formula for g yields

$$(\mathrm{I}+\mathcal{R})^\gamma(fg)(x) = \int_{\widehat{G}} \mathrm{Tr}\big(\pi(g)(\mathrm{I}+\mathcal{R})^\gamma_x\{f(x)\pi(x)\}\big)d\mu(\pi).$$

We also have $\pi(g) = E_\pi(I_j)\pi(g)$. By the Cauchy-Schwartz inequality and the Plancherel formula, we obtain

$$|(\mathrm{I}+\mathcal{R})^\gamma(fg)(x)|^2 \leq \|g\|^2_{L^2(G)} \int_{\widehat{G}} \|E_\pi(I_j)(\mathrm{I}+\mathcal{R})^\gamma_x\{f(x)\pi(x)\}\|^2_{\mathtt{HS}} d\mu(\pi).$$

Integrating on both side over $x \in G$, we have

$$\|(\mathrm{I}+\mathcal{R})^\gamma(fg)\|^2_{L^2(G)} \leq \|g\|^2_{L^2} \int_{\widehat{G}}\int_G \|E_\pi(I_j)(\mathrm{I}+\mathcal{R})^\gamma_x\{f(x)\pi(x)\}\|^2_{\mathtt{HS}} dx d\mu(\pi).$$

For each $\pi \in \widehat{G}$, we fix an orthonormal basis of $\mathcal{H}_\pi$, so that we can write the Hilbert-Schmidt norm as the square of the coefficients of a (possibly infinite dimensional) matrix. The Plancherel formula then yields

$$\begin{aligned}
&\int_G \|E_\pi(I_j)(\mathrm{I}+\mathcal{R})_x^\gamma\{f(x)\pi(x)\}\|_{\mathtt{HS}}^2 dx \\
&\quad = \sum_{kl} \int_G |[E_\pi(I_j)(\mathrm{I}+\mathcal{R})_x^\gamma\{f(x)\pi(x)\}]_{kl}|^2 dx \\
&\quad = \sum_{kl} \int_{\widehat{G}} \|\mathcal{F}\left[E_\pi(I_j)(\mathrm{I}+\mathcal{R})^\gamma f\pi\right]_{kl}(\tau)\|_{\mathtt{HS}(\mathcal{H}_\tau)}^2 d\mu(\tau),
\end{aligned}$$

where

$$\begin{aligned}
&\mathcal{F}\left[E_\pi(I_j)(\mathrm{I}+\mathcal{R})^\gamma f\pi\right]_{kl}(\tau) \\
&\quad = \int_G (\mathrm{I}+\mathcal{R})_x^\gamma \left\{f(x)[E_\pi(I_j)\pi(x)]_{kl}\right\} \ \tau(x)^* dx \\
&\quad = \tau(\mathrm{I}+\mathcal{R})^\gamma \int_G f(x)[E_\pi(I_j)\pi(x)]_{kl} \ \tau(x)^* dx \\
&\quad = \left[E_\pi(I_j) \otimes \tau(\mathrm{I}+\mathcal{R})^\gamma \int_G f(x)(\pi\otimes\tau^*)(x) dx\right]_{kl,\cdot}.
\end{aligned}$$

Here the notation $[\cdot]_{kl,\cdot}$ means considering the (kl)-coefficients in $\mathcal{H}_\pi$ in the tensor product over $\mathcal{H}_\pi \otimes \mathcal{H}_\tau$. We recognise

$$\int_G f(x)(\pi\otimes\tau^*)(x)dx = (\pi^*\otimes\tau)(f)$$

thus

$$\begin{aligned}
&\sum_{kl} \|\mathcal{F}[E_\pi(I_j)(\mathrm{I}+\mathcal{R})^\gamma f\pi]_{kl}(\tau)\|_{\mathtt{HS}(\mathcal{H}_\tau)}^2 \\
&\quad = \|\left(E_\pi(I_j)\otimes\tau(\mathrm{I}+\mathcal{R})^\gamma\right)\left((\pi^*\otimes\tau)(f)\right)\|_{\mathtt{HS}(\mathcal{H}_\pi\otimes\mathcal{H}_\tau)}^2.
\end{aligned}$$

So far, we have obtained

$$\begin{aligned}
&\int_{\widehat{G}}\int_G \|E_\pi(I_j)(\mathrm{I}+\mathcal{R})_x^\gamma\{f(x)\pi(x)\}\|_{\mathtt{HS}}^2 dx d\mu(\pi) \\
&= \int_{\widehat{G}}\int_{\widehat{G}} \|\left(E_\pi(I_j)\otimes\tau(\mathrm{I}+\mathcal{R})^\gamma\right)\left((\pi^*\otimes\tau)(f)\right)\|_{\mathtt{HS}(\mathcal{H}_\pi\otimes\mathcal{H}_\tau)}^2 d\mu(\tau)d\mu(\pi) \\
&= \|\|\left(E_\pi(I_j)\otimes\tau(\mathrm{I}+\mathcal{R})^\gamma\right)\left((\pi^*\otimes\tau)(f)\right)\|_{\mathtt{HS}(\mathcal{H}_\pi\otimes\mathcal{H}_\tau)}\|_{L^2(d\mu(\tau),d\mu(\pi))}^2.
\end{aligned}$$

We fix a dyadic decomposition, that is, we fix $\psi_0, \psi_1 \in \mathcal{D}(\mathbb{R})$ supported in $(-1,1)$ and $(1/2,2)$, respectively, valued in $[0,1]$ and such that

$$\forall \lambda \geq 0 \quad \sum_{k=0}^\infty \psi_k(\lambda) = 1 \quad \text{with } \psi_k(\lambda) = \psi_1(2^{-(k-1)}\lambda) \text{ if } k \in \mathbb{N}.$$

The series $\sum_k \psi_k(\tau(\mathcal{R}))$ converges to $\mathrm{I}_{\mathcal{H}_\tau}$ in the strong operator topology and we can apply the following general property:

$$\begin{aligned}&\|(B \otimes C)A\|_{\mathtt{HS}(\mathcal{H}_\pi \otimes \mathcal{H}_\tau)}\\ &\qquad \leq \sum_{k=0}^{\infty} \|E_\tau(I_k)C\|_{\mathscr{L}(\mathcal{H}_\tau)} \|(B \otimes \psi_k(\tau(\mathcal{R})))A\|_{\mathtt{HS}(\mathcal{H}_\pi \otimes \mathcal{H}_\tau)},\end{aligned}$$

to $B = E_\pi(I_j)$, $C = \tau(\mathrm{I} + \mathcal{R})^\gamma$, and

$$A = (\pi^* \otimes \tau)(f).$$

We keep momentarily this notation for A and C. As $\|E_\tau(I_k)C\|_{\mathscr{L}(\mathcal{H}_\tau)} \lesssim 2^{\gamma k}$, we have obtained

$$\begin{aligned}&\|\|\left(E_\pi(I_j) \otimes \tau(\mathrm{I} + \mathcal{R})^\gamma\right) A\|_{\mathtt{HS}(\mathcal{H}_\pi \otimes \mathcal{H}_\tau)}\|_{L^2(d\mu(\tau), d\mu(\pi))}\\ &\quad \lesssim \sum_{k=0}^{\infty} 2^{\gamma k} \|\|\left(E_\pi(I_j) \otimes \psi_k(\tau(\mathcal{R}))\right) A\|_{\mathtt{HS}(\mathcal{H}_\pi \otimes \mathcal{H}_\tau)}\|_{L^2(d\mu(\tau), d\mu(\pi))}.\end{aligned}$$

Now

$$A = ((\pi^* \otimes \tau)(f)) = E_{\pi^* \otimes \tau}(I_i)\left((\pi^* \otimes \tau)(f)\right),$$

thus we can apply Lemma 5.7.13 and the sum over k above is in fact from $k \geq j - a$. We claim that

$$\|\|\left(E_\pi(I_j) \otimes \psi_k(\tau(\mathcal{R}))\right) A\|_{\mathtt{HS}(\mathcal{H}_\pi \otimes \mathcal{H}_\tau)}\|_{L^2(d\mu(\tau), d\mu(\pi))} \lesssim \|f\|_{L^2(G)} 2^{k\frac{Q}{2\nu}}. \tag{5.74}$$

Collecting the equalities and estimates above, (5.74) would then imply

$$\|(\mathrm{I} + \mathcal{R})^\gamma (fg)\|^2_{L^2(G)} \lesssim \|g\|^2_{L^2} \|f\|^2_{L^2(G)} \sum_{k=j-a}^{\infty} 2^{k(\gamma + \frac{Q}{2\nu})},$$

and would conclude the proof of Proposition 5.7.12.

Hence it just remains to prove (5.74). Natural properties of tensor product and functional calculus yield

$$\begin{aligned}&\|\left(E_\pi(I_j) \otimes \psi_k(\tau(\mathcal{R}))\right) A\|_{\mathtt{HS}(\mathcal{H}_\pi \otimes \mathcal{H}_\tau)}\\ &\quad \leq \|E_\pi(I_j)\|_{\mathscr{L}(\mathcal{H}_\pi)} \|\left(\mathrm{I}_{\mathcal{H}_\pi} \otimes \psi_k(\tau(\mathcal{R}))\right) A\|_{\mathtt{HS}(\mathcal{H}_\pi \otimes \mathcal{H}_\tau)}\\ &\quad \leq \|\left(\mathrm{I}_{\mathcal{H}_\pi} \otimes \psi_k(\tau(\mathcal{R}))\right) A\|_{\mathtt{HS}(\mathcal{H}_\pi \otimes \mathcal{H}_\tau)}.\end{aligned}$$

We notice that

$$\left(\mathrm{I}_{\mathcal{H}_\pi} \otimes \psi_k(\tau(\mathcal{R}))\right) A = \int_G f(x)(\pi \otimes \psi_k(\tau(\mathcal{R}))\tau^*)(x) dx,$$

and introducing an orthonormal basis on $\mathcal{H}_\tau$,

$$\begin{aligned}[(\mathrm{I}_{\mathcal{H}_\pi} \otimes \psi_k(\tau(\mathcal{R})))\,A]_{\cdot,l'k'} &= \int_G f(x)\,[\psi_k(\tau(\mathcal{R}))]_{l'k'}\,\pi(x)dx \\ &= \mathcal{F}\,[f\psi_k(\tau(\mathcal{R}))]_{l'k'}\,(\pi^*) = \mathcal{F}\{[f\psi_k(\tau(\mathcal{R}))]_{l'k'}\,(\cdot^{-1})\}(\pi).\end{aligned}$$

Therefore we have

$$\begin{aligned}&\|\|\,(\mathrm{I}_{\mathcal{H}_\pi} \otimes \psi_k(\tau(\mathcal{R})))\,A\|_{\mathtt{HS}(\mathcal{H}_\pi\otimes\mathcal{H}_\tau)}\|^2_{L^2(d\mu(\tau),d\mu(\pi))} \\ &= \int_{\widehat{G}}\sum_{k'l'}\int_{\widehat{G}} \|\mathcal{F}\,[f\psi_k(\tau(\mathcal{R}))]_{l'k'}\,(\pi^*)\|^2_{\mathtt{HS}(\mathcal{H}_\tau)}d\mu(\pi)d\mu(\tau) \\ &= \int_{\widehat{G}}\sum_{k'l'} \|[f\psi_k(\tau(\mathcal{R}))]_{l'k'}\,(\cdot^{-1})\|^2_{L^2(G)}\,d\mu(\tau),\end{aligned}$$

having applied the Plancherel formula in π. Simple manipulations yield

$$\begin{aligned}\sum_{k'l'}\|[f\psi_k(\tau(\mathcal{R}))]_{l'k'}\,(\cdot^{-1})\|^2_{L^2(G)} &= \sum_{k'l'}\|[f\psi_k(\tau(\mathcal{R}))]_{l'k'}\|^2_{L^2(G)} \\ &= \sum_{k'l'}\int_G |f(x)\,[\psi_k(\tau(\mathcal{R}))]_{l'k'}\,|^2dx \\ &= \int_G |f(x)|^2dx\sum_{k'l'}|\,[\psi_k(\tau(\mathcal{R}))]_{l'k'}\,|^2 \\ &= \|f\|^2_{L^2(G)}\|\psi_k(\tau(\mathcal{R}))\|^2_{\mathtt{HS}(\mathcal{H}_\tau)}.\end{aligned}$$

Integrating over $\tau \in \widehat{G}$, we can apply the Plancherel formula and obtain

$$\int_{\widehat{G}}\sum_{k'l'}\|[f\psi_k(\tau(\mathcal{R}))]_{l'k'}\,(\cdot^{-1})\|^2_{L^2(G)}\,d\mu(\tau) = \|f\|^2_{L^2(G)}\|\psi_k(\mathcal{R})\delta_0\|^2_{L^2(G)}.$$

Using the properties of dilations, we have for any $k \in \mathbb{N}$:

$$\|\psi_k(\mathcal{R})\delta_0\|_{L^2(G)} = 2^{\frac{Q}{2}\frac{k-1}{\nu}}\|\psi_1(\mathcal{R})\delta_0\|_{L^2(G)}.$$

Collecting the equalities and inequalities above yields that the left-hand side of (5.74) is

$$\begin{aligned}&\|\|\,(E_\pi(I_j) \otimes \psi_k(\tau(\mathcal{R})))\,A\|_{\mathtt{HS}(\mathcal{H}_\pi\otimes\mathcal{H}_\tau)}\|_{L^2(d\mu(\tau),d\mu(\pi))} \\ &\le \|f\|_{L^2(G)}2^{\frac{Q}{2}\frac{k-1}{\nu}}\|\psi_1(\mathcal{R})\delta_0\|_{L^2(G)}.\end{aligned}$$

By Hulanicki's theorem, see Corollary 4.5.2, $\|\psi_1(\mathcal{R})\delta_0\|_{L^2(G)}$ is a finite constant. This shows (5.74) and concludes the proof of Proposition 5.7.12. □

5.7.4 Proof of the case $S^0_{\rho,\rho}$

In this section, we prove the L^2-boundedness of operators in $\Psi^0_{\rho,\rho}$ with $\rho \in (0,1)$:

Proposition 5.7.14. *Let $\sigma \in S^0_{\rho,\rho}$ with $\rho \in (0,1)$. Then $\mathrm{Op}(\sigma)$ is bounded on $L^2(G)$ and the operator norm is, up to a constant, less than a seminorm of $\sigma \in S^0_{\rho,\rho}$; the parameters of the seminorm depend on ρ but not on σ and could be computed explicitly.*

The rest of this section is devoted to the proof of Proposition 5.7.14. The strategy is broadly similar to the one in [Ste93, ch VII §2.5] for the Euclidean case. Technically, this means using analogous rescaling arguments but also replacing certain integrations by parts on the (Euclidean) Fourier side with the bilinear estimate obtained in Proposition 5.7.12.

Strategy of the proof

We fix a dyadic decomposition, that is, we fix $\psi_0, \psi_1 \in \mathcal{D}(\mathbb{R})$ supported in $(-1,1)$ and $(1/2,2)$, respectively, valued in $[0,1]$ and such that

$$\forall \lambda \geq 0 \quad \sum_{j=0}^{\infty} \psi_j(\lambda) = 1 \quad \text{with } \psi_j(\lambda) = \psi_1(2^{-(j-1)}\lambda) \text{ if } j \in \mathbb{N}.$$

Let $\sigma \in S^0_{\rho,\rho}$. We define

$$\sigma_j(x,\pi) := \sigma(x,\pi)\psi_j(\pi(\mathcal{R})) \quad \text{and} \quad T_j := \mathrm{Op}(\sigma_j) = T\psi_j(\mathcal{R}),$$

where $T = \mathrm{Op}(\sigma)$.

It is clear that $T_j T_i^* = T(\psi_j\psi_i)(\mathcal{R})T^*$ is zero if $|j-i| > 1$ and the strategy of the proof is to apply the crude version of the Cotlar-Stein Lemma, see Proposition A.5.3. We will first prove that the operator norms of the T_j's are uniformly bounded in j by a $S^0_{\rho,\rho}$-seminorm, see Lemma 5.7.15. Then we will show that there exist a constant $C > 0$ and a $S^0_{\rho,\rho}$-seminorm such that

$$\sum_{|i-j|>3} \|T_j^* T_i\|_{\mathscr{L}(L^2(G))} \leq C\|\sigma\|^2_{S^0_{\rho,\rho},a,b,c}. \tag{5.75}$$

These two claims together with Proposition A.5.3 and Remark A.5.4 imply that the series $\sum_j T_j \in \mathscr{L}(L^2(G))$ converges in the strong operator topology of $\mathscr{L}(L^2(G))$ and that the operator norm of the sum is $\lesssim \|\sigma\|_{S^0_{\rho,\rho},a,b,c}$. As $\mathrm{Op}(\sigma) = \sum_j T_j$ in the strong operator topology, this will conclude the proof of Proposition 5.7.14.

Step 1

Let us show that the operator norms of the T_j's are uniformly bounded with respect to j:

Lemma 5.7.15. *The operator $T_j = \mathrm{Op}(\sigma_j)$ is bounded on $L^2(G)$ with operator norm $\leq C\|\sigma\|_{S^0_{\rho,\rho},a,b,c}$ with a, b, c as in Proposition 5.7.7.*

The proof of Lemma 5.7.15 uses the following result which is of interest on its own. In particular, it describes the action of the dilations on $\widehat{G}$.

Lemma 5.7.16. *Let σ be a symbol with kernel κ_x and operator $T = \mathrm{Op}(\sigma)$. Let $r > 0$. We define the operator*

$$T_r : \mathcal{S}(G) \ni \phi \longmapsto (T\phi(r\,\cdot))\,(r^{-1}\cdot).$$

Then (with operator norm possibly infinite)

$$\|T\|_{\mathscr{L}(L^2(G))} = \|T_r\|_{\mathscr{L}(L^2(G))}.$$

Furthermore, the symbol of T_r is

$$\sigma_r := \mathrm{Op}^{-1}(T_r) \quad \text{given by} \quad \sigma_r(x,\pi) := \sigma\left(r^{-1}x, \pi^{(r)}\right),$$

where the representation $\pi^{(r)}$ is defined by

$$\pi^{(r)}(y) := \pi(ry).$$

The kernel of σ_r is $r^{-Q}\kappa_{r^{-1}x}(r^{-1}\cdot)$. Moreover, we have

$$\begin{aligned} \mathcal{F}_G(\kappa)(\pi^{(r)}) &= \mathcal{F}_G\left(r^{-Q}\kappa(r^{-1}\cdot)\right)(\pi), \\ \Delta^\alpha\left\{\mathcal{F}_G(\kappa)(\pi^{(r)})\right\} &= r^{[\alpha]}\left\{\Delta^\alpha\mathcal{F}_G(\kappa)\right\}(\pi^{(r)}), \\ f(\pi^{(r)}(\mathcal{R})) &= f(r^\nu\pi(\mathcal{R})), \end{aligned}$$

for any $\alpha \in \mathbb{N}_0^n$, any positive Rockland operator $\mathcal{R}$ of homogeneous degree ν, and any reasonable functions f and κ (for instance f measurable bounded and κ in some $\mathcal{K}_{a,b}$).

Proof of Lemma 5.7.16. We keep the notation of the statement. The property $\|T\|_{\mathscr{L}(L^2(G))} = \|T_r\|_{\mathscr{L}(L^2(G))}$ follows easily from $\|\phi(r\cdot)\|_2 = r^{-Q/2}\|\phi\|_2$. We compute

$$\begin{aligned} (T\phi(r\,\cdot))\,(r^{-1}x) &= \int_G \phi(ry)\;\kappa_{r^{-1}x}(y^{-1}r^{-1}x)dy \\ &= \int_G \phi(z)\;\kappa_{r^{-1}x}(r^{-1}z^{-1}r^{-1}x)r^{-Q}dz \\ &= \phi * \left(r^{-Q}\kappa_{r^{-1}x}(r^{-1}\cdot)\right)(x). \end{aligned}$$

Therefore, the kernel of the operator T_r is $r^{-Q}\kappa_{r^{-1}x}(r^{-1}\cdot)$. The computation of its symbol follows from

$$\begin{aligned} \mathcal{F}_G\left(r^{-Q}\kappa(r^{-1}\cdot)\right)(\pi) &= \int_G r^{-Q}\kappa(r^{-1}x)\pi(x)^*dx \\ &= \int_G \kappa(y)\pi(ry)^*dx \;=\; \mathcal{F}_G(\kappa)(\pi^{(r)}). \end{aligned}$$

The difference operator applied to the above expression is

$$\begin{aligned}\Delta^\alpha \left\{\mathcal{F}_G(\kappa)(\pi^{(r)})\right\} &= \Delta^\alpha \left\{\mathcal{F}_G\left(r^{-Q}\kappa(r^{-1}\cdot)\right)(\pi)\right\} \\ &= \mathcal{F}_G\left(\tilde{q}_\alpha(\cdot)\, r^{-Q}\kappa(r^{-1}\cdot)\right)(\pi) \\ &= r^{[\alpha]}\left\{\mathcal{F}_G\left(r^{-Q}(\tilde{q}_\alpha\kappa)(r^{-1}\cdot)\right)(\pi)\right\} \\ &= r^{[\alpha]}\left\{\Delta^\alpha\mathcal{F}_G(\kappa)\right\}(\pi^{(r)}).\end{aligned}$$

The kernels of the operators $f(\mathcal{R})$ and $f(r^\nu\mathcal{R})$ are respectively $f(\mathcal{R})\delta_o$ and $r^{-Q}f(\mathcal{R})\delta_o(r^{-1}\cdot)$ (see (4.3) in Corollary 4.1.16, and Example 3.1.20 for the homogeneity of δ_o). Since the group Fourier transform of the former is $f(\pi(\mathcal{R}))$, the group Fourier transform of the latter is $f(r^\nu\pi(\mathcal{R})) = f(\pi^{(r)}(\mathcal{R}))$. □

We can now show Lemma 5.7.15 using the rescaling arguments (together with the lemma above) and the case $\rho = \delta = 0$.

Proof of Lemma 5.7.15. Using the Leibniz formula in Proposition 5.2.10, we first estimate

$$\begin{aligned}&\|\pi(\mathrm{I}+\mathcal{R})^{\frac{\gamma}{\nu}}X_x^{\beta_o}\Delta^{\alpha_o}\sigma_j(x,\pi)\pi(\mathrm{I}+\mathcal{R})^{-\frac{\gamma}{\nu}}\|_{\mathscr{L}(\mathcal{H}_\pi)} \\ &\le C_{\alpha_o}\sum_{[\alpha_1]+[\alpha_2]=[\alpha_o]}\|\pi(\mathrm{I}+\mathcal{R})^{\frac{\gamma}{\nu}}X_x^{\beta_o}\Delta^{\alpha_1}\sigma(x,\pi)\pi(\mathrm{I}+\mathcal{R})^{\frac{\rho([\alpha_1]-[\beta_o])-\gamma}{\nu}}\|_{\mathscr{L}(\mathcal{H}_\pi)} \\ &qquad \quad \|\pi(\mathrm{I}+\mathcal{R})^{-\frac{\rho([\alpha_1]-[\beta_o])-\gamma}{\nu}}\Delta^{\alpha_2}\psi_j(\pi(\mathcal{R}))\pi(\mathrm{I}+\mathcal{R})^{-\frac{\gamma}{\nu}}\|_{\mathscr{L}(\mathcal{H}_\pi)} \\ &\le C_{\alpha_o}\|\sigma\|_{S^0_{\rho,\rho},[\alpha_o],[\beta_o],|\gamma|}\sum_{[\alpha_1]+[\alpha_2]=[\alpha_o]}2^{-j\frac{\nu}{\rho}\frac{[\alpha_2]+\rho([\alpha_1]-[\beta_o])}{\nu}} \\ &\le C_{\alpha_o}\|\sigma\|_{S^0_{\rho,\rho},[\alpha_o],[\beta_o],|\gamma|}2^{-j([\alpha_o]-[\beta_o])}, \end{aligned} \tag{5.76}$$

by Lemma 5.4.7.

For each $j\in\mathbb{N}_0$, we define the symbol σ'_j given by setting

$$\sigma'_j(x,\pi) := \sigma_j\left(2^{-j\rho}x, \pi^{(2^{j\rho})}\right).$$

By Lemma 5.7.16, the corresponding operator $T'_j := \mathrm{Op}(\sigma'_j)$ satisfies

$$(T'_j\phi)(x) = \left(T_j\phi(2^{j\rho}\cdot)\right)(2^{-j\rho}x).$$

Lemma 5.7.16 and Proposition 5.7.7 imply that

$$\|T_j\|_{\mathscr{L}(L^2(G))} = \|T'_j\|_{\mathscr{L}(L^2(G))} \le C\|\sigma'_j\|_{S^0_{0,0},a,b,c}, \tag{5.77}$$

with a,b,c as in Proposition 5.7.7. So we are led to compute $\|\sigma'_j\|_{S^0_{0,0},a,b,c}$. By Lemma 5.7.16, we have

$$\begin{aligned}&X_x^{\beta_o}\Delta^{\alpha_o}\sigma'_j(x,\pi) = 2^{-j\rho[\beta_o]}2^{j\rho[\alpha_o]}X^{\beta_o}_{x_o=2^{-j\rho}x}\Delta^{\alpha_o}_{\pi_o=\pi^{(2^{j\rho})}}\sigma_j(x_o,\pi_o) \\ &= 2^{j\rho([\alpha_o]-[\beta_o])}\pi(\mathrm{I}+2^{j\rho}\mathcal{R})^{-\frac{\gamma}{\nu}} \\ &\left(\pi_o(\mathrm{I}+\mathcal{R})^{\frac{\gamma}{\nu}}X^{\beta_o}_{x_o=2^{-j\rho}x}\Delta^{\alpha_o}\sigma_j(x_o,\pi_o)\pi_o(\mathrm{I}+\mathcal{R})^{-\frac{\gamma}{\nu}}\right)_{\pi_o=\pi^{(2^{j\rho})}}\pi(\mathrm{I}+2^{j\rho}\mathcal{R})^{\frac{\gamma}{\nu}},\end{aligned}$$

so that

$$\begin{aligned}&\|\pi(\mathrm{I}+\mathcal{R})^{\frac{\gamma}{\nu}}X_x^{\beta_o}\Delta^{\alpha_o}\sigma_j'(x,\pi)\pi(\mathrm{I}+\mathcal{R})^{-\frac{\gamma}{\nu}}\|_{\mathscr{L}(\mathcal{H}_\pi)}\\ &\quad\leq 2^{j\rho([\alpha_o]-[\beta_o])}\|\pi(\mathrm{I}+\mathcal{R})^{\frac{\gamma}{\nu}}\pi(\mathrm{I}+2^{j\rho}\mathcal{R})^{-\frac{\gamma}{\nu}}\|_{\mathscr{L}(\mathcal{H}_\pi)}\\ &\qquad\|\left(\pi_o(\mathrm{I}+\mathcal{R})^{\frac{\gamma}{\nu}}X_{x_o=2^{-j\rho}x}^{\beta_o}\Delta^{\alpha_o}\sigma_j(x_o,\pi_o)\pi_o(\mathrm{I}+\mathcal{R})^{-\frac{\gamma}{\nu}}\right)_{\pi_o=\pi(2^{j\rho})}\|_{\mathscr{L}(\mathcal{H}_\pi)}\\ &\qquad\|\pi(\mathrm{I}+2^{j\rho}\mathcal{R})^{\frac{\gamma}{\nu}}\pi(\mathrm{I}+\mathcal{R})^{-\frac{\gamma}{\nu}}\|_{\mathscr{L}(\mathcal{H}_\pi)}.\end{aligned}$$

By the functional calculus (Corollary 4.1.16),

$$\begin{aligned}\|\pi(\mathrm{I}+\mathcal{R})^{\frac{\gamma}{\nu}}\pi(\mathrm{I}+2^{j\rho}\mathcal{R})^{-\frac{\gamma}{\nu}}\|_{\mathscr{L}(\mathcal{H}_\pi)} &\leq \sup_{\lambda\geq 0}\left(\frac{1+\lambda}{1+2^{j\rho}\lambda}\right)^{\frac{\gamma}{\nu}}\leq C2^{-j\rho\frac{\gamma}{\nu}},\\ \|\pi(\mathrm{I}+2^{j\rho}\mathcal{R})^{\frac{\gamma}{\nu}}\pi(\mathrm{I}+\mathcal{R})^{-\frac{\gamma}{\nu}}\|_{\mathscr{L}(\mathcal{H}_\pi)} &\leq \sup_{\lambda\geq 0}\left(\frac{1+2^{j\rho}\lambda}{1+\lambda}\right)^{\gamma\nu}\leq C2^{j\rho\frac{\gamma}{\nu}},\end{aligned}$$

for any $j\in\mathbb{N}_0$. Thus, we have obtained

$$\begin{aligned}&\|\pi(\mathrm{I}+\mathcal{R})^{\frac{\gamma}{\nu}}X_x^{\beta_o}\Delta^{\alpha_o}\sigma_j'(x,\pi)\pi(\mathrm{I}+\mathcal{R})^{-\frac{\gamma}{\nu}}\|_{\mathscr{L}(\mathcal{H}_\pi)}\\ &\leq C2^{j\rho([\alpha_o]-[\beta_o])}\sup_{x_o\in G,\,\pi_o\in\widehat{G}}\|\pi_o(\mathrm{I}+\mathcal{R})^{\frac{\gamma}{\nu}}X_{x_o}^{\beta_o}\Delta^{\alpha_o}\sigma_j(x_o,\pi_o)\pi_o(\mathrm{I}+\mathcal{R})^{-\frac{\gamma}{\nu}}\|_{\mathscr{L}(\mathcal{H}_\pi)}\\ &\leq C\|\sigma\|_{S^0_{\rho,\rho},[\alpha_o],[\beta_o],|\gamma|},\end{aligned}$$

because of (5.76). Taking the supremum over $\pi\in\widehat{G}$, $x\in G$, $[\alpha_o]\leq a$, $[\beta_o]\leq b$ and $|\gamma|\leq c$ yields

$$\|\sigma_j'\|_{S^0_{0,0},a,b,c}\leq C\|\sigma\|_{S^0_{\rho,\rho},a,b,c}.$$

With (5.77), we conclude that $\|T_j\|_{\mathscr{L}(L^2(G))}\leq C\|\sigma\|_{S^0_{\rho,\rho},a,b,c}$. □

Step 2

Now let us prove Claim (5.75). This relies on the bilinear estimate obtained in Proposition 5.7.12.

Proof of Claim (5.75). For each $i\in\mathbb{N}_0$, we denote by $\kappa_{i,x}$ the kernel associated with σ_i. Then one computes easily the integral kernel $K_{ji}(x,y)$ of the operator $T_j^*T_i$, that is,

$$(T_j^*T_i)f(x)=\int_G K_{ji}(x,y)f(y)dy,\quad f\in\mathcal{S}(G),$$

with

$$K_{ji}(x,y)=\int_G\bar{\kappa}_{j,z}(x^{-1}z)\kappa_{i,z}(y^{-1}z)dz.$$

By Schur's lemma [Ste93, §2.4.1], we have

$$\begin{aligned}\|T_j^*T_i\|_{\mathscr{L}(L^2(G))} &\leq \max\left(\sup_{x\in G}\int_G |K_{ji}(x,y)|dy, \sup_{y\in G}\int_G |K_{ji}(x,y)|dx\right), \\ &\lesssim \|T_j^*T_i\|_{\Psi^2_{\rho,\rho},a,b,c} + \max_{|y^{-1}x|\leq 1}|K_{ji}(x,y)|,\end{aligned}$$

since the estimates at infinity for the kernels of a pseudo-differential operator obtained in Theorem 5.4.1 for $\rho\neq 0$ yield

$$|K_{ji}(x,y)| \lesssim \|T_j^*T_i\|_{\Psi^2_{\rho,\rho},a_1,b_1,c_1}|y^{-1}x|^{-N}$$

for any $N\in\mathbb{N}_0$. (We have assumed that a quasi-norm $|\cdot|$ has been fixed on G.) The properties of composition and of taking the adjoint of pseudo-differential operators (see Theorems 5.5.3 and 5.5.12) together with Lemma 5.4.7 yield

$$\|T_j^*T_i\|_{\Psi^2_{\rho,\rho},a_1,b_1,c_1} \lesssim \|\sigma_j\|_{S^1_{\rho,\rho},a_2,b_2,c_2}\|\sigma_i\|_{S^1_{\rho,\rho},a_3,b_3,c_3} \lesssim \|\sigma\|^2_{S^0_{\rho,\rho},a_4,b_4,c_4}2^{-\frac{i+j}{\nu}}.$$

We now analyse $\max_{|y^{-1}x|\leq 1}|K_{ji}(x,y)|$. So let $x,y\in G$ with $|y^{-1}x|\leq 1$. We fix a function $\chi\in\mathcal{D}(G)$ which is a smooth version of the indicatrix function of the ball $B(0,10)=\{z\in G:|x^{-1}z|<10\}$ about 0 with radius 10, that is, we assume that $\chi\equiv 1$ on $B(0,10)$ and $\chi\equiv 0$ on $B(0,11)$. Let us assume that the quasi-norm is in fact a norm, that is, it satisfies the triangle inequality 'with constant 1' (although we could give a proof without this restriction, it simplifies the choice of constants and therefore avoids dwelling on unimportant technical points). We can always decompose

$$\begin{aligned}K_{ji}(x,y) &= \int_{z\in G}\bar{\kappa}_{j,z}(x^{-1}z)\kappa_{i,z}(y^{-1}z)\left(\chi(x^{-1}z)+(1-\chi(x^{-1}z))\right)dz \\ &= I_1+I_2.\end{aligned}$$

We first estimate the second integral via

$$|I_2| \lesssim \|\sigma_j\|_{S^1_{\rho,\rho},a_5,b_5,c_5}\|\sigma_i\|_{S^1_{\rho,\rho},a_6,b_6,c_6}\int_{|x^{-1}z|>10}|x^{-1}z|^{-N_1}|y^{-1}z|^{-N_1}dz.$$

having used the estimates at infinity for the kernels of a pseudo-differential operator obtained in Theorem 5.4.1 for $\rho\neq 0$. As $|y^{-1}x|\leq 1$, the last integral is just a finite constant if we choose $N_1=Q+1$ for instance. We estimate the $S^1_{\rho,\rho}$-seminorms with Lemma 5.4.7 and we obtain then

$$|I_2| \lesssim \|\sigma\|^2_{S^0_{\rho,\rho},a_7,b_7,c_7}2^{-\frac{i+j}{\nu}}.$$

We now estimate the integral I_1:

$$I_1 = \int_G \bar{\kappa}_{j,z}(x^{-1}z)\kappa_{i,z}(y^{-1}z)\chi(x^{-1}z)dz.$$

It is of the form $\int_G f(z,z)dz$ for a given function f on $G\times G$. Simple formal manipulations yield for any $N\in\mathbb{N}_0$

$$\begin{aligned}\int_G f(z,z)dz &= \int_G (\mathrm{I}+\mathcal{R})^N_{z_2=z}(\mathrm{I}+\mathcal{R})^{-N}_{z_2} f(z,z_2)dz\\ &= \int_G (\mathrm{I}+\bar{\mathcal{R}})^N_{z_1=z}(\mathrm{I}+\mathcal{R})^{-N}_{z_2=z} f(z_1,z_2)dz,\end{aligned}$$

having used integration by parts or equivalently $\mathcal{R}^t=\bar{\mathcal{R}}$, since $\mathcal{R}$ is essentially self-adjoint. Hence, we obtain formally in our case

$$I_1=\int_G (\mathrm{I}+\bar{\mathcal{R}})^N_{z_1=z}(\mathrm{I}+\mathcal{R})^{-N}_{z_2=z}\left\{\bar{\kappa}_{j,z_1}(x^{-1}z_2)\kappa_{i,z_1}(y^{-1}z_2)\chi(x^{-1}z_1)\right\}dz,$$

where $N\in\mathbb{N}_0$ is to be fixed later. Note that the expression in z_1 is supported in $B(x_1,11)$, hence so is the integrand in z. This produces the following estimate

$$|I_1|\le\int_{|x^{-1}z_2|\le 11} S(z_2)dz_2$$

where $S(z_2)$ is the supremum

$$\begin{aligned}S(z_2) &= \sup_{z_1\in G}\left|(\mathrm{I}+\bar{\mathcal{R}})^N_{z_1}(\mathrm{I}+\mathcal{R})^{-N}_{z_2}\left\{\bar{\kappa}_{j,z_1}(x^{-1}z_2)\kappa_{i,z_1}(y^{-1}z_2)\chi(x^{-1}z_1)\right\}\right|\\ &\lesssim \left\|(\mathrm{I}+\bar{\mathcal{R}})^{N+\frac{s_0}{\nu}}_{z_1}(\mathrm{I}+\mathcal{R})^{-N}_{z_2}\bar{\kappa}_{j,z_1}(x^{-1}z_2)\kappa_{i,z_1}(y^{-1}z_2)\chi(x^{-1}z_1)\right\|_{L^2(dz_1)}\\ &\lesssim \sum_{\substack{[\beta_{01}]+[\beta_{02}]\\ \le\nu N+s_0}}\left\|(\mathrm{I}+\mathcal{R})^{-N}_{z_2}\{X^{\beta_{01}}_{z_1}\bar{\kappa}_{j,z_1}(x^{-1}z_2)\ X^{\beta_{02}}_{z_1}\kappa_{i,z_1}(y^{-1}z_2)\}\right\|_{L^2(B(x,11),dz_1)},\end{aligned}$$

by the properties of the Sobolev spaces, see Theorem 4.4.28, especially the Sobolev embedding in Part (5). Here $s_0\in\nu\mathbb{N}$ denotes the smallest integer multiple of ν such that $\frac{s_0}{\nu}>Q/2$. By the Cauchy-Schwartz inequality, as $B(x,11)$ has finite volume independent of x, we obtain

$$\begin{aligned}|I_1| &\lesssim \sum_{\substack{[\beta_{01}]+[\beta_{02}]\\ \le\nu N+s_0}}\left\|(\mathrm{I}+\mathcal{R})^{-N}_{z_2}\{X^{\beta_{01}}_{z_1}\bar{\kappa}_{j,z_1}(x^{-1}z_2)\ X^{\beta_{02}}_{z_1}\kappa_{i,z_1}(y^{-1}z_2)\}\right\|_{L^2(B(x,11)^2,dz_1dz_2)}\\ &\lesssim \sup_{\substack{z_1\in B(x,11)\\ [\beta_{01}]+[\beta_{02}]\le\nu N+s_0}}\left\|(\mathrm{I}+\mathcal{R})^{-N}_{z_2}\{X^{\beta_{01}}_{z_1}\bar{\kappa}_{j,z_1}(x^{-1}z_2)\ X^{\beta_{02}}_{z_1}\kappa_{i,z_1}(y^{-1}z_2)\}\right\|_{L^2(dz_2)}.\end{aligned}$$

Choosing $N>\frac{Q}{2\nu}$, we can apply Proposition 5.7.12 to the L^2-norm above, so that

$$\begin{aligned}&\left\|(\mathrm{I}+\mathcal{R})^{-N}_{z_2}\{X^{\beta_{01}}_{z_1}\bar{\kappa}_{j,z_1}(x^{-1}z_2)\ X^{\beta_{02}}_{z_1}\kappa_{i,z_1}(y^{-1}z_2)\}\right\|_{L^2(dz_2)}\\ &\qquad\lesssim \left\|X^{\beta_{01}}_{z_1}\bar{\kappa}_{j,z_1}(z_2)\right\|_{L^2(dz_2)}\left\|X^{\beta_{02}}_{z_1}\kappa_{i,z_1}(z_2)\right\|_{L^2(dz_2)}2^{(-N+\frac{Q}{2\nu})\max(i,j)}.\end{aligned}$$

By Corollary 5.4.3, we have

$$\left\|X_{z_1}^{\beta_{01}}\bar{\kappa}_{j,z_1}(z_2)\right\|_{L^2(dz_2)} \lesssim \|X_x^{\beta_{01}}\sigma_j\|_{S^{m'}_{\rho,\rho},a_7,b_7,c_7},$$

where m' is a number such that $m' < -Q/2$, for instance $m' := -1 - Q/2$. By Lemma 5.4.7, we have (with $\rho = \delta$)

$$\|X_x^{\beta_{01}}\sigma_j\|_{S^{m'}_{\rho,\rho},a_7,b_7,c_7} \lesssim \|\sigma\|_{S^0_{\rho,\rho},a_8,b_8,c_8} 2^{-j\frac{m'-\delta[\beta_{01}]}{\nu}}.$$

We have similar estimates for $\left\|X_{z_1}^{\beta_{02}}\kappa_{i,z_1}(z_2)\right\|_{L^2(dz_2)}$, thus

$$\begin{aligned}
&\max_{\substack{[\beta_{01}]+[\beta_{02}]\\ \leq \nu N+s_0}} \left\|X_{z_1}^{\beta_{01}}\bar{\kappa}_{j,z_1}(z_2)\right\|_{L^2(dz_2)} \left\|X_{z_1}^{\beta_{02}}\kappa_{i,z_1}(z_2)\right\|_{L^2(dz_2)} \\
&\qquad \lesssim \|\sigma\|^2_{S^0_{\rho,\rho},a_9,b_9,c_9} \max_{\substack{[\beta_{01}]+[\beta_{02}]\\ \leq \nu N+s_0}} 2^{-j\frac{m'-\delta[\beta_{01}]}{\nu}} 2^{-i\frac{m'-\delta[\beta_{02}]}{\nu}} \\
&\qquad \lesssim \|\sigma\|^2_{S^0_{\rho,\rho},a_9,b_9,c_9} 2^{\max(i,j)(-2m'+\delta(N+s_0))}.
\end{aligned}$$

The estimates above show that the first formal manipulations on I_1 are justified and we obtain

$$|I_1| \lesssim \|\sigma\|^2_{S^0_{\rho,\rho},a_9,b_9,c_9} 2^{\max(i,j)(-(1-\delta)N-2m'+s_0+\frac{Q}{2\nu})}.$$

Consequently, we have

$$\max_{|y^{-1}x|\leq 1} |K_{ji}(x,y)| \lesssim \|\sigma\|^2_{S^0_{\rho,\rho},a,b,c} \left(2^{-\frac{i+j}{\nu}} + 2^{\max(i,j)(-(1-\delta)N-2m'+s_0+\frac{Q}{2\nu})}\right),$$

thus

$$\|T_j^* T_i\|_{\mathscr{L}(L^2(G))} \lesssim \|\sigma\|^2_{S^0_{\rho,\rho},a,b,c} \left(2^{-\frac{i+j}{\nu}} + 2^{\max(i,j)(-(1-\delta)N-2m'+s_0+\frac{Q}{2\nu})}\right).$$

As $\delta = \rho \in (0,1)$, we can choose N such that $-(1-\delta)N - 2m' + s_0 + \frac{Q}{2\nu} < -1$. Summing over $i > j+3$ and using the symmetry of the rôle played by i and j yield (5.75). □

Hence we have shown Proposition 5.7.14 and this concludes the proof of Theorem 5.7.1.

5.8 Parametrices, ellipticity and hypoellipticity

In this section, we obtain statements regarding ellipticity and hypoellipticity which are similar to the compact case presented in Section 2.2.3 where the Laplacian has the role of the positive Rockland operator. However, on nilpotent Lie groups, since $\widehat{G}$ is not discrete and the representations are often not (and can be almost never) finite dimensional, the precise hypotheses become more technical to present.

5.8.1 Ellipticity

Roughly speaking, we define the ellipticity by requiring that the symbol is invertible for 'high frequencies'. These 'high frequencies' are determined with respect to the spectral projection E of a positive Rockland operator $\mathcal{R}$, and its group Fourier transform E_π, see Corollary 4.1.16.

We will use the following shorthand notation:

$$\mathcal{H}^\infty_{\pi,\Lambda} := E_\pi(\Lambda, +\infty)\mathcal{H}^\infty_\pi. \tag{5.78}$$

Since $E_\pi(\Lambda,\infty) = \mathcal{F}_G(1_{(\Lambda,\infty)}(\mathcal{R})\delta_0)$ yields a symbol acting on smooth vectors (see Examples 5.1.27 and 5.1.38), $\mathcal{H}^\infty_{\pi,\Lambda}$ is a subspace of $\mathcal{H}^\infty_\pi$.

We can now define our notion of ellipticity:

Definition 5.8.1. Let $\mathcal{R}$ be a positive Rockland operator of homogeneous degree ν. Let σ be a symbol given by fields of operators acting on smooth vectors, i.e. $\sigma(x,\cdot) = \{\sigma(x,\cdot) : \mathcal{H}^\infty_\pi \to \mathcal{H}^\infty_\pi, \pi \in \widehat{G}\}$ is in some $L^\infty_{a,b}(\widehat{G})$ for each $x \in G$.

The symbol σ is said to be *elliptic* with respect to $\mathcal{R}$ of *elliptic order* m_o if there is $\Lambda \in \mathbb{R}$ such that for any $\gamma \in \mathbb{R}$, $x \in G$, μ-almost all $\pi \in \widehat{G}$, and any $u \in \mathcal{H}^\infty_{\pi,\Lambda}$ we have

$$\forall\gamma \in \mathbb{R} \qquad \|\pi(\mathrm{I}+\mathcal{R})^{\frac{\gamma}{\nu}}\sigma(x,\pi)u\|_{\mathcal{H}_\pi} \geq C_\gamma\|\pi(\mathrm{I}+\mathcal{R})^{\frac{\gamma}{\nu}}\pi(\mathrm{I}+\mathcal{R})^{\frac{m_o}{\nu}}u\|_{\mathcal{H}_\pi}. \tag{5.79}$$

with $C_\gamma = C_{\sigma,\mathcal{R},m_o,\Lambda,\gamma}$ independent of $(x,\pi) \in G \times \widehat{G}$ and $u \in \mathcal{H}^\infty_{\pi,\Lambda}$.

We will say that the symbol σ or the corresponding operator $\mathrm{Op}(\sigma)$ is $(\mathcal{R},\Lambda,m_o)$-*elliptic*, or elliptic of elliptic order m_o, or just elliptic.

The notation $\mathcal{H}^\infty_{\pi,\Lambda}$ was defined in (5.78). As $\mathcal{H}^\infty_{\pi,\Lambda}$ is a subspace of $\mathcal{H}^\infty_\pi$ and since $\pi(\mathrm{I}+\mathcal{R})^{\frac{\gamma}{\nu}}$ and $\sigma(x,\cdot)$ are fields of operators acting on smooth vectors, the expression in the norm of the left-hand side of (5.79) makes sense.

In our elliptic condition in Definition 5.8.1, σ is a symbol in the sense of Definition 5.1.33 which is given by fields of operators acting on smooth vectors. It will be natural to consider symbols in the classes $S^m_{\rho,\delta}$ to construct parametrices, see Proposition 5.8.5 and Theorem 5.8.7.

Our definition of ellipticity requires a property of 'x-uniform partial injectivity'. Of course, we note that $\pi(\mathrm{I}+\mathcal{R})^{\frac{\gamma}{\nu}}\pi(\mathrm{I}+\mathcal{R})^{\frac{m_o}{\nu}} = \pi(\mathrm{I}+\mathcal{R})^{\frac{\gamma+m_o}{\nu}}$.

Naturally, we will see shortly in Corollary 5.8.4 that it suffices to check (5.79) for a sequence of real numbers $\{\gamma_\ell, \ell \in \mathbb{Z}\}$ which tends to $\pm\infty$ as $\ell \to \pm\infty$.

Our first examples of elliptic operators are provided by positive Rockland operators:

Proposition 5.8.2. *Let $\mathcal{R}$ be a positive Rockland operator of homogeneous degree ν. Then we have the following properties.*

1. *The operator $(\mathrm{I}+\mathcal{R})^{\frac{m_o}{\nu}}$, for any $m_o \in \mathbb{R}$, is elliptic with respect to $\mathcal{R}$ of elliptic order m_o.*

2. *If f_1 and f_2 are complex-valued (smooth) functions on G such that*

$$\inf_{x\in G,\lambda\geq\Lambda}\frac{|f_1(x)+f_2(x)\lambda|}{1+\lambda}>0 \quad \text{for some } \Lambda\geq 0,$$

then the differential operator $f_1(x)+f_2(x)\mathcal{R}$ is $(\mathcal{R},\Lambda,\nu)$-elliptic.

3. *The operator $E(\Lambda,\infty)\mathcal{R}$, for any $\Lambda>0$, is $(\mathcal{R},\Lambda,\nu)$-elliptic.*

   More generally, if f is a complex-valued function on G such that $\inf_G|f|>0$, then $f(x)E(\Lambda,\infty)\mathcal{R}$ is $(\mathcal{R},\Lambda,\nu)$-elliptic.

4. *Let $\psi\in C^\infty(\mathbb{R})$ be such that*

$$\psi_{|(-\infty,\Lambda_1]}=0 \quad \text{and} \quad \psi_{|[\Lambda_2,\infty)}=1,$$

for some real numbers Λ_1,Λ_2 satisfying $0<\Lambda_1<\Lambda_2$, Then the operator $\psi(\mathcal{R})\mathcal{R}$ is $(\mathcal{R},\Lambda_2,\nu)$-elliptic.

   More generally, if f is a complex-valued function on G such that $\inf_G|f|>0$, then $f(x)\psi(\mathcal{R})\mathcal{R}$ is $(\mathcal{R},\Lambda_2,\nu)$-elliptic.

Proof. The symbols involved in the statement are multipliers in $\mathcal{R}$. By Example 5.1.27 and Corollary 5.1.30, the corresponding symbols are symbols in the sense of Definition 5.1.33 which are given by fields of operators acting on smooth vectors. Hence it remains just to check the condition in (5.79).

Part (1) is easy to check using the functional calculus of $\pi(\mathcal{R})$.

Let us prove Part (2). Let Λ, f_1, f_2, and m be as in the statement. The properties of the functional calculus for $\pi(\mathcal{R})$ yield that, for each $x\in G$ fixed and $u\in\mathcal{H}^\infty_{\pi,\Lambda}$ we have

$$\pi(\mathrm{I}+\mathcal{R})^{\frac{2}{\nu}}\pi(\mathrm{I}+\mathcal{R})u=\phi_x(\pi(\mathcal{R}))\pi(\mathrm{I}+\mathcal{R})^{\frac{2}{\nu}}(f_1(x)+f_2(x)\pi(\mathcal{R}))u,$$

where $\phi_x\in L^\infty[0,\infty)$ is given by

$$\phi_x(\lambda)=\frac{1+\lambda}{f_1(x)+f_2(x)\lambda}1_{\lambda\geq\Lambda}.$$

Our assumption implies that ϕ_x is bounded on $[0,\infty)$ with

$$C:=\sup_{x\in G}\|\phi_x\|_\infty=\left(\inf_{x\in G,\lambda\geq\Lambda}\frac{|f_1(x)+f_2(x)\lambda|}{1+\lambda}\right)^{-1}<\infty.$$

The property of the functional calculus for $\pi(\mathcal{R})$ yields

$$\forall x\in G\quad \|\phi_x(\pi(\mathcal{R}))\|_{\mathscr{L}(\mathcal{H}_\pi)}\leq C.$$

Thus we have

$$\begin{aligned}\|\pi(\mathrm{I}+\mathcal{R})^{\frac{\gamma}{\nu}}\pi(\mathrm{I}+\mathcal{R})u\|_{\mathcal{H}_\pi} &= \|\phi_x(\pi(\mathcal{R}))\pi(\mathrm{I}+\mathcal{R})^{\frac{\gamma}{\nu}}(f_1(x)+f_2(x)\pi(\mathcal{R}))u\|_{\mathcal{H}_\pi}\\ &\leq C\|\pi(\mathrm{I}+\mathcal{R})^{\frac{\gamma}{\nu}}(f_1(x)+f_2(x)\pi(\mathcal{R}))u\|_{\mathcal{H}_\pi}.\end{aligned}$$

This proves Part (2).

Let us prove Part (3). The properties of the functional calculus for $\pi(\mathcal{R})$ yield

$$\pi(\mathrm{I}+\mathcal{R})u = \phi(\pi(\mathcal{R}))E_\pi(\Lambda,\infty)\pi(\mathcal{R})u,$$

where $\phi \in L^\infty[0,\infty)$ is given by

$$\phi(\lambda) = \frac{1+\lambda}{\lambda}1_{(\Lambda,\infty)}(\lambda).$$

Moreover,

$$\begin{aligned}\|\pi(\mathrm{I}+\mathcal{R})^{1+\frac{\gamma}{\nu}}u\|_{\mathcal{H}_\pi} &= \|\phi(\pi(\mathcal{R}))\pi(\mathrm{I}+\mathcal{R})^{\frac{\gamma}{\nu}}E_\pi(\Lambda,\infty)\pi(\mathcal{R})u\|_{\mathcal{H}_\pi}\\ &\leq \|\phi\|_\infty\|\pi(\mathrm{I}+\mathcal{R})^{\frac{\gamma}{\nu}}E_\pi(\Lambda,\infty)\pi(\mathcal{R})u\|_{\mathcal{H}_\pi}.\end{aligned}$$

Since $C=\|\phi\|_\infty^{-1}$ is a finite positive constant, we have obtained

$$C\|\pi(\mathrm{I}+\mathcal{R})^{1+\frac{\gamma}{\nu}}u\|_{\mathcal{H}_\pi} \leq \|\pi(\mathrm{I}+\mathcal{R})^{\frac{\gamma}{\nu}}E_\pi(\Lambda,\infty)\pi(\mathcal{R})u\|_{\mathcal{H}_\pi}.$$

This shows that $E(\Lambda,\infty)\mathcal{R}$, is elliptic.

If f is as in the statement, we proceed as above, replacing ϕ by

$$\phi_x(\lambda) = \frac{1+\lambda}{f(x)\lambda}1_{(\Lambda,\infty)}(\lambda),$$

and C such that C^{-1} is equal to the right-hand side of the estimate

$$\|\phi_x\|_\infty \leq \frac{1}{\inf_G|f|}\sup_{\lambda\geq\Lambda}\frac{1+\lambda}{\lambda} := C^{-1}.$$

This shows Part (3).

For Part (4), we proceed as in Part (3) replacing $1_{(\Lambda,\infty)}$ by $\psi(\lambda)$ and Λ by Λ_2. □

The next lemma is technical. It states that we can construct a partial inverse of an elliptic symbol. The analogue for scalar-valued symbols would be obvious: if $|a(x,\xi)|$ does not vanish for $|\xi|>\Lambda$ then we can consider $1_{|\xi|>\Lambda}1/a(x,\xi)$. However, in the context of operator-valued symbols, we need to proceed with caution.

Lemma 5.8.3. *Let σ be a symbol $(\mathcal{R}, \Lambda, m_o)$-elliptic as in Definition 5.8.1.*

For any $v \in \mathcal{H}_\pi^\infty$, if there is a vector $u \in \mathcal{H}_{\pi,\Lambda}^\infty$ such that $\sigma(x,\pi)u = v$ then this u is necessarily unique. In this sense $\sigma(x,\pi)$ is invertible on $\mathcal{H}_{\pi,\Lambda}^\infty$ and we can set

$$E_\pi(\Lambda,\infty)\sigma(x,\pi)^{-1}(v) := \begin{cases} u & \text{if } v = \sigma(x,\pi)u, \ u \in \mathcal{H}_{\pi,\Lambda}^\infty, \\ 0 & \text{if } \mathcal{H}_\pi^\infty \ni v \perp \sigma(x,\pi)\mathcal{H}_{\pi,\Lambda}^\infty. \end{cases} \tag{5.80}$$

This yields the symbol (in the sense of Definition 5.1.33) given by fields of operators acting on smooth vectors

$$\{E_\pi(\Lambda,\infty)\sigma(x,\pi)^{-1} : \mathcal{H}_\pi^\infty \to \mathcal{H}_\pi^\infty, (x,\pi) \in G \times \widehat{G}\}. \tag{5.81}$$

Furthermore, for every γ,

$$\|E_\pi(\Lambda,\infty)\sigma(x,\pi)^{-1}\|_{L^\infty_{\gamma,\gamma+m_o}(\widehat{G})} \leq C_\gamma^{-1}, \tag{5.82}$$

where C_γ is the constant appearing in (5.79) of Definition 5.8.1.

If σ is continuous in the sense of Definition 5.1.34, then the symbol in (5.81) is continuous in the sense of Definition 5.1.34. If σ is smooth, then the symbol in (5.81) is continuous and depends smoothly on $x \in G$ in the sense of Remark 1.8.16.

Proof. Recall that $E_\pi(\Lambda,\infty) = \mathcal{F}_G(1_{(\Lambda,\infty)}(\mathcal{R})\delta_0)$ yields a symbol acting on smooth vectors, see Examples 5.1.27 and 5.1.38.

If $v = \sigma(x,\pi)u$ where $u \in \mathcal{H}_{\pi,\Lambda}^\infty$, then, using (5.79), we have

$$\|\pi(\mathrm{I}+\mathcal{R})^{\frac{m_o+\gamma}{\nu}}u\|_{\mathcal{H}_\pi} \leq C_\gamma^{-1}\|\pi(\mathrm{I}+\mathcal{R})^{\frac{\gamma}{\nu}}\sigma(x,\pi)u\|_{\mathcal{H}_\pi} = C_\gamma^{-1}\|\pi(\mathrm{I}+\mathcal{R})^{\frac{\gamma}{\nu}}v\|_{\mathcal{H}_\pi}.$$

It is now easy to check $\{E_\pi(\Lambda,\infty)\sigma(x,\pi)^{-1}, (x,\pi) \in G \times \widehat{G}\}$ is a symbol in the sense of Definition 5.1.33 and that the estimates in (5.82) hold.

If σ is continuous, then one checks easily that the map

$$G \ni x \mapsto E_\pi(\Lambda,\infty)\sigma(x,\pi)^{-1} \in L^\infty_{\gamma,\gamma+m_o}(\widehat{G})$$

is continuous. Consequently $\{E_\pi(\Lambda,\infty)\sigma(x,\pi)^{-1}, (x,\pi) \in G \times \widehat{G}\}$ is continuous.

If σ is smooth, then $\{E_\pi(\Lambda,\infty)\sigma(x,\pi)^{-1}, (x,\pi) \in G \times \widehat{G}\}$ depends smoothly in $x \in G$, see Remark 1.8.16. □

Corollary 5.8.4. *Let $\mathcal{R}$ be a positive Rockland operator of homogeneous degree ν. The symbol σ satisfies (5.79) for each $\gamma \in \mathbb{R}$ if and only if σ satisfies (5.79) for a sequence of real numbers $\{\gamma_\ell, \ell \in \mathbb{Z}\}$ which tends to $\pm\infty$ as $\ell \to \pm\infty$.*

We may choose the constants C_γ such that $\max_{|\gamma|\leq c} C_\gamma$ in (5.79) is finite for any $c \geq 0$.

Proof. From the proof of Lemma 5.8.3, we see that σ satisfies (5.79) for γ if and only if

$$\sup_{x\in G} \|E_\pi(\Lambda,\infty)\sigma(x,\pi)^{-1}\|_{L^\infty_{\gamma,\gamma+m_o}(\widehat{G})} < \infty$$

is finite. The conclusion follows from Corollary 4.4.10. □

The next statement says that if a symbol in some $S^m_{\rho,\delta}$ is elliptic and if the elliptic order is equal to the order m of the symbol, then we can define a symbol in $S^{-m}_{\rho,\delta}$ using the operator $E_\pi(\Lambda,\infty)\sigma(x,\pi)^{-1}$ defined via (5.80). This will be the main ingredient in the construction of a parametrix, see the proof of Theorem 5.8.7.

Proposition 5.8.5. *Assume $1\geq\rho\geq\delta\geq 0$. Let $\sigma\in S^m_{\rho,\delta}$ be a symbol which is $(\mathcal{R},\Lambda,m)$-elliptic with respect to a positive Rockland operator $\mathcal{R}$. If $\psi\in C^\infty(\mathbb{R})$ is such that*

$$\psi_{|(-\infty,\Lambda_1]} = 0 \quad \textit{and} \quad \psi_{|[\Lambda_2,\infty)} = 1,$$

for some real numbers Λ_1,Λ_2 satisfying $\Lambda<\Lambda_1<\Lambda_2$, then the symbol

$$\{\psi(\pi(\mathcal{R}))\sigma^{-1}(x,\pi)\ ,\ (x,\pi)\in G\times\widehat{G}\},$$

given by

$$\psi(\pi(\mathcal{R}))\sigma^{-1}(x,\pi) := \psi(\pi(\mathcal{R}))E_\pi(\Lambda_1,\infty)\sigma(x,\pi)^{-1},$$

is in $S^{-m}_{\rho,\delta}$. Moreover, for any $a_o,b_o\in\mathbb{N}_0$, we have

$$\begin{aligned}&\|\psi(\pi(\mathcal{R}))\sigma^{-1}(x,\pi)\|_{S^{-m}_{\rho,\delta},a_o,b_o,0}\\ &\leq C\sum_{\substack{a_1',a_2'\leq a_o\\ b_1',b_2'\leq b_o}}\ \max_{|\gamma|\leq\rho a_o+\delta b_o} C^{a_1'+b_1'+1}_{\gamma,\sigma,\Lambda_1}\|\sigma(x,\pi)\|^{a_2'+b_2'}_{S^m_{\rho,\delta},a_o,b_o,|m|},\end{aligned}$$

where $C>0$ is a positive constant depending on a_o,b_o,ψ, and where the constant $C_{\gamma,\sigma,\Lambda_1}$ was given in (5.79).

The following lemma is helpful in the proof of Proposition 5.8.5. Indeed, in the case of $\mathbb{R}^n$, if a cut-off function $\psi(\xi)$ on the Fourier side is constant for $|\xi|>\Lambda$ (Λ large enough), then its derivatives are $\partial^\alpha_\xi\psi(\xi)=0$ if $|\xi|>\Lambda$. In our case, we can not say anything in general. If we use $\psi(\pi(\mathcal{R}))$ as 'a cut-off in frequency' with ψ as in Proposition 5.8.5 for example, it is not true in general that its (Δ^α-)derivatives will vanish on $E_\pi(\Lambda,\infty)$ or will be of the form $\psi_1(\pi(\mathcal{R}))$. However, we can show that these derivatives are smoothing:

Lemma 5.8.6. *Let $\psi\in C^\infty(\mathbb{R})$ satisfy $\psi_{|[\Lambda,+\infty)}=1$ for some $\Lambda\in\mathbb{R}$. Then for any $\alpha\in\mathbb{N}_0^n\backslash\{0\}$, the symbol given by $\Delta^\alpha\psi(\pi(\mathcal{R}))$ is smoothing, i.e. is in $S^{-\infty}$.*

Proof of Lemma 5.8.6. Let $\alpha \in \mathbb{N}_0^n \backslash \{0\}$. Then $\Delta^\alpha \mathrm{I} = 0$ by Example 5.2.8. Therefore

$$\Delta^\alpha \psi(\pi(\mathcal{R})) = -\Delta^\alpha (1-\psi)(\pi(\mathcal{R})).$$

As $1-\psi$ is a smooth function such that $\mathrm{supp}(1-\psi) \cap [0,\infty)$ is compact, the symbol $(1-\psi)(\pi(\mathcal{R}))$ is smoothing. Hence so is $\Delta^\alpha(1-\psi)(\pi(\mathcal{R}))$ and $\Delta^\alpha\psi(\pi(\mathcal{R}))$. □

Proof of Proposition 5.8.5. Recall that by the Leibniz formula (Proposition 5.2.10), we have

$$\Delta^{\alpha_o}(\sigma_1\sigma_2) = \sum_{[\alpha_1]+[\alpha_2]=[\alpha_o]} c_{\alpha_1,\alpha_2} \Delta^{\alpha_1}\sigma_1 \, \Delta^{\alpha_2}\sigma_2,$$

with

$$c_{\alpha_1,0} = \begin{cases} 1 & \text{if } \alpha_1 = \alpha_o \\ 0 & \text{otherwise} \end{cases}, \quad c_{0,\alpha_2} = \begin{cases} 1 & \text{if } \alpha_2 = \alpha_o \\ 0 & \text{otherwise} \end{cases}.$$

It is also easy to see that

$$X^{\beta_o}(f_1 f_2) = \sum_{[\beta_1]+[\beta_2]=[\beta_o]} c'_{\beta_1,\beta_2} X^{\beta_1} f_1 \, X^{\beta_2} f_2,$$

with

$$c'_{\beta_1,0} = \begin{cases} 1 & \text{if } \beta_1 = \alpha_o \\ 0 & \text{otherwise} \end{cases}, \quad c'_{0,\beta_2} = \begin{cases} 1 & \text{if } \beta_2 = \beta_o \\ 0 & \text{otherwise} \end{cases}.$$

Let $\sigma = \sigma(x,\pi) \in S^m_{\rho,\delta}$ and $\psi \in C^\infty(\mathbb{R})$ as in the statement. By Lemma 5.8.3, the continuous symbol

$$\{E_\pi(\Lambda,\infty)\sigma(x,\pi)^{-1} : \mathcal{H}_\pi^\infty \to \mathcal{H}_\pi^\infty, (x,\pi) \in G \times \widehat{G}\},$$

depends smoothly on $x \in G$. Hence so does the continuous symbol σ_o defined via

$$\sigma_o(x,\pi) := \psi(\pi(\mathcal{R}))\sigma^{-1}(x,\pi).$$

Since $\psi(\pi(\mathcal{R}))$ commutes with powers of $\pi(\mathrm{I}+\mathcal{R})$ and

$$\|\psi(\pi(\mathcal{R}))\|_{\mathscr{L}(\mathcal{H}_\pi)} \le \|\psi\|_\infty,$$

we have

$$\begin{aligned} &\|\pi(\mathrm{I}+\mathcal{R})^{\frac{m}{\nu}}\sigma_o(x,\pi)\|_{\mathscr{L}(\mathcal{H}_\pi)} \\ &\quad \le \|\psi\|_\infty \|\pi(\mathrm{I}+\mathcal{R})^{\frac{m}{\nu}} \left\{E_\pi(\Lambda,\infty)\sigma(x,\pi)^{-1}\right\}\|_{\mathscr{L}(\mathcal{H}_\pi)} \\ &\quad = \|\psi\|_\infty C_0^{-1}, \end{aligned}$$

where by Lemma 5.8.3, C_0 is the finite constant intervening in the ellipticity condition for $\gamma = 0$ in (5.79). More generally, in this proof, C_γ denotes the constant depending on γ in (5.79), see also Corollary 5.8.4.

By Proposition 5.3.4, $\psi(\pi(\mathcal{R})) \in S^0$. We also see that

$$\psi(\pi(\mathcal{R})) = \sigma_o(x,\pi)\sigma(x,\pi). \tag{5.83}$$

Hence for any left-invariant vector field X we have

$$\begin{aligned} 0 &= X_x\psi(\pi(\mathcal{R})) \\ &= X_x\sigma_o(x,\pi)\ \sigma(x,\pi) + \sigma_o(x,\pi)\ X_x\sigma(x,\pi). \end{aligned}$$

Thus

$$X_x\sigma_o(x,\pi)\sigma(x,\pi) = -\sigma_o(x,\pi)\ X_x\sigma(x,\pi),$$

and since $\sigma(x,\pi)$ is invertible on $E_\pi(\Lambda_1,\infty)\mathcal{H}_\pi^\infty$,

$$X_x\sigma_o(x,\pi) = -\sigma_o(x,\pi)\ \{X_x\sigma(x,\pi)\}\ \ E(\Lambda_1,\infty)\sigma^{-1}(x,\pi).$$

Assuming that X is homogeneous of degree d, we can take the operator norm and estimate

$$\begin{aligned} &\|\pi(\mathrm{I}+\mathcal{R})^{\frac{m-\delta d}{\nu}} X_x\sigma_o(x,\pi)\|_{\mathscr{L}(\mathcal{H}_\pi)} \\ &\quad\leq \|\pi(\mathrm{I}+\mathcal{R})^{\frac{m-\delta d}{\nu}}\sigma_o(x,\pi)\pi(\mathrm{I}+\mathcal{R})^{\frac{\delta d}{\nu}}\|_{\mathscr{L}(\mathcal{H}_\pi)} \\ &\qquad \|\pi(\mathrm{I}+\mathcal{R})^{-\frac{\delta d}{\nu}} X_x\sigma(x,\pi)\pi(\mathrm{I}+\mathcal{R})^{-\frac{m}{\nu}}\|_{\mathscr{L}(\mathcal{H}_\pi)} \\ &\qquad \|\pi(\mathrm{I}+\mathcal{R})^{\frac{m}{\nu}}\left\{E_\pi(\Lambda_1,\infty)\sigma(x,\pi)^{-1}\right\}\|_{\mathscr{L}(\mathcal{H}_\pi)} \\ &\quad\leq \|\psi\|_\infty C_{-\delta d}^{-1} C_0^{-1}\|\sigma(x,\pi)\|_{S^m_{\rho,\delta},0,d,|-m|}. \end{aligned}$$

Recursively on $d=[\beta_o]$, we can show similar properties for $X_x^{\beta_o}\left\{\psi(\pi(\mathcal{R}))\sigma(x,\pi)^{-1}\right\}$, and obtain

$$\begin{aligned} &\|\psi(\pi(\mathcal{R}))\sigma(x,\pi)^{-1}\|_{S^{-m}_{\rho,\delta},0,b_o,0} \\ &\quad\leq C_{b_o,\|\psi\|_\infty}\sum_{b_1',b_2'\leq b_o}\max_{|\gamma|\leq\delta b_o} C_\gamma^{-(b_1'+1)}\|\sigma(x,\pi)\|^{b_2'}_{S^m_{\rho,\delta},0,b_o,|m|}. \end{aligned}$$

We can proceed in a parallel way for difference operators. Indeed, for any $\alpha_o\in\mathbb{N}_0^n$ with $|\alpha_o|=1$, we apply Δ^{α_o} to both sides of (5.83) and obtain

$$\Delta^{\alpha_o}\{\psi(\pi(\mathcal{R}))\} = \Delta^{\alpha_o}\sigma_o(x,\pi)\ \sigma(x,\pi) + \sigma_o(x,\pi)\ \Delta^{\alpha_o}\{\sigma(x,\pi)\},$$

thus

$$\begin{aligned} \Delta^{\alpha_o}\sigma_o(x,\pi) &= \Delta^{\alpha_o}\{\psi(\pi(\mathcal{R}))\}E(\Lambda_1,\infty)\sigma^{-1}(x,\pi) \\ &\qquad -\sigma_o(x,\pi)\ \{\Delta^{\alpha_o}\sigma(x,\pi)\}\ E(\Lambda_1,\infty)\ \sigma^{-1}(x,\pi). \end{aligned}$$

Then

$$\|\pi(\mathrm{I}+\mathcal{R})^{\frac{\rho[\alpha_o]+m}{\nu}}\Delta^{\alpha_o}\sigma_o(x,\pi)\|_{\mathscr{L}(\mathcal{H}_\pi)} \leq N_1+N_2,$$

with

$$\begin{aligned}N_1 &= \|\pi(\mathrm{I}+\mathcal{R})^{\frac{\rho[\alpha_o]+m}{\nu}}\Delta^{\alpha_o}\{\psi(\pi(\mathcal{R}))\}E(\Lambda_1,\infty)\sigma^{-1}(x,\pi)\|_{\mathscr{L}(\mathcal{H}_\pi)},\\ N_2 &= \|\pi(\mathrm{I}+\mathcal{R})^{\frac{\rho[\alpha_o]+m}{\nu}}\sigma_o(x,\pi)\ \{\Delta^{\alpha_o}\sigma(x,\pi)\}\ E(\Lambda_1,\infty)\ \sigma^{-1}(x,\pi)\|_{\mathscr{L}(\mathcal{H}_\pi)}.\end{aligned}$$

For the first norm, we see that

$$\begin{aligned}N_1 &\le \|\pi(\mathrm{I}+\mathcal{R})^{\frac{\rho[\alpha_o]+m}{\nu}}\Delta^{\alpha_o}\{\psi(\pi(\mathcal{R}))\}\pi(\mathrm{I}+\mathcal{R})^{-\frac{m}{\nu}}\|_{\mathscr{L}(\mathcal{H}_\pi)}\\ &\qquad \|\pi(\mathrm{I}+\mathcal{R})^{\frac{m}{\nu}}E(\Lambda_1,\infty)\sigma^{-1}(x,\pi)\|_{\mathscr{L}(\mathcal{H}_\pi)}\\ &\le C_\psi C_0^{-1},\end{aligned}$$

since $\Delta^{\alpha_o}\{\psi(\pi(\mathcal{R}))\}\in S^{-\infty}$ by Lemma 5.8.6. For the second norm, we see that

$$\begin{aligned}N_2 &\le \|\pi(\mathrm{I}+\mathcal{R})^{\frac{\rho[\alpha_o]+m}{\nu}}\sigma_o(x,\pi)\pi(\mathrm{I}+\mathcal{R})^{-\frac{\rho[\alpha_o]}{\nu}}\|_{\mathscr{L}(\mathcal{H}_\pi)}\\ &\qquad \|\pi(\mathrm{I}+\mathcal{R})^{\frac{\rho[\alpha_o]}{\nu}}\Delta^{\alpha_o}\sigma(x,\pi)\pi(\mathrm{I}+\mathcal{R})^{-\frac{m}{\nu}}\|_{\mathscr{L}(\mathcal{H}_\pi)}\\ &\qquad \|\pi(\mathrm{I}+\mathcal{R})^{\frac{m}{\nu}}\ E(\Lambda_1,\infty)\ \sigma^{-1}(x,\pi)\|_{\mathscr{L}(\mathcal{H}_\pi)}\\ &\le \|\psi\|_\infty C_{\rho[\alpha_o]}^{-1}C_0^{-1}\|\sigma\|_{S^m_{\rho,\delta},[\alpha_o],0,|m|}.\end{aligned}$$

Recursively on $[\alpha_o]$, we can show similar properties for $\Delta^{\alpha_o}\left\{\psi(\pi(\mathcal{R}))\sigma(x,\pi)^{-1}\right\}$, and obtain

$$\begin{aligned}&\|\sigma_o(x,\pi)\|_{S^{-m}_{\rho,\delta},a_o,0,0}\\ &\qquad \le C_{a_o,\psi}\sum_{a_1',a_2'\le a_o}\max_{|\gamma|\le\rho a_o}C_\gamma^{-(a_1'+1)}\|\sigma(x,\pi)\|^{a_2'}_{S^m_{\rho,\delta},a_o,0,|m|}.\end{aligned}$$

More generally, we have

$$\begin{aligned}X_x^{\beta_o}\Delta^{\alpha_o}\left\{\psi(\pi(\mathcal{R}))\right\} &= \sum_{\substack{[\alpha_1]+[\alpha_2]=[\alpha_o]\\ [\beta_1]+[\beta_2]=[\beta_o]}} c'_{\beta_1,\beta_2}c_{\alpha_1,\alpha_2}\ X_x^{\beta_1}\Delta^{\alpha_1}\sigma_o(x,\pi)\\ &\qquad\qquad X_x^{\beta_2}\Delta^{\alpha_2}\sigma(x,\pi).\end{aligned}$$

Because of the very first remark of this proof, we obtain $X^{\beta_o}\Delta^{\alpha_o}\sigma_o$ in terms of $X^{\beta'}\Delta^{\alpha'}\sigma_o$ with $[\beta']<[\beta_o]$ and $[\alpha']<[\alpha_o]$ and of some derivatives of $\psi(\pi(\mathcal{R}))$ and σ. If we assume that we can control all the seminorms $\|\sigma_o\|_{S^{-m}_{\rho,\delta},a,b,c}$ with $a<[\alpha_o]$, $b<[\beta_o]$ and any $c\in\mathbb{R}$, then we can proceed as above introducing powers of $\mathrm{I}+\mathcal{R}$ to obtain the estimate for the seminorms of $\psi(\pi(\mathcal{R}))\sigma(x,\pi)^{-1}$. Recursively this shows Proposition 5.8.5. □

5.8.2 Parametrix

In the next theorem, we show that our notion of ellipticity implies the construction of a parametrix.

Theorem 5.8.7. *Let $\sigma \in S^m_{\rho,\delta}$ be elliptic of elliptic order m with $1 \geq \rho > \delta \geq 0$. We can construct a left parametrix $B \in \Psi^{-m}_{\rho,\delta}$ for the operator $A = \text{Op}(\sigma)$, that is, there exists $B \in \Psi^{-m}_{\rho,\delta}$ such that*

$$BA - \text{I} \in \Psi^{-\infty}.$$

Comparing with two-sided parametrices in the case of compact Lie groups (Theorem 2.2.17), this parametrix is one-sided. It was also the case in [CGGP92].

Proof. We can adapt the proof in [Tay81, §0.4] to our setting. Let $\psi \in C^\infty(\mathbb{R})$ be such that $\psi_{|(-\infty,\Lambda_1]} = 0$ and $\psi_{|[\Lambda_2,\infty)} = 1$ for some $\Lambda_1, \Lambda_2 \in \mathbb{R}$ with $\Lambda < \Lambda_1 < \Lambda_2$. By Proposition 5.8.5,

$$\psi(\pi(\mathcal{R}))\sigma^{-1}(x,\pi) \in S^{-m}_{\rho,\delta}.$$

Since $\psi(\pi(\mathcal{R})) = \psi(\pi(\mathcal{R}))\sigma^{-1}(x,\pi)\sigma(x,\pi)$, by Corollary 5.5.8,

$$\text{Op}\big(\psi(\pi(\mathcal{R}))\sigma^{-1}(x,\pi)\big)\ A = \psi(\mathcal{R}) \ \text{mod}\Psi^{-(\rho-\delta)}_{\rho,\delta} \ ;$$

now $\psi(\mathcal{R}) = \text{I} - (1-\psi)(\mathcal{R})$ and $(1-\psi) \in \mathcal{D}([0,\infty))$ so $(1-\psi)(\mathcal{R}) \in \Psi^{-\infty}$. This shows

$$\text{Op}\big(\psi(\pi(\mathcal{R}))\sigma^{-1}(x,\pi)\big)\ A \quad = \text{I} \ \text{mod}\Psi^{-(\rho-\delta)}_{\rho,\delta}.$$

So we have

$$\text{Op}\big(\psi(\pi(\mathcal{R}))\sigma^{-1}(x,\pi)\big)\ A \quad = \text{I} - U \quad \text{with} \quad U \in \Psi^{-(\rho-\delta)}_{\rho,\delta}.$$

By Theorem 5.5.1, there exists $T \in \Psi^0_{\rho,\delta}$ such that

$$T \sim \text{I} + U + U^2 + \ldots + U^j + \ldots$$

By Theorem 5.5.3,

$$B := T\ \text{Op}\big(\psi(\pi(\mathcal{R}))\sigma^{-1}\big) \ \in \Psi^{-m}_{\rho,\delta}.$$

Therefore, we obtain

$$BA = T(\text{I} - U) = \text{I} \ \text{mod}\Psi^{-\infty},$$

completing the proof. □

It is not difficult to construct the following examples of elliptic operators satisfying Theorem 5.8.7 out of any Rockland operator. Indeed, combining Proposition 5.3.4 or Corollary 5.3.8 together with Proposition 5.8.2 yield

Example 5.8.8. Let $\mathcal{R}$ be a positive Rockland operator of homogeneous degree ν.

1. For any $m \in \mathbb{R}$, the operator $(\text{I} + \mathcal{R})^{\frac{m}{\nu}} \in \Psi^m$ is elliptic with respect to $\mathcal{R}$ of elliptic order m.

2. If f_1 and f_2 are complex-valued smooth functions on G such that
$$\inf_{x\in G,\lambda\geq\Lambda}\frac{|f_1(x)+f_2(x)\lambda|}{1+\lambda}>0 \quad \text{for some } \Lambda\geq 0,$$
and such that $X^{\alpha_1}f_1$, $X^{\alpha_2}f_2$ are bounded for each $\alpha_1,\alpha_2\in\mathbb{N}_0^n$, then the differential operator
$$f_1(x)+f_2(x)\mathcal{R}\in\Psi^\nu$$
is $(\mathcal{R},\Lambda,\nu)$-elliptic.

3. Let $\psi\in C^\infty(\mathbb{R})$ be such that
$$\psi_{|(-\infty,\Lambda_1]}=0 \quad \text{and} \quad \psi_{|[\Lambda_2,\infty)}=1,$$
for some real numbers Λ_1,Λ_2 satisfying $0<\Lambda_1<\Lambda_2$, Then the operator $\psi(\mathcal{R})\mathcal{R}\in\Psi^\nu$ is $(\mathcal{R},\Lambda_2,\nu)$-elliptic.

   More generally, if f is a smooth complex-valued function on G such that $\inf_G|f|>0$ and that $X^\alpha f$ is bounded on G for every $\alpha\in\mathbb{N}_0^n$, then
$$f(x)\psi(\mathcal{R})\mathcal{R}\in\Psi^\nu$$
is elliptic with respect to $\mathcal{R}$ of elliptic order ν.

Hence all the operators in Example 5.8.8 admit a left parametrix.

We will see other concrete examples of elliptic differential operators on the Heisenberg group in Section 6.6.1, see Example 6.6.2.

In fact we can prove the existence of left parametrices for symbols which are elliptic with an elliptic order lower than their order. Indeed, we can modify the hypothesis of the ellipticity in Section 5.8.1 to obtain the analogue of Hörmander's theorem about hypoellipticity involving lower order terms, similar to Theorem 2.2.18 in the compact case.

Theorem 5.8.9. *Let $\sigma\in S^m_{\rho,\delta}$ with $1\geq\rho>\delta\geq 0$. We assume that σ is elliptic with respect to a positive Rockland operator $\mathcal{R}$ in the sense of Definition 5.8.1, and that its elliptic order is $m_o\leq m$.*

We also assume that the following hypothesis on the lower order terms holds: there is $\Lambda\in\mathbb{R}$ such that for any $\gamma\in\mathbb{R}$, $x\in G$, μ-almost all $\pi\in\widehat{G}$, and any $u\in\mathcal{H}^\infty_{\pi,\Lambda}$, we have
$$\begin{aligned}\|\pi(\mathrm{I}+\mathcal{R})^{\frac{\rho[\alpha]-\delta[\beta]+\gamma}{\nu}}\{\Delta^\alpha X^\beta\sigma(x,\pi)\}\,\pi(\mathrm{I}+\mathcal{R})^{-\frac{\gamma}{\nu}}u\|_{\mathcal{H}_\pi}&\\ \leq C'_{\alpha,\beta,\gamma}\|\sigma(x,\pi)u\|_{\mathcal{H}_\pi},&\end{aligned}\tag{5.84}$$
with $C'_{\alpha,\beta,\gamma}=C'_{\alpha,\beta,\gamma,\sigma,\mathcal{R},m_o,\Lambda,\gamma}$ independent of $(x,\pi)\in G\times\widehat{G}$ and $u\in\mathcal{H}^\infty_{\pi,\Lambda}$.

Then we can construct a left parametrix $B\in\Psi^{-m_o}_{\rho,\delta}$ for the operator $A=\mathrm{Op}(\sigma)$, that is, there exists $B\in\Psi^{-m_o}_{\rho,\delta}$ such that
$$BA-\mathrm{I}\in\Psi^{-\infty}.$$

Proceeding as in Corollary 5.8.4, we can show easily that it suffices to assume (5.79) and (5.84) for a countable sequence γ which goes to $+\infty$ and $-\infty$.

Proof. Let $\psi \in C^\infty(\mathbb{R})$ be such that $\psi_{|(-\infty,\Lambda_1]} = 0$ and $\psi_{|[\Lambda_2,\infty)} = 1$ for some $\Lambda_1, \Lambda_2 \in \mathbb{R}$ with $\Lambda < \Lambda_1 < \Lambda_2$. Proceeding as in the proof of Proposition 5.8.5, we see that

$$\sigma_o(x,\pi) := \psi(\pi(\mathcal{R}))\sigma^{-1}(x,\pi) \in S^{-m_o}_{\rho,\delta},$$

with similar estimates for the seminorms of σ_o and σ.

With similar ideas, using (5.84), we claim that, for any multi-index $\beta_o \in \mathbb{N}_0^n$, we have

$$X^{\beta_o}\sigma(x,\pi)\ \sigma_o(x,\pi) \in S^{\delta[\beta_o]}_{\rho,\delta}.$$

Indeed, from the proof of Proposition 5.8.5, we know that

$$X\sigma_o = -\sigma_o\ X\sigma\ E(\Lambda,\infty)\sigma^{-1},$$

hence

$$\begin{aligned} X\left(X^{\beta_o}\sigma(x,\pi)\ \sigma_o(x,\pi)\right) &= XX^{\beta_o}\sigma(x,\pi)\ \sigma_o(x,\pi) + X^{\beta_o}\sigma(x,\pi)\ X\sigma_o(x,\pi) \\ &= XX^{\beta_o}\sigma(x,\pi)\ \sigma_o(x,\pi) - X^{\beta_o}\sigma(x,\pi)\ \sigma_o\ X\sigma\ E(\Lambda,\infty)\sigma^{-1}, \end{aligned}$$

and we can use the hypothesis (5.84) on each term to control the $S^m_{\rho,\delta}$-seminorms of the expression on the right-hand side. For the difference operators, from the proof of Proposition 5.8.5, we know with $|\alpha_o| = 1$, that

$$\Delta^{\alpha_o}\sigma_o = \Delta^{\alpha_o}\psi(\pi(\mathcal{R}))\ E(\Lambda,\infty)\sigma^{-1} - \sigma_o\ \Delta^{\alpha_o}\sigma\ E(\Lambda,\infty)\sigma^{-1}.$$

Hence

$$\begin{aligned} &\Delta^{\alpha_o}\left\{X^{\beta_o}\sigma(x,\pi)\ \sigma_o(x,\pi)\right\} \\ &\quad = X^{\beta_o}\Delta^{\alpha_o}\sigma(x,\pi)\ \sigma_o(x,\pi) + X^{\beta_o}\sigma(x,\pi)\ \Delta^{\alpha_o}\sigma_o(x,\pi) \\ &\quad = X^{\beta_o}\Delta^{\alpha_o}\sigma(x,\pi)\ \sigma_o(x,\pi) - X^{\beta_o}\sigma(x,\pi)\ \sigma_o\ \Delta^{\alpha_o}\sigma\ E(\Lambda,\infty)\sigma^{-1} \\ &\qquad + X^{\beta_o}\sigma(x,\pi)\ \Delta^{\alpha_o}\psi(\pi(\mathcal{R}))\ \psi_o(\pi(\mathcal{R}))\sigma^{-1}, \end{aligned}$$

where $\psi_o \in C^\infty(\mathbb{R})$ is a fixed smooth function such that $\psi_{o|[\Lambda_1,\infty)} = 1$ and $\psi_{o|(-\infty,\Lambda_1/2)} = 0$. While we can use the hypothesis (5.84) on the first two terms, we use Lemma 5.8.6 for the last term which is then smoothing. Proceeding recursively as in the proof of Proposition 5.8.5, we obtain the estimates for the sum on the right-hand side.

We now define recursively

$$\sigma_n(x,\pi) := \left(\sum_{0<[\alpha]\leq n} \Delta^\alpha\sigma_{n-[\alpha]}X^\alpha\sigma\right)\sigma_o, \quad n = 1,2,\ldots$$

It is easy to check that each symbol $\sigma_n(x,\pi)$ is in $S^{-m_o-n(\rho-\delta)}_{\rho,\delta}$ and that as in the compact case,

$$\mathrm{Op}(\sigma_o)\mathrm{Op}(\sigma) - I - \mathrm{Op}(\sigma_1)\mathrm{Op}(\sigma) - \ldots - \mathrm{Op}(\sigma_n)\mathrm{Op}(\sigma) \in \Psi^{m-m_0-n}_{\rho,\delta}.$$

Therefore, the operator $B \in \Psi^{-m_o}_{\rho,\delta}$ whose symbol is given by the asymptotic sum $\sigma_o - \sum_{j=1}^{\infty} \sigma_j$ is a left parametrix for $A = \mathrm{Op}(\sigma)$. □

We will see a concrete example of hypoelliptic differential operators on the Heisenberg group in Section 6.6.2, see Example 6.6.4.

We now note the following generalisation of Proposition 5.8.5 that we have already used in the proof of Theorem 5.8.9.

Proposition 5.8.10. *Assume $1 \geq \rho \geq \delta \geq 0$. Let $\sigma \in S^m_{\rho,\delta}$ be a symbol which is $(\mathcal{R}, \Lambda, m_o)$-elliptic with respect to a positive Rockland operator $\mathcal{R}$. If $\psi \in C^\infty(\mathbb{R})$ is such that*

$$\psi_{|(-\infty,\Lambda_1]} = 0 \quad \textit{and} \quad \psi_{|[\Lambda_2,\infty)} = 1,$$

for some real numbers Λ_1, Λ_2 satisfying $\Lambda < \Lambda_1 < \Lambda_2$, then the symbol

$$\{\psi(\pi(\mathcal{R}))\sigma^{-1}(x,\pi)\ ,\ (x,\pi) \in G \times \widehat{G}\},$$

given by

$$\psi(\pi(\mathcal{R}))\sigma(x,\pi)^{-1} := \psi(\pi(\mathcal{R}))E_\pi(\Lambda_1,\infty)\sigma^{-1}(x,\pi),$$

is in $S^{-m_o}_{\rho,\delta}$. Moreover, for any $a_o, b_o \in \mathbb{N}_0$, we have

$$\begin{aligned}&\|\psi(\pi(\mathcal{R}))\sigma^{-1}(x,\pi)\|_{S^{-m_o}_{\rho,\delta},a_o,b_o,0}\\ &\quad \leq C \sum_{\substack{a_1',a_2'\leq a_o\\ b_1',b_2'\leq b_o}} \max_{|\gamma|\leq \rho a_o+\delta b_o} C^{a_1'+b_1'+1}_{\gamma,\sigma,\Lambda_1} \|\sigma(x,\pi)\|^{a_2'+b_2'}_{S^m_{\rho,\delta},a_o,b_o,|m|},\end{aligned}$$

where $C > 0$ is a positive constant depending on a_o, b_o, ψ, and where the constant $C_{\gamma,\sigma,\Lambda_1}$ was given in (5.79).

Here the elliptic order m_o and the symbol order m are different but the same results holds: one can construct a symbol $\psi(\pi(\mathcal{R}))\sigma^{-1}(x,\pi) \in S^{-m_o}_{\rho,\delta}$. The proof is easily obtained by generalising the proof of Proposition 5.8.5.

We now show that Theorem 5.8.7 has a partial inverse.

Proposition 5.8.11. *Suppose that the operator $A = \mathrm{Op}(\sigma) \in \Psi^m_{\rho,\delta}$, with $1 \geq \rho > \delta \geq 0$, admits a left parametrix $B \in \Psi^{-m}_{\rho,\delta}$, i.e. $BA - \mathrm{I} \in \Psi^{-\infty}$. Then σ is elliptic of order m, that is, there exist a positive Rockland operator $\mathcal{R}$ of homogeneous degree ν, and $\Lambda \in \mathbb{R}$ such that for any $\gamma \in \mathbb{R}$, $x \in G$, μ-almost all $\pi \in \widehat{G}$, and any $u \in \mathcal{H}^\infty_{\pi,\Lambda}$ we have*

$$\|\pi(\mathrm{I}+\mathcal{R})^{\frac{\gamma}{\nu}}\sigma(x,\pi)u\|_{\mathcal{H}_\pi} \geq C_\gamma \|\pi(\mathrm{I}+\mathcal{R})^{\frac{\gamma}{\nu}}\pi(\mathrm{I}+\mathcal{R})^{\frac{m}{\nu}}u\|_{\mathcal{H}_\pi}.$$

Moreover, if this property holds for one positive Rockland operator then it holds for any Rockland operator.

Proof. Let A and B be as in the statement. Let σ and τ be their respective symbols. Then the symbol

$$\begin{aligned}\varepsilon &:= \tau\sigma - \mathrm{I} \\ &= (\tau\sigma - \mathrm{Op}^{-1}(BA)) - (\mathrm{I} - \mathrm{Op}^{-1}(BA)),\end{aligned}$$

is in $S^{-(\rho-\delta)}_{\rho,\delta}$, and we can write

$$\pi(\mathrm{I}+\mathcal{R})^{\frac{m+\gamma}{\nu}}\tau\sigma = \pi(\mathrm{I}+\mathcal{R})^{\frac{m+\gamma}{\nu}} + \epsilon_0\pi(\mathrm{I}+\mathcal{R})^{-\frac{\rho-\delta}{\nu}}\pi(\mathrm{I}+\mathcal{R})^{\frac{m+\gamma}{\nu}},$$

where

$$\varepsilon_0 := \pi(\mathrm{I}+\mathcal{R})^{\frac{m+\gamma}{\nu}}\varepsilon\pi(\mathrm{I}+\mathcal{R})^{\frac{\rho-\delta}{\nu}-\frac{m+\gamma}{\nu}} \in S^0_{\rho,\delta}.$$

For any $u \in \mathcal{H}_\pi^\infty$, $(x,\pi) \in G \times \widehat{G}$, we thus have

$$\begin{aligned}&\|\pi(\mathrm{I}+\mathcal{R})^{\frac{m+\gamma}{\nu}}\tau(x,\pi)\sigma(x,\pi)u\|_{\mathcal{H}_\pi} \\ &\quad = \|\left(\pi(\mathrm{I}+\mathcal{R})^{\frac{m+\gamma}{\nu}} + \epsilon_0(x,\pi)\pi(\mathrm{I}+\mathcal{R})^{-\frac{\rho-\delta}{\nu}}\pi(\mathrm{I}+\mathcal{R})^{\frac{m+\gamma}{\nu}}\right)u\|_{\mathcal{H}_\pi}.\end{aligned}$$

We can bound the left hand side by

$$\begin{aligned}&\|\pi(\mathrm{I}+\mathcal{R})^{\frac{m+\gamma}{\nu}}\tau(x,\pi)\sigma(x,\pi)u\|_{\mathcal{H}_\pi} \\ &\quad \le \|\pi(\mathrm{I}+\mathcal{R})^{\frac{m+\gamma}{\nu}}\tau(x,\pi)\pi(\mathrm{I}+\mathcal{R})^{-\frac{\gamma}{\nu}}\|_{\mathscr{L}(\mathcal{H}_\pi)}\|\pi(\mathrm{I}+\mathcal{R})^{\frac{\gamma}{\nu}}\sigma(x,\pi)u\|_{\mathcal{H}_\pi} \\ &\quad \le \|\tau\|_{S^{-m}_{0,0,|\gamma|}}\|\pi(\mathrm{I}+\mathcal{R})^{\frac{\gamma}{\nu}}\sigma(x,\pi)u\|_{\mathcal{H}_\pi},\end{aligned}$$

and the right hand side below by

$$\begin{aligned}&\|\left(\pi(\mathrm{I}+\mathcal{R})^{\frac{m+\gamma}{\nu}} + \epsilon_0(x,\pi)\pi(\mathrm{I}+\mathcal{R})^{-\frac{\rho-\delta}{\nu}}\pi(\mathrm{I}+\mathcal{R})^{\frac{m+\gamma}{\nu}}\right)u\|_{\mathcal{H}_\pi} \\ &\quad \ge \|\pi(\mathrm{I}+\mathcal{R})^{\frac{m+\gamma}{\nu}}u\|_{\mathcal{H}_\pi} - \|\epsilon_0(x,\pi)\pi(\mathrm{I}+\mathcal{R})^{-\frac{\rho-\delta}{\nu}}\pi(\mathrm{I}+\mathcal{R})^{\frac{m+\gamma}{\nu}}u\|_{\mathcal{H}_\pi} \\ &\quad \ge \|\pi(\mathrm{I}+\mathcal{R})^{\frac{m+\gamma}{\nu}}u\|_{\mathcal{H}_\pi} \\ &\qquad - \|\epsilon_0(x,\pi)\|_{\mathscr{L}(\mathcal{H}_\pi)}\|\pi(\mathrm{I}+\mathcal{R})^{-\frac{\rho-\delta}{\nu}}\pi(\mathrm{I}+\mathcal{R})^{\frac{m+\gamma}{\nu}}u\|_{\mathcal{H}_\pi}.\end{aligned}$$

Hence if $u \in E(\Lambda,\infty)\mathcal{H}_\pi^\infty$ where $\Lambda \ge 0$ then

$$\begin{aligned}&\|\tau\|_{S^{-m}_{0,0,|\gamma|}}\|\pi(\mathrm{I}+\mathcal{R})^{\frac{\gamma}{\nu}}\sigma(x,\pi)u\|_{\mathcal{H}_\pi} \\ &\quad \ge \|\pi(\mathrm{I}+\mathcal{R})^{\frac{m+\gamma}{\nu}}u\|_{\mathcal{H}_\pi} \\ &\qquad - \|\epsilon_0(x,\pi)\|_{\mathscr{L}(\mathcal{H}_\pi)}(1+\Lambda)^{-\frac{\rho-\delta}{\nu}}\|\pi(\mathrm{I}+\mathcal{R})^{\frac{m+\gamma}{\nu}}u\|_{\mathcal{H}_\pi}.\end{aligned}$$

Clearly $\tau \not\equiv 0$ and $\|\tau\|_{S^{-m}_{0,0,|\gamma|}} \neq 0$. Furthermore

$$\|\epsilon_0(x,\pi)\|_{\mathscr{L}(\mathcal{H}_\pi)} \le \|\epsilon_0\|_{S^0_{\rho,\delta},0,0,0} < \infty,$$

hence we can choose $\Lambda \geq 0$ such that

$$\|\epsilon_0(x,\pi)\|_{\mathscr{L}(\mathcal{H}_\pi)}(1+\Lambda)^{-\frac{\rho-\delta}{\nu}} \leq \|\epsilon_0\|_{S^0_{\rho,\delta},0,0,0}(1+\Lambda)^{-\frac{\rho-\delta}{\nu}} \leq \frac{1}{2},$$

in view of $\rho > \delta$. We have therefore obtained for $u \in E(\Lambda,\infty)\mathcal{H}_\pi^\infty$ with the chosen Λ, that

$$\|\pi(\mathrm{I}+\mathcal{R})^{\frac{\gamma}{\nu}}\sigma(x,\pi)u\|_{\mathcal{H}_\pi} \geq \frac{1}{2\|\tau\|_{S^{-m}_{0,0,|\gamma|}}}\|\pi(\mathrm{I}+\mathcal{R})^{\frac{m+\gamma}{\nu}}u\|_{\mathcal{H}_\pi},$$

which is the required statement. □

5.8.3 Subelliptic estimates and hypoellipticity

The existence of a parametrix yields subelliptic estimates:

Corollary 5.8.12. *Let $m \in \mathbb{R}$ and $1 \geq \rho > \delta \geq 0$. If $A \in \Psi^m_{\rho,\delta}$ is elliptic of order m, then A satisfies the following subelliptic estimates*

$$\forall s \in \mathbb{R} \quad \forall N \in \mathbb{R} \quad \exists C>0 \quad \forall f \in \mathcal{S}(G) \qquad \|f\|_{L^2_{s+m}} \leq C\Big(\|Af\|_{L^2_s} + \|f\|_{L^2_{-N}}\Big).$$

If $A \in \Psi^m_{\rho,\delta}$ is elliptic of order m_o and satisfies the hypotheses of Theorem 5.8.9, then A satisfies the subelliptic estimates

$$\forall s \in \mathbb{R} \quad \forall N \in \mathbb{R} \quad \exists C>0 \quad \forall f \in \mathcal{S}(G) \qquad \|f\|_{L^2_{s+m_o}} \leq C\Big(\|Af\|_{L^2_s} + \|f\|_{L^2_{-N}}\Big).$$

In the case $(\rho,\delta)=(1,0)$, assume that $A \in \Psi^m$ is either elliptic of order $m_0 = m$ or is elliptic of some order m_0 and satisfies the hypotheses of Theorem 5.8.9. Then A satisfies the subelliptic estimates

$$\forall s \in \mathbb{R} \quad \forall N \in \mathbb{R} \quad \forall p \in (1,\infty) \quad \exists C>0 \quad \forall f \in \mathcal{S}(G)$$
$$\|f\|_{L^p_{s+m_o}} \leq C\Big(\|Af\|_{L^p_s} + \|f\|_{L^p_{-N}}\Big).$$

In the estimates above, $\|\cdot\|_{L^p_s}$ denotes any (fixed) Sobolev norm, for example obtained from a (fixed) positive Rockland operator.

Proof. By Theorem 5.8.7 or Theorem 5.8.9, A admits a left parametrix B, i.e. $BA - \mathrm{I} = R \in \Psi^{-\infty}$. By using the boundedness on Sobolev spaces from Corollary 5.7.2, we get

$$\|f\|_{L^2_{s+m_o}} \leq \|BAf\|_{L^2_{s+m_o}} + \|Rf\|_{L^2_{s+m_o}} \leq C(\|Af\|_{L^2_s} + \|f\|_{L^2_{-N}}).$$

In the case $(\rho,\delta)=(1,0)$, the last statement follows from Corollary 5.7.4 with Sobolev L^p-boundedness instead. □

Local hypoelliptic properties

Our construction of parametrices implies the following local property:

Proposition 5.8.13. *Let $A \in \Psi^m_{\rho,\delta}$ with $m \in \mathbb{R}$, $1 \geq \rho > \delta \geq 0$. We assume that the operator A is elliptic of order m_0 and that*

- *either $m = m_0$,*
- *or $m > m_0$ and in this case A satisfies the hypotheses of Theorem 5.8.9.*

Then the singular support of any $f \in \mathcal{S}'(G)$ is contained the singular support of Af,

$$\operatorname{sing\,supp} f \subset \operatorname{sing\,supp} Af,$$

that is, if Af coincides with a smooth function on any open subset of G, then f is also smooth there.

Consequently, if A is a differential operator, then it is hypoelliptic.

The notion of hypoellipticity for a differential operator with smooth coefficients is explained in Appendix A.1.

Proposition 5.8.13 follows easily from the following property:

Lemma 5.8.14. *Let $A \in \Psi^m_{\rho,\delta}$ with $m \in \mathbb{R}$, $1 \geq \rho > \delta \geq 0$. We assume that there exists an open set Ω such that the symbol of A satisfies the elliptic condition in (5.79) for any $x \in \Omega$ only. We also assume that*

- *either $m = m_0$,*
- *or $m > m_0$ and in this case A satisfies the hypotheses of Theorem 5.8.9 with $x \in \Omega$.*

If $f \in \mathcal{S}'(G)$ and if Ω' is an open subset of Ω where Af is smooth, i.e. $Af \in C^\infty(\Omega')$, then $f \in C^\infty(\Omega')$.

The proof requires to revisit the construction of parametrices 'to make it local'.

Proof of Lemma 5.8.14. We keep the hypotheses and notation of the statement. As the properties are essentially local, we may assume that the open subsets Ω, Ω' are open bounded and that there exists an open subset Ω_1 such that $\bar{\Omega}' \subset \Omega_1$ and $\bar{\Omega}_1 \subset \Omega$. Let $\chi \in \mathcal{D}(G)$ be such that $\chi \equiv 1$ on Ω' and $\chi \equiv 0$ outside Ω_1. The symbol of the operator $A' := \chi(x)A$ is given via $\chi(x)\sigma(x,\pi)$. An easy modification of the proof of Proposition 5.8.5 implies that the symbol given by

$$\chi(x)\psi(\pi(\mathcal{R}))\sigma(x,\tau)^{-1}$$

is in $S^{-m_0}_{\rho,\delta}$ (here ψ is a function as in Proposition 5.8.5). Adapting the proof of Theorem 5.8.7 or Theorem 5.8.9, we construct an operator $B \in \Psi^{-m_0}_{\rho,\delta}$ such that $BA' = \chi(x) + R$ with $R \in \Psi^{-\infty}$.

Let $\chi_1 \in \mathcal{D}(G)$ be such that $\chi_1 \equiv 1$ on Ω_1 and $\chi_1 \equiv 0$ outside Ω. Let $f \in \mathcal{S}'(G)$. As A admits a singular integral representation, see Lemma 5.4.15 and its proof, the function $x \mapsto \chi(x)\ A\{(1-\chi_1)f\}(x)$ is smooth and compactly supported. Let us assume that Af is smooth on Ω'. Since we have for any $x \in G$

$$A'\{\chi_1 f\}(x) = \chi(x)\ Af(x) - \chi(x)\ A\{(1-\chi_1)f\}(x),$$

the function $A'\{\chi_1 f\}$ is necessarily smooth and compactly supported on G, i.e. $A'\{\chi_1 f\} \in \mathcal{D}(G)$. Applying B, we have $BA'\{\chi_1 f\} \in \mathcal{S}(G)$ by Theorem 5.2.15. By Corollary 5.5.13. $R\{\chi_1 f\} \in \mathcal{S}(G)$ since the distribution $\chi_1 f \in \mathcal{E}'(G)$ has compact support. Hence $\chi_1 f = BA'\{\chi_1 f\} - R\{\chi_1 f\}$ must be in $\mathcal{S}(G)$. This shows that f is smooth on Ω'. □

Global hypoelliptic-type properties

Our construction of parametrix is global. Hence we also obtain the following global property:

Proposition 5.8.15. *Let $A \in \Psi^m_{\rho,\delta}$ with $m \in \mathbb{R}$, $1 \geq \rho > \delta \geq 0$. We assume that the operator A is elliptic of order m_0 and that*

- *either $m = m_0$,*
- *or $m > m_0$ and in this case A satisfies the hypotheses of Theorem 5.8.9.*

If $f \in \mathcal{S}'(G)$ and $Af \in \mathcal{S}(G)$, then f is smooth and all its left-derivatives (hence also right-derivatives and abelian derivatives) have polynomial growth. More precisely, for any multi-index $\beta \in \mathbb{N}_0^n$, there exists a constant $C > 0$, an integer $M \in \mathbb{N}_0$ and seminorms $\|\cdot\|_{\mathcal{S}'(G),N_1}$, $\|\cdot\|_{\mathcal{S}(G),N_2}$ such that for any $f \in \mathcal{S}'(G)$ with $Af \in \mathcal{S}(G)$, we have

$$|X^\beta f(x)| \leq C\left((1+|x|)^M \|f\|_{\mathcal{S}'(G),N_1} + \|Af\|_{\mathcal{S}(G),N_2}\right), \qquad x \in G.$$

Proof. We keep the hypotheses and notation of the statement. By Theorem 5.8.7 or Theorem 5.8.9, A admits a left parametrix B, i.e. $BA - \mathrm{I} \in \Psi^{-\infty}$. By Corollary 5.4.10, $(BA - \mathrm{I})f$ is smooth with polynomial growth. As $Af \in \mathcal{S}(G)$, $B(Af) \in \mathcal{S}(G)$ by Theorem 5.2.15. Thus

$$f = -(BA - \mathrm{I})f + B(Af)$$

is smooth with polynomial growth. The estimate follows easily from the ones in Corollary 5.4.10 and Theorem 5.2.15. □

Examples

Hence we have obtained hypoellipticity and subelliptic estimates for the operators in Examples 5.8.8.

Corollary 5.8.16. *Let $\mathcal{R}$ be a positive Rockland operator of homogeneous degree ν and let $p \in (1,\infty)$.*

1. *If f_1 and f_2 are complex-valued smooth functions on G such that*
$$\inf_{x\in G,\lambda\geq\Lambda} \frac{|f_1(x)+f_2(x)\lambda|}{1+\lambda} > 0 \quad \text{for some } \Lambda \geq 0,$$
and such that $X^{\alpha_1} f_1$, $X^{\alpha_2} f_2$ are bounded for each $\alpha_1,\alpha_2 \in \mathbb{N}_0^n$, then the differential operator
$$f_1(x) + f_2(x)\mathcal{R}$$
satisfies the following subelliptic estimates
$$\forall p\in(1,\infty) \quad \forall s\in\mathbb{R} \quad \forall N\in\mathbb{R} \quad \exists C>0 \quad \forall\varphi\in\mathcal{S}(G)$$
$$\|\varphi\|_{L^p_{s+\nu}} \leq C\Big(\|(f_1+f_2\mathcal{R})\varphi\|_{L^p_s} + \|\varphi\|_{L^p_{-N}}\Big),$$
and is (locally) hypoelliptic. It is also globally hypoelliptic in the sense of Proposition 5.8.15.

2. *Let $\psi \in C^\infty(\mathbb{R})$ be such that*
$$\psi_{|(-\infty,\Lambda_1]} = 0 \quad \text{and} \quad \psi_{|[\Lambda_2,\infty)} = 1,$$
for some real numbers Λ_1, Λ_2 satisfying $0 < \Lambda_1 < \Lambda_2$. Let also f_1 be a smooth complex-valued function on G such that
$$\inf_G |f_1| > 0$$
and that $X^\alpha f_1$ is bounded on G for each $\alpha \in \mathbb{N}_0^n$. Then the operator
$$f_1(x)\psi(\mathcal{R})\mathcal{R} \in \Psi^\nu$$
satisfies the following subelliptic estimates
$$\forall p\in(1,\infty) \quad \forall s\in\mathbb{R} \quad \exists C>0 \quad \forall N\in\mathbb{R} \quad \forall\varphi\in\mathcal{S}(G)$$
$$\|\varphi\|_{L^p_{s+\nu}} \leq C\Big(\|f_1\psi(\mathcal{R})\mathcal{R}\varphi\|_{L^p_s} + \|\varphi\|_{L^p_{-N}}\Big),$$
and is (locally) hypoelliptic. It is also globally hypoelliptic in the sense of Proposition 5.8.15.

Chapter 6

Pseudo-differential operators on the Heisenberg group

The Heisenberg group was introduced in Example 1.6.4. It was our primal example of a stratified Lie group, see Section 3.1.1. Due to the importance of the Heisenberg group and of its many realisations, we start this chapter by sketching various descriptions of the Heisenberg group. We also describe its dual via the well known Schrödinger representations. Eventually, we particularise our general approach given in Chapter 5 to the Heisenberg group. Among other things, we show that using the (Euclidean) Weyl quantization, the analysis of pseudo-differential operators on the Heisenberg group can be reduced to considering scalar-valued symbols parametrised not only by the elements of the Heisenberg group but also by a parameter $\lambda \in \mathbb{R}\backslash\{0\}$; such symbols will be called λ-symbols. The corresponding classes of symbols are of Shubin-type but with an interesting dependence on λ which we explore in detail in this chapter; such classes will be called λ-Shubin classes. Some results of this chapter have been announced in the authors' paper [FR14b], this chapter contains their proofs.

In [BFKG12a], a pseudo-differential calculus on the Heisenberg group was developed with a different approach (but related results) from our work presented here.

There is an important change of notation concerning the Heisenberg group in this chapter. In Example 1.6.4, where the Heisenberg group $\mathbb{H}_{n_o}$ was introduced, we used the index n_o as its subscript because the index n was already used to denote quantities associated with the homogeneous groups. However, throughout Chapter 6, general groups will hardly appear, so we can simplify the notation by denoting the Heisenberg group by $\mathbb{H}_n$ instead of $\mathbb{H}_{n_o}$, so that the notation change is

$$\boxed{\mathbb{H}_{n_o} \longrightarrow \mathbb{H}_n}$$

V. Fischer, M. Ruzhansky, *Quantization on Nilpotent Lie Groups*, Progress in Mathematics 314,
DOI 10.1007/978-3-319-29558-9_6

We emphasise that n is the index here (not the dimension): the topological dimension on $\mathbb{H}_n$ is $2n+1$, and its homogeneous dimension is $2n+2$.

6.1 Preliminaries

In this section, we discuss several aspects of the Heisenberg group, hopefully shedding some light on its importance and general structure.

6.1.1 Descriptions of the Heisenberg group

We remind the reader that the Heisenberg group $\mathbb{H}_n$ was defined in Example 1.6.4 in the following way: the *Heisenberg group* $\mathbb{H}_n$ is the manifold $\mathbb{R}^{2n+1}$ endowed with the law

$$(x,y,t)(x',y',t') := (x+x', y+y', t+t'+\frac{1}{2}(xy'-x'y)), \tag{6.1}$$

where (x,y,t) and (x',y',t') are in $\mathbb{R}^n \times \mathbb{R}^n \times \mathbb{R} \sim \mathbb{H}_n$.

In the formula above as in the whole chapter, we adopt the following convention: if x and y are two vectors in $\mathbb{R}^n$ for some $n \in \mathbb{N}$, then xy denotes their standard scalar product

$$xy = \sum_{j=1}^{n} x_j y_j \quad \text{if} \quad x = (x_1, \ldots, x_n), \; y = (y_1, \ldots, y_n).$$

First we remark that the factor $\frac{1}{2}$ in the group law given by (6.1) is irrelevant in the following sense. Let $\alpha \in \mathbb{R}^* = \mathbb{R}\backslash\{0\}$. Consider the group $\mathbb{H}_n^{(\alpha)}$ endowed with the law

$$(x,y,t)(x',y',t') := (x+x', y+y', t+t'+\frac{1}{\alpha}(xy'-x'y)).$$

Then the groups $\mathbb{H}_n^{(\alpha)}$ and $\mathbb{H}_n = \mathbb{H}_n^{(2)}$ are isomorphic via

$$\left\{ \begin{array}{ccc} \mathbb{H}_n & \longrightarrow & \mathbb{H}_n^{(\alpha)} \\ (x,y,t) & \longmapsto & (x,y,\frac{2}{\alpha}t) \end{array} \right. .$$

In the same way, consider the *polarised Heisenberg group* $\tilde{\mathbb{H}}_n$ (or $\mathbb{H}_n^{pol}$) endowed with the law

$$(x,y,t)(x',y',t') := (x+x', y+y', t+t'+xy').$$

Then the groups $\tilde{\mathbb{H}}_n$ and $\mathbb{H}_n$ are isomorphic via

$$\left\{ \begin{array}{ccc} \mathbb{H}_n & \longrightarrow & \tilde{\mathbb{H}}_n \\ (x,y,t) & \longmapsto & (x,y,t+\frac{1}{2}xy) \end{array} \right. .$$

Note that the Heisenberg group $\mathbb{H}_n$ can be also viewed as a matrix group. For simplicity, we consider $n = 1$, in which case the group $\tilde{\mathbb{H}}_1$ is isomorphic to T_3, the group of 3-by-3 upper triangular real matrices with 1 on the diagonal:

$$\begin{cases} \tilde{\mathbb{H}}_1 & \longrightarrow & T_3 \\ (x,y,t) & \longmapsto & \begin{bmatrix} 1 & x & t \\ 0 & 1 & y \\ 0 & 0 & 1 \end{bmatrix} \end{cases}.$$

All the statements above can be readily checked by a straightforward computation. Combining two isomorphisms above, we obtain the identification $\mathbb{H}_1 \longrightarrow \tilde{\mathbb{H}}_1 \longrightarrow T_3$ given by

$$\begin{cases} \mathbb{H}_1 & \longrightarrow & T_3 \\ (x,y,t) & \longmapsto & \begin{bmatrix} 1 & x & t+\frac{1}{2}xy \\ 0 & 1 & y \\ 0 & 0 & 1 \end{bmatrix} \end{cases}.$$

Although we will not use it, let us mention a couple of other important appearances of the Heisenberg group. The Heisenberg group can be also realised as a group of transformations; for example, for each

$$h = (x,y,t) \in \mathbb{H}_1,$$

the affine (holomorphic) map given by

$$\phi_h : \mathbb{C} \times \mathbb{C} \ni (z_1, z_2) \longmapsto (z_1 + x + iy, z_2 + t + 2iz_1(x - iy) + i(x^2 + y^2)) \in \mathbb{C} \times \mathbb{C},$$

sends the (Siegel) domain

$$\mathscr{U} := \{(z_1, z_2) \in \mathbb{C} \times \mathbb{C} \ : \ \operatorname{Im} z_2 > |z_1|^2\} \quad (= \mathrm{SU}(2,1)/\mathrm{U}(2))$$

to itself, and the (Shilov) boundary of $\mathscr{U}$,

$$b\mathscr{U} := \{(z_1, z_2) \in \mathbb{C} \times \mathbb{C} \ : \ \operatorname{Im} z_2 = |z_1|^2\},$$

also to itself. One can check that $\mathbb{H}_1 \ni h \mapsto \phi_h$ defines an action of $\mathbb{H}_1$ on $\mathscr{U}$ and on $b\mathscr{U}$. Furthermore, the action of $\mathbb{H}_1$ on $b\mathscr{U}$ is simply transitive. A Cayley type transform

$$(w_1, w_2) \longmapsto (z_1, z_2) \quad \text{with} \quad z_1 = \frac{w_1}{1 + w_2}, \qquad z_2 = i\frac{1 - w_2}{1 + w_2},$$

is a biholomorphic bijective mapping which sends $\mathscr{U}$ onto the unit complex ball of $\mathbb{C}^2$. It also send $b\mathscr{U}$ to the unit complex sphere $\mathbb{S}^3$, more precisely onto $\mathbb{S}^3 \backslash \{S\}$ where $S = (0, -1)$ is the south pole (which may be viewed as the image of ∞). Hence the Heisenberg group acts simply transitively on $\mathbb{S}^3 \backslash \{S\}$.

We can also mention here that the group $\mathrm{U}(n)$ acts naturally by automorphisms on $\mathbb{H}_n$ leading to the interpretation of $(\mathrm{U}(n), \mathbb{H}_n)$ as a nilpotent Gelfand pair with strong relation to the theory of commutative convolution algebras. For example, such analysis can be used to characterise Gelfand (spherical) transforms of K-invariant Schwartz functions on $\mathbb{H}_n$ for a group $K \subset \mathrm{U}(n)$ ([BJR98]), or view them as Schwartz functions on the Gelfand spectrum ([ADBR09]).

6.1.2 Heisenberg Lie algebra and the stratified structure

The Lie algebra $\mathfrak{h}_n$ of $\mathbb{H}_n$ is identified with the vector space of left-invariant vector fields. Its canonical basis is given by the left-invariant vector fields

$$X_j = \partial_{x_j} - \frac{y_j}{2}\partial_t, \quad Y_j = \partial_{y_j} + \frac{x_j}{2}\partial_t, \ j = 1, \dots, n, \quad \text{and } T = \partial_t. \tag{6.2}$$

For comparison, the corresponding right-invariant vector fields are

$$\tilde{X}_j = \partial_{x_j} + \frac{y_j}{2}\partial_t, \ \tilde{Y}_j = \partial_{y_j} - \frac{x_j}{2}\partial_t, \ j = 1, \dots, n, \text{ and } \tilde{T} = \partial_t. \tag{6.3}$$

The canonical commutation relations are

$$[X_j, Y_j] = T, \quad j = 1, \dots, n,$$

and T is the centre of $\mathfrak{h}_n$. This shows that the Lie algebra $\mathfrak{h}_n$ and the Lie group $\mathbb{H}_n$ are nilpotent of step 2. Hence the Heisenberg group $\mathbb{H}_n$ described above in Section 6.1.1, that is, $\mathbb{R}^{2n+1}$ endowed with the group law given in (6.1), is the connected simply connected (step-two nilpotent) Lie group whose Lie algebra is $\mathfrak{h}_n$ and which is realised via the exponential mapping together with the canonical basis. This means that the element $(x, y, t) = (x_1, \dots, x_n, y_1, \dots, y_n, t)$ of $\mathbb{H}_n$ can be written as

$$(x, y, t) = \exp_{\mathbb{H}_n}(x_1 X_1 + \dots + x_n X_n + y_1 Y_1 + \dots + y_n Y_n + tT).$$

We fix

$$dxdydt = dx_1 \dots dx_n dy_1 \dots dy_n dt$$

as the Lebesgue measure on $\mathbb{H}_n$, see Proposition 1.6.6. Therefore, we may be free to write formulae like

$$\int_{\mathbb{H}_n} \cdots \, dxdydt = \int_{\mathbb{R}^{2n+1}} \cdots \, dxdydt.$$

The Heisenberg Lie algebra is stratified via $\mathfrak{h}_n = V_1 \oplus V_2$, where V_1 is linearly spanned by the X_j's and Y_j's, while $V_2 = \mathbb{R}T$. Since the Heisenberg Lie algebra is stratified via $\mathfrak{h}_n = V_1 \oplus V_2$, the natural dilations on the Lie algebra are given by

$$D_r(X_j) = rX_j \quad \text{and} \quad D_r(Y_j) = rY_j, \quad j = 1, \dots, n, \quad \text{and} \quad D_r(T) = r^2 T, \tag{6.4}$$

see Section 3.1.2. We keep the same notation D_r for the dilations on the group $\mathbb{H}_n$. They are therefore given by

$$D_r(x,y,t) = r(x,y,t) = (rx, ry, r^2t), \quad (x,y,t) \in \mathbb{H}_n, \ r > 0.$$

We also keep the same notation D_r for the dilations on the universal enveloping algebra $\mathfrak{U}(\mathfrak{h}_n)$ induced by Property (6.4).

Note that the homogeneous dimension of $\mathbb{H}_n$ is $Q = 2n + 2$. This is also the homogeneous degree of the Lebesgue measure $dxdydt$.

Example 6.1.1. The sub-Laplacian

$$\begin{aligned} \mathcal{L} &:= \sum_{j=1}^{n} (X_j^2 + Y_j^2) \qquad (6.5) \\ &= \sum_{j=1}^{n} \left(\partial_{x_j} - \frac{y_j}{2}\partial_t\right)^2 + \left(\partial_{y_j} + \frac{x_j}{2}\partial_t\right)^2, \end{aligned}$$

is homogeneous of degree 2 since

$$D_r(\mathcal{L}) = r^2\mathcal{L}.$$

Remark 6.1.2. The 'canonical' positive Rockland operator in this setting is

$$\mathcal{R} = -\mathcal{L}.$$

We will also use the mapping $\Theta : \mathbb{H}_n \to \mathbb{H}_n$ given by

$$\Theta(x,y,t) := (x, -y, -t).$$

One checks easily that for any $(x,y,t), (x',y',t') \in \mathbb{H}_n$, we have

$$\Theta\big((x,y,t)(x',y',t')\big) = \Theta(x,y,t)\ \Theta(x',y',t') \quad \text{and} \quad \Theta\big(\Theta(x,y,t)\big) = (x,y,t).$$

Therefore, Θ is a group automorphism and an involution. Furthermore, it is clear that it commutes with the dilations:

$$\forall r > 0 \qquad \Theta \circ D_r = D_r \circ \Theta.$$

We keep the same notation for the corresponding Lie algebra morphism and we have

$$\Theta(X_j) = X_j, \ \Theta(Y_j) = -Y_j, \ j = 1, \dots, n, \ \Theta(T) = -T. \qquad (6.6)$$

6.2 Dual of the Heisenberg group

In this section we will analyse the unitary dual of the Heisenberg group $\mathbb{H}_n$. For our purposes, it will be more convenient to work with the Schrödinger representations. This will lead to the group Fourier transform parametrised by λ in (6.19). Such group Fourier transforms yield operators acting on the representation space $L^2(\mathbb{R}^n)$. The latter can be, in turn, analysed using the Weyl quantization on $\mathbb{R}^n$ that appears naturally.

6.2.1 Schrödinger representations π_λ

The Schrödinger representations of the Heisenberg group $\mathbb{H}_n$ are the infinite dimensional unitary representations of $\mathbb{H}_n$, where, as usual, we allow ourselves to identify unitary representations with their unitary equivalence classes. They are parametrised by the co-adjoint orbits (see Section 1.8.1) and more concretely by $\lambda \in \mathbb{R}\backslash\{0\}$. We denote these representations π_λ. Each π_λ acts on the Hilbert space

$$\mathcal{H}_{\pi_\lambda} = L^2(\mathbb{R}^n)$$

in the way we now describe. An element of $L^2(\mathbb{R}^n)$ will very often be denoted as a function h of the variable $u = (u_1, \dots, u_n) \in \mathbb{R}^n$.

First let us define π_1 corresponding to $\lambda = 1$. It is the representation of the group $\mathbb{H}_n$ acting on $L^2(\mathbb{R}^n)$ via

$$\pi_1(x, y, t)h(u) := e^{i(t+\frac{1}{2}xy)} e^{iyu} h(u + x),$$

for $h \in L^2(\mathbb{R}^n)$ and $(x, y, t) \in \mathbb{H}_n$. Here xy denotes the scalar product in $\mathbb{R}^n$ of x and y, and similarly for yu. Consequently its infinitesimal representation (see Section 1.7) is given by

$$\left\{ \begin{array}{rcll} \pi_1(X_j) & = & \partial_{u_j} & \text{(differentiate with respect to } u_j), \quad j = 1, \dots, n, \\ \pi_1(Y_j) & = & iu_j, & \text{(multiplication by } iu_j), \quad j = 1, \dots, n, \\ \pi_1(T) & = & i\mathrm{I}, & \text{(multiplication by } i). \end{array} \right. \tag{6.7}$$

The Schrödinger representations π_λ on the group are realised in this monograph using

$$\pi_\lambda := \left\{ \begin{array}{ll} \pi_1 \circ D_{\sqrt{\lambda}} & \text{if } \lambda > 0, \\ \pi_{-\lambda} \circ \Theta & \text{if } \lambda < 0, \end{array} \right.$$

that is,

$$\pi_\lambda(x, y, t)h(u) = e^{i\lambda(t+\frac{1}{2}xy)} e^{i\sqrt{\lambda} yu} h(u + \sqrt{|\lambda|}x), \tag{6.8}$$

for $h \in L^2(\mathbb{R}^n)$ and $(x, y, t) \in \mathbb{H}_n$ where we use the following convention:

$$\sqrt{\lambda} := \operatorname{sgn}(\lambda)\sqrt{|\lambda|} = \left\{ \begin{array}{ll} \sqrt{\lambda} & \text{if } \lambda > 0, \\ -\sqrt{|\lambda|} & \text{if } \lambda < 0. \end{array} \right. \tag{6.9}$$

We observe that for any $\lambda \in \mathbb{R}\backslash\{0\}$ and $r > 0$,

$$\pi_\lambda \circ \Theta = \pi_{-\lambda} \quad \text{and} \quad \pi_\lambda \circ D_r = \pi_{r^2\lambda}, \tag{6.10}$$

and this is true for the group representation π_λ on $\mathbb{H}_n$ and for its corresponding infinitesimal representation on the Lie algebra $\mathfrak{h}_n$ and on the universal enveloping algebra $\mathfrak{U}(\mathfrak{h}_n)$. As usual we keep the same notation, here π_λ for the corresponding infinitesimal representation.

Lemma 6.2.1. *The infinitesimal representation of* π_λ *acts on the canonical basis of* $\mathfrak{h}_n$ *via*

$$\pi_\lambda(X_j) = \sqrt{|\lambda|}\partial_{u_j}, \ \pi_\lambda(Y_j) = i\sqrt{\lambda}u_j, \ j = 1, \dots, n, \quad \text{and} \quad \pi_\lambda(T) = i\lambda \mathrm{I}, \quad (6.11)$$

using the convention in (6.9).

Proof. Formulae (6.11) can be computed easily from (6.8). Here we show that they also follow from Properties (6.7) and (6.10). Indeed we have for $\lambda > 0$

$$\begin{cases} \pi_\lambda(X_j) &= \pi_1(D_{\sqrt{\lambda}}(X_j)) = \sqrt{\lambda}\pi_1(X_j) = \sqrt{\lambda}\partial_{u_j} \quad j = 1, \dots, n, \\ \pi_\lambda(Y_j) &= \pi_1(D_{\sqrt{\lambda}}(Y_j)) = \sqrt{\lambda}\pi_1(Y_j) = \sqrt{\lambda}iu_j, \quad j = 1, \dots, n, \\ \pi_\lambda(T) &= \pi_1(D_{\sqrt{\lambda}}(T)) = \lambda\pi_1(T) = i\lambda, \end{cases}$$

and thus for $\lambda < 0$

$$\begin{cases} \pi_\lambda(X_j) &= \pi_{-\lambda}(\Theta(X_j)) = \pi_{-\lambda}(X_j) = \sqrt{|\lambda|}\partial_{u_j} \quad j = 1, \dots, n, \\ \pi_\lambda(Y_j) &= \pi_{-\lambda}(\Theta(Y_j)) = -\pi_{-\lambda}(Y_j) = -\sqrt{|\lambda|}iu_j, \quad j = 1, \dots, n, \\ \pi_\lambda(T) &= \pi_{-\lambda}(\Theta(T)) = -\pi_{-\lambda}(T) = -(-\lambda)i = i\lambda, \end{cases}$$

proving (6.11) in both cases. □

Consequently, the group Fourier transform of the sub-Laplacian

$$\mathcal{L} = \sum_{j=1}^{n}(X_j^2 + Y_j^2)$$

is

$$\pi_\lambda(\mathcal{L}) = |\lambda| \sum_{j=1}^{n}(\partial_{u_j}^2 - u_j^2). \quad (6.12)$$

A direct characterisation implies that the space of smooth vectors of π_λ is

$$\mathcal{H}_{\pi_\lambda}^\infty = \mathcal{S}(\mathbb{R}^n).$$

This is true more generally for any representation of a connected simply connected nilpotent Lie group realised on some $L^2(\mathbb{R}^m)$ via the orbit method, see [CG90, Corollary 4.1.2].

6.2.2 Group Fourier transform on the Heisenberg group

We could have realised the equivalence classes $[\pi_\lambda]$ of Schrödinger representations in various ways. For instance by composing with the unitary operator $U_\lambda : L^2(\mathbb{R}^n) \to L^2(\mathbb{R}^n)$ given by $Uf(x) = |\lambda|^{\frac{n}{2}} f(\sqrt{\lambda}x)$, one would have obtained a slightly different, although equivalent, representation. Another realisation is with the Bargmann representations, see, e.g., [Tay86]. Our choice of representation π_λ

to represent its equivalence class will prove useful in relation with the Weyl-Shubin calculus on $\mathbb{R}^n$ later, see Section 6.5.

The group Fourier transform of a function $\kappa \in L^1(\mathbb{H}_n)$ at π_1 is

$$\mathcal{F}_{\mathbb{H}_n}(\kappa)(\pi_1) = \pi_1(\kappa) = \int_{\mathbb{H}_n} \kappa(x,y,t)\pi_1(x,y,t)^* dxdydt,$$

that is, the operator on $L^2(\mathbb{R}^n)$ given by

$$\pi_1(\kappa)h(u) = \int_{\mathbb{H}_n} \kappa(x,y,t)e^{i(-t+\frac{1}{2}xy)}e^{-iyu}h(u-x)dxdydt.$$

We now fix the notation concerning the Euclidean Fourier transform and recall some facts about the Weyl quantization on $\mathbb{R}^n$.

The Euclidean Fourier transform

In order to give a nicer expression for the operator $\mathcal{F}_{\mathbb{H}_n}(\kappa)(\pi_1)$, we adopt here the following notation for the Euclidean Fourier transform on $\mathbb{R}^N$:

$$\mathcal{F}_{\mathbb{R}^N} f(\xi) = (2\pi)^{-\frac{N}{2}} \int_{\mathbb{R}^N} f(x)e^{-ix\xi}dx, \tag{6.13}$$

where $\xi \in \mathbb{R}^N$ and $f : \mathbb{R}^N \to \mathbb{C}$ is for instance integrable. With our choice of notation and normalisation, the mapping $\mathcal{F}_{\mathbb{R}^N}$ extends unitarily to a mapping on $L^2(\mathbb{R}^N)$ and

$$\mathcal{F}_{\mathbb{R}^N}(f)(x) = \mathcal{F}^{-1}_{\mathbb{R}^N}(f)(-x).$$

Let us also recall the Fourier inversion formula for a (e.g. Schwartz) function $f : \mathbb{R}^n \to \mathbb{C}$:

$$\int_{\mathbb{R}^N}\int_{\mathbb{R}^N} e^{i(u-v)\xi} f(v)dvd\xi = (2\pi)^N f(u). \tag{6.14}$$

In our context N will be equal to $2n+1$.

Unfortunately, due to our choice of notation π for the representations, in the formulae in the sequel π will appear both as a representation and as the constant $\pi = 3.1415926...$ However, as powers of this 2π will appear mostly as constants in front of integrals it should not lead to major confusion.

The (Euclidean) Weyl quantization

Let us also set some notation regarding the *Weyl quantization* on $\mathbb{R}^n$. If a is a symbol, that is, a reasonable function on $\mathbb{R}^n \times \mathbb{R}^n$, then the Weyl quantization associates to a the operator

$$\text{Op}^W(a) \equiv a(D,X)$$

given by

$$\mathrm{Op}^W(a)f(u) = (2\pi)^{-n} \int_{\mathbb{R}^n} \int_{\mathbb{R}^n} e^{i(u-v)\xi} a(\xi, \frac{u+v}{2}) f(v) dv d\xi, \tag{6.15}$$

where $f \in \mathcal{S}(\mathbb{R}^n)$ and $u \in \mathbb{R}^n$.

Example 6.2.2. Particular examples are

$$\mathrm{Op}^W(1) = \mathrm{I}, \quad \mathrm{Op}^W(\xi_j) = \frac{1}{i}\partial_{u_j}, \quad \mathrm{Op}^W(u_j) = u_j,$$

and

$$\mathrm{Op}^W(\xi_k u_j) = \frac{1}{2i}(\partial_{u_k} u_j + u_j \partial_{u_k}).$$

The composition of two Weyl-quantized operators is

$$\mathrm{Op}^W(a) \circ \mathrm{Op}^W(b) = \mathrm{Op}^W(a \star b), \tag{6.16}$$

where (see, e.g., [Ler10])

$$a \star b(\zeta, u) = (2\pi)^{-2n} 4^n \int_{\mathbb{R}^n} \int_{\mathbb{R}^n} \int_{\mathbb{R}^n} \int_{\mathbb{R}^n} e^{-2i\{(\xi-\zeta)(y-u)-(\eta-\zeta)(x-u)\}} a(\xi, x)\, b(\eta, y)\, d\xi d\eta dx dy,$$

and asymptotically

$$a \star b \sim \sum_{m'=0}^{\infty} c_{m',n} \sum_{|\alpha_1|+|\alpha_2|=m'} \frac{(-1)^{|\alpha_2|}}{\alpha_1! \alpha_2!} \left(\left(\frac{1}{i}\partial_\xi\right)^{\alpha_1} \partial_x^{\alpha_2} a \right) \left(\left(\frac{1}{i}\partial_\xi\right)^{\alpha_2} \partial_x^{\alpha_1} b \right), \tag{6.17}$$

with $c_{0,n_0} = 1$ and, in fact,

$$a \star b \sim ab + \frac{1}{2i}\{a, b\} + \dots \quad \text{where} \quad \{a, b\} = \sum_{j=1}^{n} \left(\frac{\partial a}{\partial \xi_j} \frac{\partial b}{\partial u_j} - \frac{\partial a}{\partial u_j} \frac{\partial b}{\partial \xi_j} \right).$$

This formula can already be checked on the basic examples given in Example 6.2.2 and on the following property:

Lemma 6.2.3. *Let a be a symbol. Then we have*

$$\begin{aligned} (\mathrm{ad} u_j)\left(\mathrm{Op}^W(a)\right) &\equiv u_j \mathrm{Op}^W(a) - \mathrm{Op}^W(a) u_j = \mathrm{Op}^W(i\partial_{\xi_j} a), \\ (\mathrm{ad} \partial_{u_j})\left(\mathrm{Op}^W(a)\right) &\equiv \partial_{u_j} \mathrm{Op}^W(a) - \mathrm{Op}^W(a)\partial_{u_j} = \mathrm{Op}^W(\partial_{u_j} a). \end{aligned}$$

Proof. Let $f \in \mathcal{S}(\mathbb{R}^n)$ and $u \in \mathbb{R}^n$. Then we have

$$\begin{aligned}
(\mathrm{ad}u_j)\left(\mathrm{Op}^W(a)\right) f(u) &= u_j \mathrm{Op}^W(a) f(u) - \mathrm{Op}^W(a)(u_j f)(u) \\
&= u_j (2\pi)^{-n} \int_{\mathbb{R}^n}\int_{\mathbb{R}^n} e^{i(u-v)\xi} a(\xi, \frac{u+v}{2}) f(v) dv d\xi \\
&\qquad -(2\pi)^{-n} \int_{\mathbb{R}^n}\int_{\mathbb{R}^n} e^{i(u-v)\xi} a(\xi, \frac{u+v}{2}) v_j f(v) dv d\xi \\
&= (2\pi)^{-n} \int_{\mathbb{R}^n}\int_{\mathbb{R}^n} e^{i(u-v)\xi} a(\xi, \frac{u+v}{2})(u_j - v_j) f(v) dv d\xi \\
&= (2\pi)^{-n} \int_{\mathbb{R}^n}\int_{\mathbb{R}^n} \frac{1}{i}\partial_{\xi_j}\left\{e^{i(u-v)\xi}\right\}\ a(\xi, \frac{u+v}{2}) f(v) dv d\xi \\
&= (2\pi)^{-n} \int_{\mathbb{R}^n}\int_{\mathbb{R}^n} e^{i(u-v)\xi} i\partial_{\xi_j}\left\{a(\xi, \frac{u+v}{2})\right\} f(v) dv d\xi,
\end{aligned}$$

after integration by parts. This shows the first equality.

For the second one, we compute

$$\partial_{u_j} \mathrm{Op}^W(a) f(u) = (2\pi)^{-n} \int_{\mathbb{R}^n}\int_{\mathbb{R}^n} \partial_{u_j}\left\{e^{i(u-v)\xi} a(\xi, \frac{u+v}{2})\right\} f(v) dv d\xi.$$

Since

$$\begin{aligned}
\partial_{u_j}\left\{e^{i(u-v)\xi}\ a(\xi, \frac{u+v}{2})\right\} \quad &= \quad -\left\{\partial_{v_j} e^{i(u-v)\xi}\right\} a(\xi, \frac{u+v}{2}) \\
&\qquad + \frac{1}{2} e^{i(u-v)\xi} \{\partial_{u_j} a\}(\xi, \frac{u+v}{2}),
\end{aligned}$$

we compute using integration by parts

$$\begin{aligned}
&\int_{\mathbb{R}^n}\int_{\mathbb{R}^n} \partial_{u_j}\left\{e^{i(u-v)\xi} a(\xi, \frac{u+v}{2})\right\} f(v) dv d\xi \\
&\quad = -\int_{\mathbb{R}^n}\int_{\mathbb{R}^n} \left\{\partial_{v_j} e^{i(u-v)\xi}\right\} a(\xi, \frac{u+v}{2}) f(v) dv d\xi \\
&\qquad\quad + \int_{\mathbb{R}^n}\int_{\mathbb{R}^n} e^{i(u-v)\xi} \frac{1}{2}\{\partial_{u_j} a\}(\xi, \frac{u+v}{2}) f(v) dv d\xi \\
&\quad = \int_{\mathbb{R}^n}\int_{\mathbb{R}^n} e^{i(u-v)\xi} \partial_{v_j}\left\{a(\xi, \frac{u+v}{2}) f(v)\right\} dv d\xi \\
&\qquad\quad + \int_{\mathbb{R}^n}\int_{\mathbb{R}^n} e^{i(u-v)\xi} \frac{1}{2}\{\partial_{u_j} a\}(\xi, \frac{u+v}{2}) f(v) dv d\xi.
\end{aligned}$$

Now

$$\partial_{v_j}\left\{a(\xi, \frac{u+v}{2}) f(v)\right\} = \frac{1}{2}\{\partial_{u_j} a\}(\xi, \frac{u+v}{2}) f(v) + a(\xi, \frac{u+v}{2}) \partial_{v_j} f(v),$$

thus

$$\begin{aligned}\int_{\mathbb{R}^n}\int_{\mathbb{R}^n} & \partial_{u_j}\left\{e^{i(u-v)\xi}a(\xi,\frac{u+v}{2})\right\}f(v)dvd\xi \\ &= \int_{\mathbb{R}^n}\int_{\mathbb{R}^n} e^{i(u-v)\xi}\{\partial_{u_j}a\}(\xi,\frac{u+v}{2})f(v)dvd\xi \\ &\qquad + \int_{\mathbb{R}^n}\int_{\mathbb{R}^n} e^{i(u-v)\xi}a(\xi,\frac{u+v}{2})\partial_{v_j}f(v)dvd\xi.\end{aligned}$$

We have obtained

$$\begin{aligned}&\partial_{u_j}\mathrm{Op}^W(a)f(u) \\ &= (2\pi)^{-n}\int_{\mathbb{R}^n}\int_{\mathbb{R}^n} e^{i(u-v)\xi}\{\partial_{u_j}a\}(\xi,\frac{u+v}{2})f(v)dvd\xi \\ &\qquad +(2\pi)^{-n}\int_{\mathbb{R}^n}\int_{\mathbb{R}^n} e^{i(u-v)\xi}a(\xi,\frac{u+v}{2})\partial_{v_j}f(v)dvd\xi.\end{aligned}$$

Therefore, we have

$$\begin{aligned}&(\mathrm{ad}\partial_{u_j})\left(\mathrm{Op}^W(a)\right)f(u) = \partial_{u_j}\mathrm{Op}^W(a)f(u) - \mathrm{Op}^W(a)(\partial_{u_j}f)(u) \\ &= (2\pi)^{-n}\int_{\mathbb{R}^n}\int_{\mathbb{R}^n} e^{i(u-v)\xi}\{\partial_{u_j}a\}(\xi,\frac{u+v}{2})f(v)dvd\xi \\ &= \mathrm{Op}^W(\partial_{u_j}a)f(u).\end{aligned}$$

This shows the second equality. □

The operator $\mathcal{F}_{\mathbb{H}_n}(\kappa)(\pi_1)$

Going back to $\pi_1(\kappa) \equiv \widehat{\kappa}(\pi_1)$ and using the well-known properties of the Euclidean Fourier transform $\mathcal{F}_{\mathbb{R}^{2n+1}}$, for instance see (6.14), it is not difficult to turn into rigorous computations the following calculations:

$$\begin{aligned}\pi_1(\kappa)h(u) &= \int_{\mathbb{R}^{2n+1}} \kappa(x,y,t)e^{i(-t+\frac{1}{2}xy)}e^{-iyu}h(u-x)dxdydt \\ &= \int_{\mathbb{R}^{2n+1}}\int_{\mathbb{R}^{2n+1}} (2\pi)^{-\frac{2n+1}{2}}\mathcal{F}_{\mathbb{R}^{2n+1}}(\kappa)(\xi,\eta,\tau)e^{it\tau}e^{iy\eta}e^{ix\xi} \\ &\qquad\qquad e^{i(-t+\frac{1}{2}xy)}e^{-iyu}h(u-x)d\xi d\eta d\tau dxdydt \\ &= \sqrt{2\pi}\int_{\mathbb{R}^n\times\mathbb{R}^n} \mathcal{F}_{\mathbb{R}^{2n+1}}(\kappa)(\xi,u-\frac{x}{2},1)e^{ix\xi}h(u-x)d\xi dx \\ &= \sqrt{2\pi}\int_{\mathbb{R}^n\times\mathbb{R}^n} \mathcal{F}_{\mathbb{R}^{2n+1}}(\kappa)(\xi,u-\frac{u-v}{2},1)e^{i\xi(u-v)}h(v)d\xi dv,\end{aligned}$$

after the change of variable $v = u - x$. Comparing this last expression with (6.15), we see that

$$\pi_1(\kappa)h(u) = \sqrt{2\pi}\int_{\mathbb{R}^n}\int_{\mathbb{R}^n} e^{i\xi(u-v)}\mathcal{F}_{\mathbb{R}^{2n+1}}(\kappa)(\xi,\frac{u+v}{2},1)h(v)d\xi dv,$$

may be written as

$$\pi_1(\kappa) = (2\pi)^{\frac{2n+1}{2}} \ \mathrm{Op}^W \left[\mathcal{F}_{\mathbb{R}^{2n+1}}(\kappa)(\cdot,\cdot,1)\right] = (2\pi)^{\frac{2n+1}{2}} \mathcal{F}_{\mathbb{R}^{2n+1}}(\kappa)(D,X,1). \quad (6.18)$$

More generally, we could compute in the same way $\pi_\lambda(\kappa)$ or use the following computational remarks.

Lemma 6.2.4. *Let $\lambda \in \mathbb{R}\backslash\{0\}$. With the convention given in (6.9) we obtain*

$$\begin{aligned} \pi_\lambda(\kappa) &= |\lambda|^{-(n+1)} \pi_{\mathrm{sgn}(\lambda)1}\left(\kappa \circ D_{1/\sqrt{|\lambda|}}\right) && (6.19) \\ &= (2\pi)^{\frac{2n+1}{2}} \ \mathrm{Op}^W \left[\mathcal{F}_{\mathbb{R}^{2n+1}}(\kappa)(\sqrt{|\lambda|}\cdot, \sqrt{\lambda}\cdot, \lambda)\right], && (6.20) \end{aligned}$$

or, equivalently,

$$\begin{aligned} &\pi_\lambda(\kappa)h(u) \\ &= \int_{\mathbb{R}^{2n+1}} \kappa(x,y,t) e^{i\lambda(-t+\frac{1}{2}xy)} e^{-i\sqrt{\lambda}yu} h(u-\sqrt{|\lambda|}x)dxdydt && (6.21) \\ &= (2\pi)^{\frac{2n+1}{2}} \int_{\mathbb{R}^n\times\mathbb{R}^n} e^{i(u-v)\xi} \mathcal{F}_{\mathbb{R}^{2n+1}}(\kappa)(\sqrt{|\lambda|}\ \xi, \sqrt{\lambda}\ \frac{u+v}{2}, \lambda)h(v)dvd\xi. && (6.22) \end{aligned}$$

We also have

$$\pi_\lambda(\kappa) = \pi_{-\lambda}(\kappa \circ \Theta), \quad (6.23)$$

and for $r > 0$, $Q = 2n+2$,

$$\pi_\lambda(r^Q \kappa \circ D_r) = \pi_{r^{-2}\lambda}(\kappa). \quad (6.24)$$

For any $X \in \mathfrak{U}(\mathfrak{h}_n)$ and $r > 0$, we have

$$\pi_\lambda(D_{r^{-1}} X) = \pi_{r^{-2}\lambda}(X). \quad (6.25)$$

Here $\mathfrak{U}(\mathfrak{h}_n)$ stands for the universal enveloping algebra of the Lie algebra $\mathfrak{h}_n$, see Section 1.3.

Proof of Lemma 6.2.4. By (6.8), we have for $h \in L^2(\mathbb{R}^n)$ and $(x,y,t) \in \mathbb{H}_n$,

$$\begin{aligned} \pi_\lambda(x,y,t)^* h(u) &= \pi_\lambda\left((x,y,t)^{-1}\right) h(u) = \pi_\lambda(-x,-y,-t)h(u) \\ &= e^{i\lambda(-t+\frac{1}{2}xy)} e^{-i\sqrt{\lambda}yu} h(u-\sqrt{|\lambda|}x). \end{aligned}$$

Thus

$$\begin{aligned} \pi_\lambda(\kappa)h(u) &= \int_{\mathbb{H}_n} \kappa(x,y,t)\ \pi_\lambda(x,y,t)^* h(u)\ dxdydt \\ &= \int_{\mathbb{R}^{2n+1}} \kappa(x,y,t) e^{i\lambda(-t+\frac{1}{2}xy)} e^{-i\sqrt{\lambda}yu} h(u-\sqrt{|\lambda|}x)dxdydt. \end{aligned}$$

This is Formula (6.21).

For Formula (6.23), since by (6.10) we have $\pi_{-\lambda} = \pi_\lambda \circ \Theta$ for any $\lambda \in \mathbb{R}\backslash\{0\}$, we see that

$$\begin{aligned}
\pi_\lambda(\kappa) &= \int_{\mathbb{H}_n} \kappa(x,y,t)\pi_\lambda(x,y,t)^* dxdydt \\
&= \int_{\mathbb{H}_n} \kappa(x,y,t)\pi_{-\lambda}\big(\Theta(x,y,t)\big)^* dxdydt \\
&= \int_{\mathbb{H}_n} \kappa\big(\Theta(x,y,t)\big)\pi_{-\lambda}(x,y,t)^* dxdydt = \pi_{-\lambda}(\kappa\circ\Theta),
\end{aligned}$$

after the change of variables given by Θ, which has the Jacobian equal to 1. We proceed in the same way for formula (6.24)

$$\begin{aligned}
\pi_\lambda(r^Q\kappa\circ D_r) &= \int_{\mathbb{H}_n} \kappa\circ D_r(x,y,t)\pi_\lambda(x,y,t)^* r^Q dxdydt \\
&= \int_{\mathbb{H}_n} \kappa(x,y,t)\pi_\lambda\big(D_r^{-1}(x,y,t)\big)^* dxdydt \\
&= \int_{\mathbb{H}_n} \kappa(x,y,t)\pi_{r^{-2}\lambda}(x,y,t)^* dxdydt = \pi_{r^{-2}\lambda}(\kappa),
\end{aligned}$$

after the change of variable given by D_r, using (6.10).

For any $X\in\mathfrak{U}(\mathfrak{h}_n)$ and $\kappa\in\mathcal{S}(G)$, recalling $D_{r^{-1}}X$ from (6.4), then using

$$(X\kappa)\circ D_r = (D_{r^{-1}}X)(\kappa\circ D_r) \tag{6.26}$$

and (6.24), we have

$$\begin{aligned}
\pi_{r^{-2}\lambda}(X)\pi_{r^{-2}\lambda}(\kappa) &= \pi_{r^{-2}\lambda}(X\kappa) \\
&= \pi_\lambda(r^Q(X\kappa)\circ D_r) \\
&= \pi_\lambda(r^Q(D_{r^{-1}}X)(\kappa\circ D_r)) \\
&= \pi_\lambda(D_{r^{-1}}X)\pi_\lambda(r^Q\kappa\circ D_r) \\
&= \pi_\lambda(D_{r^{-1}}X)\pi_{r^{-2}\lambda}(\kappa),
\end{aligned}$$

and this shows (6.25).

Thus Formulae (6.25), (6.24) and (6.23) hold for any $\lambda\in\mathbb{R}\backslash\{0\}$.

Let us assume $\lambda>0$. Using $\pi_\lambda = \pi_1\circ D_{\sqrt{\lambda}}$ we see that

$$\begin{aligned}
\pi_\lambda(\kappa) &= \int_{\mathbb{H}_n} \kappa(x,y,t)\pi_1\big(D_{\sqrt{\lambda}}(x,y,t)\big)^* dxdydt \\
&= \int_{\mathbb{H}_n} \kappa\big(D_{1/\sqrt{\lambda}}(x,y,t)\big)\pi_1(x,y,t)^*\lambda^{-(n+1)} dxdydt \\
&= \lambda^{-(n+1)}\pi_1\left(\kappa\circ D_{1/\sqrt{\lambda}}\right),
\end{aligned}$$

and this gives Formula (6.19) for $\lambda > 0$. But Formula (6.18) gives here

$$\pi_1\left(\kappa \circ D_{1/\sqrt{\lambda}}\right) = (2\pi)^{n+\frac{1}{2}} \mathrm{Op}^W \left[\mathcal{F}_{\mathbb{R}^{2n+1}}(\kappa \circ D_{1/\sqrt{\lambda}})(\cdot,\cdot,1)\right].$$

Since a simple change of variable in $\mathbb{R}^{2n+1}$ yields

$$\mathcal{F}_{\mathbb{R}^{2n+1}}\left(\kappa \circ D_{1/\sqrt{\lambda}}\right) = \lambda^{n+1}\left(\mathcal{F}_{\mathbb{R}^{2n+1}}(\kappa)\right) \circ D_{\sqrt{\lambda}}, \tag{6.27}$$

we obtain Formula (6.20) for any $\lambda > 0$.

For $\lambda < 0$, we use Formula (6.23) and the case $\lambda > 0$, that is,

$$\begin{aligned}\pi_\lambda(\kappa) &= \pi_{-\lambda}(\kappa \circ \Theta)\\ &= (-\lambda)^{-(n+1)}\pi_1\left(\kappa \circ \Theta \circ D_{1/\sqrt{-\lambda}}\right)\\ &= (-\lambda)^{-(n+1)}\pi_1\left(\kappa \circ D_{1/\sqrt{-\lambda}} \circ \Theta\right)\\ &= (-\lambda)^{-(n+1)}\pi_{-1}\left(\kappa \circ D_{1/\sqrt{-\lambda}}\right).\end{aligned}$$

Hence Formula (6.19) is proved for any $\lambda < 0$. Here, Formula (6.18) and the relation $\mathcal{F}_{\mathbb{R}^{2n+1}}(\kappa \circ \Theta) = \mathcal{F}_{\mathbb{R}^{2n+1}}(\kappa) \circ \Theta$ with (6.27) give

$$\begin{aligned}\pi_1\left(\kappa \circ \Theta \circ D_{1/\sqrt{-\lambda}}\right) &= (2\pi)^{n+\frac{1}{2}} \mathrm{Op}^W \left[\mathcal{F}_{\mathbb{R}^{2n+1}}(\kappa \circ \Theta \circ D_{1/\sqrt{-\lambda}})(\cdot,\cdot,1)\right]\\ &= (2\pi)^{n+\frac{1}{2}}(-\lambda)^{n+1}\left(\mathcal{F}_{\mathbb{R}^{2n+1}}(\kappa)\right) \circ \Theta \circ D_{\sqrt{-\lambda}}(\cdot,\cdot,1),\end{aligned}$$

we obtain Formula (6.20) for any $\lambda < 0$. □

From Lemma 6.2.4 or from (6.11), we see that

$$\pi_\lambda(X_j) = \mathrm{Op}^W(i\sqrt{|\lambda|}\xi_j) \quad \text{and} \quad \pi_\lambda(Y_j) = \mathrm{Op}^W(i\sqrt{\lambda}u_j). \tag{6.28}$$

Remark 6.2.5. This was already noted in [Tay84, BFKG12a]. However in [Tay84], the Fourier transform on $\mathbb{R}^n$ is chosen to be non-unitarily defined by

$$\xi \longmapsto \int_{\mathbb{R}^n} f(x)e^{-ix\xi}dx, \quad f \in \mathcal{S}(\mathbb{R}^n).$$

Remark 6.2.6. The Schwartz space on the Heisenberg group $\mathbb{H}_n$, realised as we have done, is defined as $\mathcal{S}(\mathbb{R}^{2n+1})$, see Section 3.1.9. The characterisation of the Fourier image of the (full) Schwartz space on $\mathbb{H}_n$ is a difficult problem analysed by Geller in [Gel80]. See also the more recent paper [ADBR13].

6.2.3 Plancherel measure

The dual $\widehat{\mathbb{H}}_n$ of the Heisenberg group $\mathbb{H}_n$ may be described together with its Plancherel measure by the orbit method, see Section 1.8.1. Here we obtain a concrete formula for the Plancherel measure μ of the Heisenberg group $\mathbb{H}_n$ using well known properties of Euclidean analysis together with our choice of representatives for the elements of $\widehat{\mathbb{H}}_n$, especially the Schrödinger representations π_λ.

Proposition 6.2.7. *Let $f \in \mathcal{S}(\mathbb{H}_n)$. Then for each $\lambda \in \mathbb{R}\backslash\{0\}$ the operator $\widehat{f}(\pi_\lambda)$ acting on $L^2(\mathbb{R}^n)$ is the Hilbert-Schmidt operator with integral kernel*

$$K_{f,\lambda} : \mathbb{R}^n \times \mathbb{R}^n \longrightarrow \mathbb{C},$$

given by

$$K_{f,\lambda}(u,v) = (2\pi)^{n+\frac{1}{2}} \int_{\mathbb{R}^n} e^{i(u-v)\xi} \mathcal{F}_{\mathbb{R}^{2n+1}}(f)(\sqrt{|\lambda|}\xi, \sqrt{\lambda}\frac{u+v}{2}, \lambda) d\xi,$$

and Hilbert-Schmidt norm

$$\begin{aligned} \|\widehat{f}(\pi_\lambda)\|_{\mathtt{HS}(L^2(\mathbb{R}^n))} &= (2\pi)^{\frac{3n+1}{2}} |\lambda|^{-\frac{n}{2}} \|\mathcal{F}_{\mathbb{R}^{2n+1}}(f)(\cdot,\cdot,\lambda)\|_{L^2(\mathbb{R}^{2n})} \\ &= (2\pi)^{\frac{3n+1}{2}} |\lambda|^{-\frac{n}{2}} \left(\int_{\mathbb{R}^n}\int_{\mathbb{R}^n} |\mathcal{F}_{\mathbb{R}^{2n+1}}(f)(\xi,w,\lambda)|^2 d\xi dw \right)^{\frac{1}{2}}. \end{aligned}$$

Furthermore, we have

$$\int_{\mathbb{H}_n} |f(x,y,t)|^2 dx dy dt = c_n \int_{\lambda \in \mathbb{R}\backslash\{0\}} \|\widehat{f}(\pi_\lambda)\|^2_{\mathtt{HS}(L^2(\mathbb{R}^n))} |\lambda|^n d\lambda,$$

where $c_n = (2\pi)^{-(3n+1)}$.

In particular, Proposition 6.2.7 implies that the Plancherel measure μ on the Heisenberg group is supported in $\{[\pi_\lambda],\ \lambda \in \mathbb{R}\backslash\{0\}\}$, see (6.29). Moreover, we have

$$d\mu(\pi_\lambda) \equiv c_n |\lambda|^n d\lambda, \quad \lambda \in \mathbb{R}\backslash\{0\}.$$

The constant c_n depends on our choice of realisation of $\pi_\lambda \in [\pi_\lambda]$.

Proof of Proposition 6.2.7. By (6.22), we have for $h \in L^2(\mathbb{R}^n)$ and $u \in \mathbb{R}^n$,

$$\begin{aligned} \widehat{f}(\pi_\lambda)h(u) &= (2\pi)^{n+\frac{1}{2}} \int_{\mathbb{R}^n}\int_{\mathbb{R}^n} e^{i(u-v)\xi} \mathcal{F}_{\mathbb{R}^{2n+1}}(f)(\sqrt{|\lambda|}\xi, \sqrt{\lambda}\frac{u+v}{2}, \lambda) h(v) dv d\xi \\ &= \int_{\mathbb{R}^n} K_{f,\lambda}(u,v) h(v) dv, \end{aligned}$$

where $K_{f,\lambda}$ is the integral kernel of $\widehat{f}(\pi_\lambda)$ hence given by

$$K_{f,\lambda}(u,v) = (2\pi)^{n+\frac{1}{2}} \int_{\mathbb{R}^n} e^{i(u-v)\xi} \mathcal{F}_{\mathbb{R}^{2n+1}}(f)(\sqrt{|\lambda|}\xi, \sqrt{\lambda}\frac{u+v}{2}, \lambda) d\xi.$$

Using the Euclidean Fourier transform (see (6.13) for our normalisation of $\mathcal{F}_{\mathbb{R}^n}$), we may rewrite this as

$$K_{f,\lambda}(u,v) = (2\pi)^{\frac{3}{2}n+\frac{1}{2}}\mathcal{F}_{\mathbb{R}^n}\left\{\mathcal{F}_{\mathbb{R}^{2n+1}}(f)(\sqrt{|\lambda|}\,\cdot, \sqrt{\lambda}\frac{u+v}{2}, \lambda)\right\}(v-u).$$

The $L^2(\mathbb{R}^n \times \mathbb{R}^n)$-norm of the integral kernel is

$$\begin{aligned}
&\int_{\mathbb{R}^n\times\mathbb{R}^n} |K_{f,\lambda}(u,v)|^2 du dv \\
&\quad = (2\pi)^{3n+1}\int_{\mathbb{R}^n\times\mathbb{R}^n} |\mathcal{F}_{\mathbb{R}^n}\left\{\mathcal{F}_{\mathbb{R}^{2n+1}}(f)(\sqrt{|\lambda|}\,\cdot, \sqrt{\lambda}\frac{u+v}{2}, \lambda)\right\}(v-u)|^2 du dv \\
&\quad = (2\pi)^{3n+1}\int_{\mathbb{R}^n}\int_{\mathbb{R}^n} |\mathcal{F}_{\mathbb{R}^n}\left\{\mathcal{F}_{\mathbb{R}^{2n+1}}(f)(\sqrt{|\lambda|}\,\cdot, w_2, \lambda)\right\}(w_1)|^2 |\lambda|^{-\frac{n}{2}} dw_1 dw_2,
\end{aligned}$$

after the change of variable $(w_1, w_2) = (v-u, \sqrt{\lambda}\frac{u+v}{2})$. The (Euclidean) Plancherel formula on $\mathbb{R}^n$ in the variable w_1 (with dual variable ξ_1) then yields

$$\begin{aligned}
&\int_{\mathbb{R}^n\times\mathbb{R}^n} |K_{f,\lambda}(u,v)|^2 du dv \\
&\quad = (2\pi)^{3n+1}\int_{\mathbb{R}^n}\int_{\mathbb{R}^n} |\mathcal{F}_{\mathbb{R}^{2n+1}}(f)(\sqrt{|\lambda|}\xi_1, w_2, \lambda)|^2 |\lambda|^{-\frac{n}{2}} d\xi_1 dw_2 \\
&\quad = (2\pi)^{3n+1}|\lambda|^{-n}\int_{\mathbb{R}^n}\int_{\mathbb{R}^n} |\mathcal{F}_{\mathbb{R}^{2n+1}}(f)(\xi, w_2, \lambda)|^2 d\xi dw_2,
\end{aligned}$$

after the change of variable $\xi = \sqrt{|\lambda|}\xi_1$. Since $f \in \mathcal{S}(\mathbb{H}_n)$, this quantity is finite. Since the integral kernel of $\widehat{f}(\pi_\lambda)$ is square integrable, the operator $\widehat{f}(\pi_\lambda)$ is Hilbert-Schmidt and its Hilbert-Schmidt norm is the L^2-norm of its integral kernel (see, e.g., [RS80, Theorem VI.23]). This shows the first part of the statement.

To finish the proof, we now integrate each side of the last equality against $|\lambda|^n d\lambda$ and then use again the (Euclidean) Plancherel formula on $\mathbb{R}^{2n+1}$ in the variable (ξ, w_2, λ). We obtain

$$\begin{aligned}
&\int_{\mathbb{R}\backslash\{0\}}\int_{\mathbb{R}^n\times\mathbb{R}^n} |K_{f,\lambda}(u,v)|^2 du dv\ |\lambda|^n d\lambda \\
&\quad = (2\pi)^{3n+1}\int_{\mathbb{R}\backslash\{0\}}\int_{\mathbb{R}^n}\int_{\mathbb{R}^n} |\mathcal{F}_{\mathbb{R}^{2n+1}}(f)(\xi, w_2, \lambda)|^2 d\xi dw_2 d\lambda \\
&\quad = (2\pi)^{3n+1}\int_{\mathbb{R}^{2n+1}} |f(x,y,t)|^2 dx dy dt.
\end{aligned}$$

This concludes the proof of Proposition 6.2.7. □

It follows from the Plancherel formula in Proposition 6.2.7 that the Schrödinger representations π_λ, $\lambda \in \mathbb{R}\backslash\{0\}$, are almost all the representations of $\mathbb{H}_n$

modulo unitary equivalence. 'Almost all' here refers to the Plancherel measure $\mu = c_n|\lambda|^n d\lambda$ on $\widehat{\mathbb{H}}_n$. The other representations are finite dimensional and in fact 1-dimensional. They are given by the unitary characters of $\mathbb{H}_n$

$$\chi_w : (x, y, t) \mapsto e^{i(xw_1 + yw_2)}, \quad w = (w_1, w_2) \in \mathbb{R}^n \times \mathbb{R}^n \sim \mathbb{R}^{2n}.$$

See also Example 1.8.1 for the link with the orbit method.

We can summarise this paragraph by writing

$$\widehat{\mathbb{H}}_n = \{[\pi_\lambda], \ \lambda \in \mathbb{R}\backslash\{0\}\} \bigcup \{[\chi_w], \ w \in \mathbb{R}^{2n}\} \overset{\mu \text{ a.e.}}{=} \{[\pi_\lambda], \ \lambda \in \mathbb{R}\backslash\{0\}\}. \quad (6.29)$$

6.3 Difference operators

In this section we compute the difference operators Δ_{x_j}, Δ_{y_j}, and Δ_t which are the operators defined via

$$\begin{aligned} \Delta_{x_j}\widehat{\kappa}(\pi_\lambda) &:= \pi_\lambda(x_j\kappa), \\ \Delta_{y_j}\widehat{\kappa}(\pi_\lambda) &:= \pi_\lambda(y_j\kappa), \\ \Delta_t\widehat{\kappa}(\pi_\lambda) &:= \pi_\lambda(t\kappa). \end{aligned}$$

General properties of such difference operators have been analysed in Section 5.2.1. Here we aim at providing explicit expressions for them in the setting of the Heisenberg group $\mathbb{H}_n$.

6.3.1 Difference operators Δ_{x_j} and Δ_{y_j}

We start with the difference operators with respect to x and y.

Lemma 6.3.1. *For any $j = 1, \ldots, n$,*

$$\begin{aligned} \Delta_{x_j}|_{\pi_\lambda} &= \frac{1}{i\lambda}\text{ad}\,(\pi_\lambda(Y_j)) = \frac{1}{\sqrt{|\lambda|}}\text{ad}u_j, \\ \Delta_{y_j}|_{\pi_\lambda} &= -\frac{1}{i\lambda}\text{ad}\,(\pi_\lambda(X_j)) = -\frac{1}{i\sqrt{\lambda}}\text{ad}\partial_{u_j}. \end{aligned}$$

By this we mean that for any κ in some $\mathcal{K}_{a,b}(\mathbb{H}_n)$ such that $x_j\kappa$ is in some $\mathcal{K}_{a',b'}(\mathbb{H}_n)$ or $y_j\kappa$ in some $\mathcal{K}_{a',b'}(\mathbb{H}_n)$ for Δ_{x_j} or Δ_{y_j}, respectively, we have for all $h \in \mathcal{S}(\mathbb{R}^n)$ that

$$\begin{aligned} \left(\Delta_{x_j}\widehat{\kappa}(\pi_\lambda)\right) h\,(u) &= \frac{1}{\sqrt{|\lambda|}}\left(u_j\left(\widehat{\kappa}(\pi_\lambda)h\right)(u) - \left(\widehat{\kappa}(\pi_\lambda)(u_jh)\right)(u)\right), \\ \left(\Delta_{y_j}\widehat{\kappa}(\pi_\lambda)\right) h\,(u) &= \frac{1}{i\sqrt{\lambda}}\left(-\partial_{u_j}\{\widehat{\kappa}(\pi_\lambda)h\}\,(u) + \widehat{\kappa}(\pi_\lambda)\{\partial_{u_j}h\}\,(u)\right). \end{aligned}$$

Proof. Although we could just use direct computations, we prefer to use the following observations. Firstly we have by (6.2) and (6.3) that

$$Y_j - \tilde{Y}_j = x_j \partial_t = \partial_t x_j \quad \text{and} \quad \tilde{X}_j - X_j = y_j \partial_t = \partial_t y_j.$$

Secondly for any κ_1 in some $\mathcal{K}_{a,b}(\mathbb{H}_n)$,

$$\pi_\lambda(\partial_t \kappa_1) = \pi_\lambda(T\kappa_1) = \pi_\lambda(T)\pi_\lambda(\kappa_1) = i\lambda \pi_\lambda(\kappa_1), \tag{6.30}$$

as $T = \partial_t$ and using (6.11). Therefore, these two observations yield

$$\begin{aligned} \pi_\lambda(x_j\kappa) &= \frac{1}{i\lambda}\pi_\lambda(\partial_t x_j \kappa) = \frac{1}{i\lambda}\pi_\lambda\left((Y_j - \tilde{Y}_j)\kappa\right) \\ &= \frac{1}{i\lambda}\left(\pi_\lambda(Y_j\kappa) - \pi_\lambda(\tilde{Y}_j\kappa)\right) \\ &= \frac{1}{i\lambda}\left(\pi_\lambda(Y_j)\pi_\lambda(\kappa) - \pi_\lambda(\kappa)\pi_\lambda(Y_j)\right), \end{aligned}$$

and

$$\begin{aligned} \pi_\lambda(y_j\kappa) &= \frac{1}{i\lambda}\pi_\lambda(\partial_t y_j \kappa) = \frac{1}{i\lambda}\pi_\lambda\left((\tilde{X}_j - X_j)\kappa\right) \\ &= \frac{1}{i\lambda}\left(\pi_\lambda(\kappa)\pi_\lambda(X_j) - \pi_\lambda(X_j)\pi_\lambda(\kappa)\right). \end{aligned}$$

Using Lemma 6.2.1, we have obtained the expressions for Δ_{y_j} and Δ_{x_j} given in the statement. □

Above and also below, we use the formula for the symbols of right derivatives, for example, $\pi_\lambda(\tilde{Y}_j\kappa) = \pi_\lambda(\kappa)\pi_\lambda(Y_j)$, see Proposition 1.7.6, (iv).

Before giving some examples of applications of the difference operators Δ_{x_j} and Δ_{y_j}, let us make a couple of remarks.

Remark 6.3.2. 1. The formulae in Lemma 6.3.1 respect the properties of the automorphism Θ. Indeed, using (6.23) we have

$$\begin{aligned} \left(\Delta_{x_j}\widehat{\kappa}(\pi)\right)|_{\pi=\pi_{-\lambda}} &= (\widehat{x_j\kappa}(\pi))|_{\pi=\pi_{-\lambda}} = \pi_{-\lambda}(x_j\kappa) = \pi_\lambda\left((x_j\kappa)\circ\Theta\right) \\ &= \pi_\lambda\left(x_j\ \kappa\circ\Theta\right) = \Delta_{x_j}\widehat{\kappa\circ\Theta}(\pi_\lambda) = \Delta_{x_j}\left(\widehat{\kappa}(\pi_{-\lambda})\right), \\ \left(\Delta_{y_j}\widehat{\kappa}(\pi)\right)|_{\pi=\pi_{-\lambda}} &= (\widehat{y_j\kappa}(\pi))|_{\pi=\pi_{-\lambda}} = \pi_{-\lambda}(y_j\kappa) = \pi_\lambda\left((y_j\kappa)\circ\Theta\right) \\ &= \pi_\lambda\left(-y_j\ \kappa\circ\Theta\right) = -\Delta_{y_j}\widehat{\kappa\circ\Theta}(\pi_\lambda) = -\Delta_{y_j}\left(\widehat{\kappa}(\pi_{-\lambda})\right). \end{aligned}$$

This can also be viewed directly from the formulae in Lemma 6.3.1:

$$\begin{aligned} \left(\Delta_{x_j}\widehat{\kappa}(\pi)\right)|_{\pi=\pi_{-\lambda}} &= \frac{1}{\sqrt{|-\lambda|}}\mathrm{ad}u_j\left(\widehat{\kappa}(\pi_{-\lambda})\right) = \Delta_{x_j}\left(\widehat{\kappa}(\pi_{-\lambda})\right), \\ \left(\Delta_{y_j}\widehat{\kappa}(\pi)\right)|_{\pi=\pi_{-\lambda}} &= -\frac{1}{i\sqrt{-\lambda}}\mathrm{ad}\partial_{u_j} = -\Delta_{y_j}\left(\widehat{\kappa}(\pi_{-\lambda})\right). \end{aligned}$$

2. The formulae in Lemma 6.3.1 respect the properties of the dilations D_r. This time using (6.24), we have

$$\begin{aligned}\left(\Delta_{x_j}\widehat{\kappa}(\pi)\right)|_{\pi=\pi_{r^{-2}\lambda}} &= (\widehat{x_j\kappa}(\pi))\,|_{\pi=\pi_{r^{-2}\lambda}} = \pi_{r^{-2}\lambda}(x_j\kappa) = \pi_\lambda\left(r^Q(x_j\kappa)\circ D_r\right)\\ &= r\;\pi_\lambda\left(r^Q x_j\;\kappa\circ D_r\right) = r\;\Delta_{x_j}\left(\widehat{\kappa}(\pi_{r^{-2}\lambda})\right).\end{aligned}$$

This can also be viewed directly from the formulae in Lemma 6.3.1:

$$\begin{aligned}\left(\Delta_{x_j}\widehat{\kappa}(\pi)\right)|_{\pi=\pi_{r^{-2}\lambda}} &= \frac{1}{\sqrt{|r^{-2}\lambda|}}\,(\mathrm{ad}u_j)\left(\widehat{\kappa}(\pi_{r^{-2}\lambda})\right)\\ &= r\times\left(\frac{1}{\sqrt{|\lambda|}}\,(\mathrm{ad}u_j)\left(\widehat{\kappa}(\pi_{r^{-2}\lambda})\right)\right)\\ &= r\;\Delta_{x_j}\left(\widehat{\kappa}(\pi_{r^{-2}\lambda})\right).\end{aligned}$$

In exactly the same two ways we obtain for Δ_{y_j} that

$$\left(\Delta_{y_j}\widehat{\kappa}(\pi)\right)|_{\pi=\pi_{r^{-2}\lambda}} = r\Delta_{y_j}\left(\widehat{\kappa}(\pi_{r^{-2}\lambda})\right).$$

Lemmata 6.3.1 and 6.2.3 imply:

Corollary 6.3.3. *If* $\widehat{\kappa}(\pi_\lambda) = \mathrm{Op}^W(a_\lambda)$ *and* $a_\lambda = \{a_\lambda(\xi,u)\}$, *then*

$$\begin{aligned}\Delta_{x_j}\widehat{\kappa}(\pi_\lambda) &= \mathrm{Op}^W\left(\frac{i}{\sqrt{|\lambda|}}\partial_{\xi_j}a_\lambda\right),\\ \Delta_{y_j}\widehat{\kappa}(\pi_\lambda) &= \mathrm{Op}^W\left(\frac{i}{\sqrt{\lambda}}\partial_{u_j}a_\lambda\right).\end{aligned}$$

If $\widehat{\kappa}(\pi_\lambda) = \mathrm{Op}^W(a_\lambda)$ and $a_\lambda = \{a_\lambda(\xi,u)\}$ as in the statement above, we will often say that a_λ is the *λ-symbol*.

Up to now, we analysed the difference operators applied to a 'general' group Fourier transform of a distribution κ (provided that the difference operators made sense, see Definition 5.2.1 and the subsequent discussion). This is equivalent to applying difference operators acting on symbols, see Section 5.1.3. In what follows, we particularise this to some known symbols, mainly to the one in Example 5.1.26, that is, to $\pi(A)$ where A is a left-invariant differential operator such as $A = X_j, Y_j$ or T.

We now give some explicit examples.

Example 6.3.4. We already know that $\Delta_{x_j}\mathrm{I} = 0$, see Example 5.2.8. We can compute

$$\Delta_{x_j}\pi_\lambda(X_k) = -\delta_{jk}\mathrm{I},\quad \Delta_{x_j}\pi_\lambda(Y_k) = 0\quad\text{and}\quad \Delta_{x_j}\pi_\lambda(T) = 0, \tag{6.31}$$

and

$$\Delta_{x_j}\pi_\lambda(\mathcal{L}) = -2\pi_\lambda(X_j). \tag{6.32}$$

Proof. By Lemma 6.3.1,

$$\begin{aligned}\Delta_{x_j}\pi_\lambda(X_k) &= \frac{1}{i\lambda}\mathrm{ad}\,(\pi_\lambda(Y_j))\,\pi_\lambda(X_k) = \frac{1}{i\lambda}[\pi_\lambda(Y_j),\pi_\lambda(X_k)]\\ &= \frac{1}{i\lambda}\pi_\lambda[Y_j,X_k],\end{aligned}$$

since π_λ is a representation of the Lie algebra $\mathfrak{g}$. Similarly,

$$\begin{aligned}\Delta_{x_j}\pi_\lambda(Y_k) &= \frac{1}{i\lambda}\mathrm{ad}\,(\pi_\lambda(Y_j))\,\pi_\lambda(Y_k) = \frac{1}{i\lambda}\pi_\lambda[Y_j,Y_k],\\ \Delta_{x_j}\pi_\lambda(T) &= \frac{1}{i\lambda}\mathrm{ad}\,(\pi_\lambda(Y_j))\,\pi_\lambda(T) = \frac{1}{i\lambda}\pi_\lambda[Y_j,T].\end{aligned}$$

By the canonical commutation relations, we have

$$[Y_j,X_k] = -\delta_{jk}T, \quad [Y_j,Y_k]=0 \quad \text{and} \quad [Y_j,T]=0.$$

Since $\pi_\lambda(T) = i\lambda\mathrm{I}$, we obtain (6.31).

In the same way, we have

$$\Delta_{x_j}\pi_\lambda(X_k)^2 = \frac{1}{i\lambda}\pi_\lambda[Y_j,X_k^2] \quad \text{and} \quad \Delta_{x_j}\pi_\lambda(Y_k^2) = \frac{1}{i\lambda}\pi_\lambda[Y_j,Y_k^2].$$

Using the canonical commutation relations, we see that Y_j and Y_k commute in the Lie algebra $\mathfrak{g}$ thus Y_j and Y_k^2 commute in the enveloping Lie algebra $\mathfrak{U}(\mathfrak{g})$: $[Y_j,Y_k^2]=0$. Again using the canonical commutation relation we compute

$$[Y_j,X_k^2] = -2\delta_{jk}X_kT,$$

since

$$\begin{aligned}Y_jX_k^2 &= Y_jX_kX_k = (-\delta_{jk}T + X_kY_j)X_k\\ &= -\delta_{jk}TX_k + X_k(-\delta_{jk}T + X_kY_j)\\ &= -2\delta_{jk}X_kT + X_k^2Y_j.\end{aligned}$$

Therefore,

$$\begin{aligned}\Delta_{x_j}\pi_\lambda(X_k)^2 &= \frac{1}{i\lambda}\pi_\lambda(-2\delta_{jk}X_kT) = \frac{-2\delta_{jk}}{i\lambda}\pi_\lambda(X_kT) = \frac{-2\delta_{jk}}{i\lambda}\pi_\lambda(X_k)\pi_\lambda(T)\\ &= \frac{-2\delta_{jk}}{i\lambda}\pi_\lambda(X_k)(i\lambda) = -2\delta_{jk}\pi_\lambda(X_k),\end{aligned}$$

and $\Delta_{x_j}\pi_\lambda(Y_k^2)=0$. This implies (6.32). □

Example 6.3.5. We already know that $\Delta_{y_j}\mathrm{I}=0$, see Example 5.2.8. We can compute

$$\Delta_{y_j}\pi_\lambda(X_k)=0, \quad \Delta_{y_j}\pi_\lambda(Y_k) = -\delta_{jk}\mathrm{I} \quad \text{and} \quad \Delta_{y_j}\pi_\lambda(T)=0, \tag{6.33}$$

and

$$\Delta_{y_j}\pi_\lambda(\mathcal{L}) = -2\pi_\lambda(Y_j). \tag{6.34}$$

Proof. Proceeding as in the proof of Example 6.3.4, we have

$$\begin{aligned}
\Delta_{y_j}\pi_\lambda(X_k) &= -\frac{1}{i\lambda}\text{ad}\,(\pi_\lambda(X_j))\,\pi_\lambda(X_k) = -\frac{1}{i\lambda}\pi_\lambda[X_j, X_k],\\
\Delta_{y_j}\pi_\lambda(Y_k) &= -\frac{1}{i\lambda}\text{ad}\,(\pi_\lambda(X_j))\,\pi_\lambda(Y_k) = -\frac{1}{i\lambda}\pi_\lambda[X_j, Y_k],\\
\Delta_{y_j}\pi_\lambda(T) &= -\frac{1}{i\lambda}\text{ad}\,(\pi_\lambda(X_j))\,\pi_\lambda(T) = -\frac{1}{i\lambda}\pi_\lambda[X_j, T],
\end{aligned}$$

and this together with the canonical commutation relations and $\pi_\lambda(T) = i\lambda\text{I}$, yield (6.33).

For the second part of Example 6.3.5, we have

$$\Delta_{y_j}\pi_\lambda(X_k)^2 = -\frac{1}{i\lambda}\pi_\lambda[X_j, X_k^2] \quad \text{and} \quad \Delta_{y_j}\pi_\lambda(Y_k^2) = -\frac{1}{i\lambda}\pi_\lambda[X_j, Y_k^2],$$

and using the canonical commutation relations we compute $[X_j, X_k^2] = 0$ whereas

$$[X_j, Y_k^2] = 2\delta_{jk}Y_k T,$$

since

$$\begin{aligned}
X_j Y_k^2 &= X_j Y_k Y_k = (\delta_{jk}T + Y_k X_j)Y_k\\
&= \delta_{jk}TY_k + Y_k(\delta_{jk}T + Y_k X_j)\\
&= 2\delta_{jk}Y_k T + Y_k^2 X_j.
\end{aligned}$$

Therefore

$$\Delta_{y_j}\pi_\lambda(Y_k)^2 = -\frac{1}{i\lambda}\pi_\lambda(2\delta_{jk}Y_k T) = -2\delta_{jk}\pi_\lambda(Y_k) \quad \text{and} \quad \Delta_{y_j}\pi_\lambda(X_k^2) = 0.$$

This implies (6.34). □

6.3.2 Difference operator Δ_t

Naturally, very important information will be contained in the difference operator corresponding to multiplication by t.

Lemma 6.3.6. *We have*

$$\Delta_t|_{\pi_\lambda} = i\partial_\lambda + \frac{1}{2}\sum_{j=1}^n \Delta_{x_j}\Delta_{y_j}|_{\pi_\lambda} + \frac{i}{2\lambda}\sum_{j=1}^n \left\{\pi_\lambda(Y_j)\Delta_{y_j}|_{\pi_\lambda} + \Delta_{x_j}|_{\pi_\lambda}\pi_\lambda(X_j)\right\}.$$

By this we mean that for any κ in some $\mathcal{K}_{a,b}(\mathbb{H}_n)$ such that $t\kappa$ is in some $\mathcal{K}_{a',b'}(\mathbb{H}_n)$, we have

$$\begin{aligned}
\Delta_t\pi_\lambda(\kappa) &= i\partial_\lambda\pi_\lambda(\kappa) + \frac{1}{2}\sum_{j=1}^n \Delta_{x_j}\Delta_{y_j}\pi_\lambda(\kappa)\\
&\quad + \frac{i}{2\lambda}\sum_{j=1}^n \left\{\pi_\lambda(Y_j)\Delta_{y_j}\pi_\lambda(\kappa) + \Delta_{x_j}\pi_\lambda(\kappa)\pi_\lambda(X_j)\right\},
\end{aligned}$$

or, rewriting this with the equivalent notation $\widehat{\kappa}(\pi_\lambda)$ as before,

$$\begin{aligned}\Delta_t\widehat{\kappa}(\pi_\lambda) &= i\partial_\lambda\widehat{\kappa}(\pi_\lambda) + \frac{1}{2}\sum_{j=1}^n \Delta_{x_j}\Delta_{y_j}\widehat{\kappa}(\pi_\lambda)\\ &\quad + \frac{i}{2\lambda}\sum_{j=1}^n \left\{\pi_\lambda(Y_j)\Delta_{y_j}\widehat{\kappa}(\pi_\lambda) + \Delta_{x_j}\widehat{\kappa}(\pi_\lambda)\pi_\lambda(X_j)\right\}.\end{aligned}$$

Before giving some examples of applications of the difference operator Δ_t, let us make a couple of remarks.

Remark 6.3.7.

1. This lemma shows that the difference operators act on the field of operators $\{\pi_\lambda(\kappa),\ \lambda\in\mathbb{R}\backslash\{0\}\}$, rather than on 'one' $\pi_\lambda(\kappa)$ for an individual λ, see Remark 5.2.2.
2. In a similar way as in Remark 6.3.2, the formula in Lemma 6.3.6 respects the properties of the automorphism Θ and the dilations D_r. Indeed, using (6.23) we have

$$\begin{aligned}(\Delta_t\widehat{\kappa}(\pi))\,|_{\pi=\pi_{-\lambda}} &= \left(\widehat{t\kappa}(\pi)\right)|_{\pi=\pi_{-\lambda}} = \pi_{-\lambda}(t\kappa) = \pi_\lambda\left((t\kappa)\circ\Theta\right)\\ &= \pi_\lambda\left(-t\ \kappa\circ\Theta\right) = -\Delta_t\widehat{\kappa\circ\Theta}(\pi_\lambda) = -\Delta_t\left(\widehat{\kappa}(\pi_{-\lambda})\right),\end{aligned}$$

that is

$$(\Delta_t\widehat{\kappa}(\pi))\,|_{\pi=\pi_{-\lambda}} = -\Delta_t\left(\widehat{\kappa}(\pi_{-\lambda})\right). \tag{6.35}$$

For the dilations, using (6.24), we have

$$\begin{aligned}(\Delta_t\widehat{\kappa}(\pi))\,|_{\pi=\pi_{r^{-2}\lambda}} &= \left(\widehat{t\kappa}(\pi)\right)|_{\pi=\pi_{r^{-2}\lambda}} = \pi_{r^{-2}\lambda}(t\kappa) = \pi_\lambda\left(r^Q(t\kappa)\circ D_r\right)\\ &= r^2\pi_\lambda\left(r^Q t\ \kappa\circ D_r\right) = r^2\Delta_t\left(\widehat{\kappa}(\pi_{r^{-2}\lambda})\right).\end{aligned}$$

that is

$$(\Delta_t\widehat{\kappa}(\pi))\,|_{\pi=\pi_{r^{-2}\lambda}} = r^2\Delta_t\left(\widehat{\kappa}(\pi_{r^{-2}\lambda})\right). \tag{6.36}$$

Formulae (6.35) and (6.36) can also be viewed directly from the formula in Lemma 6.3.6:

$$\begin{aligned}(\Delta_t\widehat{\kappa}(\pi))\,|_{\pi=\pi_{-\lambda}} &= i\partial_{\lambda_1=-\lambda}\{\pi_{\lambda_1}(\kappa)\} + \frac{1}{2}\sum_{j=1}^n\{\Delta_{x_j}\Delta_{y_j}\pi(\kappa)\}_{\pi=\pi_{-\lambda}}\\ &\quad + \frac{i}{-2\lambda}\sum_{j=1}^n\{\pi(Y_j)\Delta_{y_j}\pi(\kappa) + \Delta_{x_j}\pi(\kappa)\pi(X_j)\}_{\pi=\pi_{-\lambda}},\end{aligned} \tag{6.37}$$

$$\begin{aligned}(\Delta_t\widehat{\kappa}(\pi))\,|_{\pi=\pi_{r^{-2}\lambda}} &= i\partial_{\lambda_1=r^{-2}\lambda}\{\pi_{\lambda_1}(\kappa)\} + \frac{1}{2}\sum_{j=1}^n\{\Delta_{x_j}\Delta_{y_j}\pi(\kappa)\}_{\pi=\pi_{r^{-2}\lambda}}\\ &\quad + \frac{i}{2r^{-2}\lambda}\sum_{j=1}^n\{\pi(Y_j)\Delta_{y_j}\pi(\kappa) + \Delta_{x_j}\pi(\kappa)\pi(X_j)\}_{\pi=\pi_{r^{-2}\lambda}}.\end{aligned} \tag{6.38}$$

For the first terms in the right hand side in (6.37) and (6.38) we have easily that

$$\begin{aligned} \partial_{\lambda_1=-\lambda}\pi_{\lambda_1}(\kappa) &= -\partial_\lambda\{\pi_{-\lambda}(\kappa)\}, \\ \partial_{\lambda_1=r^{-2}\lambda}\pi_{\lambda_1}(\kappa) &= r^2\partial_\lambda\{\pi_{r^{-2}\lambda}(\kappa)\}. \end{aligned}$$

From Remark 6.3.2 we know that

$$\left\{ \begin{array}{rcl} \left(\Delta_{x_j}\widehat{\kappa}(\pi)\right)|_{\pi=\pi_{-\lambda}} &=& \Delta_{x_j}\left(\widehat{\kappa}(\pi_{-\lambda})\right) \\ \left(\Delta_{y_j}\widehat{\kappa}(\pi)\right)|_{\pi=\pi_{-\lambda}} &=& -\Delta_{y_j}\left(\widehat{\kappa}(\pi_{-\lambda})\right) \\ \left(\Delta_{x_j}\widehat{\kappa}(\pi)\right)|_{\pi=\pi_{r^{-2}\lambda}} &=& r\Delta_{x_j}\left(\widehat{\kappa}(\pi_{r^{-2}\lambda})\right) \\ \left(\Delta_{y_j}\widehat{\kappa}(\pi)\right)|_{\pi=\pi_{r^{-2}\lambda}} &=& r\Delta_{y_j}\left(\widehat{\kappa}(\pi_{r^{-2}\lambda})\right) \end{array} \right. \tag{6.39}$$

so we have for the second term of the right hand side in (6.37) and (6.38) respectively:

$$\begin{aligned} \sum_{j=1}^n\{\Delta_{x_j}\Delta_{y_j}\pi(\kappa)\}_{\pi=\pi_{-\lambda}} &= -\sum_{j=1}^n\Delta_{x_j}\Delta_{y_j}\left(\widehat{\kappa}(\pi_{-\lambda})\right), \\ \sum_{j=1}^n\{\Delta_{x_j}\Delta_{y_j}\pi(\kappa)\}_{\pi=\pi_{r^{-2}\lambda}} &= r^2\sum_{j=1}^n\Delta_{x_j}\Delta_{y_j}\left(\widehat{\kappa}(\pi_{r^{-2}\lambda})\right). \end{aligned}$$

Now viewing X_j and Y_j as elements of the Lie algebra and left invariant vector fields, we see using (6.23) and (6.6) that

$$\begin{aligned} \pi_{-\lambda}(X_j) &= \pi_{-\lambda}(\Theta(X_j)) = \pi_{-\lambda}(X_j\circ\Theta) = \pi_\lambda(X_j), \\ \pi_{-\lambda}(Y_j) &= -\pi_{-\lambda}(\Theta(Y_j)) = -\pi_{-\lambda}(Y_j\circ\Theta) = -\pi_\lambda(Y_j), \end{aligned}$$

and, using (6.25) and (6.4), we obtain

$$\begin{aligned} \pi_{r^{-2}\lambda}(X_j) &= \pi_\lambda(D_{r^{-1}}X_j) = r^{-1}\pi_\lambda(X_j), \\ \pi_{r^{-2}\lambda}(Y_j) &= \pi_\lambda(D_{r^{-1}}Y_j) = r^{-1}\pi_\lambda(Y_j). \end{aligned}$$

So from this and (6.39) we obtain for the third terms of the right hand side in (6.35) and in (6.36) that

$$\begin{aligned} &\frac{i}{-2\lambda}\sum_{j=1}^n\{\pi(Y_j)\Delta_{y_j}\pi(\kappa)+\Delta_{x_j}\pi(\kappa)\pi(X_j)\}_{\pi=\pi_{-\lambda}} \\ &\qquad = -\frac{i}{2\lambda}\sum_{j=1}^n\pi_{-\lambda}(Y_j)\Delta_{y_j}\pi_{-\lambda}(\kappa)+\Delta_{x_j}\pi_{-\lambda}(\kappa)\pi_{-\lambda}(X_j), \\ &\frac{i}{2r^{-2}\lambda}\sum_{j=1}^n\{\pi(Y_j)\Delta_{y_j}\pi(\kappa)+\Delta_{x_j}\pi(\kappa)\pi(X_j)\}_{\pi=\pi_{r^{-2}\lambda}} \\ &\qquad = r^2\frac{i}{2\lambda}\sum_{j=1}^n\pi_{r^{-2}\lambda}(Y_j)\Delta_{y_j}\pi_{r^{-2}\lambda}(\kappa)+\Delta_{x_j}\pi_{r^{-2}\lambda}(\kappa)\pi_{-\lambda}(X_j). \end{aligned}$$

Collecting the new expressions for the three terms of the right hand sides in (6.35) and in (6.36) we obtain a new proof for Equalities (6.35) and (6.36).

Proof of Lemma 6.3.6. Let κ be in some $\mathcal{K}_{a,b}(\mathbb{H}_n)$ and $h \in \mathcal{S}(\mathbb{R}^n)$. We start by differentiating with respect to λ the expression from Lemma 6.2.4:

$$\pi_\lambda(\kappa)h(u) = \int_{\mathbb{H}_n} \kappa(x,y,t)e^{i\lambda(-t+\frac{1}{2}xy)}e^{-i\sqrt{\lambda}yu}h(u-\sqrt{|\lambda|}x)dxdydt,$$

and obtain

$$\partial_\lambda\left\{\pi_\lambda(\kappa)h(u)\right\} = \int_{\mathbb{H}_n} \kappa(x,y,t)e^{i\lambda(-t+\frac{1}{2}xy)}e^{-i\sqrt{\lambda}yu}$$
$$\left(\left[i(-t+\frac{1}{2}xy) - i\frac{yu}{2\sqrt{|\lambda|}}\right]h(u-\sqrt{|\lambda|}x) - \frac{1}{2\sqrt{\lambda}}x\nabla h(u-\sqrt{|\lambda|}x)\right)dxdydt;$$

indeed with our convention we have

$$x\nabla h = \sum_{j=1}^n x_j\partial_{u_j}h, \quad \text{and} \quad \partial_\lambda\{\sqrt{\lambda}\} = \frac{1}{2\sqrt{|\lambda|}}, \quad \partial_\lambda\{\sqrt{|\lambda|}\} = \frac{1}{2\sqrt{\lambda}}.$$

We can now interpret the formula above in the light of difference operators as

$$\begin{aligned}
\partial_\lambda\pi_\lambda(\kappa) &= i\pi_\lambda((-t+\frac{1}{2}xy)\kappa) + \sum_{j=1}^n\left\{-\frac{iu_j}{2\sqrt{|\lambda|}}\pi_\lambda(y_j\kappa) - \frac{1}{2\sqrt{\lambda}}\pi_\lambda(x_j\kappa)\partial_{u_j}\right\} \\
&= -i\Delta_t\pi_\lambda(\kappa) + \frac{i}{2}\sum_{j=1}^n\Delta_{x_j}\Delta_{y_j}\pi_\lambda(\kappa) \\
&\quad -\frac{1}{2\lambda}\sum_{j=1}^n\left\{\pi_\lambda(Y_j)\left(\Delta_{y_j}\pi_\lambda(\kappa)\right) + \left(\Delta_{x_j}\pi_\lambda(\kappa)\right)\pi_\lambda(X_j)\right\},
\end{aligned}$$

using (6.11). □

We already know that

$$\Delta_t \mathrm{I} = 0 \quad \text{and} \quad \Delta_t\pi_\lambda(X_k) = \Delta_t\pi_\lambda(Y_k) = 0, \tag{6.40}$$

see Example 5.2.8 and Lemma 5.2.9, but we can also test it with the formula given in Lemma 6.3.6. We also obtain the following (more substantial) examples:

Example 6.3.8. We can compute

$$\Delta_t\pi_\lambda(T) = -\mathrm{I}, \tag{6.41}$$

and

$$\Delta_t\pi_\lambda(\mathcal{L}) = 0. \tag{6.42}$$

Proof. Since

$$\pi_\lambda(T) = i\lambda\mathrm{I}$$

(see Lemma 6.2.1), we compute directly $\partial_\lambda \pi_\lambda(T) = i\mathrm{I}$. By (6.31) and (6.33), we know

$$\Delta_{y_j}\pi_\lambda(T) = \Delta_{x_j}\pi_\lambda(T) = 0,$$

thus we have obtained (6.41) by Lemma 6.3.6. Furthermore, by (6.12), we have

$$\partial_\lambda \pi_\lambda(\mathcal{L}) = \mathrm{sgn}(\lambda)\sum_{j=1}^n \left(\partial_{u_j}^2 - u_j^2\right) = \frac{1}{\lambda}\pi_\lambda(\mathcal{L})$$

and by (6.32) and (6.34)

$$\sum_{j=1}^n \{\pi_\lambda(Y_j)\Delta_{y_j}\pi_\lambda(\mathcal{L}) + \Delta_{x_j}\pi_\lambda(\mathcal{L})\pi_\lambda(X_j)\}$$
$$= -\sum_{j=1}^n \{\pi_\lambda(Y_j)2\pi_\lambda(Y_j) + 2\pi_\lambda(X_j)\pi_\lambda(X_j) = -2\pi_\lambda(\mathcal{L})\},$$

and also by Example 6.3.4, we get

$$\Delta_{x_j}\Delta_{y_j}\pi_\lambda(\mathcal{L}) = -\Delta_{x_j}2\pi_\lambda(Y_j) = 0.$$

Combining all these equalities together with Lemma 6.3.6 yields (6.42). □

Note that (6.42) can also be obtained from (6.40) and the Leibniz formula (in the sense of (5.28)) for Δ_t.

In terms of λ-symbols, we obtain

Corollary 6.3.9. *If $\widehat{\kappa}(\pi_\lambda) \equiv \pi_\lambda(\kappa) = \mathrm{Op}^W(a_\lambda)$ with $a_\lambda = \{a_\lambda(\xi, u)\}$, then*

$$\Delta_t\widehat{\kappa}\ (\pi_\lambda) = i\mathrm{Op}^W\left(\tilde{\partial}_{\lambda,\xi,u}a_\lambda\right),$$

where

$$\tilde{\partial}_{\lambda,\xi,u} := \partial_\lambda - \frac{1}{2\lambda}\sum_{j=1}^n \left(u_j\partial_{u_j} + \xi_j\partial_{\xi_j}\right). \tag{6.43}$$

Proof. Using formulae (6.28), Corollary 6.3.3 and the properties of the Weyl calculus (see especially the composition formula in (6.16)), we obtain easily that

$$\begin{aligned}
\pi_\lambda(Y_j)\Delta_{y_j}\pi_\lambda(\kappa) &= \mathrm{Op}^W\left(i\sqrt{\lambda}u_j\right)\mathrm{Op}^W\left(\frac{-1}{i\sqrt{\lambda}}\partial_{u_j}a_\lambda\right)\\
&= -\mathrm{Op}^W(u_j)\,\mathrm{Op}^W\left(\partial_{u_j}a_\lambda\right)\\
&= -\mathrm{Op}^W\left(u_j\partial_{u_j}a_\lambda - \frac{1}{2i}\partial_{\xi_j}\partial_{u_j}a_\lambda\right),
\end{aligned}$$

and

$$\begin{aligned}
\Delta_{x_j}\pi_\lambda(\kappa)\pi_\lambda(X_j) &= \mathrm{Op}^W\left(\frac{-1}{i\sqrt{|\lambda|}}\partial_{\xi_j}a_\lambda\right)\mathrm{Op}^W\left(i\sqrt{|\lambda|}\xi_j\right)\\
&= -\mathrm{Op}^W\left(\partial_{\xi_j}a_\lambda\right)\mathrm{Op}^W\left(\xi_j\right)\\
&= -\mathrm{Op}^W\left((\partial_{\xi_j}a_\lambda)\xi_j - \frac{1}{2i}\partial_{u_j}\partial_{\xi_j}a_\lambda\right),
\end{aligned}$$

thus

$$\begin{aligned}
&\pi_\lambda(Y_j)\Delta_{y_j}\pi_\lambda(\kappa) + \Delta_{x_j}\pi_\lambda(\kappa)\pi_\lambda(X_j)\\
&= -\mathrm{Op}^W\left(u_j\partial_{u_j}a_\lambda - \frac{1}{2i}\partial_{\xi_j}\partial_{u_j}a_\lambda\right) - \mathrm{Op}^W\left((\partial_{\xi_j}a_\lambda)\xi_j - \frac{1}{2i}\partial_{u_j}\partial_{\xi_j}a_\lambda\right).\\
&= \mathrm{Op}^W\left(-u_j\partial_{u_j}a_\lambda - \xi_j\partial_{\xi_j}a_\lambda + \frac{1}{i}\partial_{\xi_j}\partial_{u_j}a_\lambda\right).
\end{aligned}$$

We also have

$$\begin{aligned}
\Delta_{x_j}\Delta_{y_j}\pi_\lambda(\kappa) &= \mathrm{Op}^W\left(\frac{-1}{i\sqrt{|\lambda|}}\partial_{\xi_j}\frac{-1}{i\sqrt{\lambda}}\partial_{u_j}a_\lambda\right)\\
&= -\frac{1}{\lambda}\mathrm{Op}^W\left(\partial_{\xi_j}\partial_{u_j}a_\lambda\right). \qquad (6.44)
\end{aligned}$$

Bringing these equalities in the formula for Δ_t in Lemma 6.3.6, we obtain

$$\begin{aligned}
\Delta_t\pi_\lambda(\kappa) &= i\partial_\lambda\pi_\lambda(\kappa) + \frac{1}{2}\sum_{j=1}^n \Delta_{x_j}\Delta_{y_j}\pi_\lambda(\kappa)\\
&\quad + \frac{i}{2\lambda}\sum_{j=1}^n\left\{\pi_\lambda(Y_j)\Delta_{y_j}\pi_\lambda(\kappa) + \Delta_{x_j}\pi_\lambda(\kappa)\pi_\lambda(X_j)\right\}\\
&= i\mathrm{Op}^W(\partial_\lambda a_\lambda) + \frac{1}{2}\sum_{j=1}^n -\frac{1}{\lambda}\mathrm{Op}^W\left(\partial_{\xi_j}\partial_{u_j}a_\lambda\right)\\
&\quad + \frac{i}{2\lambda}\sum_{j=1}^n \mathrm{Op}^W\left(-u_j\partial_{u_j}a_\lambda - \xi_j\partial_{\xi_j}a_\lambda + \frac{1}{i}\partial_{\xi_j}\partial_{u_j}a_\lambda\right)\\
&= \mathrm{Op}^W\left(i\partial_\lambda a_\lambda - \frac{i}{2\lambda}\sum_{j=1}^n\left(u_j\partial_{u_j}a_\lambda + \xi_j\partial_{\xi_j}a_\lambda\right)\right).
\end{aligned}$$

This completes the proof. □

6.3.3 Formulae

Here we summarise the formulae obtained so far in Sections 6.3.1 and 6.3.2. Let us recall our convention regarding square roots (6.9) setting

$$\sqrt{\lambda} := \operatorname{sgn}(\lambda)\sqrt{|\lambda|} = \begin{cases} \sqrt{\lambda} & \text{if } \lambda > 0 \\ -\sqrt{|\lambda|} & \text{if } \lambda < 0 \end{cases}.$$

For the Schrödinger infinitesimal representation we have obtained (see (6.11), (6.12) and (6.28)) that

$$\boxed{\begin{array}{rclcl} \pi_\lambda(X_j) & = & \sqrt{|\lambda|}\partial_{u_j} & = & \mathrm{Op}^W\left(i\sqrt{|\lambda|}\xi_j\right) \\ \pi_\lambda(Y_j) & = & i\sqrt{\lambda}u_j & = & \mathrm{Op}^W\left(i\sqrt{\lambda}u_j\right) \\ \pi_\lambda(T) & = & i\lambda\mathrm{I} & = & \mathrm{Op}^W(i\lambda) \\ \pi_\lambda(\mathcal{L}) & = & |\lambda|\sum_j(\partial_{u_j}^2 - u_j^2) & = & \mathrm{Op}^W\left(|\lambda|\sum_j(-\xi_j^2 - u_j^2)\right) \end{array}}$$

while for difference operators (cf. Lemmata 6.3.1 and 6.3.6) we have

$$\boxed{\begin{array}{rclcl} \Delta_{x_j}|_{\pi_\lambda} & = & \frac{1}{i\lambda}\mathrm{ad}\,(\pi_\lambda(Y_j)) & = & \frac{1}{\sqrt{|\lambda|}}\mathrm{ad}u_j \\ \Delta_{y_j}|_{\pi_\lambda} & = & -\frac{1}{i\lambda}\mathrm{ad}\,(\pi_\lambda(X_j)) & = & -\frac{1}{i\sqrt{\lambda}}\mathrm{ad}\partial_{u_j} \\ \Delta_t|_{\pi_\lambda} & = & \multicolumn{3}{l}{i\partial_\lambda + \frac{1}{2}\sum_{j=1}^n \Delta_{x_j}\Delta_{y_j}|_{\pi_\lambda} + \frac{i}{2\lambda}\sum_{j=1}^n\{\pi_\lambda(Y_j)|_{\pi_\lambda}\Delta_{y_j} + \Delta_{x_j}|_{\pi_\lambda}\pi_\lambda(X_j)\}} \end{array}}$$

and in terms of λ-symbols, that is, with

$$\widehat{\kappa}(\pi_\lambda) \equiv \pi_\lambda(\kappa) = \mathrm{Op}^W(a_\lambda) \text{ and } a_\lambda = \{a_\lambda(\xi, u)\},$$

(cf. Corollaries 6.3.3 and 6.3.9):

$$\boxed{\begin{array}{rcl} \Delta_{x_j}\pi_\lambda(\kappa) & = & i\mathrm{Op}^W\left(\frac{1}{\sqrt{|\lambda|}}\partial_{\xi_j}a_\lambda\right) \\ \Delta_{y_j}\pi_\lambda(\kappa) & = & i\mathrm{Op}^W\left(\frac{1}{\sqrt{\lambda}}\partial_{u_j}a_\lambda\right) \\ \Delta_t\pi_\lambda(\kappa) & = & i\mathrm{Op}^W\left(\tilde{\partial}_{\lambda,\xi,u}a_\lambda\right) \\ & = & i\mathrm{Op}^W\left((\partial_\lambda - \frac{1}{2\lambda}\sum_{j=1}^n\{u_j\partial_{u_j} + \xi_j\partial_{\xi_j}\})a_\lambda\right) \end{array}} \tag{6.45}$$

In Examples 6.3.4, 6.3.5, 6.3.8 together with (6.40), we have also obtained

	$\pi_\lambda(X_k)$	$\pi_\lambda(Y_k)$	$\pi_\lambda(T)$	$\pi_\lambda(\mathcal{L})$
Δ_{x_j}	$-\delta_{j=k}$	0	0	$-2\pi_\lambda(X_j)$
Δ_{y_j}	0	$-\delta_{j=k}$	0	$-2\pi_\lambda(Y_j)$
Δ_t	0	0	$-\mathrm{I}$	0

The equalities given in the following lemma concern another normalisation of the Weyl symbol which is motivated by (6.20) and by the fact that the expressions of the right-hand sides in (6.45), in particular for the operator $\tilde{\partial}_{\lambda,\xi,u}$, become then very simple:

Lemma 6.3.10. *Let $a_\lambda = \{a_\lambda(\xi, u)\}$ be a family of Weyl symbols depending smoothly on $\lambda \neq 0$. If $\tilde{a}_\lambda$ is the renormalisation obtained via*

$$a_\lambda(\xi, u) = \tilde{a}_\lambda(\sqrt{|\lambda|}\xi, \sqrt{\lambda}u), \tag{6.46}$$

then

$$\begin{aligned}
\left\{\tilde{\partial}_{\lambda,\xi,u} a_\lambda\right\}(\xi, u) &= \{\partial_\lambda \tilde{a}_\lambda\}(\sqrt{|\lambda|}\xi, \sqrt{\lambda}u), \\
\frac{1}{\sqrt{|\lambda|}}\left\{\partial_{\xi_j} a_\lambda\right\}(\xi, u) &= \{\partial_{\xi_j} \tilde{a}_\lambda\}(\sqrt{|\lambda|}\xi, \sqrt{\lambda}u), \\
\frac{1}{\sqrt{\lambda}}\left\{\partial_{u_j} a_\lambda\right\}(\xi, u) &= \{\partial_{u_j} \tilde{a}_\lambda\}(\sqrt{|\lambda|}\xi, \sqrt{\lambda}u).
\end{aligned}$$

Proof. We see that

$$\tilde{a}_\lambda(\xi, u) = a_\lambda\left(\frac{1}{\sqrt{|\lambda|}}\xi, \frac{1}{\sqrt{\lambda}}u\right),$$

thus

$$\begin{aligned}
\partial_\lambda \tilde{a}_\lambda(\xi, u) &= (\partial_\lambda a_\lambda)\left(\frac{1}{\sqrt{|\lambda|}}\xi, \frac{1}{\sqrt{\lambda}}u\right) \\
&\quad - \sum_{j=1}^n \frac{\xi_j}{2\lambda\sqrt{|\lambda|}}\left(\partial_{\xi_j} a_\lambda\right)\left(\frac{1}{\sqrt{|\lambda|}}\xi, \frac{1}{\sqrt{\lambda}}u\right) \\
&\quad - \sum_{j=1}^n \frac{u_j}{2|\lambda|\sqrt{|\lambda|}}\left(\partial_{u_j} a_\lambda\right)\left(\frac{1}{\sqrt{|\lambda|}}\xi, \frac{1}{\sqrt{\lambda}}u\right),
\end{aligned}$$

and

$$\begin{aligned}
\{\partial_\lambda \tilde{a}_\lambda\}\left(\sqrt{|\lambda|}\xi, \sqrt{\lambda}u\right) &= (\partial_\lambda a_\lambda)(\xi, u) \\
&\quad - \sum_{j=1}^n \left(\frac{\sqrt{|\lambda|}\xi_j}{2\lambda\sqrt{|\lambda|}}\partial_{\xi_j} a_\lambda(\xi, u) + \frac{\sqrt{\lambda}u_j}{2|\lambda|\sqrt{|\lambda|}}\partial_{u_j} a_\lambda(\xi, u)\right) \\
&= \partial_\lambda a_\lambda(\xi, u) - \frac{1}{2\lambda}\sum_{j=1}^n \left(\xi_j \partial_{\xi_j} a_\lambda(\xi, u) + u_j \partial_{u_j} a_\lambda(\xi, u)\right) \\
&= \tilde{\partial}_{\lambda,\xi,u} a_\lambda(\xi, u).
\end{aligned}$$

This shows the first stated equality. The other two are easy. ☐

Lemma 6.3.10 and the formulae already obtained yield

$$\begin{aligned}
\Delta_{x_j}\pi_\lambda(\kappa) &= i\mathrm{Op}^W\left(\partial_{\xi_j}\tilde{a}_\lambda\right), \\
\Delta_{y_j}\pi_\lambda(\kappa) &= i\mathrm{Op}^W\left(\partial_{u_j}\tilde{a}_\lambda\right), \\
\Delta_t\pi_\lambda(\kappa) &= i\mathrm{Op}^W\left(\partial_\lambda\tilde{a}_\lambda\right),
\end{aligned}$$

where the λ-symbol a_λ of $\pi_\lambda(\kappa)$, that is, $\pi_\lambda(\kappa) = \mathrm{Op}^W(a_\lambda)$, has been rescaled via (6.46), i.e.

$$a_\lambda(\xi, u) = \tilde{a}_\lambda(\sqrt{|\lambda|}\xi, \sqrt{\lambda}u).$$

Recall that

$$a_\lambda(\xi, u) = (2\pi)^{\frac{2n+1}{2}} \mathcal{F}_{\mathbb{R}^{2n+1}}(\kappa)(\sqrt{|\lambda|}\xi, \sqrt{\lambda}u, \lambda),$$

see (6.20), so

$$\tilde{a}_\lambda(\xi, u) = (2\pi)^{\frac{2n+1}{2}} \mathcal{F}_{\mathbb{R}^{2n+1}}(\kappa)(\xi, u, \lambda).$$

The above formulae in terms of the rescaled λ-symbols look neat. The drawback of using this rescaling is that one rescales the Weyl quantization:

$$\widehat{\kappa}(\pi_\lambda) = \mathrm{Op}^W(a_\lambda) = \mathrm{Op}^W\left(\tilde{a}_\lambda\left(\sqrt{|\lambda|}\,\cdot, \sqrt{\lambda}\,\cdot\right)\right).$$

Since our aim is to study the group Fourier transform on $\mathbb{H}_n$, it is more natural to study the Weyl-symbol a_λ without any rescaling.

In fact, the following two sections are devoted to understanding $\widehat{\kappa} \equiv \{\pi_\lambda(\kappa)\}$ as a family of Weyl pseudo-differential operators parametrised by $\lambda \in \mathbb{R}\backslash\{0\}$. The Weyl quantization will force us to work on the λ-symbol a_λ directly, and not on its rescaling $\tilde{a}_\lambda$.

This will lead to defining a family of symbol classes parametrised by $\lambda \in \mathbb{R}\backslash\{0\}$ for the λ-symbols a_λ. This will be done via a family of Hörmander metrics parametrised by $\lambda \in \mathbb{R}\backslash\{0\}$. Importantly the structural bounds of these metrics will be uniform with respect to λ. The resulting symbol classes will be called λ-Shubin classes.

6.4 Shubin classes

In this Section, we recall elements of the Weyl-Hörmander pseudo-differential calculus and the associated Sobolev spaces, and we apply this to obtain the Shubin classes of symbols and the associated Sobolev spaces. The dependence in a parameter λ will be of particular importance to us. We will call the resulting symbol classes the λ-Shubin classes.

6.4.1 Weyl-Hörmander calculus

Here we present the main elements of the Weyl-Hörmander calculus that will be relevant for our analysis. For more details on the underlying general theory, we can refer, for instance, to [Ler10].

We consider $\mathbb{R}^n$ and identify its cotangent bundle $T^*\mathbb{R}^n$ with $\mathbb{R}^{2n}$. The canonical symplectic form on $\mathbb{R}^{2n}$ is ω defined by

$$\omega(T, T') = x \cdot \xi' - x' \cdot \xi, \quad T = (\xi, x),\ T' = (\xi', x') \in \mathbb{R}^{2n}.$$

Definition 6.4.1. If q is a positive quadratic form on $\mathbb{R}^{2n}$, then we define its *conjugate* q^ω by

$$\forall T \in \mathbb{R}^{2n} \quad q^\omega(T) := \sup_{T' \in \mathbb{R}^{2n} \setminus \{0\}} \frac{|\omega(T,T')|^2}{q(T')},$$

and its *gain factor* by

$$\Lambda_q := \inf_{T \in \mathbb{R}^{2n} \setminus \{0\}} \frac{q^\omega(T)}{q(T)}.$$

Definition 6.4.2. A *metric* is a family of positive quadratic forms

$$g = \{g_X, X \in \mathbb{R}^{2n}\}$$

depending smoothly on $X \in \mathbb{R}^{2n}$.

- The metric g is *uncertain* when $\forall X \in \mathbb{R}^{2n}$, $\Lambda_{g_X} \geq 1$.
- The metric g is *slowly varying* when there exists a constant $\bar{C} > 0$ such that we have for any $X, X' \in \mathbb{R}^{2n}$:

$$g_X(X - X') \leq \bar{C}^{-1} \Longrightarrow \sup_{T \in \mathbb{R}^{2n} \setminus \{0\}} \left(\frac{g_X(T)}{g_{X'}(T)} + \frac{g_{X'}(T)}{g_X(T)} \right) \leq \bar{C}.$$

- The metric g is *temperate* when there are constants $\bar{C} > 0$ and $\bar{N} > 0$ such that we have for any $X, X' \in \mathbb{R}^{2n}$ and $T \in \mathbb{R}^{2n} \setminus \{0\}$:

$$\frac{g_X(T)}{g_{X'}(T)} \leq \bar{C}(1 + g_X^\omega(X - X'))^{\bar{N}}.$$

A metric g is of *Hörmander type* if it is uncertain, slowly varying and temperate. In this case the constants $\bar{C}$ and $\bar{N}$ appearing above and any constant depending only on them are called *structural.*

Proposition 6.4.3. *A metric* $g = \{g_X, X \in \mathbb{R}^{2n}\}$ *is* slowly varying *if and only if there exist constants* $C, r > 0$ *such that we have for any* $X, Y \in \mathbb{R}^{2n}$ *that*

$$g_X(Y - X) \leq r^2 \Longrightarrow \forall T \quad g_Y(T) \leq C g_X(T). \tag{6.47}$$

Proof. If g is slowly varying then it satisfies (6.47). Conversely, let us assume (6.47). Necessarily $C \geq 1$ since we can take $X = Y$ in (6.47). If $g_X(Y - X) \leq C^{-1} r^2$, then $g_X(Y - X) \leq r^2$ and, applying (6.47) with $T = Y - X$, we obtain

$$g_Y(Y - X) \leq C g_X(Y - X) \leq r^2,$$

thus re-applying (6.47) (but at g_Y), we have $g_X(T) \leq C g_Y(T)$ for all T. This shows that g is slowly varying. ◻

Remark 6.4.4. If g satisfies (6.47) with constant $C > 1$ and $r > 0$ then g is slowly varying with a constant $\bar{C} = \min(C^{-1}r^2, 2C)$.

Example 6.4.5. Let ϕ be a positive smooth function on $\mathbb{R}^{2n}$ which is Lipschitz on $\mathbb{R}^{2n}$. We denote by $T \mapsto |T|^2$ the canonical (Euclidean) quadratic form on $\mathbb{R}^{2n}$. The metric g given by

$$g_X(T) = \phi(X)^{-2}|T|^2$$

is slowly varying.

Proof. Let us assume $g_X(Y - X) \leq r^2$ for a constant $r > 0$ to be determined. This means $|Y - X| \leq r\phi(X)$. Since ϕ is Lipschitz on $\mathbb{R}^{2n}$, denoting by L its Lipschitz constant, we have

$$\phi(X) \leq \phi(Y) + L|X - Y| \leq \phi(Y) + Lr\phi(X),$$

thus

$$(1 - Lr)\phi(X) \leq \phi(Y).$$

Hence if we choose $r > 0$ so that $1 - Lr > 0$, we have obtained

$$\forall T \qquad g_Y(T) \leq C g_X(T),$$

with $C = (1 - Lr)^{-1}$. This shows that g_X satisfies (6.47) and is therefore slowly varying. □

Remark 6.4.6. If ϕ is L-Lipschitz then g given in Example 6.4.5 satisfies (6.47) with any $r \in (0, L^{-1})$ and a corresponding $C = (1 - Lr)^{-1}$.

Definition 6.4.7. Let g be a metric of Hörmander type. A positive function M defined on $\mathbb{R}^{2n}$ is a *g-weight* when there are structural constants $\bar{C}'$ and $\bar{N}'$ satisfying for any $X, Y \in \mathbb{R}^{2n}$:

$$g_X(X - Y) \leq \bar{C}'^{-1} \Longrightarrow \frac{M(X)}{M(Y)} + \frac{M(Y)}{M(X)} \leq \bar{C}',$$

and

$$\frac{M(X)}{M(Y)} \leq \bar{C}(1 + g_X^{\omega}(X - Y))^{\bar{N}'}.$$

It is easy to check that the set of g-weights forms a group for the usual multiplication of positive functions.

Definition 6.4.8 (Hörmander symbol class $S(M, g)$). Let g be a metric of Hörmander type and M a g-weight on $\mathbb{R}^{2n}$. The symbol class $S(M, g)$ is the set of functions $a \in C^\infty(\mathbb{R}^{2n})$ such that for each integer $\ell \in \mathbb{N}_0$, the quantity

$$\|a\|_{S(M,g),\ell} := \sup_{\substack{\ell' \leq \ell, X \in \mathbb{R}^{2n} \\ g_X(T_{\ell'}) \leq 1}} \frac{|\partial_{T_1} \dots \partial_{T_{\ell'}} a(X)|}{M(X)}$$

is finite.

Here $\partial_T a$ denotes the quantity (da, T).

The following properties are well known [Ler10, Chapters 1 and 2]:

Theorem 6.4.9. *Let g be a metric of Hörmander type and let M, M_1, M_2 be g-weights.*

1. *The symbol class $S(M,g)$ is a vector space endowed with a Fréchet topology via the family of seminorms $\|\cdot\|_{S(M,g),\ell}$, $\ell \in \mathbb{N}_0$.*
2. *If $a \in S(M,g)$ then the symbol b defined by*
$$\mathrm{Op}^W b = \left(\mathrm{Op}^W a\right)^*$$
is in $S(M,g)$ as well. Furthermore, for any $\ell \in \mathbb{N}_0$ there exist a constant $C > 0$ and a integer $\ell' \in \mathbb{N}_0$ such that
$$\|b\|_{S(M,g),\ell} \leq C\|a\|_{S(M,g),\ell'}.$$
The constant C and the integer ℓ' may be chosen to depend on ℓ and on the structural constants and to be independent of g, M and a.
3. *If $a_1 \in S(M_1, g)$ and $a_2 \in S(M_2, g)$ then the symbol b defined by*
$$\mathrm{Op}^W b = \left(\mathrm{Op}^W a_1\right)\left(\mathrm{Op}^W a_2\right),$$
is in $S(M_1M_2, g)$. Furthermore, for any $\ell \in \mathbb{N}_0$ there exist a constant $C > 0$ and two integers $\ell_1, \ell_2 \in \mathbb{N}_0$ such that
$$\|b\|_{S(M_1M_2,g),\ell} \leq C\|a_1\|_{S(M_1,g),\ell_1}\|a_2\|_{S(M_2,g),\ell_2}.$$
The constant C and the integers ℓ_1, ℓ_2 may be chosen to depend on ℓ and on the structural constants and to be independent of g, M_1, M_2 and a_1, a_2.

Definition 6.4.10 (Sobolev spaces $H(M,g)$)**.** Let g be a metric of Hörmander type and M a g-weight on $\mathbb{R}^{2n}$. We denote by $H(M,g)$ the set of all tempered distributions f on $\mathbb{R}^n$ such that for any symbol $a \in S(M,g)$ we have $\mathrm{Op}^W(a)f \in L^2(\mathbb{R}^n)$.

Theorem 6.4.11. *Let g be a metric of Hörmander type on $\mathbb{R}^{2n}$.*

1. *The space $H(1,g)$ coincides with $L^2(\mathbb{R}^n)$. Furthermore, there exist a structural constant $C > 0$ and a structural integer $\ell \in \mathbb{N}_0$ such that for any symbol $a \in S(1,g)$, we have*
$$\|\mathrm{Op}^W(a)\|_{\mathscr{L}(L^2(\mathbb{R}^n))} \leq C\|a\|_{S(1,g),\ell}.$$
2. *Let M_1, M_2 be g-weights. For any $a \in S(M_1, g)$, the operator $\mathrm{Op}^W(a)$ maps continuously $H(M_2, g)$ to $H(M_2M_1^{-1}, g)$. Furthermore, there exist a constant $C > 0$ and an integer $\ell \in \mathbb{N}_0$ such that*
$$\|\mathrm{Op}^W(a)\|_{\mathscr{L}(H(M_2,g),H(M_2M_1^{-1},g))} \leq C\|a\|_{S(M_1,g),\ell}.$$
The constant C and the integers ℓ may be chosen to depend only on the structural constants of g, M_1, M_2 and to be independent of g, M and a.

6.4.2 Shubin classes $\Sigma_\rho^m(\mathbb{R}^n)$ and the harmonic oscillator

It is well known (and can be readily checked) that the metric

$$\frac{d\xi^2 + du^2}{(1 + |u|^2 + |\xi|^2)^\rho},$$

is of Hörmander type with corresponding weights $(1 + |u|^2 + |\xi|^2)^{m/2}$ for $m \in \mathbb{R}$. This will be also shown later in the proof of Proposition 6.4.21. For $m \in \mathbb{R}$ and $\rho \in (0, 1]$, we denote by $\Sigma_\rho^m(\mathbb{R}^n)$ the corresponding symbol class, often called the Shubin classes of symbols on $\mathbb{R}^n$:

$$\Sigma_\rho^m(\mathbb{R}^n) := S\left((1 + |u|^2 + |\xi|^2)^{m/2}, \frac{d\xi^2 + du^2}{(1 + |u|^2 + |\xi|^2)^\rho}\right).$$

This means that a symbol $a \in C^\infty(\mathbb{R}^{2n})$ is in $\Sigma_\rho^m(\mathbb{R}^n)$ if and only if for any $\alpha, \beta \in \mathbb{N}_0^n$ there exists a constant $C = C_{\alpha,\beta} > 0$ such that

$$\forall (\xi, u) \in \mathbb{R}^{2n} \qquad |\partial_\xi^\alpha \partial_u^\beta a(\xi, u)| \leq C \left(1 + |\xi|^2 + |u|^2\right)^{\frac{m - \rho(|\alpha| + |\beta|)}{2}}.$$

The class $\Sigma_\rho^m(\mathbb{R}^n)$ is a vector subspace of $C^\infty(\mathbb{R}^n \times \mathbb{R}^n)$ which becomes a Fréchet space when endowed with the family of seminorms

$$\|a\|_{\Sigma_\rho^m, N} = \sup_{\substack{(\xi,u) \in \mathbb{R}^n \times \mathbb{R}^n \\ |\alpha|, |\beta| \leq N}} \left(1 + |\xi|^2 + |u|^2\right)^{-\frac{m - \rho(|\alpha| + |\beta|)}{2}} |\partial_\xi^\alpha \partial_u^\beta a(\xi, u)|,$$

where $N \in \mathbb{N}_0$. We denote by

$$\Psi\Sigma_\rho^m(\mathbb{R}^n) := \mathrm{Op}^W(\Sigma_\rho^m(\mathbb{R}^n))$$

the corresponding class of operators and by $\|\cdot\|_{\Psi\Sigma_\rho^m, N}$ the corresponding seminorms.

We have the inclusions

$$\rho_1 \geq \rho_2 \quad \text{and} \quad m_1 \leq m_2 \Longrightarrow \Psi\Sigma_{\rho_1}^{m_1}(\mathbb{R}^n) \subset \Psi\Sigma_{\rho_2}^{m_2}(\mathbb{R}^n).$$

Example 6.4.12. The operators $\partial_{u_j} = \mathrm{Op}^W(i\xi_j)$, $j = 1, \ldots, n$, or multiplication by $u_k = \mathrm{Op}^W(u_k)$, $k = 1, \ldots, n$, are two operators in $\Psi\Sigma_1^1(\mathbb{R}^n)$.

Standard computations also show:

Example 6.4.13. For each $m \in \mathbb{R}$, the symbol b^m, where

$$b(\xi, u) = \sqrt{1 + |u|^2 + |\xi|^2},$$

is in $\Sigma_1^m(\mathbb{R}^n)$.

The following is well known and can be viewed more generally as a consequence of the Weyl-Hörmander calculus (see Theorem 6.4.9)

Theorem 6.4.14. • *The class of operators $\cup_{m\in\mathbb{R}}\Psi\Sigma_\rho^m(\mathbb{R}^n)$ forms an algebra of operators stable by taking the adjoint. Furthermore, the operations*

$$\begin{array}{rcl}\Psi\Sigma_\rho^m(\mathbb{R}^n) & \longrightarrow & \Psi\Sigma_\rho^m(\mathbb{R}^n)\\ A & \longmapsto & A^*\end{array}$$

and

$$\begin{array}{rcl}\Psi\Sigma_\rho^{m_1}(\mathbb{R}^n)\times\Psi\Sigma_\rho^{m_2}(\mathbb{R}^n) & \longrightarrow & \Psi\Sigma_\rho^{m_1+m_2}(\mathbb{R}^n)\\ (A,B) & \longmapsto & AB\end{array}$$

are continuous.

- *The operators in $\Psi\Sigma_\rho^0(\mathbb{R}^n)$ extend boundedly to $L^2(\mathbb{R}^n)$. Furthermore, there exist $C>0$ and $N\in\mathbb{N}$ such that if $A\in\Psi\Sigma_\rho^0(\mathbb{R}^n)$ then*

$$\|A\|_{\mathscr{L}(L^2(\mathbb{R}^n))}\leq C\|A\|_{\Psi\Sigma_\rho^m,N}.$$

From Example 6.4.12, it follows that the (positive) *harmonic oscillator*

$$Q:=\sum_{j=1}^n(-\partial_{u_j}^2+u_j^2), \tag{6.48}$$

is in $\Psi\Sigma_1^2(\mathbb{R}^n)$.

Note that from now on Q denotes the harmonic oscillator and not the homogeneous dimension as in all previous chapters.

We keep the same notation for Q and for its self-adjoint extension as an unbounded operator on $L^2(\mathbb{R}^n)$. The harmonic oscillator Q is a positive (unbounded) operator on $L^2(\mathbb{R}^n)$. Its spectrum is

$$\{2|\ell|+n,\ell\in\mathbb{N}_0^n\},$$

where $|\ell|=\ell_1+\ldots+\ell_n$. The eigenfunctions associated with the eigenvalues $2|\ell|+n$ are

$$h_\ell : x=(x_1,\ldots,x_n)\longmapsto h_{\ell_1}(x_1)\ldots h_{\ell_n}(x_n),$$

where each h_j, $j=0,1,2\ldots$, is a Hermite function, that is,

$$h_j(\tau)=(-1)^j\frac{e^{\frac{\tau^2}{2}}}{\sqrt{2^j j!\sqrt{\pi}}}\frac{d^j}{d\tau^j}e^{-\tau^2},\qquad \tau\in\mathbb{R}.$$

The Hermite functions are Schwartz, i.e. $h_j\in\mathcal{S}(\mathbb{R})$. With our choice of normalisation, the functions h_j, $j=0,1,\ldots$, form an orthonormal basis of $L^2(\mathbb{R})$. Therefore, the functions h_ℓ form an orthonormal basis of $L^2(\mathbb{R}^n)$. For each $s\in\mathbb{R}$, we define the operator $(\mathrm{I}+Q)^{s/2}$ using the functional calculus, that is, in this case, the domain of $(\mathrm{I}+Q)^{s/2}$ is the space of functions

$$\mathrm{Dom}(\mathrm{I}+Q)^{s/2} = \{h \in L^2(\mathbb{R}^n) : \sum_{\ell\in\mathbb{N}_0^n} (2|\ell|+n)^s |(h_\ell, h)_{L^2(\mathbb{R}^n)}|^2 < \infty\},$$

and if $h \in \mathrm{Dom}(\mathrm{I}+Q)^{s/2}$ then

$$(\mathrm{I}+Q)^{s/2} h = \sum_{\ell\in\mathbb{N}_0^n} (2|\ell|+n)^{s/2} (h_\ell, h)_{L^2(\mathbb{R}^n)} h_\ell.$$

6.4.3 Shubin Sobolev spaces

In this section, we study Shubin Sobolev spaces. Many of their properties, especially their equivalent characterisations, are well known. Their proofs are quite easy but often omitted in the literature. Thus we have chosen to sketch their demonstrations.

The Shubin Sobolev spaces below are a special case of Sobolev spaces for measurable fields on representation spaces, see Definition 5.1.6.

Our starting point will be the following definition for the Shubin Sobolev spaces:

Definition 6.4.15. Let $s \in \mathbb{R}$. The *Shubin Sobolev space* $\mathcal{Q}_s(\mathbb{R}^n)$ is the subspace of $\mathcal{S}'(\mathbb{R}^n)$ which is the completion of $\mathrm{Dom}(\mathrm{I}+Q)^{s/2}$ for the norm

$$\|h\|_{\mathcal{Q}_s} := \|(\mathrm{I}+Q)^{s/2} h\|_{L^2(\mathbb{R}^n)}.$$

They satisfy the following properties:

Theorem 6.4.16. *1. The space $\mathcal{Q}_s(\mathbb{R}^n)$ is a Hilbert space endowed with the sesquilinear form*

$$(g,h)_{\mathcal{Q}_s} = \left((\mathrm{I}+Q)^{s/2} g, (\mathrm{I}+Q)^{s/2} h\right)_{L^2(\mathbb{R}^n)}.$$

We have the inclusions

$$\mathcal{S}(\mathbb{R}^n) \subset \mathcal{Q}_{s_1}(\mathbb{R}^n) \subset \mathcal{Q}_{s_2}(\mathbb{R}^n) \subset \mathcal{S}'(\mathbb{R}^n), \quad s_1 > s_2.$$

We also have

$$L^2(\mathbb{R}^n) = \mathcal{Q}_0(\mathbb{R}^n) \quad \text{and} \quad \mathcal{S}(\mathbb{R}^n) = \bigcap_{s\in\mathbb{R}} \mathcal{Q}_s(\mathbb{R}^n).$$

2. The dual of $\mathcal{Q}_s(\mathbb{R}^n)$ may be identified with $\mathcal{Q}_{-s}(\mathbb{R}^n)$ via the distributional duality form $\langle g,h\rangle = \int_{\mathbb{R}^n} gh$.

3. *If* $s \in \mathbb{N}_0$, $\mathcal{Q}_s(\mathbb{R}^n)$ *coincides with*

$$\mathcal{Q}_s(\mathbb{R}^n) = \{h \in L^2(\mathbb{R}^n) : u^\alpha \partial_u^\beta h \in L^2(\mathbb{R}^n) \quad \forall \alpha, \beta \in \mathbb{N}_0^n, \ |\alpha| + |\beta| \le s\}.$$

Furthermore, the norm given by

$$\|h\|_{\mathcal{Q}_s}^{(int)} = \sum_{|\alpha|+|\beta|\le s} \|u^\alpha \partial_u^\beta h\|_{L^2(\mathbb{R}^n)},$$

is equivalent to $\|\cdot\|_{\mathcal{Q}_s}$.

4. *For any* $s \in \mathbb{R}$, $\mathcal{Q}_s(\mathbb{R}^n)$ *coincides with the completion (in* $\mathcal{S}'(\mathbb{R}^n)$*) of the Schwartz space* $\mathcal{S}(\mathbb{R}^n)$ *for the norm*

$$\|h\|_{\mathcal{Q}_s}^{(b)} = \|\mathrm{Op}^W(b^s)\, h\|_{L^2(\mathbb{R}^n)},$$

where b *was given in Example 6.4.13. The norm* $\|\cdot\|_{\mathcal{Q}_s}^{(b)}$ *extended to* $\mathcal{Q}_s(\mathbb{R}^n)$ *is equivalent to* $\|\cdot\|_{\mathcal{Q}_s}$.

5. *For any* $s \in \mathbb{R}$, *the Shubin Sobolev space* $\mathcal{Q}_s(\mathbb{R}^n)$ *coincides with the Sobolev space associated with the following metric weight (see Definition 6.4.10)*

$$\mathcal{Q}_s(\mathbb{R}^n) = H\left((1 + |u|^2 + |\xi|^2)^{s/2}, \frac{d\xi^2 + du^2}{1 + |u|^2 + |\xi|^2}\right).$$

6. *For any* $s \in \mathbb{R}$, *the operators* $\mathrm{Op}^W(b^{-s})(\mathrm{I} + Q)^{s/2}$ *and* $(\mathrm{I} + Q)^{s/2}\mathrm{Op}^W(b^{-s})$ *are bounded and invertible on* $L^2(\mathbb{R}^n)$.

7. *The complex interpolation between the spaces* $\mathcal{Q}_{s_0}(\mathbb{R}^n)$ *and* $\mathcal{Q}_{s_1}(\mathbb{R}^n)$ *is*

$$(\mathcal{Q}_{s_0}(\mathbb{R}^n), \mathcal{Q}_{s_1}(\mathbb{R}^n))_\theta = \mathcal{Q}_{s_\theta}(\mathbb{R}^n), \qquad s_\theta = (1-\theta)s_0 + \theta s_1, \ \theta \in (0,1).$$

Before giving the proof of Theorem 6.4.16, let us recall the definition of complex interpolation:

Definition 6.4.17 (Complex interpolation)**.** Let X_0 and X_1 be two subspaces of a vector space Z. We assume that X_0 and X_1 are Banach spaces with norms denoted by $|\cdot|_j$, $j = 0, 1$.

Let $\mathscr{Z}$ be the space of the functions f defined on the strip $\bar{S} = \{0 \le \mathrm{Re}\, z \le 1\}$ and valued in $X_0 + X_1$ such that f is continuous on $\bar{S}$ and holomorphic in $S = \{0 < \mathrm{Re}\, z < 1\}$. For $f \in \mathscr{Z}$ we define the quantity (possibly infinite)

$$\|f\|_{\mathscr{Z}} := \sup_{y\in\mathbb{R}}\{|f(iy)|_0, |f(1+iy)|_1\}.$$

The complex interpolation space of exponent $\theta \in (0,1)$ is the space $(X_0, X_1)_\theta$ of vectors $v \in X_0 + X_1$ such that there exists $f \in \mathscr{Z}$ satisfying $f(\theta) = v$ and $\|f\|_{\mathscr{Z}} < \infty$.

The space $(X_0, X_1)_\theta$ is a subspace of Z; it is a Banach space when endowed with the norm given by

$$|v|_\theta := \inf\{\|f\|_{\mathscr{Z}} : f \in \mathscr{Z} \quad \text{and} \quad f(\theta) = v\}.$$

We also refer to Appendix A.6 for the notion of analytic interpolation.

Proof of Theorem 6.4.16. From Definition 6.4.15, it is easy to prove that the space $\mathcal{Q}_s(\mathbb{R}^n)$ is a Hilbert space, that it is included in $\mathcal{S}'(\mathbb{R}^n)$ and that $\mathcal{Q}_0(\mathbb{R}^n) = L^2(\mathbb{R}^n)$. It is a routine exercise left to the reader that the dual of $\mathcal{Q}_s(\mathbb{R}^n)$ is $\mathcal{Q}_{-s}(\mathbb{R}^n)$ via the distributional duality (Part (2)) and that the spaces $\mathcal{Q}_s(\mathbb{R}^n)$ decrease with $s \in \mathbb{R}$.

Let us prove the complex interpolation property of Part (7). We may assume $s_1 > s_0$. For $h \in \mathcal{Q}_{s_\theta}$, we consider the function

$$f(z) := (\mathrm{I} + Q)^{\frac{-(zs_1+(1-z)s_0)+s_\theta}{2}} h,$$

and we check easily that

$$f(\theta) = h, \quad \|f(iy)\|_{\mathcal{Q}_{s_0}} = \|f(1+iy)\|_{\mathcal{Q}_{s_1}} = \|h\|_{\mathcal{Q}_{s_\theta}} \quad \forall y \in \mathbb{R}.$$

This shows that $\mathcal{Q}_{s_\theta}$ is continuously included in $(\mathcal{Q}_{s_0}(\mathbb{R}^n), \mathcal{Q}_{s_1}(\mathbb{R}^n))_\theta$. By duality of the complex interpolation and of the $\mathcal{Q}_s(\mathbb{R}^n)$, we obtain the reverse inclusion and Part (7) is proved.

Let us prove Part (4). For any $s \in \mathbb{R}$, the operator $\mathrm{Op}^W(b^s)$ maps $\mathcal{S}(\mathbb{R}^n)$ to itself and the mapping $\|\cdot\|^{(b)}_{\mathcal{Q}_s}$ as defined in Part (4) is a norm on $\mathcal{S}(\mathbb{R}^n)$. We denote its completion in $\mathcal{S}'(\mathbb{R}^n)$ by $\mathcal{Q}^{(b)}_s(\mathbb{R}^n)$. From the properties of the calculus it is again a routine exercise left to the reader that the dual of $\mathcal{Q}^{(b)}_s(\mathbb{R}^n)$ is $\mathcal{Q}^{(b)}_{-s}(\mathbb{R}^n)$ via the distributional duality and that the spaces $\mathcal{Q}^{(b)}_s(\mathbb{R}^n)$ decrease with $s \in \mathbb{R}$.

We can prove the following property about interpolation between the $\mathcal{Q}^{(b)}(\mathbb{R}^n)$ spaces which is analogous to Part (7):

$$(\mathcal{Q}^{(b)}_{s_0}(\mathbb{R}^n), \mathcal{Q}^{(b)}_{s_1}(\mathbb{R}^n))_\theta = \mathcal{Q}^{(b)}_{s_\theta}(\mathbb{R}^n), \qquad s_\theta = (1-\theta)s_0 + \theta s_1, \ \theta \in (0,1). \tag{6.49}$$

Indeed we may assume $s_1 > s_0$. For $h \in \mathcal{Q}^{(b)}_{s_\theta}$, we consider the function

$$f(z) = e^{z(s_z - s_\theta)} \mathrm{Op}^W\left(b^{-s_z+s_\theta}\right) h \quad \text{where} \quad s_z = (1-z)s_0 + zs_1.$$

Clearly $f(\theta) = h$. Furthermore,

$$\begin{aligned} \|f(iy)\|^{(b)}_{\mathcal{Q}_{s_1}} &= |e^{iy(s_{iy}-s_\theta)}| \|\mathrm{Op}^W(b^{s_1}) \mathrm{Op}^W\left(b^{-s_{iy}+s_\theta}\right) h\|_{L^2(\mathbb{R}^n)} \\ &\leq e^{-y^2(s_1-s_0)} \|\mathrm{Op}^W(b^{s_1}) \mathrm{Op}^W\left(b^{-s_{iy}+s_\theta}\right) \mathrm{Op}^W(b^{-s_\theta})\|_{\mathscr{L}(L^2(\mathbb{R}^n))} \\ &\qquad \|h\|^{(b)}_{\mathcal{Q}_{s_\theta}}, \end{aligned} \tag{6.50}$$

and

$$\begin{aligned} \|f(1+iy)\|^{(b)}_{\mathcal{Q}_{s_0}} &= |e^{(1+iy)(s_{1+iy}-s_\theta)}| \|\mathrm{Op}^W(b^{s_0}) \mathrm{Op}^W\left(b^{-s_{1+iy}+s_\theta}\right) h\|_{L^2(\mathbb{R}^n)} \\ &\leq e^{s_1 - s_\theta - y^2(s_1-s_0)} \|\mathrm{Op}^W(b^{s_0}) \mathrm{Op}^W\left(b^{-s_{1+iy}+s_\theta}\right) \mathrm{Op}^W(b^{-s_\theta})\|_{\mathscr{L}(L^2(\mathbb{R}^n))} \\ &\qquad \|h\|^{(b)}_{\mathcal{Q}_{s_\theta}}. \end{aligned} \tag{6.51}$$

From the calculus we obtain that the two operator norms on $L^2(\mathbb{R}^n)$ in (6.50) and (6.51) are bounded by a constant of the form $C(1+|y|)^N$ where $C>0$ and $N \in \mathbb{N}_0$ are independent of y. This shows that $\mathcal{Q}^{(b)}_{s_\theta}$ is continuously included in $(\mathcal{Q}^{(b)}_{s_0}(\mathbb{R}^n), \mathcal{Q}^{(b)}_{s_1}(\mathbb{R}^n))_\theta$. By duality of the complex interpolation and of the spaces $\mathcal{Q}_s(\mathbb{R}^n)$, we obtain the reverse inclusion and (6.49) is proved.

Let us show that the spaces $\mathcal{Q}^{(b)}_s(\mathbb{R}^n)$ and $\mathcal{Q}_s(\mathbb{R}^n)$ coincide. First let us assume $s \in 2\mathbb{N}_0$. We have for any $h \in \mathcal{Q}^{(b)}_s(\mathbb{R}^n)$:

$$\|h\|_{\mathcal{Q}_s} \le \|(\mathrm{I}+\mathcal{Q})^{s/2}\mathrm{Op}^W(b^{-s})\|_{\mathcal{L}(L^2(\mathbb{R}^n))}\|h\|^{(b)}_{\mathcal{Q}_s}.$$

As $\mathcal{Q} \in \Psi\Sigma^2_1(\mathbb{R}^n)$, by Theorem 6.4.14, the operator $(\mathrm{I}+\mathcal{Q})^{s/2}\mathrm{Op}^W(b^{-s})$ is in $\Psi\Sigma^0_1$ and thus is bounded on $L^2(\mathbb{R}^n)$. We have obtained a continuous inclusion of $\mathcal{Q}^{(b)}_s(\mathbb{R}^n)$ into $\mathcal{Q}_s(\mathbb{R}^n)$. Conversely, we have for any $h \in \mathcal{Q}_s(\mathbb{R}^n)$ that

$$\|h\|^{(b)}_{\mathcal{Q}_s} \le \|\mathrm{Op}^W(b^s)(\mathrm{I}+\mathcal{Q})^{-s/2}\|_{\mathcal{L}(L^2(\mathbb{R}^n))}\|h\|_{\mathcal{Q}_s}.$$

The inverse of $\mathrm{Op}^W(b^s)(\mathrm{I}+\mathcal{Q})^{-s/2}$ is $(\mathrm{I}+\mathcal{Q})^{s/2}(\mathrm{Op}^W(b^s))^{-1}$ since the operators $\mathrm{I}+\mathcal{Q}$ and $\mathrm{Op}^W(b^s)$ are invertible. Moreover, for the same reason as above, $(\mathrm{I}+\mathcal{Q})^{s/2}(\mathrm{Op}^W(b^s))^{-1}$ is bounded on $L^2(\mathbb{R}^n)$. By the inverse mapping theorem, $\mathrm{Op}^W(b^s)(\mathrm{I}+\mathcal{Q})^{-s/2}$ is bounded on $L^2(\mathbb{R}^n)$. This shows the reverse continuous inclusion. We have proved

$$\mathcal{Q}^{(b)}_s(\mathbb{R}^n) = \mathcal{Q}_s(\mathbb{R}^n)$$

with equivalence of norms for $s \in 2\mathbb{N}_0$ and this implies that this is true for any $s \in \mathbb{R}$ by the properties of duality and interpolation for $\mathcal{Q}^{(b)}_s(\mathbb{R}^n)$ and $\mathcal{Q}_s(\mathbb{R}^n)$. This shows Part (4) and implies Parts (5) and (6).

Let us show that, for each $s \in \mathbb{N}_0$, the space $\mathcal{Q}_s(\mathbb{R}^n)$ coincides with the space $\mathcal{Q}^{(int)}_s(\mathbb{R}^n)$ of functions $h \in L^2(\mathbb{R}^n)$ such that the tempered distributions $u^\alpha\partial_u^\beta h$ are in $L^2(\mathbb{R}^n)$ for every $\alpha, \beta \in \mathbb{N}_0^n$ such that $|\alpha|+|\beta| \le s$. Endowed with the norm $\|\cdot\|^{(int)}_{\mathcal{Q}_s}$ defined in Part (3), $\mathcal{Q}^{(int)}_s(\mathbb{R}^n)$ is a Banach space. We have for any $h \in \mathcal{Q}_s(\mathbb{R}^n) = \mathcal{Q}^{(b)}_s(\mathbb{R}^n)$

$$\|h\|^{(int)}_{\mathcal{Q}_s} \le \sum_{|\alpha|+|\beta|\le s} \|u^\alpha\partial_u^\beta \mathrm{Op}^W(b^{-s})\|_{\mathcal{L}(L^2(\mathbb{R}^n))}\|h\|^{(b)}_{\mathcal{Q}_s}.$$

Since the operators $u^\alpha\partial_u^\beta\mathrm{Op}^W(b^{-s})$ are in $\Psi\Sigma_1^{|\alpha|+|\beta|-s}(\mathbb{R}^n)$ thus continuous on $L^2(\mathbb{R}^n)$ when $|\alpha|+|\beta| \le s$, we see that $\mathcal{Q}_s(\mathbb{R}^n)$ is continuously included in $\mathcal{Q}^{(int)}_s(\mathbb{R}^n)$. For the converse, we separate the cases s even and odd. If $s \in 2\mathbb{N}_0$ then we have easily that

$$\begin{aligned} \|h\|_{\mathcal{Q}_s} &= \|\big(\mathrm{I}+\sum_j(-\partial_{u_j}^2+u_j^2)\big)^{s/2}h\|_{L^2(\mathbb{R}^n)} \\ &\le C_s \sum_{|\alpha|+|\beta|\le s} \|u^\alpha\partial_u^\beta h\|_{L^2(\mathbb{R}^n)} = C_s\|h\|^{(int)}_{\mathcal{Q}_s}. \end{aligned}$$

Now if $s \in 2\mathbb{N}_0 + 1$, we have, since $\mathrm{Op}^W(b^{-1})(\mathrm{I}+Q)^{1/2}$ is bounded and invertible (see Part (6) already proven),

$$\begin{aligned} \|h\|_{\mathcal{Q}_s} &= \|(\mathrm{I}+Q)^{s/2}h\|_{L^2(\mathbb{R}^n)} \le C\|\mathrm{Op}^W(b^{-1})(\mathrm{I}+Q)^{1/2}(\mathrm{I}+Q)^{s/2}h\|_{L^2(\mathbb{R}^n)} \\ &\le C\|\mathrm{Op}^W(b^{-1})(\mathrm{I}+\sum_j -\partial_{u_j}^2 + u_j^2)^{(s+1)/2}h\|_{L^2(\mathbb{R}^n)} \\ &\le C_s \sum_{|\alpha|+|\beta|\le s+1} \|\mathrm{Op}^W(b^{-1})x^\alpha\partial_x^\beta h\|_{L^2(\mathbb{R}^n)} \\ &\le C_s \sum_{|\alpha'|+|\beta'|\le s} \|u^{\alpha'}\partial_u^{\beta'}h\|_{L^2(\mathbb{R}^n)} = C_s\|h\|_{\mathcal{Q}_s}^{(int)}, \end{aligned}$$

by the property of the calculus. Therefore, for s even and odd, $\mathcal{Q}_s^{(int)}(\mathbb{R}^n)$ is continuously included in $\mathcal{Q}_s(\mathbb{R}^n)$. As we have already proven the reverse inclusion, the equality holds and Part (3) is proved. This implies

$$\bigcap_{s\in\mathbb{R}} \mathcal{Q}_s(\mathbb{R}^n) = \mathcal{S}(\mathbb{R}^n)$$

and Part (1) is now completely proved. □

These Sobolev spaces enable us to characterise the operators in the calculus. We allow ourselves to use the shorthand notation

$$(\mathrm{ad}u)^{\alpha_1} := (\mathrm{ad}u_1)^{\alpha_{11}} \dots (\mathrm{ad}u_n)^{\alpha_{1n}},$$

and

$$(\mathrm{ad}\partial_u)^{\alpha_2} := (\mathrm{ad}\partial_{u_1})^{\alpha_{21}} \dots (\mathrm{ad}\partial_{u_n})^{\alpha_{2n}}.$$

Theorem 6.4.18. *We assume that $\rho \in (0,1]$. Let $A : \mathcal{S}(\mathbb{R}^n) \to \mathcal{S}'(\mathbb{R}^n)$ be a linear continuous operator such that all the operators*

$$(\mathrm{ad}u)^{\alpha_1}(\mathrm{ad}\partial_u)^{\alpha_2}A, \quad \alpha_1, \alpha_2 \in \mathbb{N}_0^n,$$

are in $\mathscr{L}(L^2(\mathbb{R}^n), \mathcal{Q}_{-m+\rho(|\alpha_1|+|\alpha_2|)})$ in the sense that they extend to continuous operators from $L^2(\mathbb{R}^n)$ to $\mathcal{Q}_{-m+\rho(|\alpha_1|+|\alpha_2|)}$. Then $A \in \Psi\Sigma_\rho^m(\mathbb{R}^n)$. Moreover, for any $\ell \in \mathbb{N}$, there exist a constant C and an integer ℓ', both independent of A, such that

$$\|A\|_{\Psi\Sigma_\rho^m,\ell} \le C \sum_{|\alpha_1|+|\alpha_2|\le\ell'} \|(\mathrm{ad}u)^{\alpha_1}(\mathrm{ad}\partial_u)^{\alpha_2}A\|_{\mathscr{L}(L^2(\mathbb{R}^n),\mathcal{Q}_{-m+\rho(|\alpha_1|+|\alpha_2|)})}.$$

Note that the converse is true, that is, given $A \in \Psi\Sigma_\rho^m$ then

$$\forall \alpha_1, \alpha_2 \in \mathbb{N}_0^n \qquad (\mathrm{ad}u)^{\alpha_1}(\mathrm{ad}\partial_u)^{\alpha_2}A \in \mathscr{L}(L^2(\mathbb{R}^n), \mathcal{Q}_{-m+\rho(|\alpha_1|+|\beta|)},).$$

This is just a consequence of the properties of the calculus.

The proof of Theorem 6.4.18 relies on the following characterisation of the class of symbols

$$\Sigma_0^0(\mathbb{R}^n) := S(1, d\xi^2 + du^2).$$

Theorem 6.4.19 (Beals' characterisation of $\Sigma_0^0(\mathbb{R}^n)$). *Let $A : \mathcal{S}(\mathbb{R}^n) \to \mathcal{S}'(\mathbb{R}^n)$ be a linear continuous operator such that all the operators*

$$(\mathrm{ad}u)^{\alpha_1}(\mathrm{ad}\partial_u)^{\alpha_2}A, \quad \alpha_1, \alpha_2 \in \mathbb{N}_0^n,$$

are in $\mathscr{L}(L^2(\mathbb{R}^n))$ in the sense that they extend to continuous operators on $L^2(\mathbb{R}^n)$. Then there exits a unique function $a = \{a(\xi, x)\} \in \Sigma_0^0(\mathbb{R}^n)$ such that $A = \mathrm{Op}^W(a)$. Moreover, for any $\ell \in \mathbb{N}$, there exist a constant C and an integer ℓ', both independent of A, such that

$$\|a\|_{\Sigma_0^0,\ell} \leq C \sum_{|\alpha_1|+|\alpha_2|\leq \ell'} \|(\mathrm{ad}u)^{\alpha_1}(\mathrm{ad}\partial_u)^{\alpha_2}A\|_{\mathscr{L}(L^2(\mathbb{R}^n))}.$$

The converse is true, that is, given $a \in \Sigma_0^0(\mathbb{R}^n)$ then $A = \mathrm{Op}^W(a)$ satisfies

$$\forall \alpha_1, \alpha_2 \in \mathbb{N}_0^n \qquad (\mathrm{ad}u)^{\alpha_1}(\mathrm{ad}\partial_u)^{\alpha_2}A \in \mathscr{L}(L^2(\mathbb{R}^n)).$$

We admit Beals' theorem stated in Theorem 6.4.19, see the original article [Bea77a] for the proof.

For the sake of completeness we prove Theorem 6.4.18. This proof can also be found in [Hel84a, Théorème 1.21.1].

Sketch of the proof of Theorem 6.4.18. Let A be as in the statement and b as in Example 6.4.13. We write

$$B_s := \mathrm{Op}^W(b^s)$$

and

$$A_{\alpha_1,\alpha_2} := (\mathrm{ad}u)^{\alpha_1}(\mathrm{ad}\partial_u)^{\alpha_2}A, \quad \alpha_1, \alpha_2 \in \mathbb{N}_0^n.$$

We set $s := m - \rho(|\alpha_1| + |\alpha_2|)$. Then $B_s^{-1}A_{\alpha_1,\alpha_2} \in \mathscr{L}(L^2(\mathbb{R}^n))$. Moreover, we have

$$\mathrm{ad}\partial_{u_1}\left(B_s^{-1}A_{\alpha_1,\alpha_2}\right) = \left(\mathrm{ad}\partial_{u_1}\left(B_s^{-1}\right)\right)A_{\alpha_1,\alpha_2} + B_s^{-1}\mathrm{ad}\partial_{u_1}\left(A_{\alpha_1,\alpha_2}\right);$$

the first operator of the right-hand side is in $\mathscr{L}(L^2(\mathbb{R}^n), \mathcal{Q}_1(\mathbb{R}^n))$ whereas the second is in $\mathscr{L}(L^2(\mathbb{R}^n), \mathcal{Q}_\rho(\mathbb{R}^n))$. Proceeding recursively, we obtain that the operator $B^{-1}_{m-\rho(|\alpha_1|+|\alpha_2|)}A_{\alpha_1,\alpha_2}$ satisfies the hypothesis of Beals' Theorem (Theorem 6.4.19). Therefore, there exists $c_{\alpha_1,\alpha_2} \in \Sigma_0^0(\mathbb{R}^n)$ such that

$$B^{-1}_{m-\rho(|\alpha_1|+|\alpha_2|)}A_{\alpha_1,\alpha_2} = \mathrm{Op}^W(c_{\alpha_1,\alpha_2})$$

or, equivalently,

$$A_{\alpha_1,\alpha_2} = \mathrm{Op}^W(a_{\alpha_1,\alpha_2}) \quad \text{with} \quad a_{\alpha_1,\alpha_2} = b_{m-\rho(|\alpha_1|+|\alpha_2|)} \star c_{\alpha_1,\alpha_2}.$$

We have $A = \mathrm{Op}^W(a_{0,0})$ and

$$\begin{aligned}\mathrm{Op}^W(a_{\alpha_1,\alpha_2}) &= A_{\alpha_1,\alpha_2} = (\mathrm{ad}u)^{\alpha_1}(\mathrm{ad}\partial_u)^{\alpha_2} A \\ &= (\mathrm{ad}u)^{\alpha_1}(\mathrm{ad}\partial_u)^{\alpha_2}\mathrm{Op}^W(a_{0,0}) \\ &= \mathrm{Op}^W\left(i^{|\alpha_1|}\partial_\xi^{\alpha_1}\partial_u^{\alpha_2} a_{0,0}\right),\end{aligned}$$

by Lemma 6.2.3, thus

$$a_{\alpha_1,\alpha_2} = i^{|\alpha_1|}\partial_\xi^{\alpha_1}\partial_u^{\alpha_2} a_{0,0}.$$

Consequently $a \in \Sigma_\rho^m$. □

Looking back at the proof, we see that it can be slightly improved in the following way:

Corollary 6.4.20. *We assume that $\rho \in (0,1]$. Let $A : \mathcal{S}(\mathbb{R}^n) \to \mathcal{S}'(\mathbb{R}^n)$ be a linear continuous operator.*

The operator A is in $\Psi\Sigma_\rho^m(\mathbb{R}^n)$ if and only if there exists $\gamma_o \in \mathbb{R}$ such that for each $\alpha_1, \alpha_2 \in \mathbb{N}_0^n$ we have

$$(\mathrm{ad}u)^{\alpha_1}(\mathrm{ad}\partial_u)^{\alpha_2} A \in \mathscr{L}(\mathcal{Q}_{\gamma_o}(\mathbb{R}^n), \mathcal{Q}_{-m+\rho(|\alpha_1|+|\alpha_2|)+\gamma_o}).$$

In this case this property is true for every $\gamma \in \mathbb{R}$, that is, for each $\gamma \in \mathbb{R}$ and $\alpha_1, \alpha_2 \in \mathbb{N}_0^n$, we have

$$(\mathrm{ad}u)^{\alpha_1}(\mathrm{ad}\partial_u)^{\alpha_2} A \in \mathscr{L}(\mathcal{Q}_{\gamma}(\mathbb{R}^n), \mathcal{Q}_{-m+\rho(|\alpha_1|+|\alpha_2|)+\gamma}).$$

Moreover, for any $\ell \in \mathbb{N}$, there exist a constant C and an integer ℓ', both independent of A, such that

$$\|A\|_{\Psi\Sigma_\rho^m,\ell} \leq C \sum_{|\alpha_1|+|\alpha_2|\leq \ell'} \|(\mathrm{ad}u)^{\alpha_1}(\mathrm{ad}\partial_u)^{\alpha_2} A\|_{\mathscr{L}(\mathcal{Q}_\gamma(\mathbb{R}^n),\mathcal{Q}_{-m+\rho(|\alpha_1|+|\alpha_2|)+\gamma})}.$$

Sketch of the proof of Corollary 6.4.20. We keep the notation of the proof of Theorem 6.4.18. Let A be as in the statement and let $s := m - \rho(|\alpha_1| + |\alpha_2|)$. Then $B_{s+\gamma_o}^{-1} A_{\alpha_1,\alpha_2} B_{\gamma_o} \in \mathscr{L}(L^2(\mathbb{R}^n))$. Moreover, we have

$$\begin{aligned}\mathrm{ad}\partial_{u_1}\left(B_{s+\gamma_o}^{-1} A_{\alpha_1,\alpha_2} B_{\gamma_o}\right) &= \left(\mathrm{ad}\partial_{u_1}\left(B_{s+\gamma_o}^{-1}\right)\right) A_{\alpha_1,\alpha_2} B_{\gamma_o} \\ &\quad + B_{s+\gamma_o}^{-1}\mathrm{ad}\partial_{u_1}\left(A_{\alpha_1,\alpha_2}\right) B_{\gamma_o} \\ &\quad + B_{s+\gamma_o}^{-1} A_{\alpha_1,\alpha_2} B_{\gamma_o}\ B_{\gamma_o}^{-1}\left(\mathrm{ad}\partial_{u_1} B_{\gamma_o}\right);\end{aligned}$$

the first operator of the right-hand side is in $\mathscr{L}(L^2(\mathbb{R}^n), \mathcal{Q}_1(\mathbb{R}^n))$, the second is in $\mathscr{L}(L^2(\mathbb{R}^n), \mathcal{Q}_\rho(\mathbb{R}^n))$ and the third is in $\mathscr{L}(L^2(\mathbb{R}^n))$. Proceeding recursively, we obtain that $B_{s+\gamma_o}^{-1} A_{\alpha_1,\alpha_2} B_{\gamma_o}$ satisfies the hypothesis of Theorem 6.4.19. We then conclude as in the proof of Theorem 6.4.18. □

6.4.4 The λ-Shubin classes $\Sigma^m_{\rho,\lambda}(\mathbb{R}^n)$

The Shubin metric depending on a parameter $\lambda \in \mathbb{R}\backslash\{0\}$ is the metric $g^{(\lambda)}$ on $\mathbb{R}^{2n}$ defined via

$$g^{(\rho,\lambda)}_{\xi,u}(d\xi, du) := \left(\frac{|\lambda|}{1+|\lambda|(1+|\xi|^2+|u|^2)}\right)^\rho (d\xi^2 + du^2).$$

The associated positive function $M^{(\lambda)}$ on $\mathbb{R}^{2n}$ is defined via

$$M^{(\lambda)}(\xi, u) := \left(1+|\lambda|(1+|\xi|^2+|u|^2)\right)^{\frac{1}{2}}.$$

These λ-families of metrics and weights were first introduced in [BFKG12a] in the case $\rho = 1$. The authors of [BFKG12a] realised that, placing λ as above, the structural constants may be chosen independently of λ:

Proposition 6.4.21. *For each $\lambda \in \mathbb{R}\backslash\{0\}$, the metric $g^{(\rho,\lambda)}$ is of Hörmander type (see Definition 6.4.2) and the function $M^{(\lambda)}$ is a $g^{(\rho,\lambda)}$-weight (see Definition 6.4.7). Furthermore, if $\rho \in (0,1]$ is fixed, then the structural constants for $g^{(\rho,\lambda)}$ and for $M^{(\lambda)}$ can be chosen independent of λ.*

The proof of Proposition 6.4.21 follows the proof of the case $\rho = 1$ given in [BFKG12a, Proposition 1.20].

Proof of Proposition 6.4.21. The conjugate of $g^{(\rho,\lambda)}_{\xi,u}$ is $(g^{(\rho,\lambda)}_{\xi,u})^\omega$ given by

$$(g^{(\rho,\lambda)}_{\xi,u})^\omega(d\xi, du) = \left(\frac{1+|\lambda|(1+|\xi|^2+|u|^2)}{|\lambda|}\right)^\rho (d\xi^2 + du^2).$$

The gain is then

$$\Lambda_{g^{(\rho,\lambda)}_{\xi,u}} = \left(\frac{1+|\lambda|(1+|\xi|^2+|u|^2)}{|\lambda|}\right)^{2\rho}.$$

We have for any ρ, λ, ξ, u:

$$\Lambda_{g^{(\rho,\lambda)}_{\xi,u}} \geq \left(\frac{1+|\lambda|}{|\lambda|}\right)^{2\rho} \geq 1.$$

This proves the uniform uncertain property in Definition 6.4.2.

To show that the metric $g^{\rho,\lambda}$ is slowly varying, we notice that it is of the form $\phi(X)^{-2}|T|^2$ as in Example 6.4.5 with

$$\phi(X) = \left(\frac{1+|\lambda|(1+|X|^2)}{|\lambda|}\right)^{\rho/2}.$$

We compute the gradient of ϕ and obtain

$$
\begin{aligned}
|\nabla_X \phi| &= \rho|\lambda|^{1-\frac{\rho}{2}}|X|(1+|\lambda|(1+|X|^2))^{\frac{\rho}{2}-1} \\
&\leq \begin{cases} \rho\left(\frac{|\lambda|}{1+|\lambda|}\right)^{1-\frac{\rho}{2}} \leq \rho & \text{if } |X| \leq 1, \\ \rho\left(\frac{|\lambda||X|^2}{1+|\lambda||X|^2}\right)^{1-\frac{\rho}{2}} |X|^{1-2(1-\frac{\rho}{2})} \leq \rho & \text{if } |X| > 1. \end{cases}
\end{aligned}
$$

So ϕ is ρ-Lipschitz on $\mathbb{R}^{2n}$. Therefore, $g^{\rho,\lambda}$ is slowly varying with a constant $\bar{C}$ independent of λ (see Example 6.4.5 as well as Remarks 6.4.4 and 6.4.6).

Let us prove that $g^{\rho,\lambda}$ is temperate. For any $X, Y \in \mathbb{R}^{2n}$ we have

$$|Y|^2 \leq 2|X|^2 + 2|X-Y|^2;$$

thus

$$\frac{1+|\lambda|(1+|Y|^2)}{1+|\lambda|(1+|X|^2)} \leq 2 + 2\frac{|\lambda|}{1+|\lambda|(1+|X|^2)}|X-Y|^2. \tag{6.52}$$

Now

$$|\lambda| \leq 1+|\lambda|(1+|X|^2) \quad \text{thus} \quad \left(\frac{|\lambda|}{1+|\lambda|(1+|X|^2)}\right)^{1+\rho} \leq 1,$$

and

$$\frac{|\lambda|}{1+|\lambda|(1+|X|^2)} \leq \left(\frac{1+|\lambda|(1+|X|^2)}{|\lambda|}\right)^{\rho}.$$

Plugging this into (6.52), we obtain

$$\frac{1+|\lambda|(1+|Y|^2)}{1+|\lambda|(1+|X|^2)} \leq 2 + 2\left(\frac{1+|\lambda|(1+|X|^2)}{|\lambda|}\right)^{\rho}|X-Y|^2.$$

Taking the ρth power yields

$$
\begin{aligned}
\frac{g_X^{(\rho,\lambda)}(T)}{g_Y^{(\rho,\lambda)}(T)} &= \left(\frac{1+|\lambda|(1+|Y|^2)}{1+|\lambda|(1+|X|^2)}\right)^{\rho} \\
&\leq 2^{\rho}\left(1+\left(\frac{1+|\lambda|(1+|X|^2)}{|\lambda|}\right)^{\rho}|X-Y|^2\right)^{\rho} \\
&= 2^{\rho}\left(1+(g_X^{(\rho,\lambda)})^{\omega}(X-Y)\right)^{\rho}.
\end{aligned}
$$

This shows that $g^{(\rho,\lambda)}$ is temperate with constant independent of λ.

So far we have shown that $g^{(\rho,\lambda)}$ is a metric of Hörmander type. Following the same computations, it is not difficult to show that $M^{(\lambda)}$ are g-weights with constants independent of λ. This concludes the proof of Proposition 6.4.21. $\square$

Let $\rho \in (0,1]$ be a fixed parameter.

For each parameter $\lambda \in \mathbb{R}\backslash\{0\}$, we define the λ-*Shubin classes* by

$$\Sigma^m_{\rho,\lambda}(\mathbb{R}^n) := S\left(\left(M^{(\lambda)}\right)^m, g^{(\rho,\lambda)}\right),$$

where we have used the Hörmander notation to define a class of symbols in terms of a metric and a weight, see Definition 6.4.8.

Here this means that $\Sigma^m_{\rho,\lambda}(\mathbb{R}^n)$ is the class of functions $a \in C^\infty(\mathbb{R}^n \times \mathbb{R}^n)$ such that for each $N \in \mathbb{N}_0$, the quantity

$$\|a\|_{\Sigma^m_{\rho,\lambda},N} := \sup_{\substack{(\xi,u)\in\mathbb{R}^n\times\mathbb{R}^n\\ |\alpha|,|\beta|\leq N}} |\lambda|^{-\rho\frac{|\alpha|+|\beta|}{2}} \left(1+|\lambda|(1+|\xi|^2+|u|^2)\right)^{-\frac{m-\rho(|\alpha|+|\beta|)}{2}} |\partial^\alpha_\xi \partial^\beta_u a(\xi,u)|,$$

is finite. This also means that a symbol $a = \{a(\xi,u)\}$ is in $\Sigma^m_{\rho,\lambda}(\mathbb{R}^n)$ if and only if it satisfies

$$\forall \alpha,\beta \in \mathbb{N}^n_0 \qquad \exists C = C_{\alpha,\beta} > 0 \qquad \forall (\xi,u) \in \mathbb{R}^n \times \mathbb{R}^n$$
$$|\partial^\alpha_\xi \partial^\beta_u a(\xi,u)| \leq C|\lambda|^{\rho\frac{|\alpha|+|\beta|}{2}} \left(1+|\lambda|(1+|\xi|^2+|u|^2)\right)^{\frac{m-\rho(|\alpha|+|\beta|)}{2}}. \qquad (6.53)$$

The class of symbols $\Sigma^m_{\rho,\lambda}(\mathbb{R}^n)$ is a vector subspace of $C^\infty(\mathbb{R}^n \times \mathbb{R}^n)$ which becomes a Fréchet space when endowed with the family of seminorms $\|\cdot\|_{\Sigma^m_{\rho,\lambda},N}$, $N \in \mathbb{N}_0$. We denote by

$$\Psi\Sigma^m_{\rho,\lambda}(\mathbb{R}^n) := \mathrm{Op}^W(\Sigma^m_{\rho,\lambda}(\mathbb{R}^n))$$

the corresponding class of operators, and by $\|\cdot\|_{\Psi\Sigma^m_{\rho,\lambda},N}$ the corresponding seminorms on the Fréchet space $\Psi\Sigma^m_{\rho,\lambda}(\mathbb{R}^n)$.

It is clear that all the spaces of the same order m and parameter ρ coincide in the sense that

$$\forall \lambda \neq 0 \qquad \Sigma^m_{\rho,\lambda}(\mathbb{R}^n) = \Sigma^m_{\rho,1}(\mathbb{R}^n) = \Sigma^m_\rho(\mathbb{R}^n), \qquad (6.54)$$

and the same is true for $\Psi\Sigma^m_{\rho,\lambda}(\mathbb{R}^n) = \Psi\Sigma^m_\rho(\mathbb{R}^n)$. However, the seminorms

$$\|\cdot\|_{\Sigma^m_{\rho,\lambda},N} \quad \text{and} \quad \|\cdot\|_{\Psi\Sigma^m_{\rho,\lambda},N}$$

carry the dependence on λ. This dependence on λ will be crucial for our purposes. From the general properties of metrics of Hörmander type (see Theorem 6.4.9 and Proposition 6.4.21), we readily obtain the following 'λ-uniform' calculus.

Proposition 6.4.22. *1. If, for each $\lambda \in \mathbb{R}\backslash\{0\}$, we are given a symbol $a_\lambda = \{a_\lambda(\xi,u)\}$ in $\Sigma^m_{\rho,\lambda}(\mathbb{R}^n)$ such that*

$$\forall N \in \mathbb{N}_0 \qquad \sup_{\lambda\neq 0} \|a_\lambda\|_{\Sigma^m_{\rho,\lambda},N} < \infty, \qquad (6.55)$$

then each symbol b_λ defined by

$$\operatorname{Op}^W b_\lambda = \left(\operatorname{Op}^W a_\lambda\right)^*$$

is in $\Sigma^m_{\rho,\lambda}(\mathbb{R}^n)$ as well. Furthermore, for any $\ell \in \mathbb{N}_0$ there exist a constant $C > 0$ and a integer $\ell' \in \mathbb{N}_0$ such that for any $\lambda \neq 0$

$$\|b_\lambda\|_{\Sigma^m_{\rho,\lambda},\ell} \leq C\|a_\lambda\|_{\Sigma^m_{\rho,\lambda},\ell'}.$$

The constant C and the integer ℓ' may be chosen to depend on ℓ, m, n and to be independent of λ and a.

2. *If, for each $\lambda \in \mathbb{R}\backslash\{0\}$, we are given two symbols $a_{1,\lambda} = \{a_{1,\lambda}(\xi, u)\}$ in $\Sigma^{m_1}_{\rho,\lambda}(\mathbb{R}^n)$ and $a_{2,\lambda} = \{a_{2,\lambda}(\xi, u)\}$ in $\Sigma^{m_2}_{\rho,\lambda}(\mathbb{R}^n)$ such that*

$$\forall N \in \mathbb{N}_0 \qquad \sup_{\lambda\neq 0} \|a_{1,\lambda}\|_{\Sigma^{m_1}_{\rho,\lambda},N} < \infty \quad \textit{and} \quad \sup_{\lambda\neq 0} \|a_{2,\lambda}\|_{\Sigma^{m_2}_{\rho,\lambda},N} < \infty,$$

then each symbol b_λ defined by

$$\operatorname{Op}^W b_\lambda = \left(\operatorname{Op}^W a_{1,\lambda}\right)\left(\operatorname{Op}^W a_{2,\lambda}\right),$$

is in $\Sigma^{m_1+m_2}_{\rho,\lambda}(\mathbb{R}^n)$. Furthermore, for any $\ell \in \mathbb{N}_0$ there exist a constant $C > 0$ and two integers $\ell_1, \ell_2 \in \mathbb{N}_0$ such that

$$\|b_\lambda\|_{\Sigma^{m_1+m_2}_{\lambda},\ell} \leq C\|a_{1,\lambda}\|_{\Sigma^{m_1}_{\rho,\lambda},\ell_1}\|a_{2,\lambda}\|_{\Sigma^{m_2}_{\rho,\lambda},\ell_2}.$$

The constant C and the integers ℓ_1, ℓ_2 may be chosen to depend on ℓ, m_1, m_2, n and to be independent of λ and $a_{1,\lambda}, a_{2,\lambda}$.

We will say that a family of symbols $a_\lambda = \{a_\lambda(\xi, u)\}$, $\lambda \in \mathbb{R}\backslash\{0\}$, which satisfies Property (6.55) is λ-*uniform* in $\Sigma^m_{\rho,\lambda}(\mathbb{R}^n)$. The corresponding family of operators via the Weyl quantization is said to be λ-uniform in $\Psi\Sigma^m_{\rho,\lambda}(\mathbb{R}^n)$.

Let us give some useful examples of such families of operators.

Example 6.4.23. The families of symbols given by

$$\pi_\lambda(X_j) = i\sqrt{|\lambda|}\xi_j, \quad \pi_\lambda(Y_j) = i\sqrt{\lambda}u_j \quad \text{and} \quad \pi_\lambda(T) = i\lambda$$

are λ-uniform in $\Sigma^1_{1,\lambda}(\mathbb{R}^n)$, $\Sigma^1_{1,\lambda}(\mathbb{R}^n)$, and $\Sigma^2_{1,\lambda}(\mathbb{R}^n)$, respectively.

In particular, the constant operator $\pi_\lambda(T) = i\lambda$ has to be considered as being of order 2 because of the dependence on λ.

Proof. We want to estimate the supremum over $\lambda \neq 0$ of each of the seminorms

$$\|\pi_\lambda(X_j)\|_{\Psi\Sigma^1_{1,\lambda},N} = \|i\sqrt{|\lambda|}\xi_j\|_{\Sigma^1_{1,\lambda},N} \quad \text{and} \quad \|\pi_\lambda(Y_j)\|_{\Psi\Sigma^1_{1,\lambda},N} = \|i\sqrt{\lambda}u_j\|_{\Sigma^1_{1,\lambda},N}.$$

We compute directly for $N=0$:

$$\sup_{\lambda\neq 0}\|i\sqrt{|\lambda|}\xi_j\|_{\Sigma^1_{1,\lambda},0} = \sup_{\lambda\neq 0,(\xi,u)\in\mathbb{R}^n\times\mathbb{R}^n}\frac{\sqrt{|\lambda|}|\xi_j|}{\sqrt{1+|\lambda|(1+|\xi|^2+|u|^2)}} < \infty,$$

$$\sup_{\lambda\neq 0}\|i\sqrt{\lambda}u_j\|_{\Sigma^1_{1,\lambda},0} = \sup_{\lambda\neq 0,(\xi,u)\in\mathbb{R}^n\times\mathbb{R}^n}\frac{\sqrt{|\lambda|}|u_j|}{\sqrt{1+|\lambda|(1+|\xi|^2+u|^2)}} < \infty,$$

and

$$\sup_{\substack{|\alpha|+|\beta|=1\\(\xi,u)\in\mathbb{R}^n\times\mathbb{R}^n}}|\partial_\xi^\alpha\partial_u^\beta\{\sqrt{|\lambda|}\xi_j\}| = \sup_{\substack{|\alpha|+|\beta|=1\\(\xi,u)\in\mathbb{R}^n\times\mathbb{R}^n}}|\partial_\xi^\alpha\partial_u^\beta\{\sqrt{\lambda}u_j\}| = \sqrt{|\lambda|},$$

therefore

$$\sup_{\lambda\neq 0}\|i\sqrt{|\lambda|}\xi_j\|_{\Sigma^1_{1,\lambda},1} < \infty \quad\text{and}\quad \sup_{\lambda\neq 0}\|i\sqrt{\lambda}u_j\|_{\Sigma^1_{1,\lambda},1} < \infty.$$

Since all the higher derivatives $\partial_\xi^\alpha\partial_u^\beta$ with $|\alpha|+|\beta|>1$ of the symbols $i\sqrt{|\lambda|}\xi_j$ and $i\sqrt{\lambda}u_j$ are zero, we obtain that the families of symbols given by $\pi_\lambda(X_j)$, $\pi_\lambda(Y_j)$, are λ-uniform in $\Sigma^1_{1,\lambda}(\mathbb{R}^n)$.

For $\pi_\lambda(T)=\mathrm{Op}^W(i\lambda)$, we see that

$$\|i\lambda\|_{\Sigma^2_{1,\lambda},0} = \sup_{(\xi,u)\in\mathbb{R}^n\times\mathbb{R}^n}\frac{|i\lambda|}{1+|\lambda|(1+|\xi|^2+|u|^2)} < \infty,$$

and since $i\lambda$ is a constant, its derivatives are zero and the family of symbols given by $\pi_\lambda(T)$, is λ-uniform in $\Sigma^2_{1,\lambda}(\mathbb{R}^n)$. $\square$

As a consequence of Example 6.4.23 and Proposition 6.4.22, we also have

Example 6.4.24. The family of operators

$$\pi_\lambda(\mathcal{L}) = \sum_{j=1}^n\left\{\pi_\lambda(X_j)^2+\pi_\lambda(Y_j)^2\right\} = -|\lambda|Q$$

is λ-uniform in $\Psi\Sigma^2_{1,\lambda}(\mathbb{R}^n)$.

Standard computations also show:

Example 6.4.25. For each $m\in\mathbb{R}$, the family of symbols b_λ^m, $\lambda\in\mathbb{R}\backslash\{0\}$, where

$$b_\lambda(\xi,u)=\sqrt{1+|\lambda|(1+|u|^2+|\xi|^2)},$$

is λ-uniform in $\Psi\Sigma^m_{1,\lambda}(\mathbb{R}^n)$.

6.4.5 Commutator characterisation of λ-Shubin classes

In this section, we characterise the λ-Shubin classes in terms of commutators and continuity on the Shubin Sobolev spaces.

First we need to understand some properties of the Sobolev spaces associated with the λ-dependent metric used to define the λ-Shubin symbols.

Proposition 6.4.26. *1. For each $\lambda \in \mathbb{R}\backslash\{0\}$ and $s \in \mathbb{R}$, the Sobolev space corresponding to $g^{(1,\lambda)}$ and $(M^{(\lambda)})^s$ coincides with the Shubin Sobolev space:*

$$H\left((M^{(\lambda)})^s, g^{(1,\lambda)}\right) = \mathcal{Q}_s(\mathbb{R}^n).$$

2. The following define norms on $\mathcal{Q}_s(\mathbb{R}^n)$ equivalent to $\|\cdot\|_{\mathcal{Q}_s}$:

$$\begin{aligned} \|h\|_{\mathcal{Q}_{s,\lambda}} &:= \|(\mathrm{I} + |\lambda| Q)^{s/2} h\|_{L^2(\mathbb{R}^n)}, \\ \|h\|^{(b_\lambda)}_{\mathcal{Q}_{s,\lambda}} &:= \|\mathrm{Op}^W(b_\lambda^s)\, h\|_{L^2(\mathbb{R}^n)}, \end{aligned}$$

where b_λ was defined in Example 6.4.25. Moreover, in the case $s \in \mathbb{N}_0$, we also have an equivalent norm

$$\|h\|^{(int)}_{\mathcal{Q}_{s,\lambda}} := \sum_{|\alpha|+|\beta| \leq s} |\lambda|^{\frac{|\alpha|+|\beta|}{2}} \|u^\alpha \partial_u^\beta h\|_{L^2(\mathbb{R}^n)}.$$

3. Furthermore, for each $s \in \mathbb{R}$ there exists a constant $C_1 = C_{1,s} > 0$ such that

$$\forall \lambda \in \mathbb{R}\backslash\{0\},\ h \in \mathcal{Q}_s(\mathbb{R}^n) \qquad C_1^{-1}\|h\|_{\mathcal{Q}_{s,\lambda}} \leq \|h\|^{(b_\lambda)}_{\mathcal{Q}_{s,\lambda}} \leq C_1\|h\|_{\mathcal{Q}_{s,\lambda}},$$

and for each $s \in \mathbb{N}_0$ there exists a constant $C_2 = C_{2,s} > 0$ such that

$$\forall \lambda \in \mathbb{R}\backslash\{0\},\ h \in \mathcal{Q}_s(\mathbb{R}^n) \qquad C_2^{-1}\|h\|_{\mathcal{Q}_{s,\lambda}} \leq \|h\|^{(int)}_{\mathcal{Q}_{s,\lambda}} \leq C_2\|h\|_{\mathcal{Q}_{s,\lambda}}.$$

Naturally, in Part (2), the constants in the equivalences between each of the norms $\|\cdot\|_{\mathcal{Q}_{s,\lambda}}$, $\|\cdot\|^{(int)}_{\mathcal{Q}_{s,\lambda}}$, $\|\cdot\|^{(b_\lambda)}_{\mathcal{Q}_{s,\lambda}}$, and the norm $\|\cdot\|_{\mathcal{Q}_s}$, depend on λ.

Proof of Proposition 6.4.26. Part (1) follows easily from (6.54), Definition 6.4.10, Theorem 6.4.16 especially Part (5).

Using the Shubin calculus $\cup_m \Psi\Sigma_1^m$, it is not difficult to see that the norms $\|\cdot\|^{(b)}_{\mathcal{Q}_s}$ and $\|\cdot\|^{(b_\lambda)}_{\mathcal{Q}_{s,\lambda}}$ are equivalent.

The fact that the norms $\|\cdot\|_{\mathcal{Q}_{s,\lambda}}$, $\|\cdot\|^{(b_\lambda)}_{\mathcal{Q}_{s,\lambda}}$ and, if $s \in \mathbb{N}_0$, $\|\cdot\|^{(int)}_{\mathcal{Q}_{s,\lambda}}$, are equivalent with λ-uniform constants comes from following the same proof as Theorem 6.4.16 but using the seminorms of $\cup_m \Sigma^m_{1,\lambda}$. This is left to the reader and concludes the proof of Proposition 6.4.26. □

Theorem 6.4.27. *We assume that $\rho \in (0,1]$. Let $A_\lambda : \mathcal{S}(\mathbb{R}^n) \to \mathcal{S}'(\mathbb{R}^n)$, $\lambda \in \mathbb{R}\backslash\{0\}$, be a family of linear continuous operators.*

We assume that for every $\alpha_1, \alpha_2 \in \mathbb{N}_0^n$ all the operators

$$|\lambda|^{-\frac{|\alpha_1|+|\alpha_2}{2}} (\mathrm{ad}u)^{\alpha_1} (\mathrm{ad}\partial_u)^{\alpha_2} A_\lambda, \quad \lambda \in \mathbb{R}\backslash\{0\},$$

are λ-uniformly in $\mathscr{L}(L^2(\mathbb{R}^n), \mathcal{Q}_{-m+\rho(|\alpha_1|+|\alpha_2|)})$. This means that

$$\sup_{\lambda\in\mathbb{R}\backslash\{0\}} |\lambda|^{-\frac{|\alpha_1|+|\alpha_2}{2}} \|(\mathrm{ad}u)^{\alpha_1} (\mathrm{ad}\partial_u)^{\alpha_2} A_\lambda\|_{\mathscr{L}(L^2(\mathbb{R}^n), \mathcal{Q}_{-m+\rho(|\alpha_1|+|\alpha_2|)})} < \infty. \quad (6.56)$$

Then $A_\lambda \in \Psi\Sigma^m_{\rho,\lambda}(\mathbb{R}^n)$. Moreover, for any $\ell \in \mathbb{N}$, there exist a constant C and an integer ℓ', both independent of $\{A_{\lambda'}\}$ and λ, such that

$$\|A_\lambda\|_{\Psi\Sigma^m_{\rho,\lambda},\ell} \le C \sum_{|\alpha_1|+|\alpha_2|\le\ell'} |\lambda|^{-\frac{|\alpha_1|+|\alpha_2}{2}} \|(\mathrm{ad}u)^{\alpha_1} (\mathrm{ad}\partial_u)^{\alpha_2} A_\lambda\|_{\mathscr{L}(L^2(\mathbb{R}^n), \mathcal{Q}_{-m+\rho(|\alpha_1|+|\alpha_2|)})}.$$

Proof. The proof follows exactly the same steps as the proof of Theorem 6.4.18 using the calculi $\cup_m \Sigma^m_{\rho,\lambda}(\mathbb{R}^n)$ to give the uniformity in λ. This is left to the reader. □

The converse is true from the λ-Shubin calculus: if $A_\lambda : \mathcal{S}(\mathbb{R}^n) \to \mathcal{S}'(\mathbb{R}^n)$, $\lambda \in \mathbb{R}\backslash\{0\}$, is uniformly in $\Psi\Sigma^m_{\rho,\lambda}(\mathbb{R}^n)$ in the sense that

$$\forall N \in \mathbb{N}_0 \quad \sup_{\lambda\in\mathbb{R}\backslash\{0\}} \|A_\lambda\|_{\Psi\Sigma^m_{\rho,\lambda},N} < \infty, \quad (6.57)$$

then (6.56) holds for every $\alpha_1, \alpha_2 \in \mathbb{N}_0^n$.

Proceeding as for Corollary 6.4.20, we obtain

Corollary 6.4.28. *We assume that $\rho \in (0,1]$. Let $A_\lambda : \mathcal{S}(\mathbb{R}^n) \to \mathcal{S}'(\mathbb{R}^n)$, $\lambda \in \mathbb{R}\backslash\{0\}$, be a family of linear continuous operators.*

The family of operators $\{A_\lambda, \lambda \in \mathbb{R}\backslash\{0\}\}$ is uniformly in $\Psi\Sigma^m_{\rho,\lambda}(\mathbb{R}^n)$ in the sense of (6.57) if and only if there exists $\gamma_o \in \mathbb{R}$ such that for each $\alpha_1, \alpha_2 \in \mathbb{N}_0^n$,

$$\sup_{\lambda\in\mathbb{R}\backslash\{0\}} |\lambda|^{-\frac{|\alpha_1|+|\alpha_2}{2}} \|(\mathrm{ad}u)^{\alpha_1} (\mathrm{ad}\partial_u)^{\alpha_2} A_\lambda\|_{\mathscr{L}(\mathcal{Q}_{\gamma_o}(\mathbb{R}^n), \mathcal{Q}_{-m+\rho(|\alpha_1|+|\alpha_2|)+\gamma_o})} < \infty.$$

In this case this property is also true for every $\gamma \in \mathbb{R}$. Moreover, for any $\gamma \in \mathbb{R}$ and $\ell \in \mathbb{N}$, there exist a constant C and an integer ℓ', both independent of $\{A_{\lambda'}\}$ and λ, such that

$$\|A_\lambda\|_{\Psi\Sigma^m_{\rho,\lambda},\ell} \le C \sum_{|\alpha|+|\alpha_2|\le\ell'} |\lambda|^{-\frac{|\alpha_1|+|\alpha_2}{2}} \|(\mathrm{ad}u)^{\alpha_1} (\mathrm{ad}\partial_u)^{\alpha_2} A_\lambda\|_{\mathscr{L}(\mathcal{Q}_{\gamma}(\mathbb{R}^n), \mathcal{Q}_{-m+\rho(|\alpha_1|+|\alpha_2|)+\gamma})}.$$

6.5 Quantization and symbol classes $S^m_{\rho,\delta}$ on the Heisenberg group

We recall that in Section 5.2.2 we have introduced symbol classes $S^m_{\rho,\delta}(G)$ for general graded Lie groups G. In particular, this yields symbol classes $S^m_{\rho,\delta}(\mathbb{H}_n)$ for the particular case of $G = \mathbb{H}_n$. In this section, working with Schrödinger representations π_λ, we obtain a characterisation of these symbol classes $S^m_{\rho,\delta}(\mathbb{H}_n)$ in terms of scalar-valued symbols which will depend on the parameter $\lambda \in \mathbb{R}\backslash\{0\}$; these symbols will be called λ-symbols. The dependence on λ will be of crucial importance here.

We start by adapting the notation of the general construction described in Chapter 5 to the case of the Heisenberg group $\mathbb{H}_n$. It will be convenient to change slightly the notation with respect to the general case. Firstly we want to keep the letter x for denoting part of the coordinates of the Heisenberg group and we choose to denote the general element of the Heisenberg group by, e.g.,

$$g = (x, y, t) \in \mathbb{H}_n.$$

Secondly we may define a symbol as parametrised by

$$\sigma(g, \lambda) := \sigma(g, \pi_\lambda), \quad (g, \lambda) \in \mathbb{H}_n \times \mathbb{R}\backslash\{0\}.$$

Thirdly we modify the indices $\alpha \in \mathbb{N}_0^{2n+1}$ in order to write them as

$$\alpha = (\alpha_1, \alpha_2, \alpha_3),$$

with

$$\alpha_1 = (\alpha_{1,1}, \ldots, \alpha_{1,n}) \in \mathbb{N}_0^n, \quad \alpha_2 = (\alpha_{2,1}, \ldots, \alpha_{2,n}) \in \mathbb{N}_0^n, \quad \alpha_3 \in \mathbb{N}_0.$$

The homogeneous degree of α is then

$$[\alpha] = |\alpha_1| + |\alpha_2| + 2\alpha_3.$$

6.5.1 Quantization on the Heisenberg group

Here we summarise the quantization formula of Section 5.1.3 and its consequences in the particular setting of the Heisenberg group $\mathbb{H}_n$.

As introduced in Definition 5.1.33, a symbol is given by a field of operators

$$\sigma = \{\sigma(g, \lambda) : \mathcal{S}(\mathbb{R}^n) \to L^2(\mathbb{R}^n), (g, \lambda) \in \mathbb{H}_n \times (\mathbb{R}\backslash\{0\})\},$$

satisfying (quite weak) properties so that the quantization makes sense. More rigorously, we require that, for each $\beta \in \mathbb{N}_0^{2n+1}$, the map $g \longmapsto \partial_g^\beta \sigma(g, \lambda)$ is continuous from $\mathbb{H}_n$ to some $L^\infty_{a,b}(\widehat{\mathbb{H}}_n)$.

Recall now, that on the Heisenberg group $\mathbb{H}_n$, the Plancherel measure is given by $c_n|\lambda|^n d\lambda$ (see Proposition 6.2.7). By Theorem 5.1.39, the quantization of a symbol σ as above is the operator

$$A = \mathrm{Op}(\sigma)$$

given by

$$A\phi(g) = c_n \int_{\mathbb{R}\backslash\{0\}} \mathrm{Tr}\left(\pi_\lambda(g)\ \sigma(g,\lambda)\ \widehat{\phi}(\pi_\lambda)\right) |\lambda|^n d\lambda, \tag{6.58}$$

for any $\phi \in \mathcal{S}(\mathbb{H}_n)$ and $g = (x, y, t) \in \mathbb{H}_n$.

Note that, by (1.5), we have

$$\widehat{\varphi}(\pi_\lambda)\,\pi_\lambda(g) = \mathcal{F}_{\mathbb{H}_n}(\varphi(g\,\cdot))(\pi_\lambda),$$

thus the properties of the trace imply that

$$\mathrm{Tr}\left(\pi_\lambda(g)\sigma(g,\lambda)\widehat{\phi}(\pi_\lambda)\right) = \mathrm{Tr}\left(\sigma(g,\lambda)\ \mathcal{F}_{\mathbb{H}_n}(\varphi(g\,\cdot))(\pi_\lambda)\right). \tag{6.59}$$

Furthermore, by (6.20), we have

$$\mathcal{F}_{\mathbb{H}_n}(\varphi(g\,\cdot))(\pi_\lambda) = (2\pi)^{\frac{2n+1}{2}}\mathrm{Op}^W\left[\mathcal{F}_{\mathbb{R}^{2n+1}}(\varphi(g\,\cdot))(\sqrt{|\lambda|}\,\cdot, \sqrt{\lambda}\,\cdot, \lambda)\right]. \tag{6.60}$$

This formula shows that the Weyl quantization is playing an important role in the quantization (6.58) due to its close relation to the group Fourier transform on the Heiseneberg group.

Now, for each $(g,\lambda) \in \mathbb{H}_n \times (\mathbb{R}\backslash\{0\})$, each operator $\sigma(g,\lambda) : \mathcal{S}(\mathbb{R}^n) \to L^2(\mathbb{R}^n)$ in the symbol σ can also be written as the Weyl quantization of some symbol on the Euclidean space $\mathbb{R}^n$, depending on (g,λ). In other words, we can think of the symbol σ as

$$\sigma(g,\lambda) = \mathrm{Op}^W(a_{g,\lambda}), \tag{6.61}$$

where $a = \{a(g,\lambda,\xi,u) = a_{g,\lambda}(\xi,u)\}$ is a function on $\mathbb{H}_n \times \mathbb{R}\backslash\{0\} \times \mathbb{R}^n \times \mathbb{R}^n$. This scalar-valued symbol a will be called the λ-*symbol* of the operator A in (6.58).

In other words, the symbol of the operator A acting on the Heisenberg group is σ, related to A by the quantization formula (6.58). For each (g,λ), the symbol $\sigma_{g,\lambda}$ is itself an operator mapping the Schwartz space $\mathcal{S}(\mathbb{R}^n)$ to $L^2(\mathbb{R}^n)$. So, the λ-symbol a of the operator A is given by the collection of the Weyl symbols $a_{g,\lambda}$ of $\sigma(g,\lambda)$.

Note that if $A \in \Psi^m_{\rho,\delta}$, then its symbol acts on smooth vectors so $\sigma_{g,\lambda}$ is itself an operator mapping the Schwartz space $\mathcal{S}(\mathbb{R}^n)$ to itself, for each (g,λ).

Consequently, using (6.59), we can rewrite our quantization given in (6.58), now using only Euclidean objects, as

$$\begin{aligned} &A\varphi(g) \qquad (6.62)\\ &= c_n' \int_{\mathbb{R}\backslash\{0\}} \operatorname{Tr}\left(\pi_\lambda(g)\operatorname{Op}^W(a_{g,\lambda})\operatorname{Op}^W\left[\mathcal{F}_{\mathbb{R}^{2n+1}}(\varphi)(\sqrt{|\lambda|}\,\cdot,\sqrt{\lambda}\,\cdot,\lambda)\right]\right)|\lambda|^n d\lambda\\ &= c_n' \int_{\mathbb{R}\backslash\{0\}} \operatorname{Tr}\left(\operatorname{Op}^W(a_{g,\lambda})\operatorname{Op}^W\left[\mathcal{F}_{\mathbb{R}^{2n+1}}(\varphi(g\,\cdot))(\sqrt{|\lambda|}\,\cdot,\sqrt{\lambda}\,\cdot,\lambda)\right]\right)|\lambda|^n d\lambda, \end{aligned}$$

with $c_n' = c_n(2\pi)^{n+\frac{1}{2}} = (2\pi)^{-2n-\frac{1}{2}}$.

In Definition 5.2.11 we have introduced the symbol classes $S^m_{\rho,\delta}(G)$ for general graded Lie groups G. Now, in the particular case $G = \mathbb{H}_n$ of the Heisenberg group, using the relation (6.61) between symbols σ and a, we can ask the following question:

what does the condition $\sigma \in S^m_{\rho,\delta}(\mathbb{H}_n)$ mean in terms of the λ-symbol $a_{g,\lambda}$?

This question will be answered in the following sections.

6.5.2 An equivalent family of seminorms on $S^m_{\rho,\delta} = S^m_{\rho,\delta}(\mathbb{H}_n)$

We now follow Definition 5.2.11 to define the symbol class

$$S^m_{\rho,\delta} = S^m_{\rho,\delta}(\mathbb{H}_n).$$

As positive Rockland operator, we will use $\mathcal{R} = -\mathcal{L}$ where $\mathcal{L}$ is the (canonical) sub-Laplacian given in (6.5). We realise almost all the elements of $\widehat{\mathbb{H}}_n$ via their representatives given by the Schrödinger representations π_λ, $\lambda \in \mathbb{R}\backslash\{0\}$, which all act on

$$\mathcal{H}_{\pi_\lambda} = L^2(\mathbb{R}^n),$$

see Section 6.2. Therefore, our symbol class on $\mathbb{H}_n$ is defined by the following family of seminorms

$$\|\sigma\|_{S^m_{\rho,\delta},a,b,c} := \sup_{\lambda\in\mathbb{R}\backslash\{0\},\, g\in\mathbb{H}_n} \|\sigma(g,\lambda)\|_{S^m_{\rho,\delta},a,b,c}, \qquad a,b,c \in \mathbb{N}_0,$$

where

$$\begin{aligned} &\|\sigma(g,\lambda)\|_{S^m_{\rho,\delta},a,b,c}\\ &:= \sup_{\substack{[\alpha]\leq a\\ [\beta]\leq b,\,|\gamma|\leq c}} \|\pi_\lambda(\mathrm{I}-\mathcal{L})^{\frac{\rho[\alpha]-m-\delta[\beta]+\gamma}{2}} X_g^\beta \Delta^\alpha \sigma(g,\lambda)\pi_\lambda(\mathrm{I}-\mathcal{L})^{-\frac{\gamma}{2}}\|_{\mathscr{L}(L^2(\mathbb{R}^n))}. \end{aligned}$$

Here the difference operators Δ^α correspond to the family of operators $\Delta_{\tilde{q}_\alpha}$ where the q_α's are the polynomials appearing in the Taylor expansion. See Example 5.2.4 for some explicit formulae.

By Remark 5.2.13 (4), we can also use the canonical basis

$$x^{\alpha_1} y^{\alpha_2} t^{\alpha_3}, \quad \alpha = (\alpha_1, \alpha_2, \alpha_3) \in \mathbb{N}^{2n+1} = \mathbb{N}_0^n \times \mathbb{N}_0^n \times \mathbb{N}_0,$$

where

$$x^{\alpha_1} = x_1^{\alpha_{11}} \dots x_n^{\alpha_{1n}}, \quad y^{\alpha_2} = y_1^{\alpha_{21}} \dots y_n^{\alpha_{2n}}.$$

We define

$$\Delta'^\alpha := \Delta_{x^{\alpha_1} y^{\alpha_2} t^{\alpha_3}}, \quad \alpha \in \mathbb{N}^{2n+1}.$$

In this case, for any $\alpha, \beta \in \mathbb{N}_0^{2n+1}$, we have

$$\Delta'^{\alpha+\beta} = \Delta'^\alpha \Delta'^\beta.$$

An equivalent family of seminorms on $S^m_{\rho,\delta}$ using the difference operators Δ'^α is given by

$$\|\sigma\|'_{S^m_{\rho,\delta},a,b,c} := \sup_{\lambda \in \mathbb{R}\setminus\{0\},\, g \in \mathbb{H}_n} \|\sigma(g,\lambda)\|'_{S^m_{\rho,\delta},a,b,c}, \quad a,b,c \in \mathbb{N}_0,$$

where

$$\|\sigma(g,\lambda)\|'_{S^m_{\rho,\delta},a,b,c} := \sup_{\substack{[\alpha]\leq a \\ [\beta]\leq b,\ |\gamma|\leq c}} \|\pi_\lambda(\mathrm{I}-\mathcal{L})^{\frac{\rho[\alpha]-m-\delta[\beta]+\gamma}{2}} X_g^\beta \Delta'^\alpha \sigma(g,\lambda) \pi_\lambda(\mathrm{I}-\mathcal{L})^{-\frac{\gamma}{2}}\|_{\mathscr{L}(L^2(\mathbb{R}^n))}.$$

Although the difference operators which intervene in the asymptotic expansions of the composition and the adjoint properties are the difference operators Δ^α, the operators Δ'_α are more handy for the computations to follow.

6.5.3 Characterisation of $S^m_{\rho,\delta}(\mathbb{H}_n)$

In this section we describe the symbol classes $S^m_{\rho,\delta}(\mathbb{H}_n)$ from Section 5.2.2 (more specifically, from Definition 5.2.11) in terms of scalar-valued λ-symbols. More precisely, we show that the symbols $\sigma = \{\sigma(g,\lambda)\}$ in $S^m_{\rho,\delta}$ are all of the form

$$\sigma(g,\lambda) = \mathrm{Op}^W(a_{g,\lambda}(\xi,u)), \tag{6.63}$$

with the λ-symbol $a_{g,\lambda}$ satisfying some properties described below in terms of the family of λ-Shubin classes described in Section 6.4.4 and of the operator $\tilde{\partial}_{\lambda,\xi,u}$ defined in (6.43).

Theorem 6.5.1. *Let $m, \rho, \delta \in \mathbb{R}$ with $1 \geq \rho \geq \delta \geq 0$, $\rho \neq 0$, $\delta \neq 1$. If $\sigma = \{\sigma(g, \lambda)\}$ is in $S^m_{\rho,\delta}$ then there exists a unique smooth function $a = \{a(g, \lambda, \xi, u) = a_{g,\lambda}(\xi, u)\}$ on $\mathbb{H}_n \times \mathbb{R}\backslash\{0\} \times \mathbb{R}^n \times \mathbb{R}^n$ such that*

$$\sigma(g, \lambda) = \mathrm{Op}^W(a_{g,\lambda}), \tag{6.64}$$

with $\tilde{\partial}^{\alpha_3}_{\lambda,\xi,u} X^\beta_g a_{g,\lambda} \in \Sigma^{m-2\rho\alpha_3+\delta[\beta]}_{\rho,\lambda}(\mathbb{R}^n)$ for each $(g, \lambda) \in \mathbb{H}_n \times \mathbb{R}\backslash\{0\}$ satisfying

$$\sup_{(g,\lambda)\in\mathbb{H}_n\times\mathbb{R}\backslash\{0\}} \|\tilde{\partial}^{\alpha_3}_{\lambda,\xi,u} X^\beta_g a_{g,\lambda}\|_{\Sigma^{m-2\rho\alpha_3+\delta[\beta]}_{\rho,\lambda}(\mathbb{R}^n),N} < \infty, \tag{6.65}$$

for every $N \in \mathbb{N}_0$. More precisely, for every $N \in \mathbb{N}_0$ there exist $C > 0$ and a, b, c such that

$$\sup_{(g,\lambda)\in\mathbb{H}_n\times\mathbb{R}\backslash\{0\}} \|\tilde{\partial}^{\alpha_3}_{\lambda,\xi,u} X^\beta_g a_{g,\lambda}\|_{\Sigma^{m-2\rho\alpha_3+\delta[\beta]}_{\rho,\lambda}(\mathbb{R}^n),N} \leq C\|\sigma\|_{S^m_{\rho,\lambda}(\mathbb{H}_n),a,b,c}.$$

Conversely, if $a = \{a(g, \lambda, \xi, u) = a_{g,\lambda}(\xi, u)\}$ is a smooth function on $\mathbb{H}_n \times \mathbb{R}\backslash\{0\} \times \mathbb{R}^n \times \mathbb{R}^n$ satisfying (6.65) for every $N \in \mathbb{N}_0$, then there exists a unique symbol $\sigma \in S^m_{\rho,\delta}$ such that (6.64) holds. Furthermore, for every a, b, c there exists $C > 0$ and $N \in \mathbb{N}_0$ such that

$$\|\sigma\|_{S^m_{\rho,\lambda}(\mathbb{H}_n),a,b,c} \leq C \sup_{(g,\lambda)\in\mathbb{H}_n\times\mathbb{R}\backslash\{0\}} \|\tilde{\partial}^{\alpha_3}_{\lambda,\xi,u} X^\beta_g a_{g,\lambda}\|_{\Sigma^{m-2\rho\alpha_3+\delta[\beta]}_{\rho,\lambda}(\mathbb{R}^n),N}.$$

In other words, Theorem 6.5.1 shows that

$$\sigma \in S^m_{\rho,\delta}(\mathbb{H}_n)$$

is equivalent to

$$\sigma(g, \lambda) = \mathrm{Op}^W(a_{g,\lambda}),$$

for each (g, λ) with $a_{g,\lambda} \in C^\infty(\mathbb{R}^{2n})$ satisfying

$$\forall\alpha \in \mathbb{N}^{2n+1}_0 \quad \exists C > 0 \quad \forall(g, \lambda) \in \mathbb{H}_n\times(\mathbb{R}\backslash\{0\}) \quad \forall(\xi, u) \in \mathbb{R}^{2n}$$
$$|\partial^{\alpha_1}_\xi \partial^{\alpha_2}_u \tilde{\partial}^{\alpha_3}_{\lambda,\xi,u} X^\beta_g a_{g,\lambda}(\xi, u)| \leq C|\lambda|^{\rho\frac{|\alpha_1|+|\alpha_2|}{2}} \left(1 + |\lambda|(1 + |\xi|^2 + |u|^2)\right)^{\frac{m-\rho[\alpha]+\delta[\beta]}{2}}.$$

Choosing a rescaled Weyl symbol as in Lemma 6.3.10, we see that

$$\sigma \in S^m_{\rho,\delta}(\mathbb{H}_n)$$

is equivalent to

$$\sigma(g, \lambda) = \mathrm{Op}^W\left(\tilde{a}_{g,\lambda}(\sqrt{|\lambda|}\xi, \sqrt{\lambda}u)\right),$$

for each (g, λ) with $\tilde{a}_{g,\lambda} \in C^\infty(\mathbb{R}^{2n})$ satisfying

$$\forall\alpha \in \mathbb{N}^{2n+1}_0 \quad \exists C > 0 \quad \forall(g, \lambda) \in \mathbb{H}_n\times(\mathbb{R}\backslash\{0\}) \quad \forall(\xi, u) \in \mathbb{R}^{2n}$$
$$|\partial^{\alpha_1}_\xi \partial^{\alpha_2}_u \partial^{\alpha_3}_\lambda X^\beta_g \tilde{a}_{g,\lambda}(\xi, u)| \leq C\left(1 + |\lambda| + |\xi|^2 + |u|^2\right)^{\frac{m-\rho[\alpha]+\delta[\beta]}{2}}.$$

Note that, by (6.20),

$$\tilde{a}_{g,\lambda}(\xi,u) = (2\pi)^{\frac{2n+1}{2}} \mathcal{F}_{\mathbb{R}^{2n+1}}(\kappa_g)(\xi,u,\lambda)$$

where $\{\kappa_g(x,y,t)\}$ is the kernel of the symbol $\{\sigma(g,\lambda)\}$, i.e.

$$\sigma(g,\lambda) = \pi_\lambda(\kappa_g),$$

(see Definition 5.1.36).

Proof of Theorem 6.5.1. Let $\sigma \in S^m_{\rho,\delta}$. This means that for each $\alpha,\beta \in \mathbb{N}_0^{2n+1}$ and $\gamma \in \mathbb{R}$ we have

$$\pi_\lambda(\mathrm{I}-\mathcal{L})^{\frac{\rho[\alpha]-m-\delta[\beta]+\gamma}{2}} X_g^\beta \Delta'^\alpha \sigma(g,\lambda) \pi_\lambda(\mathrm{I}-\mathcal{L})^{-\frac{\gamma}{2}} \in \mathscr{L}(L^2(\mathbb{R}^n)),$$

with operator norm uniformly bounded with respect to λ, or equivalently, (see the formulae in Section 6.3.3),

$$|\lambda|^{-\frac{|\alpha_1|+|\alpha_2}{2}} \|(\mathrm{ad}u)^{\alpha_1}(\mathrm{ad}\partial_u)^{\alpha_2} X_g^\beta \Delta_3'^{\alpha_3} \sigma(g,\lambda) h\|_{\mathcal{Q}_{\rho[\alpha]-m-\delta[\beta]+\gamma,\lambda}} \leq C\|h\|_{\mathcal{Q}_{\gamma,\lambda}}$$

with $C = C_{\alpha,\beta,\gamma}$ independent of λ. Taking $\gamma = 0$, we see that the λ-family of $X_g^\beta \Delta_3'^{\alpha_3} \sigma(g,\lambda)$ satisfies the hypotheses of Theorem 6.4.27. For $\beta = \alpha_3 = 0$, this shows that $\sigma(g,\lambda) = \mathrm{Op}^W(a_{g,\lambda})$ with $a_{g,\lambda} \in \Sigma^m_{\rho,\lambda}$ uniformly in λ. For any β and α_3, this shows that the λ-family of

$$X_g^\beta \Delta_3'^{\alpha_3} \sigma(g,\lambda) = i^{\alpha_3} \mathrm{Op}^W(X_g^\beta \tilde{\partial}^{\alpha_3}_{\lambda,\xi,u} a_{g,\lambda})$$

(see the formulae in Section 6.3.3, or equivalently Corollary 6.3.9) also satisfies the hypotheses of Theorem 6.4.27. Therefore, $X_g^\beta \tilde{\partial}^{\alpha_3}_{\lambda,\xi,u} a_{g,\lambda}$ is in $\Sigma^{m-2\rho[\alpha_3]+\delta[\beta]}_{\rho,\lambda}$ uniformly in λ. This proves the first part of the statement.

The converse follows from the Shubin calculi depending on λ. □

The proof above shows that we can always assume $\gamma = 0$ in the definition of a class of symbols. But we could have fixed any γ and use Corollary 6.4.28 instead of Theorem 6.4.27 in the proof above. This shows:

Corollary 6.5.2. *A symbol $\sigma = \{\sigma(g,\lambda)\}$ is in $S^m_{\rho,\delta}$ if and only if there exists* <u>*one*</u> *$\gamma \in \mathbb{R}$ such that for every $\alpha,\beta \in \mathbb{N}_0^{2n+1}$ the quantity*

$$\sup_{\lambda\in\mathbb{R}\setminus\{0\}} \|\pi_\lambda(\mathrm{I}-\mathcal{L})^{\frac{\rho[\alpha]-m-\delta[\beta]+\gamma}{2}} X_g^\beta \Delta'^\alpha \sigma(g,\lambda) \pi_\lambda(\mathrm{I}-\mathcal{L})^{-\frac{\gamma}{2}}\|_{\mathscr{L}(L^2(\mathbb{R}^n))} \tag{6.66}$$

is finite.

In this case the quantity (6.66) is finite for every $\gamma \in \mathbb{R}$ and $\alpha,\beta \in \mathbb{N}_0^n$.

Furthermore, for any $\gamma_o \in \mathbb{R}$ fixed, an equivalent family of seminorms for $S^m_{\rho,\delta}$ is given by

$$\sigma \longmapsto \sup_{\lambda\in\mathbb{R}\setminus\{0\},[\alpha]\leq a,\,[\beta]\leq b} \|\pi_\lambda(\mathrm{I}-\mathcal{L})^{\frac{\rho[\alpha]-m-\delta[\beta]+\gamma_o}{2}} X_g^\beta \Delta'^\alpha \sigma(g,\lambda) \pi_\lambda(\mathrm{I}-\mathcal{L})^{-\frac{\gamma_o}{2}}\|_{\mathscr{L}(L^2(\mathbb{R}^n))}$$

with $a,b \in \mathbb{N}_0$.

6.6 Parametrices

In this section, we present conditions for the ellipticity and hypoellipticity in the setting of the Heisenberg group as a special case of those presented in Sections 5.8.1 and 5.8.3. In particular, we can also derive conditions in terms of the λ-symbols discussed in Section 6.5.3.

6.6.1 Condition for ellipticity

We start by providing conditions on the λ-symbol ensuring that the assumptions for the ellipticity in Definition 5.8.1 and in Theorem 5.8.7 are satisfied.

Theorem 6.6.1. *Let $m \in \mathbb{R}$ and $1 \geq \rho > \delta \geq 0$. Let $\sigma = \{\sigma(g, \lambda)\}$ be in $S^m_{\rho,\delta}(\mathbb{H}_n)$ with*

$$\sigma(g, \lambda) = \mathrm{Op}^W (a_{g,\lambda})$$

as in Theorem 6.5.1. Assume that there are $R \in \mathbb{R}$ and $C > 0$ such that for any $(\xi, u) \in \mathbb{R}^{2n}$ and $\lambda \neq 0$ satisfying $|\lambda|(|\xi|^2 + |u|^2) \geq R$ we have

$$|a_{g,\lambda}(\xi, u)| \geq C \left(1 + |\lambda|(1 + |\xi|^2 + |u|^2)\right)^{\frac{m}{2}}. \tag{6.67}$$

Then there exists Λ such that σ is $(-\mathcal{L}, \Lambda, m)$-elliptic in the sense of Definition 5.8.1. Thus it satisfies the hypotheses of Theorem 5.8.7 and we can construct a left parametrix $B \in \Psi^{-m}_{\rho,\delta}$ for the operator $A = \mathrm{Op}(\sigma)$, that is, there exists $B \in \Psi^{-m}_{\rho,\delta}$ such that

$$BA - \mathrm{I} \in \Psi^{-\infty}.$$

Proof. Let $\chi \in C^\infty(\mathbb{R})$ be such that $0 \leq \chi \leq 1$ with $\chi = 0$ on $(-\infty, R)$ and $\chi = 1$ on $[2R, +\infty)$. We set for any $(\xi, u) \in \mathbb{R}^{2n}$ and $\lambda \neq 0$

$$b_{\lambda,g}(\xi, u) := \frac{\chi(|\lambda|(|\xi|^2 + |u|^2))}{a_{g,\lambda}(\xi, u)}.$$

Using the properties of a, one check easily that this defines a symbol $b_{\lambda,g}$ with $b_{\lambda,g} \in \Sigma^{-m}_{\rho,\lambda}$, and more precisely for every $N \in \mathbb{N}_0$ there exist $C > 0$ and a, b, c all independent on λ or g such that

$$\sup_{(g,\lambda)\in\mathbb{H}_n\times\mathbb{R}\setminus\{0\}} \|\tilde{\partial}^{\alpha_3}_{\lambda,\xi,u} X^\beta_g b_{g,\lambda}\|_{\Sigma^{-m-2\rho\alpha_3+\delta[\beta]}_{\rho,\lambda}(\mathbb{R}^n),N} \leq C\|\sigma\|_{S^m_{\rho,\lambda},a,b,c} := C'.$$

By the properties of uniform families of Weyl-Hörmander metrics (see Proposition 6.4.22), we have

$$\mathrm{Op}^W(b_{\lambda,g})\mathrm{Op}^W(a_{g,\lambda}) = \mathrm{Op}^W(\chi(|\lambda|(|\xi|^2 + |u|^2))) + E_{\lambda,g} = \mathrm{I} + \tilde{E}_{\lambda,g} \tag{6.68}$$

with

$$\|\tilde{\partial}^{\alpha_3}_{\lambda,\xi,u} X^\beta_g E_{\lambda,g}\|_{\Psi\Sigma^{-\rho-2\rho\alpha_3+\delta[\beta]}_{\rho,\lambda}(\mathbb{R}^n),N} \leq C_1\|\sigma\|_{S^m,a_1,b_1,c_1} := C_1', \tag{6.69}$$

and similarly for the 'error' term $\tilde{E}_{\lambda,g}$. Also we have

$$\|\pi_\lambda(\mathrm{I}-\mathcal{L})^{\frac{m}{2}}\mathrm{Op}^W(b_{\lambda,g})\|_{\mathscr{L}(L^2(\mathbb{R}^n))} \leq C_2\|\sigma\|_{S^m,a_2,b_2,c_2} := C_2'. \tag{6.70}$$

In Estimates (6.69) and (6.70), the constants C_1 and C_2, and the parameters $a_1, b_1, c_1, a_2, b_2, c_2$ do not depend on λ, g or σ. By (6.70), we have

$$C_2'\|\mathrm{Op}^W(a_{g,\lambda})u\|_{L^2(\mathbb{R}^n)} \geq \|\pi_\lambda(\mathrm{I}-\mathcal{L})^{\frac{m}{2}}\mathrm{Op}^W(b_{\lambda,g})\mathrm{Op}^W(a_{g,\lambda})u\|_{L^2(\mathbb{R}^n)}.$$

We now use (6.68) on the right hand-side and the reverse triangle inequality to obtain

$$\begin{aligned}
&\|\pi_\lambda(\mathrm{I}-\mathcal{L})^{\frac{m}{2}}\mathrm{Op}^W(b_{\lambda,g})\mathrm{Op}^W(a_{g,\lambda})u\|_{L^2(\mathbb{R}^n)}\\
&\quad= \|\pi_\lambda(\mathrm{I}-\mathcal{L})^{\frac{m}{2}}\left(\mathrm{I}+\tilde{E}_{\lambda,g}\right)u\|_{L^2(\mathbb{R}^n)}\\
&\quad\geq \|\pi_\lambda(\mathrm{I}-\mathcal{L})^{\frac{m}{2}}u\|_{L^2(\mathbb{R}^n)} - \|\pi_\lambda(\mathrm{I}-\mathcal{L})^{\frac{m}{2}}\tilde{E}_{\lambda,g}u\|_{L^2(\mathbb{R}^n)}.
\end{aligned}$$

We can write the last term as

$$\|\pi_\lambda(\mathrm{I}-\mathcal{L})^{\frac{m}{2}}\tilde{E}_{\lambda,g}u\|_{L^2(\mathbb{R}^n)} = \|U_{\lambda,g}\pi_\lambda(\mathrm{I}-\mathcal{L})^{\frac{m-\rho}{2}}u\|_{L^2(\mathbb{R}^n)}.$$

with $U_{\lambda,g} := \pi_\lambda(\mathrm{I}-\mathcal{L})^{\frac{m}{2}}\tilde{E}_{\lambda,g}\pi_\lambda(\mathrm{I}-\mathcal{L})^{\frac{-m+\rho}{2}}$ of order 0 and, therefore, bounded on $L^2(\mathbb{R}^n)$ satisfying

$$\|U_{\lambda,g}\|_{\mathscr{L}(L^2(\mathbb{R}^n))} \leq C_3\|\sigma\|_{S^m_{\rho,\delta},a_3,b_3,c_3} := C_3'.$$

Let us consider $\Lambda \in \mathbb{R}$ and $u \in \mathcal{S}(\mathbb{R}^n)$ with $u \in E_{\pi_\lambda}(\Lambda,\infty)L^2(\mathbb{R}^n)$, then

$$\|\pi_\lambda(\mathrm{I}-\mathcal{L})^{\frac{m-\rho}{2}}u\|_{L^2(\mathbb{R}^n)} \leq (1+\max(\Lambda,0))^{-\frac{\rho}{2}}\|\pi_\lambda(\mathrm{I}-\mathcal{L})^{\frac{m}{2}}u\|_{L^2(\mathbb{R}^n)},$$

thus

$$\begin{aligned}
&\|U_{\lambda,g}\pi_\lambda(\mathrm{I}-\mathcal{L})^{\frac{m-\rho}{2}}u\|_{L^2(\mathbb{R}^n)}\\
&\quad\leq C_3'\|\pi_\lambda(\mathrm{I}-\mathcal{L})^{\frac{m-\rho}{2}}u\|_{L^2(\mathbb{R}^n)}\\
&\quad\leq C_3'(1+\max(\Lambda,0))^{-\frac{\rho}{2}}\|\pi_\lambda(\mathrm{I}-\mathcal{L})^{\frac{m}{2}}u\|_{L^2(\mathbb{R}^n)}.
\end{aligned}$$

We choose $\Lambda \in \mathbb{R}$ such that

$$C_3'(1+\max(\Lambda,0))^{-\frac{\rho}{2}} \leq \frac{1}{2},$$

for example for $\Lambda > 0$, the smallest Λ satisfying the equality. We have obtained

$$\|\pi_\lambda(\mathrm{I}-\mathcal{L})^{\frac{m}{2}}\tilde{E}_{\lambda,g}u\|_2 = \|U_{\lambda,g}\pi_\lambda(\mathrm{I}-\mathcal{L})^{\frac{m-\rho}{2}}u\|_2 \leq \frac{1}{2}\|\pi_\lambda(\mathrm{I}-\mathcal{L})^{\frac{m}{2}}u\|_2.$$

Collecting the estimates, we obtain

$$\begin{aligned}
C_2'\|\mathrm{Op}^W(a_{g,\lambda})u\|_2 &\geq \|\pi_\lambda(\mathrm{I}-\mathcal{L})^{\frac{m}{2}}\mathrm{Op}^W(b_{\lambda,g})\mathrm{Op}^W(a_{g,\lambda})u\|_2\\
&\geq \|\pi_\lambda(\mathrm{I}-\mathcal{L})^{\frac{m}{2}}u\|_2 - \|\pi_\lambda(\mathrm{I}-\mathcal{L})^{\frac{m}{2}}\tilde{E}_{\lambda,g}u\|_2 \geq \frac{1}{2}\|\pi_\lambda(\mathrm{I}-\mathcal{L})^{\frac{m}{2}}u\|_2.
\end{aligned}$$

This shows that σ satisfies (5.79) for $-\mathcal{L}$, Λ and m. □

From the proof, it follows that the choice of Λ depends on ρ, δ, and a bound for a (computable) seminorm of σ in $S^m_{\rho,\delta}$.

We have already proved that, for instance, $\mathrm{I} - \mathcal{L}$ is elliptic for $-\mathcal{L}$, see Proposition 5.8.2.

Here is another example.

Example 6.6.2. On $\mathbb{H}_1$, if $m \in 2\mathbb{N}$ is an even integer, then the operator $X^m + iY^m + T^{m/2} \in \Psi^m$ is elliptic with respect to $-\mathcal{L}$ and of elliptic order m.

Proof. The symbol of $X^m + iY^m + T^{m/2}$ is

$$\begin{aligned} \sigma(\lambda) &= \pi_\lambda(X)^m + i\pi_\lambda(Y)^m + \pi_\lambda(T)^{\frac{m}{2}} \\ &= \left(\mathrm{Op}^W(i\sqrt{|\lambda|}\xi)\right)^m + i\left(\mathrm{Op}^W(i\sqrt{\lambda}u)\right)^m + (i\lambda)^{\frac{m}{2}}, \end{aligned}$$

by (6.28) and (6.11). Hence its λ-symbol is

$$\begin{aligned} a_\lambda(\xi, u) &= \left(i\sqrt{|\lambda|}\xi\right)^m + i\left(i\sqrt{\lambda}u\right)^m + (i\lambda)^{\frac{m}{2}} \\ &= (-1)^{\frac{m}{2}}|\lambda|^{\frac{m}{2}}\left(\xi^m + iu^m + (-(\mathrm{sgn}\lambda)i)^{\frac{m}{2}}\right). \end{aligned}$$

Clearly a_λ satisfies the condition of Theorem 6.6.1. □

6.6.2 Condition for hypoellipticity

We have also proved a general result regarding hypoellipticity in Theorem 5.8.9 (in the sense of the existence of a left parametrix). In the case of the Heisenberg group, we obtain the following sufficient condition on the scalar-valued symbol:

Theorem 6.6.3. *Let $m \in \mathbb{R}$ and $1 \geq \rho > \delta \geq 0$. Let $\sigma = \{\sigma(g,\lambda)\}$ be in $S^m_{\rho,\delta}(\mathbb{H}_n)$ with*

$$\sigma(g,\lambda) = \mathrm{Op}^W(a_{g,\lambda})$$

as in Theorem 6.5.1.

We assume that there is $m_o < m$ such that σ satisfies for a given R, for any $(\xi,u) \in \mathbb{R}^{2n}$ such that $|\lambda|(|\xi|^2 + |u|^2) \geq R$, the inequalities

$$|a_{g,\lambda}(\xi,u)| \geq C\left(1 + |\lambda|(1 + |\xi|^2 + |u|^2)\right)^{\frac{m_o}{2}} \tag{6.71}$$

and

$$\begin{aligned} &|\partial_\xi^{\alpha_1}\partial_u^{\alpha_2}\tilde{\partial}^{\alpha_3}_{\lambda,\xi,u}X_g^\beta a_{g,\lambda}(\xi,u)| \\ &\qquad \leq C_{\alpha,\beta}|\lambda|^{\rho\frac{|\alpha_1|+|\alpha_2|}{2}}\left(1 + |\lambda|(1 + |\xi|^2 + |u|^2)\right)^{\frac{-\rho[\alpha]+\delta[\beta]}{2}}|a_{g,\lambda}(\xi,u)|. \qquad (6.72) \end{aligned}$$

Then $\sigma(g,\lambda)$ satisfies the hypotheses of Theorem 5.8.9 for $-\mathcal{L}$ and m_o. Therefore, we can construct a left parametrix $B \in \Psi^{-m_o}_{\rho,\delta}$ for the operator $A = \mathrm{Op}(\sigma)$, that is, there exists $B \in \Psi^{-m_o}_{\rho,\delta}$ such that

$$BA - \mathrm{I} \in \Psi^{-\infty}.$$

In (6.71) and (6.72), the constants C and $C_{\alpha,\beta}$ are assumed to be independent of λ, ξ, u or g.

For each fixed $\lambda \in \mathbb{R}\backslash\{0\}$, the conditions (6.71) and (6.72) are very close to Shubin's in [Shu87, §25.1]. However Theorem 6.6.3 asks for these conditions to be satisfied uniformly in $\lambda \in \mathbb{R}\backslash\{0\}$.

The proof is in essence an adaptation of the proof of Theorem 6.6.1.

Proof. We choose χ and define $b_{\lambda,g}$ as in the proof of Theorem 6.6.1. This time, $b_{\lambda,g}$ is in $\Sigma^{-m_o}_{\rho,\delta}$, with

$$\sup_{(g,\lambda)\in\mathbb{H}_n\times\mathbb{R}\backslash\{0\}} \|\tilde{\partial}^{\alpha_3}_{\lambda,\xi,u} X^\beta_g b_{g,\lambda}\|_{\Sigma^{-m_o-2\rho\alpha_3+\delta[\beta]}_{\rho,\lambda}(\mathbb{R}^n),N} \le C\|\sigma\|_{S^m_{\rho,\lambda},a,b,c},$$

and

$$\|\pi_\lambda(\mathrm{I}-\mathcal{L})^{\frac{m_o}{2}}\mathrm{Op}^W(b_{\lambda,g})\|_{\mathscr{L}(L^2(\mathbb{R}^n))} \le C_2\|\sigma\|_{S^m_{\rho,\lambda},a_2,b_2,c_2} := C_2'. \tag{6.73}$$

In the proof of Theorem 6.6.1, we developed the product $\mathrm{Op}^W(b_{\lambda,g})\mathrm{Op}^W(a_{g,\lambda})$ at order 0, but here we now develop it up to order M such that the error term is of strictly negative order:

$$\mathrm{Op}^W(b_{\lambda,g})\mathrm{Op}^W(a_{g,\lambda}) = \sum_{m'=0}^{M} \mathrm{Op}^W(d_{m',\lambda,g}) + E_{\lambda,g}, \tag{6.74}$$

where (see (6.17))

$$d_{m',\lambda,g} := c_{m',n} \sum_{|\alpha_1|+|\alpha_2|=m'} \frac{(-1)^{|\alpha_2|}}{\alpha_1!\alpha_2!} \left(\left(\frac{1}{i}\partial_\xi\right)^{\alpha_1} \partial_x^{\alpha_2} b_{\lambda,g}\right)\left(\left(\frac{1}{i}\partial_\xi\right)^{\alpha_2} \partial_x^{\alpha_1} a_{g,\lambda}\right).$$

To fix the idea, we choose $M \in \mathbb{N}_0$ the smallest integer such that

$$m - m_o - 2(M+1)\rho \le -\rho.$$

Using (6.17) and the properties of uniform families of Weyl-Hörmander metrics (see Proposition 6.4.22), the error term satisfies

$$\|\tilde{\partial}^{\alpha_3}_{\lambda,\xi,u} X^\beta_g E_{g,\lambda}\|_{\Psi\Sigma^{-\rho-2\rho\alpha_3+\delta[\beta]}_{\rho,\lambda}(\mathbb{R}^n),N} \le C_1\|\sigma\|_{S^m,a_1,b_1,c_1} := C_1'. \tag{6.75}$$

For the term of order 0, we see that

$$d_{0,\lambda,g} = \chi(|\lambda|(|\xi|^2+|u|^2)) = 1 + (\chi-1)(|\lambda|(|\xi|^2+|u|^2),$$

and clearly the symbol $(\chi-1)(|\lambda|(|\xi|^2+|u|^2)$ is smoothing. For the term of positive order $m'>0$, we can write

$$d_{m',\lambda,g} = c_{m',n}\tilde{d}_{m',\lambda,g} + r_{m',\lambda,g},$$

where

$$\tilde{d}_{m',\lambda,g} := \chi(|\lambda|(|\xi|^2+|u|^2) \sum_{|\alpha_1|+|\alpha_2|=m'} \frac{(-1)^{|\alpha_2|}}{\alpha_1!\alpha_2!} \left(\left(\frac{1}{i}\partial_\xi\right)^{\alpha_1} \partial_x^{\alpha_2}\left\{\frac{1}{a_{g,\lambda}}\right\}\right)\left(\left(\frac{1}{i}\partial_\xi\right)^{\alpha_2}\partial_x^{\alpha_1} a_{g,\lambda}\right),$$

and the small reminder contains all the χ-derivatives, that is, is of the form

$$r_{m',\lambda,g} = \sum_{\substack{\alpha_1'',\alpha_2'' \\ 0<\alpha_1''+\alpha_2''\leq 2M}} \left(\left(\partial_\xi^{\alpha_1''}\partial_x^{\alpha_2''}\right)\chi(|\lambda|(|\xi|^2+|u|^2)\right)\left(\dots\right).$$

Clearly the derivatives of the χ's are smoothing. One can check that the conditions on the symbol a imply that $\tilde{d}_{m',\lambda,g}$ is of order $-2m'\rho$. For example,

$$\left|\partial_{\xi_1} a_{g,\lambda}\partial_{\xi_1}\frac{1}{a_{g,\lambda}}\right| = \left|\frac{\partial_{\xi_1} a_{g,\lambda}}{a_{g,\lambda}}\right|^2 \leq C_{1,0}|\lambda|^\rho\left(1+|\lambda|(1+|\xi|^2+|u|^2)\right)^{-\rho}.$$

We also write

$$\chi(|\lambda|(|\xi|^2+|u|^2) = 1 + (\chi-1)(|\lambda|(|\xi|^2+|u|^2),$$

and the symbol $(\chi-1)(|\lambda|(|\xi|^2+|u|^2)$ is smoothing.

We now incorporate all the terms of order $\leq -\rho$ in a new error term. Indeed, the considerations above show that we can now write

$$\mathrm{Op}^W(b_{\lambda,g})\mathrm{Op}^W(a_{g,\lambda}) = \mathrm{I} + \tilde{E}_{\lambda,g},$$

with $\tilde{E}_{\lambda,g}$ satisfying similar estimates to (6.69).

The end of the proof is now identical to the one of Theorem 6.6.1 with m replaced by m_o. □

Modifying Example 6.6.2, we have the following example of hypoelliptic operators in the sense that they satisfy the hypotheses of Theorem 5.8.9, and therefore admit a left parametrix.

Example 6.6.4. On $\mathbb{H}_1$, if $m, m_o \in 2\mathbb{N}$ are two even integers such that $m \geq m_0$, then the operators

$$X^m + iY^{m_o} + T^{m_o/2} \in \Psi^m \text{ and } X^{m_o} + iY^m + T^{m_o/2} \in \Psi^m$$

satisfy the hypotheses of Theorem 5.8.9 for $-\mathcal{L}$ and m_o.

Proof. The symbols of

$$A_1 := X^m + iY^{m_o} + T^{m_o/2} \quad \text{and} \quad A_2 := X^{m_o} + iY^m + T^{m_o/2},$$

are

$$\begin{aligned}\sigma_{A_1}(\lambda) &= \pi_\lambda(X)^m + i\pi_\lambda(Y)^{m_o} + \pi_\lambda(T)^{\frac{m_o}{2}} \\ &= \left(\mathrm{Op}^W(i\sqrt{|\lambda|}\xi)\right)^m + i\left(\mathrm{Op}^W(i\sqrt{\lambda}u)\right)^{m_o} + (i\lambda)^{\frac{m_o}{2}}, \\ \sigma_{A_2}(\lambda) &= \pi_\lambda(X)^{m_o} + i\pi_\lambda(Y)^m + \pi_\lambda(T)^{\frac{m_o}{2}} \\ &= \left(\mathrm{Op}^W(i\sqrt{|\lambda|}\xi)\right)^{m_o} + i\left(\mathrm{Op}^W(i\sqrt{\lambda}u)\right)^m - (i\lambda)^{\frac{m_o}{2}},\end{aligned}$$

by (6.28) and (6.11). Hence their λ-symbols are

$$\begin{aligned}a_{A_1,\lambda}(\xi,x) &= \left(i\sqrt{|\lambda|}\xi\right)^m + i\left(i\sqrt{\lambda}u\right)^{m_o} + (i\lambda)^{\frac{m_o}{2}}, \\ a_{A_2,\lambda}(\xi,x) &= \left(i\sqrt{|\lambda|}\xi\right)^{m_o} + i\left(i\sqrt{\lambda}u\right)^m + (i\lambda)^{\frac{m_o}{2}}.\end{aligned}$$

From this, it is not difficult to see that $a_{A_j,\lambda}$, $j=1,2$ satisfy

$$|\lambda|\max(|\xi|,|u|) \geq 1 \Longrightarrow |a_{A_j,\lambda}(\xi,u)| \geq C|\lambda|^{m_o}\left(\max(|\xi|,|u|)^{m_o}+1\right),$$

thus they also satisfy (6.71). The other condition in (6.72) of Theorem 6.6.3 is easy to check. □

6.6.3 Subelliptic estimates and hypoellipticity

The sufficient conditions for ellipticity in Theorem 6.6.1, or at least the existence of left parametrix (see Theorem 6.6.3) yield sufficient conditions for subelliptic estimates and hypoellipticity. More precisely, Corollary 5.8.12 and Propositions 5.8.13 and 5.8.15 imply:

Corollary 6.6.5. *Let $m \in \mathbb{R}$ and $1 \geq \rho > \delta \geq 0$. Let $\sigma = \{\sigma(g,\lambda)\}$ be in $S^m_{\rho,\delta}(\mathbb{H}_n)$ with $\sigma(g,\lambda) = \mathrm{Op}^W(a_{g,\lambda})$ as in Theorem 6.5.1.*

(i) Assume that there are $R \in \mathbb{R}$ and $C > 0$ such that for any $(\xi,u) \in \mathbb{R}^{2n}$ and $\lambda \neq 0$ satisfying $|\lambda|(|\xi|^2+|u|^2) \geq R$, we have (6.67), that is,

$$|a_{g,\lambda}(\xi,u)| \geq C\left(1+|\lambda|(1+|\xi|^2+|u|^2)\right)^{\frac{m}{2}}.$$

Then $A = \mathrm{Op}(\sigma) = \mathrm{Op}(\mathrm{Op}^W(a_{g,\lambda}))$ is (locally) hypoelliptic. It is also globally hypoelliptic in the sense of Proposition 5.8.15. The operator A also satisfies the following subelliptic estimates

$$\forall s \in \mathbb{R} \quad \forall N \in \mathbb{R} \quad \exists C > 0 \quad \forall f \in \mathcal{S}(\mathbb{H}_n)$$
$$\|f\|_{L^2_{s+m}} \leq C\left(\|Af\|_{L^2_s} + \|f\|_{L^2_{-N}}\right).$$

(ii) *We assume that there is $m_o < m$ such that σ satisfies for a given R, for any $(\xi, u) \in \mathbb{R}^{2n}$ such that $|\lambda|(|\xi|^2 + |u|^2) \geq R$, the inequalities (6.71) and (6.72), that is,*

$$|a_{g,\lambda}(\xi, u)| \geq C \left(1 + |\lambda|(1 + |\xi|^2 + |u|^2)\right)^{\frac{m_o}{2}},$$

and

$$|\partial_\xi^{\alpha_1} \partial_u^{\alpha_2} \tilde{\partial}_{\lambda,\xi,u}^{\alpha_3} X_g^\beta a_{g,\lambda}(\xi, u)|$$
$$\leq C_{\alpha,\beta} |\lambda|^{\rho \frac{|\alpha_1|+|\alpha_2|}{2}} \left(1 + |\lambda|(1 + |\xi|^2 + |u|^2)\right)^{\frac{-\rho[\alpha]+\delta[\beta]}{2}} |a_{g,\lambda}(\xi, u)|.$$

Then $A = \text{Op}(\sigma) = \text{Op}(\text{Op}^W(a_{g,\lambda}))$ is (locally) hypoelliptic. It is also globally hypoelliptic in the sense of Proposition 5.8.15. The operator A also satisfies the following subelliptic estimates

$$\forall s \in \mathbb{R} \quad \forall N \in \mathbb{R} \quad \exists C > 0 \quad \forall f \in \mathcal{S}(\mathbb{H}_n)$$
$$\|f\|_{L^2_{s+m_o}} \leq C\Big(\|Af\|_{L^2_s} + \|f\|_{L^2_{-N}}\Big).$$

(iii) *In the case $(\rho, \delta) = (1, 0)$, assume that $A \in \Psi^m$ is either elliptic of order $m_0 = m$ or is elliptic of some order m_0 and satisfies the hypotheses of Parts (i) or (ii), respectively. Then A satisfies the subelliptic estimates*

$$\forall s \in \mathbb{R} \quad \forall N \in \mathbb{R} \quad \forall p \in (1, \infty) \quad \exists C > 0 \quad \forall f \in \mathcal{S}(\mathbb{H}_n)$$
$$\|f\|_{L^p_{s+m_o}} \leq C\Big(\|Af\|_{L^p_s} + \|f\|_{L^p_{-N}}\Big).$$

In the estimates above, $\|\cdot\|_{L^p_s}$ denotes any (fixed) Sobolev norm, for example obtained from a (fixed) positive Rockland operator $\mathcal{R}$, such as $\mathcal{R} = -\mathcal{L}$.

Examples

We proceed by giving examples, applying Corollary 5.8.12 to obtain subelliptic estimates for some of the examples of operators encountered in previous sections. First, naturally, we can apply Corollary 6.6.5 to Examples 6.6.2 and 6.6.4, which we now continue.

Example 6.6.2, continued: On $\mathbb{H}_1$, if $m \in 2\mathbb{N}$ is an even integer, then the operator $X^m + iY^m + T^{m/2}$ is hypoelliptic and satisfies the following estimate

$$\forall p \in (1, \infty) \quad \forall s \in \mathbb{R} \quad \forall N \in \mathbb{R} \quad \exists C > 0 \quad \forall f \in \mathcal{S}(\mathbb{H}_1)$$
$$\|f\|_{L^p_{s+m}} \leq C \left(\|(X^m + iY^m + T^{m/2})f\|_{L^p_s} + \|f\|_{L^p_{-N}}\right).$$

Example 6.6.4, continued: Let $m, m_o \in 2\mathbb{N}$ be two even integers such that $m \geq m_0$. Then the differential operators $X^m + iY^{m_o} + T^{m_o/2}$ and $X^{m_o} + iY^m + T^{m_o/2}$ on

$\mathbb{H}_1$ are hypoelliptic and satisfy the following subelliptic estimates

$$\forall p \in (1,\infty) \quad \forall s \in \mathbb{R} \quad \forall N \in \mathbb{R} \quad \exists C>0 \quad \forall f \in \mathcal{S}(\mathbb{H}_1)$$
$$\|f\|_{L^p_{s+m}} \leq C\left(\|(X^m + iY^{m_o} + T^{m_o/2})f\|_{L^p_s} + \|f\|_{L^p_{-N}}\right).$$

and

$$\forall p \in (1,\infty) \quad \forall s \in \mathbb{R} \quad \forall N \in \mathbb{R} \quad \exists C>0 \quad \forall f \in \mathcal{S}(\mathbb{H}_1)$$
$$\|f\|_{L^p_{s+m}} \leq C\left(\|(X^{m_o} + iY^m + T^{m_o/2})f\|_{L^p_s} - \|f\|_{L^p_{-N}}\right).$$

We can also obtain the hypoellipticity and subelliptic estimates for the elliptic operators in Corollary 5.8.16 choosing first the Rockland operator $\mathcal{R} = -\mathcal{L}$:

Corollary 6.6.6. *As usual, $\mathcal{L}$ denotes the canonical sub-Laplacian on the Heisenberg group $\mathbb{H}_n$ (see (6.5)).*

1. *If f_1 and f_2 are complex-valued smooth functions on $\mathbb{H}_n$ such that*

$$\inf_{x\in\mathbb{H}_n,\lambda\geq\Lambda} \frac{|f_1(x)+f_2(x)\lambda|}{1+\lambda} > 0 \quad \textit{for some } \Lambda \geq 0,$$

and such that $X^{\alpha_1}f_1$, $X^{\alpha_2}f_2$ are bounded on $\mathbb{H}_n$ for each $\alpha_1,\alpha_2 \in \mathbb{N}_0^n$, then the differential operator $f_1(x) - f_2(x)\mathcal{L}$ is (locally) hypoelliptic. It is also globally hypoelliptic in the sense of Proposition 5.8.15. This operator also satisfies the following subelliptic estimates

$$\forall p \in (1,\infty) \quad \forall s \in \mathbb{R} \quad \forall N \in \mathbb{R} \quad \exists C>0 \quad \forall \varphi \in \mathcal{S}(\mathbb{H}_n)$$
$$\|\varphi\|_{L^p_{s+2}} \leq C\Big(\|f_1\varphi - f_2\mathcal{L}\varphi\|_{L^p_s} + \|\varphi\|_{L^p_{-N}}\Big).$$

2. *Let $\psi \in C^\infty(\mathbb{R})$ be such that*

$$\psi_{|(-\infty,\Lambda_1]} = 0 \quad \textit{and} \quad \psi_{|[\Lambda_2,\infty)} = 1,$$

for some real numbers Λ_1,Λ_2 satisfying $0 < \Lambda_1 < \Lambda_2$. Let also f_1 be a continuous complex-valued function on $\mathbb{H}_n$ such that $\inf_{\mathbb{H}_n}|f_1| > 0$ and that $X^\alpha f_1$ is bounded on $\mathbb{H}_n$ for each $\alpha \in \mathbb{N}_0^n$. Then the operator $f_1(x)\psi(-\mathcal{L})\mathcal{L}$ is (locally) hypoelliptic. It is also globally hypoelliptic in the sense of Proposition 5.8.15. This operator also satisfies the following subelliptic estimates

$$\forall p \in (1,\infty) \quad \forall s \in \mathbb{R} \quad \forall N \in \mathbb{R} \quad \exists C>0 \quad \forall \varphi \in \mathcal{S}(\mathbb{H}_n)$$
$$\|\varphi\|_{L^p_{s+2}} \leq C\Big(\|f_1\psi(-\mathcal{L})\mathcal{L}\varphi\|_{L^p_s} + \|\varphi\|_{L^p_{-N}}\Big).$$

We could also use Corollary 5.8.16 with other Rockland operators, such as $\mathcal{R} = \mathcal{L}^2$ or $\mathcal{R} = \mathcal{L}^2 + T^2$. In this case, it would yield:

Corollary 6.6.7. *Let $\mathcal{R} = \mathcal{L}^2$ or $\mathcal{R} = \mathcal{L}^2 + T^2$ where $\mathcal{L}$ denotes the canonical sub-Laplacian on the Heisenberg group $\mathbb{H}_n$ and T is the central derivative.*

1. *If f_1 and f_2 are complex-valued smooth functions on $\mathbb{H}_n$ such that*

$$\inf_{x\in\mathbb{H}_n,\lambda\geq\Lambda} \frac{|f_1(x) + f_2(x)\lambda|}{1+\lambda} > 0 \quad \textit{for some } \Lambda \geq 0,$$

and such that $X^{\alpha_1} f_1$, $X^{\alpha_2} f_2$ are bounded on $\mathbb{H}_n$ for each $\alpha_1, \alpha_2 \in \mathbb{N}_0^n$, then the differential operator $f_1(x) + f_2(x)\mathcal{R}$ is (locally) hypoelliptic. It is also globally hypoelliptic in the sense of Proposition 5.8.15. This operator also satisfies the following subelliptic estimates

$$\forall p \in (1,\infty) \quad \forall s \in \mathbb{R} \quad \forall N \in \mathbb{R} \quad \exists C > 0 \quad \forall \varphi \in \mathcal{S}(\mathbb{H}_n)$$
$$\|\varphi\|_{L^p_{s+4}} \leq C\Big(\|f_1\varphi + f_2\mathcal{R}\varphi\|_{L^p_s} + \|\varphi\|_{L^p_{-N}}\Big).$$

2. *Let $\psi \in C^\infty(\mathbb{R})$ be such that*

$$\psi_{|(-\infty,\Lambda_1]} = 0 \quad \textit{and} \quad \psi_{|[\Lambda_2,\infty)} = 1,$$

for some real numbers Λ_1, Λ_2 satisfying $0 < \Lambda_1 < \Lambda_2$. Let also f_1 be a continuous complex-valued function on $\mathbb{H}_n$ such that $\inf_{\mathbb{H}_n} |f_1| > 0$ and that $X^\alpha f_1$ is bounded on $\mathbb{H}_n$ for each $\alpha \in \mathbb{N}_0^n$. Then the operator $f_1(x)\psi(\mathcal{R})\mathcal{R} \in \Psi^4$ is (locally) hypoelliptic. It is also globally hypoelliptic in the sense of Proposition 5.8.15. This operator also satisfies the following subelliptic estimates

$$\forall p \in (1,\infty) \quad \forall s \in \mathbb{R} \quad \forall N \in \mathbb{R} \quad \exists C > 0 \quad \forall \varphi \in \mathcal{S}(\mathbb{H}_n)$$
$$\|\varphi\|_{L^p_{s+4}} \leq C\Big(\|f_1\psi(\mathcal{R})\mathcal{R}\varphi\|_{L^p_s} + \|\varphi\|_{L^p_{-N}}\Big).$$

Appendix A

Miscellaneous

In this chapter we collect a number of analytic tools that are used at some point in the monograph. These are all well-known, and we present them without proofs providing references to relevant sources when needed. Thus, here we make short expositions of topics including local hypoellipticity and solvability, operator semigroups, fractional powers of operators, singular integrals, almost orthogonality, and the analytic interpolation.

A.1 General properties of hypoelliptic operators

In this section, we recall the definition and first properties of locally hypoelliptic operators. We will also point out the useful duality between local solvability and local hypoellipticity in Theorem A.1.3.

Roughly speaking, a differential operator L is (locally) *hypoelliptic* if whenever u and f are distributions satisfying $Lu = f$, u must be smooth where f is smooth. Usually, we omit the word 'local' and just speak of hypoellipticity. More precisely:

Definition A.1.1. Let Ω be an open subset of $\mathbb{R}^n$ and let L be a differential operator on Ω with smooth coefficients. Then L is said to be *hypoelliptic* if, for any distribution $u \in \mathcal{D}'(\Omega)$ and any open subset Ω' of Ω, the condition $Lu \in C^\infty(\Omega')$ implies that $u \in C^\infty(\Omega')$.

This definition extends to an open subset of a smooth manifold.

Of course elliptic operators such as Laplace operators are hypoelliptic. Less obvious examples are provided by the celebrated Hörmander's Theorem on sums of squares of vector fields [Hör67a] which we recall here even if we will not use it in this monograph:

Theorem A.1.2 (Hörmander sum of squares). *Let $X_o, X_1, \ldots, X_p$ be smooth real-valued vector fields on an open set $\Omega \subset \mathbb{R}^n$, and let $c_o \in C^\infty(\Omega)$. We assume*

V. Fischer, M. Ruzhansky, *Quantization on Nilpotent Lie Groups*, Progress in Mathematics 314,
DOI 10.1007/978-3-319-29558-9

that the vector fields $X_o, X_1, \ldots, X_p$ satisfy Hörmander's condition, that is, the Lie algebra generated by $\{X_o, X_1, \ldots, X_p\}$ is of dimension n at every point of Ω. Then the operator $X_1^2 + \ldots + X_p^2 + X_o + c$ is hypoelliptic on Ω.

This extends to smooth manifolds.

Consequently any sub-Laplacian (see Definition 4.1.6) on a stratified Lie group is hypoelliptic on the whole group since any basis of the first stratum satisfies Hörmander's condition.

Hörmander's condition in Theorem A.1.2 is sufficient but not necessary for the hypoellipticity of sums of squares, thus allowing for sharper versions, see e.g. [BM95].

In the following sense, local hypoellipticity is dual to local solvability:

Theorem A.1.3. *Let L be hypoelliptic on Ω. Then L^t is locally solvable at every point of Ω.*

Let us briefly recall the definitions of the local solvability and of transpose:

Definition A.1.4. Let L be a linear differential operator with smooth coefficients on Ω. We say that L is *locally solvable* at $x \in \Omega$ if x has an open neighbourhood V in Ω such that, for every function $f \in \mathcal{D}(V)$ there is a distribution $u \in \mathcal{D}'(V)$ satisfying $Lu = f$ on V.

Definition A.1.5. The *transpose* of a differential operator L with smooth coefficients on an open subset Ω of $\mathbb{R}^n$ is the operator, denoted by L^t, given by

$$\forall \phi, \psi \in \mathcal{D}(\Omega) \qquad \langle L\phi, \psi\rangle = \langle \phi, L^t\psi\rangle.$$

This extends to manifolds.

Note that if

$$Lf(x) = \sum_{|\alpha|\leq m} a_\alpha(x)\partial^\alpha f(x),$$

then

$$L^t f(x) = \sum_{|\alpha|\leq m} \partial^\alpha\big(a_\alpha(x)f(x)\big) = \sum_{|\alpha|\leq m} b_\alpha(x)\partial^\alpha f(x),$$

where the b_α's are linear combinations of derivatives of the a_α's, in particular they are smooth functions.

We will need the following property:

Theorem A.1.6 (Schwartz-Trèves). *Let L be a differential operator with smooth coefficients on an open subset Ω of $\mathbb{R}^n$. We assume that L and L^t are hypoelliptic on $\Omega \subset \mathbb{R}^n$. Then the $\mathcal{D}'(\Omega)$ and $C^\infty(\Omega)$ topologies agree on*

$$N_L(\Omega) = \{f \in \mathcal{D}'(\Omega) \ : \ Lf = 0\}.$$

For its proof, we refer to [Tre67, Corollary 1 in Ch. 52].

A.2 Semi-groups of operators

In this section we discuss operator semi-groups and their infinitesimal generators.

Definition A.2.1. Suppose that for every $t \in (0, \infty)$, there is an associated bounded linear operator $Q(t)$ on a Banach space $\mathcal{X}$ in such a way that

$$\forall s, t > 0 \qquad Q(s+t) = Q(s)Q(t).$$

Then the family $\{Q(t)\}_{t>0}$ is called a *semi-group* of operators on $\mathcal{X}$.

If we have for every $x \in \mathcal{X}$, that

$$\|Q(t)x - x\|_{\mathcal{X}} \underset{t\to 0}{\longrightarrow} 0,$$

then the semi-group is said to be *strongly continuous*.

If the operator norm of each $Q(t)$ is less or equal to one, $\|Q(t)\|_{\mathscr{L}(\mathcal{X})} \leq 1$, then the semi-group is called a *contraction* semi-group.

Let $\{Q(t)\}_{t>0}$ be a semi-group of operators on $\mathcal{X}$. If $x \in \mathcal{X}$ is such that $\frac{1}{\epsilon}(Q(\epsilon)x - x)$ converges in the norm topology of $\mathcal{X}$ as $\epsilon \to 0$, then we denote its limit by Ax and we say that x is in the domain $\mathrm{Dom}(A)$ of A. Clearly $\mathrm{Dom}(A)$ is a linear subspace of $\mathcal{X}$ and A is a linear operator on $\mathrm{Dom}(A) \subset X$. This operator is essentially $A = Q'(0)$.

Definition A.2.2. The operator A defined just above is called the *infinitesimal generator* of the semi-group $\{Q(t)\}_{t>0}$.

We now collect some properties of semi-groups and their generators.

Proposition A.2.3. *Let $\{Q(t)\}_{t>0}$ be a strongly continuous semi-group with infinitesimal generator A. We also set $Q(0) := \mathrm{I}$, the identity operator. Then*

1. *there are constants C, γ such that for all $t \in [0, \infty)$,*

$$\|Q(t)\|_{\mathscr{L}(\mathcal{X})} \leq Ce^{\gamma t};$$

2. *for every $x \in \mathcal{X}$, the map $[0, \infty) \ni t \mapsto Q(t)x \in \mathcal{X}$ is continuous;*
3. *the operator A is closed with dense domain;*
4. *the differential equation*

$$\partial_t Q(t)x = Q(t)A\,x = AQ(t)x,$$

holds for every $x \in \mathrm{Dom}(A)$ and $t \geq 0$;

5. *for every $x \in \mathcal{X}$ and $t > 0$,*

$$Q(t)x = \lim_{\epsilon\to 0} \exp(tA_\epsilon)x,$$

where

$$A_\epsilon = \frac{1}{\epsilon}\left(Q(\epsilon) - \mathrm{I}\right) \quad \text{and} \quad \exp(tA_\epsilon) = \sum_{k=0}^{\infty} \frac{1}{k!}(tA_\epsilon)^k;$$

furthermore the convergence is uniform on every compact subset of $[0, \infty)$*;*

6. *if* $\lambda \in \mathbb{C}$ *and* $\operatorname{Re}\lambda > \gamma$ *(where* γ *is any constant such that (1) holds), the integral*

$$R(\lambda)x = \int_0^\infty e^{-\lambda t} Q(t)x\, dt,$$

defines a bounded linear operator $R(\lambda)$ *on* $\mathcal{X}$ *(*often called the resolvent of the semi-group $\{Q(t)\}$*) whose range is* $\mathrm{Dom}(A)$ *and which inverts* $\lambda\mathrm{I} - A$*. In particular, the spectrum of* A *lies in the half plane* $\{\lambda : \operatorname{Re}\lambda \leq \gamma\}$*.*

For the proof, see e.g. Rudin [Rud91, §13.35].

Theorem A.2.4 (Hille-Yosida). *A densely defined operator* A *on a Banach space* $\mathcal{X}$ *is the infinitesimal generator of a strongly continuous semi-group* $\{Q(t)\}_{t>0}$ *if and only if there are constants* C, γ *such that*

$$\forall \lambda > \gamma,\ m \in \mathbb{N} \qquad \|(\lambda\mathrm{I} - A)^{-m}\| \leq C(\lambda - \gamma)^{-m}.$$

The constant γ *can be taken as in Proposition A.2.3.*

For the proof of the Hille-Yosida Theorem, see e.g [Rud91, §13.37].

In this case the operators of the semi-group $\{Q(t)\}_{t>0}$ generated by A are denoted by

$$Q(t) = e^{tA}.$$

Theorem A.2.5 (Lumer-Phillips). *A densely defined operator* A *on a Banach space* $\mathcal{X}$ *is the infinitesimal generator of a strongly continuous contraction semi-group* $\{Q(t)\}_{t>0}$ *if and only if*

- A *is dissipative, i.e.*

$$\forall \lambda > 0,\ x \in \mathrm{Dom}(A) \qquad \|(\lambda\mathrm{I} - A)x\| \geq \lambda\|x\|;$$

- *there is at least one* λ_o *such that* $A - \lambda_o\mathrm{I}$ *is surjective.*

For the proof of the Lumer-Phillips Theorem, see [LP61].

For this monograph, the facts given in this section will be enough. We refer for the general theory of semi-groups to the fundamental work of Hille and Phillips [HP57], or to later expositions e.g. by Davies [Dav80] or Pazy [Paz83].

A.3 Fractional powers of operators

Here we summarise the definition of fractional powers for certain operators. We refer the interested reader to the monograph of Martinez and Sanz [MCSA01] and all the explanations and historical discussions therein.

Let $A : \mathrm{Dom}(A) \subset \mathcal{X} \to \mathcal{X}$ be a linear operator on a Banach space $\mathcal{X}$. In order to present only the part of the theory that we use in this monograph, we make the following assumptions

(i) The operator A is closed and densely defined.

(ii) The operator A is injective, that is, A is one-to-one on its domain.

(iii) The operator A is Komatsu-non-negative, that is, $(-\infty, 0)$ is included in the resolvent $\rho(A)$ of A and

$$\exists M > 0 \quad \forall \lambda > 0 \quad \|(\lambda + A)^{-1}\| \leq M\lambda^{-1}.$$

Remark A.3.1. This implies (cf. [MCSA01, Proposition 1.1.3 (iii)]) that for all $n, m \in \mathbb{N}$, $\mathrm{Dom}(A^n)$ is dense in $\mathcal{X}$, and $\mathrm{Range}(A^m)$ as well as $\mathrm{Dom}(A^n) \cap \mathrm{Range}(A^m)$ are dense in the closure of $\mathrm{Range}(A)$.

The powers A^n, $n \in \mathbb{N}$, are defined using iteratively the following definition:

Definition A.3.2. The product of two (possibly) unbounded operators A and B acting on the same Banach space $\mathcal{X}$ is as follows. A vector x is in the domain of the operator AB whenever x is in the domain of B and Bx is in the domain of A. In this case $(AB)(x) = A(Bx)$.

Remark A.3.3. Note that if an operator A satisfies (i), (ii) and (iii), then it is also the case for $\mathrm{I} + A$.

Following Balakrishnan (cf. [MCSA01, Section 3.1]), the (Balakrishnan) operators J^α, $\alpha \in \mathbb{C}_+ := \{z \in \mathbb{C}, \operatorname{Re} z > 0\}$, are (densely) defined by the following:

- If $0 < \operatorname{Re}\alpha < 1$, $\mathrm{Dom}(J^\alpha) := \mathrm{Dom}(A)$ and for $\phi \in \mathrm{Dom}(A)$,

$$J^\alpha \phi := \frac{\sin \alpha\pi}{\pi} \int_0^\infty \lambda^{\alpha-1} (\lambda \mathrm{I} + A)^{-1} A\phi \, d\lambda.$$

- If $\operatorname{Re}\alpha = 1$, $\mathrm{Dom}(J^\alpha) := \mathrm{Dom}(A^2)$ and for $\phi \in \mathrm{Dom}(A^2)$,

$$J^\alpha \phi := \frac{\sin \alpha\pi}{\pi} \int_0^\infty \lambda^{\alpha-1} \left[(\lambda \mathrm{I} + A)^{-1} - \frac{\lambda}{\lambda^2 + 1} \right] A\phi \, d\lambda + \sin \frac{\alpha\pi}{2} A\phi.$$

- If $n < \operatorname{Re}\alpha < n+1$, $n \in \mathbb{N}$, $\mathrm{Dom}(J^\alpha) := \mathrm{Dom}(A^{n+1})$ and for $\phi \in \mathrm{Dom}(A)$,

$$J^\alpha \phi := J^{\alpha - n} A^n \phi.$$

- If $\operatorname{Re}\alpha = n+1$, $n \in \mathbb{N}$, $\operatorname{Dom}(J^\alpha) := \operatorname{Dom}(A^{n+2})$ and for $\phi \in \operatorname{Dom}(A^{n+2})$,

$$J^\alpha \phi := J^{\alpha-n} A^n \phi.$$

We now define fractional powers distinguishing between three different cases:

Case 0: A is bounded.

Case I: A is unbounded and $0 \in \rho(A)$, that is, the resolvent of A contains zero; in other words, A^{-1} is bounded.

Case II: A is unbounded and $0 \in \sigma(A)$, that is, the spectrum of A contains zero.

The fractional powers A^α, $\alpha \in \mathbb{C}_+$, are defined in the following way (cf. [MCSA01, Section 5.1]):

Case 0: A being bounded, J^α is bounded and we define $A^\alpha := J^\alpha$, $\alpha \in \mathbb{C}_+$.

Case I: A^{-1} being bounded, we can use Case 0 to define $(A^{-1})^\alpha$ which is injective; then we define

$$A^\alpha := \left[(A^{-1})^\alpha\right]^{-1} \quad (\alpha \in \mathbb{C}_+).$$

Case II: Using Case I for $A + \epsilon \mathrm{I}$, $\epsilon > 0$, we define

$$A^\alpha := \lim_{\epsilon \to 0} (A + \epsilon \mathrm{I})^\alpha \quad (\alpha \in \mathbb{C}_+);$$

that is, the domain of A^α is composed of all the elements $\phi \in \operatorname{Dom}\left[(A+\epsilon\mathrm{I})^\alpha\right]$, $\epsilon > 0$ close to zero, and such that $(A+\epsilon\mathrm{I})\phi$ is convergent for the norm topology of $\mathcal{X}$ as $\epsilon \to 0$; the limit defines $A^\alpha\phi$.

In all cases, J^α is closable and we have (cf. [MCSA01, Theorem 5.2.1]):

$$A^\alpha = (A+\lambda\mathrm{I})^n \overline{J^\alpha} (A+\lambda\mathrm{I})^{-n} \qquad (\alpha \in \mathbb{C}_+,\ \lambda \in \rho(-A),\ n \in \mathbb{N}).$$

Hence A^α, $\alpha \in \mathbb{C}_+$, can be understood as the maximal domain operator which extends J^α and commutes with the resolvent of A (in other words *commutes strongly with* A).

We can now define the powers for complex numbers also with non-positive real parts (cf. [MCSA01, Section 7.1]):

- Given $\alpha \in \mathbb{C}_+$, the operators A^α, $\alpha \in \mathbb{C}_+$, are injective, and we can define

$$A^{-\alpha} := (A^\alpha)^{-1}.$$

- Given $\tau \in \mathbb{R}$, we define

$$A^{i\tau} := (A+\mathrm{I})^2 A^{-1} A^{1+i\tau} (A+\mathrm{I})^{-2}.$$

We now collect properties of fractional powers.

Theorem A.3.4. *Let $A : \mathrm{Dom}(A) \subset \mathcal{X} \to \mathcal{X}$ be a linear operator on a Banach space $\mathcal{X}$. Assume that the operator A satisfies Properties (i), (ii) and (iii), and define its fractional powers A^α as above.*

1. *For every $\alpha \in \mathbb{C}$, the operator A^α is closed and injective with $(A^\alpha)^{-1} = A^{-\alpha}$. In particular, $A^0 = \mathrm{I}$.*
2. *For $\alpha \in \mathbb{C}_+$, the operator A^α coincides with the closure of J^α.*
3. *If A has dense range and for all $\tau \in \mathbb{R}$, $A^{i\tau}$ is bounded, then there exist $C > 0$ and $\theta \in (0, \pi)$ such that*
$$\forall \tau \in \mathbb{R} \qquad \|A^{i\tau}\|_{\mathscr{L}(\mathcal{X})} \leq Ce^{\theta\tau}.$$
Given $\tau \in \mathbb{R}\backslash\{0\}$, if $A^{i\tau}$ is bounded then $\mathrm{Dom}(A^\alpha) \subset \mathrm{Dom}(A^{\alpha+i\tau})$ for all $\alpha \in \mathbb{R}$. Conversely, if $\mathrm{Dom}(A^\alpha) \subset \mathrm{Dom}(A^{\alpha+i\tau})$ for all $\alpha \in \mathbb{R}\backslash\{0\}$, then $A^{i\tau}$ is bounded.
4. *For any $\alpha, \beta \in \mathbb{C}$, we have $A^\alpha A^\beta \subset A^{\alpha+\beta}$, and if $\mathrm{Range}(A)$ is dense in $\mathcal{X}$ then the closure of $A^\alpha A^\beta$ is $A^{\alpha+\beta}$.*
5. *Let $\alpha_o \in \mathbb{C}_+$.*
   - *If $\phi \in \mathrm{Range}(A^{\alpha_o})$ then $\phi \in \mathrm{Dom}(A^\alpha)$ for all $\alpha \in \mathbb{C}$ with $0 < -\mathrm{Re}\,\alpha < \mathrm{Re}\,\alpha_o$ and the function $\alpha \mapsto A^\alpha\phi$ is holomorphic in $\{\alpha \in \mathbb{C} : -\mathrm{Re}\,\alpha_o < \mathrm{Re}\,\alpha < 0\}$.*
   - *If $\phi \in \mathrm{Dom}(A^{\alpha_o})$ then $\phi \in \mathrm{Dom}(A^\alpha)$ for all $\alpha \in \mathbb{C}$ with $0 < \mathrm{Re}\,\alpha < \mathrm{Re}\,\alpha_o$ and the function $\alpha \mapsto A^\alpha\phi$ is holomorphic in $\{\alpha \in \mathbb{C} : 0 < \mathrm{Re}\,\alpha < \mathrm{Re}\,\alpha_o\}$.*
   - *If $\phi \in \mathrm{Dom}(A^{\alpha_o}) \cap \mathrm{Range}(A^{\alpha_o})$ then $\phi \in \mathrm{Dom}(A^\alpha)$ for all $\alpha \in \mathbb{C}$ with $|\mathrm{Re}\,\alpha| < \mathrm{Re}\,\alpha_o$ and the function $\alpha \mapsto A^\alpha\phi$ is holomorphic in $\{\alpha \in \mathbb{C} : -\mathrm{Re}\,\alpha_o < \mathrm{Re}\,\alpha < \mathrm{Re}\,\alpha_o\}$.*
6. *If $\alpha, \beta \in \mathbb{C}_+$ with $\mathrm{Re}\,\beta > \mathrm{Re}\,\alpha$, then*
$$\exists C = C_{A,\alpha,\beta} > 0 \quad \forall \phi \in \mathrm{Dom}(A^\beta) \quad \|A^\alpha\phi\|_{\mathcal{X}} \leq C\|\phi\|_{\mathcal{X}}^{1-\frac{\mathrm{Re}\,\alpha}{\mathrm{Re}\,\beta}} \|A^\beta\phi\|_{\mathcal{X}}^{\frac{\mathrm{Re}\,\alpha}{\mathrm{Re}\,\beta}}.$$
7. *If B^* denotes the dual of an operator B on $\mathcal{X}$, then $(A^\alpha)^* = (A^*)^\alpha$.*
8. *For $\alpha \in \mathbb{C}_+$ and $\epsilon > 0$, $\mathrm{Dom}\left[(A + \epsilon\mathrm{I})^\alpha\right] = \mathrm{Dom}(A^\alpha)$.*
9. *Let $\tau \in \mathbb{R}$. Let $S_{i\tau}$ be the strong limit of $(A + \epsilon\mathrm{I})^{i\tau}$ as $\epsilon \to 0^+$, with domain $\mathrm{Dom}(S_{i\tau}) = \{\phi \in \mathrm{Dom}\left[(A+\epsilon)^{i\tau}\right] : \exists \lim_{\epsilon\to 0^+}(A+\epsilon)^{i\tau}\phi\}$. Then $S_{i\tau}$ is closable and the closure of (the graph of) $J^{i\tau}$ is included in the closure of (the graph of) $S_{i\tau}$ which is included in (the graph of) $A^{i\tau}$.*

   In particular, if A has dense domain and range, then the closure of $S_{i\tau}$ is $A^{i\tau}$.

10. *Let us assume that A generates an equibounded semi-group $\{e^{-tA}\}_{t>0}$ on $\mathcal{X}$, that is,*

$$\exists M \qquad \forall t > 0 \qquad \|e^{-tA}\|_{\mathcal{X}} \leq M. \tag{A.1}$$

If $0 < \operatorname{Re} \alpha < 1$ and $\phi \in \operatorname{Range}(A)$ then

$$A^{-\alpha}\phi = \frac{1}{\Gamma(\alpha)} \int_0^\infty t^{\alpha-1} e^{-tA}\phi \, dt, \tag{A.2}$$

in the sense that $\lim_{N\to\infty} \int_0^N$ converges in the $\mathcal{X}$-norm.

Moreover, if $\{e^{-tA}\}_{t>0}$ is exponentially stable, that is,

$$\exists M, \mu > 0 \quad \forall t > 0 \qquad \|e^{-tA}\|_{\mathscr{L}(\mathcal{X})} \leq Me^{-t\mu},$$

then Formula (A.2) holds for all $\alpha \in \mathbb{C}_+$ and $\phi \in \mathcal{X}$, and the integral converges absolutely: $\int_0^\infty \|t^{\alpha-1}e^{-tA}\phi\|_{\mathcal{X}} dt < \infty$.

References for these results are in [MCSA01] as follows:

(1) Corollary 5.2.4 and Section 7.1;

(2) Corollary 5.1.12;

(3) Proposition 8.1.1, Section 7.1 and Corollary 7.1.2;

(4) Theorem 7.1.1;

(5) Proposition 7.1.5 with its proof, and Corollary 5.1.13;

(6) Corollary 5.1.13;

(7) Corollary 5.2.4 for $\alpha \in \mathbb{C}_+$, consequently for any $\alpha \in \mathbb{C}$;

(8) Theorem 5.1.7;

(9) Theorem 7.4.6;

(10) Lemma 6.1.5.

In Theorem A.3.4 Part (10), Γ denotes the Gamma function. Let us recall briefly its definition. For each $\alpha \in \mathbb{C}_+$, it is defined by the convergent integral

$$\Gamma(\alpha) := \int_0^\infty t^{\alpha-1} e^{-t} dt.$$

A direct computation gives $\Gamma(1) = \int_0^\infty e^{-t}dt = 1$ and an integration by parts yields the functional equation $\alpha\Gamma(\alpha) = \Gamma(\alpha + 1)$. Hence the Gamma function coincides with the factorial in the sense that if $\alpha \in \mathbb{N}$, then the equality $\Gamma(\alpha) = (\alpha - 1)!$ holds. It is easy to see that Γ is analytic on the half plane $\{\operatorname{Re} \alpha > 0\}$. Because of the functional equation, it admits a unique analytic continuation to the whole complex plane except for non-positive integers where it has simple pole. We keep the same notation Γ for its analytic continuation.

For $\operatorname{Re} z > 0$, we have the Sterling estimate

$$\Gamma(z) = \sqrt{\frac{2\pi}{z}} \left(\frac{z}{e}\right)^z (1 + O(\frac{1}{z})). \tag{A.3}$$

Also, the following known relation will be of use to us,

$$\int_{t=0}^{1} t^{x-1}(1-t)^{y-1} dt = \frac{\Gamma(x)\Gamma(y)}{\Gamma(x+y)}, \quad \operatorname{Re} x > 0, \ \operatorname{Re} y > 0. \tag{A.4}$$

We will use Part (6) also in the following form: let $\alpha, \beta, \gamma \in \mathbb{C}$ with $\operatorname{Re}\alpha < \operatorname{Re}\beta$ and $\operatorname{Re}\alpha \le \operatorname{Re}\gamma \le \operatorname{Re}\beta$; then there exists $C = C_{\alpha,\beta,\gamma,A} > 0$ such that for any $f \in \operatorname{Dom}(A^\alpha)$ with $A^\alpha f \in \operatorname{Dom}(A^{\beta-\alpha})$, we have

$$\|A^\gamma f\|_{\mathcal{X}} \le C \|A^\alpha f\|_{\mathcal{X}}^{1-\theta} \|A^\beta f\|_{\mathcal{X}}^\theta \qquad \text{where} \quad \theta := \frac{\operatorname{Re}(\gamma - \alpha)}{\operatorname{Re}(\beta - \alpha)}.$$

A.4 Singular integrals (according to Coifman-Weiss)

The operators appearing 'in practice' in the theory of partial differential equations on $\mathbb{R}^n$ often have kernels κ satisfying the following properties:

1. the restriction of $\kappa(x,y)$ to $(\mathbb{R}^n_x \times \mathbb{R}^n_y)\backslash\{x=y\}$ coincides with a smooth function $\kappa_o = \kappa_o(x,y) \in C^\infty((\mathbb{R}^n_x \times \mathbb{R}^n_y)\backslash\{x=y\})$;
2. away from the diagonal $x = y$, the function κ_o decays rapidly;
3. at the diagonal, κ_o is singular but not completely wild: κ_o and some of its first derivatives admit a control of the form $|\kappa_o(x,y)| \le C_x |x-y|^k$ for some power $k \in (-\infty, \infty)$ with C_x varying slowly in x.

These types of operators include all the (Hörmander, Shubin, semi-classical, ...) pseudo-differential operators, and these types of operators appear when looking for fundamental solutions or parametrices of differential operators.

In general, we want our operator T to map continuously some well-known functional space to another. For example, we are looking for conditions to ensure that our operator extends to a bounded operator from L^p to L^q. This is the subject of the theory of singular integrals on $\mathbb{R}^n$, especially when the power k above equals $-n$. In the classical Euclidean case, we refer to the monograph [Ste93] by Stein for a detailed presentation of this theory.

Here, let us present the main lines of the generalisation of the theory of singular integrals to the setting of 'spaces of homogeneous type' where there is no (apparent) trace of a group structure. This generalisation is relevant for us since examples of such spaces are compact manifolds and homogeneous nilpotent Lie groups. We omit the proofs, referring to [CW71a, Chapitre III] for details.

Definition A.4.1. A *quasi-distance* on a set X is a function $d : X \times X \to [0, \infty)$ such that

1. $d(x, y) > 0$ if and only if $x \neq y$;
2. $d(x, y) = d(y, x)$;
3. there exists a constant $K > 0$ such that

$$\forall x, y, z \in X \qquad d(x, z) \leq K\left(d(x, y) + d(y, z)\right).$$

We call

$$B(x, r) := \{y \in G \ : \ d(x, y) < r\},$$

the *quasi-ball of radius r around x.*

Definition A.4.2. A *space of homogeneous type* is a topological space X endowed with a quasi-distance d such that

1. The quasi-balls $B(x, r)$ form a basis of open neighbourhood at x;
2. homogeneity property

   there exists $N \in \mathbb{N}$ such that for every $x \in X$ and every $r > 0$ the ball $B(x, r)$ contains at most N points x_i such that $d(x_i, x_j) > r/2$.

The constants K in Definition A.4.1 and N in Definition A.4.2 are called the constants of the space of homogeneous type X.

Some authors (like in the original text of [CW71a]) prefer using the vocabulary pseudo-norms, pseudo-distance, etc. instead of quasi-norms, quasi-distance, etc. In this monograph, following e.g. both Stein [Ste93] and Wikipedia, we choose the perhaps more widely adapted convention of the term quasi-norm.

Examples of spaces of homogeneous type:

1. A homogeneous Lie group endowed with the quasi-distance associated to any homogeneous quasi-norm (see Lemma 3.2.12).
2. The unit sphere $\mathbb{S}^{n-1}$ in $\mathbb{R}^n$ with the quasi-distance

$$d(x, y) = |1 - x \cdot y|^\alpha,$$

   where $\alpha > 0$ and $x \cdot y = \sum_{j=1}^n x_j y_j$ is the real scalar product of $x, y \in \mathbb{R}^n$.
3. The unit sphere $\mathbb{S}^{2n-1}$ embedded in $\mathbb{C}^n$ with the quasi-distance

$$d(z, w) = |1 - (z, w)|^\alpha,$$

   where $\alpha > 0$ and $(z, w) = \sum_{j=1}^n z_j \bar{w}_j$.
4. Any compact Riemannian manifold.

The proof that these spaces are effectively of homogeneous type comes easily from the following lemma:

Lemma A.4.3. *Let X be a topological set endowed with a quasi-distance d satisfying (1) of Definition A.4.2.*

Assume that there exist a Borel measure μ on X satisfying

$$0 < \mu\left(B(x,r)\right) \leq C\mu\left(B(x,\frac{r}{2})\right) < \infty. \tag{A.5}$$

Then X is a space of homogeneous type.

The condition (A.5) is called the *doubling condition.* For instance, the Riemannian measure of a Riemannian compact manifold or the Haar measure of a homogeneous Lie group satisfy the doubling condition; we omit the proof of these facts, as well as the proof of Lemma A.4.3.

Let (X,d) be a space of homogeneous type. The hypotheses are 'just right' to obtain a covering lemma. We assume now that X is also equipped with a measure μ satisfying the doubling condition (A.5). A maximal function with respect to the quasi-balls may be defined. Then given a level, any function f can be decomposed 'in the usual way' into good and bad functions $f = g+\sum_j b_j$. The Euclidean proof of the Singular Integral Theorem can be adapted to obtain

Theorem A.4.4 (Singular integrals)**.** *Let (X,d) be a space of homogeneous type equipped with a measure μ satisfying the doubling condition given in (A.5).*

Let T be an operator which is bounded on $L^2(X)$:

$$\exists C_o \qquad \forall f \in L^2 \quad \|Tf\|_2 \leq C_o\|f\|_2. \tag{A.6}$$

We assume that there exists a locally integrable function κ on $(X\times X)\setminus \{(x,y)\in X\times X : x=y\}$ such that for any compactly supported function $f\in L^2(X)$, we have

$$\forall x \notin \text{supp} f \qquad Tf(x) = \int_X \kappa(x,y)f(y)d\mu(y).$$

We also assume that there exist $C_1, C_2 > 0$ such that

$$\forall y, y_o \in X \qquad \int_{d(x,y_o)>C_1 d(y,y_o)} |\kappa(x,y) - \kappa(x,y_o)|d\mu(x) \leq C_2. \tag{A.7}$$

Then for all p, $1 < p \leq 2$, T extends to a bounded operator on L^p because

$$\exists A_p \qquad \forall f \in L^2 \cap L^p \quad \|Tf\|_p \leq A_p\|f\|_p;$$

for $p=1$, the operator T extends to a weak-type (1,1) operator since

$$\exists A_1 \qquad \forall f \in L^2 \cap L^1 \quad \mu\{x \ : \ |Tf(x)| > \alpha\} \leq A_1\frac{\|f\|_1}{\alpha};$$

the constants A_p, $1 \leq p \leq 2$, depend only on C_o, C_1 and C_2.

Remark A.4.5. 1. In the statement of the fundamental theorem of singular integrals on spaces of homogeneous types, cf. [CW71a, Théorème 2.4 Chapitre III], the kernel κ is assumed to be square integrable in $L^2(X \times X)$. However, the proof requires only that the kernel κ is locally integrable away from the diagonal, beside the L^2-boundedness of the operator T. We have therefore chosen to state it in the form given above.

2. Following the constants in the proof of [CW71a, Théorème 2.4 Chapitre III], we find

$$A_2 = C_1 \qquad \text{and} \qquad A_1 = C(C_1^2 + C_3),$$

where C is a constant which depends only on the constants of the space of homogeneous type. The constants A_p for $p \in (1, 2)$ are obtained via the constants appearing in the Marcinkiewicz interpolation theorem (see e.g. [DiB02, Theorem 9.1]):

$$A_p = \frac{2p}{(2-p)(1-p)} A_1^\delta A_2^{1-\delta} \quad \text{with } \delta = 2(\frac{1}{p} - \frac{1}{2}).$$

Let us discuss the two main hypotheses of Theorem A.4.4.

About Condition (A.7) in the Euclidean case. As explained at the beginning of this section, we are interested in 'nice' kernels $\kappa_o(x, y)$ with a control of the form $|\kappa_o(x, y)| \le C_x |x - y|^k$ with a particular interest for $k = -n$, and similar estimates for their derivatives with power $-n-1$. Hence they should satisfy Condition (A.7). They are called Calderón-Zygmund kernels, which we now briefly recall:

Calderón-Zygmund kernels on $\mathbb{R}^n$

A *Calderón-Zygmund kernel* on $\mathbb{R}^n$ is a measurable function κ_o defined on $(\mathbb{R}^n_x \times \mathbb{R}^n_y)\backslash\{x = y\}$ satisfying for some γ, $0 < \gamma \le 1$, the inequalities

$$\begin{aligned}
|\kappa_o(x, y)| &\le A|x - y|^{-n}, \\
|\kappa_o(x, y) - \kappa_o(x', y)| &\le A\frac{|x - x'|^\gamma}{|x - y|^{n+\gamma}} \qquad \text{if } |x - x'| \le \frac{|x - y|}{2}, \\
|\kappa_o(x, y) - \kappa_o(x, y')| &\le A\frac{|y - y'|^\gamma}{|x - y|^{n+\gamma}} \qquad \text{if } |y - y'| \le \frac{|x - y|}{2}.
\end{aligned}$$

Sometimes the condition of Calderón-Zygmund kernels refers to a smooth function κ_o defined on $(\mathbb{R}^n_x \times \mathbb{R}^n_y)\backslash\{x = y\}$ satisfying

$$\forall \alpha, \beta \quad \exists C_{\alpha,\beta} \qquad \left|\partial_x^\alpha \partial_y^\beta \kappa_o(x, y)\right| \le C_{\alpha,\beta}|x - y|^{-n-\alpha-\beta}.$$

For a detailed discussion, the reader is directed to [Ste93, ch.VII].

A *Calderón-Zygmund operator* on $\mathbb{R}^n$ is an operator $T : \mathcal{S}(\mathbb{R}^n) \to \mathcal{S}'(\mathbb{R}^n)$ such that the restriction of its kernel κ to $(\mathbb{R}^n_x \times \mathbb{R}^n_y)\backslash\{x = y\}$ is a Calderón-Zygmund kernel κ_o. In other words, $T : \mathcal{S}(\mathbb{R}^n) \to \mathcal{S}'(\mathbb{R}^n)$ is a Calderón-Zygmund operator if there exists a Calderón-Zygmund kernel κ_o satisfying

$$Tf(x) = \int_{\mathbb{R}^n} \kappa_o(x, y) f(y) dy,$$

for $f \in \mathcal{S}(\mathbb{R}^n)$ with compact support and $x \in \mathbb{R}^n$ outside the support of f.

The Calderón-Zygmund conditions imply Condition (A.7) for the operator T and its formal adjoint T^* but they are not sufficient to imply the L^2-boundedness for which some additional 'cancellation' conditions are needed.

About Condition (A.6). The difficulty with applying the main theorem of singular integrals (i.e. Theorem A.4.4) is often to know that the operator is L^2-bounded. The next section explains the Cotlar-Stein lemma which may help to prove the L^2-boundedness in many cases.

A.5 Almost orthogonality

On $\mathbb{R}^n$, a convolution operator (for the usual convolution) is bounded on $L^2(\mathbb{R}^n)$ if and only if the Fourier transform of its kernel is bounded. Similar result is valid on compact Lie groups, see (2.23), and more generally on any Hausdorff locally compact separable group, see the decomposition of group von Neumann algebras in the abstract Plancherel theorem in Theorem B.2.32. For operators on spaces without readily available Fourier transform or with no control on the Fourier transform of its kernel, or for non-convolution operators this becomes more complicated (however, see Theorem 2.2.5 for the case of non-invariant operators on compact Lie groups).

Fortunately, the space L^2 is a Hilbert space and to prove that an operator is bounded on L^2, it suffices to do the same for TT^* (or T^*T). The reason that this observation is useful in practice is that if T is formally representable by a kernel κ (see Schwartz kernel theorem, Theorem 1.4.1), then T^*T is representable by the kernel

$$\int \overline{\kappa(z, x)} \kappa(z, y)\, dz;$$

the latter kernel is often better than κ because the integration can have a smoothing effect and/or can take into account the cancellation properties of κ. This remark alone does not always suffice to prove the L^2-boundedness. Sometimes some 'smart' decomposition $T = \sum_k T_k$ of the operator is needed and again the properties of a Hilbert space may help.

The next statement is an easy case of 'exact' orthogonality:

Proposition A.5.1. *Let $\mathcal{H}$ be a Hilbert space and let $\{T_k,\ k \in \mathbb{Z}\}$ be a sequence of linear operators on $\mathcal{H}$. We assume that the operators $\{T_k\}$ are uniformly bounded:*

$$\exists C > 0 \qquad \forall k \in \mathbb{Z} \qquad \|T_k\|_{\mathscr{L}(\mathcal{H})} \leq C,$$

and that

$$\forall j \neq k \qquad T_j^* T_k = 0 \quad \text{and} \quad T_j T_k^* = 0. \tag{A.8}$$

Then the series $\sum_{k\in\mathbb{Z}} T_k$ converges in the strong operator norm topology to an operator S satisfying $\|S\|_{\mathscr{L}(\mathcal{H})} \leq C$.

Note that (A.8) is equivalent to

$$\forall j \neq k \qquad (\ker T_j)^\perp \perp (\ker T_k)^\perp \quad \text{and} \quad \operatorname{Im} T_j \perp \operatorname{Im} T_k.$$

Proof. Let $v \in \mathcal{H}$ and $N \in \mathbb{N}$. Since the images of the T_j's are orthogonal, the Pythagoras equality implies

$$\| \sum_{|j|\leq N} T_j v\|^2 = \sum_{|j|\leq N} \|T_j v\|^2.$$

Denoting by P_j the orthogonal projection onto $(\ker T_j)^\perp$, we have

$$\|T_j v\| = \|T_j P_j v\| \leq C\|P_j v\|,$$

since $\|T_j\|_{\mathscr{L}(\mathcal{H})} \leq C$. Thus

$$\| \sum_{|j|\leq N} T_j v\|^2 \leq C^2 \sum_{|j|\leq N} \|P_j v\|^2.$$

As the kernels of the T_j's are mutually orthogonal, we have

$$\sum_{|j|\leq N} \|P_j v\|^2 \leq \|v\|^2.$$

We have obtained that

$$\| \sum_{|j|\leq N} T_j v\|^2 \leq C^2 \|v\|^2,$$

for any $N \in \mathbb{N}$ and $v \in \mathcal{H}$. The constant C here is the uniform bound of the operator norms of the T_j's and is independent of v or N. The same proof shows that the sequence $(\sum_{|j|\leq N} T_j v)_{N\in\mathbb{N}}$ is Cauchy when v is in a finite number of $(\ker T_j)^\perp$. This allows us to define the operator S on the dense subspace $\sum_j (\ker T_j)^\perp$. The conclusion follows. □

In practice, the orthogonality assumption above is rather demanding, and is often substituted by a condition of 'almost' orthogonality:

Theorem A.5.2 (Cotlar-Stein lemma). *Let $\mathcal{H}$ be a Hilbert space and $\{T_k,\ k \in \mathbb{Z}\}$ be a sequence of linear operators on $\mathcal{H}$. We assume that we are given a sequence of positive constants $\{\gamma_j\}_{j=-\infty}^{\infty}$ with*

$$A = \sum_{j=-\infty}^{\infty} \gamma_j < \infty.$$

If for any $i, j \in \mathbb{Z}$,

$$\max \left(\|T_i^* T_j\|_{\mathscr{L}(\mathcal{H})}, \|T_i T_j^*\|_{\mathscr{L}(\mathcal{H})} \right) \leq \gamma_{i-j}^2,$$

then the series $\sum_{k \in \mathbb{Z}} T_k$ converges in the strong operator topology to an operator S satisfying $\|S\|_{\mathscr{L}(\mathcal{H})} \leq A$.

For the proof of the Cotlar-Stein lemma, see e.g. [Ste93, Ch. VII §2], and for its history see Knapp and Stein [KS69].

When working on groups, one sometimes has to deal with operators mapping the L^2-space on the group to the L^2-space on its unitary dual. This requires one to use the version of Cotlar's lemma for operators mapping between two different Hilbert spaces. In this case, the statement of Theorem A.5.2 still holds, for an operator $T : \mathcal{H} \to \mathcal{G}$, provided we take the operator norms $T_i^* T_j$ and $T_i T_j^*$ in appropriate spaces. For details, we refer to [RT10a, Theorem 4.14.1].

The following crude version of the Cotlar lemma will be also useful to us:

Proposition A.5.3 (Cotlar-Stein lemma; crude version). *Let $\mathcal{H}$ be a Hilbert space and $\{T_k,\ k \in \mathbb{Z}\}$ be a sequence of linear operators on $\mathcal{H}$. We assume that*

$$T_i T_j^* = 0 \qquad \text{if } i \neq j. \tag{A.9}$$

We also assume that the operators T_k, $k \in \mathbb{Z}$, are uniformly bounded,

$$\text{i.e.} \quad \sup_{k \in \mathbb{Z}} \|T_k\|_{\mathscr{L}(\mathcal{H})} < \infty, \tag{A.10}$$

and that the following sum is finite

$$\sum_{i \neq j} \|T_i^* T_j\|_{\mathscr{L}(\mathcal{H})} < \infty. \tag{A.11}$$

Then the series $\sum_{k \in \mathbb{Z}} T_k$ converges in the strong operator topology to an operator S satisfying

$$\|S\|_{\mathscr{L}(\mathcal{H})}^2 \leq 2 \max \left(\sup_{k \in \mathbb{Z}} \|T_k\|_{\mathscr{L}(\mathcal{H})}^2, \sum_{i \neq j} \|T_i^* T_j\|_{\mathscr{L}(\mathcal{H})} \right).$$

For the proof of this statement, see [Ste93, Ch. VII §2.3].

Remark A.5.4. The condition (A.9) can can be relaxed slightly with the following modifications.

For instance, (A.9) can be replaced with

$$T_i^* T_j = 0 \quad \text{if} \quad i \neq j \text{ have the same parity.}$$

(This condition appears often when considering dyadic decomposition.) Indeed, applying Proposition A.5.3 to $\{T_{2k+1}\}_{k\in\mathbb{Z}}$ and to $\{T_{2k}\}_{k\in\mathbb{Z}}$, we obtain that the series $\sum_k T_k = \sum_k T_{2k} + \sum_k T_{2k+1}$ converges in the strong operator norm topology to an operator S satisfying

$$\|S\|_{\mathscr{L}(\mathcal{H})} \leq 2^{1/2} \times 2 \times \max\left(\sup_{k\in\mathbb{Z}} \|T_k\|_{\mathscr{L}(\mathcal{H})}, \big(2 \sum_{i-j\in 2\mathbb{N}} \|T_i^* T_j\|_{\mathscr{L}(\mathcal{H})}\big)^{1/2} \right).$$

More generally, (A.9) can be replaced with

$$T_i^* T_j = 0 \quad \text{for} \quad |i-j| > a,$$

where $a \in \mathbb{N}$ is a fixed positive integer. It suffices to apply Proposition A.5.3 to each $\{T_{ak+b}\}_{k\in\mathbb{Z}}$ for $b = 0, \ldots, a-1$. Then the series $\sum T_k = \sum_{0\leq b<a} T_{ak+b}$ converges in the strong operator norm topology to an operator S satisfying

$$\|S\|_{\mathscr{L}(\mathcal{H})} \leq 2^{1/2} \times a \times \max\left(\sup_{k} \|T_k\|_{\mathscr{L}(\mathcal{H})}, \big(2 \sum_{i-j>a} \|T_i^* T_j\|_{\mathscr{L}(\mathcal{H})}\big)^{1/2} \right).$$

A.6 Interpolation of analytic families of operators

Let $(M, \mathcal{M}, \mu)$ and $(N, \mathcal{N}, \nu)$ be measure spaces. We suppose that to each $z \in \mathbb{C}$ in the strip

$$S := \{z \in \mathbb{C} \ : \ 0 \leq \operatorname{Re} z \leq 1\},$$

there corresponds a linear operator T_z from the space of simple functions in $L^1(M)$ to measurable functions on N, in such a way that $(T_z f)g$ is integrable on N whenever f is a simple function in $L^1(M)$ and g is a simple function in $L^1(N)$. (Recall that a simple function is a measurable function which takes only a finite number of values.)

We assume that the family $\{T_z\}_{z\in S}$ is admissible in the sense that the mapping

$$z \mapsto \int_N (T_z f) g \, d\nu$$

is analytic in the interior of S, continuous on S, and there exists a constant $a < \pi$ such that

$$e^{-a|\operatorname{Im} z|} \ln \left| \int_N (T_z f) g \, d\nu \right|,$$

is uniformly bounded from above in the strip S.

Theorem A.6.1. *Let $\{T_z\}_{z\in S}$ be an admissible family as above. We assume that*

$$\|T_{iy}f\|_{q_0} \leq M_0(y)\|f\|_{p_0} \quad \text{and} \quad \|T_{1+iy}f\|_{q_1} \leq M_1(y)\|f\|_{p_1},$$

for all simple functions in $L^1(M)$ where $1 \leq p_j, q_j \leq \infty$, and functions $M_j(y)$, $j = 1, 2$ are independent of f and satisfy

$$\sup_{y\in\mathbb{R}} e^{-b|y|} \ln M_j(y) < \infty,$$

for some $b < \pi$. Then if $0 \leq t \leq 1$, there exists a constant M_t such that

$$\|T_t f\|_{q_t} \leq M_t \|f\|_{p_t},$$

for all simple functions f in $L^1(M)$, provided that

$$\frac{1}{p_t} = (1-t)\frac{1}{p_0} + t\frac{1}{p_1} \quad \text{and} \quad \frac{1}{q_t} = (1-t)\frac{1}{q_0} + t\frac{1}{q_1}.$$

For the proof of this theorem, we refer e.g. to [SW71, ch. V §4].

Remark A.6.2. The following remarks are useful.

- The constant M_t depends only on t and on $a, b, M_0(y), M_1(y)$, but not on T.
- From the proof, it appears that, if $N = M = \mathbb{R}^n$ is endowed with the usual Borel structure and the Lebesgue measures, one can require the assumptions and the conclusion to be on simple functions f with compact support.

We also refer to Definition 6.4.17 for the notion of the complex interpolation (which requires stronger estimates).

Appendix B

Group C^* and von Neumann algebras

In this chapter we make a short review of the machinery related to group von Neumann algebras that will be useful for setting up the Fourier analysis in other parts of book, in particular in Section 1.8.2. We try to make a short and concise presentation of notions and ideas without proofs trying to make the presentation as informal as possible. All the material presented in this chapter is well known but is often scattered over the literature in different languages and with different notation. Here we collect what is necessary for us giving references along the exposition. The final aim of this chapter is to introduce the notion of the von Neumann algebra of the group (or the group von Neumann algebra) and describe its main properties.

B.1 Direct integral of Hilbert spaces

We start by describing direct integrals of Hilbert spaces. For more details and overall proofs we can refer to more classical literature such as Bruhat [Bru68] or to more modern exposition of Folland [Fol95, p. 219].

B.1.1 Convention: Hilbert spaces are assumed separable

All the Hilbert spaces considered in this chapter are separable, unless stated otherwise. Let us recall the definition and some properties of separable spaces.

Definition B.1.1. A topological space is *separable* if its topology admits a countable basis of neighbourhoods.

When a topological space is metrisable, being separable is equivalent to having a (countable) sequence which is dense in the space.

V. Fischer, M. Ruzhansky, *Quantization on Nilpotent Lie Groups*, Progress in Mathematics 314,
DOI 10.1007/978-3-319-29558-9

Moreover, a separable Hilbert space of infinite dimension is unitarily equivalent to the Hilbert space of square integrable complex sequences: that is, to

$$\ell^2(\mathbb{N}_0) = \{(x_j)_{j\in\mathbb{N}_0}, \sum_{j=0}^{\infty} |x_j|^2 < \infty\}.$$

Naturally a separable Hilbert space of finite dimension n is unitarily equivalent to $\mathbb{C}^n$.

We can refer e.g. to Rudin [Rud91] for different topological implications of the separability.

B.1.2 Measurable fields of vectors

Here we recall the definitions of measurable fields of Hilbert spaces, of vectors and of operators.

Definition B.1.2. Let Z be a set and let $(\mathcal{H}_\zeta)_{\zeta\in Z}$ is a family of vector spaces (on the same field) indexed by Z. Then $\prod_{\zeta\in Z}\mathcal{H}_\zeta$ denotes the *direct product* of $(\mathcal{H}_\zeta)_{\zeta\in Z}$, that is, the set of all tuples $v = (v(\zeta))_{\zeta\in Z}$ with $v(\zeta) \in \mathcal{H}_\zeta$ for each $\zeta \in Z$. It is naturally endowed with a structure of a vector space with addition and scalar multiplication being performed componentwise.

An element of $\prod_{\zeta\in Z}\mathcal{H}_\zeta$, that is, a tuple $v = (v(\zeta))_{\zeta\in Z}$, may be called a field of vectors parametrised by Z, or, when no confusion is possible, a *vector field.*

We will use this definition for a measurable space Z. In practice, for the set Γ in the following definition, we may also choose $\Gamma \subset \prod_{\zeta\in Z}\mathcal{H}_\zeta^\infty$ in view of Gårding's theorem (see Proposition 1.7.7).

Definition B.1.3. Let Z be a measurable space and μ a positive sigma-finite measure on Z. A *μ-measurable field of Hilbert spaces* over Z is a pair $\mathcal{E} = \big((\mathcal{H}_\zeta)_{\zeta\in Z}, \Gamma\big)$ where $(\mathcal{H}_\zeta)_{\zeta\in Z}$ is a family of (separable) Hilbert spaces indexed by Z and where $\Gamma \subset \prod_{\zeta\in Z}\mathcal{H}_\zeta$ satisfies the following conditions:

(i) Γ is a vector subspace of $\prod_{\zeta\in Z}\mathcal{H}_\zeta$;

(ii) there exists a sequence $(x_\ell)_{\ell\in\mathbb{N}}$ of elements of Γ such that for every $\zeta \in Z$, the sequence $(x_\ell(\zeta))_{\ell\in\mathbb{N}}$ spans $\mathcal{H}_\zeta$ (in the sense that the subspace formed by the finite linear combination of the $x_\ell(\zeta), \ell \in \mathbb{N}$, is dense in $\mathcal{H}_\zeta$);

(iii) for every $x \in \Gamma$, the function $\zeta \mapsto \|x(\zeta)\|_{\mathcal{H}_\zeta}$ is μ-measurable;

(iv) if $x \in \prod_{\zeta\in Z}\mathcal{H}_\zeta$ is such that for every $y \in \Gamma$, the function

$$Z \ni \zeta \mapsto (x(\zeta), y(\zeta))_{\mathcal{H}_\zeta}$$

is measurable, then $x \in \Gamma$.

Under these conditions, the elements of Γ are called the *measurable vector fields* of $\mathcal{E}$. We always identify two vector fields which are equal almost everywhere. This means that we identify two elements x and x' of Γ when, for every $y \in \Gamma$, the two mappings

$$Z \ni \zeta \mapsto (x(\zeta), y(\zeta))_{\mathcal{H}_\zeta} \quad \text{and} \quad Z \ni \zeta \mapsto (x'(\zeta), y(\zeta))_{\mathcal{H}_\zeta},$$

can be identified as measurable functions.

A vector field x is *square integrable* if $x \in \Gamma$ and $\int_Z \|x(\zeta)\|^2_{\mathcal{H}_\zeta} d\mu(\zeta) < \infty$. One may write then

$$x = \int_Z^{\oplus} x(\zeta) d\mu(\zeta).$$

The set of square integrable vector fields form a (possibly non-separable) Hilbert space denoted by

$$\mathcal{H} := \int_Z^{\oplus} \mathcal{H}_\zeta d\mu(\zeta),$$

and called the *direct integral* of the $\mathcal{H}_\zeta$. The inner product is given via

$$(x|y)_{\mathcal{H}} = \int_Z^{\oplus} (x(\zeta)|y(\zeta)_{\mathcal{H}_\zeta} d\mu(\zeta), \quad x, y \in \mathcal{H}.$$

B.1.3 Direct integral of tensor products of Hilbert spaces

After a brief recollection of the definitions of tensor products, we will be able to analyse the direct integral of tensor products of Hilbert spaces, as well as their decomposable operators.

Definition of tensor products

Here we define firstly the algebraic tensor product of two vector spaces, and secondly the tensor products of Hilbert spaces.

Definition B.1.4. Let V and W be two complex vector spaces.

The free space generated by V and W is the vector space $\mathbb{F}(V \times W)$ linearly spanned by $V \times W$, that is, the space of finite $\mathbb{C}$-linear combinations of elements of $V \times W$.

The *algebraic tensor product* of V and W is the quotient of $\mathbb{F}(V \times W)$ by its subspace generated by the following elements

$$\begin{aligned} (v_1, w) + (v_2, w) - (v_1 + v_2, w), &\qquad (v, w_1) + (v, w_2) - (v, w_1 + w_2), \\ c(v, w) - (cv, w), &\qquad c(v, w) - (v, cw), \end{aligned}$$

where v, v_1, v_2 are arbitrary elements of V, w, w_1, w_2 are arbitrary elements of W, and c is an arbitrary complex number.

The equivalence class of an element $(v, w) \in V \times W \subset \mathbb{F}(V \times W)$ is denoted $v \otimes w$.

The algebraic tensor product of V and W is naturally a complex vector space which we will denote in this monograph by

$$V \overset{alg}{\otimes} W.$$

The algebraic tensor product has the following universal property (which may be given as an alternate definition):

Proposition B.1.5 (Universal property). *Let V, W and X be (complex) vector spaces and let $\Psi : V \times W \to X$ be a bilinear mapping. Then there exists a unique map $\tilde{\Psi} : V \overset{alg}{\otimes} W \to X$ such that*

$$\Psi = \tilde{\Psi} \circ \pi$$

where $\pi : V \times W \to V \overset{alg}{\otimes} W$ is the map defined by $\pi(v, w) = v \otimes w$.

More can be said when the complex vector spaces are also Hilbert spaces. Indeed one checks easily:

Lemma B.1.6. *Let $\mathcal{H}_1$ and $\mathcal{H}_2$ be Hilbert spaces. Then the mapping defined on $\mathcal{H}_1 \overset{alg}{\otimes} \mathcal{H}_2$ via*

$$(u_1 \otimes v_1, u_2 \otimes v_2) := (u_1, u_2)(v_1, v_2), \quad u_1, u_2 \in \mathcal{H}_1,\ v_1, v_2 \in \mathcal{H}_2,$$

is a complex inner product on $\mathcal{H}_1 \overset{alg}{\otimes} \mathcal{H}_2$.

This shows that $\mathcal{H}_1 \overset{alg}{\otimes} \mathcal{H}_2$ is a pre-Hilbert space.

Definition B.1.7. The *tensor product of the Hilbert spaces* $\mathcal{H}_1$ and $\mathcal{H}_2$ is the completion of $\mathcal{H}_1 \overset{alg}{\otimes} \mathcal{H}_2$ for the natural sesquilinear form from Lemma B.1.6. It is denoted by $\mathcal{H}_1 \otimes \mathcal{H}_2$.

Naturally we have the universal property of tensor products of Hilbert spaces:

Proposition B.1.8 (Universal property). *Let $\mathcal{H}_1$, $\mathcal{H}_2$ and $\mathcal{H}$ be Hilbert spaces and let $\Psi : \mathcal{H}_1 \times \mathcal{H}_2 \to \mathcal{H}$ be a continuous bilinear mapping. Then there exists a unique continuous map $\tilde{\Psi} : \mathcal{H}_1 \otimes \mathcal{H}_2 \to \mathcal{H}$ such that*

$$\Psi = \tilde{\Psi} \circ \pi$$

where $\pi : \mathcal{H}_1 \times \mathcal{H}_2 \to \mathcal{H}_1 \otimes \mathcal{H}_2$ is the map defined by $\pi(v, w) = v \otimes w$.

Tensor products of Hilbert spaces as Hilbert-Schmidt spaces

The tensor product of two Hilbert spaces may be identified with a space of Hilbert Schmidt operators in the following way. To any vector $w \in \mathcal{H}_2$, we associate the continuous linear form on $\mathcal{H}_2$

$$w^* : v \longmapsto (v, w)_{\mathcal{H}_2}.$$

Conversely any element of $\mathcal{H}_2^*$, that is, any continuous linear form on $\mathcal{H}_2$, is of this form. To any $u \in \mathcal{H}_1$ and $v \in \mathcal{H}_2$, we associate the rank-one operator

$$\Psi_{u,v} : \begin{cases} \mathcal{H}_2^* & \longrightarrow & \mathcal{H}_1 \\ w^* & \longmapsto & w^*(v)u \end{cases}$$

Lemma B.1.9. *With the notation above, the continuous bilinear mapping*

$$\Psi : \mathcal{H}_1 \times \mathcal{H}_2 \to \mathtt{HS}(\mathcal{H}_2^*, \mathcal{H}_1)$$

extends to an isometric isomorphism of Hilbert spaces

$$\tilde{\Psi} : \mathcal{H}_1 \otimes \mathcal{H}_2 \to \mathtt{HS}(\mathcal{H}_2^*, \mathcal{H}_1).$$

Moreover, if $T_1 \in \mathscr{L}(\mathcal{H}_1)$ and $T_2 \in \mathscr{L}(\mathcal{H}_2)$, then the operator $T_1 \otimes T_2$ defined via

$$(T_1 \otimes T_2)(v_1 \otimes v_2) := (T_1 v_1) \otimes (T_2 v_2), \quad v_1 \in \mathcal{H}_1, v_2 \in \mathcal{H}_2,$$

is in $\mathscr{L}(\mathcal{H}_1 \otimes \mathcal{H}_2)$ and corresponds to the bounded operator

$$\tilde{\Psi}(T_1 \otimes T_2)\tilde{\Psi}^{-1} : \begin{cases} \mathtt{HS}(\mathcal{H}_2^*, \mathcal{H}_1) & \longrightarrow & \mathtt{HS}(\mathcal{H}_2^*, \mathcal{H}_1) \\ A & \longmapsto & T_1 A T_2 \end{cases}.$$

Recall that the scalar product of $\mathtt{HS}(\mathcal{H}_2^*, \mathcal{H}_1)$ is given by

$$(T_1, T_2)_{\mathtt{HS}(\mathcal{H}_2^*, \mathcal{H}_1)} = \sum_j (T_1 f_j^*, T_2 f_j^*)_{\mathcal{H}_1}.$$

where $(f_j^*)_{j \in \mathbb{N}}$ is any orthonormal basis of $\mathcal{H}_2^*$.

Proof. By Proposition B.1.8, Ψ leads to a continuous linear mapping $\tilde{\Psi} : \mathcal{H}_1 \otimes \mathcal{H}_2 \to \mathtt{HS}(\mathcal{H}_2^*, \mathcal{H}_1)$. The image of $\tilde{\Psi}$ contains the rank-one operators, thus all the finite ranked operators which form a dense subset of $\mathtt{HS}(\mathcal{H}_2^*, \mathcal{H}_1)$. Thus $\tilde{\Psi}$ is surjective.

If $(f_j^*)_{j \in \mathbb{N}}$ is an orthonormal basis of $\mathcal{H}_2^*$, we can compute easily the scalar product between Ψ_{u_1,v_1} and Ψ_{u_2,v_2}:

$$\begin{aligned}(\Psi_{u_1,v_1}, \Psi_{u_2,v_2})_{\mathtt{HS}(\mathcal{H}_2^*, \mathcal{H}_1)} &= \sum_j (\Psi_{u_1,v_1} f_j^*, \Psi_{u_2,v_2} f_j^*)_{\mathcal{H}_1} \\ &= \sum_j (f_j^*(v_1)u_1, f_j^*(v_2)u_2)_{\mathcal{H}_1} = (u_1, u_2)_{\mathcal{H}_1} \sum_j f_j^*(v_1)\overline{f_j^*(v_2)} \\ &= (u_1, u_2)_{\mathcal{H}_1} \sum_j (v_1, f_j)\overline{(v_2, f_j)} = (u_1, u_2)_{\mathcal{H}_1} (v_1, v_2)_{\mathcal{H}_2}.\end{aligned}$$

This implies that the mapping $\tilde{\Psi} : \mathcal{H}_1 \otimes \mathcal{H}_2 \to \mathtt{HS}(\mathcal{H}_2^*, \mathcal{H}_1)$ is an isometry.

For the last part of the statement, one checks easily that

$$(T_1 \Psi_{u,v} T_2)(w^*) = w^*(T_2 v)\, T_1 u,$$

concluding the proof. □

Let us apply this to $\mathcal{H}_1 = \mathcal{H}$ and $\mathcal{H}_2 = \mathcal{H}^*$.

Corollary B.1.10. *Let $\mathcal{H}$ be a Hilbert space. The Hilbert space given by the tensor product $\mathcal{H} \otimes \mathcal{H}^*$ of Hilbert spaces is isomorphic to $\mathtt{HS}(\mathcal{H})$ via*

$$u \otimes v^* \longleftrightarrow \Psi_{u,v}, \quad \Psi_{u,v}(w) = (w, v)_{\mathcal{H}} u.$$

Via this isomorphism, the bounded operator $T_1 \otimes T_2^*$ where $T_1, T_2 \in \mathscr{L}(\mathcal{H})$, corresponds to the bounded operator

$$\tilde{\Psi}(T_1 \otimes T_2)\tilde{\Psi}^{-1} : \left\{ \begin{array}{rcl} \mathtt{HS}(\mathcal{H}) & \longrightarrow & \mathtt{HS}(\mathcal{H}) \\ A & \longmapsto & T_1 A T_2^* \end{array} \right. .$$

Direct integral of tensor products of Hilbert spaces

Let μ be a positive sigma-finite measure on a measurable space Z and $\mathcal{E} = \big((\mathcal{H}_\zeta)_{\zeta \in Z}, \Gamma\big)$ a *μ-measurable field of Hilbert spaces* over Z. Then

$$\mathcal{E}^{\otimes} := \big((\mathcal{H}_\zeta \otimes \mathcal{H}_\zeta^*)_{\zeta \in Z}, \Gamma \otimes \Gamma^*\big)$$

is a *μ-measurable field of Hilbert spaces* over Z.

Identifying each tensor product $\mathcal{H}_\zeta \otimes \mathcal{H}_\zeta^*$ with $\mathtt{HS}(\mathcal{H}_\zeta)$, see Corollary B.1.10, we may write

$$\int_Z^{\oplus} \mathcal{H}_\zeta \otimes \mathcal{H}_\zeta^* d\mu(\zeta) \equiv \int_Z^{\oplus} \mathtt{HS}(\mathcal{H}_\zeta) d\mu(\zeta).$$

Furthermore if $x \in \int_Z^{\oplus} \mathcal{H}_\zeta \otimes \mathcal{H}_\zeta^* d\mu(\zeta)$ then

$$\|x\|^2 = \int_Z \|x(\zeta)\|^2_{\mathtt{HS}(\mathcal{H}_\zeta)} d\mu(\zeta).$$

B.1.4 Separability of a direct integral of Hilbert spaces

In this chapter, we are always concerned with separable Hilbert spaces (see Section B.1.1). A sufficient condition to ensure the separability of a direct integral is that the measured space is standard (the definition of this notion is recalled below):

Proposition B.1.11. *Keeping the setting of Definition B.1.3, if (Z, μ) is a standard space, then $\int_Z^{\oplus} \mathcal{H}_\zeta d\mu(\zeta)$ is a separable Hilbert space.*

For the proof we refer to Dixmier [Dix96, §II.1.6].

Definition B.1.12. A measurable space Z is a *standard Borel space* if Z is a Polish space (i.e. a separable complete metrisable topological space) and the considered sigma-algebra is the Borel sigma-algebra of Z (i.e. the smallest sigma-algebra containing the open sets of Z).

These Borel spaces have a simple classification: they are isomorphic (as Borel spaces) either to a (finite or infinite) countable set, or to $[0,1]$. For these and other details see, for instance, Kechris [Kec95, Chapter II, Theorem 15.6] and its proof.

Definition B.1.13. A positive measure μ on a measure space Z is a *standard measure* if μ is sigma-finite, (i.e. there exists a sequence of mutually disjoint measurable sets $Y_1, Y_2, \ldots$ such that $\mu(Y_j) < \infty$ and $Z = Y_1 \cup Y_2 \cup \ldots$) and there exists a null set E such that $Z \backslash E$ is a standard Borel space.

In this monograph, we consider only the setting described in Proposition B.2.24 which is standard.

B.1.5 Measurable fields of operators

Let Z be a measurable space and μ a positive sigma-finite measure on Z. The main application for our analysis of these constructions will be in Section 1.8.3 dealing with measurable fields of operators over $\widehat{G}$.

Definition B.1.14. Let $\mathcal{E} = \big((\mathcal{H}_\zeta)_{\zeta\in Z}, \Gamma\big)$ be a μ-measurable field of Hilbert spaces over Z. A *μ-measurable field of operators* over Z is a collection of operators $(T(\zeta))_{\zeta\in Z}$ such that $T(\zeta) \in \mathscr{L}(\mathcal{H}_\zeta)$ and for any $x \in \Gamma$, the field $(T(\zeta)x(\zeta))_{\zeta\in Z}$ is measurable. If furthermore the function $\zeta \mapsto \|T(\zeta)\|_{\mathscr{L}(\mathcal{H}_\zeta)}$ is μ-essentially bounded, then the field of operators $(T(\zeta))_{\zeta\in Z}$ is *essentially bounded.*

Let us continue with the notation of Definition B.1.14. Let $(T(\zeta))_{\zeta\in Z}$ be an essentially bounded field of operators. Then we can define the operator T on the Hilbert space $\mathcal{H} = \int_Z^\oplus \mathcal{H}_\zeta d\mu(\zeta)$ via $(Tx)(\zeta) := T(\zeta)x(\zeta)$. Clearly the operator T is linear and bounded. It is often denoted by

$$T := \int_Z^\oplus T(\zeta)d\mu(\zeta).$$

Naturally two fields of operators which are equal up to a μ-negligible set yield the same operator on $\mathcal{H}$ and may be identified. Furthermore the operator norm of $T \in \mathscr{L}(\mathcal{H})$ is

$$\|T\|_{\mathscr{L}(\mathcal{H})} = \sup_{\zeta\in Z} \|T(\zeta)\|_{\mathscr{L}(\mathcal{H}_\zeta)},$$

where sup denotes here the essential supremum with respect to μ.

Definition B.1.15. An operator on $\mathcal{H}$ as above, that is, obtained via

$$T := \int_Z^\oplus T(\zeta)d\mu(\zeta)$$

where $(T(\zeta))_{\zeta\in Z}$ is an essentially bounded field of operators, is said to be *decomposable.*

The set of decomposable operators form a subspace of $\mathscr{L}(\mathcal{H})$ stable by composition and taking the adjoint.

B.1.6 Integral of representations

In the following definition, μ is a positive sigma-finite measure on a measurable space Z, $\mathcal{A}$ is a separable C^*-algebra, and G is a (Hausdorff) locally compact separable group. For further details on the constructions of this section we refer to Dixmier [Dix77, §8]. For the definition of representations of C^*-algebras see Definition B.2.16.

Definition B.1.16. Let $\mathcal{E} = \big((\mathcal{H}_\zeta)_{\zeta\in Z}, \Gamma\big)$ be a μ-measurable field of Hilbert spaces over Z. A *μ-measurable field of representations* of $\mathcal{A}$, resp. G, is a μ-measurable field of operator $(T(\zeta))_{\zeta\in Z}$ (see Definition B.1.14) such that for each $\zeta \in Z$, $T(\zeta) = \pi_\zeta$ is a representation of $\mathcal{A}$, resp. a unitary continuous representation of G, in $\mathcal{H}_\zeta$.

In this case, for each $x \in G$, we can define the operator

$$\pi(x) := \int_Z^\oplus \pi_\zeta(x) d\mu(\zeta) \quad \text{acting on } \mathcal{H} := \int_Z^\oplus \mathcal{H}_\zeta d\mu(\zeta).$$

One checks easily that this yields a representation π of $\mathcal{A}$, resp. a unitary continuous representation of G, on $\mathcal{H}$ denoted by

$$\pi := \int_Z^\oplus \pi_\zeta d\mu(\zeta),$$

often called *the integral of the representations* $(\pi_\zeta)_{\zeta\in Z}$.

The following technical properties give sufficient conditions for two integrals of representations to yield equivalent representations. Again $\mathcal{A}$ is a separable C^*-algebra and G a (Hausdorff) locally compact separable group.

Proposition B.1.17. *Let μ_1 and μ_2 be two positive sigma-finite measures on measurable spaces Z_1 and Z_2 respectively. For $j = 1, 2$, let $\mathcal{E}_j = \big((\mathcal{H}^{(j)}_{\zeta_j})_{\zeta_j\in Z_j}, \Gamma_j\big)$ be a μ_j-measurable field of Hilbert spaces over Z_j and let $(\pi^{(j)}_{\zeta_j})$ be a measurable field of representations of $\mathcal{A}$, resp. of unitary continuous representations of G.*

We assume that μ_1 and μ_2 are standard. We also assume that there exist a Borel μ_1-negligible part $E_1 \subset Z_1$, a Borel μ_2-negligible part $E_2 \subset Z_2$ and a Borel isomorphism $\eta : Z_1\backslash E_1 \to Z_2\backslash E_2$ which transforms μ_1 to μ_2 and such that $\pi^{(1)}_{\zeta_1}$ and $\pi^{(2)}_{\eta(\zeta_1)}$ are equivalent for any $\zeta_1 \in Z_1\backslash E_1$. Then there exists a unitary mapping from $\mathcal{H}^{(1)} := \int_{Z_1}^\oplus \mathcal{H}^{(1)}_{\zeta_1} d\mu_1(\zeta_1)$ onto $\mathcal{H}^{(2)} := \int_{Z_2}^\oplus \mathcal{H}^{(2)}_{\zeta_2} d\mu_2(\zeta_2)$ which intertwines the representations of $\mathcal{A}$, resp. the unitary continuous representations of G,

$$\pi^{(1)} := \int_{Z_1}^\oplus \pi^{(1)}_{\zeta_1} d\mu_1(\zeta_1) \quad \text{and} \quad \pi^{(2)} := \int_{Z_2}^\oplus \pi^{(2)}_{\zeta_2} d\mu_2(\zeta_2).$$

B.2 C^*- and von Neumann algebras

The main reference for this section are Dixmier's books [Dix81, Dix77], Arveson [Arv76] or Blackadar [Bla06]. For a more basic introduction to C^*-algebras and elements of the Gelfand theory see also Ruzhansky and Turunen [RT10a, Chapter D].

B.2.1 Generalities on algebras

Here we recall the definitions of an algebra, together with its possible additional structures (involution, norm) and sets usually associated with it (spectrum, bi-commutant).

Algebra

Let us start with the definition of an algebra over a field.

Let $\mathcal{A}$ be a vector space over a field $\mathbb{K}$ equipped with an additional binary operation

$$\begin{array}{ccc} \mathcal{A} \times \mathcal{A} & \longrightarrow & \mathcal{A}, \\ (x, y) & \longmapsto & x \cdot y. \end{array}$$

It is an *algebra* over $\mathbb{K}$ when the binary operation (then often called the product) satisfies:

- left distributivity: $(x + y) \cdot z = x \cdot z + y \cdot z$ for any $x, y, z \in \mathcal{A}$,
- right distributivity: $z \cdot (x + y) = z \cdot x + z \cdot y$ for any $x, y, z \in \mathcal{A}$,
- compatibility with scalars: $(ax) \cdot (by) = (ab)(x \cdot y)$ for any $x, y \in \mathcal{A}$ and $a, b \in \mathbb{K}$.

The algebra $\mathcal{A}$ is said to be *unital* when there exists a unit, that is, an element $1 \in \mathcal{A}$ such that $x \cdot 1 = 1 \cdot x = x$ for every $x \in \mathcal{A}$.

A subspace $\mathcal{Y} \subset \mathcal{A}$ is a *sub-algebra* of $\mathcal{A}$ whenever $y_1 \cdot y_2 \in \mathcal{Y}$ for any $y_1, y_2 \in \mathcal{Y}$.

Commutant and bi-commutant

We will need the notion of commutant:

Definition B.2.1. Let $\mathcal{M}$ be a subset of the algebra $\mathcal{A}$. The *commutant* of $\mathcal{M}$ is the set denoted by $\mathcal{M}'$ of the elements which commute with all the elements of $\mathcal{M}$, that is,

$$\mathcal{M}' := \{x \in \mathcal{A} \ : \ xm = mx \ \text{ forall } \ m \in \mathcal{M}\}.$$

The *bi-commutant* of $\mathcal{M}$ is the commutant of the commutant of $\mathcal{M}$, that is,

$$\mathcal{M}'' := (\mathcal{M}')'.$$

Keeping the notation of Definition B.2.1, one checks easily that a commutant $\mathcal{M}'$ is a sub-algebra of $\mathcal{A}$. It contains the unit if $\mathcal{A}$ is unital. Furthermore, in any case, $\mathcal{M} \subset \mathcal{M}''$.

Involution and norms

We consider now algebras endowed with an involution:

Definition B.2.2. Let $\mathcal{A}$ be an algebra over the complex numbers $\mathbb{C}$. It is called an *involutive algebra* or a $*$*-algebra* when there exists a map $* : \mathcal{A} \to \mathcal{A}$ which is

- sesquilinear (that is, $(ax+by)^* = \bar{a}x^* + \bar{b}y^*$ for every $x, y \in \mathcal{A}$ and $a, b \in \mathbb{C}$),
- involutive (that is, $(x^*)^* = x$ for every $x \in \mathcal{A}$).

In this case, x^* may be called the *adjoint* of $x \in \mathcal{A}$. An element $x \in \mathcal{A}$ is *hermitian* if $x^* = x$. An element $x \in \mathcal{A}$ is *unitary* if $xx^* = x^*x = 1$.

Example B.2.3. Let $\mathcal{A}$ be a $*$-algebra. If $\mathcal{M}$ is a subset of $\mathcal{A}$ stable under the involution (that is, $m^* \in \mathcal{M}$ for every $m \in \mathcal{M}$), then its commutant $\mathcal{M}'$ is a $*$-subalgebra of $\mathcal{A}$.

Definition B.2.4. A *normed involutive algebra* is an involutive algebra $\mathcal{A}$ endowed with a norm $\|\cdot\|$ such that

$$\|x^*\| = \|x\|$$

for each $x \in \mathcal{A}$. If, in addition, $\mathcal{A}$ is $\|\cdot\|$-complete, then $\mathcal{A}$ is called an *involutive Banach algebra.*

The notions of (involutive, normed involutive / involutive Banach) sub-algebra and morphism between (involutive / normed involutive / involutive Banach) algebras follow naturally. Furthermore if $\mathcal{A}$ is a (involutive / normed involutive / involutive Banach) non unital algebra, then there exists a unique (involutive / normed involutive / involutive Banach) unital algebra $\tilde{\mathcal{A}} = \mathcal{A} \oplus \mathbb{C}1$, up to isomorphism, which contains $\mathcal{A}$ as a (involutive, normed involutive / involutive Banach) sub-algebra.

Examples

Example B.2.5. The complex field $\mathcal{A} = \mathbb{C}$ is naturally a unital commutative involutive Banach algebra.

Example B.2.6. Let X be a locally compact space and let $\mathcal{A} = C_o(X)$ be the space of continuous functions $f : X \to \mathbb{C}$ vanishing at infinity, that is, for every $\epsilon > 0$, there exists a compact neighbourhood out of which $|f| < \epsilon$. Then $\mathcal{A}$ is a commutative involutive Banach algebra when endowed with pointwise multiplication and involution $f \mapsto \bar{f}$. When X is a singleton, this reduces to Example B.2.5.

Example B.2.7. If η is a positive measure on a measurable space X and if $\mathcal{A}$ is the space of η-essentially bounded functions $f : X \to \mathbb{C}$, that is, $\mathcal{A} = L^\infty(X, \eta)$, then $\mathcal{A}$ is a unital commutative involutive Banach algebra when endowed with pointwise multiplication and involution $f \mapsto \bar{f}$. When X is a singleton, this reduces to Example B.2.5.

Recall that all the Hilbert spaces we consider are separable.

Example B.2.8. The space $\mathscr{L}(\mathcal{H})$ of continuous linear operators on a Hilbert space $\mathcal{H}$ is naturally a unital involutive Banach algebra for the usual structure. This means that the product is given by the composition of operators $(A, B) \mapsto AB$, the involution by the adjoint and the norm by the operator norm. The unit is the identity mapping $\mathrm{I}_{\mathcal{H}} = \mathrm{I} : v \mapsto v$.

Example B.2.9. If G is a locally compact (Hausdorff) group which is unimodular, then $L^1(G)$ is naturally an involutive Banach algebra where the product is given by the convolution and the involution $f \mapsto f^*$ by $f^*(x) = \bar{f}(x^{-1})$. If G is separable then $L^1(G)$ is separable.

Example B.2.9 can be generalised to locally compact groups which are not necessarily unimodular. First, let us recall the following definitions:

Definition B.2.10. Let G be a locally compact (Hausdorff) group. Let us fix a left Haar measure dx. We also denote by $|E|$ the volume of a Borel set for this measure. Then there exists a unique function Δ such that

$$|Ex| = \Delta(x)|E|$$

for any Borel set E and $x \in G$. It is called the *modular* function of G and is independent of the chosen left Haar measure. It is a group homomorphism $G \to (\mathbb{R}^+, \times)$.

If the modular function is constant then $\Delta \equiv 1$ and G is said to be *unimodular.*

Remark B.2.11. Any Lie group is a separable locally compact (Hausdorff) group. Any compact (Hausdorff) group is necessarily a locally compact (Hausdorff) group and it is also unimodular. Any abelian locally compact (Hausdorff) group is unimodular. Any nilpotent or semi-simple Lie group is unimodular.

Example B.2.12. If G is a locally compact (Hausdorff) group then $L^1(G)$ is naturally an involutive Banach algebra often called the *group algebra.* The product is given by the convolution and the involution $f \mapsto f^*$ by

$$f^*(x) = \bar{f}(x^{-1})\Delta(x)^{-1},$$

where Δ is the modular function (see Definition B.2.10).

The space $M(G)$ of complex measures on G is also naturally an involutive Banach algebra and $L^1(G)$ may be viewed as a closed involutive sub-algebra. The algebra $M(G)$ always admits the Dirac measure δ_e at the neutral element of the group as unit.

Note that $L^1(G)$ is unital if and only if G is discrete and in this case $L^1(G) = M(G)$.

B.2.2 C^*-algebras

In this subsection we briefly review the notion of C^*-algebra and its main properties. We can refer to Ruzhansky and Turunen [RT10a, Chapter D] for a longer exposition.

Definition B.2.13. A C^**-algebra* is an involutive Banach algebra $\mathcal{A}$ such that

$$\|x\|^2 = \|x^*x\|$$

for every $x \in \mathcal{A}$.

Example B.2.14. Examples B.2.5, B.2.6, B.2.7, and B.2.8 are C^*-algebras.

Remark B.2.15.

1. If we choose a Hilbert space $\mathcal{H}$ of finite dimension n in Example B.2.8, the Banach algebra $\mathscr{L}(\mathcal{H}) \sim \mathscr{L}(\mathbb{C}^n) \sim \mathbb{C}^{n\times n}$ is a C^*-algebra if endowed with the operator norm, but is not a C^*-algebra when equipped with the Euclidean norm of $\mathbb{C}^{n^2}$ for instance.

2. Example B.2.6 is fundamental in the sense that one can show that any commutative C^*-algebra $\mathcal{A}$ is isomorphic to $C_o(X)$, where X is the spectrum of $\mathcal{A}$, that is, the set of non-zero complex homomorphisms with its usual topology. Moreover the isomorphism often called the Gelfand-Fourier transform is $*$-isometric. For further details see e.g. Rudin [Rud91] but with a different vocabulary.

3. In the non-commutative setting, the previous point may be generalised via the Gelfand-Naimark theorem: this theorem states that any C^*-algebra is $*$-isometric to a closed sub-$*$-algebra of $\mathscr{L}(\mathcal{H})$ for a suitable Hilbert space $\mathcal{H}$. Note that Example B.2.8 give the precise structure of $\mathscr{L}(\mathcal{H})$ and shows that a closed sub-$*$-algebra of $\mathscr{L}(\mathcal{H})$ is indeed a C^*-algebra. The proof is based on the Gelfand-Naimark-Segal construction, see e.g. Arveson [Arv76] for more precise statements.

The general definition of the spectrum of a (not necessarily commutative) C^* algebra is more involved than in the commutative case (Remark B.2.15 (2)):

Definition B.2.16 (Representations of C^*-algebras)**.** Let $\mathcal{A}$ be a C^*-algebra.

A *representation* of $\mathcal{A}$ is a continuous mapping $\mathcal{A} \to \mathscr{L}(\mathcal{H})$ for some Hilbert space $\mathcal{H}$, this mapping being a homomorphism of involutive algebras. Two representations $\pi_j : \mathcal{A} \to \mathscr{L}(\mathcal{H}_j)$, $j = 1, 2$, of $\mathcal{A}$, are *unitarily equivalent* if there exists a unitary operator $U : \mathcal{H}_1 \to \mathcal{H}_2$ such that $U\pi_1(x) = \pi_2(x)U$ for every $x \in \mathcal{A}$. A representation $\pi : \mathcal{A} \to \mathscr{L}(\mathcal{H})$ is *irreducible* if the only subspaces of $\mathcal{H}$ which are invariant under π, that is, under every $\pi(x)$, $x \in \mathcal{A}$, are trivial: $\{0\}$ and $\mathcal{H}$.

The *dual* (or spectrum) of $\mathcal{A}$ is the set of unitary irreducible representations of $\mathcal{A}$ modulo unitary equivalence. It is denoted by $\widehat{\mathcal{A}}$.

Remark B.2.17. The dual of a C^*-algebra is equipped with the hull-kernel topology due to Jacobson, and, if it is separable, with a structure of measurable space due to Mackey, see Dixmier [Dix77, §3].

B.2.3 Group C^*-algebras

In general, the group algebra of a locally compact (Hausdorff) group G, that is, the involutive Banach algebra $L^1(G)$ in Example B.2.12, is not a C^* algebra (see Remark B.2.26 below). The group C^* algebra is the C^*-enveloping algebra of $L^1(G)$, meaning that it is a 'small' C^* algebra containing $L^1(G)$ and built in the following way.

First, let us mention that many authors, for instance Jacques Dixmier, prefer to use for the Fourier transform

$$\pi_{\mathscr{D}}(f) := \int_G \pi(x) f(x) dx, \quad f \in L^1(G), \tag{B.1}$$

instead of $\pi(f)$ defined via

$$\pi(f) = \int_G \pi(x)^* f(x) dx, \quad f \in L^1(G), \tag{B.2}$$

which we adopt in this monograph, starting from (1.2), see Remark 1.1.4 for the explanation of this choice.

An advantage of using $\pi_{\mathscr{D}}$ would be that it yields a morphism of involutive Banach algebras from $L^1(G)$ to $\mathscr{L}(\mathcal{H}_\pi)$ as one checks readily:

Lemma B.2.18. *Let π be a unitary continuous representation of G. Then $\pi_{\mathscr{D}}$ is a (non-degenerate) representation of the involutive Banach algebra $L^1(G)$:*

$$\forall f, g \in L^1(G) \qquad \pi_{\mathscr{D}}(f * g) = \pi_{\mathscr{D}}(f)\pi_{\mathscr{D}}(g), \quad \pi_{\mathscr{D}}(f)^* = \pi_{\mathscr{D}}(f^*),$$

and

$$\|\pi_{\mathscr{D}}(f)\|_{\mathscr{L}(\mathcal{H}_\pi)} \leq \|f\|_{L^1(G)}.$$

For the proof, see Dixmier [Dix77, Proposition 13.3.1].

The choice of the Fourier transform in (B.2) made throughout this monograph, yields in contrast

$$\forall f, g \in L^1(G) \qquad \pi(f * g) = \pi(g)\pi(f)$$

and still

$$\pi(f)^* = \pi(f^*), \quad \|\pi(f)\|_{\mathscr{L}(\mathcal{H}_\pi)} \le \|f\|_{L^1(G)}.$$

The main advantage of our choice of Fourier transform is the fact that the Fourier transform of left-invariant operators will act on the left, as is customary in harmonic analysis, see our presentation of the abstract Plancherel theorem in Section 1.8.2.

Definition B.2.19. On $L^1(G)$, we can define $\|\cdot\|_*$ via

$$\|f\|_* := \sup_\pi \|\pi_{\mathcal{D}}(f)\|_{\mathscr{L}(\mathcal{H}_\pi)}, \quad f \in L^1(G),$$

where the supremum runs over all continuous unitary irreducible representations π of the group G.

One checks easily that $\|\cdot\|_*$ is a seminorm on $L^1(G)$ which satisfies

$$\|f\|_* \le \|f\|_{L^1} < \infty.$$

One can show that it is in fact also a norm on $L^1(G)$, see Dixmier [Dix77, §13.9.1].

Definition B.2.20. The *group C^*-algebra* is the Banach space obtained by completion of $L^1(G)$ for the norm $\|\cdot\|_*$. It is often denoted by $C^*(G)$.

Remark B.2.21. Choosing the definition of $\|\cdot\|_*$ using $\pi_{\mathcal{D}}$ as above or using our usual Fourier transform leads to the same C^*-algebra of the group. Indeed one checks easily that the adjoint of the operator $\pi(f)$ acting on $\mathcal{H}_\pi$ is $\pi_{\mathscr{D}}(\bar{f})$:

$$\pi(f) = \pi_{\mathscr{D}}(\bar{f})^* = \pi_{\mathscr{D}}(\bar{f}^*) \quad \text{and} \quad \|\pi(f)\|_{\mathscr{L}(\mathcal{H}_\pi)} = \|\pi_{\mathscr{D}}(\bar{f})\|_{\mathscr{L}(\mathcal{H}_\pi)}, \tag{B.3}$$

for all $f \in L^1(G)$.

Naturally $C^*(G)$ is a C^*-algebra and there are natural one-to-one correspondences between the representation theories of the group G, of the involutive Banach algebra $L^1(G)$, and of the C^*-algebra $C^*(G)$ in the following sense:

Lemma B.2.22. *If π is a continuous unitary representation of G, then $f \mapsto \pi_{\mathscr{D}}(f)$ defined via (B.1) is a non-degenerate $*$-representation of $L^1(G)$ which extends naturally to $C^*(G)$. Conversely any non-degenerate $*$-representation of $L^1(G)$ or $C^*(G)$ arise in this way.*

Hence

$$\|f\|_* = \sup_\pi \|\pi(f)\|_{\mathscr{L}(\mathcal{H}_\pi)}, \quad f \in L^1(G),$$

where the supremum runs over all representations π of the involutive Banach algebra $L^1(G)$ or over all representations π of the C^*-algebra $C^*(G)$.

For the proof see Dixmier [Dix77, §13.3.5 and §13.9.1].

Definition B.2.23. The dual of the group G is the set $\widehat{G}$ of (continuous) irreducible unitary representations of G modulo equivalence, see (1.1).

Given the correspondence explained in Lemma B.2.22, $\widehat{G}$ can be identified with the dual of $C^*(G)$ and inherit the structure that may occur on $\widehat{C^*(G)}$, see Remark B.2.17.

In particular, $\widehat{G}$ inherits a topology, called the *Fell topology*, corresponding to the hull-kernel (Jacobson) topology on $C^*(G)$, see e.g. Folland [Fol95, §7.2], Dixmier [Dix77, §18.1 and §3]. If G is separable, then $C^*(G)$ is separable, see [Dix77, §13.9.2], and $\widehat{G}$ also inherits the Mackey structure of measurable space.

Proposition B.2.24. *Let G be a separable locally compact group of type I. Then its dual $\widehat{G}$ is a standard Borel space. Moreover the Mackey structure coincides with the sigma-algebra associated with the Fell topology.*

For the definition of groups of type I, see Dixmier [Dix77, §13.9.4] or Folland [Fol95, §7.2]. See also hypothesis (H) in Section 1.8.2 for a relevant discussion. For the definition of the Plancherel measure, see (1.28), as well as Dixmier [Dix77, Definition 8.8.3] or Folland [Fol95, §7.5].

References for the proof of Proposition B.2.24. As G is of type I and separable, its group C^*-algebra $C^*(G)$ is of type I, postliminar and separable, see Dixmier [Dix77, §13.9]. Hence the Mackey Borel structure on the spectrum of this C^*-algebra (cf. [Dix77, §3.8]) is a standard Borel space by Dixmier [Dix77, Proposition 4.6.1]. □

Reduced group C^*-algebra

Although we do not use the following in this monograph, let us mention that one can also define another 'small' C^* algebra which contains $L^1(G)$.

Let us recall that the left regular representation π_L is defined on the group via

$$\pi_L(x)\phi(y) := \phi(x^{-1}y), \quad x, y \in G, \ \phi \in L^2(G). \tag{B.4}$$

This leads to the representation of $L^1(G)$ given by

$$(\pi_L)_{\mathscr{D}}(f)\phi = \int_G f(x)\pi_L(x)\phi\, dx = \int_G f(x)\phi(x^{-1}\cdot)\, dx = f * \phi, \tag{B.5}$$

which may be extended onto the closure $\overline{(\pi_L)_{\mathcal{D}}(L^1(G))}$ of $(\pi_L)_{\mathcal{D}}(L^1(G))$ for the operator norm, see Lemma B.2.22. This closure is naturally a C^*-algebra, often called the *reduced* C^*-algebra of the group and denoted by $C^*_r(G)$. Equivalently, $C^*_r(G)$ may be realised as the closure of $L^1(G)$ for the norm given by

$$\|f\|_{C^*_r} = \|(\pi_L)_{\mathscr{D}}(f)\|_{\mathscr{L}(L^2(G))} = \{\|f * \phi\|_{L^2}, \phi \in L^2(G) \text{ with } \|\phi\|_{L^2} = 1\}.$$

The 'full' and reduced C^* algebras of a group may be different. When they are equal, that is, $C^*_r(G) = C^*(G)$, then the group G is said to be *amenable*. Amenability can be described in many other ways. The advantage of considering the 'full'

C^*-algebra of a group is the one-to-one correspondence between the representations theories of G, $L^1(G)$, and $C^*(G)$.

The groups considered in this monograph, that is, compact groups and nilpotent Lie groups, are amenable.

Pontryagin duality

Although we do not use it in this monograph, let us recall briefly the Pontryagin duality, as this may be viewed as one of the historical motivation to develop the theory of (noncommutative) C^*-algebras.

The case of a locally compact (Hausdorff) abelian ($\equiv$ commutative) group G is described by the Pontryagin duality, see Section 1.1. In this case, the group algebra $L^1(G)$ (see Example B.2.9) is an abelian involutive Banach algebra. Its spectrum $\widehat{G}$ may be identified with the set of the continuous characters of G and is naturally equipped with the structure of a locally compact (Hausdorff) abelian group. The group G is amenable, that is, the full and reduced group C^*-algebras coincide: $C^*(G) = C^*_r(G)$. Moreover, the Fourier-Gelfand transform (see Remark B.2.15 (2)) extends into an isometry of C^*-algebra from $C^*(G)$ onto $C_o(\widehat{G})$.

Example B.2.25. In the particular example of the abelian group $G = \mathbb{R}^n$, the dual $\widehat{G}$ may also be identified with $\mathbb{R}^n$ and the Fourier-Gelfand transform in this case is the (usual) Euclidean Fourier transform $\mathcal{F}_{\mathbb{R}^n}$.

The group C^*-algebra $C^*(\mathbb{R}^n) = C^*_r(\mathbb{R}^n)$ may be viewed as a subspace of $\mathcal{S}'(\mathbb{R}^n)$ which contains $L^1(\mathbb{R}^n)$. Recall that, by the Riemann-Lebesgue Theorem (see e.g. [RT10a, Theorem 1.1.8]), the Euclidean Fourier transform $\mathcal{F}_{\mathbb{R}^n}$ maps $L^1(\mathbb{R}^n)$ to $C_o(\mathbb{R}^n)$, and one can show that

$$C^*(\mathbb{R}^n) = \mathcal{F}^{-1}_{\mathbb{R}^n} C_o(\mathbb{R}^n).$$

Remark B.2.26. Note that the inclusion $\mathcal{F}_{\mathbb{R}^n}(L^1(\mathbb{R}^n)) \subset C_o(\mathbb{R}^n)$ is strict. Indeed for $n > 1$, the kernel of the Bochner Riesz means $\mathcal{F}^{-1}_{\mathbb{R}^n}\{\sqrt{1-|\xi|^2} 1_{|\xi|\leq 1}\}$ is not in $L^1(\mathbb{R}^n)$ but its Fourier transform is in $C_o(\mathbb{R}^n)$. For $n = 1$, see e.g. Stein and Weiss [SW71, Ch 1, §4.1].

B.2.4 Von Neumann algebras

Let us recall the von Neumann bi-commutant theorem:

Theorem B.2.27. *Let $\mathscr{L}(\mathcal{H})$ be the space of continuous linear operators on a Hilbert space $\mathcal{H}$ with its natural structure (see Example B.2.8). Let $\mathcal{M}$ be a $*$-subalgebra of $\mathscr{L}(\mathcal{H})$ containing the identity mapping* I. *Then the following are equivalent:*

(i) $\mathcal{M}$ is equal to its bi-commutant (in the sense of Definition B.2.1):

$$\mathcal{M} = \mathcal{M}''.$$

(ii) $\mathcal{M}$ *is closed in the weak-operator topology, i.e. the topology given by the family of seminorms* $\{T \mapsto (Tv, w)_{\mathcal{H}},\ v, w \in \mathcal{H}\}$.

(iii) $\mathcal{M}$ *is closed in the strong-operator topology, i.e. the topology on* $\mathscr{L}(\mathcal{H})$ *given by the family of seminorms* $\{T \mapsto \|Tv\|_{\mathcal{H}},\ v \in \mathcal{H}\}$.

This leads to the notion of a von Neumann algebra where we take the above equivalent properties as its definition:

Definition B.2.28. We keep the notation of Theorem B.2.27. A *von Neumann algebra in* $\mathcal{H}$ *is a* $*$*-subalgebra* $\mathcal{M}$ *of* $\mathscr{L}(\mathcal{H})$ *which satisfies any of the equivalent properties (i), (ii), or (iii) in Theorem B.2.27.*

Note that the operator-norm topology on $\mathscr{L}(\mathcal{H})$ is stronger than the strong-operator topology, which in turn is stronger than the weak-operator topology. Thus a von Neumann algebra in $\mathcal{H}$ is a $*$-subalgebra of $\mathscr{L}(\mathcal{H})$ closed for the operator-norm topology, hence is a C^*-subalgebra of $\mathscr{L}(\mathcal{H})$ and a C^*-algebra itself. Among C^*-algebras, the von Neumann algebras are the C^*-algebras which are realised as a closed $*$-subalgebra of $\mathscr{L}(\mathcal{H})$ and furthermore satisfy any of the equivalent properties (*i*), (*ii*), or (*iii*) in Theorem B.2.27.

It is also possible to define the von Neumann algebras abstractly as the C^*-algebras having a predual, see e.g. Sakai [Sak98].

Example B.2.29. Naturally $\mathscr{L}(\mathcal{H})$ and $\mathbb{C}\mathrm{I}_{\mathcal{H}}$ are von Neumann algebras in $\mathcal{H}$.

Example B.2.30. If η is a positive and sigma-finite measure on a locally compact space X, then $\mathcal{A} = L^\infty(X, \eta)$ is a commutative unital C^*-algebra (see Example B.2.7). The operator of pointwise multiplication

$$L^\infty(X,\eta) \ni f \mapsto T_f \in \mathscr{L}(L^2(X,\mu)), \quad T_f(\phi) = f\phi,$$

is an isometric ($*$-algebra) morphism. This yields a C^*-algebra isomorphism from $\mathcal{A} = L^\infty(X, \eta)$ onto an abelian von Neumann algebra acting on the separable Hilbert space $L^2(X, \mu)$.

Conversely any abelian von Neumann algebra on a separable Hilbert space may be realised in the way described in Example B.2.30, see Dixmier [Dix96, §I.7.3].

The main example of von Neumann algebras of interest for us is the one associated with a group. This is explained in the next subsection.

B.2.5 Group von Neumann algebra

In this section we follow Dixmier [Dix77, §13]. The main application of these constructions are in Section 1.8.2, see Definition 1.8.7 and the subsequent discussion.

Now, first let us define the (isomorphic) left and right von Neumann algebras of a (Hausdorff) locally compact group G.

The left, resp. right, von Neumann algebra of G is the von Neumann algebra $\mathrm{VN}_L(G)$, resp. $\mathrm{VN}_R(G)$, in $L^2(G)$ generated by the left, resp. right, regular representation. This means that $\mathrm{VN}_L(G)$ *is the smallest von Neumann algebra containing all the operators* $\pi_L(x)$, $x \in G$, where π_L is defined in (B.4), i.e.

$$\pi_L(x)\phi(y) := \phi(x^{-1}y), \quad x, y \in G, \ \phi \in L^2(G).$$

Let us recall that the right regular representation π_R is given by

$$\pi_R(x)\phi(y) = \Delta(x)^{\frac{1}{2}}\phi(yx).$$

Here Δ denotes the modular function (see Definition B.2.10).

One checks easily that the isomorphism U of $L^2(G)$ given by

$$U\phi(y) = \Delta(y)^{\frac{1}{2}}\phi(y^{-1}), \quad \phi \in L^2(G), \ y \in G,$$

intertwines π_L and π_R:

$$\forall x \in G \qquad U\pi_L(x) = \pi_R(x)U.$$

Thus one is sometimes allowed to speak of 'the regular representation' and 'the group von Neumann algebra'. However, in this subsection, we will keep making the distinction between left and right regular representations.

Let us assume that the group G is also separable. In this case, the group von Neumann algebra can be described further.

Clearly $\mathrm{VN}_L(G)$, resp. $\mathrm{VN}_R(G)$, is the smallest von Neumann algebra containing all the operators $(\pi_L)_{\mathscr{D}}(f)$, $f \in C_c(G)$, resp. $(\pi_R)_{\mathscr{D}}(f)$, $f \in C_c(G)$, see [Dix77, §13.10.2]. Here $C_c(G)$ denotes the space of continuous functions with compact support on G. For the definitions of $(\pi_L)_{\mathscr{D}}(f)$ and $(\pi_R)_{\mathscr{D}}$, see (B.5) and (B.1). This easily implies that $\mathrm{VN}_L(G)$, resp. $\mathrm{VN}_R(G)$, is the smallest von Neumann algebra containing all the operators $(\pi_L)_{\mathscr{D}}(f)$, resp. $(\pi_R)_{\mathscr{D}}(f)$, where f runs over $L^1(G)$ or $C^*(G)$.

Applying the commutation theorem (cf. Dixmier [Dix96, Ch 1, §5.2]) to the quasi-Hilbertian algebra $C_c(G)$ ([Eym72, p. 210]) we see that

$$\mathrm{VN}_L(G) = (\mathrm{VN}_R(G))' \quad \text{and} \quad \mathrm{VN}_R(G) = (\mathrm{VN}_L(G))'.$$

See Definition B.2.1 for the definition of the commutant. This implies

Proposition B.2.31. *The group von Neumann algebra coincides with the invariant bounded operators in the following sense:*

- $\mathrm{VN}_L(G)$ *is the space* $\mathscr{L}_R(L^2(G))$ *of operators in* $\mathscr{L}(L^2(G))$ *which commute with* $\pi_R(x)$, *for all* $x \in G$,
- $\mathrm{VN}_R(G)$ *is the space* $\mathscr{L}_L(L^2(G))$ *of operators in* $\mathscr{L}(L^2(G))$ *which commute with* $\pi_L(x)$, *for all* $x \in G$:

$$\mathrm{VN}_L(G) = \mathscr{L}_R(L^2(G)) \quad and \quad \mathrm{VN}_R(G) = \mathscr{L}_L(L^2(G)).$$

Denoting by J the involutive anti-automorphism on $L^2(G)$ given by

$$J(\phi)(x) := \bar{\phi}(x^{-1})\Delta(x)^{-\frac{1}{2}}, \quad \phi \in L^2(G),\ x \in G,$$

we also have

$$J\ \mathrm{VN}_L(G)\ J = \mathrm{VN}_R(G) \quad and \quad J\ \mathrm{VN}_R(G)\ J = \mathrm{VN}_L(G).$$

Under our hypotheses, it is possible to describe the group von Neumann algebra as a space of convolution operators, see Eymard [Eym72, Theorem 3.10 and Proposition 3.27]. In the special case of Lie groups, this is a consequence of the Schwartz kernel theorem, see Corollary 3.2.1 and its right-invariant version.

B.2.6 Decomposition of group von Neumann algebras and abstract Plancherel theorem

The full abstract version of the Plancherel theorem allows us to decompose not only the Hilbert space $L^2(G)$ (thus obtaining the Plancherel formula) but also the operators in $\mathrm{VN}_R(G)$ and $\mathrm{VN}_L(G)$:

Theorem B.2.32 (Plancherel theorem). *We assume that the (Hausdorff locally compact separable) group G is also unimodular and of type I and that a (left) Haar measure has been fixed.*

Then there exist

- *a positive sigma-finite measure μ on $\widehat{G}$,*
- *a μ-measurable field of unitary continuous representations $(\pi_\zeta)_{\zeta\in\widehat{G}}$ of G on the μ-measurable field of Hilbert spaces $(\mathcal{H}_\zeta)_{\zeta\in\widehat{G}}$,*
- *and a unitary map W from $L^2(G)$ onto*

$$\int_{\widehat{G}}^{\oplus} (\mathcal{H}_\zeta \otimes \mathcal{H}_\zeta^*)\, d\mu(\zeta) \equiv \int_{\widehat{G}}^{\oplus} \mathtt{HS}(\mathcal{H}_\zeta)\, d\mu(\zeta),$$

(see Subsection B.1.3)

such that W satisfies the following properties:

1. *If $\phi \in L^2(G)$, then $W\phi = \int_{\widehat{G}}^{\oplus} v_\zeta d\mu(\zeta)$ where each v_ζ is a Hilbert-Schmidt operator on $\mathcal{H}_\zeta$ and we have*

$$WJ\phi = \int_{\widehat{G}}^{\oplus} v_\zeta^*\, d\mu(\zeta), \quad where \quad (J\phi)(x) = \bar{\phi}(x^{-1}).$$

2. *For any $f \in L^1(G)$ (or $C^*(G)$), the operators $(\pi_R)_{\mathcal{D}}(f)$ and $(\pi_L)_{\mathcal{D}}(f)$ acting on $L^2(G)$ are transformed via W into the decomposable operators (in the sense of Definition B.1.15) on $\int_{\widehat{G}}^{\oplus}(\mathcal{H}_\zeta \otimes \mathcal{H}_\zeta^*)d\mu(\zeta)$,*

$$W\{(\pi_L)_{\mathcal{D}}(f)\}W^{-1} = \int_{\widehat{G}}^{\oplus} (\pi_\zeta)_{\mathcal{D}}(f) \otimes \mathrm{I}_{\mathcal{H}_\zeta^*}\, d\mu(\zeta),$$

and

$$W\{(\pi_R)_{\mathcal{D}}(f)\}W^{-1} = \int_{\widehat{G}}^{\oplus} \mathrm{I}_{\mathcal{H}_\zeta} \otimes (\pi_\zeta^{dual})_{\mathcal{D}}(f)\, d\mu(\zeta).$$

See (B.1) for the notation $(\pi)_{\mathcal{D}}$, and here π_ζ^{dual} denotes the dual representation to π_ζ which acts on $\mathcal{H}_\zeta^*$ via

$$(\pi_\zeta^{dual}(x))v^* : w \mapsto (\pi_\zeta(x^{-1})w, v)_{\mathcal{H}_\zeta}.$$

3. *If T is a bounded operator on $L^2(G)$ which commutes with $\pi_L(x)$, for all $x \in G$, that is, $T \in \mathrm{VN}_R(G) = \mathscr{L}_L(L^2(G))$, then T is transformed via W into a decomposable operator (in the sense of Definition B.1.15) on the Hilbert space $\int_{\widehat{G}}^{\oplus}(\mathcal{H}_\zeta \otimes \mathcal{H}_\zeta^*)d\mu(\zeta)$ of the form*

$$WTW^{-1} = \int_{\widehat{G}}^{\oplus} T_\zeta \otimes \mathrm{I}_{\mathcal{H}_\zeta^*}\, d\mu(\zeta).$$

Conversely any decomposable operator of this type yields an operator in $\mathscr{L}_L(L^2(G))$. Hence we may summarise this by writing

$$\mathrm{VN}_R(G) = \mathscr{L}_L(L^2(G)) = W^{-1} \int_{\widehat{G}}^{\oplus} \mathscr{L}(\mathcal{H}_\zeta) \otimes \mathbb{C}\, d\mu(\zeta)\, W.$$

Similarly

$$\mathrm{VN}_L(G) = \mathscr{L}_R(L^2(G)) = W^{-1} \int_{\widehat{G}}^{\oplus} \mathbb{C} \otimes \mathscr{L}(\mathcal{H}_\zeta^*)\, d\mu(\zeta)\, W.$$

A consequence of Points 1. and 2. is that if $f \in L^1(G) \cap L^2(G)$, then $(\pi_\zeta)_{\mathcal{D}}(f) \in \mathtt{HS}(\mathcal{H}_\zeta)$ for almost every $\zeta \in \widehat{G}$ and

$$Wf = \int_{\widehat{G}}^{\oplus} (\pi_\zeta)_{\mathcal{D}}(f) d\mu(\zeta) \quad \text{thus} \quad \|f\|_{L^2(G)}^2 = \int_{\widehat{G}} \|(\pi_\zeta)_{\mathcal{D}}(f)\|_{\mathtt{HS}(\mathcal{H}_\zeta)}^2 d\mu(\zeta).$$

The measure μ is standard (in the sense of Definition B.1.13, see also Proposition B.2.24) and unique modulo equivalence (see Proposition B.1.17).

Reference for the proof of Theorem B.2.32. For the Plancherel measure being standard, see Dixmier [Dix77, Proposition 18.7.7 and Theorems 8.8.1 and 8.8.2]. For the Plancherel theorem expressed in terms of the canonical fields, see [Dix77, 18.8.1 and 18.8.2]. □

The main application of the above theorem for us is Theorem 1.8.11.

Definition B.2.33. The measure μ is called the *Plancherel measure* (associated to the fixed Haar measure).

A different choice of the Haar measure would lead to a different Plancherel measure. Up to this choice, the Plancherel measure is unique. Proposition B.1.17 then implies that we do not need to specify the choice of a measurable field of continuous representations.

In our monograph, our group Fourier transform and Dixmier's defined in (B.2) and (B.1) respectively, are related via (B.3). This implies that the statement of Theorem B.2.32 remains valid if we replace firstly $(\pi)_{\mathcal{D}}$ with our definition of the group Fourier transform and, secondly, W with the isometric isomorphism

$$\tilde{W} : L^2(G) \to \int_{\widehat{G}}^{\oplus} \mathtt{HS}(\mathcal{H}_\zeta) d\mu(\zeta)$$

given by

$$\tilde{W}\phi := W(\phi \circ \mathrm{inv}) \quad \text{where} \quad \mathrm{inv}(x) = x^{-1}.$$

In particular, if $\phi \in L^2(G)$ then

$$\tilde{W}\phi = \int_{\widehat{G}}^{\oplus} \phi_\zeta d\mu(\zeta), \tag{B.6}$$

and we understand $(\phi_\zeta)_{\zeta \in \widehat{G}}$ as the group Fourier transform of ϕ. If $T \in \mathscr{L}_L(L^2(G))$ then it may be decomposed by

$$\tilde{W} T \tilde{W}^{-1} = \int_{\widehat{G}}^{\oplus} T_\zeta \otimes \mathrm{I}_{\mathcal{H}_\zeta^*} \, d\mu(\zeta),$$

which means that if $\phi \in L^2(G)$ with (B.6), then

$$\tilde{W}(T\phi) = \int_{\widehat{G}}^{\oplus} T_\zeta \phi_\zeta d\mu(\zeta).$$

Theorem B.2.32 is reformulated in Theorem 1.8.11 with our choice of group Fourier transform.

We end this appendix with the following observation. Comparing closely the contents of Chapter 1 and Chapter B, there is a small discrepancy about the separability of Hilbert spaces. Indeed, in Chapter B, all the Hilbert spaces on which the representations act are assumed separable, see Section B.1.1, whereas the separability of the Hilbert spaces is not mentioned in Chapter 1. This leeds however to no contradiction when considering a continuous irreducible unitary representation π of a Hausdorff locally compact separable group G on a Hilbert space $\mathcal{H}_\pi$. Indeed, in this case, this yields a continuous non-degenerate representation of $L^1(G)$ on $\mathcal{H}_\pi$ as in Lemma B.2.18. As $L^1(G)$ is separable [Dix77, §13.2.4] and π is irreducible, one can easily adapt the arguments in [Dix77, §2.3.3] to show that $\mathcal{H}_\pi$ is separable. Consequently, the dual $\widehat{G}$ of a Hausdorff locally compact separable group G may be defined as in Section 1.1 as the equivalence classes of the continuous unitary representations, without stating the hypothesis of separability on the representation spaces.

Schrödinger representations and Weyl quantization

Here we summarise the choices of normalisations and give some relations between the Schrödinger representations π_λ, $\lambda \in \mathbb{R}\backslash 0$, of the Heisenberg group $\mathbb{H}_n$ and the Weyl quantization on $L^2(\mathbb{R}^n)$. Detailed justifications and some proofs are given in Section 6.2.

Euclidean Fourier transform (for $f \in \mathcal{S}(\mathbb{R}^N)$ and $\xi \in \mathbb{R}^N$)

$$\boxed{\mathcal{F}_{\mathbb{R}^N} f(\xi) = (2\pi)^{-\frac{N}{2}} \int_{\mathbb{R}^N} f(x) e^{-ix\xi} dx}$$

Weyl quantization (for $f \in \mathcal{S}(\mathbb{R}^N)$ and $u \in \mathbb{R}^N$)

$$\boxed{\mathrm{Op}^W(a) f(u) = (2\pi)^{-N} \int_{\mathbb{R}^N} \int_{\mathbb{R}^N} e^{i(u-v)\xi} a(\xi, \frac{u+v}{2}) f(v) dv d\xi}$$

The useful convention for abbreviating the expressions below is

$$\sqrt{\lambda} := \mathrm{sgn}(\lambda)\sqrt{|\lambda|} = \begin{cases} \sqrt{\lambda} & \text{if } \lambda > 0, \\ -\sqrt{|\lambda|} & \text{if } \lambda < 0. \end{cases} \tag{B.7}$$

Schrödinger representations (for $(x, y, t) \in \mathbb{H}_n$, $h \in L^2(\mathbb{R}^n)$, and $u \in \mathbb{R}^n$)

$$\boxed{\pi_\lambda(x, y, t) h(u) = e^{i\lambda(t + \frac{1}{2}xy)} e^{i\sqrt{\lambda} yu} h(u + \sqrt{|\lambda|} x)}$$

Notation for the group Fourier transform

$$\boxed{\pi_\lambda(\kappa) \equiv \widehat{\kappa}(\pi_\lambda) = \int_{\mathbb{H}_n} \kappa(x, y, t)\ \pi_\lambda(x, y, t)^*\ dx dy dt}$$

Relation between Schrödinger representation and Weyl quantization

$$\boxed{\pi_\lambda(\kappa) = (2\pi)^{\frac{2n+1}{2}} \mathrm{Op}^W \left[\mathcal{F}_{\mathbb{R}^{2n+1}}(\kappa)(\sqrt{|\lambda|}\,\cdot, \sqrt{\lambda}\,\cdot, \lambda)\right]}$$

or, with more details,

$$\begin{aligned} \pi_\lambda(\kappa) h(u) &= \int_{\mathbb{H}_n} \kappa(x, y, t)\ \pi_\lambda(x, y, t)^* h(u)\ dx dy dt \\ &= \int_{\mathbb{R}^{2n+1}} \kappa(x, y, t) e^{i\lambda(-t + \frac{1}{2}xy)} e^{-i\sqrt{\lambda} yu} h(u - \sqrt{|\lambda|} x) dx dy dt \\ &= (2\pi)^{\frac{2n+1}{2}} \int_{\mathbb{R}^n} \int_{\mathbb{R}^n} e^{i(u-v)\xi} \mathcal{F}_{\mathbb{R}^{2n+1}}(\kappa)(\sqrt{|\lambda|}\xi, \sqrt{\lambda}\frac{u+v}{2}, \lambda) h(v) dv d\xi. \end{aligned}$$

Plancherel formula

$$\boxed{\int_{\mathbb{H}_n} |f(x, y, t)|^2 dx dy dt = c_n \int_{\lambda \in \mathbb{R}\backslash\{0\}} \|\widehat{f}(\pi_\lambda)\|^2_{\mathrm{HS}(L^2(\mathbb{R}^n))} |\lambda|^n d\lambda}$$

Explicit symbolic calculus on the Heisenberg group

Here we give a summary of some explicit formulae for symbolic analysis of concrete operators on the Heisenberg group $\mathbb{H}_n$. We refer to Section 6.3.3 for more details. We always employ the convention in (B.7) for $\sqrt{\lambda}$.

Symbols of left-invariant vector fields and the sub-Laplacian

$$\begin{array}{rclcl}
\pi_\lambda(X_j) &=& \sqrt{|\lambda|}\partial_{u_j} &=& \mathrm{Op}^W\left(i\sqrt{|\lambda|}\xi_j\right)\\
\pi_\lambda(Y_j) &=& i\sqrt{\lambda}u_j &=& \mathrm{Op}^W\left(i\sqrt{\lambda}u_j\right)\\
\pi_\lambda(T) &=& i\lambda \mathrm{I} &=& \mathrm{Op}^W(i\lambda)\\
\pi_\lambda(\mathcal{L}) &=& |\lambda|\sum_j(\partial_{u_j}^2-u_j^2) &=& \mathrm{Op}^W\left(|\lambda|\sum_j(-\xi_j^2-u_j^2)\right)
\end{array}$$

Difference operators

$$\begin{array}{rclcl}
\Delta_{x_j}|_{\pi_\lambda} &=& \frac{1}{i\lambda}\mathrm{ad}\,(\pi_\lambda(Y_j)) &=& \frac{1}{\sqrt{|\lambda|}}\mathrm{ad}u_j\\
\Delta_{y_j}|_{\pi_\lambda} &=& -\frac{1}{i\lambda}\mathrm{ad}\,(\pi_\lambda(X_j)) &=& -\frac{1}{i\sqrt{\lambda}}\mathrm{ad}\partial_{u_j}\\
\Delta_t|_{\pi_\lambda} &=& \multicolumn{3}{l}{i\partial_\lambda+\frac{1}{2}\sum_{j=1}^n\Delta_{x_j}\Delta_{y_j}|_{\pi_\lambda}+\frac{i}{2\lambda}\sum_{j=1}^n\{\pi_\lambda(Y_j)|_{\pi_\lambda}\Delta_{y_j}+\Delta_{x_j}|_{\pi_\lambda}\pi_\lambda(X_j)\}}
\end{array}$$

Difference operators acting on symbols of left-invariant vector fields

	$\pi_\lambda(X_k)$	$\pi_\lambda(Y_k)$	$\pi_\lambda(T)$	$\pi_\lambda(\mathcal{L})$
Δ_{x_j}	$-\delta_{j=k}$	0	0	$-2\pi_\lambda(X_j)$
Δ_{y_j}	0	$-\delta_{j=k}$	0	$-2\pi_\lambda(Y_j)$
Δ_t	0	0	$-\mathrm{I}$	0

Relation between the group Fourier transform and the λ-symbols

$$\widehat{\kappa}(\pi_\lambda)\equiv\pi_\lambda(\kappa)=\mathrm{Op}^W(a_\lambda)=\mathrm{Op}^W(\tilde{a}_\lambda(\sqrt{|\lambda|}\cdot,\sqrt{\lambda}\cdot))$$

with
$$\begin{array}{rcl}
a_\lambda &=& \{a_\lambda(\xi,u)=\sqrt{2\pi}\mathcal{F}_{\mathbb{R}^{2n+1}}(\kappa)(\sqrt{|\lambda|}\xi,\sqrt{\lambda}u,\lambda)\}\\
\tilde{a}_\lambda &=& \{\tilde{a}_\lambda(\xi,u)=\sqrt{2\pi}\mathcal{F}_{\mathbb{R}^{2n+1}}(\kappa)(\xi,u,\lambda)\}
\end{array}$$

Difference operators in terms of the Weyl quantization of λ-symbols

$$\begin{array}{rclcl}
\Delta_{x_j}\pi_\lambda(\kappa) &=& i\mathrm{Op}^W\left(\frac{1}{\sqrt{|\lambda|}}\partial_{\xi_j}a_\lambda\right) &=& i\mathrm{Op}^W\left(\partial_{\xi_j}\tilde{a}_\lambda\right)\\
\Delta_{y_j}\pi_\lambda(\kappa) &=& i\mathrm{Op}^W\left(\frac{1}{\sqrt{\lambda}}\partial_{u_j}a_\lambda\right) &=& i\mathrm{Op}^W\left(\partial_{u_j}\tilde{a}_\lambda\right)\\
\Delta_t\pi_\lambda(\kappa) &=& i\mathrm{Op}^W\left(\tilde{\partial}_{\lambda,\xi,u}a_\lambda\right) &=& i\mathrm{Op}^W\left(\partial_\lambda\tilde{a}_\lambda\right)
\end{array}$$

$$\left(\text{with } \tilde{\partial}_{\lambda,\xi,u}=\partial_\lambda-\frac{1}{2\lambda}\sum_{j=1}^n\{u_j\partial_{u_j}+\xi_j\partial_{\xi_j}\}\right)$$

List of quantizations

We refer to Sections 2.2, 5.1.3 and 6.5.1 for the cases of compact, graded, and Heisenberg groups, respectively.

Quantization on compact Lie groups (for $\varphi \in C^\infty(G)$ and $x \in G$)

$$\boxed{A\varphi(x) = \sum_{\pi \in \widehat{G}} d_\pi \operatorname{Tr}\left(\pi(x)\ \sigma_A(x,\pi)\ \widehat{\varphi}(\pi)\right)}$$

with the formula for the symbol

$$\boxed{\sigma_A(x,\pi) = \pi(x)^*(A\pi)(x)}$$

Quantization on general graded Lie groups (for $\varphi \in \mathcal{S}(G)$ and $x \in G$)

$$\boxed{A\varphi(x) = \int_{\widehat{G}} \operatorname{Tr}\left(\pi(x)\ \sigma_A(x,\pi)\ \widehat{\varphi}(\pi)\right) d\mu(\pi)}$$

Symbols of vector fields $\sigma_X(\pi) \equiv d\pi(X) = X\pi(e)$, see (1.22)

In the compact and graded cases, relation with the right-convolution kernel

$$\boxed{A\varphi(x) = \varphi * \kappa_x(x) = \int_G \varphi(y)\kappa_x(y^{-1}x)dy \quad \text{with} \quad \widehat{\kappa_x}(\pi) = \sigma_A(x,\pi)}$$

Quantization on the Heisenberg group (for $\varphi \in \mathcal{S}(\mathbb{H}_n)$ and $g = (x,y,t) \in \mathbb{H}_n$)

$$\boxed{A\varphi(g) = c_n \int_{\mathbb{R}\setminus\{0\}} \operatorname{Tr}\left(\pi_\lambda(g)\ \sigma_A(g,\lambda)\ \widehat{\varphi}(\pi_\lambda)\right) |\lambda|^n d\lambda}$$

and in terms of λ-symbols $a_{g,\lambda} : \mathbb{R}^n \times \mathbb{R}^n \to \mathbb{C}$,

$$\boxed{\sigma_A(g,\lambda) = \operatorname{Op}^W(a_{g,\lambda})} \quad (g \in \mathbb{H}_n,\ \lambda \in \mathbb{R}\setminus\{0\})$$

$$\begin{aligned}
&A\varphi(g)\\
&= c_n' \int_{\mathbb{R}\setminus\{0\}} \operatorname{Tr}\left(\pi_\lambda(g)\operatorname{Op}^W(a_{g,\lambda})\ \operatorname{Op}^W\left[\mathcal{F}_{\mathbb{R}^{2n+1}}(\varphi)(\sqrt{|\lambda|}\,\cdot, \sqrt{\lambda}\,\cdot, \lambda)\right]\right) |\lambda|^n d\lambda\\
&= c_n' \int_{\mathbb{R}\setminus\{0\}} \operatorname{Tr}\left(\operatorname{Op}^W(a_{g,\lambda})\ \operatorname{Op}^W\left[\mathcal{F}_{\mathbb{R}^{2n+1}}(\varphi(g\,\cdot))(\sqrt{|\lambda|}\,\cdot, \sqrt{\lambda}\,\cdot, \lambda)\right]\right) |\lambda|^n d\lambda
\end{aligned}$$

Bibliography

[ACC05] G. Arena, A. O. Caruso, and A. Causa. Taylor formula for Carnot groups and applications. *Matematiche (Catania)*, 60(2):375–383 (2006), 2005.

[ADBR09] F. Astengo, B. Di Blasio, and F. Ricci. Gelfand pairs on the Heisenberg group and Schwartz functions. *J. Funct. Anal.*, 256(5):1565–1587, 2009.

[ADBR13] F. Astengo, B. Di Blasio, and F. Ricci. Fourier transform of Schwartz functions on the Heisenberg group. *Studia Math.*, 214(3):201–222, 2013.

[Ale94] G. Alexopoulos. Spectral multipliers on Lie groups of polynomial growth. *Proc. Amer. Math. Soc.*, 120(3):973–979, 1994.

[AR15] R. Akylzhanov and M. Ruzhansky. Hausdorff-Young-Paley inequalities and L^p-L^q Fourier multipliers on locally compact groups. *arXiv:1510.06321*, 2015.

[Arv76] W. Arveson. *An invitation to C^*-algebras*. Springer-Verlag, New York-Heidelberg, 1976. Graduate Texts in Mathematics, No. 39.

[AtER94] P. Auscher, A. F. M. ter Elst, and D. W. Robinson. On positive Rockland operators. *Colloq. Math.*, 67(2):197–216, 1994.

[Bea77a] R. Beals. Characterization of pseudodifferential operators and applications. *Duke Math. J.*, 44(1):45–57, 1977.

[Bea77b] R. Beals. Opérateurs invariants hypoelliptiques sur un groupe de Lie nilpotent. In *Séminaire Goulaouic-Schwartz 1976/1977: Équations aux dérivées partielles et analyse fonctionnelle, Exp. No. 19*, page 8. Centre Math., École Polytech., Palaiseau, 1977.

[BFKG12a] H. Bahouri, C. Fermanian-Kammerer, and I. Gallagher. Phase-space analysis and pseudodifferential calculus on the Heisenberg group. *Astérisque*, (342):vi+127, 2012.

V. Fischer, M. Ruzhansky, *Quantization on Nilpotent Lie Groups*, Progress in Mathematics 314,
DOI 10.1007/978-3-319-29558-9

[BFKG12b] H. Bahouri, C. Fermanian-Kammerer, and I. Gallagher. Refined inequalities on graded Lie groups. *C. R. Math. Acad. Sci. Paris*, 350(7-8):393–397, 2012.

[BG88] R. Beals and P. Greiner. *Calculus on Heisenberg manifolds*, volume 119 of *Annals of Mathematics Studies*. Princeton University Press, Princeton, NJ, 1988.

[BGGV86] R. W. Beals, B. Gaveau, P. C. Greiner, and J. Vauthier. The Laguerre calculus on the Heisenberg group. II. *Bull. Sci. Math. (2)*, 110(3):225–288, 1986.

[BGJR89] O. Bratteli, F. Goodman, P. Jorgensen, and D. W. Robinson. Unitary representations of Lie groups and Gårding's inequality. *Proc. Amer. Math. Soc.*, 107(3):627–632, 1989.

[BGX00] H. Bahouri, P. Gérard, and C.-J. Xu. Espaces de Besov et estimations de Strichartz généralisées sur le groupe de Heisenberg. *J. Anal. Math.*, 82:93–118, 2000.

[BJR98] C. Benson, J. Jenkins, and G. Ratcliff. The spherical transform of a Schwartz function on the Heisenberg group. *J. Funct. Anal.*, 154(2):379–423, 1998.

[BL76] J. Bergh and J. Löfström. *Interpolation spaces. An introduction.* Springer-Verlag, Berlin, 1976. Grundlehren der Mathematischen Wissenschaften, No. 223.

[Bla06] B. Blackadar. *Operator algebras*, volume 122 of *Encyclopaedia of Mathematical Sciences*. Springer-Verlag, Berlin, 2006. Theory of C^*-algebras and von Neumann algebras, Operator Algebras and Non-commutative Geometry, III.

[BLU07] A. Bonfiglioli, E. Lanconelli, and F. Uguzzoni. *Stratified Lie groups and potential theory for their sub-Laplacians*. Springer Monographs in Mathematics. Springer, Berlin, 2007.

[BM95] D. R. Bell and S. E. A. Mohammed. An extension of Hörmander's theorem for infinitely degenerate second-order operators. *Duke Math. J.*, 78(3):453–475, 1995.

[Bon09] A. Bonfiglioli. Taylor formula for homogeneous groups and applications. *Math. Z.*, 262(2):255–279, 2009.

[Bou98] N. Bourbaki. *Lie groups and Lie algebras. Chapters 1–3.* Elements of Mathematics (Berlin). Springer-Verlag, Berlin, 1998. Translated from the French, Reprint of the 1989 English translation.

[BP08] E. Binz and S. Pods. *The geometry of Heisenberg groups*, volume 151 of *Mathematical Surveys and Monographs*. American Mathematical

Society, Providence, RI, 2008. With applications in signal theory, optics, quantization, and field quantization, With an appendix by Serge Preston.

[Bru68] F. Bruhat. *Lectures on Lie groups and representations of locally compact groups.* Tata Institute of Fundamental Research, Bombay, 1968. Notes by S. Ramanan, Tata Institute of Fundamental Research Lectures on Mathematics, No. 14.

[BV08] I. Birindelli and E. Valdinoci. The Ginzburg-Landau equation in the Heisenberg group. *Commun. Contemp. Math.*, 10(5):671–719, 2008.

[Cap99] L. Capogna. Regularity for quasilinear equations and 1-quasiconformal maps in Carnot groups. *Math. Ann.*, 313(2):263–295, 1999.

[CCG07] O. Calin, D.-C. Chang, and P. Greiner. *Geometric analysis on the Heisenberg group and its generalizations*, volume 40 of *AMS/IP Studies in Advanced Mathematics.* American Mathematical Society, Providence, RI, 2007.

[CdG71] R. R. Coifman and M. de Guzmán. Singular integrals and multipliers on homogeneous spaces. *Rev. Un. Mat. Argentina*, 25:137–143, 1970/71. Collection of articles dedicated to Alberto González Domínguez on his sixty-fifth birthday.

[CDPT07] L. Capogna, D. Danielli, S. D. Pauls, and J. T. Tyson. *An introduction to the Heisenberg group and the sub-Riemannian isoperimetric problem*, volume 259 of *Progress in Mathematics.* Birkhäuser Verlag, Basel, 2007.

[CG84] M. Christ and D. Geller. Singular integral characterizations of Hardy spaces on homogeneous groups. *Duke Math. J.*, 51(3):547–598, 1984.

[CG90] L. J. Corwin and F. P. Greenleaf. *Representations of nilpotent Lie groups and their applications. Part I*, volume 18 of *Cambridge Studies in Advanced Mathematics.* Cambridge University Press, Cambridge, 1990. Basic theory and examples.

[CGGP92] M. Christ, D. Geller, P. Głowacki, and L. Polin. Pseudodifferential operators on groups with dilations. *Duke Math. J.*, 68(1):31–65, 1992.

[Che99] C. Chevalley. *Theory of Lie groups. I*, volume 8 of *Princeton Mathematical Series.* Princeton University Press, Princeton, NJ, 1999. Fifteenth printing, Princeton Landmarks in Mathematics.

[Chr84] M. Christ. Characterization of H^1 by singular integrals: necessary conditions. *Duke Math. J.*, 51(3):599–609, 1984.

[Cow83] M. G. Cowling. Harmonic analysis on semigroups. *Ann. of Math. (2)*, 117(2):267–283, 1983.

[CR81] L. Corwin and L. P. Rothschild. Necessary conditions for local solvability of homogeneous left invariant differential operators on nilpotent Lie groups. *Acta Math.*, 147(3-4):265–288, 1981.

[CS01] M. Cowling and A. Sikora. A spectral multiplier theorem for a sublaplacian on SU(2). *Math. Z.*, 238(1):1–36, 2001.

[CW71a] R. R. Coifman and G. Weiss. *Analyse harmonique non-commutative sur certains espaces homogènes.* Lecture Notes in Mathematics, Vol. 242. Springer-Verlag, Berlin, 1971. Étude de certaines intégrales singulières.

[CW71b] R. R. Coifman and G. Weiss. Multiplier transformations of functions on SU(2) and $\sum_2$. *Rev. Un. Mat. Argentina*, 25:145–166, 1971. Collection of articles dedicated to Alberto González Domínguez on his sixty-fifth birthday.

[CW74] R. R. Coifman and G. Weiss. Central multiplier theorems for compact Lie groups. *Bull. Amer. Math. Soc.*, 80:124–126, 1974.

[Cyg81] J. Cygan. Subadditivity of homogeneous norms on certain nilpotent Lie groups. *Proc. Amer. Math. Soc.*, 83(1):69–70, 1981.

[Dav80] E. B. Davies. *One-parameter semigroups*, volume 15 of *London Mathematical Society Monographs.* Academic Press Inc. [Harcourt Brace Jovanovich Publishers], London, 1980.

[DHZ94] J. Dziubański, W. Hebisch, and J. Zienkiewicz. Note on semigroups generated by positive Rockland operators on graded homogeneous groups. *Studia Math.*, 110(2):115–126, 1994.

[DiB02] E. DiBenedetto. *Real analysis.* Birkhäuser Advanced Texts: Basler Lehrbücher. [Birkhäuser Advanced Texts: Basel Textbooks]. Birkhäuser Boston, Inc., Boston, MA, 2002.

[Dix53] J. Dixmier. Formes linéaires sur un anneau d'opérateurs. *Bull. Soc. Math. France*, 81:9–39, 1953.

[Dix77] J. Dixmier. *C^*-algebras.* North-Holland Publishing Co., Amsterdam, 1977. Translated from the French by Francis Jellett, North-Holland Mathematical Library, Vol. 15.

[Dix81] J. Dixmier. *von Neumann algebras*, volume 27 of *North-Holland Mathematical Library.* North-Holland Publishing Co., Amsterdam, 1981. With a preface by E. C. Lance, Translated from the second French edition by F. Jellett.

[Dix96] J. Dixmier. *Les algèbres d'opérateurs dans l'espace hilbertien (algèbres de von Neumann).* Les Grands Classiques Gauthier-Villars. [Gauthier-Villars Great Classics]. Éditions Jacques Gabay, Paris, 1996. Reprint of the second (1969) edition.

[DM78] J. Dixmier and P. Malliavin. Factorisations de fonctions et de vecteurs indéfiniment différentiables. *Bull. Sci. Math. (2)*, 102(4):307–330, 1978.

[DR14a] A. Dasgupta and M. Ruzhansky. Gevrey functions and ultradistributions on compact Lie groups and homogeneous spaces. *Bull. Sci. Math.*, 138(6):756–782, 2014.

[DR14b] J. Delgado and M. Ruzhansky. L^p-nuclearity, traces, and Grothendieck-Lidskii formula on compact Lie groups. *J. Math. Pures Appl. (9)*, 102(1):153–172, 2014.

[DR16] A. Dasgupta and M. Ruzhansky. Eigenfunction expansions of ultradifferentiable functions and ultradistributions. *arXiv:1410.2637. To appear in Trans. Amer. Math. Soc.*, 2016.

[DtER03] N. Dungey, A. F. M. ter Elst, and D. W. Robinson. *Analysis on Lie groups with polynomial growth*, volume 214 of *Progress in Mathematics.* Birkhäuser Boston Inc., Boston, MA, 2003.

[Dye70] J. L. Dyer. A nilpotent Lie algebra with nilpotent automorphism group. *Bull. Amer. Math. Soc.*, 76:52–56, 1970.

[Dyn76] A. S. Dynin. An algebra of pseudodifferential operators on the Heisenberg groups. Symbolic calculus. *Dokl. Akad. Nauk SSSR*, 227(4):792–795, 1976.

[Dyn78] A. Dynin. Pseudodifferential operators on Heisenberg groups. In *Pseudodifferential operator with applications (Bressanone, 1977)*, pages 5–18. Liguori, Naples, 1978.

[Dzi93] J. Dziubański. On semigroups generated by subelliptic operators on homogeneous groups. *Colloq. Math.*, 64(2):215–231, 1993.

[Edw72] R. E. Edwards. *Integration and harmonic analysis on compact groups.* Cambridge Univ. Press, London, 1972. London Mathematical Society Lecture Note Series, No. 8.

[Eym72] P. Eymard. *Moyennes invariantes et représentations unitaires.* Lecture Notes in Mathematics, Vol. 300. Springer-Verlag, Berlin-New York, 1972.

[Feg91] H. D. Fegan. *Introduction to compact Lie groups*, volume 13 of *Series in Pure Mathematics.* World Scientific Publishing Co. Inc., River Edge, NJ, 1991.

[Fis15] V. Fischer. Intrinsic pseudo-differential calculi on any compact Lie group. *J. Funct. Anal.*, 268(11):3404–3477, 2015.

[FMV07] G. Furioli, C. Melzi, and A. Veneruso. Strichartz inequalities for the wave equation with the full Laplacian on the Heisenberg group. *Canad. J. Math.*, 59(6):1301–1322, 2007.

[Fol75] G. B. Folland. Subelliptic estimates and function spaces on nilpotent Lie groups. *Ark. Mat.*, 13(2):161–207, 1975.

[Fol77a] G. B. Folland. Applications of analysis on nilpotent groups to partial differential equations. *Bull. Amer. Math. Soc.*, 83(5):912–930, 1977.

[Fol77b] G. B. Folland. On the Rothschild-Stein lifting theorem. *Comm. Partial Differential Equations*, 2(2):165–191, 1977.

[Fol89] G. B. Folland. *Harmonic analysis in phase space*, volume 122 of *Annals of Mathematics Studies*. Princeton University Press, Princeton, NJ, 1989.

[Fol94] G. B. Folland. Meta-heisenberg groups. In *Fourier analysis (Orono, ME, 1992)*, volume 157 of *Lecture Notes in Pure and Appl. Math.*, pages 121–147. Dekker, New York, 1994.

[Fol95] G. B. Folland. *A course in abstract harmonic analysis*. Studies in Advanced Mathematics. CRC Press, Boca Raton, FL, 1995.

[Fol99] G. B. Folland. *Real analysis*. Pure and Applied Mathematics (New York). John Wiley & Sons Inc., New York, second edition, 1999. Modern techniques and their applications, A Wiley-Interscience Publication.

[FP78] C. Fefferman and D. H. Phong. On positivity of pseudo-differential operators. *Proc. Nat. Acad. Sci. U.S.A.*, 75(10):4673–4674, 1978.

[FR66] E. B. Fabes and N. M. Rivière. Singular integrals with mixed homogeneity. *Studia Math.*, 27:19–38, 1966.

[FR13] V. Fischer and M. Ruzhansky. Lower bounds for operators on graded Lie groups. *C. R. Math. Acad. Sci. Paris*, 351(1-2):13–18, 2013.

[FR14a] V. Fischer and M. Ruzhansky. A pseudo-differential calculus on graded nilpotent Lie groups. In *Fourier analysis*, Trends Math., pages 107–132. Birkhäuser/Springer, Cham, 2014.

[FR14b] V. Fischer and M. Ruzhansky. A pseudo-differential calculus on the Heisenberg group. *C. R. Math. Acad. Sci. Paris*, 352(3):197–204, 2014.

[Fri70] K. O. Friedrichs. *Pseudo-diferential operators. An introduction*. Notes prepared with the assistance of R. Vaillancourt. Revised edition. Courant Institute of Mathematical Sciences New York University, New York, 1970.

[FS74] G. B. Folland and E. M. Stein. Estimates for the $\bar{\partial}_b$ complex and analysis on the Heisenberg group. *Comm. Pure Appl. Math.*, 27:429–522, 1974.

[FS82] G. B. Folland and E. M. Stein. *Hardy spaces on homogeneous groups*, volume 28 of *Mathematical Notes*. Princeton University Press, Princeton, N.J., 1982.

[Går53] L. Gårding. Dirichlet's problem for linear elliptic partial differential equations. *Math. Scand.*, 1:55–72, 1953.

[Gel80] D. Geller. Fourier analysis on the Heisenberg group. I. Schwartz space. *J. Funct. Anal.*, 36(2):205–254, 1980.

[Gel83] D. Geller. Liouville's theorem for homogeneous groups. *Comm. Partial Differential Equations*, 8(15):1665–1677, 1983.

[Gel90] D. Geller. *Analytic pseudodifferential operators for the Heisenberg group and local solvability*, volume 37 of *Mathematical Notes*. Princeton University Press, Princeton, NJ, 1990.

[GGV86] B. Gaveau, P. Greiner, and J. Vauthier. Intégrales de Fourier quadratiques et calcul symbolique exact sur le groupe d'Heisenberg. *J. Funct. Anal.*, 68(2):248–272, 1986.

[GKS75] P. C. Greiner, J. J. Kohn, and E. M. Stein. Necessary and sufficient conditions for solvability of the Lewy equation. *Proc. Nat. Acad. Sci. U.S.A.*, 72(9):3287–3289, 1975.

[Gło89] P. Głowacki. The Rockland condition for nondifferential convolution operators. *Duke Math. J.*, 58(2):371–395, 1989.

[Gło91] P. Głowacki. The Rockland condition for nondifferential convolution operators. II. *Studia Math.*, 98(2):99–114, 1991.

[Gło04] P. Głowacki. A symbolic calculus and L^2-boundedness on nilpotent Lie groups. *J. Funct. Anal.*, 206(1):233–251, 2004.

[Gło07] P. Głowacki. The Melin calculus for general homogeneous groups. *Ark. Mat.*, 45(1):31–48, 2007.

[Gło12] P. Głowacki. Invertibility of convolution operators on homogeneous groups. *Rev. Mat. Iberoam.*, 28(1):141–156, 2012.

[Goo76] R. W. Goodman. *Nilpotent Lie groups: structure and applications to analysis*. Lecture Notes in Mathematics, Vol. 562. Springer-Verlag, Berlin, 1976.

[Goo80] R. Goodman. Singular integral operators on nilpotent Lie groups. *Ark. Mat.*, 18(1):1–11, 1980.

[Gro96] M. Gromov. Carnot-Carathéodory spaces seen from within. In *Sub-Riemannian geometry*, volume 144 of *Progr. Math.*, pages 79–323. Birkhäuser, Basel, 1996.

[GS85] A. Grigis and J. Sjöstrand. Front d'onde analytique et sommes de carrés de champs de vecteurs. *Duke Math. J.*, 52(1):35–51, 1985.

[Hel82] B. Helffer. Conditions nécessaires d'hypoanalyticité pour des opérateurs invariants à gauche homogènes sur un groupe nilpotent gradué. *J. Differential Equations*, 44(3):460–481, 1982.

[Hel84a] B. Helffer. *Théorie spectrale pour des opérateurs globalement elliptiques*, volume 112 of *Astérisque*. Société Mathématique de France, Paris, 1984. With an English summary.

[Hel84b] S. Helgason. *Groups and geometric analysis*, volume 113 of *Pure and Applied Mathematics*. Academic Press Inc., Orlando, FL, 1984. Integral geometry, invariant differential operators, and spherical functions.

[Hel01] S. Helgason. *Differential geometry, Lie groups, and symmetric spaces*, volume 34 of *Graduate Studies in Mathematics*. American Mathematical Society, Providence, RI, 2001. Corrected reprint of the 1978 original.

[HJL85] A. Hulanicki, J. W. Jenkins, and J. Ludwig. Minimum eigenvalues for positive, Rockland operators. *Proc. Amer. Math. Soc.*, 94(4):718–720, 1985.

[HN79] B. Helffer and J. Nourrigat. Caracterisation des opérateurs hypoelliptiques homogènes invariants à gauche sur un groupe de Lie nilpotent gradué. *Comm. Partial Differential Equations*, 4(8):899–958, 1979.

[HN05] B. Helffer and F. Nier. *Hypoelliptic estimates and spectral theory for Fokker-Planck operators and Witten Laplacians*, volume 1862 of *Lecture Notes in Mathematics*. Springer-Verlag, Berlin, 2005.

[Hör60] L. Hörmander. Estimates for translation invariant operators in L^p spaces. *Acta Math.*, 104:93–140, 1960.

[Hör66] L. Hörmander. Pseudo-differential operators and non-elliptic boundary problems. *Ann. of Math. (2)*, 83:129–209, 1966.

[Hör67a] L. Hörmander. Hypoelliptic second order differential equations. *Acta Math.*, 119:147–171, 1967.

[Hör67b] L. Hörmander. Pseudo-differential operators and hypoelliptic equations. In *Singular integrals (Proc. Sympos. Pure Math., Vol. X, Chicago, Ill., 1966)*, pages 138–183. Amer. Math. Soc., Providence, R.I., 1967.

[Hör77] L. Hörmander. The Cauchy problem for differential equations with double characteristics. *J. Analyse Math.*, 32:118–196, 1977.

[Hör03] L. Hörmander. *The analysis of linear partial differential operators. I.* Classics in Mathematics. Springer-Verlag, Berlin, 2003. Distribution theory and Fourier analysis, Reprint of the second (1990) edition [Springer, Berlin; MR1065993 (91m:35001a)].

[How80] R. Howe. On the role of the Heisenberg group in harmonic analysis. *Bull. Amer. Math. Soc. (N.S.)*, 3(2):821–843, 1980.

[How84] R. Howe. A symbolic calculus for nilpotent groups. In *Operator algebras and group representations, Vol. I (Neptun, 1980)*, volume 17 of *Monogr. Stud. Math.*, pages 254–277. Pitman, Boston, MA, 1984.

[HP57] E. Hille and R. S. Phillips. *Functional analysis and semi-groups.* American Mathematical Society Colloquium Publications, vol. 31. American Mathematical Society, Providence, R. I., 1957. rev. ed.

[HR70] E. Hewitt and K. A. Ross. *Abstract harmonic analysis. Vol. II: Structure and analysis for compact groups. Analysis on locally compact Abelian groups.* Die Grundlehren der mathematischen Wissenschaften, Band 152. Springer-Verlag, New York, 1970.

[HS90] W. Hebisch and A. Sikora. A smooth subadditive homogeneous norm on a homogeneous group. *Studia Math.*, 96(3):231–236, 1990.

[Hul84] A. Hulanicki. A functional calculus for Rockland operators on nilpotent Lie groups. *Studia Math.*, 78(3):253–266, 1984.

[Hun56] G. A. Hunt. Semi-groups of measures on Lie groups. *Trans. Amer. Math. Soc.*, 81:264–293, 1956.

[Kec95] A. S. Kechris. *Classical descriptive set theory*, volume 156 of *Graduate Texts in Mathematics.* Springer-Verlag, New York, 1995.

[Kg81] H. Kumano-go. *Pseudodifferential operators.* MIT Press, Cambridge, Mass., 1981. Translated from the Japanese by the author, Rémi Vaillancourt and Michihiro Nagase.

[Kir04] A. A. Kirillov. *Lectures on the orbit method*, volume 64 of *Graduate Studies in Mathematics.* American Mathematical Society, Providence, RI, 2004.

[Kna01] A. W. Knapp. *Representation theory of semisimple groups.* Princeton Landmarks in Mathematics. Princeton University Press, Princeton, NJ, 2001. An overview based on examples, Reprint of the 1986 original.

[Koh73] J. J. Kohn. Pseudo-differential operators and hypoellipticity. In *Partial differential equations (Proc. Sympos. Pure Math., Vol. XXIII, Univ. California, Berkeley, Calif., 1971)*, pages 61–69. Amer. Math. Soc., Providence, R.I., 1973.

[Kol34] A. Kolmogoroff. Zufällige Bewegungen (zur Theorie der Brownschen Bewegung). *Ann. of Math. (2)*, 35(1):116–117, 1934.

[Kor72] A. Koranyi. Harmonic functions on symmetric spaces. In *Symmetric spaces (Short Courses, Washington Univ., St. Louis, Mo., 1969–1970)*, pages 379–412. Pure and Appl. Math., Vol. 8. Dekker, New York, 1972.

[Kra09] S. G. Krantz. *Explorations in harmonic analysis*. Applied and Numerical Harmonic Analysis. Birkhäuser Boston Inc., Boston, MA, 2009. With applications to complex function theory and the Heisenberg group, With the assistance of Lina Lee.

[KS69] A. W. Knapp and E. M. Stein. Singular integrals and the principal series. I, II. *Proc. Nat. Acad. Sci. U.S.A. 63 (1969), 281–284; ibid.*, 66:13–17, 1969.

[Kun58] R. A. Kunze. L_p Fourier transforms on locally compact unimodular groups. *Trans. Amer. Math. Soc.*, 89:519–540, 1958.

[KV71] A. Korányi and S. Vági. Singular integrals on homogeneous spaces and some problems of classical analysis. *Ann. Scuola Norm. Sup. Pisa (3)*, 25:575–648 (1972), 1971.

[Lan60] R. P. Langlands. Some holomorphic semi-groups. *Proc. Nat. Acad. Sci. U.S.A.*, 46:361–363, 1960.

[Ler10] N. Lerner. *Metrics on the phase space and non-selfadjoint pseudo-differential operators*, volume 3 of *Pseudo-Differential Operators. Theory and Applications*. Birkhäuser Verlag, Basel, 2010.

[Liz75] P. I. Lizorkin. Interpolation of weighted L_p spaces. *Dokl. Akad. Nauk SSSR*, 222(1):32–35, 1975.

[LN66] P. D. Lax and L. Nirenberg. On stability for difference schemes: A sharp form of Gårding's inequality. *Comm. Pure Appl. Math.*, 19:473–492, 1966.

[LP61] G. Lumer and R. S. Phillips. Dissipative operators in a Banach space. *Pacific J. Math.*, 11:679–698, 1961.

[LP94] E. Lanconelli and S. Polidoro. On a class of hypoelliptic evolution operators. *Rend. Sem. Mat. Univ. Politec. Torino*, 52(1):29–63, 1994. Partial differential equations, II (Turin, 1993).

[Man91] D. Manchon. Formule de Weyl pour les groupes de Lie nilpotents. *J. Reine Angew. Math.*, 418:77–129, 1991.

[MCSA01] C. Martínez Carracedo and M. Sanz Alix. *The theory of fractional powers of operators*, volume 187 of *North-Holland Mathematics Studies*. North-Holland Publishing Co., Amsterdam, 2001.

[Mel71] A. Melin. Lower bounds for pseudo-differential operators. *Ark. Mat.*, 9:117–140, 1971.

[Mel81] A. Melin. Parametrix constructions for some classes of right-invariant differential operators on the Heisenberg group. *Comm. Partial Differential Equations*, 6(12):1363–1405, 1981.

[Mel83] A. Melin. Parametrix constructions for right invariant differential operators on nilpotent groups. *Ann. Global Anal. Geom.*, 1(1):79–130, 1983.

[Mét80] G. Métivier. Hypoellipticité analytique sur des groupes nilpotents de rang 2. *Duke Math. J.*, 47(1):195–221, 1980.

[Mih56] S. G. Mihlin. On the theory of multidimensional singular integral equations. *Vestnik Leningrad. Univ.*, 11(1):3–24, 1956.

[Mih57] S. G. Mihlin. Singular integrals in L_p spaces. *Dokl. Akad. Nauk SSSR (N.S.)*, 117:28–31, 1957.

[Mil80] K. G. Miller. Parametrices for hypoelliptic operators on step two nilpotent Lie groups. *Comm. Partial Differential Equations*, 5(11):1153–1184, 1980.

[MPP07] M. Mughetti, C. Parenti, and A. Parmeggiani. Lower bound estimates without transversal ellipticity. *Comm. Partial Differential Equations*, 32(7-9):1399–1438, 2007.

[MPR99] D. Müller, M. M. Peloso, and F. Ricci. On local solvability for complex coefficient differential operators on the Heisenberg group. *J. Reine Angew. Math.*, 513:181–234, 1999.

[MR15] M. Mantoiu and M. Ruzhansky. Pseudo-differential operators, Wigner transform and Weyl systems on type I locally compact groups. *arXiv:1506.05854*, 2015.

[MS87] G. A. Meladze and M. A. Shubin. A functional calculus of pseudodifferential operators on unimodular Lie groups. *Trudy Sem. Petrovsk.*, (12):164–200, 245, 1987.

[MS99] D. Müller and E. M. Stein. L^p-estimates for the wave equation on the Heisenberg group. *Rev. Mat. Iberoamericana*, 15(2):297–334, 1999.

[MS11] P. McKeag and Y. Safarov. Pseudodifferential operators on manifolds: a coordinate-free approach. In *Partial differential equations and spectral theory*, volume 211 of *Oper. Theory Adv. Appl.*, pages 321–341. Birkhäuser Basel AG, Basel, 2011.

[Nac82] A. I. Nachman. The wave equation on the Heisenberg group. *Comm. Partial Differential Equations*, 7(6):675–714, 1982.

[Nag77] M. Nagase. A new proof of sharp Gårding inequality. *Funkcial. Ekvac.*, 20(3):259–271, 1977.

[Nik77] S. M. Nikolskii. *Priblizhenie funktsii mnogikh peremennykh i teoremy vlozheniya (Russian) [Approximation of functions of several variables and imbedding theorems].* "Nauka", Moscow, 1977. Second edition, revised and supplemented.

[Nom56] K. Nomizu. *Lie groups and differential geometry.* The Mathematical Society of Japan, 1956.

[NS59] E. Nelson and W. F. Stinespring. Representation of elliptic operators in an enveloping algebra. *Amer. J. Math.*, 81:547–560, 1959.

[NSW85] A. Nagel, E. M. Stein, and S. Wainger. Balls and metrics defined by vector fields. I. Basic properties. *Acta Math.*, 155(1-2):103–147, 1985.

[OR73] O. A. Oleĭnik and E. V. Radkevič. *Second order equations with nonnegative characteristic form.* Plenum Press, New York, 1973. Translated from the Russian by Paul C. Fife.

[Pan89] P. Pansu. Métriques de Carnot-Carathéodory et quasiisométries des espaces symétriques de rang un. *Ann. of Math. (2)*, 129(1):1–60, 1989.

[Paz83] A. Pazy. *Semigroups of linear operators and applications to partial differential equations*, volume 44 of *Applied Mathematical Sciences.* Springer-Verlag, New York, 1983.

[Pon66] L. S. Pontryagin. *Topological groups.* Translated from the second Russian edition by Arlen Brown. Gordon and Breach Science Publishers, Inc., New York, 1966.

[Pon08] R. S. Ponge. Heisenberg calculus and spectral theory of hypoelliptic operators on Heisenberg manifolds. *Mem. Amer. Math. Soc.*, 194(906):viii+ 134, 2008.

[Puk67] L. Pukánszky. *Leçons sur les représentations des groupes.* Monographies de la Société Mathématique de France, No. 2. Dunod, Paris, 1967.

[PW27] F. Peter and H. Weyl. Die Vollständigkeit der primitiven Darstellungen einer geschlossenen kontinuierlichen Gruppe. *Math. Ann.*, 97(1):737–755, 1927.

[Ric] F. Ricci. Sub-Laplacians on nilpotent Lie groups. Unpublished lecture notes accessible on webpage http://homepage.sns.it/fricci/corsi.html.

[Roc78] C. Rockland. Hypoellipticity on the Heisenberg group-representation-theoretic criteria. *Trans. Amer. Math. Soc.*, 240:1–52, 1978.

[Rot83] L. P. Rothschild. A remark on hypoellipticity of homogeneous invariant differential operators on nilpotent Lie groups. *Comm. Partial Differential Equations*, 8(15):1679–1682, 1983.

[RS75] M. Reed and B. Simon. *Methods of modern mathematical physics. II. Fourier analysis, self-adjointness.* Academic Press [Harcourt Brace Jovanovich Publishers], New York, 1975.

[RS76] L. P. Rothschild and E. M. Stein. Hypoelliptic differential operators and nilpotent groups. *Acta Math.*, 137(3-4):247–320, 1976.

[RS80] M. Reed and B. Simon. *Methods of modern mathematical physics. I.* Academic Press Inc. [Harcourt Brace Jovanovich Publishers], New York, second edition, 1980. Functional analysis.

[RT10a] M. Ruzhansky and V. Turunen. *Pseudo-differential operators and symmetries. Background analysis and advanced topics*, volume 2 of *Pseudo-Differential Operators. Theory and Applications.* Birkhäuser Verlag, Basel, 2010.

[RT10b] M. Ruzhansky and V. Turunen. Quantization of pseudo-differential operators on the torus. *J. Fourier Anal. Appl.*, 16(6):943–982, 2010.

[RT11] M. Ruzhansky and V. Turunen. Sharp Gårding inequality on compact Lie groups. *J. Funct. Anal.*, 260(10):2881–2901, 2011.

[RT13] M. Ruzhansky and V. Turunen. Global quantization of pseudo-differential operators on compact Lie groups, SU(2), 3-sphere, and homogeneous spaces. *Int. Math. Res. Not. IMRN*, (11):2439–2496, 2013.

[RTW14] M. Ruzhansky, V. Turunen, and J. Wirth. Hörmander class of pseudo-differential operators on compact Lie groups and global hypoellipticity. *J. Fourier Anal. Appl.*, 20(3):476–499, 2014.

[Rud87] W. Rudin. *Real and complex analysis.* McGraw-Hill Book Co., New York, third edition, 1987.

[Rud91] W. Rudin. *Functional analysis.* International Series in Pure and Applied Mathematics. McGraw-Hill Inc., New York, second edition, 1991.

[RW13] M. Ruzhansky and J. Wirth. On multipliers on compact Lie groups. *Funct. Anal. Appl.*, 47(1):87–91, 2013.

[RW14] M. Ruzhansky and J. Wirth. Global functional calculus for operators on compact Lie groups. *J. Funct. Anal.*, 267(1):144–172, 2014.

[RW15] M. Ruzhansky and J. Wirth. L^p Fourier multipliers on compact Lie groups. *Math. Z.*, 280(3-4):621–642, 2015.

[Saf97] Y. Safarov. Pseudodifferential operators and linear connections. *Proc. London Math. Soc. (3)*, 74(2):379–416, 1997.

[Sak79] K. Saka. Besov spaces and Sobolev spaces on a nilpotent Lie group. *Tôhoku Math. J. (2)*, 31(4):383–437, 1979.

[Sak98] S. Sakai. *C^*-algebras and W^*-algebras.* Classics in Mathematics. Springer-Verlag, Berlin, 1998. Reprint of the 1971 edition.

[See69] R. T. Seeley. Eigenfunction expansions of analytic functions. *Proc. Amer. Math. Soc.*, 21:734–738, 1969.

[Seg50] I. E. Segal. An extension of Plancherel's formula to separable unimodular groups. *Ann. of Math. (2)*, 52:272–292, 1950.

[Seg53] I. E. Segal. A non-commutative extension of abstract integration. *Ann. of Math. (2)*, 57:401–457, 1953.

[Sem03] S. Semmes. An introduction to Heisenberg groups in analysis and geometry. *Notices Amer. Math. Soc.*, 50(6):640–646, 2003.

[Sha05] V. A. Sharafutdinov. Geometric symbol calculus for pseudodifferential operators. I [Translation of Mat. Tr. **7** (2004), no. 2, 159–206]. *Siberian Adv. Math.*, 15(3):81–125, 2005.

[Shu87] M. A. Shubin. *Pseudodifferential operators and spectral theory.* Springer Series in Soviet Mathematics. Springer-Verlag, Berlin, 1987. Translated from the Russian by Stig I. Andersson.

[Sjö82] J. Sjöstrand. Singularités analytiques microlocales. In *Astérisque, 95*, volume 95 of *Astérisque*, pages 1–166. Soc. Math. France, Paris, 1982.

[Ste70a] E. M. Stein. *Singular integrals and differentiability properties of functions.* Princeton Mathematical Series, No. 30. Princeton University Press, Princeton, N.J., 1970.

[Ste70b] E. M. Stein. *Topics in harmonic analysis related to the Littlewood-Paley theory.* Annals of Mathematics Studies, No. 63. Princeton University Press, Princeton, N.J., 1970.

[Ste93] E. M. Stein. *Harmonic analysis: real-variable methods, orthogonality, and oscillatory integrals*, volume 43 of *Princeton Mathematical Series.* Princeton University Press, Princeton, NJ, 1993. With the assistance of Timothy S. Murphy, Monographs in Harmonic Analysis, III.

[Str72] R. S. Strichartz. Invariant pseudo-differential operators on a Lie group. *Ann. Scuola Norm. Sup. Pisa (3)*, 26:587–611, 1972.

[SW71] E. M. Stein and G. Weiss. *Introduction to Fourier analysis on Euclidean spaces.* Princeton University Press, Princeton, N.J., 1971. Princeton Mathematical Series, No. 32.

[SZ02] A. Sikora and J. Zienkiewicz. A note on the heat kernel on the Heisenberg group. *Bull. Austral. Math. Soc.*, 65(1):115–120, 2002.

[Tar78] D. S. Tartakoff. Local analytic hypoellipticity for $\Box_b$ on nondegenerate Cauchy-Riemann manifolds. *Proc. Nat. Acad. Sci. U.S.A.*, 75(7):3027–3028, 1978.

[Tar80] D. S. Tartakoff. The local real analyticity of solutions to $\Box_b$ and the $\bar{\partial}$-Neumann problem. *Acta Math.*, 145(3-4):177–204, 1980.

[Tay81] M. E. Taylor. *Pseudodifferential operators*, volume 34 of *Princeton Mathematical Series*. Princeton University Press, Princeton, N.J., 1981.

[Tay84] M. E. Taylor. Noncommutative microlocal analysis. I. *Mem. Amer. Math. Soc.*, 52(313):iv+182, (Revised version accessible at http://math.unc.edu/Faculty/met/ncmlms.pdf) 1984.

[Tay86] M. E. Taylor. *Noncommutative harmonic analysis*, volume 22 of *Mathematical Surveys and Monographs*. American Mathematical Society, Providence, RI, 1986.

[tER97] A. F. M. ter Elst and D. W. Robinson. Spectral estimates for positive Rockland operators. In *Algebraic groups and Lie groups*, volume 9 of *Austral. Math. Soc. Lect. Ser.*, pages 195–213. Cambridge Univ. Press, Cambridge, 1997.

[Tha98] S. Thangavelu. *Harmonic analysis on the Heisenberg group*, volume 159 of *Progress in Mathematics*. Birkhäuser Boston Inc., Boston, MA, 1998.

[Tre67] F. Treves. *Topological vector spaces, distributions and kernels*. Academic Press, New York, 1967.

[Trè78] F. Trèves. Analytic hypo-ellipticity of a class of pseudodifferential operators with double characteristics and applications to the $\bar{\partial}$-Neumann problem. *Comm. Partial Differential Equations*, 3(6-7):475–642, 1978.

[Vai70] R. Vaillancourt. A simple proof of Lax-Nirenberg theorems. *Comm. Pure Appl. Math.*, 23:151–163, 1970.

[vE10a] E. van Erp. The Atiyah-Singer index formula for subelliptic operators on contact manifolds. Part I. *Ann. of Math. (2)*, 171(3):1647–1681, 2010.

[vE10b] E. van Erp. The Atiyah-Singer index formula for subelliptic operators on contact manifolds. Part II. *Ann. of Math. (2)*, 171(3):1683–1706, 2010.

[Vil68] N. J. Vilenkin. *Special functions and the theory of group representations.* Translated from the Russian by V. N. Singh. Translations of Mathematical Monographs, Vol. 22. American Mathematical Society, Providence, R. I., 1968.

[VK91] N. J. Vilenkin and A. U. Klimyk. *Representation of Lie groups and special functions. Vol. 1*, volume 72 of *Mathematics and its Applications (Soviet Series).* Kluwer Academic Publishers Group, Dordrecht, 1991. Simplest Lie groups, special functions and integral transforms, Translated from the Russian by V. A. Groza and A. A. Groza.

[VK93] N. J. Vilenkin and A. U. Klimyk. *Representation of Lie groups and special functions. Vol. 2*, volume 74 of *Mathematics and its Applications (Soviet Series).* Kluwer Academic Publishers Group, Dordrecht, 1993. Class I representations, special functions, and integral transforms, Translated from the Russian by V. A. Groza and A. A. Groza.

[VSCC92] N. T. Varopoulos, L. Saloff-Coste, and T. Coulhon. *Analysis and geometry on groups*, volume 100 of *Cambridge Tracts in Mathematics.* Cambridge University Press, Cambridge, 1992.

[Wal73] N. R. Wallach. *Harmonic analysis on homogeneous spaces.* Marcel Dekker Inc., New York, 1973. Pure and Applied Mathematics, No. 19.

[Wal92] N. R. Wallach. *Real reductive groups. II*, volume 132 of *Pure and Applied Mathematics.* Academic Press Inc., Boston, MA, 1992.

[Wei72] N. J. Weiss. L^p estimates for bi-invariant operators on compact Lie groups. *Amer. J. Math.*, 94:103–118, 1972.

[Wid80] H. Widom. A complete symbolic calculus for pseudodifferential operators. *Bull. Sci. Math. (2)*, 104(1):19–63, 1980.

[Žel73] D. P. Želobenko. *Compact Lie groups and their representations.* American Mathematical Society, Providence, R.I., 1973. Translated from the Russian by Israel Program for Scientific Translations, Translations of Mathematical Monographs, Vol. 40.

[Zie04] J. Zienkiewicz. Schrödinger equation on the Heisenberg group. *Studia Math.*, 161(2):99–111, 2004.

[Zui93] C. Zuily. Existence globale de solutions régulières pour l'équation des ondes non linéaire amortie sur le groupe de Heisenberg. *Indiana Univ. Math. J.*, 42(2):323–360, 1993.

Index

V. Fischer, M. Ruzhansky, *Quantization on Nilpotent Lie Groups*, Progress in Mathematics 314,
DOI 10.1007/978-3-319-29558-9

www.ingramcontent.com/pod-product-compliance
Lightning Source LLC
LaVergne TN
LVHW062153200926
848484LV00007B/1911
* 9 7 8 1 0 1 3 2 6 7 3 1 4 *